Insect Development and Evolution

Insect Development and Evolution

Bruce S. Heming
University of Alberta

COMSTOCK PUBLISHING ASSOCIATES
a division of
CORNELL UNIVERSITY PRESS
Ithaca and London

First published 2003 by Cornell University Press

Printed in the United States of America

Library of Congress Cataloging-in-Publication Data

Heming, B. S.
   Insect development and evolution / Bruce S. Heming.
      p. cm.
Includes bibliographical references (p.   ).
   ISBN 0-8014-3933-7 (cloth : alk. paper)
   1. Insects—Development. 2. Insects—Evolution. 3. Insects—Reproduction. I. Title.
   QL495.5 .H46 2003
   571.8'157—dc21

                          2002011087

Cornell University Press strives to use environmentally responsible suppliers and materials to the fullest extent possible in the publishing of its books. Such materials include vegetable-based, low-VOC inks and acid-free papers that are recycled, totally chlorine-free, or partly composed of nonwood fibers. For further information, visit our website at www.cornellpress.cornell.edu.

Cloth printing   10 9 8 7 6 5 4 3 2 1

To those students of insect structure and development whose elegant and ingenious contributions in the past contributed to our present spectacular success

Contents

Preface   xiii

Acknowledgments   xv

Introduction
Development and Biology   1
Reductionist versus Evolutionary Approaches to the Study of Insect Development   1
Insect Ontogeny: A Summary   1
A Note on Phylogeny   4

Chapter 1. The Male Reproductive System and Spermatogenesis
1.1 Male Reproductive Systems   6
   1.1.1 Testis   6
   1.1.2 Primary Exit System   7
   1.1.3 Secondary Exit System   7
   1.1.4 External Genitalia   7
1.2 Insect Spermatozoa   8
   1.2.1 Structure   8
   1.2.2 Motility   11
   1.2.3 Phylogenetic Aspects   11
1.3 Spermatogenesis   13
   1.3.1 Primordial Development   13
   1.3.2 Premeiotic Events   13
   1.3.3 Meiosis   14
   1.3.4 Spermiogenesis   18
   1.3.5 Rate of Spermatogenesis   18
   1.3.6 Productivity of Spermatogenesis   20
   1.3.7 Control of Spermatogenesis   21

Chapter 2. The Female Reproductive System and Oogenesis
2.1 Female Reproductive Systems   29
   2.1.1 Ovaries   29
   2.1.2 Primary Exit System   30
   2.1.3 Secondary Exit System   30
   2.1.4 External Genitalia   32
2.2 Oogenesis   32
   2.2.1 Panoistic Ovarioles   34
   2.2.2 Meroistic Ovarioles   41
   2.2.3 Staging Oogenesis   52
   2.2.4 Duration of Oogenesis   52
   2.2.5 Fecundity   54
   2.2.6 The Adaptive Significance of Conspecific Differences in Egg Size   54
   2.2.7 Phylogenetic Aspects   54
2.3 Endocrine Control of Oogenesis   58
   2.3.1 Previtellogenesis   58
   2.3.2 Vitellogenesis   59
   2.3.3 Choriogenesis   61

2.4 Influence of Extrinsic Factors on Vitellogenesis   61
2.5 Genes and Oogenesis   63

Chapter 3. Sperm Transfer, Allocation, and Use

3.1 Sperm Passage from Sperm Tubes into Seminal Vesicles and Spermatophore   65
3.2 Sperm Activation   66
3.3 Intromission and Insemination   66
    3.3.1 Marine Arthropods   67
    3.3.2 Terrestrial Arthropods   67
3.4 Ovulation   77
3.5 Passage of Sperm from Spermatheca to Micropyle of Egg   77
3.6 Fertilization   78
    3.6.1 Molecular Aspects   80
    3.6.2 Contributions of the Sperm to Successful Fertilization   81
3.7 Sperm Competition and Sexual Selection   82
    3.7.1 Influence of Bacterial Parasites on Fertilization and Sperm Competition   86
    3.7.2 Copulatory Inhibitors in Drosophila melanogaster   86
3.8 Copulatory Courtship   86
3.9 Oviposition   86
    3.9.1 Ovipositor Function   88
    3.9.2 Adaptive Significance of the Ovipositor   89
3.10 Viviparity   89

Chapter 4. Sex Determination

4.1 Sex Chromosomes   91
    4.1.1 Evolution   91
    4.1.2 Discovery   92
4.2 Sex Determination in *Drosophila melanogaster*   92
    4.2.1 Somatic Sex Determination   92
    4.2.2 Sex Determination of Germ Cells   93
4.3 Sex Determination in Haplodiploid (Arrhenotokous) Insects   93
4.4 Chromosome Elimination and Sex Determination in Symphypleon Collembola   93
4.5 Environmental Effects on Sex Determination   94
    4.5.1 Temperature   94
    4.5.2 Nutrition and Inhibitory Pheromones   95
    4.5.3 Androgenic Hormones   95
    4.5.4 Maternal Hemolymph   96
4.6 Hermaphrodites   96
4.7 Gynandromorphs   97
4.8 Intersexes   97
4.9 Parasite Effects on Sex Determination   98
4.10 Why Are There Only Two Sexes?   99

Chapter 5. Parthenogenesis

5.1 Arrhenotoky   101
5.2 Thelytoky   103
    5.2.1 Automictic   103
    5.2.2 Apomictic   103
5.3 Amphitoky   104
5.4 Evolutionary Considerations   104
    5.4.1 Arrhenotoky   104
    5.4.2 Thelytoky   104
5.5 Cyclical Parthenogenesis   105
    5.5.1 Gall Wasps   105
    5.5.2 Aphids   105
    5.5.3 Gall Midges   107
    5.5.4 Micromalthus debilis   108
5.6 Microorganisms and Parthenogenesis   109

Chapter 6. Early Embryogenesis
6.1 Egg Structure 111
6.2 Cleavage 111
   6.2.1 Control of Cleavage Rate 112
   6.2.2 Movement of Cleavage Energids to the Periphery 113
   6.2.3 Chromosome Elimination in Embryos of Lower Diptera 113
   6.2.4 Vitellophages 115
6.3 Blastoderm Formation 115
6.4 DNA, RNA, and Protein Synthesis during Cleavage and Blastoderm Formation 117
6.5 Mitotic Domains in the Blastoderm of Flies 118
6.6 Pole Cell Formation 119
   6.6.1 Genes and Pole Cell Formation in Drosophila melanogaster *Embryos* 120
   6.6.2 Substitutive Embryogeny and the Origin of Germ Cells in Orthopteroid Embryos 121
   6.6.3 Germ Cell Origin in Polyembryonic Insects 121
6.7 Germ Band Formation 122
   6.7.1 Factors Affecting Orientation of the Germ Band in Pentatomid (Heteroptera) Eggs 122
   6.7.2 Polyembryony 123
6.8 Gastrulation and Germ Layer Formation 125
   6.8.1 Genes and Gastrulation in Drosophila melanogaster *Embryos* 126
6.9 Blastokinesis and Embryonic Envelopes 126
   6.9.1 Anatrepsis 127
   6.9.2 Katatrepsis 128
   6.9.3 Blastodermal and Serosal Cuticles 129
   6.9.4 Blastokinesis in Long Germ Embryos and Specification of the Amnioserosa 129
   6.9.5 Yolk Extrusion and Consumption in Long Germ Embryos of the Tephritid Fly
     Anastrepha fraterculus 131
6.10 Segmentation and Appendage Formation 134
   6.10.1 Short Germ Embryos 134
   6.10.2 Long Germ Embryos 135

Chapter 7. Specification of the Body Plan in Insect Embryos
7.1 Early Experiments 136
   7.1.1 Determinants and Mosaic Development 136
   7.1.2 Physiological Centers 136
   7.1.3 Gradients and Pattern Formation 137
   7.1.4 Fate Maps 143
   7.1.5 Developmental Genetics 144
   7.1.6 The Compartment Hypothesis 144
7.2 Molecular Genetics of Pattern Formation in *Drosophila melanogaster* Embryos 146
   7.2.1 Pattern-Control Genes 147
   7.2.2 Evolutionary Considerations 174

Chapter 8. Organogenesis
8.1 Mesoderm 176
   8.1.1 Genes and Mesodermal Fate in Drosophila *Embryos* 176
   8.1.2 Muscle Pioneers and Muscle Formation 179
   8.1.3 Hemocytes 182
   8.1.4 Body Cavity (Hemocoel) 182
   8.1.5 Dorsal Vessel 183
   8.1.6 Fat Body 184
   8.1.7 Subesophageal Body 184
8.2 Endoderm and Ectoderm 184
   8.2.1 Midgut 184
   8.2.2 Foregut and Hindgut 185
   8.2.3 Malpighian Tubules 185
8.3 Ectoderm 186
   8.3.1 Central Nervous System 186

8.3.2 *Peripheral Nervous System* — 202
8.3.3 *Stomatogastric (Enteric) Nervous System* — 209
8.3.4 *Tracheal System* — 212
8.3.5 *Salivary Glands* — 215
8.3.6 *Epidermis and Imaginal Discs* — 215
8.3.7 *Head Involution in Embryos of Higher Flies* — 218
8.3.8 *Embryogenesis of Piercing and Sucking Mouthparts* — 220
8.3.9 *Skeletal Apodemes* — 225
8.4 Primordia of Mixed Germ Layer Origin — 225
8.4.1 *Reproductive System* — 225
8.5 Other Tissues — 228
8.6 Miscellaneous Events — 228
8.6.1 *The Role of Apoptosis in Insect Embryogenesis* — 228
8.6.2 *Induction* — 228
8.6.3 *Hormones and Embryogenesis* — 229
8.6.4 *Environmental Factors* — 231
8.7 Overview — 233
8.8 Computer Simulation of Gene Expression during Embryogenesis of *Drosophila melanogaster* — 233

Chapter 9. Postembryonic Development and Life History
9.1 Life History Theory — 235
9.2 Hatching — 235
9.3 Stage and Instar — 236
9.4 Metamorphosis — 236
9.4.1 *Ametabola* — 236
9.4.2 *Hemimetabola* (sensu latu) *(Exopterygota)* — 237
9.4.3 *Holometabola* — 239
9.4.4 *Recent Concepts* — 240
9.5 The Concept of Stase — 242
9.6 Larvae and Pupae in Holometabola — 243
9.6.1 *Larvae* — 243
9.6.2 *Pupae* — 246
9.7 Growth — 256
9.7.1 *Patterns of Growth* — 256
9.7.2 *Growth and Molting* — 256
9.7.3 *Effects of Sclerotization* — 258
9.7.4 *Why Don't Caterpillars Grow Short and Fat?* — 259
9.7.5 *Disproportionate (Allometric) Growth* — 259
9.7.6 *Diet and Developmental Polymorphism* — 262
9.7.7 *Fluctuating and Directional Asymmetry* — 262
9.7.8 *Number of Molts* — 265

Chapter 10. Molting and Metamorphosis
10.1 Molting — 266
10.1.1 *Sensory Continuity during a Molt* — 269
10.1.2 *Muscular Continuity during a Molt* — 269
10.2 Structural Change during Metamorphosis — 270
10.2.1 *Metamorphosis in Higher Flies* — 270
10.2.2 *Mesodermal Structures* — 273
10.2.3 *Structures of Mixed Germ Layer Origin* — 275
10.2.4 *Ectodermal Structures* — 278

Chapter 11. Specification of the Adult Body Pattern
11.1 Wound Healing — 302
11.2 Regeneration — 302
11.2.1 *The Role of Morphogens in Regeneration* — 302
11.2.2 *The Polar Coordinate Hypothesis* — 304
11.2.3 *The Reaction-Diffusion Hypothesis* — 304

11.2.4 *The Cartesian Coordinate Hypothesis*   306
11.3 Genetic and Molecular Basis of Leg and Wing Formation   306
    11.3.1 *Antigenic Differences around the Circumference and between the Base and*
    Apex of Leg Segments   307
    11.3.2 *Genetic Specification of Circumferential and Radial (Proximodistal) Axes in Leg and*
    Wing Discs   307
11.4 Genetic Specification of Other External Imaginal Primordia during Development of
Drosophila melanogaster   316
    11.4.1 *Antennal versus Leg Development*   316
    11.4.2 *Antennae, Tarsi, Maxillary Palpus, and Proboscis*   316
    11.4.3 *Abdominal Terga, Sterna, and Pleura*   316
    11.4.4 *The Genital Disc*   317
11.5 Effects of Symbiotic Microorganisms   318

Chapter 12. Hormones, Molting, and Metamorphosis
12.1 Classic Experiments Revealing How Hormones Induce Molting   319
12.2 Classic Experiments Revealing How Hormones Induce Metamorphosis   321
12.3 Neurosecretory Cells and Endocrine Organs   323
12.4 The Developmental Hormones   324
    12.4.1 *20-OH Ecdysone*   324
    12.4.2 *Prothoracicotropic Hormone*   336
    12.4.3 *Other Ecdysiotropins*   336
    12.4.4 *Allatotropin*   337
    12.4.5 *Juvenile Hormones*   337
    12.4.6 *Allatostatins*   340
    12.4.7 *Precocenes as Antiallatotropins*   341
    12.4.8 *Eclosion Hormone*   341
    12.4.9 Manduca sexta *Pre-ecdysis– and Ecdysis–Triggering Hormones*   341
    12.4.10 *Bursicon*   342
    12.4.11 *Diapause Hormone*   343
12.5 The Interaction of Hormones during Molting and Metamorphosis   343
12.6 The Role of Critical Size in Molting and Metamorphosis   343
    12.6.1 *Molting*   343
    12.6.2 *Size, Growth, and Differentiation of Wing Imaginal Discs in Lepidopterans*   344
    12.6.3 *Metamorphosis*   344
12.7 Endocrine Control of Molting and Metamorphosis in Other Arthropods   345
12.8 Hormones, Polymorphism, and Polyphenism   345
    12.8.1 *Polyphenism in Nonsocial Insects*   345
    12.8.2 *Polymorphism in Eusocial Insects*   347
12.9 Future Developments   351

Chapter 13. Ontogeny and Hexapod Evolution
13.1 The Origin of Taxonomic Units or Taxa   352
    13.1.1 *The Role of Variation and Natural Selection*   352
    13.1.2 *Speciation*   352
    13.1.3 *The Origin of Higher Taxa: Saltationism versus Gradualism*   353
    13.1.4 *The Origin of Higher Taxa: Adaptive Zones and Key Innovations*   354
13.2 Key Innovations and the Insect Fossil Record   355
    13.2.1 *Key Innovations and the Adaptive Radiation of the Hexapods*   355
    13.2.2 *Earth History and the Origin of the Hexapod Orders*   357
13.3 Ontogeny and Hexapod Evolution   363
    13.3.1 *The "Biogenetic Law" of Müller-Haeckel*   363
    13.3.2 *The Influence of von Baer and Darwin*   364
    13.3.3 *The Role of Heterochrony in Ontogeny and Phylogeny*   365
    13.3.4 *Hormones, Abnormal Metamorphosis, and Macroevolution*   366
    13.3.5 *The Effects of Phyletic Change in Size*   367
    13.3.6 *The Influence of Conserved Cellular Processes*   367
    13.3.7 *The Role of Developmental Constraint*   368
    13.3.8 *Modularity*   369

13.3.9 *Expression of Homeotic and* Distal-less *Genes and the Evolution of Arthropod Limbs*   373
13.3.10 *Fossils, Pattern-Control Genes, Key Innovations, and Hexapod Evolution*   375

References   385

Index   429

Preface

The last three decades have witnessed a revolution in our understanding of insect development, primarily due to effective application of the methods of developmental genetics and molecular biology to embryos and larvae of the well-known laboratory fly *Drosophila melanogaster* and culminating in the publication of its genome in March of 2000 (Adams et al., 2000). Resulting discoveries on the identity and mode of action of genes specifying the development of larval and adult body plans not only have enhanced our understanding of ontogeny in this insect but also have revitalized comparative study and the realization that the progenitors of these genes date back to the Precambrian, their descendant genes serving similar functions in other animals including humans. The fundamental significance of these discoveries to biology was recognized in 1995 by award of the Nobel Prize in medicine or physiology to three key contributors: E. B. Lewis, Christiane Nüsslein-Volhard, and Eric Wieschaus. Along with these revelations in insect ontogeny, the conception of phylogenetic systematics (now called cladistics) by the dipterist Willi Hennig and the invention and use of automated methods for sequencing genes and inferring phylogeny have stimulated vigorous and increasingly successful attempts to synthesize discoveries in animal ontogeny with those in paleontology and evolution. This book is my attempt to do so for insects.

For over 30 years I have taught a one-semester course in insect development to senior undergraduates and graduate students at the University of Alberta and have tried to introduce these discoveries as they were made. Since there was no single source available on insect ontogeny, I have had to use reprints, review articles, and more recently, the Web for supplementary reading. Students have found this frustrating because they had nowhere to go for the "big picture" and because most sources are relentlessly reductionist, limit their coverage to work on widely used model organisms such as *D. melanogaster*, and generally ignore the wonderful diversity of hexapods and their development. Also, because of the increasing emphasis on molecular biology in the modern biology curriculum, more and more students are coming to the course with a sound background in genetics, cell biology, and biochemistry but with a diminished understanding of normal insect structure, development, function, histology, and evolution.

My goal in this book is to synthesize knowledge from the elegant, descriptive, comparative, and experimental studies of the past with the recent exciting discoveries on *Drosophila* and *Manduca* development. I write not only for insect developmental biologists but also for those intrigued by insects' structure, function, behavior, systematics, paleontology, evolution, and pathology and for developmental biologists investigating other organisms. I address the topics of reproductive systems, gametogenesis, sperm transfer and use, fertilization, sex determination, parthenogenesis, embryogenesis, postembryogenesis and metamorphosis, hormones, the quantitative analysis of growth, and the contributions of ontogeny to insect diversification. For each one, I summarize structural events, comparative aspects, and results of experimental analysis, the last presented in chronological order to provide some understanding of how we got to where we are today in our knowledge of the subject. To achieve this breadth of coverage I have had to sacrifice depth. Thus, my treatment of genetic and molecular aspects, though reasonably up-to-date, may seem superficial and naive to the ingenious practitioners of these dark arts.

Committed readers with an introductory knowledge of entomology, cell biology, genetics, and evolution should have no difficulty understanding the text, for I have tried to write it as clearly and simply as possible. Because I have organized the book sequentially, it can be read either from beginning to end or dipped into here and there for information on particular topics, most of which are extensively cross-referenced.

The book can be used as the text for a one- or two-semester course in insect development. In my course, I have laboratory sessions on insect reproductive systems, spermatogenesis, oogenesis, early and late embryogenesis, and imaginal discs and early and late metamorphosis in higher flies. The labs for each topic begin with a study of living material and finish with close examination of histological sections of representatives of several species. Such course work requires access to insect-rearing facilities, a variety of species in culture, and personnel to care for them, to prepare sections, and to ensure that particular, living stages of development are available at appropriate times. The descriptive emphasis of the labs results from my interest in insect functional morphology and development and from my strongly held belief that correct interpretation of experimental results requires detailed understanding of normal development. Since causal analysis is fully covered in the text, an experimental component could easily be introduced into such laboratories by instructors having interests, expertise, and facilities to do so.

Acknowledgments

I am indebted to the enthusiasm, stimulation, and insight provided by two generations of students in my insect development course at the University of Alberta and to the legions of entomologists, paleontologists, and developmental biologists who have provided me with reprints over the years. The book would not have been written without the constant support of my wife Karin and of present and former colleagues at the University of Alberta, particularly George Ball (who constructively criticized the first draft of chapter 13), Doug Craig, Kris Justus, Chris Klingenberg, John Spence, Andy Keddie, and Felix Sperling—all of whom, for months, admonished me to "submit the manuscript now, Bruce; it will *never* be up-to-date!" I thank the late René Cobben, Erwin Huebner, Conrad Labandeira, Jim Nardi, the late Ellis MacLeod, Niels Kristensen, Laurence Mound (a fellow thysanopterist), Klaus Sander, and Petr Svácha for providing inspiration by their example and for sharing their enthusiasm for and encyclopedic knowledge of insects and their development. The late Jake Rempel got me off to a good start early in my career by providing me with an exquisite collection of histological slides recording the complete embryogenesis of the Caragana blister beetle *Lytta viridana* and the black widow spider *Lactrodectus mactans*. I am particularly indebted to John Ewer, Jim Nardi, and Fred Nijhout, whose detailed and constructive criticism of the entire manuscript provided me with loads of additional work that resulted in a better book. And for their numerous suggestions and infinite patience during the endless process of editing and producing the book, I thank Peter Prescott, Louise E. Robbins, Brianna Burke, and Bob Tombs of Cornell University Press and Mary Babcock, copyeditor extraordinaire.

The Natural Science and Engineering Research Council of Canada (NSERC) has supported my work financially for many years, including during preparation of the manuscript, and I am most grateful for it. Finally, I thank Jack Scott of the Department of Biological Sciences of the University of Alberta for the many contributions he has made to my teaching and research in securing specimens, preparing histological sections, and producing graphics and photomicrographs. He scanned, relabeled, and reformatted the illustrations for this book with the eye for detail and rigorous standards he has exhibited in all his work.

Insect Development and Evolution

Introduction

Development and Biology

Little is more captivating to a naturalist than watching a tiny translucent cell, a newly fertilized insect egg, develop first into a satiny green caterpillar that feeds, grows, and molts, and then into an unprepossessing chrysalis from which emerges an iridescent butterfly that spreads its wings, glides across a meadow, visits a flower, meets a conspecific of the opposite sex, and mates to propagate its kind. Study of such change in any organism comprises developmental biology, a discipline that seeks to describe and explain the acquisition of its form and function. Development has a central place in biology since it generates the organisms that fascinate us and encompasses the proximate causes of evolutionary change: for a descendant to differ from its parents, it must alter the way it develops.

Reductionist versus Evolutionary Approaches to the Study of Insect Development

Two kinds of development occur in organisms: ontogenetic and phylogenetic. *Ontogeny* encompasses the processes by which a single, fertilized egg transforms into a sexually mature adult, and *phylogeny*, the origin and diversification of lineages through time. For too long, these phenomena have been examined by biologists totally oblivious of each other: reductionists committed to unravelling the proximate causes of individual ontogenies, and evolutionary biologists investigating the roles of variation, selection, and ontogeny in generating organismal diversity (i.e., the ultimate causes). Key differences in the questions, thinking, and approach of the members of each camp to the study of ontogeny are summarized in Table 0.1.

A reductionist seeks to identify and explain the underlying similarities in organismal development, whereas an evolutionary biologist is fascinated by both similarities and differences and tries to explain how such differences might have arisen. Thus, a reductionist chooses a suitable model organism such as *Drosophila melanogaster* on which to work and considers variation to interfere with experiments and to obscure fundamental elements of ontogeny. An evolutionary biologist, however, delights in the differences and more likely than not is an expert on the systematics, ecology, behavior, zoogeography, paleontology, and/or phylogeny of the organisms on which he or she works.

Today, most research on insect development focuses on some dozen model species, with an emphasis on *D. melanogaster*, and is carried out by thousands of ingenious experimentalists who hope to discover and explain developmental mechanisms at the lowest levels of organization and to uncover information of use in solving the problems of cancer and of inherited abnormalities in humans. Happily, the last decade has seen increasing numbers of developmental biologists expanding their focus to other species in hopes of discovering how universal are the molecular and genetic rules governing animal development. This change in focus was perceived in 1994 in a survey in *Science* in which the world's top developmental biologists rated the connection between development and evolution to be the second greatest unsolved mystery in developmental biology—an observation reflected since in the emergence of evolutionary developmental biology, which has grown rapidly, though not without some teething pains.

Insect Ontogeny: A Summary

Development of an organism following fertilization is usually continuous but, to facilitate description, is generally discussed as if it occurred in stages, in arbitrary slices through the time axis of ontogeny. For an insect these stages are as follows (Fig. 0.1):

- Gametogenesis: Formation of eggs and sperm creates the basis for subsequent development.
- Fertilization: Fusion of sperm and egg nuclei to form a zygote nucleus combines the reshuffled genes of mother and father and activates the egg to commence embryogenesis.
- Cleavage: Following fertilization, several synchronous mitotic divisions follow each other within the yolk. During cleavage, the nuclei of the zygote proliferate and become relatively smaller, and the nuclear-cytoplasmic ratio increases. These divisions are nuclear rather than cellular because most insect eggs are yolk rich. Such eggs are said to be *centrolecithal* because a central yolk mass is surrounded by the blastoderm later in embryogenesis.

Table 0.1 Differences between Evolutionary and Reductionist Biologists in Their Perception and Study of Ontogeny

Topic	Evolutionary biologist	Reductionist
Principal goal	Explanation of the origin, diversity, and relationships of living organisms.	Explanation of the proximate causes of individual ontogenies.
Type of questions	Why questions.	How questions.
Principal causes	Selection acting on genetic and phenotypic variation in members of populations (*externalism*).	Action of proximate genetic and developmental mechanisms on individuals (*internalism*).
Scientific approach	Description and comparison between members of naturally occurring species to discover ontogenetic characters of use in inferring homology and phylogeny. Experimentation secondary.	Testing of hypotheses of cause by experimenting on genetically uniform model organisms reared under controlled conditions in the laboratory.
Hypothesis making	Based on historical scenarios (effects of long-term selection, environmental change, speciation, and extinction on evolution).	Based on mechanistic details of ontogeny.
Hypothesis testing	By tracing predicted effects in the fossil record or in the distribution of new characters or character states on an inferred phylogeny. Experimentation secondary.	By experiment.
Role of genes	Mutation, recombination, and introgression are sources of genetic variation to be acted on by selection (displacement of genetic alleles by others to yield higher fitness).	Controllers and executors of ontogenetic processes.
Importance of variation	Substrate on which selection acts to generate diversity and evolutionary change.	A "pain": variation interferes with the execution of clean, reproducible experiments.
Time scale	Tens of thousands to hundreds of millions of years.	Seconds to months.
History	Phylogeny.	Cell lineages within individual embryos.

Source: Modified from Raff, 1996: 21.

- Blastoderm formation: In most insect eggs, cellularization does not occur until the cleavage nuclei and their surrounding cytoplasm, collectively, the *cleavage energids*, enter a thin superficial layer of cytoplasm, the *periplasm*, beneath the egg's oolemma. The cell membrane then invaginates between adjacent nuclei to form an epithelium around the yolk called the *blastoderm*, the hexapod equivalent of the blastula or blastosphere in echinoderm and amphibian embryos.
- Germ band formation: Blastodermal cells on either side of the egg converge posteroventrally to form the germ band (embryo), while those remaining persist about the yolk as a thin, extraembryonic serosa.
- Anatrepsis, gastrulation, germ layer formation, and germ band elongation: The embryo rapidly elongates through cellular proliferation and movement and is pulled posteriorly into the yolk during *anatrepsis* so that its anteroposterior and dorsoventral axes become temporarily reversed from those of the egg, in a posi-

tion called *intertrepsis*. Simultaneously, cells in its ventral longitudinal midline invaginate (*gastrulate*) to form an inner layer of mesoderm, with cells at its two ends becoming endoderm while those remaining on the surface become ectoderm.
- Segmentation and appendage formation: Ectoderm and mesoderm of the embryo now begin to segment transversely into 20 body segments. Soon after, four cephalic, three thoracic, and one or more abdominal segments each develop a pair of appendages, which subsequently become the antennae, mouthparts, legs, abdominal appendages (if present in the juveniles that hatch from the eggs), and cerci of the insect.
- Organogenesis: Simultaneously, cells of the three germ layers begin to assemble into organ primordia.
- Katatrepsis, dorsal closure, and differentiation: The embryo re-emerges from the yolk in a reverse movement called *katatrepsis* to resume its former position on the ventral side of the egg. With subsequent development, the sides of the body grow upward around the diminishing yolk on either side and meet on the dorsal

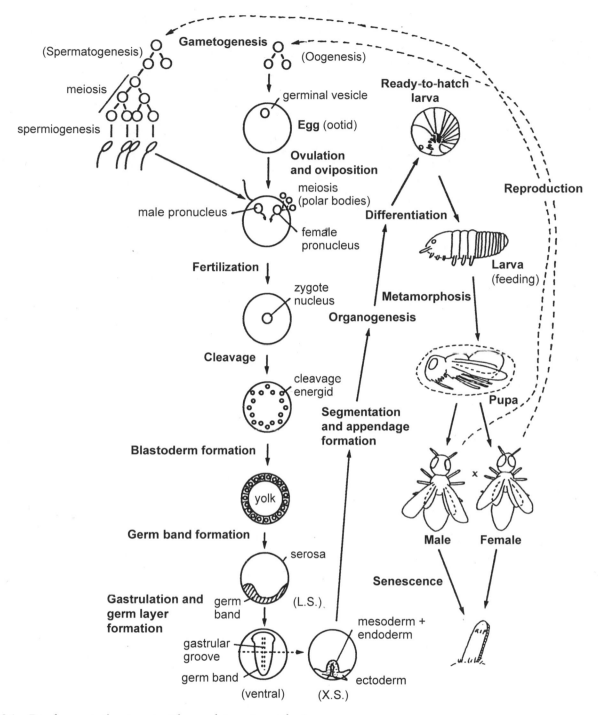

Fig. 0.1. Development of an insect with complete metamorphosis.

midline, a process called *dorsal closure*. The cells of each primordium continue to differentiate and gradually acquire the structure and function they will have at the time of hatch.

• Hatching, postembryogenesis, and metamorphosis: The young insect hatches when most of the yolk in the egg has been mobilized and when it is capable of fending for itself. At hatch, all insects are small and sexually immature, lack wings, and may

differ more or less from their adults in form, food, habitat, and behavior. As it grows, a young insect periodically sheds and replaces its exoskeleton in a series of molts, each mediated by hormones. Because juvenile and adult have more or less diverged from each other in form and function during evolution—the juvenile for more efficient feeding, the adult for dispersal and reproduction—the juvenile must change its form (i.e., metamorphose) to that of the

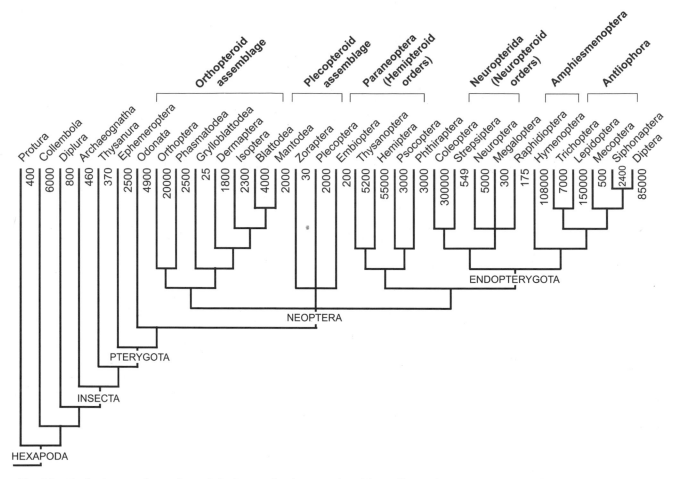

Fig. 0.2. Evolutionary relationships of the hexapod orders as inferred by Gullan and Cranston (1994), indicating their known diversity (from various sources). (Adapted with permission from P. J. Gullan, and P. S. Cranston. *The insects. An outline of entomology*, p. 193, Fig. 7.5. 1st ed. Chapman and Hall, London. © 1994 Blackwell Science, Ltd.)

adult, the amount of change varying with the species. These events are also under the control of hormones.

- Senescence: After the adults have mated and reproduced, they are redundant so far as subsequent evolution is concerned, and they die.

Most insects are developmentally plastic and can replace or repair certain body parts that have been injured or removed. This capacity is greatest early in embryogenesis and is steadily lost as the insect ages. Detailed consideration of these topics, throughout the hexapods, and from both historical and causal perspectives, constitutes the contents of this book.

A Note on Phylogeny

Throughout this book I use a phylogenetic diagram to illustrate similarities and differences in the ontogeny of investigated representatives distributed throughout the hexapod orders. The tree shown in Fig. 0.2 was recon-

structed by Gullan and Cranston (1994) using cladistic techniques in which those insects sharing derived characters or character states are grouped together into the same order. For example, juveniles and adults of all 55,000 known species of the order Hemiptera share a unique, complex, and highly specialized feeding mechanism that consists of long, co-adapted mandibular and maxillary stylets, efficient cibarial and salivary pumps, and a complex of salivary glands and enables them to feed on the vascular tissues of plants or on the blood of other animals (Fig. 8.49). And adults of most of the 85,000 known species of true flies (order Diptera) bear a pair of functional mesothoracic wings and two metathoracic halteres derived from the hind wings of a four-winged, mecopteroid ancestor (Fig. 7.19). Both the mouthparts of bugs and the wings and halteres of flies are sources of structural disparity between the members of these and other orders and important components of the body plans of bugs and flies.

If such an inferred phylogeny represents the true course evolution has taken to generate the orders we recognize today (there is still major argument concerning the placement of the apterygote, paleopterous, orthopteroid, and neuropteroid orders and of the Hymenoptera and Strepsiptera), and if a sufficiently broad selection of species has been examined within each order to encompass its diversity and to allow one to generalize (unfortunately, yet to be realized for any order for any developmental event), then we can use it to trace the origin and direction of change in the developmental processes occurring in insects, their underlying genetic and molecular bases, and the influence these may have had on diversification. And if the phylogeny is supported by a rich fossil record, as is true for some groups of insects (Fig. 13.1), then it is possible to pin down the minimum time of origin of those features defining extant body plans and to trace their modification within the members of each lineage.

Unfortunately, problems persist in the phylogenetic analysis of living and fossil taxa. They have to do with some of the assumptions of the methods currently used to infer relationship, to polarize character states (i.e., to decide whether they are primitive or derived), and to reconstruct the distribution of structural and molecular characters (Jenner and Schram, 1999). Published cladistic studies differ in their methods of character selection, coding, scoring, and weighting; in ground plan reconstruction; and in the selection of study taxa. These differences are rarely made explicit, making such studies difficult to compare. Such weaknesses should be kept in mind when examining comparative information on the trees.

1/ The Male Reproductive System and Spermatogenesis

It seems appropriate to begin an analysis of how insects develop by considering the organ systems responsible for generating, nurturing, and bringing together for fusion the gametes linking one generation to the next. In insects, the reproductive system is unique among organ systems in the body because it

- is sexually dimorphic;
- usually functions only in adults;
- is not required for survival of the individual;
- generates and nurtures the gametes (germ plasm) from which members of subsequent generations develop; and, at least in males,
- is the site of meiosis, the two cell divisions so important in generating genetic variability in bisexual species.

The basic organization of the reproductive system is practically identical in males and females (Matsuda, 1976; Kaulenas, 1992). In both it comprises a pair of gonads, a pair of primary exit ducts, a common secondary exit duct, and external genitalia that function to transfer sperm to females (male) or to deposit eggs (female). The gonads and primary exit ducts originate before hatch from mesoderm; the secondary system and external genitalia usually develop after hatch, either by invagination (secondary exit system) or evagination (genitalia) from surface ectoderm. Structures of ectodermal origin can be recognized in adults because they are lined with or covered by cuticle, a characteristic one can evaluate even in pinned specimens decades after they were collected.

1.1 Male Reproductive Systems

The principal components of the reproductive system in male insects are the testes, vasa deferentia, ejaculatory duct, and external genitalia (Fig. 1.1A). Though all components are usually present in adults, they differ greatly in relative position, size, shape, and number between males of different species (Matsuda, 1976; Kaulenas, 1992).

1.1.1 TESTIS
Each testis contains one or more sperm tubes or follicles, each equivalent to an egg tube or ovariole of the female's ovary (Fig. 1.1B, C). These range in number

from one in adult male thrips and higher flies, to hundreds in certain mayflies, stoneflies, termites, grasshoppers, beetles, and bees (Fig. 1.2). The tubes can lie loose in the body cavity (as in some stoneflies), can be bound together tightly by fat body cells and tracheoles (as in grasshoppers and cockroaches), or are invested in a common epithelium bounded inside and out by basal laminae (as in lepidopterans and some bugs) (Matsuda, 1976). The outer cellular layer of the testis usually contains pigment granules of characteristic color (e.g., yellow, orange, red, brown, green, or violet), which are thought to protect the developing gametes from ultraviolet (UV) radiation. Though usually paired, the testes of each side fuse together during metamorphosis in male lepidopterans and in some stoneflies and grasshoppers.

Each sperm tube (Fig. 1.1B, C) is round to elongate, is bounded externally by one or two layers of epithelial cells, opens proximally into the male primary exit duct by a short exit duct or *vas efferens* (plural: vasa efferentia), and may be divided internally by septa, as in certain trogine and geotrupine scarab beetles (Friedländer et al., 1999; Martínez and Cruz, 1999). In locusts, an outer perifollicular layer is sandwiched between basal laminae while an inner parietal layer of cells with complex, overlapping processes is well developed proximally (Szöllosi, 1982). Walls of the testes and sperm tubes have been shown by experiment to be differentially permeable to blood-borne factors in certain moths and locusts, this being influenced by the presence and concentration of the molting hormone, 20-OH ecdysone (20E) (1.3.7.2).

Within a fully mature sperm tube can usually be distinguished successive regions in which the germ cells are at progressively more advanced stages of development (Fig. 1.3). At the apex is a germinal proliferation center or *germarium* containing stem spermatogonia and cyst progenitor cells surrounding, in some insects, an apical cell or apical complex (Roosen-Runge, 1977; Szöllosi, 1982). A zone of multiplication follows containing clones of interconnected, synchronously dividing secondary spermatogonia, each within a sheath of mesodermal cyst cells. In a zone of growth, the spermatogonia within a clone cease dividing, enlarge into primary spermatocytes, and enter meiotic prophase. These undergo the two meiotic divisions in a zone of maturation, producing secondary spermatocytes after

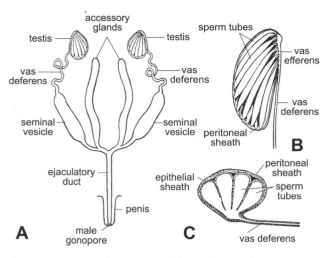

Fig. 1.1. A. General structure of the male reproductive system in an insect. B. Structure of a testis. C. Sagittal section of a testis with four sperm tubes, showing their relationship to the vas deferens. (Reproduced from R. E. Snodgrass. 1935. *Principles of insect morphology*, p. 568, Fig. 7.5. McGraw-Hill. Figures owned by U.S. Department of Agriculture, Bureau of Entomology and Plant Quarantine.)

the first division and spermatids after the second. The clones of spermatids transform into spermatozoa within a proximal zone of transformation, the sperms in each cyst accumulating as a sperm bundle at the base of the tube in a zone of spermatozoa.

In long-lived males of larger anautogenous (1.3.7.1) insects (Fig. 1.3), all zones are usually present in each sperm tube after adult emergence so that all stages of spermatogenesis can be examined within one individual. In testes of the tiny short-lived males of terebrantian thrips, only mature sperm are present, and all stages of the life cycle must be examined since each testis has only one sperm tube containing a single clone of germ cells (Fig. 1.2) (Heming, 1995). Other insects range between these two extremes.

1.1.2 PRIMARY EXIT SYSTEM

The primary exit duct of each testis, the *vas deferens* (plural: vasa deferentia) (Fig. 1.1), lacks a cuticular lining in most species, is thick walled, and is surrounded by circular muscle fibers. In members of some species, it is coiled along its length into an epididymis and in most, is expanded into a seminal vesicle (vesicula seminales) in which the sperm are stored after their release from the testis. Mesodermal accessory glands (*mesadenes*), if present, generally develop postembryologically by evagination from rudiments of the vasa in juveniles of most species (10.2.3.1) but in some are reduced to secretory cells within their walls (Matsuda, 1976; Kaulenas, 1992). In larvae and pupae of higher Diptera, a primary exit system seems not to develop, and the testes are serviced by distal outgrowth of the secondary exit system from a median, ectodermal genital disc.

1.1.3 SECONDARY EXIT SYSTEM

The vasa deferentia open posteriorly into a common ejaculatory duct (*ductus ejaculatorius*) (Fig. 1.1). This is lined with cuticle continuous posteriorly with that of the body wall and is surrounded by powerful circular and sometimes longitudinal muscles. Its posterior opening, the *male gonopore*, is usually located within the male intromittent organ or *phallus* contained within the ninth abdominal segment (Snodgrass, 1935, 1957). A set of ectodermal accessory glands, the *ectadenes*, opens into the anterior end of the ejaculatory duct of males of some species. They can be differentiated from mesadenes by their lining of cuticle. Both mesadenes and ectadenes are active prior to copulation (Kaulenas, 1992) and discharge a liquid or viscous substance with the sperm that sustains them or forms a sperm sac or *spermatophore* about them (3.3.2.1). In addition, their products have been shown to assist in sperm transfer and activation, to prevent just-mated females from mating with conspecific males, to influence subsequent copulatory behavior of females, or to stimulate egg maturation and oviposition (3.2, 3.3, 3.6, 3.7).

1.1.4 EXTERNAL GENITALIA

The external genitalia of male insects function to transfer sperm to the female during copulation. The intromittent organ or phallus differentiates from a pair of ectodermal, primary phallic lobes at the back of abdominal sternum 9 and, in adults, is usually concealed within an ectodermal invagination from the back of the ninth segment, the genital chamber (Fig. 10.12). In males of other species, periphallic organs that function to hold or stimulate the female during copulation originate close enough to the phallus to be considered part of the external genitalia (Snodgrass, 1957).

Male mayflies (Ephemeroptera) and some earwigs (Dermaptera) have paired penes, each with its own ejaculatory duct, but in those of other pterygotes the phallus comprises a complex of heavily sclerotized lobes or phallomeres (e.g., cockroaches, mantids) or forms a median, partially sclerotized telescoping tube. In males of most species, the phallus consists of a proximal, sclerotized phallobase and a more slender, terminal aedeagus (Snodgrass, 1935, 1957; Tuxen, 1970). The external walls of both comprise the ectophallus and enclose a membranous endophallus invaginated from the tip of the organ. The male gonopore opens into the lumen of the endophallus and is either eversible or not, depending on species. In males of most species, the phallobase bears a pair of sensory parameres laterally whose sensory neurons project to the terminal abdominal ganglion of the ventral nerve cord. These parts, and others not mentioned, differ enormously in complexity, shape, and size and are an important source of characters for discriminating between the males of different species (Snodgrass, 1957; Tuxen, 1970; Matsuda, 1976; Eberhard, 1985).

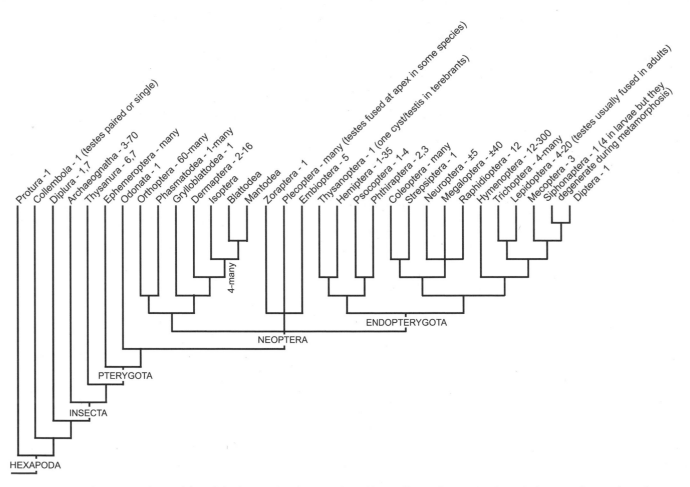

Fig. 1.2. Evolutionary relationships of the hexapod orders as inferred by Gullan and Cranston (1994), showing the number of sperm tubes per testis. (Data from Matsuda, 1976, and Roosen-Runge, 1977.) Cladogram adapted with permission from P. J. Gullan and P. S. Cranston. *The insects. An outline of entomology*, p. 193, Fig. 7.5. 1st ed. Chapman and Hall, London. © 1994 Blackwell Science, Ltd.

1.2 Insect Spermatozoa

Animal sperm are characteristically "stripped down" cells with their structure adapted

- to their mode of transmission from male to female;
- to their duration of survival in the male and female reproductive tracts or within the spermatophore or spermatheca;
- to the nature and viscosity of the fluid they swim in;
- to the complexity of the egg envelopes they must penetrate; and
- as a result of selection for competitiveness in sperm competition systems.

Most insect sperm have evolved in the direction of increased motility.

1.2.1 STRUCTURE

In males of most insect species, living sperm resemble long, delicate, rippling hairs when observed in vitro with a phase contrast microscope (Fig. 1.4). From front to back they generally contain the following elements (Fig.

1.5), whose details are resolvable only by transmission electron microscopy:

- Acrosome: Though yet to be demonstrated experimentally in insects, the acrosome is known to contain hydrolytic enzymes (lysins) that enable the sperm to penetrate the egg's envelope (vitelline membrane), a process understood in great detail in the eggs and sperm of some echinoderms and mammals (Alberts et al., 1994).
- Head (nucleus): The nucleus contains the genetically inactive DNA of the meiotically reshuffled chromosomes complexed with lysine-rich (not arginine-rich as in somatic chromatin) histone proteins, and in sperm of most species, has a paracrystalline structure when fully differentiated (Phillips, 1970). The head is long, slender (diameter 0.2–0.5 μm), and stiff in the sperm of most species (Fig. 1.4).
- Centriole: In most animal sperm there are two centrioles: proximal (C1) and distal (C2), with C1 at right angles to C2 and to the long axis of the sperm. Fully differentiated insect sperm, however, usually have

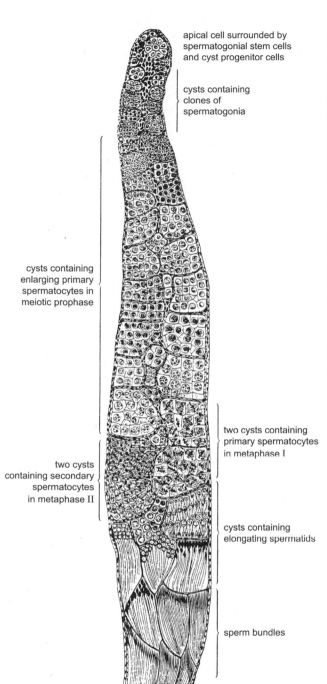

apical cell surrounded by
spermatogonial stem cells
and cyst progenitor cells

cysts containing
clones of
spermatogonia

cysts containing
enlarging primary
spermatocytes in
meiotic prophase

two cysts containing
primary spermatocytes
in metaphase I

two cysts
containing secondary
spermatocytes
in metaphase II

cysts containing
elongating spermatids

sperm bundles

Fig. 1.3. Sagittal section of a sperm tube in the testis of a grasshopper (Orthoptera), showing clones of synchronously developing male germ cells. There is a general progression from early stages at its apex to fully differentiated sperm bundles at its base. (Reprinted with permission from H. Weber. *Grundriss der Insektenkunde*, p. 126, Fig. 85. 2nd ed. © 1966 G. Fischer Verlag.

only centriole C2 (Fig. 1.11E) because of the way these organelles are distributed during meiosis (Fig. 1.10), and even this is usually visible only in developing spermatids (Fig. 1.11A–D).

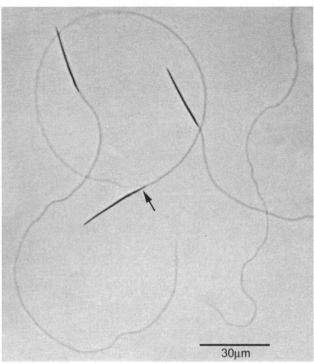

30μm

Fig. 1.4. Phase contrast photomicrograph of three Feulgen-stained spermatozoa of the fly *Megaselia scalaris* (Diptera: Phoridae), showing their characteristic threadlike shape. (Ultrastructure of the spermatozoon of *Megaselia scalaris* Loew (Diptera: Brachycera, Cyclorrhapha: Phoridea: Phoridae), by S. K. Curtis, D. B. Benner and G. Musil, *J. Morphol.*, 200: 47–61, Copyright © 1989 Wiley-Liss, Inc. Reprinted by permission of Wiley-Liss, Inc., a subsidiary of John Wiley & Sons, Inc.)

- Centriolar adjunct: An electron-dense material surrounds the centriole in some insect sperm and has been suggested to anchor the axoneme to the head (Fig. 1.11C–E). It is not present in all species.
- Mitochondrial derivatives: One (Figs. 1.5, 1.11) or two (Fig. 1.12) long, highly ordered mitochondrial derivatives extend the length of the axoneme from the base of the head to near the end of the flagellum with a structure characteristic for the species (Jamieson, 1987; Jamieson et al., 1999). The cristae characteristic of mitochondria are much reduced in the sperm of most species and are ordered in a regular array along its long axis, with most of the derivative filled with a paracrystalline material containing the protein crystallomitin (Baccetti, 1979, 1998). This protein seems to be devoid of enzymatic activity and may assist in maintaining and stabilizing flagellar beat during swimming. In spite of their highly specialized structure, the derivatives still function in oxidation, electron transport, and phosphorylation and supply the ATP required for sperm motility (Alberts et al., 1994).
- Axoneme: The axoneme in insect sperm flagella is very long and commonly has a 9 + 9 + 2 microtubular

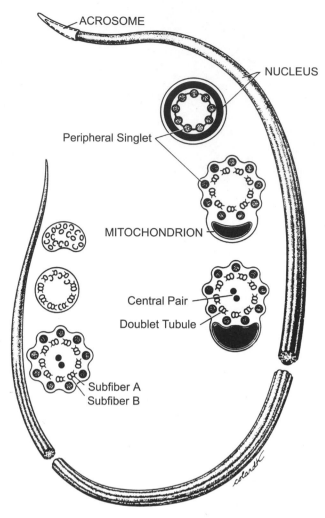

Fig. 1.5. Spermatozoon of the caddisfly *Neuronice* sp. (Trichoptera), showing the principal features common to insect sperm. The breaks indicate where long sections of the cell have been omitted. (Reproduced with permission from D. M. Phillips. *Spermiogenesis*. © 1974 Academic Press, Inc.)

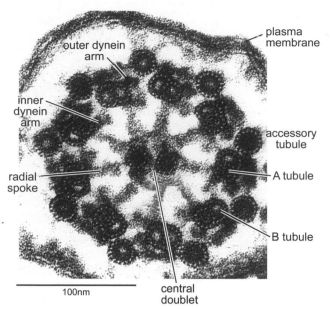

Fig. 1.6. Transmission electron micrograph of a transverse section of the sperm flagellum of *Lepismodes inquilinus* (Zygentoma: Lepismatidae), showing the 9 + 9 + 2 arrangement of microtubules within its axoneme. (Reproduced with permission from R. Dallai and B. A. Afzelius. Microtubular diversity in insect spermatozoa: results obtained with a new fixative. *J. Struct. Biol.* 103: 164–179. © 1990 Academic Press, Inc.)

formula as seen in transverse section: nine single, outer accessory tubules, nine peripheral doublets, and a central doublet (Figs. 1.5, 1.6). Only collembolans, japygid diplurans, scorpionflies, fleas, and some fungus gnats (Diptera: Mycetophilidae) are known to have 9 + 2 axonemes (i.e., with accessory tubules absent) (Fig. 1.8). These microtubules consist of the proteins alpha- and beta-tubulin typically organized into either 13 (A tubule) or 10 (B tubule) protofilaments (Fig. 1.6). However, the number of protofilaments per accessory tubule in insect sperm is known to vary from 13 to 19 depending on taxon, and to be useful in phylogenetic analysis (Dallai and Afzelius, 1993; Baccetti, 1998).

Each tubulin molecule consists of 450 amino acids and, in the fly *Drosophila melanogaster*, is encoded by a family of closely related tubulin genes, four each for each protein. The enzyme ATPase occurs in the dynein arms of the A tubule of each peripheral doublet (Figs.

1.6, 1.7), and much of the energy for sperm motility arises from action of this enzyme on ATP supplied by oxidation from the mitochondrial derivatives (Fig. 1.7) (Alberts et al., 1994). Other proteins, including tektin, nexin, and radial spoke proteins, are present in various additional structures of the axoneme.

Axonemes with the 9 + 2 arrangement are typical of sperm in animals reproducing by external fertilization in seawater (Wolfe, 1981). Accessory tubules or fibers (as in mammalian sperm) occur almost exclusively in animals having internal fertilization as do insects, suggesting that they provide additional power for sperm movement through viscous fluids in the reproductive tracts of females (Wolfe, 1981).

- Plasma membrane: Insect sperm are surrounded by a plasma membrane that is usually unembellished (Fig. 1.6). However, the normal, nucleate eupyrene sperm of lepidopterans bear an elaborate radial array of short laciniate appendages on their surfaces that are characteristic for the species (Fig. 3.4A), while those of certain grasshoppers bear radially striated fibrils (Jamieson, 1987).

The characteristic threadlike shape of insect sperm (Fig. 1.4) probably results from the terrestrial lifestyle of most insects (Davey, 1965). Because of surface-volume relationships and the generally small size of insect eggs, they are well waterproofed, usually with a wax layer between the vitelline membrane and chorion (shell), and

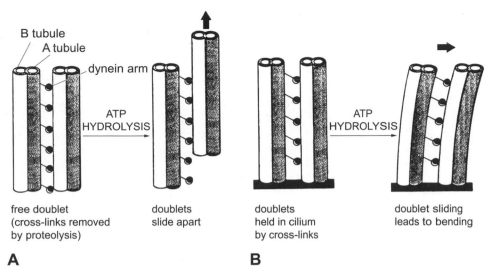

Fig. 1.7. The sliding of outer microtubule doublets against each other (A) causes bending if the doublets are tied to each other at one end (B). The base of the dynein molecule attaches only to the A tubule, leaving the heads free to interact with the adjacent B tubule. Apparently, the different structure of the B tubule prevents the base of a dynein molecule from binding to it. The resulting asymmetry in arrangement of dynein molecules is required to prevent a fruitless tug-of-war between neighboring microtubules, which probably explains why each of the nine outer tubules is in an A/B doublet. (© 1989 from *Molecular biology of the cell*, 3rd ed., by B. Alberts et al. Reproduced by permission of Routledge, Inc., part of The Taylor & Francis Group.)

have very small (0.3–0.6-μm-diameter) holes (*micropyles*) in their shells for sperm entry (the chorion is deposited before fertilization while the egg is still within its follicle in an ovariole [2.2.1.6]). Sperm must be very slender (0.2–0.5-μm diameter) if they are to pass through these holes. That sperm of the American cockroach *Periplaneta americana* are relatively short and have short, thick heads correlates with the eggs of this species being enclosed within an ootheca, affording partial protection from desiccation. The micropyles of eggs of this insect are probably larger.

1.2.2 MOTILITY

Flagellar movement during swimming results from action of a sliding microtubule mechanism (Brokaw, 1989), similar to the sliding filament mechanism functioning in muscle contraction and first proposed by Afzelius in 1959 and Satir in 1965. Satir (1968), studying various echinoderm and mammalian sperm, demonstrated that adjacent peripheral doublets of the 9 + 2 axoneme have fixed lengths and slide past each other when the axoneme bends. Sliding is induced through interaction of ATP with the dynein arms, which form transient cross-bridges between adjacent doublets (Fig. 1.7) and function as oscillating force generators moving the adjacent B tubule and its doublet toward its plus (growing) end (Shingyogi et al., 1998). ATP concentration, temperature, and number of dynein arms on the A tubules contribute to beat frequency (Wolfe, 1981).

In living *Tenebrio molitor* (Coleoptera: Tenebrionidae) sperm, the 9 + 9 + 2 axoneme has a cylindrical wave type consisting of two superimposed, helical waves (Baccetti and Afzelius, 1976). The large wave is 20–30 μm long, has an amplitude of 9–15 μm, beats 0.9–2.8 times/sec, and is propagated posteriorly at a speed of 20–90 μm/sec. The small wave is 6–12 μm long, has an amplitude of 3–4 μm, beats 7–28 times/sec, and moves posteriorly at 40–300 μm/sec. These waves propel the spermatozoon forward at speeds of 16–100 μm/sec depending on temperature and fluid viscosity. And sperm of the phorid fly *Megasalia scalaris* can move both forward and backward and can adjust their speed within wide limits (Curtis and Benner, 1991). They also swim up to ninefold faster in natural fluids of the male and female reproductive tracts than in artificial ones such as physiological saline.

1.2.3 PHYLOGENETIC ASPECTS

Figure 1.8 is a reconstructed phylogeny of the insect orders, with the known sperm form and axonemal microtubular formula indicated for members of each order (Jamieson, 1987; Baccetti, 1998; Jamieson et al., 1999). In most orders, sperm of only a few species have been examined (noted), and additional forms undoubtedly exist. Note that in members of some taxa (Collembola, and certain Protura, Isoptera, Thysanoptera, Hemiptera, Trichoptera, and Diptera), sperm motility has been lost (indicated by J′–J‴, H, and I; in figure), while in others (Hemiptera: Coccoidea; Diptera: Cecidomyiidae), it has been lost and secondarily regained (indicated by L). If dynein arms are absent from A microtubules of the peripheral doublets, the sperm

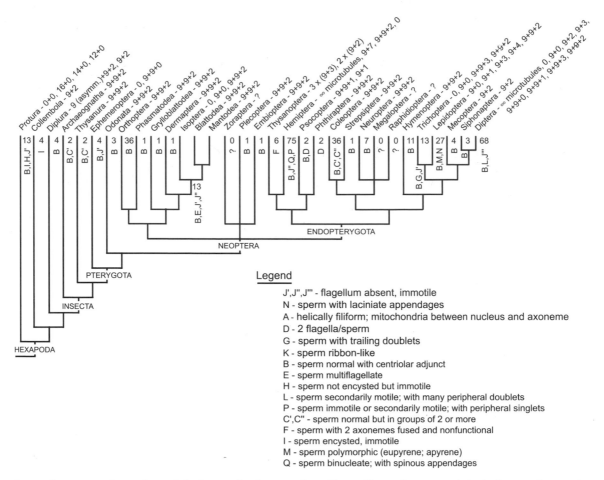

Fig. 1.8. Evolutionary relationships of the hexapod orders as inferred by Gullan and Cranston (1994), showing known diversity in sperm structure. Known axonemal microtubular arrangements are presented above each order name. Numbers below these names indicate the number of species examined in each order by 1987. Other sperm characteristics and their distribution are explained in the legend. (Data from Jamieson, 1987, and other sources. Cladogram adapted with permission from P. J. Gullan and P. S. Cranston. *The insects. An outline of entomology*, p. 193, Fig. 7.5. 1st ed. Chapman and Hall, London. © 1994 Blackwell Science, Ltd.)

are immotile. Sperm of other insects (some Psocoptera, Phthiraptera, Thysanoptera [two or three fused axonemes], and Heteroptera) characteristically have two or more axonemes in their flagella (indicated by D and F), while those of others (Archaeognatha, Thysanura, some Coleoptera) are released in pairs (C′ and C″ in figure).

Investigated lepidopterans (M in figure) produce two kinds of sperm (discovered by Meves in 1903): *apyrene* (anucleate) and *eupyrene* (normal, nucleate). Significantly, adult males of the phylogenetically basal, mandibulate micropterygid moths *Micropteryx calthella* and *M. aruncella* produce only eupyrene sperm (Sonnenschein and Hauser, 1990).

The lengths of insect sperm and sperm heads can vary within and between species (Bircher and Hauschteck-Jungen, 1997; Presgraves et al., 1999; Morrow and Gage, 2000). In known *Drosophila* spp., for example, they are from 55 μm to 58 mm in length, and single males of

some species can generate up to three size classes (*polymegaly*) (Jamieson, 1987; Pitnick et al., 1995b, 1999; Karr and Pitnick, 1996). Males of the drosophilid *D. bifurca* produce sperm over 58 mm long (20-fold longer than the male's body), the longest recorded in the animal kingdom, and the testes in which they develop constitute 11% of the fly's body mass (Pitnick et al., 1995a). The adaptive significance of such giant sperm is unknown, but Pitnick and coworkers (1995) speculated that their longer flagella may confer a competitive advantage in fertilizing eggs or contribute to nutritional support of the embryo (the latter suggestion now seems unlikely [Karr and Pitnick, 1996]). And at least in hebrid bugs (Heming-van Battum and Heming, 1989), ptiliid beetles (Dybas and Dybas, 1981), moths (Morrow and Gage, 2000), *Drosophila* spp. (Pitnick et al., 1999), and diopsid flies (Presgraves et al., 1999), sperm length is known to correlate with the structure of the female reproductive system, particularly the size or length of

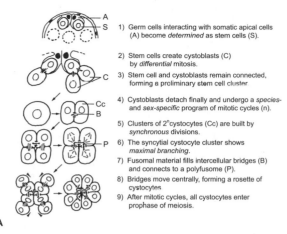

1) Germ cells interacting with somatic apical cells (A) become *determined* as stem cells (S).

2) Stem cells create cystoblasts (C) by *differential* mitosis.

3) Stem cell and cystoblasts remain connected, forming a proliminary stem cell cluster.

4) Cystoblasts detach finally and undergo a *species-* and *sex-specific* program of mitotic cycles (n).

5) Clusters of 2^n cystocytes (Cc) are built by *synchronous* divisions.

6) The syncytial cystocyte cluster shows *maximal branching.*

7) Fusomal material fills intercellular bridges (B) and connects to a polyfusome (P).

8) Bridges move centrally, forming a rosette of cystocytes.

9) After mitotic cycles, all cystocytes enter prophase of meiosis.

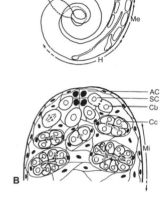

Fig. 1.9. A. Early events in spermatogonial cluster formation within an insect sperm tube. B. Top: Diagram of the mature, coiled testis of a male *Drosophila melanogaster* (Diptera: Drosophilidae). Cysts containing growing primary spermatocytes (G), spermatocytes in meiosis (Me), and spermatids undergoing spermiogenesis (H) are outlined. Bottom: Enlargement of the testis apex showing apical cells (AC), a stem cell (SC) still connected to a cystoblast (Cb), and cystocytes (Cc; definitive spermatogonia) within cysts, each enclosed by two cyst cells. Polyfusomes (stippled) connect all spermatogonia within each cyst. Mi, spermatogonial mitosis. (Reprinted from *Int. J. Insect Morphol. Embryol.*, 22, J. Büning, Germ cell cluster formation in insect ovaries, pp. 237–253, Copyright 1993, with permission from Elsevier Science.)

the spermathecae and/or seminal receptacle or, in moths, the length of the spermathecal duct (Morrow and Gage, 2000).

Males of certain tropical pentatomid bugs have a specialized sperm tube in each testis that typically generates aberrant spermatozoa (Schrader, 1960a, 1960b). These sperm and the meiotic divisions producing them are so bizarre that the aberrant sperm tube is referred to as the "harlequin lobe." The sperm produced by this lobe can vary in size and in number of chromosomes and are thus unlikely to generate viable offspring with fertilization. Schrader suggested that they may provide

additional nutrient to the egg if they succeed in entering it.

Figure 1.8 surveys known sperm axonemal microtubular configurations throughout the Hexapoda. Most insects produce sperm with a $9 + 9 + 2$ axoneme, as summarized earlier. However, representatives of some orders have additional patterns, particularly the Hemiptera, Lepidoptera, and Diptera (but note that more species have been examined in these three orders). Among the most unusual are the multiflagellate sperm of the basal Australian termite *Mastotermes darwiniensis*, the aflagellate sperm of certain leptophlebiid mayflies (Ephemeroptera) (Gaino and Mazzini, 1991), and the bizarre flagellum of the the western flower thrips *Frankliniella occidentalis* and other thysanopterans (Dallai et al., 1991). In male scale insects (Hemiptera: Coccoidea), the sperm have no recognizable head, middle piece, or tail; lack centrioles and mitochondrial deriviatives; have 20–410 microtubules arranged into longitudinally disposed, concentric whorls or spirals about the central, threadlike nucleus; and yet are still motile.

Comparative knowledge of sperm structure in animals indicates that they are among the most rapidly evolving of known cell types and that their characteristics are as likely to be related to function as to phylogeny.

1.3 Spermatogenesis

Cellular processes generating the spermatozoa occur in three stages: primordial development, prespermatogenesis, and spermatogenesis. Spermatogenesis, in turn, includes premeiotic events, meiosis, and spermiogenesis.

1.3.1 PRIMORDIAL DEVELOPMENT

Sperm cell progenitors originate within the embryo as pole cells or *primordial germ cells* at various sites and times, depending on species; are asynchronously mitotic; and are usually unrecognizable as to sex (6.6). With subsequent development, they and their progeny separate laterally into two groups, become surrounded by mesodermal cells to form the testicular rudiments, and become spermatogonial stem cells (8.4.1).

1.3.2 PREMEIOTIC EVENTS

1.3.2.1 GERMINAL PROLIFERATION CENTER

At the apex of each sperm tube or testis (if only one sperm tube is present, as in higher Diptera) are one or more, nondividing apical cells surrounded by stem spermatogonia and cyst progenitor cells (Figs. 1.3, 1.9) (Hardy et al., 1979; Szöllosi and Marcaillou, 1979; Büning, 1993). The apical cells, in males of *D. melanogaster*, are located within a hub of mesodermal origin, with the other two cell types alternating around it (Fig. 1.9B: bottom). Each stem spermatogonium and cyst progenitor cell maintains contact with the apical complex throughout the cell cycle in such a way that with each asymmetrical mitosis, one daughter cell remains associated with the hub as a stem cell and the other becomes a spermatogonium or a cyst cell. Somatic cells surrounding the stem cells ensure their renewal as stem cells after each mitosis (Kiger et al., 2000; Tran et al., 2000).

Within each stem spermatogonium is a spherical organelle called a *spectrosome* containing the membrane-associated proteins spectrin, ankyrin, and Hu-li tai shao (Hts) (compare Fig. 2.11) (Deng and Lin, 1997; Lin and Schagat, 1997). During each stem cell division, the spectrosome anchors one pole of the spindle in such a way that the axis of each mitosis is parallel to the long axis of the tube. In stocks of *Drosophila* in which the germ cells are missing (*chickadee* and *diaphanous* mutants), the daughter cyst cells continue to proliferate and some become apical cells, showing that signaling from spermatogonia regulates proliferation and the fate of cyst cells and has some role in directing the size and position of the hub (Gönczy et al., 1992; Gönczy and DiNardo, 1996).

1.3.2.2 SPERMATOGONIAL PROLIFERATION

Indefinitive spermatogonia divide asynchronously and comprise the generations of spermatogonia between the stem cells and the definitive spermatogonia occurring within cysts; they are absent in males of some species (King et al., 1982). Eventually, each indefinitive spermatogonium becomes surrounded by two or more cyst cells (Fig. 1.9B: bottom), becomes a definitive spermatogonium (a cystoblast or gonialblast), and begins to divide synchronously but incompletely a species-specific number of times, usually in some multiple of two (2^n where n is the number of mitotic divisions). The number of spermatogonial divisions known to occur in cysts of male insects is 2^9–2^{14} in odonates, 2^4–2^9 in orthopterans, 2^3–2^8 in hemipterans, 2^3–2^6 in neuropterans, 2^5–2^7 in beetles, 2^5 in caddisflies, 2^5–2^7 in lepidopterans, and 2^3–2^6 in flies (Virkki, 1969; Phillips, 1970; Roosen-Runge, 1977). These divisions ultimately generate clones of interconnected primary spermatocytes of characteristic number (8–16, 384/cyst). Exceptions to this mitotic synchrony are known in several *Drosophila* species (Liebrich et al., 1982; Mojica et al., 2000). Primary spermatocytes are easy to recognize in sections because they are the largest cells in the sperm tube, have a growth phase just like the primary oocytes of females but of lesser extent (usually 2–25 but up to 78-fold [Heming-van Battum and Heming, 1989]), and have nuclei that are large relative to their cytoplasm and that are pushed apart within a cyst as the spermatocytes grow (Figs. 1.3, 1.19).

Genes such as *bag-of-marbles* (*bam*) and *benign gonial cell neoplasm* (*bgcn*) are known in *D. melanogaster* males whose normal expression limits the number of spermatogonial divisions occurring (Fig. 1.19) (Gönczy et al., 1997), while expression of *punt* (*put*) and *schnuuri* (*shn*) in surrounding cyst cells regulates pro-

duction of a signal inhibiting such division (Matunes et al., 1997).

1.3.2.2.1 MITOTIC SYNCHRONY

Mitotic synchrony within a cyst is maintained by incomplete cytokinesis during each division cycle. Following telophase of each division, the spindle microtubules persist and are compressed by the cleavage furrow to form a tubelike gel, the *fusome*, extending through the ring canals between adjacent spermatogonia and serving to prevent complete cytokinesis (King et al., 1982; Marec et al., 1993; McKearin, 1997). With subsequent proliferation, new fusomes are linked to preexisting ones to form a *polyfusome* (Fig. 1.9A), an interconnection not breaking until completion of spermiogenesis.

Formation and function of the fusome in both male and female gametogenesis in *Drosophila* require the presence and action of various membrane-associated skeletal proteins including alpha-spectrin, ankyrin, and Hts, all of which are found in fusomes (Lin et al., 1994; Deng and Lin, 1997; McKearin, 1997). Males with the *Hts* mutation that lack fusomes form abnormal cysts containing a reduced number of spermatogonia.

1.3.2.2.2 THE ROLE OF CHROMOSOMES AND MICROTUBULES IN SPERMATOGONIAL MITOSIS

The process of mitosis, including that occurring during spermatogenesis and oogenesis, has long been thought to require the presence of chromosomes, kinetochores, microtubules, and centrosomes (Hyams, 1996; Sharp et al., 2000). However, Zhang and Nicklas (1996), by removing the chromosomes from dividing primary spermatocytes of the grasshoppers *Chortophaga australior* and *Melanoplus sanguinipes*, showed that both anaphase and cytokinesis occur normally in their absence, and Heald and coworkers (1996), using tiny magnetic beads covered with plasmid DNA, induced formation of functional nuclei and mitotic spindles in incubated egg extract of *Xenopus* (Amphibia: Anura) lacking both centrosomes and kinetochores. These results indicate that the formation of bipolar spindles in dividing cells is an intrinsic property of the microtubules able to assemble around chromatin, and that it requires dynein-dependent translocation of microtubules across one another.

1.3.3 MEIOSIS

Meiosis is the occurrence of two successive nuclear divisions accompanied by a single replication of chromosomes (Fig. 1.10) (White, 1973; John, 1990). Its complete details were not fully elucidated until the 1930s.

1.3.3.1 PROPHASE

Meiotic prophase is much longer than the prophase of mitosis and can be divided into five stages based on a changing configuration of chromosomes within the nucleus of primary spermatocytes: *leptotene* (or leptonema), *zygotene* (or zygonema), *pachytene* (or pachynema), *diplotene* (or diplonema), and *diakinesis* (words ending

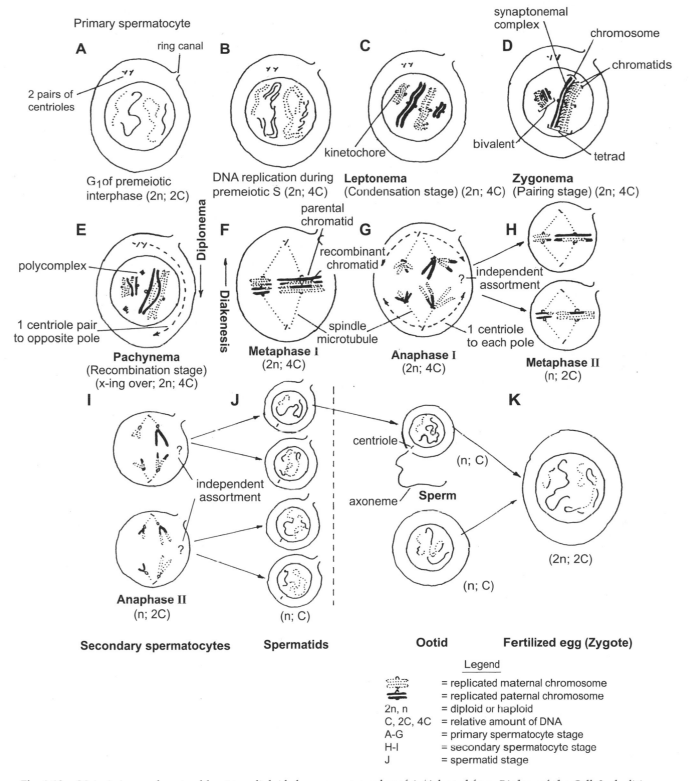

Primary spermatocyte

A

2 pairs of centrioles

ring canal

G_1 of premeiotic interphase (2n; 2C)

B

DNA replication during premeiotic S (2n; 4C)

C

kinetochore

Leptonema (Condensation stage) (2n; 4C)

D

synaptonemal complex

chromosome

chromatids

bivalent

tetrad

Zygonema (Pairing stage) (2n; 4C)

E

polycomplex

1 centriole pair to opposite pole

Pachynema (Recombination stage) (x-ing over; 2n; 4C)

Diplonema

Diakenesis

F

parental chromatid

recombinant chromatid

spindle microtubule

Metaphase I (2n; 4C)

G

?

Anaphase I (2n; 4C)

H

independent assortment

1 centriole to each pole

Metaphase II (n; 2C)

I

?

?

Anaphase II (n; 2C)

independent assortment

J

(n; C)

K

centriole

axoneme

Sperm

(n; C)

(n; C)

(n; C)

(2n; 2C)

Secondary spermatocytes

Spermatids

Ootid

Fertilized egg (Zygote)

Legend

= replicated maternal chromosome
= replicated paternal chromosome
2n, n = diploid or haploid
C, 2C, 4C = relative amount of DNA
A-G = primary spermatocyte stage
H-I = secondary spermatocyte stage
J = spermatid stage

Fig. 1.10. Meiosis in a male animal having a diploid chromosome number of 4. (Adapted from *Biology of the Cell*, 2nd edition, by S. L. Wolfe © 1981. With permission of Wadsworth, an imprint of the Wadsworth Group, a division of Thomson Learning. Fax 800 730-2215.)

with -tene are the adjectival forms, and though incorrect, are preferred to the noun form, ending with -nema ["thread"]). Autoradiography using [³H]thymidine indicates that DNA replication occurs in the premeiotic S phase of primary spermatocytes (Fig. 1.10A, B).

1.3.3.1.1 LEPTOTENE (THIN-THREAD OR CONDENSATION STAGE)

Chromatin of the chromosomes begins to coil, and they reappear as thin threads in the nucleus often in a polarized "bouquet" configuration, with their ends (telomeres) attached to the nuclear membrane adjacent to a centrosome containing two centriolar pairs. The remainder of each chromosome extends as a loop into the nucleoplasm. At this time each chromosome already has two chromatids (Fig. 1.10C).

1.3.3.1.2 ZYGOTENE ("MATING-THREAD" OR PAIRING STAGE)

The homologous chromosomes contributed by mother and father come together in pairs and become closely juxtaposed along their lengths (Fig. 1.10D). This synapsis starts at the telomeres of the chromosomes where they attach to the nuclear membrane, and proceeds zipper-like. Pairing is intimate; is between homologous regions of paternal and maternal chromosomes, perhaps by direct DNA-DNA interaction (Marec, 1996); and is accompanied by the formation of a three-part synaptonemal complex between the synapsed chromosomes. This proteinaceous structure, discovered independently by Moses and Fawcett in 1956, is of protein, is always present at this time in premeiotic cells of both sexes and in all investigated eukaryotes, and seems to have a role in crossing-over (Marec, 1996).

1.3.3.1.3 PACHYTENE ("THICK-THREAD" OR RECOMBINATION STAGE)

Pachytene begins as soon as synapsis is complete and lasts a long time, particularly in females. Shortening and coiling of the chromosomes continues throughout this stage so that at its end, they are one-fourth to one-sixth their leptotene lengths (Wolfe, 1981). Each pair of homologous chromosomes is a bivalent of two chromosomes and a tetrad of four chromatids (Fig. 1.10E).

After synapsis, the two chromatids of each chromosome become partly independent of each other but remain together because of their common kinetochores. While chromatids of the two chromosomes are juxtaposed, breakages occur, and the resulting parts join up in new combinations so that sections belonging to different partners of the pair of chromosomes become joined in the same chromatid (Fig. 1.10E–G). This exchange is mediated by a recombination nodule at each cross-over point, first described by Adelaide Carpenter in 1975 in *Drosophila* oocytes and now known to constitute a multienzyme complex (Marec, 1996: 225–226).

This exchange between chromosomes is referred to as *crossing-over*, the genes from the parental chromosomes being reshuffled and redistributed to the four chromatids (Fig. 1.10E, F). At any cross-over point, breaks and crossing-over occur between only one of the two chromatids (the *recombinant* chromatid) of each chromosome; the other, *parental*, chromatid remains intact (Fig. 1.10F). Since the two chromosomes exchange corresponding parts, each recombinant chromatid emerges with a different though full complement of genes, except in males of higher flies and of some lepidopterans in which crossing-over does not occur. Of course, errors—inversions, deletions, duplications, translocations, or the perambulations of transposable elements—can occur during crossing-over such that the chromatids end up with a different mix of genes.

1.3.3.1.4 DIPLOTENE (SYNTHESIS STAGE)

The homologous chromosomes begin to separate but are held together at cross-over points called *chiasmata* (singular: chiasma), the first visible sign of crossing-over (Fig. 4.1). At this time, the two homologous chromosomes are still elongate but can have a hairy or fuzzy outline similar to that of the "lampbrush" chromosomes of amphibian oocytes. Such loops are particularly prominent in the Y-chromosome of the giant primary spermatocytes of *Drosophila hydei* (Hackstein et al., 1990) and other species with long sperm. Their chromatin unwinds and becomes diffuse, and they become active in ribosomal, messenger, and transfer RNA synthesis as shown by autoradiography using [³H]uridine.

Formation of chiasmata is usually necessary for normal segregation of homologous chromosomes at anaphase I, as those lacking these cross-over points fail to segregate normally and the resulting sperm have either more or fewer chromosomes than normal (*aneuploidy*). An exception is the "achiasmatic disjunction" of chromosome 4 in *D. melanogaster*, which is too short to undergo recombination. Using artificially synthesized "minichromosomes," Karpen and colleagues (1996) showed that efficient disjunction requires at least 1000 kilobases (kb) of overlap in heterochromatin around the centromeres of the two homologs.

1.3.3.1.5 DIAKINESIS ("MOVEMENT-TO-PERIPHERY" OR RECONDENSATION STAGE)

The now recombined homologous chromosomes become more condensed and darker staining, move about within the nucleus, and eventually adhere throughout their lengths to the nuclear membrane. Simultaneously, the spermatocyte nucleolus disappears, the synaptonemal complexes degenerate or are cast off and accumulate as polycomplexes in the nucleoplasm or cytoplasm, and the chiasmata move toward the ends of the chromosome arms (*terminalization*). At this stage, the chromosomes look like X's, O's, or 8's in sections (Fig. 1.3, two cysts at base of primary spermatocytes section) and squash preparations of the spermatocytes, depending on their number of chiasmata.

In premetaphase I, the spindle microtubules assemble, the nuclear membrane disappears, and the bivalents move to the equator of the spindle (Fig. 1.10F). The four kinetochores of each tetrad (one for each chromatid) attach to spindle microtubules, the two centromeres of each homolog attaching to only one pole of the spindle (Fig. 1.10G).

1.3.3.2 MEIOSIS I AND II

The separated bivalents assume an equatorial position between the centrosomes (Fig. 1.10F), separate from one another (anaphase I, Fig. 1.10G), move to opposite poles of the spindle while casting off their chiasmata, and end up within two secondary spermatocytes (Fig. 1.10H). In male insects, this phase is usually very short, telophase I leading directly into metaphase II, so that secondary spermatocytes are either absent (as in most insects) or of short duration and thus difficult to spot in sections or squash preparations. If they do form, secondary spermatocytes can be recognized within a cyst because they are half the size of and twice as numerous as primary spermatocytes and larger than and half as numerous as early spermatids. Meiosis I is usually reductional (Fig. 1.10F–H, but see 5.2.1). In each chromosome, the two chromatids are still joined together and the kinetochore has duplicated, but both kinetochores are still on the same side of the spindle equator (Fig. 1.10G).

In metaphase and anaphase II, the daughter chromatids of each chromosome, now chromosomes in their own right, segregate to opposite poles (Figs. 1.10I). This division is equational and can be differentiated from the first meiotic division by its smaller size and greater frequency within a cyst (Fig. 1.3) and by the longer, more slender, and often kinked chromosomes.

As a result of meiosis, four spermatids are formed from each primary spermatocyte, each with a different complement of chromosomes (Fig. 1.10J). The amount of DNA (and the number of chromosomes) in male germ cells is 4C in interphase spermatogonia (2n chromosomes), 4C in primary spermatocytes (2n chromosomes), 2C in secondary spermatocytes (n chromosomes), and 1C in spermatids (n chromosomes) (Fig. 1.10).

1.3.3.2.1 PROPER ALLOCATION OF CHROMOSOMES TO DAUGHTER SPERMATOCYTES DURING MEIOSIS

The mechanism by which the chromosomes segregate and move to opposite poles during mitosis and meiosis is under intense scrutiny and is becoming well understood at the molecular level, as summarized by Pluta and coauthors (1995), McKim and Hawley (1995), Nicklas (1997), and Sharp and colleagues (2000). Error-free allocation of daughter chromosomes to each secondary spermatocyte and spermatid during meiosis I and II is critical since missing or extra chromosomes (aneuploidy) result in abnormal zygotes following fertilization (Nicklas, 1997). However, it is up to chance whether the two attachment sites (kinetochores) of a particular tetrad attach to microtubules emanating from centrosomes at the same or at opposite poles of a dividing primary spermatocyte prior to metaphase I. If they attach normally to microtubules from opposite poles, tension is exerted on the bivalents at anaphase I and they sepa-

rate. If they attach to the same pole, no tension is generated, they fail to separate, and both chromosomes end up within the same secondary spermatocyte. Thus, there must be mechanisms acting to correct such errors.

To examine this, Nicklas (1997) and colleagues investigated anaphase I in primary spermatocytes of mantid testes cultured in vitro. When the two kinetochores of a particular tetrad attached to the same centrosome, they hooked it with a micromanipulation needle, artificially created tension, and caused the division to proceed by activating a cell cycle checkpoint.

This tension seems to act by dephosphorylating certain kinetochore proteins. Prior to correct attachment to microtubules, these proteins are phosphorylated (when immunostained with a tagged antibody specific for the phosphorylated state, such kinetochores appear larger and brighter with fluorescent microscopy). When the correct microtubules attach and begin to pull the chromosomes apart, these proteins are dephosphorylated (such kinetochores appear smaller and less bright), and this chemical change seems to activate the cell cycle checkpoint so that anaphase can proceed. These responses can be duplicated in misattached chromosomes stretched with a micromanipulation needle.

1.3.3.3 CENTRIOLES DURING MEIOSIS

Primary spermatocytes in insects usually contain four centrioles in two pairs located next to each other within a centrosome at one side of the nucleus. Each member of a pair is at right angles to the other (Fig. 1.10A–D) (Shi et al., 1982; Chapman et al., 2000). Late in diakinesis, one pair migrates to the opposite side of the nucleus (Fig. 1.10E) so that at metaphase and anaphase I, each pole of the spindle contains one pair (Fig. 1.10F, G). Each centriole of the pair then separates from the other to form the poles of the spindles for meiosis II so that each spermatid receives only one centriole (Fig. 1.10G–J). Centriolar behavior is similar in males of *Ephestia kühniella* and probably of other lepidopterans, even though they are known to produce both normal (eupyrene) and anucleate (apyrene) spermatozoa (Wolf, 1997).

1.3.3.4 ADAPTIVE SIGNIFICANCE OF MEIOSIS

The reduction in chromosome number from diploid to haploid at meiosis I prevents a continuous increase in chromosome number, which doubles in each generation at fertilization. During both metaphase I and II, the chromosomes line up at random on the equator of the spindle so that at anaphase I and II, the paternal and maternal chromosomes independently assort, though in the correct number, to secondary spermatocytes and spermatids (Fig. 1.10G, I). Therefore, the spermatids and sperm resulting from meiosis will have the correct number but any combination of paternal and maternal chromosomes, the number of possibilities equaling 2^n, where n is the haploid chromosome number. Thus, in our own species, *Homo sapiens*, where n = 23, there are 2^{23} or 8×10^6 possibilities. This, in conjunction with crossing-over, provides for an infinite variety of combinations of paternal and maternal genes in any sperm and constitutes *genetic recombination*, a principal source of genetic variation in bisexual organisms (the same occurs during meiosis in females). The amount of recombination occurring in a species depends on its haploid chromosome number and on the frequency and distribution of chiasmata (White, 1973).

1.3.4 SPERMIOGENESIS

Although they have a haploid set of chromosomes, the spermatids resulting from meiosis are not equipped to function as male gametes. They must reach the egg, penetrate it, activate it, and pass parental genes into it. All these roles require the action of organelles that are not present in newly formed spermatids but that differentiate during *spermiogenesis* (Phillips, 1974; Szöllosi, 1975; Hamilton and Waites, 1996). This process is summarized in Fig. 1.11 for a staphylinid beetle and in Fig. 1.12 for *D. melanogaster*. Recall that all spermatids within a cyst differentiate simultaneously because they are still joined to each other by ring canals, a connection only severed as a final event of spermiogenesis (Fig. 1.11) (Fabrizio et al., 1998).

Formation of each organelle of a mature sperm is summarized below:

- Acrosome: The acrosome forms within Golgi cisternae of the spermatid (Fig. 1.11A). Small pro-acrosomal granules appear within the vesicles of this system that fuse to form a single, large acrosomal vesicle (Fig. 1.11B). The acroblast and its granule then apply themselves to the side of the elongating nucleus and spread over it, eventually assuming a form characteristic for the species (Fig. 1.11C–E). Later, all Golgi material is discarded in the residual cytoplasm (Fig. 1.11D, E). Both the mode of formation and the function of the acrosome in the sperm of certain echinoderms and mammals suggest that it is a highly specialized lysosome, with its digestive activity directed toward components of the egg membranes (Wolfe, 1981).
- Head: Chromatin is usually diffuse in early spermatids and highly condensed in mature spermatozoa (Fig. 1.11A, E). During spermiogenesis, the DNA in spermatid nuclei is coiled by arginine-rich proteins into countless toroidal structures, each containing up to 60 kb of DNA, a process that inactivates the entire genome (Brewer et al., 1999). The protamines responsible contain a series of arginine-rich anchoring domains that bind to the phosphodiester backbone of DNA in a base sequence–independent fashion. One protamine molecule is bound to each turn of DNA, and adjacent arginines in the anchoring domains interlock both strands of the helix. This change accompanies a change in configuration of the nuclear material from a loose to a tightly fused paracrystalline configuration, a process that is reversed within the egg after sperm entry (3.6).

Judging from analysis of mutant genes in *Drosophila* spermiogenesis (Fig. 1.19B), shaping of the sperm head is determined both from within, by a genetically

determined pattern of DNA and protein aggregation during chromatin condensation, and from without, by development of a species-specific array of longitudinally disposed microtubules about the head prior to elongation (Kessel, 1966; Fuller, 1993).

- Mitochondrial derivatives: At anaphase and telophase II, the mitochondria of each newly formed spermatid cluster near the ring canal and begin a complex series of rearrangements and fusions that results in their forming a large, spherical Nebenkern (plural: Nebenkernen) next to the nucleus (Fig. 1.11A) (Pratt, 1968; first described by Retzius in 1904). This structure undergoes a complicated series of topographical changes and either divides into two separate bodies, the mitochondrial derivatives, or remains as one. These lengthen along with the axoneme and on either side of it (Fig. 1.11B–E). In males of *D. melanogaster*, the Nebenkern measures 7 μm across in early spermatids, whereas fully differentiated mitochondrial derivatives are 1.8 mm long, even though their volume remains the same (Fig. 1.12).
- Flagellum: Growth of the axoneme usually begins at anaphase or telophase II. It grows out from centriole C2, either adjacent to the nuclear membrane (Fig. 1.11A) or next to the plasma membrane, through the addition of tubulin molecules to the distal, or plus, ends of its microtubules at the apex of the elongating flagellum.
- Cytoplasm: Since most spermatid cytoplasm is redundant in the sperm, it is sloughed off during the separation (*individualization*) of individual spermatozoa at the end of spermiogenesis as droplets of residual cytoplasm, which soon degenerate (Figs. 1.11D, E; 1.13) (Fabrizio et al., 1998). When viewed by phase or interference contrast microscopy, these are easily seen in living, older spermatids as rows of droplets about the flagellum.

When spermiogenesis is complete within a cyst, its sperm are held together in a sperm bundle surrounded by the cyst membrane. The number of sperms per bundle is characteristic for the species, depends on the number of spermatogonial mitoses occurring after enclosure of the definitive spermatogonium by cyst cells, and is known to range from a low of 16 in certain hemipterans (with 2 synchronous, spermatogonial mitoses) to 65,536 in certain odonates (with 14) (Virkki, 1969).

1.3.5 RATE OF SPERMATOGENESIS

Spermatogenesis usually begins soon after the insect hatches from the egg. To calculate the time required for each phase of spermatogenesis, one usually fixes animals for autoradiographic analysis at progressive intervals or chase periods following injection of [³H]thymidine. Several testes are exposed to the radiolabeled nucleotide during the S phase of a spermatogonial or primary spermatocyte cell cycle, and are then fixed, sectioned, and exposed to film at various time intervals afterward as the labeled cells differentiate. Cells labeled with silver grains are found at progressively later stages

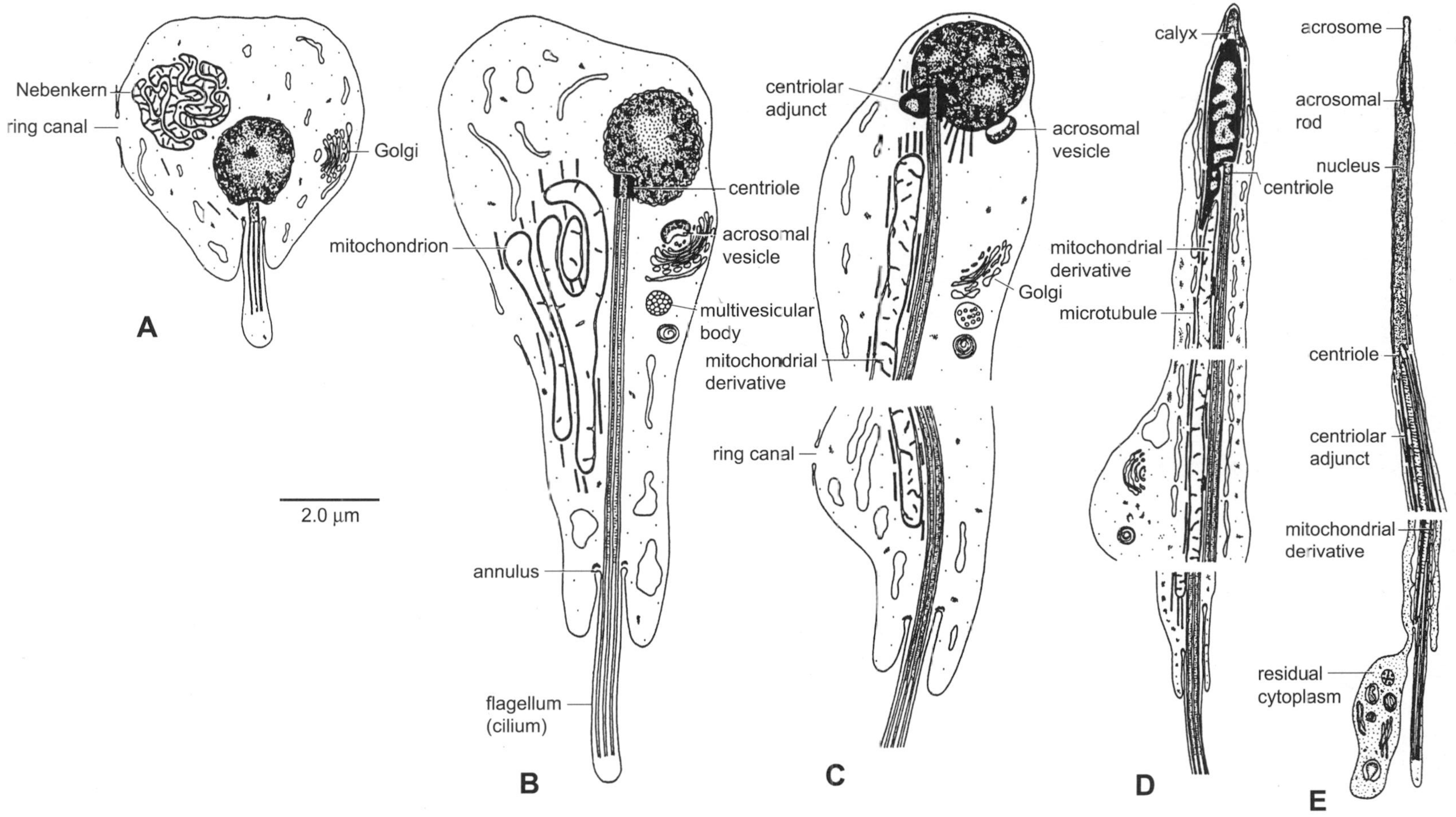
Nebenkern
ring canal
Golgi
A
2.0 μm
mitochondrion
centriole
acrosomal
vesicle
multivesicular
body
mitochondrial
derivative
annulus
flagellum
(cilium)
B
centriolar
adjunct
acrosomal
vesicle
Golgi
mitochondrial
derivative
ring canal
C
calyx
centriole
mitochondrial
derivative
microtubule
D
acrosome
acrosomal
rod
nucleus
centriole
centriolar
adjunct
mitochondrial
derivative
residual
cytoplasm
E

Fig. 1.11. Spermiogenesis in a bathysciine staphylinid beetle (Coleoptera). See text for details. (Adapted with permission from L. Juberthie-Jupeau, J. Durand, and M. Cazals. Spermatogenèse comparée chez les coléoptères Bathysciinae souterrains. *Cytobios* 37: 187–208. © 1983 Faculty Press.)

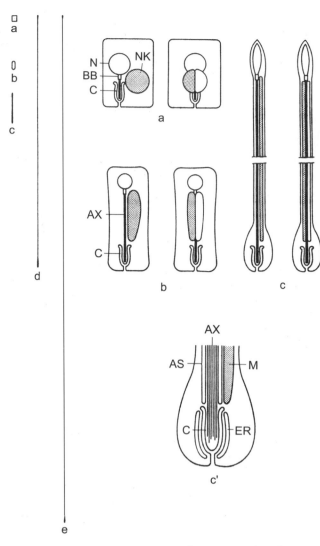

Fig. 1.12. Spermiogenesis in *D. melanogaster*. The relative dimensions of cysts before (a) and during (b–e) elongation are shown on the left. On the right the relative positions of the nucleus (N), Nebenkern (NK), centriole (BB), cilium (C), and axoneme (AX) are shown (lettered stages on left and right are comparable). The Nebenkern and mitochondrial derivatives are bipartite, as indicated by stippled and nonstippled components. In c′, the bulbous posterior tip of an elongating spermatid is illustrated to show the mitochondrial derivatives (M), axonemal sheath (AS), and the tip of the axoneme within the cilium that is invaginated into the tip of the flagellum. Endoplasmic reticulum (ER) surrounds this invagination. (Reprinted with permission from D. L. Lindsley and K. T. Tokuyasu. Spermatogenesis. In M. Ashburner and T. R. F. Wright (eds.), *The genetics and biology of* Drosophila. Vol. 2D, p. 250, Fig. 6. © 1980 Academic Press, Ltd.)

of development and at more proximal positions within the sperm tube as time passes (Roosen-Runge, 1977). The time required for the labeled cells to become mature spermatozoa can thus be determined with great accuracy.

In adult males of *D. melanogaster* at 25°C, transit time from stem spermatogonium to mature sperm is 250 h, with 25 developing cysts, as descendants of each stem

cell, present at any one time within the testis (this constitutes a family of cysts: Fig. 1.13). Since there are 5–8 stem cells/testis, this means there are 5–8 such families in each testis at any one time (Lindsley and Tokuyasu, 1980). Each definitive spermatogonium gives rise to 64 sperm (Figs. 1.13, 1.19). The growth phase of primary spermatocytes lasts 90 h and the cells enlarge 25-fold. Spermiogenesis lasts about 134 h, with the spermatid length-width ratio increasing from 1 to over 1000 (Figs. 1.12, 1.13). And during spermiogenesis, the spermatid nucleus undergoes a 30-fold reduction in volume and a 5-fold reduction in surface area.

In 42 species of *Drosophila* producing sperm from 0.32 to 58.3 mm long, Pitnick and coworkers (1995a) found the premating period of males (the time from adult eclosion to reproductive maturity) to correlate with sperm length and to range from 0 to 19 d; those producing the longest sperm required the most time. In their "sperm production hypothesis," they suggested that the machinery necessary to manufacture longer sperm requires more resources and production time than does that for shorter sperm (the mature testes constitute 1%–5% of total dry body mass in males producing short sperm and 8%–11% in those producing long sperm).

In other insects, spermatogenesis is generally slower. For example, the male onion maggot *Hylemya antiqua* (Diptera: Anthomyiidae) at 22°C (Theunissen, 1976) requires 9–11 d for definitive spermatogonia to develop into spermatozoa; the mosquito *Culex pipiens* (Diptera: Culicidae), 9 d; the locust *Melanoplus differentialis*, 28 d; and the boll weevil *Anthonomus grandis* (Coleoptera: Curculionidae), 10 d.

All stages of spermatogenesis in *D. melanogaster* and other insects can be maintained for short time intervals within individual living cysts cultured in vitro in appropriate physiological saline solution, and can be easily observed by phase contrast or interference contrast microscopy and photographed or videotaped (Fuller, 1993). Gönczy and colleagues (1992) developed a series of p-element enhancer detectors in *D. melanogaster* by which they could probe the distribution and ontogeny of each cell type in the testes from embryo to adult.

1.3.6 PRODUCTIVITY OF SPERMATOGENESIS

Total sperm production is part of the "male gametic strategy" of a species (Roosen-Runge, 1977; Dumser, 1980). This is characterized by the amount of energy invested per gamete, the total amount of energy invested in the production of gametes, and the pattern of gamete allocation to successive females; it has been investigated thoroughly in few species (Pitnick and Markow, 1994a; Pitnick, 1996).

In *D. melanogaster*, 9–13 primordial germ cells are encapsulated within each testicular rudiment within the embryo (Tihen, 1946). Each of these becomes a spermatogonial stem cell and produces a definitive spermatogonium each time it divides, with each definitive spermatogonium generating 64 spermatozoa (Figs. 1.13, 1.19). Tihen (1946) calculated the total number of cell

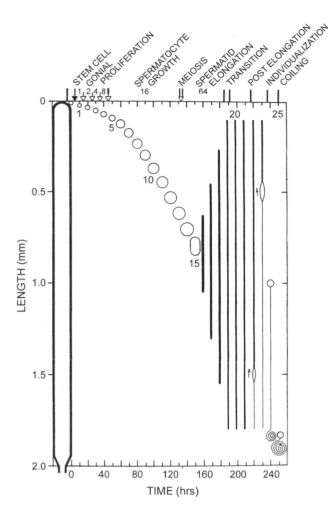

Fig. 1.13. Quantitative and temporal aspects of spermatogenesis in *D. melanogaster*. Twenty-five cysts descend from a single stem spermatogonium in the steady-state testis. The ordinate contains a diagram of the testis and indicates the longitudinal position of each cyst within it. The abscissa indicates the temporal position, at 25°C, of each cyst in the developmental sequence from stem cell to spermatozoon. Each testis normally contains 5–8 such cyst families. Arrows at the top represent cell divisions: solid arrow, stem cell mitosis; single-shafted open arrows, four synchronous spermatogonial mitoses; open arrow, two meiotic divisions. Numbers between the arrows indicate the number of germ cells within cysts of corresponding stages. Numbers adjacent to the cysts indicate their rank within the family. Vertical lines at the top divide the process of spermatogenesis into a series of recognizable stages: gonial proliferation (cysts 1–5)—the cystoblast, after four rounds of mitosis, produces 16 primary spermatocytes; spermatocyte growth (cysts 5–12)—these undergo a 25-fold increase in volume; meiosis (cyst 13)—the two meiotic divisions produce 64 spermatids; pre-elongation (cyst 14—not labeled)—Nebenkern is formed; elongation (cysts 15–18)—spermatids elongate along longitudinal axis of testis, and axoneme and mitochondrial derivatives elongate; transition (cyst 19)—head and tail align; postelongation (cysts 20–21)—spermatids mature; individualization (cysts 22–23)—a cystic bulge traverses the spermatid bundle from head to tail, removes excess cytoplasm, causes individual spermatids to become invested in their own membranes, and liberates them from the syncytium; coiling (cysts 24–25)—the extended bundle of spermatids is withdrawn and coils into the base of the testis, and the spherical waste bag, the final form of the cystic bulge, follows. (Reprinted with permission from D. L. Lindsley and K. T. Tokuyasu. Spermatogenesis. In M. Ashburner and T. R. F. Wright (eds.), *The genetics and biology of* Drosophila. Vol. 2D, p. 228, Fig. 2. © 1980 Academic Press, Ltd.)

divisions required to reach the average spermatozoon to be 121: 8 cleavage divisions in the embryo, 107 primordial germ cell and stem cell divisions, 4 definitive spermatogonial divisions, and 2 meiotic divisions. However, since there are actually 13 cleavage divisions in the eggs of this insect (Fig. 6.3), the total should be 126. From these figures, Roosen-Runge (1977) concluded that in most animals the critical proliferation, in terms of overall lifetime production of sperm, occurs in prespermatogonial cell generations.

Pitnick (1996) examined the relationships between dry body mass (176–704 μg), dry testis mass (6–75 μg), testis length (1.7–67.0 mm), sperm length (1.4–58.3 mm), number of sperm bundles in a midtestis cross section (24.6–94.0), and number of sperms per bundle (24–64) in 11 *Drosophila* species. After controlling for phylogenetic affinity, he found that the males of larger-bodied species invested relatively more in testis development than did those of smaller-bodied species, with the variation in testis mass being a function of variation in sperm length—not in number of sperm produced and transferred per copulation.

Males of *D. hydei* produce sperm 23.3 mm long

regardless of adult body size (Pitnick et al., 1995a). However, smaller males of this species were found to make a greater relative investment in testicular growth, to mature fewer sperm bundles within each testis at any point in time, to require a longer premating period to become reproductively mature, to mate with fewer females, to transfer fewer sperm per copulation, and to produce fewer progeny (Pitnick and Markow, 1994b).

Actual counts of sperm numbers in adult male insects are scarce. However, it is known that virgin males of the sarcophagid fly *Neobellieria bullata* examined by Berrigan and Locke (1991) and weighing from 15 to 76 mg contained an average of 28,000—581,000 sperm and produced 17,000–27,000 mature sperm/d after adult emergence.

1.3.7 CONTROL OF SPERMATOGENESIS
The control of gametogenesis is much less studied in male insects than in females (2.3). This is probably because the germ cells of males are so small and because the females of pest species are thought to contribute more to the success of succeeding generations than males and thus to be more worthy of study (Dumser,

1980). Also, vitellogenesis, the most spectacular aspect of egg production (2.2.1.2), often occurs after adult emergence, when results of experimental manipulation are easier to evaluate.

1.3.7.1 AUTOGENOUS VERSUS ANAUTOGENOUS INSECTS
In *autogenous* (self-generating) insects, adult males become sexually active immediately upon emergence, mature sperm are already present within their seminal vesicles, and the contents of their accessory glands have been synthesized and stored ready for use. Such males usually do not feed (often they lack functional mouthparts), copulate immediately and generally only once, and die shortly thereafter. In these, spermatogenesis and production of accessory gland secretions are completed while the insects are still juvenile.

Most insects, however, are *anautogenous* (non-self-generating). In these, newly emerged males require a premating period of varying length in which they feed, spermatogenesis is completed, sperm move into the seminal vesicles, and synthesis and storage of accessory gland products occur. These insects usually copulate repeatedly, remain sexually active for a long time, and have continuous spermatogenesis in the adult.

1.3.7.2 ENDOCRINE CONTROL OF SPERMATOGENESIS
In autogenous and anautogenous males, reproduction appears to be, at least partly, under endocrine control via the following sequence:

environmental cues → sense organs → central nervous system → endocrine system → hormone synthesis and release → spermatogenesis

At least two hormones are involved: juvenile hormone (JH) produced in cells of the corpora allata (CA) and 20-OH ecdysone (20E) produced in those of the prothoracic glands (PG) (12.4).

1.3.7.2.1 EARLY EXPERIMENTS
Bugs of the blood-feeding species *Rhodnius prolixus* (Hemiptera: Reduviidae) have five juvenile instars and require a blood meal for each molt. V. B. Wigglesworth (1936) observed that if the corpora allata were removed from fourth instar males, they molted to precocious adults having fully differentiated sperm bundles, whereas the control larvae (now fifth instars) had testes still containing only spermatogonial cysts. This suggested to him that JH inhibits meiosis and spermiogenesis (JH is absent or at very low concentrations in fifth instars; see 12.5).

In 1953, Schmidt and Williams observed that testes of overwintering, diapausing, silkworm pupae, *Hyalophora cecropia* and *Samia walkeri*, contained only spermatogonial cysts (these moths are autogenous). Since 20E is absent from the blood during winter, their observation suggested that 20E stimulates meiosis and spermiogenesis in the spring when its production is renewed. Kambysellis and Williams (1971a, 1971b) confirmed this

by culturing intact testes of diapausing silkworm pupae in hanging drops of culture media on the underside of a coverslip placed over the well of a well slide. When ecdysteroids were added to the medium, meiosis and spermiogenesis ensued; if not, the cysts remained spermatogonial.

The above-mentioned work would suggest that ecdysteroids stimulate differentiation of male gametes and that JH inhibits it. However, there is a problem: most adult male insects at that time were thought to lack ecdysteroids (cells of the prothoracic glands, then thought to be the only source of this hormone, degenerate during or just after adult emergence) but synthesize JH (which stimulates accessory gland synthesis and mating behavior either directly or indirectly), yet spermatogenesis still continues—a seeming paradox.

In fact, the story is more complicated than this. In a series of experiments using both intact testes and isolated spermatogonial cysts from diapausing pupae of the giant silkworm *H. cecropia*, cultured in hanging drop cultures, Kambysellis and Williams (1971a, 1971b) showed that the function of ecdysteroids is to render the wall of the testis permeable to passage of a "macromolecular factor" (MF) produced elsewhere in the body, perhaps by the hemocytes, with this second factor actually stimulating meiosis and spermiogenesis (Fig. 1.14).

1.3.7.2.2 BLOOD-TESTIS BARRIER
An additional complicating factor is that, at least in certain locusts, each sperm tube in the adult testis is divided into apical and basal compartments that are differentially permeable to molecules placed into the blood such as horseradish peroxidase or lanthanum chloride and presumably to hormones (Fig. 1.15) (Szöllosi and Marcaillou, 1977). In the apical compartment, tracer molecules reach the innermost spermatogonia (Fig. 1.15A, B). In the basal compartment, an additional, inner parietal layer of cells in the walls of each sperm tube, having extensive septate desmosomes between them, prevents tracer molecules from entering this part of the follicle and contacting the cysts (Fig. 1.15A, C). It is thus a blood-testis barrier similar to that found in the seminiferous tubules of mammalian testes (Fig. 1.16). (But why can't tracer move from apical to basal compartments once it is inside the sperm tube or seminiferous tubule?) This layer provides a specific tight environment in which the germ cells can undergo meiosis and differentiate into spermatozoa seemingly free of hormonal influence. In locusts, this barrier develops in third instar larvae prior to the onset of meiosis (Jones, 1978). Similar septate desmosome barriers reside between the cyst cells in testicular cysts of the moth *Anagasta kühniella* (Lepidoptera: Pyralidae) containing later-generation spermatogonia (Szöllosi et al., 1980).

1.3.7.2.3 INFLUENCE OF HORMONES ON MITOSIS, MEIOSIS, AND SPERMIOGENESIS
The male germ cells of insects proliferate, grow, and differentiate in clones enclosed within a series of meso-

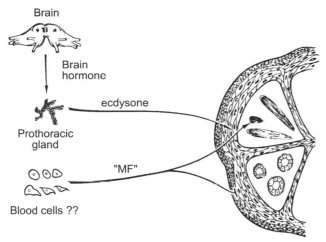

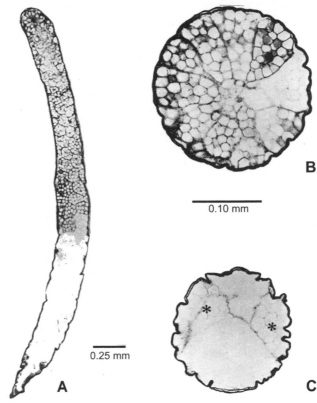

Fig. 1.14. Summary of the control of spermatogenesis in males of *Hyalophora cecropia* (Lepidoptera: Saturniidae). Results of hanging drop experiments suggest that ecdysteroids from the prothoracic glands render the testis wall permeable to macromolecular factor (MF) from the hemocytes (?), which is thought to induce meiosis and spermiogenesis. (Reproduced with permission from M. P. Kambysellis and C. M. Williams. *In vitro* development of insect tissues. II. The role of ecdysone in the spermatogenesis of silkworms. *Biol. Bull.* 141: 541–552. © 1971 The Marine Biological Laboratory.)

Fig. 1.15. The blood-testis barrier in a sperm tube of *Locusta migratoria* (Orthoptera: Acrididae). The tube was removed and incubated in horseradish peroxidase. A. Sagittal section. B. Transverse section through apical compartment. C. Transverse section through basal compartment. Two cysts in C are slightly penetrated by tracer (asterisks). Two others contain more developed germ cells and are tight. (Reproduced with permission from A. Szöllosi and C. Marcaillou. Electron microscope study of the blood-testis barrier in an insect: *Locusta migratoria. J. Ultrastruct. Res.* 59: 158–172. © 1977 Academic Press, Inc.)

dermal membranes (Figs. 1.1, 1.3). These membranes form barriers between the immediate environment of a clone, the primary compartment or cyst, and its wider environment, the secondary compartment or sperm tube, the tertiary compartment or testis, and a fourth compartment, the body cavity (Roosen-Runge, 1977). These structural relationships suggest that developing male gametes both influence each other and can be influenced by differential passage of molecules from their surroundings.

The number of synchronous spermatogonial divisions occurring between isolation of a definitive spermatogonium within a cyst membrane and appearance of primary spermatocytes is characteristic for the species and does not appear to be influenced by hormones (1.3.2.2). In some insects, primary spermatocyte-containing cysts accumulate in the late larval or pupal testes; Dumser (1980) suggested that this accumulation could result from differences in cell cycle duration between spermatogonia and primary spermatocytes. For example, in testes of the grasshopper *Melanoplus differentialis* (Orthoptera: Acrididae), the seven, synchronous spermatogonial mitoses occurring within a cyst have an average duration of 28 h, whereas that of primary spermatocytes is 10 d, principally because meiotic prophase is much longer than mitotic prophase.

Dumser and Davey (1974, 1975a, 1975b), working on larvae of *R. prolixus*, provided evidence to suggest that hormones affect only the rate of spermatogonial mitosis, not meiosis or spermiogenesis. In testes of this insect

there is a basal endogenous rate of spermatogonial mitosis that occurs in the absence of hormones. This rate is characteristic of the species, is inherited, and may be affected by interactions with the cyst cells. Application of 20E doubles this rate while JH depresses it, though never below the basal rate (20E is known to accelerate passage of spermatogonia of the grasshopper *Locusta migratoria* from G1 into S of the cell cycle and from G2 into M [Dumser, 1980]). In addition, meiosis and spermiogenesis follow the spermatogonial divisions only after a species-specific number of spermatogonial mitoses have occurred within each cyst (2^8 to yield 256 primary spermatocytes/cyst in *R. prolixus*). In order for cysts containing 256 primary spermatocytes to undergo meiosis and spermiogenesis, a blood meal is required. If the bug cannot feed, it enters diapause (a state of arrested development and low metabolic rate) and any cysts already differentiated abort. It is also known that

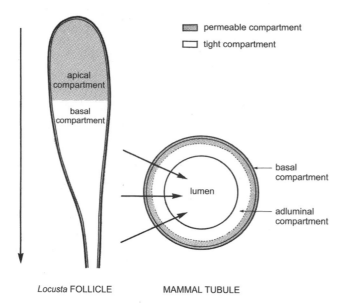

apical
compartment

basal
compartment

basal
compartment

lumen

adluminal
compartment

Locusta FOLLICLE   MAMMAL TUBULE

Fig. 1.16. Schematic diagrams of the blood-testis barrier in a sagittal section of a locust sperm tube and in a transverse section of a seminiferous tubule of a mammal. In both, the young germ cells proliferate in a compartment that is permeable to substances in the blood but differentiate within a tight compartment. Arrows indicate the direction of germ cell differentiation. (Reproduced with permission from A. Szöllosi and C. Marcaillou. Electron microscope study of the blood-testis barrier in an insect: *Locusta migratoria. J. Ultrastruct. Res.* 59: 158–172. © 1977 Academic Press, Inc.)

injection of ecdysteroids alone can induce differentiation in a bug that has not had a blood meal. Dumser and Davey thus proposed the following scenerio to explain testis growth in this insect (Fig. 1.17).

In instars 1–4, the spermatogonial mitotic rate is at the basal rate because of high relative concentrations of both ecdysteroids and JH. In the fifth instar, when no JH is present (this is necessary if the fifth instar is to molt to an adult), 20E doubles the spermatogonial mitotic rate. This results in rapid enlargement of the sperm tubes and in accumulation of cysts containing 256 primary spermatocytes. If the insect then takes a blood meal, the cells of these cysts undergo meiosis and differentiate into sperm bundles, and the insect molts to an adult. In adults, spermatogenesis continues at the basal rate again because JH is again present in the blood.

In 1979, Koolman and coworkers discovered that testes of adult males of the blowfly *Calliphora vicina* (Diptera: Calliphoridae) contain large quantities of ecdysteroids, particularly 20E, which could not have been synthesized in the prothoracic glands since these degenerate following adult emergence. Loeb and colleagues (1982), working with fifth (last) instars of *Heliothis virescens* (Lepidoptera: Noctuidae), showed these hormones to be contained in both the cysts and the testis sheaths and suggested that they were synthesized in cells of the sheaths. They also demonstrated that ecdysteroids appeared in three peaks during this instar: first at onset of meiosis I, then at onset of spermiogenesis, and finally with fusion of the paired larval testes to form the single composite testis of the adult, observations all suggesting a direct stimulatory role for ecdysteroids in meiosis and spermiogenesis.

Finally, Friedländer and Reynolds (1988) showed that meiosis in testes of mature (fifth instar), postwandering larvae (wandering larvae are looking for a place to spin up and pupate) of the tobacco hornworm *Manduca sexta* (Lepidoptera: Sphingidae) is induced, probably indirectly, by release of 20E. In these insects, a peak of 20E is released from the prothoracic glands 2 d after wandering is complete. Abdomens of caterpillars isolated after this peak displayed meiotic metaphases, whereas those isolated before did not. Instead, in the latter the primary spermatocytes were blocked in meiotic prophase. This block could be removed by implanting active prothoracic glands into these isolated abdomens or by injecting 20E. However, meiosis could not be induced in this way in abdomens isolated from caterpillars that had yet to wander (their testes were not yet able to respond to 20E, perhaps because the required number of ecdysteroid receptors [12.4.1.5.1] was not yet present in their cells). Friedländer (1989) then showed, in larvae of the codling moth *Cydia pomonella* (Lepidoptera: Tortricidae), that 20E may induce meiosis by restoring sheath permeability, perhaps by affecting the septate desmosomes between cyst cells, to other factors in the blood that actually stimulate meiosis, such as the macromolecular factor mentioned earlier (Fig. 1.14).

In last instar larval testes of the gypsy moth *Lymantria dispar* (Lepidoptera: Lymantriidae) cultured in vitro, ecdysteroid synthesis can be induced by application of a brain peptide, *Lymantria* testis ecdysiotropin (LTE) (Loeb et al., 1998). Testes of midpupae can do so without such stimulation, though application of LTE can boost the basal rate of ecdysteroid production. All these findings are summarized in Fig. 1.17.

1.3.7.2.4 CONTROL OF APYRENE AND EUPYRENE SPERMATOGENESIS IN LEPIDOPTERANS

As mentioned earlier (1.2.3), most adult male lepidopterans produce both nucleate (eupyrene) and anucleate (apyrene) spermatozoa, in a ratio of 1:9 in the pyralid moth *Plodia interpunctella* (Gage and Cook, 1994) and the pierid *Pieris napi* (Cook and Wedell, 1999). Both sperm types develop within cysts containing similar primary spermatocytes (but see Tschudi-Rein and Benz [1990b]) except that meiosis is aberrant in apyrene-producing cysts, resulting in the formation of spermatids having one or more micronuclei with unequal chromosome numbers that are discarded during spermiogenesis (Friedländer and Wahrman, 1972; Friedländer and Miesel, 1977; Lai-Fook, 1982; Friedländer, 1997; Kawamura et al., 2000). In addition, eupyrene spermatogenesis is regular and highly sensitive to genetic and experimental manipulation, while that of apyrene sperm is irregular and is able to withstand such insult (Friedländer, 1997).

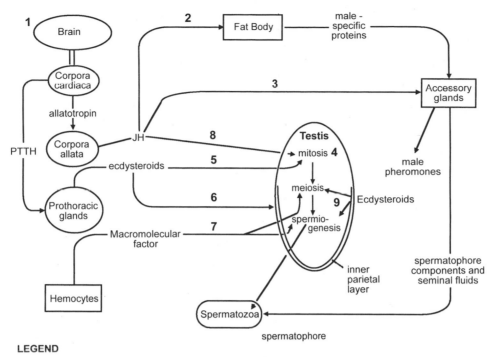

LEGEND
1. Feeding, mating, etc. induces release of prothoracicotropic hormone (PTTH) and allatotropin.
2. Induces synthesis of male-specific proteins in fat body.
3. Induces synthesis and release of male pheromones and spermatophore proteins.
4. Species-specific number and rate of spermatogonial mitoses.
5. Doubles basal spermatogonial mitotic rate.
6. Makes testis wall permeable to macromolecular factor.
7. Stimulates meiosis and spermiogenesis probably indirectly.
8. Inhibits 20E-induced doubling of spermatogonial mitotic rate.
9. Sheath cell ecdysteroids stimulate meiosis and spermiogenesis probably indirectly.

Fig. 1.17. Summary of the endocrine control of spermatogenesis in male insects based mostly on study of *Rhodnius prolixus* and *Hyalophora cecropia*.

The shift from eupyrene to apyrene spermiogenesis correlates with a shortening of meiotic prophase and with a halt in the synthesis of a meiotic, lysine-rich protein fraction in those spermatocytes generating apyrene sperm (Friedländer, 1997). In developing testes of the codling moth *C. pomonella*, eupyrene precedes apyrene spermatogenesis (Fig. 1.18A) (Friedländer and Benz, 1981): at 26°C, eupyrene meiosis begins in fourth instar larvae; apyrene meiosis, only on day 4 of fifth instars. Friedländer and colleagues concluded that an apyrene spermatogenesis inducing factor becomes active on day 4 of the fifth instar and that commitment to apyrene spermatogenesis and to pupal development of somatic tissues coincide (Friedländer and Benz, 1982; Friedländer, 1997).

1.3.7.2.5 DIAPAUSE AND SPERMATOGENESIS
IN LEPIDOPTERA
During diapause in male insects, specific degeneration of the most differentiated cohorts of germ cells occurs, usually in clones containing primary or secondary spermatocytes or spermatids (Dumser, 1980). As described earlier (1.3.7.2.3), injection of 20E can induce meiosis

and spermiogenesis in diapausing insects, and it is likely that absence of this hormone during diapause indirectly causes this degeneration to occur.

Friedländer and Brown (1995) showed that injection of tebufenozide (Mimic), a nonecdysteroidal ecdysone inhibitor and insecticide (Fig. 12.11C), can reinitiate spermatogenesis in isolated abdomens of diapausing larvae of *C. pomonella* in which spermatogenesis is in arrest. Apparently, molecules of this compound attach to and saturate a specific number of ecdysone receptors, causing an "all-or-none" response, the germ cells themselves being more susceptible than support tissues of the testes.

1.3.7.2.6 INFLUENCE OF PARASITOIDS ON APYRENE
AND EUPYRENE SPERMATOGENESIS IN LEPIDOPTERA
Codling moth larvae developing from embryos parasitized by the braconid *Ascogaster quadridentata* after dorsal closure (i.e., late in embryogenesis; 8.1.4) contain both a first instar parasitoid larva and host testes, whose development can be studied together (larvae developing from eggs parasitized before dorsal closure contain only the parasitoid larva) (Brown and Friedländer, 1995).

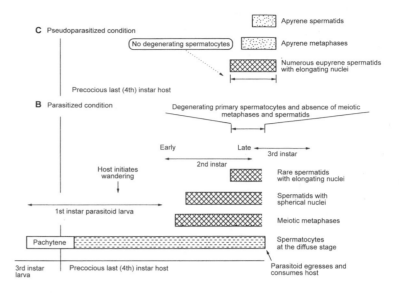

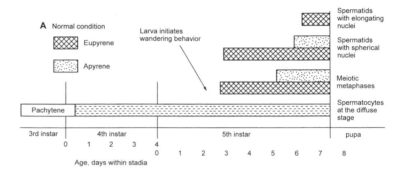

Fig. 1.18. Influence of parasitism by larvae of *Ascogaster quadridentata* (Hymenoptera: Braconidae) on spermatogenesis in larvae of their host, the codling moth *Cydia pomonella* (Lepidoptera: Pyralidae). A. Developmental sequence of eupyrene and apyrene sperm in unparasitized larvae. B. The sequence in parasitized larvae. C. The sequence in pseudoparasitized larvae. (Adapted from *J. Insect Physiol.*, 41, J. J. Brown and M. Friedländer, Influence of parasitism on spermatogenesis in the codling moth, *Cydia pomonella* L., pp. 957–963, Copyright 1995, with permission from Elsevier Science.)

Spermatogenesis in parasitized host larvae is identical to that of nonparasitized larvae up to the beginning of the fourth instar; the latest stage observed is primary spermatocytes in the diffuse stage of meiotic prophase (Fig. 1.18A, B). However, 2 d after the molt into this instar, parasitized hosts initiate precocious wandering and cocoon spinning, and the parasitoid larva molts to a second instar (Fig. 1.18B) (normal host larvae molt to the fifth instar [Fig. 1.18A]). In hosts with a late second instar parasitoid present, primary spermatocytes undergo precocious meiosis, and eupyrene spermatids commence spermiogenesis on day 5 of the fourth instar, but both are accompanied by degeneration of primary spermatocytes. Spermatogenesis ends when the parasitoid larva molts to a third instar, exits the host, and consumes it (Fig. 1.18B). In normal host larvae, meiosis and spermiogenesis of eupyrene spermatocytes do not begin until day 3 of the fifth instar, and those of apyrene spermatocytes do not begin until day 5, the larva pupating on day 7 (Fig. 1.18A).

Other host eggs, chilled after late exposure to a parasitoid egg (chilling kills the parasitoid), develop into pseudoparasitized larvae having testes but no parasitoid. Spermatogenesis in these is identical in timing to that of late parasitized larvae except that apyrene meiosis and

spermiogenesis also occur (Fig. 1.18C). Host wandering, cocoon spinning, and presumably endocrine changes associated with this behavior in both parasitized and pseudoparasitized host larvae cause both precocious spermatogenesis and the parasitoid larva to molt to instar 2. When both parasitoid larva and testes are present, they compete for available ecdysteroids so that some primary spermatocytes degenerate (Fig. 1.18B), exactly as occurs when normal host males enter diapause (Fig. 1.18A). However, in pseudoparasitized larvae where only testes are present, spermatogenesis continues because more ecdysteroids are available (Fig. 1.18C). These observations show that the timetable for spermatogenesis in insects is flexible, is correlated with somatic development, and is induced by changes in ecdysteroid titer.

1.3.7.3 GENES AND SPERMATOGENESIS
As usual, the role of genes in spermatogenesis is best studied in *D. melanogaster* males, where action of both *pleiotropic genes* (those having many effects throughout the body) located on chromosomes 2 and 3 and male-specific genes on the Y-chromosome are involved (female *D. melanogaster* is 2n = 6A + XX; male is 2n = 6A + XY). The latter genes are expressed only during diplotene

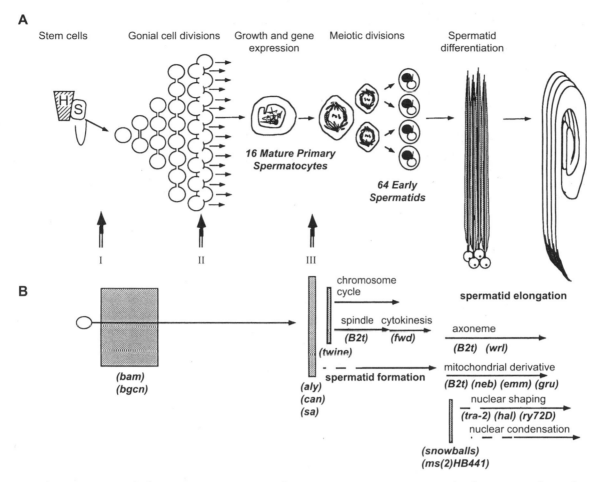

Fig. 1.19. The genetic control of spermatogenesis in *D. melanogaster*. A. Major events, I–III, the three principal switch points in the development of male germ cells. I. A stem cell (S) produces a single primary spermatogonium (cystoblast) by asymmetrical mitosis. This cell divides synchronously to produce a clone of 16 interconnected primary spermatocytes (cystocytes). II. Premeiotic DNA sythesis occurs early in the 16-cell stage, marking the onset of meiosis, gene expression, and growth in the primary spermatocytes. III. The primary spermatocytes exit meiotic prophase (extended G2) and divide twice in the two meiotic divisions to generate 64 spermatids, which then undergo spermiogenesis. H, hub. B. Regulatory circuit diagram summarizing the genetic control of spermatogenesis. Horizontal arrows, independent differentiation pathways; shaded vertical bars, control points. Some genes thought to function at control points or in specific morphogenetic pathways are shown in parentheses. Genes: *aly*, always early; *bam*, bag-of-marbles; *bcgn*, benign gonial cell neoplasm; *B2t*; *can*, cannonball; *emm*, emmenthal; *fwd*, four wheel drive; *gru*, gruyère; *hal*, halted; *ms(2)HB441*, male sterile(2)HB441; *neb*, nebbish; *ry72D*; *sa*, spermatocyte arrest; *tra-2*, transformer-2; *wrl*, whirlygig. (Reprinted with permission from M. T. Fuller. Spermatogenesis. In M. Bate and A. Martinez Arias (eds.), *The development of* Drosophila melanogaster. Vol. 1, p. 112, Fig. 9. © 1993 Cold Spring Harbor Laboratory Press.)

of meiotic prophase when long-lived mRNAs functioning in spermiogenesis are synthesized, as shown by [³H]uridine autoradiography (Lifschytz and Hareven, 1977; White-Cooper et al., 1998). Differentiation of sperm from haploid spermatids is thus governed by the diploid genome. In fact, in *disjunction-defective* mutants of *Drosophila*, the chromosomes are allocated unequally to secondary spermatocytes and spermatids during meiosis. As a result, some spermatids contain too few chromosomes, some too many, and others none at all. Yet all these spermatids, including those without chromosomes, differentiate into normally behaving sperm during spermiogenesis that can fertilize eggs (though

not successfully). The genetic products of the missing chromosomes thus must have been passed into the cytoplasm of the primary spermatocytes before meiosis.

Many other mutant loci also interfere with spermatogenesis in *Drosophila* males (Fuller, 1993), and the nature of the defects they cause enables one to infer what the functions of the wild-type alleles of these genes might be in normal spermatogenesis. However, it has been difficult to separate different phenotypic effects genetically in differentiating spermatids (Fuller, 1993).

Fig. 1.19, summarizing the principal events of spermatogenesis in *D. melanogaster*, shows the three switch points, I, II, and III, believed to act during male germ cell development, and indicates some of the genes involved. Note that many of the events of meiosis (cell cycle, spindle formation, cytokinesis) and spermiogenesis (DNA condensation, shaping of the sperm head, elongation of axoneme, and mitochondrial derivatives) appear to be under separate genetic control and to still proceed when other aspects have been interrupted by action of mutant alleles. In fact, I have commonly observed one or more giant spermatids with two or four axonemes and centriolar adjuncts in otherwise normal locust cysts, evidence for disruptions in the meiotic divisions but not in differentiation of spermatids. Table 1 in Fuller, 1993, summarizes the phenotypes induced by most mutations known to affect spermatogenesis in this insect.

2/ The Female Reproductive System and Oogenesis

2.1 Female Reproductive Systems

The eggs of insects are generated by, fertilized in, and deposited by the female's reproductive system, the principal components of which are the ovaries, lateral oviducts, common oviduct, genital chamber, spermatheca, accessory glands, and external genitalia (Fig. 2.1A). All these elements are present in females of most species but vary in relative number, size, shape, and position (Matsuda, 1976; Martoja, 1977; Kaulenas, 1992).

2.1.1 OVARIES

Each ovary contains from 1 (in certain apterygote hexapods) to more than 2000 (in certain queen termites) egg tubes or *ovarioles* (Fig. 2.2). The number is constant in females of some orders (e.g., 4 in female thrips) but can vary in others, within females of a single species, depending on how well they feed as larvae. For example, numbers ranged from 10 to 58 in females of the sarcophagid fly *Neobellieria bullata*, from 51 to 165 in those of the blowfly *Phormia regina* (Bennettova and Fraenkel, 1981), and from 36 to 46 in *Drosophila melanogaster* (Hodin and Riddiford, 2000a) (in females of the latter species, differences result from alterations in cell differentiation during the wandering stage of third instar larvae).

Each ovariole is surrounded by a loose-fitting outer sheath that is well supplied with tracheoles and, in some insects, with ramifying muscle fibers and *mycetocytes* (bacteria-containing cells [Fig. 2.6]) and within by a tightly fitting basal lamina or tunica propria (e.g., Bonhag and Arnold, 1961). In heteropterans, a second, tightly fitting inner sheath encloses the ovariole within the tunica (Huebner, 1984).

From apex to base, a typical ovariole comprises a terminal filament, an egg tube, and a short exit duct or pedicel (Fig. 2.1B). The terminal filaments of adjacent ovarioles in each ovary fuse together distally into a suspensory ligament, which may or may not fuse with that of the other ovary to form a median ligament. These end anteriorly in the fat body or are attached to the dorsal diaphragm, accessory glands, or body wall, depending on species (Matsuda, 1976; Martoja, 1977). Each terminal filament is surrounded by a basal lamina continuous proximally with that of the egg tube and may contain muscle fibers. The filaments function to suspend the ovarioles in the body cavity.

The apex of each ovariole is more or less expanded into a *germarium* originating during embryogenesis and containing proliferating female germ cells or oogonia, at least in juveniles; primary oocytes in early stages of meiotic prophase; and mesodermal prefollicular cells (Fig. 2.6).

The long proximal part of each ovariole is a zone of growth, the *vitellarium* (Fig. 2.1B). As each primary oocyte begins to enlarge at the base of the germarium, it pushes itself posteriorly into the apex of the vitellarium and becomes surrounded by mesodermal prefollicular cells, which eventually form a follicular epithelium about it, the whole being called an egg chamber or *follicle* (Figs. 2.3A, 2.6). While in the vitellarium, each primary oocyte completes meiotic prophase, rapidly enlarges primarily through uptake of yolk (*vitellogenesis*), and has a vitelline membrane and *chorion* (shell) deposited about it by its follicular cells after its growth is complete (Fig. 2.5). Since these events occur sequentially in each oocyte following adult emergence in anautogenous insects (those in which imaginal feeding is required for vitellogenesis and membrane deposition), the vitellarium in each ovariole of an ovipositing female usually contains a sequence of progressively larger follicles, with the largest, terminal (T) follicle at its base (Fig. 2.3A).

Insect ovarioles are classified according to how their eggs are nourished during oogenesis (Goss, 1903). In *panoistic* (all egg) ovarioles, each oocyte is surrounded by a follicular epithelium as described earlier (Fig. 2.3A) (follicles like these occur in the ovaries of many arthropods). This type is believed to be ancestral and that from which the other types evolved (Büning, 1994, 1998b). In *meroistic* (part egg) ovarioles, each oocyte, in addition, is serviced by one or more nurse cells (trophocytes) derived by incomplete cytokinesis of a common parental oogonium or *cystoblast*. The number of nurse cells per oocyte ranges from 1 to 31 and is characteristic of the species.

Two types of meroistic ovarioles exist. *Polytrophic* ovarioles contain an alternating succession of nurse cells and oocytes in the same (Fig. 2.3C) or in adjacent follicles. In *telotrophic* (acrotrophic) ovarioles, the nurse cells remain within the germarium (*tropharium*) after

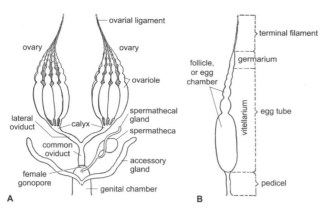

Fig. 2.1. Structure of the female reproductive system in an insect. A. The ovaries, exit ducts, and associated structures. B. An ovariole. (Reproduced from R. E. Snodgrass. 1935. *Principles of insect morphology*, p. 553, Fig. 284. McGraw-Hill. Figure owned by U.S. Department of Agriculture, Bureau of Entomology and Plant Quarantine.)

they are generated, but are attached to each oocyte by way of a *trophic core* (the microtubule-rich center of the tropharium) and individual *trophic (nutritive) cords*, the latter lengthening and thickening as the oocytes enlarge and move out of the germarium and down the vitellar-

ium (Fig. 2.3B). The adaptive significance, phylogenetic distribution, and evolution of these ovariolar types are discussed in 2.2.7.

2.1.2 PRIMARY EXIT SYSTEM
Each ovariole empties proximally into its own pedicel, and each pedicel into a lateral oviduct, which may be more or less expanded at its tip into a calyx (Fig. 2.1). Both ducts lack a cuticular intima, indicating their mesodermal origin, but in higher flies can be lined with cuticle because they derive from the ectodermal, secondary exit system (10.2.3.1.2) (Matsuda, 1976). Each lateral oviduct is a simple epithelial tube having external circular and longitudinal muscles and serves to convey fully developed eggs to the secondary exit system following ovulation, to store eggs prior to fertilizing them, or to produce secretions about them (Kaulenas, 1992). In female mayflies (Ephemeroptera), only this system is present.

2.1.3 SECONDARY EXIT SYSTEM
The secondary exit system develops after hatch by invagination of ventral abdominal epidermis in genital segments 7–9 (Fig. 10.13), is lined throughout with cuticle continuous posteriorly with that of the body wall, and comprises the common oviduct, genital chamber,

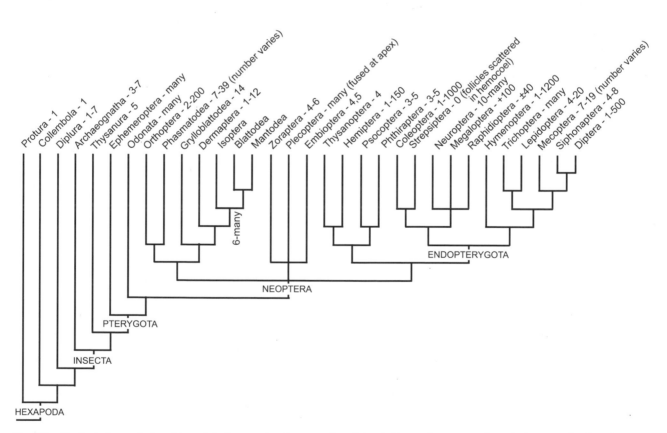

Fig. 2.2. Evolutionary relationships of the hexapod orders as inferred by Gullan and Cranston (1994), showing the distribution of ovariolar number per ovary. (Data from Matsuda [1976] and Martoja [1977]. Cladogram adapted with permission from P. J. Gullan and P. S. Cranston. *The insects. An outline of entomology*, p. 193, Fig. 7.5. 1st ed. Chapman and Hall, London. © 1994 Blackwell Science, Ltd.)

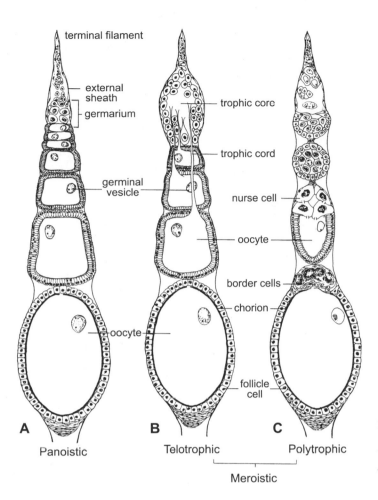

terminal filament

external
sheath
germarium

germinal
vesicle

oocyte

A
Panoistic

trophic core

trophic cord

nurse cell

oocyte

border cells

chorion

oocyte

follicle
cell

B
Telotrophic

C
Polytrophic

Meroistic

Fig. 2.3. The principal types of insect ovarioles. A. Panoistic. B. Telotrophic. C. Polytrophic. (Reprinted with permission from F. E. Schwalm. *Insect morphogenesis*, p. 27, Fig. 8. Monogr. Dev. Biol. 20. © 1988 S. Karger.)

and its two derivatives, the spermatheca and accessory glands (Fig. 2.1A). The posterior ends of the lateral oviducts open into a common oviduct, which varies in length and complexity depending on species and is usually surrounded by longitudinal and circular muscles. Its posterior opening, the *female gonopore*, can be at the posterior margin of either the seventh (Dermaptera) or eighth (other insects) abdominal sternum but is more often concealed within a genital chamber.

The genital chamber is a cuticle-lined inflexion of body wall from behind the eighth or ninth sternum that has the common oviduct and spermathecal duct opening dorsally into it at its anterior end and the accessory gland(s) more posteriorly. Since it is usually occupied by the tip of the male's intromittent organ or by a spermatophore during copulation (3.3.2), it is often called the *bursa copulatrix* or *vagina*. It opens posteriorly to the outside or into the base of the ovipositor through an opening called the *vulva* at the posterior margin of the eighth sternum in most insects but at the back of the ninth in female cicadas, scorpionflies, caddisflies, lepidopterans, and beetles (Snodgrass, 1933, 1935; Matsuda, 1976). In females of Lepidoptera-Ditrysia and of certain Hemiptera-Fulgoromorpha, openings on both the eighth (vulva or *ostium bursae*) and ninth (*oviporus*) sterna persist, the former functioning during copulation and the

latter during oviposition. And, in some tachinid and pupiparous flies (e.g., *Glossina* spp.), the genital chamber is greatly enlarged to store eggs prior to oviposition or to hold the larvae after they hatch; in the latter instance it is called a *uterus*.

Since females of most insects fertilize and release their eggs some time after insemination, most have one or more *spermathecae* (sperm receptacles, receptacula seminis) in which the sperm are stored prior to use. These open anteriorly by ducts into the dorsal wall of the genital chamber near its juncture with the common oviduct. This part of the chamber is more or less expanded into a fertilization chamber in which actual transfer of sperm from spermatheca to egg occurs (Fig. 3.11). In most species the spermatheca is single (although there can be up to four in flies) and has glands opening into it that secrete products maintaining the stored sperm, and in some species it has a muscular pump in its duct that is innervated from the terminal ganglion of the ventral nerve cord and that functions to release sperm prior to fertilization of eggs (Fig. 3.7C).

Accessory glands associated with the reproductive system in female insects generally assist in the deposition of eggs (Snodgrass, 1935; Kaulenas, 1992). They open either dorsally into the genital chamber behind the spermatheca or into the base of the ovipositor between

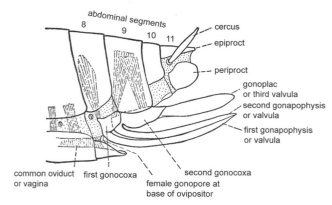

Fig. 2.4. Structure of the valvular ovipositor in a pterygote insect, showing the segmental relations of its parts. (Reproduced from R. E. Snodgrass. 1935. *Principles of insect morphology*, p. 611, Fig. 314A. McGraw-Hill. Figure owned by U.S. Department of Agriculture, Bureau of Entomology and Plant Quarantine.)

the upper valves. Depending on species, these glands produce an adhesive used to stick eggs to the substrate and secretions that lubricate the movement of her ovipositor valves upon one another during oviposition or that may accumulate about an egg batch as a frothy mass that hardens in air to form a protective egg case or *ootheca* (in some mantids and cockroaches). In aculeate (sting-bearing) hymenopterans, a modified accessory (Dufour's) gland produces venoms used in defense or in paralyzing prey, and in pupiparous flies the glands synthesize a "milk" used to nourish the larva until its deposition prior to pupariation.

2.1.4 EXTERNAL GENITALIA

The external genitalia in female insects comprise an ovipositor for depositing eggs into soil, leaves, stems, or wood or on or into the bodies of other insects (Snodgrass, 1933; Scudder, 1971; Matsuda, 1976; Hinton, 1981; Zeh et al., 1989). It has been secondarily reduced or lost in females of many insects (e.g., Coleoptera, Lepidoptera, Diptera), and in these, the terminal abdominal segments, the *oviscapt*, can be slender, telescoping, and prehensile and used for depositing eggs into fruit, seeds, dung, or carrion or into crevices or under leaves (Fig. 3.20) (Zeh et al., 1989).

In most female pterygotes, the ovipositor is valvular and consists of a shaft and basal apparatus (Fig. 2.4). The shaft usually comprises two or three pairs of long valves, the first and second *gonapophyses* (valvulae) (Scudder, 1971). The first pair are borne by the eighth segment and are ventral, while the second pair arise from the ninth sternum, are dorsal, and are often fused to each other medially into a single functional unit. The lower margin of each upper valve is fashioned into a ridge (*rhachis*), which fits into a corresponding groove (*aulax*) in the upper margin of each lower valve, the fully engaged interlock being called the *olistheter* (Smith, 1969). Thus, the upper and lower valves can slide along each other in

a longitudinal direction without separating, owing to alternate contraction and relaxation of tractor muscles inserting into their basal plates, the first and second *gonocoxae* or valvifers (Snodgrass, 1935; Scudder, 1971). The valves generally bear characteristic teeth on their dorsal and ventral margins, which slice the substrate during insertion, and posteriorly pointing microtrichia (*pectines*) on their inner surfaces, which function to move an egg down the shaft of the ovipositor during oviposition (Austin and Browning, 1981) (3.9.1).

A third pair of valves, the *gonoplacs* (third valvulae), borne by the posterior ends of the second gonocoxae, usually ensheath the distal part of the shaft but form a third pair of valves, either short or long, in orthopterans (Snodgrass, 1933, 1935). All three pairs of valves and the inner surface of the secondary exit system are richly supplied with sense organs or *sensilla* whose sensory neurons project to the terminal ganglion of the ventral nerve cord (Fig. 3.12). These sensilla are receptive to stimuli emanating from the oviposition substrate, to passage of the egg during oviposition (Sugawara, 1993), or to the intromittent organ or spermatophore of the male during copulation (Scudder, 1971). The genital chamber opens posteriorly into the base of the ovipositor between its upper and lower valves (e.g., in many hemipterans), or at its base at the back of the eighth sternum (in terebrantian thrips) (compare Fig. 2.4).

Zeh and colleagues (1989) consider evolution and adaptive radiation of the ovipositor, along with associated development of complex egg coverings and embryonic membranes, to have contributed to the diversification of ectognathous insects by enabling them to enter previously inaccessible habitats (3.9.2).

2.2 Oogenesis

A newly fertilized insect egg contains a blueprint for embryogenesis and postembryogenesis encoded in the DNA of its zygote nucleus and all resources required for development of the embryo until it can produce them itself or obtain them from its environment. It is surrounded by a vitelline membrane and chorion that enable it to respire but more or less protect it from desiccation, mechanical damage, predation, or parasitism (Browder, 1980; Zeh et al., 1989; Gerhart and Kirschner, 1997).

The early history of oogonia is identical to that of spermatogonia (1.3.1, 8.4.1) except that they eventually lodge in two ovary rudiments. After hatch, the oogonia begin to proliferate through asynchronous mitosis, a process continuing, in larger anautogenous females, until after adult emergence within germaria of the ovarioles of each ovary. Eventually, each oogonium ceases dividing, becomes a primary oocyte, and begins to grow just like a primary spermatocyte. This growth involves the synthesis of new cytoplasm followed by yolk uptake and continues until the vitelline membrane and chorion are deposited about the oocyte by the follicular cells shortly before it is ovulated.

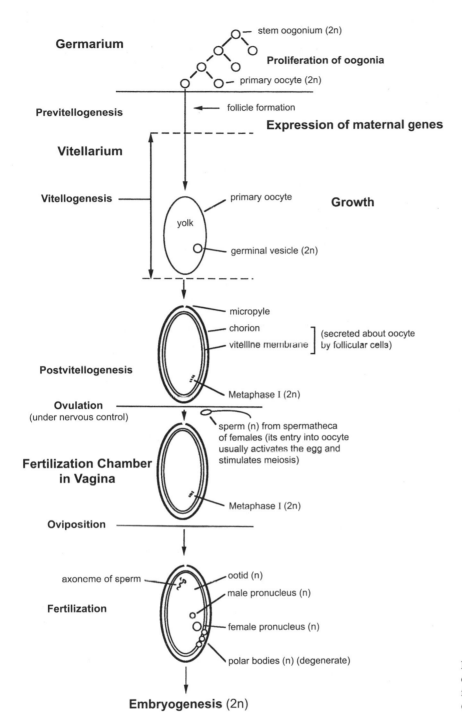

Germarium

stem oogonium (2n)

Proliferation of oogonia

primary oocyte (2n)

Previtellogenesis — follicle formation

Expression of maternal genes

Vitellarium

Vitellogenesis — primary oocyte

Growth

yolk

germinal vesicle (2n)

micropyle

chorion

vitelline membrane] (secreted about oocyte by follicular cells)

Postvitellogenesis

Metaphase I (2n)

Ovulation
(under nervous control)

sperm (n) from spermatheca of females (its entry into oocyte usually activates the egg and stimulates meiosis)

Fertilization Chamber in Vagina

Metaphase I (2n)

Oviposition

axoneme of sperm — ootid (n)

male pronucleus (n)

Fertilization

female pronucleus (n)

polar bodies (n) (degenerate)

Embryogenesis (2n)

Fig. 2.5. Summary of oogenesis, oviposition, and fertilization in a female insect having panoistic ovarioles.

In insects, oogenesis involves three phases—previtellogenesis, vitellogenesis, and postvitellogenesis (Fig. 2.5)—and differs from spermatogenesis in seven ways:
- Fully differentiated eggs are not motile.
- Differentiation of insect oocytes precedes meiosis rather than follows it.
- Primary oocytes enlarge much more than primary spermatocytes (up to 90,000-fold in oocytes of *D. melanogaster* [King, 1970] and up to 2.5 million-fold

in those of the American cockroach *Periplaneta americana* [Bonhag, 1959], compared with a maximum of 78-fold in primary spermatocytes).
- Fully developed oocytes usually have two external envelopes deposited about them by the follicular cells, the vitelline membrane, and *chorion*.
- Meiosis is usually stimulated by sperm entry and occurs after ovulation and often after oviposition (in males meiosis occurs within the sperm tubes of each testis).

- Division of the cytoplasm during meiosis is unequal, resulting in one large *ootid* (ovum, mature egg) and two or three small polar bodies, which later degenerate (four functional spermatids descend from each primary spermatocyte in males).
- Meiosis is generally blocked at metaphase I in females but not in males (it continues when the egg is activated by the sperm at fertilization).

2.2.1 PANOISTIC OVARIOLES

Oogenesis is similar in most female animals (Matova and Cooley, 2001). In insects, the simplest sequence occurs in females having panoistic ovarioles in which each division of an oogonium is followed by complete cytokinesis, and in which all types of RNA accumulating in the ooplasm are synthesized on chromosomes of the oocyte's own nucleus. As shown recently, the females of some insects, long thought to have typical panoistic ovarioles, have evolved this condition secondarily from polytrophic ancestors; such insects are said to have "neopanoistic" ovarioles (2.2.7) (Stys and Bilinski, 1990; Büning, 1994, 1998b).

2.2.1.1 PREVITELLOGENESIS

2.2.1.1.1 OOGONIAL PROLIFERATION

Proliferation of oogonia usually occurs briefly in the embryo, just after the germ cells are segregated; ceases for the rest of embryogenesis; and begins again in each ovary rudiment just after hatch. All oogonia have the potential to become oocytes, and in anautogenous species, they continue to proliferate throughout postembryogenesis and into the adult. In those of autogenous females, proliferation generally ends well before adult eclosion.

2.2.1.1.2 GENERATIVE PHASE

Eventually, each oogonium ceases dividing and becomes a primary oocyte at the base of the germarium (Fig. 2.6) (Bonhag, 1959). The movements of its chromosomes during meiotic prophase (leptotene, zygotene, pachytene, diplotene, and diakenesis) are identical to those of primary spermatocytes, and the cells resemble them in having relatively large nuclei and little cytoplasm.

2.2.1.1.3 VEGETATIVE (PREVITELLOGENIC) PHASE

During the vegetative phase of development, the nucleus of each primary oocyte enlarges through uptake of water to become a germinal vesicle, while its cytoplasm begins to grow and to increase its uptake of basic dyes. Because of this growth, each vegetative oocyte "pushes itself" out of the germarium into the apex of the vitellarium, where it is surrounded by mesodermal prefollicular cells (Fig. 2.6). These subsequently proliferate to form a follicular epithelium about each oocyte (Fig. 2.7C).

Meiotic prophase is temporarily arrested early in diplotene at a time following crossing-over when the homologous chromosomes are just beginning to separate from one another. Thereafter, the chromatin of each chromosome uncoils to such an extent that it is difficult to identify, even in cells stained with DNA-specific stains such as the Feulgen reagent, a stage thus often referred to as the *diffuse stage* (Fig. 2.6) (Telfer, 1975).

Amplification of rDNA

In insects having panoistic (and in some having polytrophic) ovarioles, the genes encoding rRNA are replicated many times (i.e., they are amplified) prior to and during meiotic prophase (Telfer, 1975: Table 1; Cave, 1982; Büning, 1994, 1998b). This amplification is correlated with development of a prominent, extrachromosomal DNA body or nucleolus within the nuclei of oogonia and primary oocytes (Fig. 2.7B, C). The genes encoding 18S and 28S rRNAs are located within the nucleolar organizer region of the chromosomes and are present in 50–300 copies per haploid genome as tandem repeat units (the number of copies varies with species and population) (Cave, 1982; Büning, 1994, 1998b).

Amplification of rDNA genes can be monitored autoradiographically by following the uptake of [3H]thymidine during pachytene. Transcription of rDNA into rRNA is usually associated with the appearance of many small "multiple nucleoli" in the germinal vesicle (Figs. 2.6, 2.7C) and usually begins in diplotene (demonstrated by autoradiographic uptake of [3H]uridine). Each of these nucleoli consists of a ring of circular DNA and associated proteins bearing adjacent pairs of fuzzy and smooth segments repeated one to six times along its axis (Fig. 2.7C) (Miller and Beatty, 1969; Trendelenberg et al., 1977; Büning, 1994, 1998b; Kubrakiewicz and Bilinski, 1995). The fuzzy portions of the ring are actively transcribing *transcription units* (rRNA genes) and the smooth portions are of nontranscribed "spacer DNA."

Ribosomal proteins synthesized on already present cytoplasmic ribosomes are transported through nuclear pores into the germinal vesicle and to multiple nucleoli where they complex with pre-rRNA and 5S rRNA to form the ribosomes (Browder, 1980). These pass in large numbers back into the ooplasm through the nuclear pores as emission bodies (EBs), where they cause it to darken (Fig. 2.6, growing oocytes at bottom). Eighty percent to 90% of the RNA synthesized during previtellogenesis in the cricket *Acheta domesticus* is stable rRNA (Hansen-Delkeskamp, 1969). Protein synthesis later occurring in preblastoderm embryos (Fig. 6.8) takes place on ribosomes synthesized primarily during the previtellogenic phase of oogenesis (monitored by autoradiographic uptake of 14C-labeled amino acids), and practically no rRNA synthesis occurs in embryos at this time (autoradiographic uptake of [3H]uridine) (Hansen-Delkeskamp, 1969).

Lampbrush Chromosomes

During meiotic prophase, in both primary spermatocytes (Fig. 4.1) and oocytes, pachytene and diplotene chromosomes (actually each of their two chromatids)

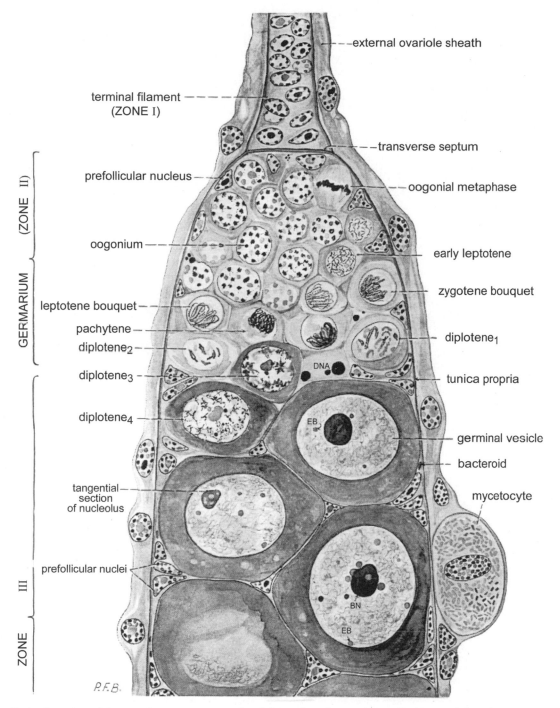

Fig. 2.6. Sagittal section of the tip of a panoistic ovariole of the cockroach *Periplaneta americana* (Blattodea), showing interphase and dividing oogonia and primary oocytes in leptotene, zygotene, pachytene, and diplotene of meiotic prophase. Notice how the chromatin becomes diffuse in diplotene stages 2, 3, and 4. BN, "budding" nucleolus (extrachromosomal DNA body); DNA, pyknotic remnants of degenerating female germ cell; EB, emission body (circular rDNA and associated proteins). (Reproduced with permission from P. F. Bonhag. Histological and histochemical studies on the ovary of the American cockroach, *Periplaneta americana* (L.). *Univ. Calif. Publ. Entomol.* 16: 109; pl. 10. © 1959 University of California Press.)

extrude thousands of lateral DNA loops, each 2.5–14 μm long and covered by a matrix of ribonucleoprotein (Kunz and Glätzer, 1980). Marec (1996) estimated that about 4600 loops develop in each haploid genome in primary oocytes of the clothes moth *Ephestia kühniella*. These chromosomes appear fuzzy under the light microscope and are referred to as *lampbrush chromosomes* (Fig. 2.7C). Although most loops show no evidence of

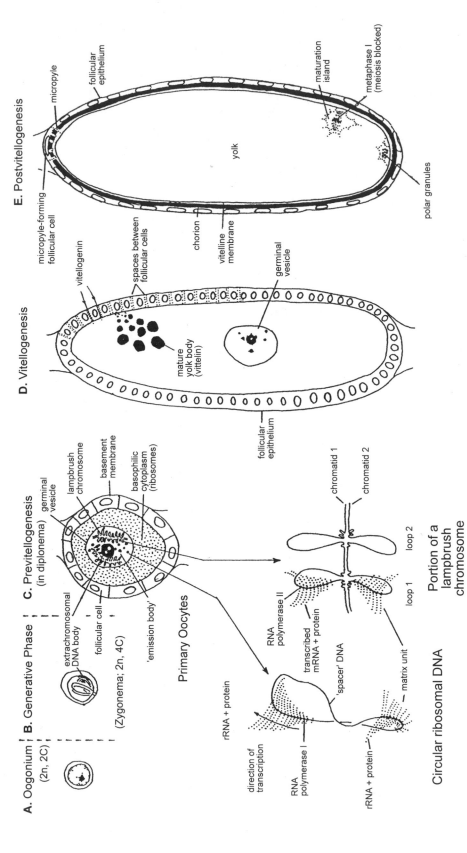

Fig. 2.7. Gene expression and vitellogenesis in, and deposition of vitelline membrane and chorion about, a primary oocyte of an insect with panoistic ovarioles. The oocyte in E is about to be released from its follicle.

gene transcription, some bear one fuzzy region or more, each representing pre-mRNA or tRNA being transcribed, which later function in protein synthesis in preblastoderm embryos. This mRNA is often referred to as "long-lived maternal mRNA" because it is synthesized in the ovaries of the mother and is present for such a long time before being translated into protein in the early embryo.

Because of the relatively low (4C) amounts of DNA (one C for each chromatid of each of the two homologous chromosomes of each pair) in the chromosomes of their germinal vesicles, panoistic oocytes are relatively slow growing (only ribosomal genes are amplified) prior to yolk uptake. Gene expression continues into vitellogenesis.

2.2.1.2 VITELLOGENESIS

The component responsible for most growth in an insect oocyte is yolk, and its uptake during vitellogenesis seems to be basically the same in insects regardless of ovariole type (Fig. 2.7D).

When an autogenous (pro-ovigenic [Papaj, 2000]) female emerges as an adult, each of her ovarioles already contains many fully developed and chorionated eggs, and additional ovulated ones are held in the primary and/or secondary exit duct ready for fertilization and release (Engelmann, 1970: Table 6). The yolk constituents for these are accumulated by larval feeding. For example, in 18-d female pupae of the giant silkworm moth *Hyalophora cecropia*, each of its eight ovarioles contains about 40 vitellogenic follicles, the yolk constituents of these being accumulated during larval feeding and yolk uptake beginning in older larvae or in early pupae (Telfer et al., 1982).

In anautogenous (synovigenic [Papaj, 2000]) females, the synthesis and accumulation of yolk do not begin until after the adult has emerged and begun to feed (Engelmann, 1970). Therefore, a "preoviposition period" of characteristic length ensues between adult emergence and appearance of the first batch of eggs (Hinton, 1981: Table 1). In such insects, vitellogenesis usually occurs only within the most proximal (T) oocyte of each ovariole, with the growth of these, in all ovarioles of both ovaries, being synchronized (Engelmann, 1970). When fully developed, all T eggs are ovulated together in an egg batch. With their deposition, a block to vitellogenesis of the next batch of eggs (in T-1 follicles) is removed and the process is repeated. Such insects thus tend to have *ovarian* (gonotrophic) *cycles*, times when eggs are being deposited and times when they are not.

2.2.1.2.1 LIPID YOLK

Most insect yolk proteins or *vitellogenins* (Vgs) are 7%–15% lipid (Raikhel and Dhadialla, 1992; Kaulenas, 1992). The remaining lipids are delivered to the yolk from their synthesis in the fat body by *lipophorins*, a group of shuttle proteins classified as low-density lipoproteins (LDLps), high-density lipoproteins (HDLps), and very high-density lipoproteins (VHDLps) based on their buoyant density. In lepidopterans such as *Bombyx mori* and *Manduca sexta*, the lipophorins shuttle lipid precursors from fat body to ovary and to other tissues, particularly the wing muscles, and also become major constituents of the egg proteins. In *M. sexta*, for example, over 90% of egg lipid is transported by LDLp, which is not itself sequestered into oocytes—it is recycled. HDLp is taken into oocytes by receptor-mediated endocytosis just like the Vgs (see 2.2.1.2.3).

2.2.1.2.2 CARBOHYDRATE YOLK

Carbohydrate yolk occurs within the egg as glycogen deposits or as mucopolysaccharide-protein complexes in the protein yolk bodies and is synthesized within the oocyte itself toward the end of vitellogenesis (monitored autoradiographically by uptake of ^{14}C-labeled glucose) (Mahowald, 1972).

2.2.1.2.3 PROTEIN YOLK

Sixty percent to 90% of the yolk proteins in insect eggs are synthesized in the fat body (the insect "liver") of female insects and are released into the blood as female-specific blood proteins, Vgs (Raikhel and Dhadialla, 1992; Izunni et al., 1994; Raikhel and Snigirevskaya, 1998; Sappington and Raikhel, 1998). These generally large glycolipophosphoproteins consist of two or more subunits and are synthesized as single or multiple precursors before being processed for secretion into the blood. In most insects examined, the large Vgs have molecular sizes of 210–652 kilodaltons (kDa) but in higher dipterans, the yolk proteins are very different and consist of three to five discrete peptides ranging from 44 to 51 kDa. In addition, the yolk proteins of such flies are synthesized in both fat body and follicle cells. After uptake by the oocyte, they are stored in crystalline form and are referred to as *vitellins* (Vns). These are identical to their precursor Vgs immunologically but differ in lipid content.

Additional yolk proteins are synthesized in the follicular cells. For example, in lepidopterans, paravitellogenin and an egg-specific protein (ESP) are secreted into the interfollicular spaces and are internalized within the oocyte.

Vitellogenin Uptake

Vgs are carried by the hemolymph to the ovaries; pass through the basal lamina surrounding each follicle and then between the follicular cells, which are separated ("patent") at this time; and are selectively concentrated within the oocyte at 20- to 100-fold their concentration in the blood by receptor-mediated pinocytosis (reviewed by Marsh and McMahon, 1999; Gillooly and Stenmark, 2001). Molecules of a particular Vg, the *ligand*, are bound by Vg-specific receptor proteins in the oolemma (Fig. 2.8). Following binding, ligand-receptor complexes concentrate in coated pits that invaginate and pinch off to form intracellular coated vesicles. These carry the Vg to an endosome, which directs subsequent intracellular movement of both ligand and receptor. Acidification of

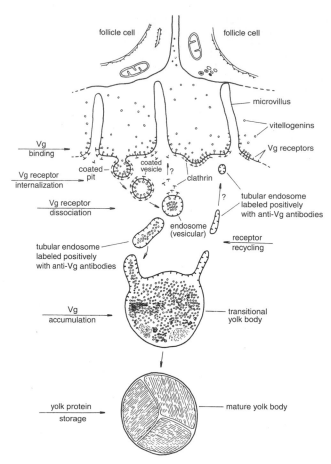

Labels in figure:
follicle cell · follicle cell
microvillus
vitellogenins
Vg receptors
Vg binding
coated pit · coated vesicle · clathrin
Vg receptor internalization
Vg receptor dissociation
tubular endosome labeled positively with anti-Vg antibodies
endosome (vesicular)
receptor recycling
tubular endosome labeled positively with anti-Vg antibodies
Vg accumulation
transitional yolk body
yolk protein storage
mature yolk body

Fig. 2.8. Internalization of vitellogenins (Vg) in a mosquito oocyte undergoing vitellogenesis. (With permission, from the *Annual Review of Entomology*, Volume 37 © 1992 by Annual Reviews www.AnnualReviews.org.)

the endosome results in disassociation of ligand-receptor complexes, followed by recycling of receptors back to the cell surface and delivery of free Vn into mature yolk bodies. Both coated pits and vesicles are surrounded by a protein lattice of clathrin. Rate of Vg uptake varies at different periods during vitellogenesis.

Vg receptors in the oolemma have been partially identified in several insects (Sappington and Raikhel, 1998). In females of the grasshopper *Locusta migratoria*, for example, one Vg receptor is an acidic, 180-kDa glycoprotein localized in the oolemma and in coated vesicles (Ferenz, 1993), while that of the mosquito *Aedes aegypti* has a molecular size of 205 kDa and exists as a 390-kDa homodimer in its native state (Sappington and Raikhel, 1998). The latter is present in the oolemma of previtellogenic oocytes on adult emergence, increases rapidly between 8 and 24 h after a blood meal, and begins to decline thereafter. Sequence homologies in the Vg receptors of several insects, of other invertebrates, and in several vertebrates suggest that the genes encoding them have been evolutionarily conserved (Wahli, 1988).

Specificity of Vitellogenin Uptake

In vitellogenic oocytes of the giant silkworm *Hyalophora cecropia*, patent follicular cells secrete a sulfated proteoglycan matrix into the interfollicular spaces and into the space between the follicle cells and the oocyte (Telfer et al., 1982). This secretion is also necessary for Vg uptake, because if it is removed by treating the follicles with the enzymes pronase or testicular hyaluronidase, vitellogenesis ceases. Removing the follicular cells has the same effect.

Uptake of Vgs by the oolemma is very selective, as has been demonstrated by cross-species transplantation of ovarioles where only oocytes of species in the same or in closely related genera will take up host Vgs (Kambysellis, 1968, 1970; Bell, 1972; Hatakeyama et al., 2000). Ovary transplants have also been placed into the hemocoels of males of the same species. For example, Ballarino and coworkers (1991) transplanted ovary rudiments from newly ecdysed last instar caterpillars of the gypsy moth *Lymantria dispar* into male caterpillars of similar age and followed their subsequent development. The follicles in such transplants detached from the germarium and increased in size, but oocyte growth proceeded more slowly than normally. Only male hemolymph proteins were taken up by the oocytes during vitellogenesis, and Vg synthesis was not stimulated in the fat bodies of the male host. Choriogenesis eventually occurred but was much delayed from that occurring in females. It was not investigated whether such eggs were viable.

Male Proteins in Vitellogenesis

In some insects, proteins that are synthesized in the male's accessory glands and allocated to the semen or spermatophore wall enter the female during insemination and are incorporated into both vitellogenic oocytes and somatic tissues in a species-specific pattern (Kaulenas, 1992; Pitnick et al., 1997). These "ejaculatory donations" may both facilitate egg production by females and benefit males by enhancing their fertilization success or the viability of their offspring (Pitnick et al., 1997). For example, in a lineage of 34 *Drosophila* species, Pitnick and colleagues (1997) found ejaculate incorporation by female somatic tissues to have arisen independently seven and, in ovaries, three times; to be relatively lineage specific, perhaps owing to phylogenetic niche conservation; to be influenced by body size, sperm length, and induction of the insemination reaction in females (3.7); and to correlate with sex-specific age of reproductive maturity.

2.2.1.2.4 TERMINATION OF VITELLOGENESIS

Termination of Vg uptake in pupae of *H. cecropia* occurs when each oocyte is two-thirds its final volume (Telfer and Anderson, 1968), and is caused by formation of an occlusion zone in the form of septate desmosomes between adjacent follicular cells (Rubenstein, 1979). This halts the entry of tracer molecules such as horseradish peroxidase from the blood to the surface of the oocyte, and presumably the entry of Vgs.

The entire vitellogenic sequence has been studied in detail in at least 13 species of insects in seven orders having panoistic, polytrophic, or telotrophic ovarioles, and has been found to be basically the same in all (although the amino acid composition and complexity of the various Vns, and sometimes their source, differ between species) (Raikhel and Dhadialla, 1992; Raikhel and Snigirevskaya, 1998).

Most research today is on the molecular biology of the Vns (amino acid sequences, molecular sizes, synthesis, routes of entry into the oocyte, mobilization during embryogenesis, etc.); on their receptor proteins on the surface of the oocyte; and on the identity, nature, control, and expression pattern of the genes encoding the Vns and Vg receptor proteins (Wahli, 1988; Raikhel and Dhadialla, 1992; Kaulenas, 1992; Raikhel and Snigirevskaya, 1998).

2.2.1.2.5 GENES AND VITELLOGENESIS
Insect Vns are usually encoded by a small gene family. In *D. melanogaster*, for example, there are three single copy genes in the family—*yp1*, *yp2*, and *yp3*—all linked on the X-chromosome (Kaulenas, 1992; Spradling, 1993). These genes are expressed only in fat body and follicle cell chromosomes and are controlled by products of the autosomal sex determination genes, to be discussed later (4.2). Separate fat body– and follicle cell–specific consensus sequences (125 base pairs [bp]) located upstream of the initiation site of each gene determine the sex (female) and tissue specificity (fat body, follicle cells) of its expression.

Expression of Vg genes in fat bodies of females of the yellow fever mosquito *Aedes aegypti* is induced by release of 20-OH ecdysone (20E) (Raikhel et al., 1999). This binds to a nuclear ecdysteroid receptor (AaEcR) together with one or the other of two isoforms of an ortholog of the *Drosophila* Ultraspiracle protein (AsUsp). The ecdysone receptor complex then binds to the ecdysone response element (EcRE) of putative early ecdysone response genes including *AaE75*, with their products possibly inducing expression of the yolk protein genes (see 12.4.1.5 for a full summary of ecdysteroid function).

2.2.1.3 POLE PLASM FORMATION IN OOCYTES
Newly deposited eggs of the females in many advanced clades, particularly of those having polytrophic ovarioles, contain a prominent area of cytoplasm at their posterior ends called the *pole plasm* (germ plasm, oöplasm, germ tract determinants) (Fig. 2.7E) (Klag and Bilinski, 1993). This part of the egg is free of yolk and contains polar granules that seem to have a germ tract–determining function: cleavage energids entering this area prior to blastoderm formation are subsequently segregated as pole cells, whose progeny eventually differentiate into sperm or eggs (6.6). This specialized cytoplasm gradually differentiates, beginning just before vitellogenesis.

Though uninvestigated in any panoistic insect having pole plasm (e.g., the thrips *Haplothrips verbasci*

[Heming, 1979] and *Bactrothrips brevitubus* [Haga, 1985]), the molecular genetics of pole plasm assembly in polytrophic ovarioles of *D. melanogaster* has been investigated in several laboratories. At least eight genes are involved—*cappuccino* (*capu*), *spire* (*spir*), *staufen* (*stau*), *oskar* (*osk*), *vasa* (*vas*), *valois* (*vls*), *mago nashi* (*mago*), and *tudor* (*tud*)—and some of these have been characterized at the molecular level (Ephrussi et al., 1991; Mahowald, 1992). Embryos derived from females that are homozygous mutant in any of these "*grandchildless*" genes lack polar granules in their pole plasm, fail to form pole cells, and develop into sterile adults.

All these genes begin to be expressed in nurse cell nuclei of previtellogenic follicles at about the time that the oocyte becomes differentiated from the nurse cells (the cystocyte clones are still in the germarium at this time), and their transcripts or proteins seem to be required for distribution, insertion, localization, or anchoring of polar granule components at the posterior pole of the oocyte and for specifying pole cells.

Similar events occur during pole plasm formation in eggs of the polytrophic ichneumonid wasps *Cosmoconus meridionator* and *Lissonota catenator* but have yet to be investigated at the molecular level (Klag and Bilinski, 1993). In these, proteinaceous nuage–mitochondrial complexes form in the nurse cells and pass through the intercellular bridges of each follicle to the oocyte and along its oolemma to the posterior pole of the egg, where they interact with nuage of medium density, originating in the ooplasm. Both disappear with the formation of the RNA-positive pole plasm.

2.2.1.4 ECDYSTEROID PRODUCTION BY FOLLICULAR CELLS
In 1977, Lagueux and others demonstrated that follicular cells surrounding the terminal oocytes in panoistic ovarioles of the grasshopper *Locusta migratoria* synthesize ecdysteroids in large amounts and pass them into the oocyte at the end of vitellogenesis at the close of each ovarian cycle. Within the yolk, the ecdysteroids are conjugated and are bound to Vn. In addition, isolated follicle cells of this insect can synthesize ecdysone in vitro from [^{3}H]cholesterol. Thus, there are now known to be at least two sources of ecdysone production in addition to the prothoracic glands of juvenile insects: testis sheath cells in adult males (1.3.7.2) and follicular cells in adult females. Ecdysteroids have been recovered from the eggs of at least 11 other insects in nine orders, are probably universal in members of this class (Gutzeit, 1985), and seem to have a role in inducing deposition of the serosal and of the first and second embryonic cuticles (6.9.3, 8.6.3).

2.2.1.5 THE GERMINAL VESICLE AND THE METAPHASE I BLOCK
Yolk uptake by the oocyte occurs while its chromosomes are in diplotene of prophase I (the synthesis stage). Toward the end of vitellogenesis, during a period of oocyte maturation, the germinal vesicle moves to the

periphery of the cell and assumes a position characteristic for the species (Figs. 2.5, 2.7E). Its membrane disassembles, indicating the end of prophase I, and its chromosomes condense and congress at the equator of a spindle that has its long axis perpendicular to the surface of the egg within an island of cytoplasm called the *maturation island*; this spindle lacks centrosomes (Sharp et al., 2000). Meiosis is blocked at metaphase I and usually does not continue until after the egg is fertilized by a sperm or deposited by the female. In eggs of the parthenogenetic stick insect *Carausius morosus*, the block is removed by uptake of oxygen following oviposition (parthenogenetic females produce offspring in the absence of males and without the stimulus of sperm entry) (Pijnacker and Ferwerda, 1976).

The nature of this block is unknown in insects but may result from production of an "ovarian maturation factor" by follicle cells adjacent to the maturation island. Evidence for this comes from the study of oogenesis in the fungus-feeding gall midge *Heteropyza pygmaea* (Diptera: Cecidomyiidae) (Went, 1979). In this fly, females may reproduce bisexually or parthenogenetically as adults or paedogenetically as larvae, adults appearing only when living conditions experienced by the larvae begin to deteriorate (5.5.3). In female adults, oocytes undergo vitellogenesis and are arrested, as usual, in metaphase I. In paedogenetic larvae, previtellogenic ovarian follicles are released into the hemolymph, where meiosis (one division in female oocytes; two in males) ensues, followed immediately by embryogenesis. If the ovaries of such larvae are cultured in vitro in male-determining hemolymph, meiosis in their oocytes is blocked. If the follicle cells are removed from the oocytes prior to culture, the block does not occur, suggesting that both hemolymph factors and follicle cells are involved.

In *D. melanogaster* ovarioles, metaphase I arrest may be based on "chiasma-based kinetochore tension" (Jang et al., 1995). In normal oocytes, meiosis arrests, as usual, at metaphase I and resumes after ovulated eggs have passed through the lateral oviduct. However, it only occurs in oocytes that have undergone at least one meiotic exchange. Cross-overs between chromatids of homologous chromosomes attached by spindle microtubules to the same kinetochore do not induce meiotic arrest, whereas those between homologs attached to different kinetochores do, suggesting the arrest signal is the resulting tension on homologous kinetochores.

In eggs of the grasshopper *L. migratoria*, the block seems to be lifted, at least partly, by release of ecdysteroids from follicle cells in the posterior fourth of the egg, as it can be induced in a dose-related manner by culturing eggs of appropriate length in a medium containing free ecdysteroids (Lanot et al., 1987). In nature this occurs 60–90 min before fertilization. The whole process is well understood in clawed toed toad *Xenopus laevis* eggs, in which the block to meiosis occurs in metaphase II and is ended by release of gonadotropins from the pituitary gland (Alberts et al., 1994).

2.2.1.6. POSTVITELLOGENESIS

Toward the end of yolk uptake, the protein-synthesizing machinery of the follicle cells is reorganized to synthesize and secrete the vitelline membrane, followed by the chorion, about each full-sized oocyte. But first a few words on the history of the follicular cells during oogenesis.

2.2.1.6.1 FOLLICULAR CELLS

As each oocyte begins to grow, it enters the apex of the vitellarium and becomes surrounded by mesodermal prefollicular cells (Fig. 2.6). These rapidly proliferate about each oocyte for the rest of previtellogenesis, separate into subgroups that move to specific positions, and cease dividing prior to yolk uptake (Fig. 2.7C, D) (Mahowald, 1972; Kaulenas, 1992). The final number of follicle cells about each oocyte is characteristic for the species and is usually quite large. For example, there are approximately 27,000 in follicles of the cockroach *Leucophaea maderae*, about 10,000 in those of the moth *Antheraea polyphemus*, and approximately 1200 in those of *D. melanogaster* (Regier and Kafatos, 1985; Kaulenas, 1992).

When vitellogenesis begins, the follicle cells begin to grow and the DNA of their chromosomes to replicate endomitotically (only S and G phases in the cell cycle) (e.g., Pijnacker and Godeke, 1984). As a result, the cells gradually become columnar and their nuclei either become large and complexly lobed or divide in two (Heteroptera). The level of ploidy reached by nuclei is up to 45n in follicular cells of *D. melanogaster*, 64n in those of the stick insect *Carausius morosus* (Pijnacker and Godeke, 1984), and 512–1024n in those of *A. polyphemus* (Regier and Kafatos, 1985).

During vitellogenesis, the follicular cells separate somewhat from each other (they become patent) to allow passage of Vgs to the surface of the oocyte from the blood (Figs. 2.7D, 2.8), and with continued growth of the oocyte, the follicle cells become cuboidal and then flattened because their growth fails to match that of the cell they surround (Fig. 2.7E) (Kaulenas, 1992).

2.2.1.6.2 DEPOSITION OF EGG ENVELOPES

The synthesis and deposition of vitelline membrane and chorionic proteins has been followed in follicles of many pterygote insects by transmission (TEM) and scanning (SEM) electron microscopy, and in all instances, both envelopes have been shown to be products of the follicular cells (Fig. 2.7E) (although microvilli of the oocyte also contribute to vitelline membrane formation and, in some entognath hexapods, it is secreted by the oocyte) (Regier and Kafatos, 1985; Kaulenas, 1992; Margaritis and Mazzini, 1998). Both layers are thus secondary (primary envelopes are produced by the oocyte itself, and tertiary envelopes by cells in walls of the exit system, as in the eggs of stick insects [Phasmatodea] and of birds and reptiles).

Vitelline Membrane

The vitelline membrane (vm) is proteinaceous and often elastic. It usually begins to be deposited before the end of vitellogenesis. TEM and autoradiography with ^{3}H-labeled amino acid precursors (leucine, tyrosine) suggest the following sequence occurs:

vm genes in follicle cell chromosomes → vm mRNA
→ polysomes on rough endoplasmic reticulum (RER)
→ vm proteins → Golgi apparatus → Golgi vesicles
→ follicular cell apex → deposition

In eggs of *D. melanogaster*, the vitelline membrane consists of a few proteins and some other minor constituents encoded by a single gene cluster, is 0.5–0.75 μm thick, and requires 11 h to deposit at 25°C (Table 2.2) (Kaulenas, 1992; Spradling, 1993). In addition, recent evidence suggests that some of the positional signals specifying dorsoventral and terminal axes of the embryo may be temporarily embedded in the membrane during oogenesis (7.2.1.1). A wax layer serving to reduce water loss is deposited by the follicle cells on the surface of the vitelline membrane in eggs of most insects (Kaulenas, 1992).

Chorion

Insect egg shells are structurally and biochemically complex, are intricately sculptured on their surfaces with the imprint of their depositing follicle cells, and are punctured here and there by holes for sperm entry (*micropyles*), gas exchange (*aeropyles*), and water uptake (*chorionic hydropyles*), all in patterns characteristic for the species (there are numerous identification keys to the eggs of insects: Hinton, 1981; Margaritis and Mazzini, 1998). These holes are formed during chorion deposition by secretion of chorionic proteins about protoplasmic extensions of specialized follicle cells (Fig. 2.7E) (Regier and Kafatos, 1985).

In spite of this complexity, the deposition of chorion about an egg can be very rapid: in *D. melanogaster*, the chorion is 1.5 μm thick and is deposited during stages 11–13 in 5 h at 25°C (Table 2.2), while in *A. polyphemus* it is 25–40 μm thick and is deposited in 51 h at this temperature (Regier and Kafatos, 1985).

The chorion of *D. melanogaster* eggs contains 20 different proteins encoded by chorion genes on the X and third chromosomes. The mRNAs of these genes are transcribed in the follicle cells in a stage-specific sequence prior to and during choriogenesis (Spradling, 1993), and their expression is presently being investigated intensely in relation to the deposition of particular chorionic structures such as the operculum (Dobeus et al., 2000) and dorsal appendages (Peri and Roth, 2000).

Regier and Kafatos (1985) examined choriogenesis in eggs of two silkworm moths (*Antheraea polyphemus* and *Bombyx mori*) using a variety of techniques (SEM, TEM, quantitative autoradiography, two-dimensional gel chromatography, and immunocytochemistry) and showed isotope-labeled chorionic proteins to have the following origin and fate:

chorion genes in follicle cell chromosomes → chorion mRNAs → RER (2 min) → chorionic proteins → Golgi apparatus (10–20 min) → follicle cell apex (10–30 min) → discharge complete over surface of oocyte

As they are deposited in the chorion, the proteins rapidly assume a characteristic distribution that varies for different developmental stages. In choria of *A. polyphemus*, about 186 proteins are present, each with characteristic developmental kinetics. In addition, 23 of these are associated with specific chorionic structures such as micropyles and aeropyles.

Chorion Gene Families

Most chorionic proteins, though not those of *D. melanogaster*, are encoded by distinct but evolutionarily related families of chorion genes. These have probably evolved through selection to facilitate the accumulation of diverse chorionic gene products in massive amounts over a short time (many replicates of each gene with amplification of each gene resulting from endopolyploidy). In moths, over 100 genes are in the chorion family (Regier and Kafatos, 1985; Kaulenas, 1992).

Such families of closely related genes are believed to result from repeated tandem duplication of an ancestral gene followed by selection for divergence in descendant genes to encode different but related proteins (Dover, 1986; Lynch and Conery, 2000). As well, the proteins governing vesicular traffic in animal cells, including the follicle cells of insect ovaries, seem to be universal (Allan and Balch, 1999; Ellgaard et al., 1999).

A six-nucleotide consensus sequence associated with each of the chorion genes of both moths and flies specifies their expression in the follicle cells, and there are also control sequences that regulate when the genes are to be active.

2.2.2 MEROISTIC OVARIOLES

In meroistic ovarioles, at least some oogonial divisions within the germaria are incomplete, such that clusters (clones) of interconnected germ cells result that are identical to the spermatogonial clones of male sperm tubes (1.3.2.2) (Telfer, 1975). The cells within each cluster likewise remain connected to each other by polyfusomes (compare to Fig. 1.9), with one or more differentiating into oocytes and the remainder into nurse cells. In addition, the nurse cells have more or less usurped the function of the oocyte's chromosomes and cytoplasm in producing RNA and protein for the egg. In fact, Büning (1996) suggested the supraordinal taxon name "Meroista" for taxa having nurse cells in their ovarioles, including dermapterans, hemipteroids, and endopterygotes (Fig. 2.22).

Two types of clones are known to be produced in hexapod ovarioles: *linear* clones in ovariolar germaria of collembolans, campodeid diplurans, and certain

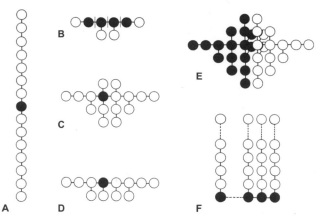

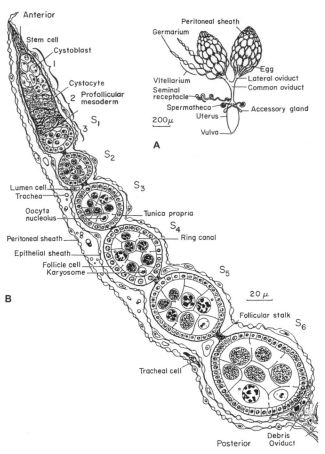

Fig. 2.9. Arrangement of sibling cells within linear (A) and branched (B–F) germ cell clusters. A. *Campodea* sp. (Diplura: Campodeidae). B. *Anisolabis maritima* (Dermaptera: Carcinophoridae). C. *Drosophila melanogaster* (Diptera: Drosophilidae). D. *Chrysopa perla* (Neuroptera: Chrysopidae). E. *Drepanosiphum platanoides* (Hemiptera: Aphididae). F. *Creophilus maxillosus* (Coleoptera: Staphylinidae). Open circles, prospective nurse cells; filled circles, prospective oocytes. (From P. Stys and S. Bilinski. 1990. Ovariole types and the phylogeny of the hexapods. *Biol. Rev.* 65: 401–429. Reprinted with the permission of Cambridge University Press.)

Fig. 2.10. A. The female reproductive system of *D. melanogaster* in dorsal aspect (two ovarioles have been teased out of the left ovary). B. Sagittal section of a single polytrophic ovariole with its investing membranes, showing the three germarial regions (1–3) and stages 1–6 of follicle development (S^1–S^6). (Reproduced with permission from R. C. King. *Ovarian development in* Drosophila melanogaster. © 1970 Academic Press, Inc.)

mayflies (Figs. 2.9A, 2.17A), and *branched* clones in those of pterygote insects having meroistic ovarioles (Fig. 2.9B–F) (Stys and Bilinski, 1990; Büning, 1994).

2.2.2.1 POLYTROPHIC OVARIOLES

In polytrophic ovarioles, the nurse cells accompany each vegetative oocyte as it enters the vitellarium (Fig. 2.3C). At the apex of each ovariole in *D. melanogaster* are two oogonial stem cells that develop from two prestem cells early in the pupal stage (Figs. 2.10B, 2.11) (King, 1970; Bhat and Schedl, 1997). Each is in contact with apical, mesodermal, terminal filament, cap, and inner germarial sheath cells (Xie and Spradling, 2000). The cap cells express the proteins Hedgehog (Hh), Fs(1)Yb, Decapentaplegic (Dpp), and Piwi (Lin and Schagat, 1997; King and Lin, 1999). Expression of the genes *decapentaplegic* (*dpp*), *fs(1)yb*, and *piwi* by the cap cells is required to maintain the stem cells and to promote their division. Mutations in *dpp*, or in the *saxaphone* gene encoding the receptor of Dpp protein in the stem cell membrane, retard their mitosis (Xie and Spradling, 1998, 2000), while receipt of the secreted Hh protein is required for proliferation of somatic (prefollicular) stem cells, ultimately generating the follicle cells that later surround each egg chamber (King and Lin, 1999).

Within each stem cell is a spherical, fusome-like organelle called a *spectrosome* (Fig. 2.11) containing the membrane-associated proteins ankyrin, alpha- and beta-spectrin (SP), and Hu-li tai shao (Hts) (Deng and Lin, 1997; Lin and Schagat, 1997; de Cuevas and Spradling, 1998). During each asymmetrical stem cell division, the spectrosome anchors one pole of the spindle to a cap cell

in such a way that the axis of each mitosis is parallel to the long axis of the ovariole (*hts* mutants lack a spectrosome and have randomized spindle orientation during stem cell mitoses). With each division, one daughter cell remains as the stem cell (S) and the other becomes a cystoblast (C_b) (Figs. 2.11, 2.12).

Each stem cell has a half-life of 4–5 wk and eventually separates from the cap cells to become a cystoblast; a new, replacement stem cell arises from division of the remaining stem cell perpendicular to the long axis of the ovariole (Fig. 2.11) (Xie and Spradling, 2000).

Just as in the sperm tubes of male insects (1.3.2.2), each cystoblast also contains a spectrosome and undergoes a characteristic number of incomplete, synchronous, mitotic divisions to yield a clone of 2^n cystocytes, with n being the number of division cycles producing the clone. The value for n is 2 in most mecopterans (to yield 3 nurse cells), 3 in most lepidopterans (7 nurse cells), 4 in most higher flies including *Drosophila* (Fig. 2.12) and in adephagan beetles (15 nurse cells), and 5 in

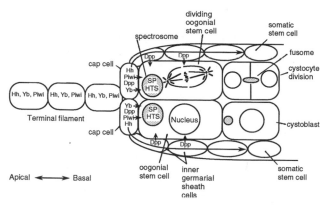

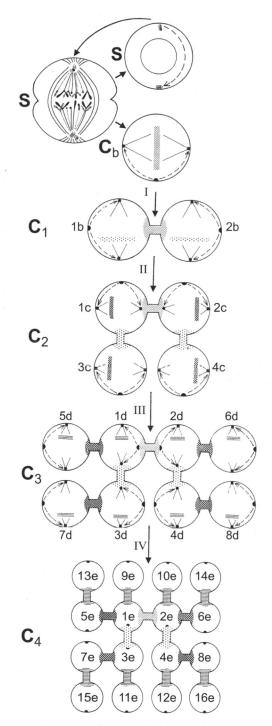

Fig. 2.11. The ovariolar apex of a female *D. melanogaster*, showing gene products involved in the maintenance and asymmetrical division of two germ-line, oogonial stem cells. The stem cells are in contact with apical cap cells expressing the Hedgehog (Hh), Fs(1)Yb (Yb), and Piwi proteins. Yb, Piwi, and Decapentaplegic (Dpp) proteins control maintenance and division of the germ-line stem cells. Within the apex of both interphase and dividing stem cells is a spectrosome (shaded) containing the spectrin (SP) and Hu-li tai shao (Hts) proteins. During stem cell mitosis, the spectrosome anchors one pole of the spindle to a cap cell so that it is parallel to the long axis of the ovariole. At mitosis, a daughter cystoblast is produced while the stem cell remains in contact with the cap cells. Signaling by Hh protein also induces proliferation in somatic (prefollicular) stem cells located 2–5 cells proximal to the signaling cap cells. (Adapted with permission from T. Xie and A. C. Spradling. A niche maintaining germ line stem cells in the *Drosophila* ovary. *Science* 290: 328–330. Copyright 2000 American Association for the Advancement of Science.)

some hymenopterans and fleas (31 nurse cells; in neuropterans, there may be 19–23 nurse cells/clone, indicating that some cystocyte divisions are asynchronous) (Kubrakiewicz, 1997). In a fully developed, 16-cell clone of *D. melanogaster* (C_4) two cells are attached to n bridges, two to n-1, four to n-2, and eight to n-3 (Fig. 2.12 [King, 1970]). The oocyte always develops from one of two pro-oocytes with n bridges and requires expression of the *spindle, egalitarian* (*egl*), and *Bicaudal-D* (*Bic-D*) genes to do so (Swan and Suter, 1996; Mach and Lehmann, 1997; González-Reyes et al., 1997); the other cystocytes become nurse cells.

Cytokinesis within the clone is synchronous and incomplete, owing to persistence of a polyfusome between adjacent cystocytes, just as in males (1.3.2.2 [McKearin, 1997; de Cuevas and Spradling, 1998]). That pro-oocyte within the cyst retaining the largest proportion of the polyfusome, and which is most posteriorly situated (it adheres more strongly to posterior follicle cells [González-Reyes and St. Johnston, 1998]), becomes the oocyte and is probably the original cystoblast (Fig. 2.10B: $S_1–S_6$).

Before the oocyte becomes visibly distinguishable within a clone (the latter is still within the germarium at this time; Fig. 2.10B: 1–3), a microtubule organizing center (MTOC) develops within it (Theurkauf et al.,

Fig. 2.12. The mitotic history of a cluster of 16 interconnected cystocytes in germarial region 1 (see Fig. 2.10B: 1) of a *D. melanogaster* ovariole. The cells are shown in a single plane, and the ring canals between them have been lengthened for clarity. Notice how the centrosomes divide and separate during each mitotic cycle and how the size of each cystocyte decreases as the number of cells in the clone increase. C_b, cystoblast; C^1–C^4, first-, second-, third-, and fourth-generation cystocytes; S, stem oogonium. (Reproduced with permission from R. C. King. *Ovarian development in Drosophila melanogaster*. © 1970 Academic Press, Inc.)

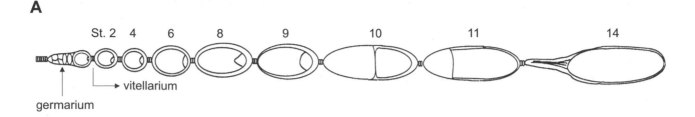

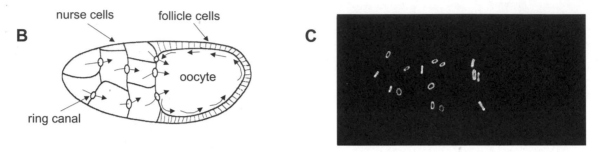

Fig. 2.13. A. A single ovariole of a mature *D. melanogaster* female, showing germarium and stages 2–14 of oogenesis (stages 1–6 are shown in Fig. 2.10B: S_1–S_6). B. In stage 10, nurse cell cytoplasm flows into the oocyte through the ring canals and is vigorously mixed within the ooplasm (arrows). C. Ring canals linking adjacent nurse cells and oocyte as illustrated by a fluorescently labeled antibody to Hu-li tai shao (Hts) protein (required for ring canal formation). (Reprinted with permission from L. Cooley and W. E. Theurkauf. Cytoskeletal functions during *Drosophila* oogenesis. *Science* 266: 590–596. Copyright 1994 American Association for the Advancement of Science.)

1993) (this fails to form in *hts* mutants [Deng and Lin, 1997]). The MTOC is a polarized network of microtubules that connects this cystocyte to all others within the clone by way of the ring canals, and later has a role in transporting transcripts, centrioles, and other organelles and proteins from the nurse cells into the oocyte (Fig. 2.13B). If colchicine, an inhibitor of microtubular assembly, is fed to a fly, the MTOC fails to form and all cystocytes of a clone become nurse cells. In *Bic-D* mutant flies, the MTOC likewise fails to form, whereas in *egl* mutants, it forms but is not stable, the end result being oocyte-free follicles in both instances. Proteins of both genes colocalize to the oocyte at all stages of oogenesis and apparently are components of a protein complex.

When mitosis is complete, each clone leaves the germarium, enters the apex of the vitellarium, and becomes surrounded by mesodermal prefollicular cells in the usual way (Fig. 2.10B: 2, 3). The oocyte in each clone then undergoes previtellogenic and vitellogenic growth, followed by vitelline membrane deposition and choriogenesis. Fourteen stages of development are recognized in an egg follicle of *D. melanogaster* after it leaves the germarium (Table 2.2; Figs. 2.10A, B: S_1–S_6; 2.13A: St. 2–14) (King, 1970). In cecidomyiid midges (Diptera), some nurse cells are mesodermal and do not belong to the germ line (Madhavan, 1973; Schüpbach and Camenzind, 1983).

The only known exception to this division sequence occurs in ovarioles of certain earwigs (Dermaptera) in which each cystoblast undergoes three synchronous mitotic divisions to generate a clone of eight interconnected cystocytes (Fig. 2.9B) (Yamauchi and Yoshitake, 1982). This subsequently divides into four 2-cell groups, each containing one oocyte and one nurse cell.

And ovaries of female twisted wings (Strepsiptera), though polytrophic, are highly modified: ovarial somatic tissues and the primary exit system are greatly reduced; oogonial stem cell mitosis occurs only in embryos; a previtellogenic growth phase for oocytes is missing; and growth of cytoplasm is shifted into the period of cystoblast mitosis before differentiation into nurse cells and oocytes occurs. The nurse cells remain diploid and their membranes degenerate at onset of vitellogenesis, which itself is greatly reduced (Büning, 1998a). Nevertheless, cystocyte proliferation follows the 2^n rule; all cystocytes form a rosette with a common polyfusome; and only one becomes an oocyte.

Each clone in ovarioles of the hystrichopsylloid flea *Hystrichopsylla talpae* (Siphonaptera) consists of 32 cystocytes: the oocyte and 31 nurse cells. However, the latter remain small, do not polyploidize their genomes endomitotically, do not exhibit streaming through their ring canals into the oocyte, show the same ultrastructural characteristics as oocytes (at least at first), and eventually end up as part of the follicular epithelium about each oocyte (Büning, 1993). Fleas in other families have secondarily panoistic (neopanoistic) ovarioles derived from the polytrophic type (Stys and Bilinski, 1990).

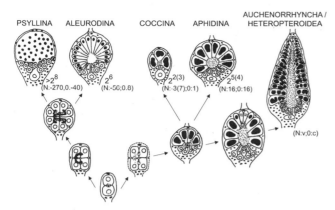

PSYLLINA   ALEURODINA   COCCINA   APHIDINA   AUCHENORRHYNCHA/HETEROPTEROIDEA

$>2^8$ 2^6 $2^{2(3)}$ $2^{5(4)}$
(N:-270,0,-40)   (N:-50;0.8)   (N: 3(7);0:1)   (N:16;0:16)

(N:v;0:c)

Fig. 2.14. The terminal ovariolar chambers of representative species of the principal subordinal taxa of Hemiptera, showing their phylogenetic relationships. Short dashes symbolize masses of microtubules. Permanent polyfusomes during cystocyte divisions exist only in the Psyllina/Aleurodina lineage. The hypothetical four-cell stage is surrounded by a broken line since, in coccines, cluster formation occurs in the ovary rudiment not in the ovariole, as occurs in females of all other taxa. Numbers represent the number of sibling cells. N, nurse cells (with dark nuclei); O, oocytes (with white nuclei); v, number varies; c, number constant. (Reprinted from *Int. J. Insect Morphol. Embryol.*, 22, J. Büning, Germ cell cluster formation in insect ovaries, pp. 237–253, copyright 1993, with permission from Elsevier Science.)

Known thrips (Thysanoptera) have long been thought to have panoistic ovarioles (Heming, 1995). However, Pritsch and Büning (1989) showed in ovarioles of the palm thrips *Parthenothrips dracaenae* that each oocyte arises by separation from a clone of 8 (possibly 16) cystocytes that originates exactly as in *D. melanogaster*. And very much the same events occur in ovarioles of investigated stoneflies (Plecoptera) and in at least some proturans, mecopterans (Boreidae), and megalopterans (Corydalidae), indicating they are all neopanoistic and derived from an ancestral polytrophic type (Stys and Bilinski, 1990).

2.2.2.2 TELOTROPHIC OVARIOLES
In telotrophic ovarioles, all or nearly all oogonial divisions are similarly incomplete and likewise generate one to several clones of interconnected cystocytes but with several to many of these developing into oocytes in each clone (Fig. 2.14) (Büning, 1993, 1994, 1998b). Primary oocytes move one by one into the apex of the vitellarium, as in other insects, while the nurse cells remain behind in the germarium to form a nutritive structure called the *tropharium* (Fig. 2.3B). As it moves into the apex of the vitellarium from the tropharium, each oocyte maintains contact with nurse cells in the tropharium by means of elongate, intercellular bridges called *trophic* (nutritive) *cords*.

Ovarioles having these characteristics appear to have independently evolved five times within the insects (once each in Ephemeroptera, Hemiptera, Coleoptera-Polyphaga, Raphidioptera, and Megaloptera: Sialidae), as

can be recognized by subtle differences in the development and basic cellular organization of each type (Büning, 1993, 1994). These differences, and their disjunct phylogenetic distribution within the Hexapoda (Fig. 2.22), indicate their independent evolution from polytrophic ancestors in all except the Ephemeroptera, which probably had a panoistic ancestor (Stys and Bilinski, 1990).

2.2.2.2.1 OVARIOLAR DIFFERENTIATION IN HEMIPTERA
In fifth (last) instars of the blood-feeding bugs *Panstrongylus megistus* and *Rhodnius prolixus* (Reduviidae), each ovary has seven ovarioles and each ovariole contains a large number of oogonia surrounding a trophic core, delimited by a meshwork of F-actin, and a proximal area of prefollicular cells (compare to Fig. 2.15: 1) (Furtado, 1979; Lutz and Huebner, 1980, 1981; Huebner and Diehl-Jones, 1998). After the larva takes a blood meal, the ovarioles metamorphose through (1) a proliferation phase (unfed to 3 d post feeding [DPF] at 27°C), (2) an early differentiation phase (9–15 DPF) and (3) a later differentiation phase (16–21 DPF when the adult molt occurs). In the proliferation phase, asynchronous proliferation of oogonia increases the size of the tropharium and is accompanied by proliferation of prefollicular cells (Fig. 2.15: 2). In the early differentiation phase, a zone of generative oocytes is established at the base of the tropharium above the prefollicular tissue while nurse cell nuclei begin to enlarge and their chromosomes begin to undergo endomitosis (Fig. 2.15: 3) (Dittman et al., 1984). In the later differentiation phase, adjacent nurse cells in the tropharium fuse into groups of polyploid nuclei in a common cytoplasm while intercellular bridges between adjacent oocytes develop into trophic cords (Lutz and Huebner, 1981); the trophic core within the tropharium mediates the passage of RNAs and other molecules from nurse cell nuclei into the oocytes (Fig. 2.15: 4, 5).

As an oocyte begins its previtellogenic growth, it pushes itself back into the vitellarium and becomes surrounded, in the usual way, by prefollicular cells (Fig. 2.15: 5). As it enlarges, it moves posteriorly down the vitellarium, and its trophic cord lengthens and thickens a corresponding amount, reaching a maximum diameter of 25—32 μm in ovaries of *R. prolixus* (compare Fig. 2.3B). In most bugs, the trophic cord breaks prior to commencement of vitellogenesis (Huebner, 1984), but in *R. prolixus* it persists until the oocyte is about 1 mm long and at midvitellogenesis. The cord then narrows, breaks just above the oocyte, and is withdrawn anteriorly into the trophic core. Division kinetics and the number of nurse cells and oocytes per clone differ within the tropharia of different bugs (Fig. 2.14) (Büning, 1993, 1994, 1998b).

2.2.2.2.2 OVARIOLAR DIFFERENTIATION IN OTHER INSECTS WITH TELOTROPHIC OVARIOLES
In ovarioles of polyphagous beetles, the tropharia are relatively larger, lack a trophic core, and contain

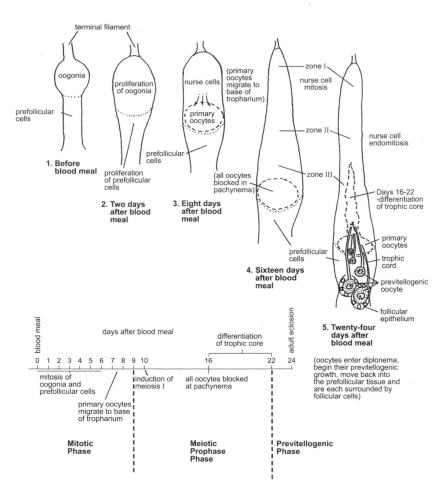

Fig. 2.15. Summary of the differentiation of a telotrophic ovariole after a blood meal in a fifth (last) instar larva of the triatomine reduviid *Panstrongylus megistus* at 27°C and 70 ± 5% relative humidity. Events are almost identical in the ovarioles of *Rhodnius prolixus* (Lutz and Huebner, 1980). (Drawn from photomicrographs in *J. Insect Physiol.*, 25, A. F. Furtado, The hormonal control of mitosis and meiosis during oogenesis in a blood-feeding bug *Panstrongylus megistus*, pp. 561–570, Copyright 1979, with permission from Elsevier Science.)

mesodermal interstitial cells in addition to nurse cells (Fig. 2.16A). The trophic cords are also relatively shorter and more delicate.

Division kinetics within the tropharium of beetles also differ. Clones of eight cystocytes arise, the inner four becoming oocytes (Fig. 2.16B) (Büning, 1993, 1994). Each of the outer four cells becomes a chordoblast that proliferates a linear chain of chordocytes anteriorly. This occurs through a series of mitoses, with the orientation of the spindle consistently parallel to the long axis of the ovariole. Because of incomplete cytokinesis, the chordocytes of each chordoblast remain linked to each other by ring canals. In each chain, the chordoblast starts to divide first. From this cell a wave of mitoses spreads anteriorly in a mitotic gradient. Several linear clusters can be arranged in parallel, linked together by transverse intercellular bridges between their chordocytes, and these divide in relative synchrony.

Similar linear clusters of chordocytes develop within the ovarioles of mayflies (Ephemeroptera), and the chordocyte most proximally located within the ovariole and becoming surrounded by prefollicular cells becomes the oocyte. Others, more anteriorly situated and contacting inner sheath cells within the tropharium, become nonfunctioning nurse cells (Fig. 2.17A) (Büning, 1993, 1994).

In snakeflies (Raphidioptera) and alderflies (Megaloptera: Sialidae), several clones of cystocytes occur in each ovariole. Each clone contains several oocytes, with these clusters fusing secondarily to form a syncytium surrounded by a layer of tapetum cells. In addition, the trophic cords are slender and devoid of microtubules (Fig. 2.17B).

2.2.2.3 PREVITELLOGENESIS IN MEROISTIC OVARIOLES

Previtellogenic growth of oocytes is similar in panoistic and meroistic ovarioles, but the germinal vesicle does not enlarge as much. As in panoistic ovarioles, the chromosomes of vegetative oocytes enter meiotic prophase and are arrested in early diplotene. The chromosomes then coil, condense, and aggregate into a *karyosphere* (karyosome), which, with some exceptions, seems to prevent them from transcribing RNA (Telfer, 1975).

2.2.2.3.1 GENE EXPRESSION DURING PREVITELLOGENESIS

Depending on species, the oocyte's chromosomes either are inactive or have low levels of transcriptional activity (Cave, 1982; Büning, 1994, 1998b). Their function is taken over by chromosomes in the nurse cell nuclei. However, ovarioles of certain basal neuropterans,

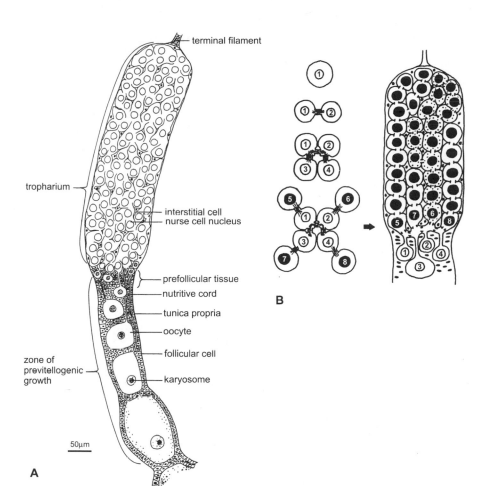

terminal filament

tropharium

interstitial cell
nurse cell nucleus

prefollicular tissue

nutritive cord

tunica propria

oocyte

follicular cell

zone of
previtellogenic
growth

karyosome

50μm

A

B

Fig. 2.16. A. Diagrammatic sagittal section of the telotrophic ovariole of a polyphage beetle. Notice the large size of the tropharium relative to the rest of the ovariole, the absence of a trophic core, the interstitial cells, and the delicate trophic (nutritive) cords. B. Model of cystocyte cluster formation in the tropharium of the telotrophic ovariole of a polyphage beetle. (Reprinted from *Int. J. Insect Morphol. Embryol.*, 22, J. Büning, Germ cell cluster formation in insect ovaries, pp. 237–253, Copyright 1993, with permission from Elsevier Science.)

beetles, and crane flies, though polytrophic, are intermediate between panoistic and meroistic ovarioles in that the chromosomes of both oocyte and nurse cells function in transcription (Cave, 1982: Table 1). In hystrichopsylloid fleas, as already mentioned, the oocytes are functionally panoistic and the nurse cells become follicular cells.

2.2.2.4 NURSE CELL ENDOMITOSIS
In most investigated meroistic ovarioles, though not in those of strepsipterans (Büning, 1998a), the nurse cell chromosomes replicate endomitotically to reach high levels of ploidy. For example, those in *D. melanogaster* follicles reach 1024–2048n and have the transcriptional potential of about 5000 diploid panoistic oocytes, while those in *Calliphora erythrocephala* follicles proliferate to 4096n (Telfer, 1975; Dittman et al., 1984, 1987; Kaulenas, 1992; Gerhart and Kirschner, 1997; Dej and Spradling, 1999). The entire genome, not just rDNA as in panoistic germinal vesicles, is replicated, the levels of ploidy reached in adjacent nurse cell nuclei of a cluster tending to differ (Dej and Spradling, 1999). High levels of transcription occur and these RNAs together with mitochondria, centrioles, ribosomes, and other materials are passed into the oocyte through the ring canals (Fig. 2.13B) or trophic cords for storage and for later use

in preblastoderm embryogenesis (6.4), as described for panoistic oogenesis.

Nurse cell polyploidy results when all cystocytes except the oocyte enter the endo cell cycle in which DNA continues to replicate in the absence of mitosis (i.e., the cycle changes from S and M to S and G) (Reed and Orr-Weaver, 1997; Dej and Spradling, 1999). In *Drosophila* females, nurse cell mitosis is inhibited by expression of the *morula (mr)* gene, and in female-sterile alleles of *mr*, the nurse cells begin endopolyploidy but revert to a mitotic-like state in which the chromosomes condense and spindles form. Expression of *mr* is also required for dividing cells to exit mitosis.

2.2.2.5 TRANSPORT OF NURSE CELL CONTENTS INTO OOCYTE
In polytrophic ovarioles, passage of materials (centrosomes, ribosomes, proteins, RNAs, etc.) from the nurse cells into the oocyte occurs in two stages: a sustained period of days or weeks while the oocyte is in the previtellogenic phase (stages 2–6 in *D. melanogaster*) (Fig. 2.13A), followed by an abrupt terminal injection of nurse cell cytoplasm into the oocyte toward the end of vitellogenesis during stages 10b–12; the nuclei stay behind and eventually degenerate (Fig. 2.3C) (Telfer, 1975). The first phase of movement results from actin- and myosin-

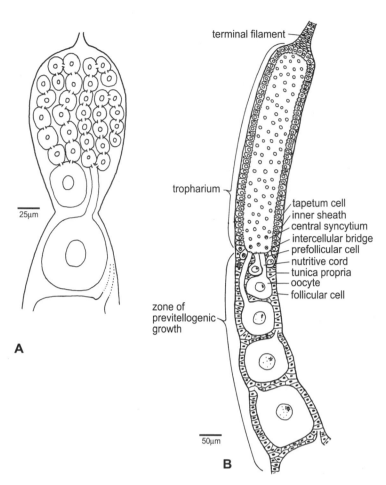

terminal filament

tropharium

tapetum cell
inner sheath
central syncytium
intercellular bridge
prefollicular cell
nutritive cord
tunica propria
oocyte
follicular cell

zone of
previtellogenic
growth

25μm

50μm

A

B

Fig. 2.17. A. Diagrammatic sagittal section of the telotrophic ovariole of the mayfly *Siphlonurus armatus* (Ephemeroptera: Siphlonuridae), showing two of five linear clones of cystocytes within the tropharium. The cystocyte most proximally situated becomes the oocyte. B. Diagrammatic sagittal section of the telotrophic ovariole characteristic of snakeflies (Raphidioptera) and hellgramites (Megaloptera). Notice the large size of the tropharium relative to the rest of the ovariole, the central syncytium of nurse cell nuclei, and the parietal layer of tapetal cells. (Reprinted from *Int. J. Insect Morphol. Embryol.*, 22, J. Büning, Germ cell cluster formation in insect ovaries, pp. 237–253, Copyright 1993, with permission from Elsevier Science.)

based contraction of nurse cell cytoplasm (McCall and Steller, 1998) and through function of the polyfusome and a complex and exceedingly active network of microtubules in the oocyte (Fig. 2.13B) (Cooley and Theurkauf, 1994; Bolvar et al., 2001).

The second phase of nurse cell transport, at least in ovarioles of *D. melanogaster*, seems to depend on the relative concentrations of juvenile hormone (JH) and 20E in the hemolymph (chapter 12). These regulate whether the oocytes will progress through stage 9 and continue vitellogenesis (note the rapid enlargement of the oocyte and shrinkage of the nurse cells between stages 9 and 11 in Fig. 2.13A) or degenerate (Soller et al., 1999). Application of the JH analog methoprene (Fig. 12.11F) to sexually mature, virgin females induces this progress, as does receipt of a sex peptide synthesized in the male's accessory glands and passed to the female during copulation (3.7.2). This observation suggests that JH release is induced directly or indirectly by the peptide. Injection of 20E alone at physiological concentrations will induce degeneration of nurse cells in stage 9 follicles and resorption of their oocytes.

In both polytrophic and telotrophic (Fig. 2.18) ovarioles, electrophoresis seems to assist in this transport, the oocyte being the anode (+) and the nurse cells the cathode (–) (Dittman et al., 1981; De Loof, 1983; Overall

and Jaffe, 1985; Kaulenas, 1992; Huebner and Diehl-Jones, 1993, 1998). This was shown by microinjecting electrically charged macromolecules into the nurse cells and following their subsequent movement into the oocyte, and by external use of a small, ultrasensitive vibrating probe for monitoring external currents caused by a changing pattern of activity in various K^+, Na^+, and Ca^{2+} channels and Na^+/K^+-ATPase in the cell membranes of nurse cells, oocytes, and follicular cells (Huebner and Diehl-Jones, 1993, 1998).

In addition, in telotrophic ovarioles of hemipterans, the trophic cords contain thousands of longitudinally disposed microtubules with their plus ends toward the tropharium, which cause the trophic cords to appear birefringent when viewed under polarized light (Fig. 2.19). The microtubules have been shown to function in cytoplasmic transport by means of kinesin-like motors (Hyams and Stebbings, 1977; Stebbings and Hunt, 1983; Dittman et al., 1987; Huebner and Diehl-Jones, 1998; Vale and Milligan, 2000). At onset of vitellogenesis, just before they withdraw into the tropharium, the microtubules lose their characteristic spacing and degenerate (Lane and Stebbings, 1994).

Development of microtubules within a trophic cord begins late in the fifth (last) instar of *R. prolixus* at the apex of the oocyte and gradually progresses distally into

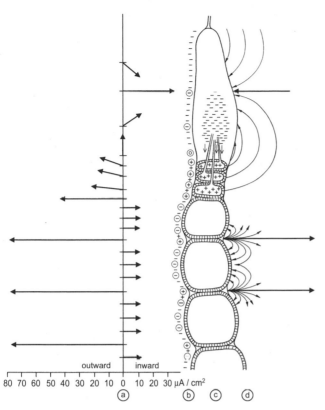

Fig. 2.18. Diagrammatic sagittal section of a telotrophic ovariole of the pyrrhocorid bug *Dysdercus intermedius*, showing the pattern of electrical current normal to it. a, current vectors: composite of measurements made on all sides of five ovarioles taken from a single female; b, surface charge: circled symbols indicate points of measurement for a; c, intracellular charge distribution; d, model of two steady current systems normal to the tropharial and vitellarial region of the ovariole. (Reprinted with permission from F. Dittman, R. Ehni, and W. Engels. Bioelectric aspects of the hemipteran telotrophic ovariole (*Dysdercus intermedius*). *W. Roux's Arch. Entwicklungsmech. Organ.* 190: 221–225. © 1981 Springer-Verlag.)

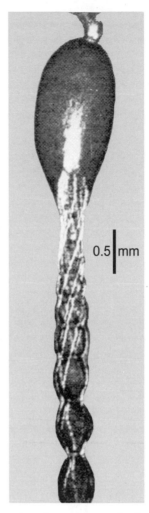

Fig. 2.19. Polarization photomicrograph of a telotrophic ovariole of the aquatic bug *Notonecta glauca* (Hemiptera: Notonectidae), showing the birefringent trophic cords. Birefringence is due to the presence of thousands of longitudinally disposed microtubules within each cord and also within the trophic core. (Reprinted from *Tissue and Cell*, 9, J. S. Hyams, and H. Stebbings, The distribution and function of microtublules in nutritive tubes, pp. 537–545, © 1977, by permission of the publisher Churchill Livingstone.)

the trophic core of the tropharium, suggesting action of a microtubule organizing center (MTOC) similar to that in polytrophic oocytes of *D. melanogaster* (Huebner and Diehl-Jones, 1998).

2.2.2.6 ESTABLISHING THE TERMINI AND THE ANTEROPOSTERIOR AND DORSOVENTRAL AXES IN EGGS OF *DROSOPHILA MELANOGASTER*

Longitudinal and dorsoventral polarity of fully developed eggs is established within the follicles of the female as they develop, their anteroposterior (A-P) and dorsoventral (D-V) axes being the same as those of the mother (Halley's law, 1886). In fact, in developing oocytes of the ichneumonid *Cosmoconus meridionator* (Hymenoptera), there are structural markers for these axes, including location of the germinal vesicle and pole plasm and graded distribution of lipid droplets, yolk bodies, and "accessory nuclei" within the ooplasm (Bilinski,

1991) (the latter bud off from the germinal vesicle and are characteristic of hymenopteran oocytes).

In spite of this generalization, Kleine-Schonnefeld and Engels (1981) observed few eggs of the house fly *Musca domestica* to have this relationship with the mother. In fact, in transverse sections of the abdomens of mature females, they found the follicles of the 60–70 ovarioles of each ovary to be arranged in four or five concentric circles, with the dorsal sides of their enclosed eggs oriented toward an imaginary center eccentrically placed within each ovary near the midline of the female's abdomen.

The molecular genetics of axis specification in eggs of *D. melanogaster* is established during stages 6–10 of oogenesis (Figs. 2.13A, 2.20) and is summarized in Table

Table 2.1 Maternal Genes of the Four Pathways Specifying the Termini and the Anterior, Posterior, and Dorsoventral Axes in the Embryo of *Drosophila melanogaster*

	Anterior	Posterior	Terminal	Ventral
Germ line				*K10* *cornichon* *gurken* ↓
Follicle cells			*torsolike*	*torpedo* *pipe* *windbeutel* *nudel* ↓
Germ line	*exuperantia* *swallow* *staufen* ↓	*staufen* *oskar* *vasa* *valois* *tudor* *mago nashi* ↓ *nanos* *pumilio* ↓	*s(1)Nasrath* *fs(1)pole hole* *trunk* ↓ *torso* ↓ *l(1)pole hole* *MAPK* ↓	*gastrulation deficient* *snake* *easter* *spätzle* ↓ *Toll* ↓ *tube* *pelle* *cactus* ↓
	bicoid	*hunchback*	*gene X*	*dorsal*

Source: Modified from Klingler and Tautz, 1999: 316.

2.1 (Theurkauf, 1994; Anderson, 1995; Morgan and Mahowald, 1996; González-Reyes et al., 1997; Newmark et al., 1997; Gerhart and Kirschner, 1997: 412–421; Huebner and Diehl-Jones, 1998; Klingler and Tautz, 1999).

2.2.2.6.1 ANTEROPOSTERIOR AXIS

A single signaling protein encoded by the *gurken* (*grk*) gene initiates polarity development in both the A-P and D-V axes of the egg (Anderson, 1995; Gerhart and Kirschner, 1997; Klingler and Tautz, 1999). First, the oocyte nucleus, together with its centrosome and aster, moves to the posterior pole of the oocyte (Fig. 2.20A: stages 1–6). Gurken protein is locally secreted from the oocyte nucleus, passes across the oolemma, and binds to the receptor protein Torpedo in adjacent polar follicular cells, activating a posterior-specific program in them (the Torpedo receptor is present in all follicle cells but only those nearest the oocyte nucleus receive Gurken). These follicle cells then signal back to the oocyte by releasing a secreted protein encoded by the *mago nashi* (*mago*) gene (Fig. 2.20A: stages 6–8) (Newmark et al., 1997). This protein causes the oocyte's centrosome and aster to degenerate, and influences assembly of the pole plasm and of a new MTOC anteriorly within the oocyte that has the minus ends of its microtubules toward the nurse cells and plus end near its posterior pole (Fig.

2.20A: stage 9). Maternal *nanos* (*nos*) mRNA is then transported to the posterior pole of the oocyte and *bicoid* (*bcd*) mRNA, to the anterior pole (Figs. 2.20A: stages 9–14). Both are transcribed from nurse cell chromosomes, pass into the oocyte through the ring canals (Fig. 2.13B, C), and are transported in opposite directions within the oocyte by different motor proteins associated with the microtubules (Brendza et al., 2000). After the egg is fertilized, the two RNAs are translated into their respective proteins, which then diffuse posteriorly (Bicoid) and anteriorly (Nanos) within the periplasm, establishing two gradients in mirror-image symmetry with their high points at either end of the egg (Fig. 7.24A–C). These gradients provide the basis for subsequent specification of the A-P axis during early embryogenesis (7.2.1.1.3).

2.2.2.6.2 DORSOVENTRAL AXIS

As microtubules of the MTOC are transporting *bcd* and *nos* RNAs to opposite ends of the oocyte, they also begin to convey its nucleus anteriorly along the dorsal oolemma to its anterodorsal margin next to the nurse cells (Fig. 2.20B: stages 6–8). On arrival, the nucleus again locally secretes Gurken protein to adjacent dorsal follicle cells (Fig. 2.20B: stage 9), which respond by repressing transcription of the genes *pipe* (*pip*), *windbeutel* (*wind*), and *nudel* (*ndl*) (these genes remain active ventrally) (Fig. 2.20B: stage 9) (Gerhart and Kirschner, 1997;

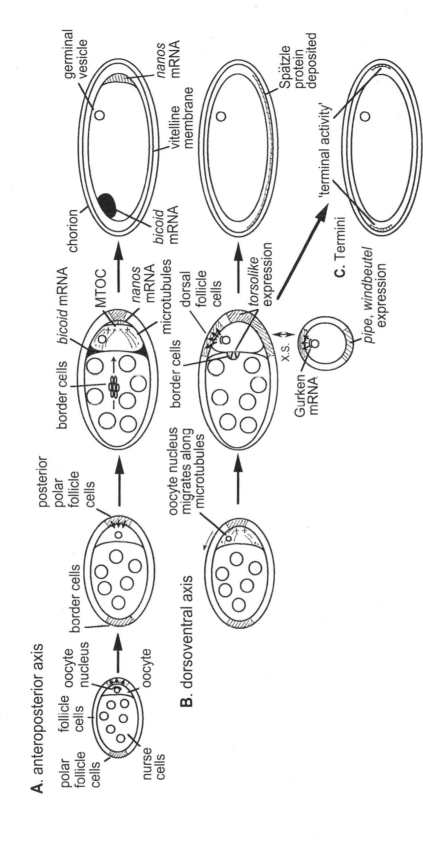

Oocyte Stage 1-6   6-8   9   10-14

A. anteroposterior axis

polar follicle cells

follicle cells

oocyte nucleus

oocyte

nurse cells

posterior polar follicle cells

border cells

border cells   *bicoid* mRNA

MTOC

nanos mRNA

microtubules

germinal vesicle

nanos mRNA

vitelline membrane

chorion

bicoid mRNA

B. dorsoventral axis

oocyte nucleus migrates along microtubules

border cells

dorsal follicle cells

torsolike expression

Gurken mRNA

x.s.

pipe, windbeutel expression

Spätzle protein deposited

'terminal activity'

C. Termini

Fig. 2.20. Anteroposterior (A), dorsoventral (B), and terminal (C) axis specification in an oocyte of *D. melanogaster*. Follicles are shown in sagittal section and contain the oocyte, 15 nurse cells, and two kinds of follicle cells. A. See text for details. (Adapted with permission from J. Gerhart and M. Kirschner. *Cells, embryos, and evolution. Toward a cellular and developmental understanding of phenotypic variation and evolutionary adaptability*, pp. 416–417, Fig. 8-14. © 1997 Blackwell Science, Ltd.)

Klingler and Tautz, 1999). Ventrally, Spätzle (Spz) protein (and Windbeutel [Wind] and Pipe [Pip] proteins according to Nilson and Schubach, 1998, and Sen et al., 1998) is secreted and is deposited in the ventral vitelline membrane and inner chorion during their deposition (Fig. 2.20B: stages 10–14). Following fertilization, Spz is released and activates the Toll signal transduction pathway (Table 2.1) (Gerhart and Kirschner, 1997: 423).

The first eight steps in this pathway occur during, embryonic cleavage within the perivitelline space between vitelline membrane and oolemma and end in production of an inducer protein, Spz, that binds to the transmembrane Toll receptor in the ventral oolemma (Table 2.1). Activation of Toll, by way of a signal transduction pathway, in turn activates a protein kinase encoded by the *pelle* (*pll*) gene that degrades Cactus (Cact) protein and releases, into the yolk, the transcription factor Dorsal (Dl), to which it was bound (the Cact/Dl complex is uniformly distributed throughout the yolk system but Dl is only released ventrally). Released Dl protein enters ventral preblastoderm nuclei in a concentration gradient, with its high point ventrally to activate the venter-specific, downstream, selector genes *twist* (*twi*) and *snail* (*sna*) that specify mesoderm, and to repress expression of dorsal-specific ones (Fig. 7.20). This regulation of downstream genes occurs at at least five different thresholds for Dl that are set by its affinity for its binding sites in the regulatory region of each target gene and by its interaction with other, already bound transcription factors (Gurdon et al., 1998). (The toll pathway is reactivated after hatch in larvae and adults, where it is an important component of the immune response to entry of disease-causing fungi and bacteria [Hoffman et al., 1999].)

The dorsal side of the egg is specified in a similar manner (Wasserman and Freeman, 1998). When Gurken protein is released dorsally from the oocyte nucleus, it binds to *Drosophila* epidermal growth factor receptors (Dm-EGFR) in the dorsal follicle cells. Binding triggers release of two other EGFR ligands—Spitz and Vein—which leads to localized expression of the diffusible inhibitor protein Argos.

The question might arise as to whether the oocyte has a previously determined dorsal side to which the oocyte nucleus is drawn or whether the nucleus itself determines this side. This question was answered in studies of stocks of *Drosophila* having binucleate oocytes (Roth et al., 1999). The two nuclei of each oocyte were found to migrate independently and to be equal in their ability to induce a D-V pattern in overlying follicle cells in any position around the anterior circumference of the egg chamber. Thus, the D-V axis of the egg is determined by the oocyte nucleus when it releases Gurken protein to dorsal follicle cells.

2.2.2.6.3 TERMINI

Specification of the two unsegmented ends of the embryo, the *acron* and *telson*, is achieved by a process related to that of A-P axis specification. Beginning in stage 9 of oogenesis, 6–10 follicle cells at the apex of the follicle, the *border cells*, migrate posteriorly between the nurse cells to the anterior pole of the oocyte to become anterior polar cells (Fig. 2.20A, B) (King, 1970). They then migrate dorsally toward the oocyte nucleus, guided by the presence of the Gurken protein (Duchek and Rørth, 2001). During choriogenesis, these and two additional follicle cells at the posterior pole of the oocyte begin to express the gene *torsolike* (*tsl*) (Gerhart and Kirschner, 1997: 418–419). Torsolike (Tsl) protein is incorporated into the vitelline membrane and inner chorion at each end of the egg during their deposition (Fig. 2.20C). After fertilization, this protein seems to mediate cleavage of the Trunk (Trk) protein into a form that can bind to a transmembrane receptor tyrosine kinase in the oolemma encoded by the *torso* (*tor*) gene (Table 2.1). Binding activates expression of the gap genes *tailless* (*tll*) and *huckebein* (*hkb*) at the two ends of the egg that specify the termini (Fig. 7.24D). Vitellogenesis and postvitellogenesis are the same as in panoistic ovarioles.

2.2.3 STAGING OOGENESIS

Individuals studying oogenesis and monitoring its rate in different insects have tended to devise their own schemes for staging, with little regard for those of other workers. In Table 2.2, I compare five systems devised for insects having panoistic (Bonhag, 1959; Mahowald, 1972), polytrophic (King, 1970; Trepte, 1979), and telotrophic (Schreiner, 1977) ovarioles. Mahowald's and Bonhag's systems can be used for insects having any type of ovariole, but lack sufficient resolution for detailed comparison. King's (1970) system of 14 stages is universally used by *Drosophila* workers and could easily be adapted for use in all polytrophic insects. Finally, it might be useful to stage oogenesis in terms of percentage of total development time (TDT) from stem oogonium to fully chorionated egg, as Bentley and colleagues (1979) did for embryos of the grasshopper *Schistocerca nitens*. This method is widely used for staging insect embryos and for comparing the embryogenesis of different species reared under differing conditions.

2.2.4 DURATION OF OOGENESIS

The time required for an oogonium to grow and differentiate into a mature egg has been followed in few insects. It is best understood in *D. melanogaster*, where it is known to be influenced by genotype, age, temperature, relative humidity, abundance of mates, degree of adult crowding, texture and odor of oviposition site, and nutritional state and water content of adults (Figs. 2.10, 2.13A; Table 2.2) (King, 1970; Büning, 1998b; Drummond-Barbosa and Spradling, 2001). Under ideal conditions at 25°C, an egg takes 8 d to develop and oocyte volume increases 90,000-fold. In stages 1–6 (50 h) the increase in volume is 40-fold with a doubling time of 9 h and in stages 8–12 (18 h) the increase is 1500-fold, with a doubling time of 2 h. During stages 8–10, the nurse chamber increases in volume at a rate half that of

Table 2.2 Staging Oogenesis

Ovariolar Type	Panoistic	All types	Telotrophic	Polytrophic	Polytrophic
Order	Blattaria *Periplaneta americana* (Bonhag, 1959)	Hexapoda (Mahowald, 1972)	Hemiptera *Oncopeltus fasciatus* (Schreiner, 1977)	Diptera *Drosophila melanogaster* (King, 1970)	Diptera *Musca domestica* (Trepte, 1979)
Terminal filament	Zone 1	Zone 1			
Germarium	Zone 2	Zone 2	Stage 1 (1° ooc. in base of tropharium)	Stem cells; Cystoblast 2^4 synchronous mitoses; Stage 1 (9.7h) 16-cell cluster	Stage 1 (70h)
Previtellogenesis	Zone 3 (vegetative ooc. not linearly arranged); Zone 4 (vegetative ooc. linearly arranged and each surrounded by follicle cells)	Zone 3	Stage 2 (vegetative ooc. side by side); Stage 3 (vegetative ooc. in single row)	Stage 2 (9.7h); Stage 3 (9.7h); Stage 4 (9.8h); Stage 5 (2.6h); Stage 6 (8.5h); Stage 7 (8.7h); Stage 8 (5.2h)	Stage 2 (12h) ooc. shorter than 4 proximal nurse cells.
Vitellogenesis	Zone 5	Zone 4	Stage 4 (yolk uptake just beginning); Stage 5 (ooc. with protein yolk bodies along sides); Stage 6 (intense vitellogenesis; follicle cells "patent")	Stage 9 (5.6h); Stage 10 (5.1h); Stage 11 (0.4h); Stage 12 (1.0h)	Stage 3 (33h) ooc. longer than 4 proximal nurse cells but shorter than nurse chamber; Stage 4 (9h) ooc. longer than nurse chamber
Choriogenesis	Zone 5	Zone 4	Stage 7	Stage 13 (.79h)	Stage 5 (11h)
Mature egg	Zone 5	Zone 5	Stage 8	Stage 14 (1.9h)	Stage 6 (6–9h)
Pedicel	Zone 6				
Growth Rate	oog. → mature egg: requires ± 100d (*Acheta domesticus*)			oog. → mature egg: requires 9–11d; vol. increase = 90,000-fold	16-cell cluster → mature egg: requires 145h; vol. increase: 5000-fold from primary ooc. and 100,000-fold from oog.

Annotations (Drosophila/Telotrophic): 3 d; Ooc. and nurse cells grow at similar rate; Follicle reaches maximum of 1200 cells; (18h) ooc. grows 4–5-fold faster than nurse cells; Vitelline membrane (11h); Nurse cells degenerate; Choriogenesis (5h)

Annotations (Musca): DNA, RNA, and protein synthesis in nurse cells; DNA, RNA, and protein transport into ooc.; Glycogen synthesis in ooc.

Note: ooc., oocyte; oog., oogonium.

the oocyte, and in stages 11–12 it shrinks, injecting its cytoplasm into the oocyte while its nuclei degenerate (Fig. 2.3C).

The number of 16-cell cystocyte clusters per ovariolar germarium is six or seven (Fig. 2.10B), and in a female ovulating two eggs per ovariole per day, each of these spends 3 d growing and migrating in the germarium before entering the vitellarium. The four division cycles generating a clone after a stem cell division (Fig. 2.12) occur every 12 h so that 2 d is required for a cystoblast to become a 16-cell clone. Thus, it takes about 8 d for all postoogonial stages in an ovariole to develop to maturity and be deposited (Fig. 11.20 in King, 1970). Since there are about 16 postoogonial stages per ovariole, a female with 30 ovarioles would deposit 500 eggs before mature oocytes, which were oogonia when oviposition started, would be deposited.

2.2.5 FECUNDITY
The number of eggs deposited by a female during her lifetime has been recorded for dozens of species in all orders (Hinton, 1981: Table 2). The largest numbers are estimates for queens of certain eusocial insects, although none of these are reliable. For example, a queen honeybee, *Apis mellifera*, may live over 5 y and deposit up to 600,000 eggs in the first three. Queen termites have been reported to deposit eggs steadily at a rate of 2–3/s (i.e., 3×10^4/d and 1×10^7/y) and to live 15 or 20 y. Most bisexual insects have a sex ratio of about 1:1, and only two eggs from any female need to develop to reproductive maturity to maintain the population (Hinton, 1981). Thus, there is probably a rough relationship between the average fecundity of females of a particular species and the hazards likely to be encountered by their offspring. The lowest average number of eggs per female Hinton recorded was 8 (an aphid, *Schizolachnus piniradiatae*) and the highest 12,500 (a parasitoid eucharitid wasp, *Kapala terminalis*). Females of many species produce 1000–4000 eggs/lifetime but the average is well below 1000.

By controlling the amount of protein-rich nutrient in the diet of *D. melanogaster* females, Drummond-Barbosa and Spradling (2001) were able to vary the rate of egg production by 60-fold, with both germ-line and somatic cells adjusting their rate of proliferation up to 4-fold in response to changes in food intake. In addition, the rate of cell death during oogenesis, occurring mostly at stages 2a/2b in the germarium (Fig. 2.10B) and at stage 8 in the vitellarium (Fig. 2.13A), varied up to 15-fold.

2.2.6 THE ADAPTIVE SIGNIFICANCE OF CONSPECIFIC DIFFERENCES IN EGG SIZE
Intuitively, one might think that differences in egg size for a given species would correlate with larval performance, those juveniles hatching from larger eggs doing better. Azevedo and colleagues (1997) tested this hypothesis by crossing four independent pairs of outbred *D. melanogaster* populations differing genetically in egg size and examined their female offspring for positive

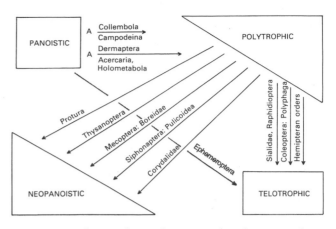

Fig. 2.21. Evolution of ovariolar type within the Hexapoda. (From P. Stys and S. Bilinski. 1990. Ovariole types and the phylogeny of the hexapods. *Biol. Rev.* 65: 401–429. Reprinted with the permission of Cambridge University Press.)

maternal genetic effects on life history traits. They showed larger egg size to correlate positively with embryonic viability and development rate, with hatching weight and larval feeding rate, and with egg-larva and egg-adult development rate, but not with larval competitive ability, adult weight, or egg size in their offspring. Similar relationships may exist in other insects, but there is too much variation to generalize (Fox and Czesak, 2000).

2.2.7 PHYLOGENETIC ASPECTS
The direction of evolution of ovariolar structure in insects appears to be as follows (Fig. 2.21):

panoistic (1) → polytrophic (2) → telotrophic (3)

Neopanoistic ovarioles (4), secondarily devoid of nurse cells and thus identical to panoistic ovarioles when viewed by light microscopy, have evolved independently at least five times from the polytrophic type.

Transition 1–2 is characterized by development of incomplete cytokinesis (or by reinstatement of it, since germ cell cluster formation may be ancestral in both male and female insects [Büning, 1993, 1994, 1996, 1998b]); by differentiation of most cystocytes into nurse cells; by amplification of the nurse cell genome; and by inhibition of transcription in oocyte chromosomes. Transition 2–3 is characterized by two or more cystocytes per clone developing into oocytes (Figs. 2.14, 2.16B) and by retention of nurse cells in the germarium, and transition 2–4, by development of all cystocytes into additional oocytes or follicle cells. Finally, transition 1–3 (in certain Ephemeroptera) is characterized by possible ancestral retention of cluster formation of the linear type (Fig. 2.9A), with the cystocyte situated most proximally in the ovariole becoming the oocyte and the remainder becoming nonfunctional nurse cells (Fig. 2.17A) (they remain diploid and microtubules are absent from the trophic cords).

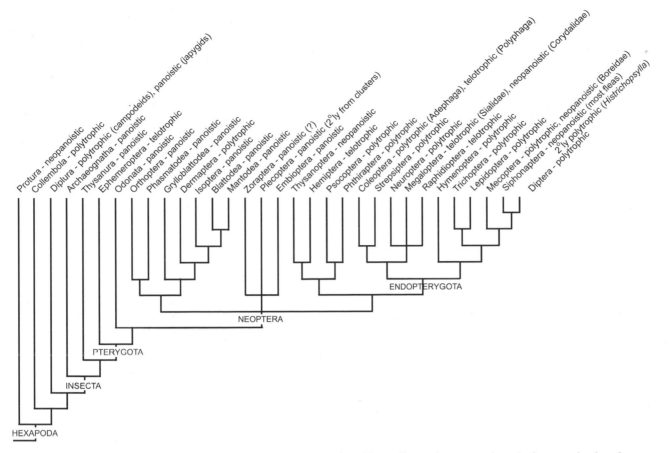

Fig. 2.22. Evolutionary relationships of the hexapod orders as inferred by Gullan and Cranston (1994), showing the distribution of ovariolar type throughout the Hexapoda. (Data from Büning, 1994.) Cladogram adapted with permission from P. J. Gullan and P. S. Cranston. *The insects. An outline of entomology*, p. 193, Fig. 7.5. 1st ed. Chapman and Hall, London. © 1994 Blackwell Science, Ltd.)

In hexapods (Fig. 2.22), transition 1–2 has evolved four times: once each in Collembola, Diplura-Campodeidae, Dermaptera, and Phalloneoptera (Acercaria + Endopterygota); transition 2–3, four times (once each in Hemiptera, Megaloptera: Sialidae, Raphidioptera, and Coleoptera: Polyphaga); transition 2–4, at least six times (once each in Protura, Plecoptera, Thysanoptera, Megaloptera: Corydalidae, Mecoptera: Boreidae, and Siphonaptera: Pulicoidea, and possibly in some other taxa)—all independently; and transition 1–3, once, in Ephemeroptera. The phylogenetic distribution of ovariolar type is presented in Figure 2.22 (Büning, 1994); a much more detailed account is provided in Büning, 1998b.

2.2.7.1 ADAPTIVE SIGNIFICANCE OF THE NURSE CELLS

The functional advantage of nurse cells is in the extreme amplification of the genome in chromosomes of their nuclei. In the germinal vesicles of panoistic oocytes, only 4C amounts of DNA are present, one C for each chromatid of each homologous chromosome pair (Fig. 2.7B, C). In meroistic ovarioles, the nurse cell nuclei are typically highly polyploid (up to 4096n in those of the blowfly *Calliphora erythrocephala*), and there can be one to 31 nurse cells/oocyte. Thus, in females of *C. erythrocephala*, with 15 nurse cells/oocyte, there are about 61,000 genome copies to satisfy the requirements of the oocyte for ribosomal, messenger, and transfer RNAs rather than four. One result is that the eggs of insects having meroistic ovarioles, under optimal conditions, can be produced much more rapidly than can those having panoistic ovarioles. For example, an oocyte in a panoistic ovariole of the house cricket *Acheta domesticus* requires about 100 d to develop (Cave, 1982; no temperature given) while that of a house fly takes only 72 h (Table 2.2 [Trepte, 1979]).

The increased speed of egg production is thus the principal selective advantage of meroistic over panoistic ovarioles, but no one has yet convinced me why the telotrophic type should have a selective advantage over the polytrophic type even though it is obvious, in three instances, that telotrophic ovarioles are derived from polytrophic ones. And in mayflies, the telotrophic ovarioles apparently have nonfunctional nurse cells and thus appear to have no adaptive advantage over the panoistic type, suggesting that mayfly telotrophy is a relict of its ancestral, clonal proliferation of oogonia.

If one examines the distribution of ovariolar type on a reconstructed phylogeny of the insect orders (Fig. 2.22) (Büning, 1998b: Fig. 7), one notices first that females of many basal lineages (on the left) are characterized by panoistic ovarioles and that females of the more advanced and diverse orders (Fig. 0.2) have meroistic ovarioles. Recall also that some beetles, flies, and neuropteroids of the more basal, subordinal lineages have polytrophic ovarioles with both panoistic and meroistic characteristics (Table 2.3), supporting derivation of the polytrophic from the panoistic type. In beetles, females of the more advanced and diverse Polyphaga have telotrophic ovarioles, while those of the basal lineages Myxophaga and Adephaga have polytrophic ovarioles (Büning, 1998b: Fig. 10). In addition, both basal (proturans, stoneflies, corydalids) and more advanced (thrips, some fleas, boreid mecopterans) hexapods have neopanoistic ovarioles derived secondarily from the polytrophic type, perhaps as an adaptation to small size (proturans, thrips, boreid mecopterans), to an ectoparasitic lifestyle (fleas), or to both. But why should large female corydalids have abandoned their polytrophic ovarioles?

Büning (1993, 1994, 1998b) suggested that germ cell cluster formation could be ancestral in both male and female insects since clusters form in the ovaries of females of certain entognath hexapods and in crustaceans and mammals. If this is true, then all genes responsible for germ cell cluster formation (those functioning in maintaining the fusomes, polyfusomes, and cell synchrony during each cell cycle) must be blocked in females having panoistic ovarioles, while in males of most orders, cluster formation is usual during spermatogenesis (1.3.2.2). Expression of the cluster-forming genes might be blocked in panoistic females because they generally produce a relatively small number of large gametes. Reactivation of these cluster genes is then required for nurse cell formation in females having meroistic ovarioles in response to selection for increased rate of egg production.

2.2.7.2 EVOLUTIONARY ORIGIN OF THE OVARIOLES
In adult females of Protura, Collembola, and Campodeina (Diplura) and in most other arthropods, the ovaries are saclike and nonmetameric (Fig. 2.23A), whereas in those of Japygina (Diplura) (Fig. 2.23B, C) and all other insects, they are either metameric (Fig. 2.23B–D) or postmetameric (Fig. 2.23E) and divided into ovarioles (Bilinski, 1993; Stys et al., 1993). The primitive saclike ovary is not homologous to a single ovariole. Instead, the ovarioles (and sperm tubes) of most insects probably descend from paired segmental groups of primordial germ cells originating from the mesoderm of several, primitively seven, abdominal segments in the embryos of a common ancestor (Fig. 2.24: ancestral state). This is exactly what occurs in embryos of the pyrrhocorid bug *Pyrrhocoris apterus* (Seidel, 1924). Such segmentally organized ovaries are said to be metameric or primary (Fig. 2.23B–D). The adaptive advantage of

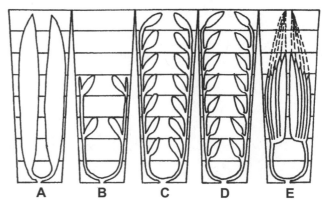

Fig. 2.23. Ovary types within the Hexapoda. A. Nonmetameric; *Campodea* sp. (Diplura: Campodeina). B–D. Metameric. B. *Anajapyx* sp. (Diplura: Japygina). C. *Japyx* sp. (Diplura: Japygina). D. *Machilis* sp. (Archaeognatha: Machilidae). E. Postmetameric: *Lepisma* sp. (Zygentoma: Lepismatidae). (From P. Stys, J. Zrzavy, and F. Weyda. 1993. Phylogeny of the Hexapoda and ovarian metamerism. *Biol. Rev.* 68: 365–379. Reprinted with the permission of Cambridge University Press.)

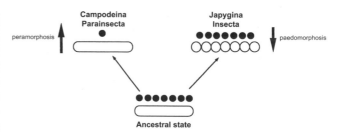

Fig. 2.24. Modifications suggested to have occurred in ovarian ontogeny during hexapod evolution. Filled circles, developing (usually embryonic) ovariolar or ovary rudiments; white areas or circles, fully developed (usually postembryonic) ovaries either saclike (Campodeina + Parainsecta) or in ovarioles (Japygina + Insecta); large arrows, possible heterochronic shifts in development (peramorphic in Campodeina + Parainsecta; paedomorphic in Japygina + Insecta s. str.) (see chapter 13 for an explanation of these terms). (From P. Stys, J. Zrzavy, and F. Weyda. 1993. Phylogeny of the Hexapoda and ovarian metamerism. *Biol. Rev.* 68: 365–379. Adapted with the permission of Cambridge University Press.)

numerous ovarioles over two saclike ovaries is in the presence of additional germaria to generate additional oocytes that increase the number of eggs produced per female per unit of time (saclike ovaries have only one germarium each).

In members of more advanced lineages, the germ cells originate just before blastoderm formation from pole cells at the posterior end of the egg (Fig. 6.3). These proliferate, move forward into the abdomen later in embryogenesis, and separate into two masses on either side, where each mass becomes surrounded by mesoderm to form an ovary rudiment (8.4.1). The apical sheath of each rudiment then invaginates from apex to base, secondarily dividing it into a characteristic number

Table 2.3 Functional Differences in Ovarioles

Ovariolar type	Panoistic	Polytrophic	Telotrophic
Stem oogonia	Not yet identified; present in insects with neopanoistic ovarioles.	One or more in each ovariole.	One or more in each ovariole.
Oocyte/nurse cell clones from individual cystoblasts?	No. But present in neopanoistic ovarioles.	Yes. One to 31 nurse cells/ oocyte.	Yes. But number of nurse cells/oocyte evaluated in few species and probably differs in mayfly, bug, beetle, and megalopteran insects.
Growth of germinal vesicle of oocyte during oogenesis	Extensive.	Usually limited but more so in certain primitive neuropterans, beetles, and flies.	Limited.
Amplification of genome	No. Chromosomes 2n in germinal vesicle of oocyte. Many nuclear pores in gv membrane.	Extensive: through endomitosis of chromosomes in nurse cell nuclei. Many nuclear pores in membranes of nurse cell nuclei but not in membrane of oocyte.	Extensive: through endomitosis of chromosomes in nurse cell nuclei. Many nuclear pores in membranes of nurse cell nuclei but not in membrane of oocyte.
Amplification of rDNA	Yes. Extrachromosomal DNA body and multiple nucleoli (circular rDNAs) in gv rDNA content/genome high. Rapid production of rRNA during previtellogenic stage of oogenesis.	Yes, but usually no more than genome as a whole (differential amplification of rDNA is known in some flies).	Yes, but usually no more than genome as a whole.
Lampbrush chromosomes in oocyte germinal vesicle	Yes. Extensive transcription of m, t, and rRNAs from diplotene lampbrush chromosomes is associated with many pores in gv membrane.	Yes, but of short duration early in previtellogenic growth and probably of little significance. Chromosomes quickly condensed into karyosome and relatively inactive in transcription, this function having been usurped by polyploid nurse cell chromosomes. Few pores are present in membrane of oocyte gv.	Yes, but of short duration early in vegetative growth and probably of little significance. Chromosomes quickly condensed into karyosome and relatively inactive in transcription, this function having been usurped by polyploid nurse cell chromosomes. Few pores are present in membrane of oocyte gv.
Rate of oogenesis	Slow (e.g., ± 100 d from oog. →to mature egg in *Acheta domesticus*).	Rapid (e.g., 9–11 d from oog. →mature egg at 25°C in *Drosophila melanogaster*.	Rapid.
Fecundity (see Table 2 in Hinton, 1981)	Depends on size of female and on species.	Depends on size of female and on species.	Depends on size of female and on species.

Note: gv, germinal vesicle; oog., oogenesis.

of ovarioles—either before or after hatch and not until the pupal stage in higher flies (Godt and Lanski, 1995). Such ovarioles are said to be postmetameric or secondary (Fig. 2.23E). For example, females of *D. melanogaster* generally have 20 ovarioles in each ovary (Fig. 2.10A) that develop de novo from each ovary rudiment beginning in the third (last) larval instar (Godt and Lanski, 1995). Formation of terminal filaments and of the basal and interfollicular stalks (the row of cells between adjacent follicles; Fig. 2.10B) of each ovariole involves recruitment, intercalation, and sorting of mesodermal cells from this mass and requires expression of Bric-à-brac (Bab) protein in their nuclei. In *bab* mutant larvae, disruption in formation of these leads to defects in ovariole formation and to emergence of sterile adults.

The principal functional differences between panoistic, polytrophic, and telotrophic ovarioles in insects are summarized in Table 2.3.

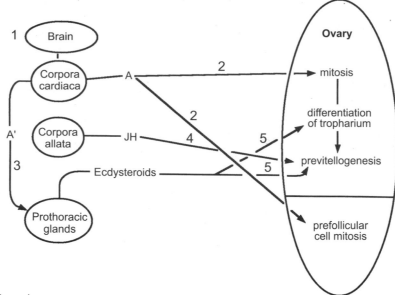

Legend

1. Blood meal.
2. Secretion of A neurosecretory cells 24–48 h after a blood meal induces mitosis in oogonia and prefollicular cells. (Destruction of A cells inhibits proliferation of both cell types.)
3. Secretion of A' neurosecretory cells (allatotropin?) induces secretion of ecdysteroids by prothoracic gland cells. (Destruction of A' cells inhibits differentiation of tropharium and prefollicular cells and previtellogenic growth of oocytes. If both A and A' cells are destroyed just after a blood meal, both ovary rudiments degenerate.)
4. Juvenile hormone (JH) stimulates previtellogenic growth of oocytes.
5. Ecdysteroids secreted 9.5–20 days after a blood meal induce meiotic prophase, differentiation of the tropharium and prefollicular cells, and previtellogenic growth of primary oocytes.

Fig. 2.25. Summary of the endocrine control of previtellogenesis in telotrophic ovarioles of the blood-feeding bug *Panstrongylus megistus* (Hemiptera: Reduviidae). The structural events occurring are illustrated in Fig. 2.15. (Drawn from information in *J. Insect Physiol.*, 25, A. F. Furtado, The hormonal control of mitosis and meiosis during oogenesis in a blood-feeding bug *Panstrongylus megistus*, pp. 561–570, Copyright 1979, with permission from Elsevier Science.)

2.3 Endocrine Control of Oogenesis

The principal responsibilities of a female insect are to produce viable eggs at an appropriate time and to deposit them in a place suitable for survival of her offspring when they hatch (Davey, 1974). She can do this because she is sensitive to diverse cues from the environment (photoperiod, temperature, relative humidity, substrate odor, food, proximity and state of mates, etc.), is able to integrate this information in her central nervous system, and via her endocrine system, can use it to develop and deposit her eggs at the right time and place (de Wilde and De Loof, 1973; Wyatt and Davey, 1996).

2.3.1 PREVITELLOGENESIS

In most investigated insects, oogonia divide by mitosis and differentiate into primary oocytes (and also into nurse cells in insects having meroistic ovarioles) during juvenile or adult life. Development of oocytes is arrested in late diplotene of prophase I when the oocytes are in previtellogenesis, and does not continue until

"called out" by appropriate hormonal commands (Dumser, 1980). The division sequence of stem cells, oogonia, and cystocyte clones is programmed in the genome just as in male insects (1.3.2.2).

However, in 1979, Furtado demonstrated that proliferation of oogonia and prefollicular cells and differentiation of the tropharium and previtellogenic oocytes in telotrophic ovarioles of the blood-feeding reduviid bug *Panstrongylus megistus* is under control of blood-borne factors produced by certain median neurosecretory cells (MNSCs) in the pars intercerebralis of the brain in fifth (last) instars (Figs. 2.15, 2.25). He showed two types of MNSCs to be involved, A and A', both of which are induced to produce their product by uptake of a blood meal. A cells release a factor into the blood that stimulates proliferation of oogonia and prefollicular cells in fifth instar ovarioles up until day 6 (Fig. 2.25). A factor from the A' cells then stimulates secretion of 20E from the prothoracic glands, which causes the ovarioles to differentiate between days 16 and 24. JH from the corpora allata then stimulates the previtellogenic growth of primary oocytes after adult emergence.

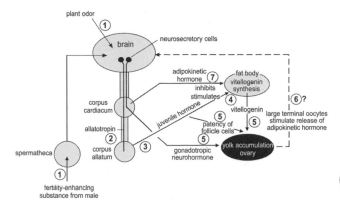

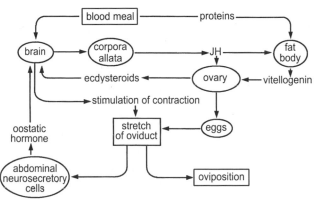

Fig. 2.26. Endocrine control of vitellogenesis in *Locusta migratoria* (Orthoptera: Acrididae). This diagram shows the sequence of events (numbers) following insemination of the female and the transfer of fertility-enhancing substances by the male. Step 6 is inferred. (From R. F. Chapman. 1998. *The insects. Structure and function*, p. 305, Fig. 13.9a. 4th ed. Reprinted with the permission of Cambridge University Press.)

Fig. 2.27. Endocrine control of vitellogenesis in the blood-feeding bug *Rhodnius prolixus* (Hemiptera: Reduviidae). JH, juvenile hormone. (Nijhout, H. F. *Insect hormones.* © 1994 by Princeton University Press. Reprinted by permission of Princeton University Press.)

2.3.2 VITELLOGENESIS

In most female insects investigated, secretions from two endocrine organs in the head control vitellogenesis: paired corpora allata (CA) below and behind the brain and certain MNSCs in its pars intercerebralis (Fig. 8.24) (Kaulenas, 1992).

2.3.2.1 CORPORA ALLATA AND JUVENILE HORMONE

Swelling and subsidence of the CA during oogenesis were first observed in 1918 by Ito in females of the commercial silkworm *Bombyx mori*, and he suggested that they might be endocrine organs. Their role in vitellogenesis was first demonstrated experimentally in 1936 independently by Weed-Pfeiffer in sixth instar larvae and adults of the grasshopper *Melanoplus differentialis* and by Wigglesworth in newly emerged adults of the blood-feeding reduviid *Rhodnius prolixus*. When Weed-Pfeiffer removed the CA from newly emerged adult females, she observed oogenesis to stop prior to yolk uptake and concluded that these glands secrete a hormone necessary for normal functioning of the follicle cells during vitellogenesis.

Changes in size of the CA (they are fused in this insect) immediately before and after the imaginal molt in *R. prolixus* suggested to Wigglesworth that the gland secreted a blood-borne factor necessary for oogenesis. He confirmed this by cutting and removing the head either in front of or behind the CA of newly emerged, unfed and just-fed females. When only the brain was removed before the end of a critical period (the time required for the factor to be released after the blood meal), terminal oocytes in the insect's ovarioles underwent vitellogenesis. However, when both brain and CA were removed, oogenesis was halted prior to yolk uptake and any oocytes already undergoing vitellogenesis degenerated and were resorbed by their follicle cells. If such an insect

was then joined in parabiosis (i.e., if their body cavities were rendered confluent by means of a glass capillary) to another adult of either sex having active CA or if active CA were implanted into it, vitellogenesis resumed. He thus concluded that cells of the CA produce a gonadotropin necessary for vitellogenesis.

This pivotal role for CA in vitellogenesis has since been demonstrated by use of similar experiments in one or more anautogenous species of Zygentoma, Orthoptera, Blattodea, Hemiptera, Coleoptera, Lepidoptera, Hymenoptera, and Diptera (Engelmann, 1970: Table 9; 1979: Table 4; Wyatt, 1997: Table 1), but not in autogenous species such as the giant silkworm *Hyalophora cecropia*, in viviparous (live-bearing) species such as tsetse flies (*Glossina* spp.), or in neotenous species such as the stick insect *Carausius morosus* (such insects are capable of producing viable eggs while still of juvenile phenotype).

In all anautogenous species tested, vitellogenesis was halted by removing the CA after adult emergence and reinstated by implanting active CA from other insects of either the same or different species, or by culturing ovaries in JH-free media (vitellogenesis stopped) and then adding JH or JH analogs (vitellogenesis resumed) (Raikhel and Dhadialla, 1992; Kaulenas, 1992).

The hormone produced by cells of the CA is now known to be JH (Fig. 12.11E). It is a primary coordinator of reproductive processes in both sexes and affects all tissues either directly or indirectly related to reproduction (Nijhout, 1994b; Wyatt, 1997). During oogenesis, JH
• stimulates Vg synthesis in the fat body of representative Orthoptera (Fig. 2.26), Blattodea, Hemiptera (Fig. 2.27), Coleoptera, Lepidoptera, and Hymenoptera via the following sequence (Chen and Hillen, 1983; Wyatt, 1997):

CA → JH → putative JH receptors in fat body cells (?)
→ JH receptor complex (?) → Vg genes → Vg mRNA
→ Vgs → hemolymph → ovary → receptor-mediated
yolk uptake by vegetative oocytes → Vns;

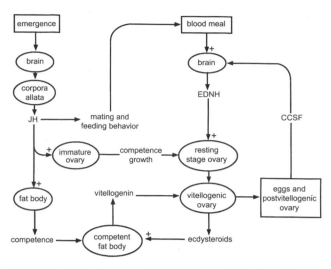

Fig. 2.28. Endocrine control of vitellogenesis in the yellow fever mosquito *Aedes aegypti* (Diptera: Culicidae). JH, juvenile hormone; EDNH, egg development neurosecretory hormone; CCSF, corpus cardiacum stimulating factor. (Nijhout, H. F. *Insect hormones.* © 1994 by Princeton University Press. Reprinted by permission of Princeton University Press.)

- primes the follicle cells to prepare for vitellogenesis (DNA and protein synthesis), to synthesize ecdysteroids (Figs. 2.27, 2.28), and in mosquitoes, to synthesize 20E in response to receipt of egg development neurosecretory hormone (EDNH) from MNSCs of the brain (Fig. 2.28);
- primes the fat body of *Aedes aegypti* (Diptera: Culicidae) to synthesize Vgs when it is subsequently exposed to 20E from the ovaries (Fig. 2.28);
- induces patency between adjacent follicular cells by binding to putative JH receptors on the follicle cell membrane, activating Na⁺/K⁺-ATPase, and influencing assembly of microtubules and microfilaments about their horizontal circumferences, thereby allowing uptake of Vgs to occur (Fig. 2.26) (Abu-Hakima and Davey, 1975, 1977; Huebner and Diehl-Jones, 1993, 1998; Wyatt and Davey, 1996);
- stimulates early vitellogenic oocytes in *D. melanogaster* to progress through a putative control point at about stage 9 (Fig. 2.13A) (Soller et al., 1999);
- influences development of oocyte competence to internalize Vgs (by inducing formation of the endocytotic complex) (Fig. 2.8); and
- stimulates protein synthesis in the female's accessory glands either directly or indirectly (Nijhout, 1994b; Wyatt, 1997; Chapman, 1998). In female moths, for example, synthesis and release of sex attractant and "calling" behavior are evoked via JH-induced release, from certain neurosecretory cells in the brain, of a pheromone biosynthesis activating neuropeptide (PBAN) (Cusson and McNeil, 1989).

Finally, Park and coworkers (1998) discovered that the accessory reproductive glands of 3-d old recently mated *males* of the tobacco budworm *Heliothis*

virescens (Lepidoptera: Noctuidae) synthesize small amounts of JH that are transferred to the female during copulation and stimulate JH secretion by her CA up to 2.5-fold the levels reached by virgin females.

2.3.2.2 MEDIAN NEUROSECRETORY CELLS

For many years, decreases in the size of certain protocerebral MNSCs (the A cells) have been observed to correlate with cycles of vitellogenesis in female insects, and their removal or destruction inhibits vitellogenesis (Raabe, 1982: Table 14; 1986). They have since been shown to influence vitellogenesis in females of various species by synthesizing and releasing the following hormones:

- *Allatotropin* (Fig. 2.26), a hormone inducing cells of the CA to synthesize and release JH. Demonstration of a direct role for the MNSCs in vitellogenesis is shown by destroying them and then implanting inactive CA into the body. If vitellogenesis fails to occur, then action of the MNSCs is probably required to stimulate JH synthesis and release from the CA.
- *Gonadotropin*, which stimulates Vg and ecdysteroid uptake by the ovaries (Fig. 2.26).
- EDNH, which is released on adult emergence of the yellow fever mosquito *A. aegypti* (Diptera) (Fig. 2.28). When the mosquito takes a blood meal, something causes release of a corpus cardiacum stimulating factor (CCSF) from the ovary, which induces release of EDNH. EDNH then evokes release of ecdysteroids from the ovary, which enter the fat body, are hydroxylated to 20E, and stimulate Vg synthesis (Raikhel et al., 1999).

In all instances examined, hormones synthesized in MNSCs of the pars intercerebralis are stored either there or in the corpora cardiaca prior to release. Thus, both the MNSCs and the corpora cardiaca must be removed in experiments demonstrating the role of the MNSCs.

2.3.2.3 NEUROSECRETORY CELLS IN THE VENTRAL NERVE CORD

Decreases in the size of A cells in ganglia of the ventral nerve cord have been observed to correlate with cycles of vitellogenesis in females of representative Orthoptera, Blattodea, and Phasmida, and these cells enlarge when the ovaries are removed (Raabe, 1982).

2.3.2.4 ANTIGONADOTROPINS AND OOSTATIC HORMONE

As mentioned in 2.2.1.2, females of many anautogenous insects deposit their eggs in batches, vitellogenesis only occurring in the most proximal (terminal or T) follicle of each ovariole. The block to vitellogenesis in T-1 follicles has various causes, depending on species:

- In females of *R. prolixus*, oostatic hormone, a neuropeptide released from four pairs of abdominal neurosecretory organs during vitellogenesis of T follicles, antagonizes JH-induced patency of follicular cells in T-1 follicles and also the release of JH from CA (Fig.

2.27) (Abu-Hakima and Davey, 1975, 1977; Chiang and Davey, 1988; Davey, 1993).

- In females of the ovoviviparous cockroach *Diploptera punctata*, 20E, released from follicle cells of T oocytes undergoing vitellogenesis, inhibits the synthesis or release of JH from CA either directly or indirectly (Stay et al., 1980).
- In females of the cockroach *Eublaberus posticus* and some other species, eggs are enclosed within an ootheca that the female carries about with her within her genital chamber (Roth, 1973). Nervous impulses from mechanoreceptors in the wall of this chamber, via the ventral nerve cord, inhibit either directly or indirectly the synthesis or release of JH by cells of the CA, probably by way of the MNSCs and allatotropin. When the ootheca is dropped, this inhibition is lifted, JH is released, and vitellogenesis of the next batch of eggs proceeds.
- In females of the house fly *Musca domestica*, oostatic hormone is released from follicle cells of the T follicles, which inhibits the synthesis or release of JH from CA (again probably by way of MNSCs and allatotropin) and hence vitellogenesis in T-1 follicles (Adams, 1970).

2.3.2.5 OVERLAPPING CYCLES OF VITELLOGENESIS

In females of the cockroach *Periplaneta americana* (Weaver et al., 1975), the grasshopper *Melanoplus sanguinipes* (McCaffery and McCaffery, 1983), and some other insects having panoistic ovarioles, the vitellogenic cycles of egg follicles T, T-1, and T-2 overlap within a single ovariole. *Melanoplus sanguinipes* females, maintained at 28°C (light) and 23°C (dark), have three phases of vitellogenesis: (1) an initial, 24-h phase of slow development (oocyte length: 1.0–1.2 mm), (2) a phase of rapid growth (1.2–3.5 mm) lasting 2 d in the first gonotrophic cycle and 3 d in later ones, and (3) a final phase of rapid growth (3.5–4.5 mm) lasting 3 d. Phases 1, 2, and 3 of oocytes T, T-1, and T-2 overlap entirely, with increased secretion of JH by the CA coinciding exactly with initiation of phase 2 in each cycle. Thus, each threesome of oocytes within an ovariole is subject to two and occasionally to three peaks of JH during vitellogenesis.

The endocrine control of vitellogenesis in *Thermobia domestica* (Zygentoma: Lepismatidae), *Diploptera punctata* (Blattodea: Blaberidae), *Locusta migratoria* (Orthoptera: Acrididae; Fig. 2.26), *R. prolixus* (Hemiptera: Reduviidae; Fig. 2.27), *A. aegypti* (Diptera: Culicidae; Fig. 2.28), and *Phormia regina* (Diptera: Calliphoridae) is discussed in detail by Nijhout (1994b) and Chapman (1998).

2.3.3 CHORIOGENESIS

In females of *Blattella germanica* (Blattodea), application of 20E will induce precocious choriogenesis by follicular cells in ovarioles cultured in vitro (in vivo there is an ecdysteroid peak during maturation of each batch of eggs just before oviposition which may be correlated with this) (Belles et al., 1993).

2.4 Influence of Extrinsic Factors on Vitellogenesis

Environmental factors usually influence vitellogenesis by modifying secretory activity of the endocrine organs, particularly that of the brain's MNSCs, via the following sequence:

environmental stimuli → sense organs → central nervous system → MNSCs of brain → allatotropin → CA → JH → Vg synthesis in fat body (and follicular cells in flies) and its uptake by the oocytes

The effects of 13 environmental factors on vitellogenesis in various insects are summarized below.

1. Flight muscle autolysis: Flight muscle proteins in female bark beetles, young ant and termite queens, and some other female insects depend on larval feeding for their synthesis and allocation. After adult emergence and following a mating or dispersal flight, the female alights to reproduce, her flight muscles degenerate, and the processed proteins from them contribute to vitellogenesis of the first batch of eggs (Kobayashi and Ishikawa, 1993; Tanaka, 1993; Stjernholm and Karlsson, 2000). This breakdown appears to be mediated by release of JH from the CA, as topical application of JH to a newly emerged female will cause flight muscle autolysis and allatectomy will prevent it (Borden and Slater, 1968):

CA → JH → flight muscle degeneration → proteins → fat body → Vg synthesis → vitellogenesis

Application of the antiallatotropin precocene II (12.4.7) prevents this degeneration (Sahota and Farris, 1980). The trade-off between the ability of a female to fly and disperse and to produce eggs is referred to as the *oogenesis-flight syndrome*.

2. Photoperiod: Adult *diapause* (arrested development and low metabolic rate) in female insects usually involves an arrest in vitellogenesis (Raabe, 1982, 1986; Nijhout, 1994b). In overwintering females it is usually induced by shorter photoperiods in autumn, and in aestivating females in midsummer, by longer photoperiods, although there is certainly a genetic component (Engelmann, 1970). These responses have been best studied in females of the Colorado potato beetle *Leptinotarsa decemlineata*, where a shorter photoperiod, in conjunction with lower temperature and changes in food quality, results in allatotropin from MNSCs in the brain not being released and thus, in no JH and no vitellogenesis.

3. Temperature: Most female insects have a characteristic range of temperatures within which they will develop eggs (often more narrow than that in which they are active, their *effective temperature*) and an *optimum temperature* at which egg production is greatest (Engelmann, 1970; Wigglesworth, 1972).

Photoperiod also interacts with temperature, as recently demonstrated in the carabid beetle *Notiophilus biguttatus*. Ermating and Isaake (2000) showed that eggs increase in number and decrease in size with increased temperature and that the effect of temperature is weaker when days are short than when they are long.

4. Relative humidity: Relative humidity is correlated with temperature, so their effects on oogenesis must be considered together (Engelmann, 1970). At low relative humidity, females lose water by evaporation or excretion and insufficient amounts may remain for oogenesis.

5. Food: In anautogenous females, vitellogenesis will not occur until a newly emerged female has begun to feed, resulting in an extended preoviposition period between emergence and deposition of the first batch of eggs (Wigglesworth, 1972). In addition, the number of eggs per female is usually heavily influenced by availability of food (Fox and Czesak, 2000). For example, in most blood-feeding insects, the number of eggs produced is proportional to the size of the blood meal (Wigglesworth, 1972; de Wilde and De Loof, 1973; Clements, 1992: chapter 27). And in locusts, removal of the frontal ganglion prevents the insects from swallowing and hence vitellogenesis. Possible ways that food could influence vitellogenesis in insects are summarized by Engelmann (1970: Fig. 47) and by Clements (1992: chapter 27) specifically for mosquitoes.

6. Host effects: In females of many insects, both oviposition and oogenesis, particularly vitellogenesis, can be influenced by the quality and availability of the host (Papaj, 2000). Oogenesis in a female can be induced either by feedback resulting from the act of depositing eggs, by host feeding, or by stimulatory sensory input from the host. Also, trade-offs between egg production and either survival or dispersal (the oogenesis-flight syndrome), uncertainty in the host environment, and host condition between the onset of oogenesis and time of egg deposition can select for development of host effects on oogenesis. Finally, some host defense factors actually inhibit egg production by the female. Papaj (2000) provided a fascinating review of these interactions.

7. Mating stimuli: In many flies and lepidopterans, the number of eggs produced is greater in females that have mated than in those that haven't (Wigglesworth, 1972). In *Thermobia domestica* (Zygentoma), *Diploptera punctata* and *Leucophae maderae* (Blattodea), *Cimex lectularius* (Hemiptera), and *Anopheles* (Fig. 2.28) and *Glossina* spp. (Diptera), mating is required before vitellogenesis will occur (see references in de Wilde and De Loof, 1973).

The apterygote *T. domestica* continues to molt after adult emergence, the gonotrophic cycle between each molt being induced by insemination (Watson, 1964). If mating does not occur, all vitellogenic oocytes degenerate and are resorbed by the follicle cells before the next molt. Mating is also required between each molt because spermatozoa within the spermatheca are shed along with its cuticular lining at each ecdysis.

In *R. prolixus*, presence of spermatozoa within the spermatheca evokes release into the hemolymph of a spermathecal factor (SF) from spermathecal cells, which in turn induces release of an ovulation hormone (myotropin) from five pairs of MNSCs in the pars inter-

cerebralis of the brain. Ovulation hormone stimulates ovulation and oviposition so that oostatic hormone is no longer released from the abdominal neurosecretory organs. This in turn removes the block to vitellogenesis of T-l follicles (Fig. 2.27).

8. Precocenes: Chromenes (Fig. 12.11I) produced by leaves of the composite bedding plant *Ageratum houstonianum,* when imbibed by a feeding female insect, cause her to become sterile by inducing degeneration of her CA, the source of JH. With no JH there is no vitellogenesis. (These compounds are discussed in more detail in 12.4.7.)

9. Male pheromones: In the locust *Schistocerca gregaria*, male pheromones affect release of allatotropin from MNSCs in the brain in the female, which stimulates vitellogenesis, probably via the CA/JH axis.

10. Inhibitory pheromones in eusocial insects: In growing colonies of the honeybee *Apis mellifera*, workers in contact with the queen and elsewhere in the hive are usually unable to produce eggs (in rare, anarchistic colonies, some 5% of workers can deposit viable eggs even when the queen is present [Oldroyd et al., 1999]). This is because inhibitory pheromones from the queen's mandibular glands and elsewhere, when taken up by workers through *trophallaxis* (food exchange), inhibit release of allatotropin from their MNSCs and thus release of JH from CA and vitellogenesis (de Wilde and Beetsma, 1982; Winston and Slessor, 1992). If the queen is removed, this inhibition is lifted, and some workers begin to develop and deposit a few unfertilized eggs, which will develop into drones (males) (hymenopterans are haplodiploid; 5.1). Simultaneously, others fashion one or more emergency queen cells from worker cells already containing young female larvae and nurse bees will rear them as queens by feeding them royal jelly (12.8.2). The same may be true in ants and in some termites (de Wilde and Beetsma, 1982), but in the bee *Frieseomelitta varia*, the workers fail to lay eggs even in queenless colonies because their ovarioles degenerate in the late pupal and pharate adult stages (Boleli et al., 1999).

11. Aging: Though older females often deposit a smaller number of smaller eggs than do younger females of the same species, with fewer of these hatching, there is too much variation between species to generalize (Fox and Czesak, 2000). In addition, females developing from such eggs often, but not always, have lower fecundity. Where the effects occur are usually unknown, but where they are, they are diverse (Fox and Czesak, 2000).

12. Effects of endoparasites: Parasitic nematodes or strepsipterans living in the body cavity of an adult female compete for nutriment with vitellogenic oocytes (de Wilde and De Loof, 1973). They can also inhibit development of CA and hence release of JH. However, females of the braconid wasp *Asobara tabida* must be infected with *Wolbachia* bacteria if normal oogenesis is to occur (Dedeins et al., 2001). If such females are treated with antibiotics, they are unable to develop and deposit eggs, while those of the related wasp *A. citri*,

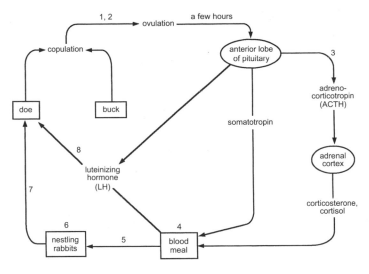

Fig. 2.29. Summary of how endocrine activity in a host rabbit affects reproductive physiology and gametogenesis in the rabbit flea *Spilopsyllus cuniculi*. (Drawn from information in Rothschild et al., 1970.)

a species not naturally infected with *Wolbachia*, are not affected by such treatment. This is the first example known in which *Wolbachia* infection is obligatory for the successful reproduction of an insect.

13. Host effects on ectoparasitic insects: Male and female rabbit fleas *Spilopsyllus cuniculi* cannot breed unless the female rabbit on whose ears they feed becomes pregnant (Rothschild, 1965). Oocytes of female fleas feeding in the ears of a pregnant doe commence vitellogenesis about 10 d prior to birth of the young rabbits (Fig. 2.29). A few hours after birth, both male and female fleas detach from the ears of the doe and move to her face and then on to the nestlings while she is tending them and eating the afterbirth. There, both sexes feed and mate, and the female deposits her eggs on the nestlings or in the nest. After about 7 d, the fleas leave the nestlings and return to the lactating doe. If she becomes pregnant again, the cycle is repeated.

The various hormones circulating in the blood of the pregnant doe and in that of the nestlings, and affecting the reproductive cycle of the flea, were identified by injecting castrated bucks, lacking most of these hormones, with the hormones in various combinations or by spraying them directly on fleas. Results of these experiments demonstrated that somatotropin induces mating in fleas, hydrocortisone and corticosterone induce vitellogenesis and egg deposition, and luteinizing hormone and progesterone induce oocyte resorption. External application of insect growth regulators had the same effect on fleas of both sexes (Rothschild et al., 1970). Thus, host hormones probably affect activity of the female flea's MNSCs and hence CA function and vitellogenesis.

A similar relationship exists between *Cediopsylla simplex* and the cottontail rabbit but does not hold for Oriental (*Xenopsylla cheopis*) and European (*Nosopsyllus fasciatus*) rat fleas on Norway rats, probably because these two species are not host-specific (Rothschild et al., 1970).

2.5 Genes and Oogenesis

As usual, the role of genes is best understood in females of *D. melanogaster*. Many *female sterile* (*fs*) mutations are known in this insect. They affect germ cell determination, differentiation of ovarioles, cystocyte divisions, nurse cell and oocyte determination, meiosis, nurse cell function, follicle cell development, vitellogenesis, vitelline membrane deposition, choriogenesis, and ovulation (Spradling, 1993: Table 2). Three types of defects that are recognized result from action of such mutant loci: females fail to deposit eggs, they deposit abnormal eggs, or they deposit apparently normal eggs which fail to develop normally after fertilization.

To identify the site of action of a particular mutant locus, the usual first step is a *reciprocal transplant*: The mutant ovaries are transplanted into a wild-type host female and the wild-type ovaries, into a mutant host. If the mutant ovaries remain defective in the wild-type host, and if the wild-type ovaries function in the mutant host, then the fault lies within the ovary itself and the mutation is *ovary autonomous*. If the wild-type ovary fails to function in the mutant host and if the mutant ovary functions in the wild-type host, then the defect lies outside the ovary and is *ovary nonautonomous*. There are many examples of both types of mutation. A full review of the genetics of oogenesis in *D. melanogaster* is provided in Spradling, 1993.

3/ Sperm Transfer, Allocation, and Use

Although the male and female of a given species of insect may have fully differentiated gametes in their gonads, they cannot breed successfully until the male has successfully introduced his sperm into the spermatheca of the female. Depending on species, males and females locate each other by chance or by use of visual, chemical, or auditory cues (Thornhill and Alcock, 1983; Bailey and Ridsdill-Smith, 1991). After the two have met, more or less complex precopulatory and copulatory behavioral sequences ensue involving the exchange of signals indicating the species, sex, quality, and receptivity of each partner. If a female is conspecific and receptive, the male then mounts her in a characteristic way (Fig. 3.1) and begins the behavioral and physiological processes culminating in insemination.

Based on the distribution of copulatory positions known at that time from throughout the pterygotes, Alexander (1964) postulated the ancestral position to be female above or male side. However, the false male above position seems to be the most widespread (Fig. 3.2) and thus probably the ancestral mode. After the external genitalia of the two sexes are linked, the insects may change their positions relative to each other prior to insemination.

3.1 Sperm Passage from Sperm Tubes into Seminal Vesicles and Spermatophore

When the male and female are *in copula*, the problem remaining is to transfer sperm from the sperm tubes of the male into the spermatheca of the female. First, the sperm must move to the seminal vesicles and then into the spermatophore or intromittent organ by some variation of the following route (numbers in parentheses correspond to those in Fig. 3.3):

- Prior to mating, individual spermatozoa are released from sperm bundles in the base of each sperm tube through rupture of cyst membranes (1, 2).
- They pass into the vasa efferentia and deferentia and then are moved to the seminal vesicles by peristaltic contractions in their walls (3).
- Sperm is stored immobile in the seminal vesicles until ejaculation (4).
- Prior to ejaculation, accessory gland (mesadene) cells

release seminal secretions about the sperm and an activator inducing motility in them (5, 6).
- Activated sperm, suspended in semen, move into the apex of the ejaculatory duct and are surrounded sequentially by various proteins from cells of the mesadenes. These may or may not form a spermatophore before or during copulation, within the ejaculatory bulb (7) or duct, the endophallus (7A), or the genital chamber of the male or in the vagina of the female, depending on species.

There are many variations in this sequence. For example, in *Locusta migratoria* (Orthoptera: Acrididae) (Szöllosi, 1975) and some odonates (Siva-Jothy, 1997), the heads of sperm released from each sperm bundle are inserted into a gelatinous cap of material secreted by the cyst cells and move to the seminal vesicles by the combined lashing of their flagella. During this passage, *L. migratoria* (Szöllosi, 1975) and mosquito (Ndiaye et al., 1997) sperm undergo structural and functional changes (capacitation) that may facilitate their readiness to fertilize eggs.

In lepidopterans, the plasma membrane of individual spermatozoa in cysts within the testes is covered by short, radially arranged laciniate appendages (Fig. 3.4A: RM). Just before sperm release from the seminal vesicles, these appendages disappear, the cyst membrane degenerates, a thick extracellular sheath develops about each spermatozoon (Fig. 3.4B), and a network of proteinaceous strands arises and binds the sperm together into spermatodesms (Riemann and Thorson, 1971; Friedländer and Gitay, 1972; Friedländer et al., 2001), changes that do not occur in males maintained in continuous light (Riemann et al., 1974). Upon their activation within the spermatophore (3.2), both apyrene and eupyrene sperm become motile, are released from the strands, and intermingle, and the eupyrene sperm "hatch" from their sheaths (Fig. 3.4C) (Riemann and Thorson, 1971; Friedländer et al., 2001). Capacitation of eupyrene sperm can be induced in vitro by application of trypsin and seems to be required for them to leave the spermatophore and to fertilize eggs (Friedländer et al., 2001).

In testes of virgin male Mediterranean flour moths *Anagasta kühniella*, gypsy moths *Lymantria dispar*, and perhaps other insects, sperm release is under endocrine control and has a precise, circadian (24-h) rhythm. The

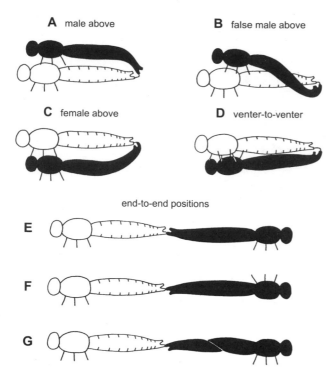

A male above

B false male above

C female above

D venter-to-venter

end-to-end positions

E

F

G

Fig. 3.1. Positions assumed by male (black) and female (white) insects during copulation. (Reprinted by permission of the publisher from R. F. Chapman. *The insects. Structure and function*, p. 358, Fig. 227. 3rd ed. Harvard University Press, Copyright by R. F. Chapman. © 1969, 1972, 1982. Based on Richards, 1927, Sexual selection and allied problems in the insects, *Biol. Rev.* 2: 298–364.)

pacemaker for this release, in *L. dispar*, is located within the testes and vasa deferentia, is sensitive to light, and can be entrained even when the system is removed from the insect (Giebultowicz et al., 1989; Riemann and Giebultowicz, 1991) (in fact, cells throughout the body of *Drosophila* flies are known to contain similar clocks driven by expression of the clock gene, *period* [Plautz et al., 1997]). Under a 16:8 light-dark cycle, release of bundles from the testis into the upper vas deferens (UVD) occurs during a 3-h period beginning just before lights out, and they remain there for 12 h before moving to the seminal vesicles (Giebultowicz et al., 1994). During release and while retained in the UVD, the sperm undergo extensive structural changes (Fig. 3.4), and several protein bands, characteristic of bundles in the testis, are missing from the sperm in the UVD and are replaced by several new protein bands, some of which are secreted by the wall cells of these ducts.

To be released from each sperm tube in the testis, the first sperm bundles must pass through a *terminal epithelium*, a layer of specialized epithelial cells separating the lumina of the sperm tubes from the lumen of the UVD (Giebultowicz et al., 1997). About 3 h before sperm first appears in the UVD, sperm bundles, still enclosed in cyst membranes, start to penetrate between cells of the membrane. The cyst cells degenerate and are phagocytosed by cells of the terminal epithelium, and individual sperm rapidly enter the lumen of the UVD.

3.2 Sperm Activation

Insect sperm are generally immotile while stored within the seminal vesicles (Leopold, 1976). This may be because they are closely packed, because they lack sufficient oxygen, because some substrate necessary for movement is lacking, or because an inhibitory substance is released from cells of the vesicle wall (Davey, 1965). Transfer of sperm from the vesicles to physiological saline solution usually activates them, but this is believed to be an artifact and the motility induced to be nonfunctional (Leopold, 1976).

Prior to, during, or just after transfer to the female, the sperm are activated by secretions from the male's accessory glands (Fig. 3.3: 5). In males of the reduviid bug *Rhodnius prolixus*, for example, a change in pH from 7 in the testes and seminal vesicles, to 5.5 in the ejaculatory bulb and in the exit system of the female, caused by products of the male's transparent accessory glands, activates the sperm (Davey, 1959), whereas in males of the human bed bug *Cimex lectularius*, exposure of sperm to oxygen and metabolizable substrates provided by the male has the same effect (Davis, 1966). And in males of the saturniid moth *Antheraea pernyi*, apyrene (anucleate) sperm become vigorously active during entry into the spermatophore by action of a polypeptide activator secreted by wall cells of the common exit duct (Shepherd, 1974). This occurs in about 90 s at 25°C, and a 30-min exposure is sufficient for 31 h of activity. In addition, enough activator is produced by a single male to induce activity of apyrene sperm in 400,000 other males. Close chemical similarity in the activators of males of different species is suggested by the observation that those from the accessory glands of six other lepidopteran and two grasshopper species also activate *A. pernyi* apyrene sperm (Shepherd, 1974).

A serine-type endopeptidase, *initiatorin*, secreted by male accessory gland cells in the domestic silkworm *Bombyx mori* both activates apyrene sperm and dissociates eupyrene sperm bundles within the spermatophore after it has been transferred to the female (compare Fig. 3.7G) (Osanai, 1996). It also activates the arginine degradation cascade, an energy-yielding system for sperm maturation.

3.3 Intromission and Insemination

The next step is to introduce sperm into the genital ducts of the female, a process accomplished in innumerable ways in different species. To understand the distribution and evolution of sperm transfer mechanisms in hexapods, it is necessary to have some appreciation for

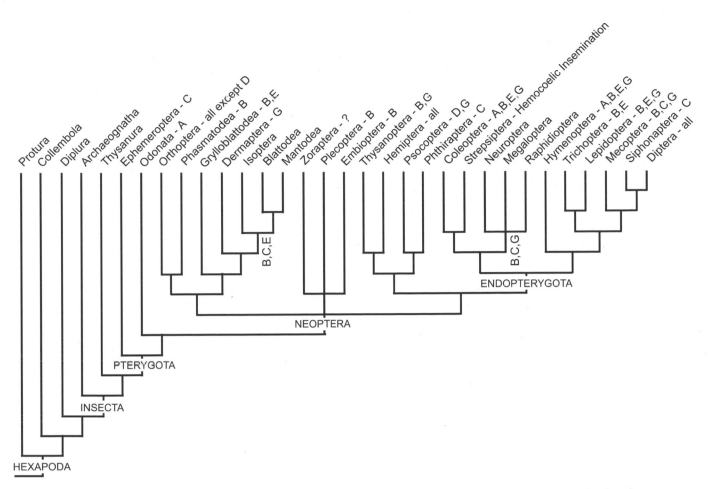

Fig. 3.2. Evolutionary relationships of the hexapod orders as inferred by Gullan and Cranston (1994), showing the distribution of known copulatory positions throughout the pterygotes (apterygotes use indirect sperm transfer via the substrate; see Fig. 3.5: 2a, 2b). The wide distribution of false male above (B) suggests that this position is plesiotypic. Letters refer to drawings in Figure 3.1. (Cladogram adapted with permission from P. J. Gullan and P. S. Cranston. *The insects. An outline of entomology*, p. 193, Fig. 7.5. 1st ed. Chapman and Hall, London. © 1994 Blackwell Science, Ltd.)

that occurring between males and females of other arthropods.

3.3.1 MARINE ARTHROPODS

In marine arthropods, desiccation of sperm during transfer from male to female is not a problem (Fig. 3.5). Nevertheless, three different mechanisms of sperm transfer have been observed in such animals (Schaller, 1979) (numbers in parentheses correspond to those in the top part of Fig. 3.5):

- release of sperm and eggs into water with external fertilization (widespread; 1a, 1b);
- internal fertilization by means of a spermatophore, either indirectly, via the substrate and with (2b) or without (2a) pairing, or directly by copulation (3); and
- free sperm directly by use of an intromittent organ (4).

External fertilization is undoubtedly ancestral in marine arthropods, judging by its wide distribution;

internal fertilization, either with or without a spermatophore, has evolved innumerable times to reduce sperm loss and to increase the probability of sperm-egg contact (Schaller, 1979).

3.3.2 TERRESTRIAL ARTHROPODS

In terrestrial arthropods, internal fertilization of eggs is usually necessary as desiccation of sperm is probable except under humid situations (in leaf litter and flowers, under bark of rotten logs, at night, at dawn or dusk, or during rainy periods) (Fig. 3.5) (Proctor, 1998). Since the ancestors of extant classes of arthropods were undoubtedly marine, males of those species transferring sperm by spermatophore or by use of an intromittent organ would have been preadapted reproductively to make the transition to land, and both modes of sperm transfer could be construed as ancestral in terrestrial hexapods (Schaller, 1979; Labandeira and Beall, 1990). However,

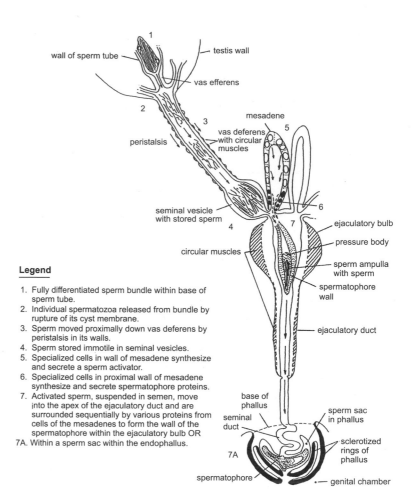

wall of sperm tube

testis wall

vas efferens

mesadene

peristalsis

vas deferens
with circular
muscles

seminal vesicle
with stored sperm

circular muscles

ejaculatory bulb

pressure body

sperm ampulla
with sperm

spermatophore
wall

ejaculatory duct

base of
phallus

sperm sac
in phallus

seminal
duct

sclerotized
rings of
phallus

genital chamber

spermatophore

Legend

1. Fully differentiated sperm bundle within base of sperm tube.
2. Individual spermatozoa released from bundle by rupture of its cyst membrane.
3. Sperm moved proximally down vas deferens by peristalsis in its walls.
4. Sperm stored immotile in seminal vesicles.
5. Specialized cells in wall of mesadene synthesize and secrete a sperm activator.
6. Specialized cells in proximal wall of mesadene synthesize and secrete spermatophore proteins.
7. Activated sperm, suspended in semen, move into the apex of the ejaculatory duct and are surrounded sequentially by various proteins from cells of the mesadenes to form the wall of the spermatophore within the ejaculatory bulb OR
7A. Within a sperm sac within the endophallus.

Fig. 3.3. Composite diagram of the reproductive system in a generalized male insect, showing the movement of spermatozoa from a sperm bundle in the base of a sperm tube into a spermatophore.

judging from the phylogenetic distribution of both methods (Fig. 3.10), use of an intromittent organ with direct sperm transfer (4 in Fig. 3.5: bottom) seems to have evolved independently in most orders from ancestral use of a spermatophore (1–3).

3.3.2.1 HEXAPODS

In extant terrestrial hexapods, sperm is passed from male to female within a spermatophore either indirectly via the substrate (2a, 2b) or directly by copulation (3a), directly into the spermatheca via an elongate intromittent organ (4), or by traumatic insemination (5 in Fig. 3.5: bottom).

3.3.2.1.1 SPERMATOPHORES

In Fig. 3.10, the known distribution of sperm transfer mechanisms in hexapods is superimposed on a reconstructed phylogeny of the hexapod orders. That at least some members of most orders use a spermatophore (1–3) suggests that this method is ancestral.

Spermatophore Formation

Proteins forming the spermatophore are synthesized within cells of the male's accessory glands (Fig. 3.3: 6), either directly or indirectly upon receipt of juvenile hormone (JH) from the corpora allata (CA), and after adult emergence in anautogenous species, via the following sequence (Mann, 1984; Kaulenas, 1992):

environmental stimuli → sense organs → median neurosecretory cells of brain → allatotropin → CA → JH → (other factors?) → male accessory glands → spermatophore proteins → spermatophore

If these glands are removed, spermatophore production and sperm transfer stop (there are exceptions) but not attempts to copulate (Leopold, 1976). Histological, histochemical, autoradiographic, and immunocytochemical methods have been used to trace the origins of spermatophore components to specific cell types within the accessory glands of males (discussion and references in Happ, 1992, and Kaulenas, 1992). Particularly fine studies of spermatophore formation are those of Linley (1981a, 1981b) on *Culicoides melleus* (Diptera: Ceratopogonidae), of Davey (1959) on *R. prolixus* (Hemiptera: Reduviidae), of Holt and North (1970) on *Trichoplusia ni* (Lepidoptera: Noctuidae), and of Gregory (1965) on *Locusta migratoria* (Orthoptera: Acrididae). In the latter, the spermatophore is 35–45 mm long, extends from the ejaculatory sac in the male's lower ejaculatory duct

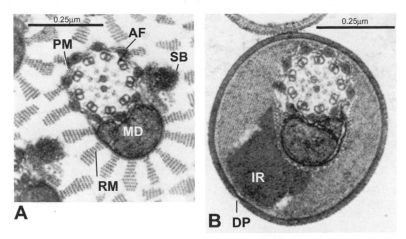

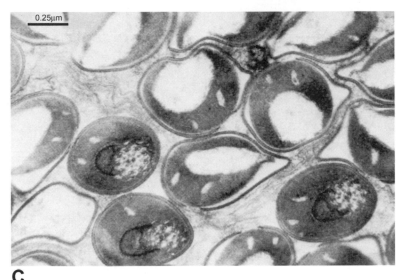

Fig. 3.4. Sperm capacitation in females of the clothes moth *Anagasta kühniella* (Lepidoptera: Pyralidae). A. Transmission electron micrograph (TEM) of transverse section through the flagellum of a eupyrene spermatozoon within a sperm bundle in the testis, showing the radial mantle (RM) of laciniate appendages and the satellite body (SB) outside the plasma membrane (PM). The latter encloses a mitochondrial deriviative (MD) and axoneme (AF). B. Same of a spermatozoon within the spermatheca of a female after capacitation, showing the sheath, dense plate (DP), and intrasheath rod (IR). C. TEM of transverse section of an area within the spermatheca of a female 9 d after insemination, showing parts of 14 empty sperm sheaths and three still containing a spermatozoon. (Reprinted from *Int. J. Insect Morphol Embryol.*, 1, J. G. Riemann and B. J. Thorson, Sperm maturation in the male and female genital tracts of *Anagasta kühniella* (Lepidoptera: Pyralidae), pp. 11–19, Copyright 1971, with permission from Elsevier Science.)

into the spermatheca of the female, and takes 8–10 h to form at 33–35°C.

Spermatophore Types

Spermatophores and their mode of formation, transfer, and use are almost as diverse as the hexapods in which they occur. The known distribution of spermatophore type throughout the insects is shown on a reconstructed phylogeny of the insect orders in Fig. 3.10.

Apterygote Hexapods. Investigated males of Collembola and Diplura deposit naked droplet spermatophores individually on stalks on the substrate (Fig. 3.5: bottom, 2a) (Schaller, 1979; Mann, 1984; Proctor et al., 1995). A receptive female must first find the drop and then take up the sperm with her genital opening. Search is facilitated by the high population densities often characteristic of these hexapods, by pheromones emanating from the spermatophore itself, or by the fact that one male can often deposit over 100 spermatophores. In intermediate and advanced clades of collembolans, the male

assists the female in finding the spermatophore by means of characteristic behavior (2b), and in some smintherids, actually manipulates the female with a pincer-like structure on his antennae by which he grasps an antenna of the female (Schaller, 1979; Proctor et al., 1995).

Male bristletails (Archaeognatha) and silverfish (Zygentoma) indulge in complex and characteristic courtship behavior and deposit one or more droplet spermatophores on or in association with a carrier thread stretched taut by the abdomen of the male (additional threads are used in silverfish). He then guides the female to pick up the droplet(s) (Sturm, 1992, and references). Most of the above-mentioned animals are confined to cryptic, humid situations, are nocturnal or crepuscular, or are active only on wet days.

Pterygote Insects. Males of many pterygote insects, including those of secondarily aquatic species, transfer sperm directly into the exit system of the female, within a spermatophore whose form can be determined by

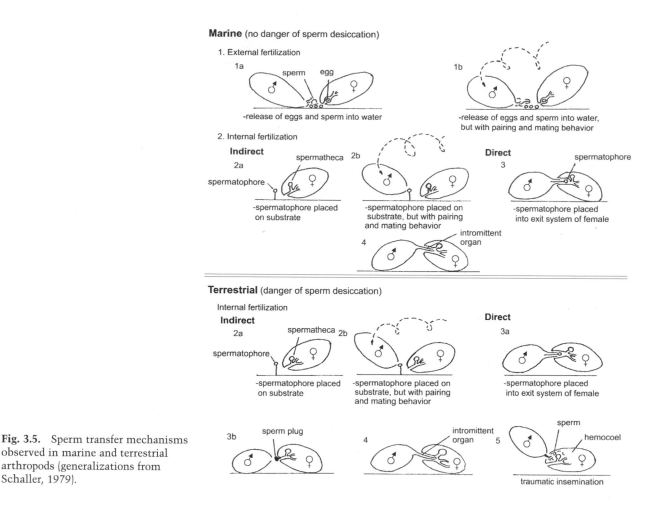

Fig. 3.5. Sperm transfer mechanisms observed in marine and terrestrial arthropods (generalizations from Schaller, 1979).

the male, the female, or both (Fig. 3.5: bottom, 3a) (Gerber, 1970; Tuzet, 1977; Mann, 1984).

There are two types of male-determined spermatophores. Depending on species, the type 1 spermatophore is formed within the lumen of the ejaculatory duct (Fig. 3.3: 7), within the endophallus (spermatophore sac) of the intromittent organ (Fig. 3.3: 7A), or within the genital chamber of the male before copulation. Such spermatophores are usually large, consist of a complex gelatinous matrix (sphraga) containing one or more sperm sacs or *ampullae*, and are individually lodged in the genital chamber of the female, with much of it remaining outside her body, only its "neck" penetrating close to or into the spermathecal duct.

A good example is the indirect sperm transfer typical of dragonflies and damselflies (Odonata; first accurately described by Jan Swammerdam in 1669), in which sperm transfer is by way of accessory genitalia on the venters of abdominal segments 2 and 3 of the male (Fig. 3.6E) (Carle, 1982; Askew, 1988; Corbet, 1999). In dragonflies, a depressed genital fossa on sternum 2 opens posteriorly into an internal vesicle derived from the anterior margin of sternum 3. The vesicle connects externally with a three-segmented "penis" and associated accessory "horns" (Fig. 3.6E). When a male is ready to mate, he transfers sperm or a small spermatophore from his genital opening at the back of abdominal sternum 9 into the vesicle of his accessory genitalia by flexing the apex of his abdomen ventrally and forward (Fig. 3.6A). Thus "loaded," the male approaches a flying female from above and grasps her thorax with his second and third pairs of legs while his forelegs touch the bases of her antennae (Fig. 3.6B). He then flexes his abdomen ventrally and forward above the body of the female, spreads the two pairs of claspers at its tip, and clasps the female. In dragonflies, the superior pair clasps the neck of the female while the inferior pair presses down on the top of her head (Fig. 3.6C), whereas in damselflies, these claspers fit into sockets in the female's pronotum. After attaching, the male releases the grip of his legs and straightens his abdomen, and the two fly off in tandem, the male pulling the female (compare Fig. 3.6F).

Either in flight or after landing, the male again flexes the tip of his abdomen ventrally and forward, so that the female's head touches his accessory genitalia. The female then flexes her abdomen ventrally and forward, her genitalia are grasped by the accessory genitalia of the male, and copulation ensues (lasting 3 s to 340 min depending on species) (Fig. 3.6D). Copulation ends when the male disengages the abdomen of the female from his

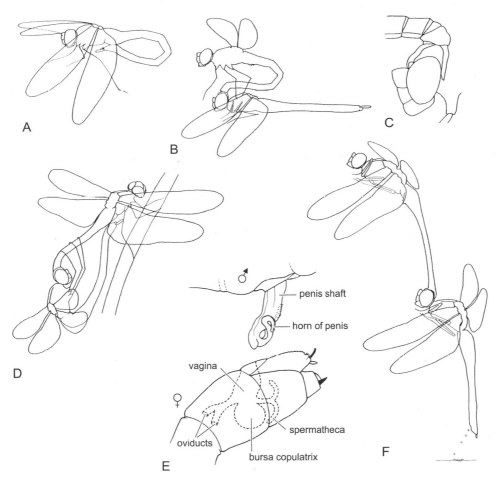

Fig. 3.6. Reproductive behavior in male and female dragonflies (Odonata). A. Male loading his accessory genitalia in flight. B. Male seizing flying female by the dorsum of her head with his claspers. C. Male claspers of *Cordulia aenea* (Corduliidae) grasping head of female. D. Female flexing her abdomen forward under that of the male to receive sperm from his accessory genitalia. E. Closeup, showing the functional relationship in *Calopteryx* sp. (Odonata: Calopteryidae) damselflies between the "penis" of the male's accessory genitalia and the secondary exit system of the female. The male spends the greater portion of his time physically removing the sperm of a previous male from the female's spermatheca and bursa copulatrix using recurved microtrichia on the horns of his penis; only at the last minute does he introduce his own. F. Female of *Sympetrum sanguineum* (Libellulidae) in tandem ovipositing onto wet ground. (A–D and F reproduced with permission from R. R. Askew, *The dragonflies of Europe.* Harley Books. © 1988 B. H. and A Harley Ltd. E reproduced with permission from P. J. Gullan and P. S. Cranston. *The insects. An outline of entomology*, p. 137, Box 5.3. 1st ed. Chapman and Hall, London. © 1994 Blackwell Science, Ltd.)

accessory genitalia with his hind legs so that she hangs passively from his claspers. The male may then separate from the female (most dragonflies), guard her (libellulids), or remain in tandem with her as she oviposits (Fig. 3.6F).

This peculiar mode of sperm transfer is believed to have evolved from that of an ancestor having indirect sperm transfer via the substrate (Fig. 3.5: bottom, 2a, 2b) (Corbet, 1999). Brinck (1962) suggested that when an ancestral protodonate first developed wings, it might have become necessary for the male to deposit sperm on his body rather than on the ground, the ventral surfaces of the second and third abdominal segments being the site most easily reached by the tip of his abdomen.

Development of accessory genitalia followed to increase the efficiency of spermatophore transfer.

The type 2 male-determined spermatophore is formed within the phallus during copulation, its size and shape being determined by the shape of the spermatophore sac within the endophallus and by the amount of room available in the vagina of the female (Fig. 3.3: 7A) (Mann, 1984). The male subsequently withdraws his phallus, leaving the spermatophore in the female.

There are also two types of female-determined spermatophores. The type 1 spermatophore is formed within the vagina of the female during copulation, and its shape is determined by the shape of this sac (Mann,

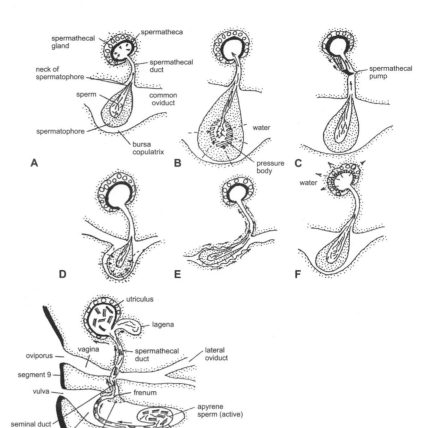

Fig. 3.7. Mechanisms that function to pass sperm from within the spermatophore into the spermatheca. See text for details in the following insects. A. *Anthonomus grandis* (Coleoptera: Curculionidae). B. Many ensiferans (Orthoptera). C. *Apis mellifera* (Hymenoptera: Apidae). D. Some lepidopterans. E. *Rhodnius prolixus* (Hemiptera: Reduviidae). F. *Culicoides melleus* (Diptera: Ceratopogonidae). G. Lepidoptera: Ditrysia. In most insects, the spermatheca is lined with sclerotized cuticle perforated by the ductules of spermathecal glandular cells (indicated in F).

1984). During or after sperm transfer, the male's accessory gland materials are ejected in a characteristic sequence into the vagina, where they encapsulate the sperm. In type 2, a sperm plug (sphragis, spermatophylax), probably derived by reduction from a spermatophore of type 1, is placed into the opening of the female gonopore just after sperm transfer (Fig. 3.5: bottom, 3b). The plug must be removed before the female can receive sperm from another male.

Transfer of Sperm from Spermatophore
to Spermatheca

Do sperm swim from the spermatophore into the spermatheca or are they passively moved there by action of the female's reproductive tract? Both mechanisms are known to occur in insects, and at least nine variations have been demonstrated by observation or experiment (Fig. 3.7) (Mann, 1984):

- Spermatozoa swim out of the spermatophore and up the spermathecal duct, attracted chemotactically by secretions of the spermathecal gland (Fig. 3.7A) (e.g., the boll weevil *Anthonomus grandis* [Coleoptera: Curculionidae]) (Grodner and Steffans, 1978).
- Osmotic uptake of water by a pressure body about the sperm ampulla, within the spermatophore causes it to

swell, collapses the ampulla, and forces sperm out of its neck and up the spermathecal duct into the spermatheca (Fig. 3.7B) (e.g., many ensiferans [Orthoptera]).

- Action of a pump within the spermathecal duct sucks sperm out of the spermatophore and into the spermatheca (Fig. 3.7C) (e.g., queens of the domestic honeybee *Apis mellifera* [Hymenoptera: Apidae]) (Gessner and Ruttner, 1977).
- Peristaltic contractions in the wall of the vagina squeeze the spermatophore and force both apyrene and eupyrene sperm up the spermathecal duct into the spermatheca (Fig. 3.7D). Active apyrene sperm may assist in freeing the inactive eupyrene sperm from sperm bundles within the spermatophore (e.g., some lepidopterans) (Tschudi-Rein and Benz, 1990a).
- Peristaltic contractions in the wall of the vagina press a spinous sclerite, the *signum*, against the wall of the spermatophore and rupture it, the sperm released into the vagina being moved into the spermatheca by rhythmic contractions in the walls of the seminal and spermathecal ducts (compare Fig. 3.7G) (e.g., some lepidopterans).
- Peristaltic contractions in walls of the female's exit ducts, induced by action of a serotonin-like secretion within the jelly coat of the spermatophore and origi-

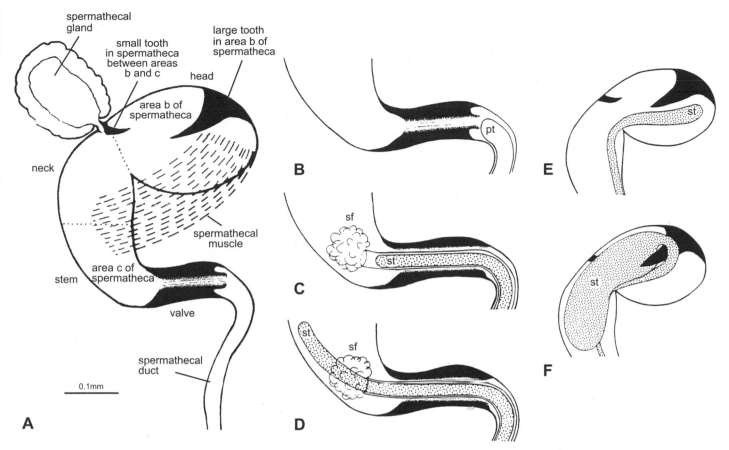

Fig. 3.8. Sperm transfer and displacement by males of *Aleochara curtula* (Coleoptera: Staphylinidae). A. Structure of the spermatheca. B. Elongation of the primary tube (pt) of the spermatophore within the spermathecal duct. C. Tip of primary tube enters stem region of spermatheca and bursts to release a fluid (sf) serving to flush out any sperm already present from a previous insemination. D. Tip of secondary tube (st) of spermatophore passes through opening of the primary tube, enters the spermatheca (E), doubles back on itself, and expands into a balloon (F). (Reprinted with permission from C. Gack and K. Peschke. Spermathecal morphology, sperm transfer and a novel mechanism of sperm displacement in the rove beetle, *Aleochara curtula* (Coleoptera, Staphylinidae). *Zoomorphology* 114: 227–237. © 1994 Springer-Verlag.)

nating in cells of the male's opaque accessory glands, move the sperm into the spermatheca (Fig. 3.7E) (e.g., *R. prolixus* [Hemiptera: Reduviidae]) (Davey, 1959).

- Osmotic removal of water from the lumen of the spermatheca, assisted or not by osmotic uptake by a pressure body within the spermatophore, sucks sperm out of the spermatophore and up the spermathecal duct into the spermatheca (Fig. 3.7F) (e.g., *Culicoides melleus* [Diptera: Ceratopogonidae]) (Linley, 1981b).
- Swimming of apyrene sperm is suggested to move inactive eupyrene sperm out of the spermatophore and up the seminal and spermathecal ducts into the utriculus of the spermatheca (Fig. 3.7G). There, the apyrene sperm are passed into the lagena, where they degenerate while the eupyrene sperm are activated (Lepidoptera: Ditrysia) (Holt and North, 1970).
- The double-walled tube of the spermatophore "grows" into the spermathecal duct, guided by the flagellum of the male's intromittent organ, and through a valve into the spermatheca (Fig. 3.8A, B). There, the outer tube bursts, and the inner tube "grows" through its

apex and farther into the apical bulb of the spermatheca (Fig. 3.8C, D), where it doubles back on itself and swells into a balloon, completely filling the spermatheca (Fig. 3.8E, F). Two sclerotized teeth within the spermatheca then pierce the balloon on contraction of a spermathecal muscle, releasing its sperm (Fig. 3.8A) (e.g., *Aleochara curtula* [Coleoptera: Staphilinidae]) (Gack and Peschke, 1994).

Fate of Empty Spermatophores

Empty spermatophores have various fates depending on species (Mann, 1984). Females of *R. prolixus* eject the spermatophore about 12 h after copulating (Davey, 1959), whereas certain female megalopterans, neuropterans, cockroaches, and orthopterans usually eat it, though not before all sperm have moved into the spermatheca (e.g., Hayashi, 1999). In other cockroaches and in haglid ensiferans, the female feeds on the notal glands or fleshy wings, respectively, of the male during copulation, which prevents her from eating the spermatophore until its sperm have entered the spermatheca. In female lepidopterans and some caddisflies, and in certain female cockroaches, crickets, and beetles, the spermatophore is partly or entirely digested and resorbed and its components used in vitellogenesis (Mann, 1984).

3.3.2.1.2 DIRECT SPERM TRANSFER
In males of higher lineages in most pterygote orders, the structure of the phallus is adapted to convey sperm directly into the female's spermatheca or spermathecal duct (Fig. 3.5: bottom, 4). This part of the phallus, variously called the endophallus, endosoma, internal sac of median lobe, and flagellum (Tuxen, 1970), is everted by hemocoelic pressure during copulation that is generated by contraction of abdominal muscles (Heming-van Battum and Heming, 1989) or by action of a fluid erection system in the ejaculatory duct (Bonhag and Wick, 1953).

Because of the often intricate structure of the phallus in these insects, the male may take a long time to maneuver its phallus into the opening of the female's spermathecal duct. For example, males of the lygaeid *Lygaeus simulans* take over 30 min to locate the entrance to the spermathecal duct and another 30 for full penetration (Micholitsch et al., 2000).

Particularly detailed studies of insemination by such insects are those of Bonhag and Wick (1953) on *Oncopeltus fasciatus* (Hemiptera: Lygaeidae), of Heming-van Battum and Heming (1989) on *Hebrus pusillus* (Hemiptera: Hebridae) (Fig. 3.17), of Günther (1961) on fleas, and of Jones and Wheeler (1965) on *Aedes aegypti* (Diptera: Culicidae).

3.3.2.1.3 TRAUMATIC (HEMOCOELIC) INSEMINATION
The most unusual form of sperm transfer known in insects is that in which the male introduces his spermatozoa directly into the body cavity of the female by puncturing her integument with his phallus (Fig. 3.5: bottom, 5) (Carayon, 1966, 1977). This occurs in members of the order Strepsiptera and of six families of the monophyletic taxon Cimiciformes of the Heteroptera—Nabidae, Plokiophilidae, Lyctocoridae, Anthocoridae, Cimicidae, and Polyctenidae (Schuh and Stys, 1991)—and has been examined experimentally in the human bed bug *Cimex lectularius* by Davis (1966) and Stutt and Siva-Jothy (2001).

In *C. lectularius*, sperm transfer occurs in three phases and involves the function of the male's phallus and of adaptations in the female's reproductive system and body, collectively referred to as the *paragenital system* (Fig. 3.9):
- In the spermalege phase, the male penetrates an ectospermalege of soft, thickened endocuticle between the fifth and sixth abdominal sternites of the female with his sickle-shaped, grooved, and heavily sclerotized left paramere; everts his endosoma along the length of the groove and through the hole; and within 1–5 min, releases a mass of ejaculate into a cellular mesospermalege, subtending it (ectospermalege and mesospermalege together are collectively referred to as the *organ of Berlese* or *Ribaga*). The sperm are

activated in 30 s by an activator from the male's accessory glands and move to a conducting lobe of the mesospermalege in 1–2 h; many of them are engulfed and digested by amoebocytes and collophages (modified hemocytes) within the mesospermalege, whose activity is also induced by entry of ejaculate into the mesospermalege.
- Three to four hours after copulation, during a hemocoelic phase, motile sperm enter the hemocoel from the conducting lobe and swim to a lateral seminal conceptacle at the base of each lateral oviduct (these bugs lack true spermathecae). They enter first, the left and then the right conceptacles through their walls via the *hemochrism*, a space between the inner and outer sheaths of the walls, the conceptacles becoming tightly packed with immotile sperm between 6 and 12 h after copulation. Additional sperm accumulate in the hemocoel adjacent to the sacs after they are full but disappear in a day or two.
- In an intragenital phase, individual spermatozoa begin to swim out of the conceptacles into the hemochrism of the lateral oviducts and enter the posterior openings of the *spermodes*, an intraepithelial network of tubules within the walls of the lateral oviducts. They swim distally within these to an epithelial plug (syncytial body) at the base of each ovariole, which they enter.
- When insemination stimulates vitellogenesis in the terminal (T) oocyte of each ovariole, sperm in the syncytial body enter its follicular epithelium and move forward to a newly differentiating syncytial body behind follicle T-1. From here, individual sperm enter the oocyte from its anterior end before chorion deposition and fertilize it; at least one-third of embryogenesis occurs before ovulation (Davis, 1966).

Evolution in Cimiciform Heteroptera
Hinton (1964) recognized six grades in the evolution of traumatic insemination in cimiciform bugs:
- Grade 1: In basal members of this lineage (e.g., the prostemmatine nabid *Allocorrhynchus plebejus*), the male deposits sperm in the usual way into the vagina of the female with his intromittent organ. These accumulate within the lateral oviducts, enter hemochrisms in their walls, and migrate to the ovarioles. A few sperm escape to the hemocoel, where they are phagocytosed.
- Grade 2: In members of intermediate prostemmatine lineages (e.g., *A. flavipes*) and in the lyctocorids *Lyctocoris campestris* and *L. dorini*, sperm are injected into the hemocoel through the walls of the vagina by the phallus and make their way to hemochrisms of the lateral oviducts or ovarioles. Absorption of sperm occurs within a more or less compact mass of mesodermal cells, the mesospermalege, of blood or fat cell origin.
- Grade 3: Sperm are introduced into the hemocoel through the integument of the abdomen, the female being pierced at a different place each time (e.g., *Primicimex cavernis* [Cimicidae]). Sperm accumulate in the dorsal vessel of the female and are dispersed

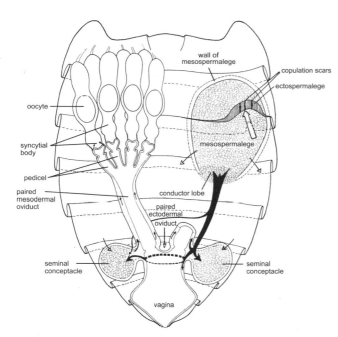

Fig. 3.9. The paragenital system and the process of insemination in bed bugs (Hemiptera: Cimicidae) based principally on that occurring in the human bed bug *Cimex lectularius*. The right ovary and most of the lateral oviduct are omitted and the mesospermalege is shown farther forward than in reality. The short, broad white arrow shows the course followed by the paramere of the male in reaching the ectospermalege (the three black bars represent copulation scars remaining from previous copulations). Black arrows indicate the normal routes of sperm migration from the mesospermalege to the base of ovarioles. Small arrows with white heads show migratory routes never or rarely used in *Cimex* spp. but occurring in other cimicids. (Reprinted with permission from J. Carayon. Traumatic insemination and the paragenital system. In R. L. Usinger (ed.), *Monograph of Cimicidae*, pp. 81–166. Entomol. Soc. Am. Vol. 7. Thomas Say Foundation. © 1966 Entomological Society of America.)

throughout the body via the blood, eventually entering a seminal conceptacle at the base of each lateral oviduct.

- Grade 4: Areas of female abdominal cuticle become specialized into one or more ectospermaleges, which receive the phallus of the male during copulation, the number, shape, and position of these varying with the species.
- Grade 5: Absorption of sperm is taken over by a more or less compact mesospermalege subtending the ectospermalege (e.g., most cimicids including *C. lectularius*) (Fig. 3.9).
- Grade 6: The mesospermalege, in addition to phagocytosing some sperm, conveys them directly to hemochrisms in walls of the lateral oviducts so that they no longer enter the hemocoel. This level of specialization has evolved independently in certain cimicids (*Stricticimex brevispinosus*) and anthocorids (e.g., *Anthocoris* and *Elatophilus* spp.).

Traumatic Insemination in Strepsiptera

Female strepsipterans are larviform, viviparous, internal parasites of hymenopterans and hemipterans (Kathirithamby, 1991). At the end of larval development, the female penetrates the cuticle of the host with her cephalothorax but does not emerge. A male copulates with the female by penetrating the wall of the female's brood canal with his intromittent organ and releases motile sperm into her hemocoel that subsequently fertilize free-floating follicles. In these insects, hemocoelic insemination is probably adapted to the greatly shortened, parasitic lifestyle of the female.

Adaptive Significance

Hemocoelic insemination usually results in the digestion of seminal fluid and of some sperm by free hemocytes, by hemocytes and mesospermalege cells, or by those of the mesospermalege alone, suggesting that the adaptive value of this mode of sperm transfer is to provide nutriment to the female (Hinton, 1974). It may be significant that cimiciform males transfer relatively greater volumes of semen and sperm per copulation than do other insects that transfer sperm directly or by spermatophore.

All cimicids are blood-feeding ectoparasites of birds or mammals, plokiophilids are on bats, and the other cimiciform bugs mostly prey on other insects. An ability to resist starvation would seem to be adaptive in insects feeding at irregular intervals, as most cimicids would do (Hinton, 1974). Chance encounters with males and receipt of their ejaculate could, perhaps, enable such females to live a little longer. In fact, males of two species of bat bugs, *Afrocimex leleupi* and *A. constrictus*, have three ectospermaleges on the left side of the abdomen, one each behind sternites 3, 4, and 5, although these lack mesospermaleges (Carayon, 1966: Fig. 12–63); Carayon found wound scars in these resembling those in mated females of the same species, indicating that males copulate with each other. Hinton (1964) observed that *Afrocimex* bugs are the only bisexual animals known in which the male is structurally modified to accommodate homosexual copulation, the only benefit possibly accruing to an inseminated male being the protein-rich "meal" of ejaculate it receives, presumably increasing its resistance to starvation (but what of the donor?).

Results of more recent experiments on *C. lectularius* by Stutt and Siva-Jothy (2001) and further speculation on the adaptive value of traumatic insemination suggest instead that its evolution has resulted from a coevolutionary race between male and female as to which sex will control fertilization of eggs (3.7) (Thornhill and Alcock, 1983: 326–327; Eberhard, 1996). To maximize their lifetime reproductive success, females of *C. lectularius* need to copulate successfully only once after every four blood meals. The normal lifetime copulation rate of 20 times significantly shortens their lives, presumably because of the cuticle repair required after each insemination (Stutt and Siva-Jothy, 2001). The true function of the mesospermalege in *C. lectularius* seems to be

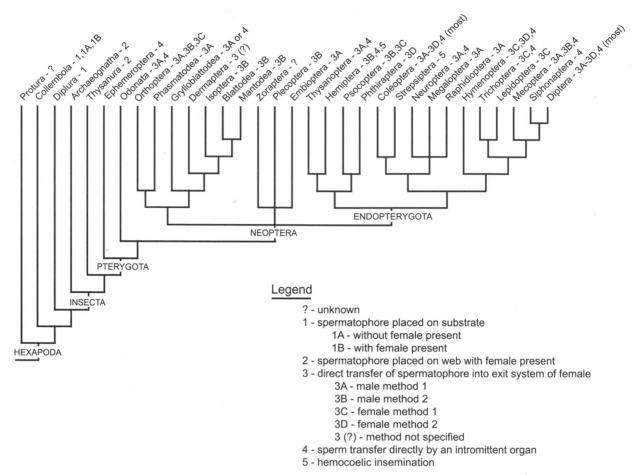

Taxa labels (top, left to right): Protura - ?, Collembola - 1,1A,1B, Diplura - 1, Archaeognatha - 2, Thysanura - 2, Ephemeroptera - 4, Odonata - 3A,4, Orthoptera - 3A,3B,3C, Phasmatodea - 3A, Grylloblattodea - 3A or 4, Dermaptera - 3 (?), Isoptera - 3B, Blattodea - 3B, Mantodea - 3B, Zoraptera - ?, Plecoptera - 3B, Embioptera - 3A, Thysanoptera - 3A,4, Hemiptera - 3B,4,5, Psocoptera - 3B,3C, Phthiraptera - 3D, Coleoptera - 3A-3D,4 (most), Strepsiptera - 5, Neuroptera - 3A,4, Megaloptera - 3A, Raphidioptera - 3A, Hymenoptera - 3C,3D,4, Trichoptera - 3C,4, Lepidoptera - 3C, Mecoptera - 3A,3B,4, Siphonaptera - 4, Diptera - 3A-3D,4 (most)

Clade labels: ENDOPTERYGOTA, NEOPTERA, PTERYGOTA, INSECTA, HEXAPODA

Legend

? - unknown
1 - spermatophore placed on substrate
    1A - without female present
    1B - with female present
2 - spermatophore placed on web with female present
3 - direct transfer of spermatophore into exit system of female
    3A - male method 1
    3B - male method 2
    3C - female method 1
    3D - female method 2
    3 (?) - method not specified
4 - sperm transfer directly by an intromittent organ
5 - hemocoelic insemination

Fig. 3.10. Evolutionary relationships of the hexapod orders as inferred by Gullan and Cranston (1994), showing the distribution of known sperm transference mechanisms. (Data from various sources. Cladogram adapted with permission from P. J. Gullan and P. S. Cranston. *The insects. An outline of entomology*, p. 193, Fig. 7.5. 1st ed. Chapman and Hall, London. © 1994 Blackwell Science, Ltd.)

to diminish the potential effects of wounding and infection caused by traumatic insemination. Traumatic insemination also results in last-male-sperm precedence in this insect.

The known distribution of sperm transfer mechanisms throughout the hexapods is summarized in Fig. 3.10.

3.3.2.1.4 NUMBER OF SPERM TRANSFERRED
The number of sperm transferred from male to female during a single insemination varies with species, body size, age, sperm length, number of previous inseminations, and the incidence of male-male competition, and has been determined by actual sperm counts in smaller insects and by various estimation methods in larger ones. A few examples follow:

• In the bush cricket *Metaplastes ornatus*, the ampulla of the spermatophore contains a mean of 1.49 million sperm, and the spermatheca of first-inseminated females, a mean of 1.23 million sperm (von Hilversen and von Hilversen, 1991).
• Males of the crickets *Acheta domesticus* and

Gryllodes supplicans increase the amount of sperm transferred as apparent male competition increases, and those of *A. domesticus* transfer more sperm when encountering larger females (Gage and Bernard, 1996).
• In the tiny (<1 mm) ptiliid beetle *Bambara invisibilis*, the sperm are complex and species-specific in structure and one- to two-thirds the length of the body. Twenty-eight are passed one at a time up the spermathecal duct into a spermatheca, whose shape is closely co-adapted to their length (Fig. 3.15A) (Dybas and Dybas, 1981).
• Virgin males of the bruchine beetle *Callosobruchus maculatus* transfer approximately 46,000 sperm during a single insemination—about 85% more than the female can store in her spermatheca (Eady, 1994). The excess sperm remain in her vagina and appear to degenerate there.
• By 6 h after copulation by females of the noctuid moth *Pseudaletia separata*, 293,900 ± 122,100 sperm are found within the spermatheca, of which 92.5% ± 2.9% are apyrene (He et al., 1995). The number of apyrene

sperm decreases rapidly within 3 d after which time the female becomes susceptible to remating.

- In the ceratopogonid fly *Culicoides melleus*, the number of sperm per spermatophore increases from 760 ± 195 within 1 h of emergence to 943 ± 199 from 4 to 8 h afterward and declines to 125 ± 110 at 176 h (Hinds and Linley, 1974).
- The mean number of sperm counted in 43 spermatophores of the black fly *Simulium decorum* was 4048 ± 230, with about 73 sperm/min being transferred into the spermatophore during the 51 min required to form it (Linley and Simmons, 1983).
- Males of the sarcophagid fly *Neobellieria bullata* transfer an average of 6965 ± 712 sperm in copulations lasting a mean of 165 ± 15 min, the number of sperm correlating with body size, with a potential fecundity in females of 3000 eggs, and with a probable sperm utilization efficiency of 2.3 sperm/egg (Berrigan and Locke, 1991).
- Males of *Drosophila acanthoptera* (sperm length, 5.8 mm) transfer 1023 ± 48 sperm/copulation, and those of *D. pachea* (sperm length, 16.5 mm), 44 ± 6 (Pitnick and Markow, 1994a).
- In males of *Drosophila bifurca*, individual sperm are 58 mm long and require 17 d to mature following adult emergence (Joly et al., 1995). The male rolls each sperm into a coil 80 μm across and lines them up single file in each of his seminal vesicles like peas in a pod. The female then hauls all 25 of them out, one after another, and passes them into her two tubular storage organs in 374 s, where they uncoil. Each sperm bundle in males of this species contains only 24 sperm, indicating a trade-off between sperm size and number.
- Queens of the fire ant *Solenopsis invicta*, mated in the laboratory at the beginning of their reproductive lives, receive an initial supply of about 7 million sperm, which they gradually allocate over a period of 6.8 y to produce about 0.5 million worker offspring with a utilization efficiency of 2.6 sperm/worker (Tschinkel and Porter, 1988).

As might be expected, nutritional stress experienced during larval development can affect male adult size and thus the number of sperm produced and transferred during insemination. A low-protein diet in developing males of the Indian meal moth *Plodia interpunctella* (Lepidoptera: Pyralidae), for example, reduced the mean number of eupyrene sperm in the first spermatophore from 10,000 to 5700 and that of apyrene sperm from 130,000 to 75,000 but without changing the length of either type or the 9 : 1 proportion of apyrene to eupyrene sperm (Gage and Cook, 1994).

3.4 Ovulation

Before an egg can be fertilized, it must exit its follicle and ovariole and position itself within the exit system in such a way as to be entered by a sperm from the spermatheca (Fig. 3.11A). When the proximal oocyte of an ovariole is fully formed and its external envelopes are complete, it ruptures an epithelial plug closing the vitellarium from the pedicel and moves through the pedicel and the lateral and common oviducts into the vagina, where it is fertilized. The follicle cells remain behind within the ovariole, collapse, degenerate, and form a corpus luteum or "yellow body." This structure is not known to have any reproductive function in insect ovarioles (in mammals its cells produce progesterone, a steroid promoting growth of the uterine lining). The yellow bodies persist for some time in the ovarioles of some insects (e.g., mosquitoes) and can be counted and used to indicate the number of ovarian cycles a particular female has had; subsequent eggs can apparently squeeze by the yellow bodies without dislodging them.

In females of the stick insect *Clitumnus extradentatus*, ovulation results from dissolution and opening of the epithelial plug that closed the ovariole and by muscular contraction of the ovariolar sheath and pedicel and is under nervous and endocrine control (Mesnier, 1981). In females of the blood-feeding bug *Rhodnius prolixus*, it is stimulated by release of an ovulation hormone from five pairs of neurosecretory cells in the pars intercerebralis of the brain (Davey, 1993).

3.5 Passage of Sperm from Spermatheca to Micropyle of Egg

Peristaltic contractions in walls of the pedicel and lateral and common oviducts propel the egg posteriorly into the vagina (Fig. 3.11A, B). Here, its apex is temporarily lodged within an anterior diverticulum of this duct, the *fertilization chamber*, in such a way that its micropyles are temporarily adjacent to the mouth of the spermathecal duct (Fig. 3.11C). A few sperm pass out of the spermatheca, down the spermathecal duct, and into the egg through these openings.

Such passage has been examined in greater or lesser detail in females of the house cricket *Acheta domesticus* (McFarlane and McFarlane, 1988), the Australian bush cricket *Teleogryllus commodus* (Pohlhammer, 1978), the fruit fly *D. melanogaster* (Fowler, 1973), and the house fly *Musca domestica* (Degrugillier, 1985; Degrugillier and Leopold, 1973, 1976; Leopold and Degrugillier, 1973; Leopold et al., 1978), and I summarize each of these sequences below.

- In females of *A. domesticus* at 30°C, one egg is laid every 2 s, with 10–250 eggs being deposited per batch. Sperm penetration, as observed after in vitro fertilization by visualization of 4-6-diamidino-2-phenylindole (DAPI)–stained sperm, seems not to be restricted to the micropyles, and an acrosomal filament is said to form during sperm entry (McFarlane and McFarlane [1988] refer to a Fig. 7, which was not included in their paper).
- In *T. commodus* females, sperm entry takes 8–10 s/egg (Fig. 3.12) and is intimately related to oviposition behavior, which involves searching, positioning, ovipositor penetration, and lift; a rest phase; egg

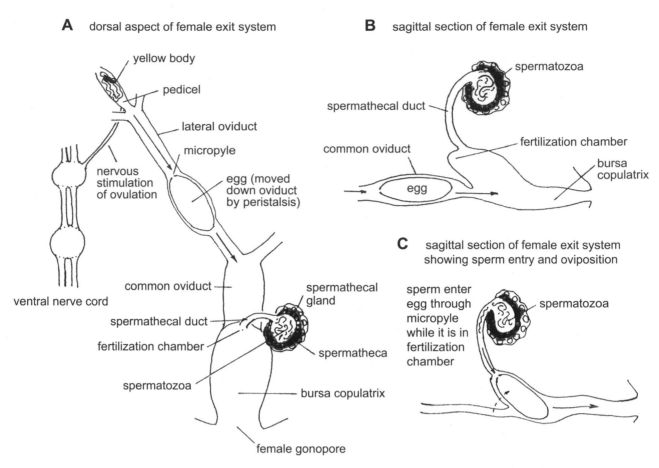

A dorsal aspect of female exit system

yellow body
pedicel
lateral oviduct
micropyle
nervous stimulation of ovulation
egg (moved down oviduct by peristalsis)
ventral nerve cord
common oviduct
spermathecal gland
spermathecal duct
fertilization chamber
spermatheca
spermatozoa
bursa copulatrix
female gonopore

B sagittal section of female exit system

spermatozoa
spermathecal duct
common oviduct
fertilization chamber
bursa copulatrix
egg

C sagittal section of female exit system showing sperm entry and oviposition

sperm enter egg through micropyle while it is in fertilization chamber
spermatozoa

Fig. 3.11. Summary of ovulation and sperm entry into an egg. Many variations on this theme have been observed.

deposition; and ovipositor withdrawal (Pohlhammer, 1978; Sugawara, 1993). During the lift phase, the posterior half of an egg enters the genital chamber from the common oviduct. Stimulation of multipolar and bipolar mechanosensory neurons in the wall of the genital chamber by egg entry during the rest phase causes it to contract, forcing the egg farther into the genital chamber and contracting the spermathecal duct. This opens into the genital chamber subterminally on a papilla above the egg, and several motile sperm are pressed out of the duct into a groove in the roof of the genital chamber and posteriorly into a micropylar region situated middorsally on the egg (Pohlhammer, 1978: Fig. 11). The heads of several sperm are caught in the micropylar openings, and their flagellar action is said to push them into the egg.

- In ovipositing females of *M. domestica*, at 27°C, each egg is held for 5.4 s in the fertilization chamber, where 2–4 sperm enter its micropyle (a fertile female deposits an egg every 8 s). Release of a secretion from the female's accessory glands is essential if sperm are to enter, for in glandless females the sperm reach the fertilization chamber but do not enter the egg. An acrosome response occurs that seems to require action of both the secretion and release of some factor in a micropylar cap substance within the fertilization chamber. In sperm still within the spermatheca (Fig. 3.13A) and within the fertilization chamber of glandless females, the acrosomal membrane is intact, but in sperm within the micropyle of the egg in normal females, it is dehisced and its contents absent (Fig. 3.13B, C).

- In females of *D. bifurca*, each egg appears to suck the anterior end of a single giant (58-mm) sperm through the micropyle and into yolk at the front of the egg, where it coils into a characteristic configuration (Boyce, 1998).

A wax layer extends across the inner end of the micropyle and above the vitelline membrane in most insect eggs, while in *M. domestica* eggs, part of the endochorion does (Kaulenas, 1992). These layers must be dissolved by action of hydrolytic enzymes within the acrosome of the sperm, as is well understood in sea urchin and mouse eggs (Alberts et al., 1994) but is just beginning to be investigated in *D. melanogaster* (Foe et al., 1993; Perotti et al., 1996; Fitch and Wakimoto, 1998).

3.6 Fertilization

The fusion of egg and sperm has been investigated in eggs of several species but is best understood in those of

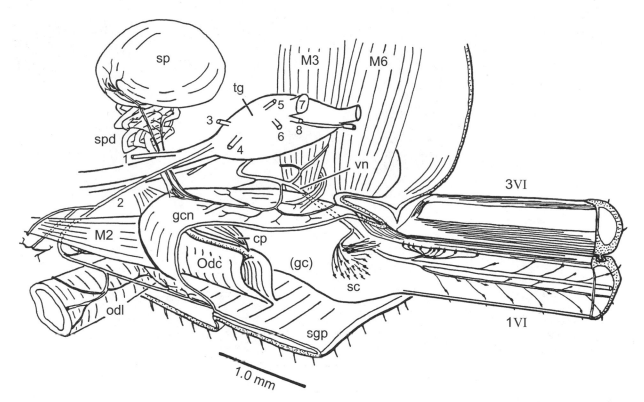

Fig. 3.12. Structure of the genital chamber (gc) and ovipositor base of a female of the Australian bush cricket *Teleogryllus commodus* (Orthoptera: Gryllidae). Eight pairs of nerves (1–8) project from the terminal ganglion (tg) among which N2 and N4 innervate the ventral and dorsal portions of the genital chamber, respectively. A branch of N4, the valve nerve (vn), enters each ventral ovipositor valve (1Vl). Mechanosensory cells (sc) are present in the posterolateral wall of the genital chamber at the bases of these valves. cp, copulatory papilla; gcn, genital chamber nerve; M2, M3, M6, ovipositor protractor and retractor muscles numbered according to Snodgrass (1933); Odc, common oviduct; odl, lateral oviduct; sgp, subgenital plate; sp, spermatheca; spd, spermathecal duct; 3Vl, third or dorsal ovipositor valve. (Reprinted from *J. Insect Physiol.*, 39, T. Sugawara, Oviposition behaviour of the cricket *Teleogryllus commodus*: mechanosenory cells in the genital chamber and their role in the switch-over of steps, pp. 335–346, Copyright, 1993, with permission from Elsevier Science.)

D. melanogaster (Retnakaran and Percy, 1985; Longo, 1997) and of the honeybee *Apis mellifera* (Yu and Omholt, 1999). Fertilization includes the following events (Figs. 3.13D, 3.14):

- A sperm penetrates the egg through a micropyle and induces structural changes in the vitelline membrane and subtending periplasm, the *fertilization response* (Fig. 3.14A).
- Its head separates from its centriole, axoneme, and mitochondrial derivatives (Fig. 3.13D); its chromatin decondenses in 5–10 min, depending on temperature, partly by exchange of sperm-specific protamines or histones for maternally provided histones (see 1.3.4) (Brewer et al., 1999; Loppin et al., 2000); and it swells into a male pronucleus (Fig. 3.14B, C). In most insects, sperm entry also activates the egg and induces its meiotic divisions (these occur without centrioles) and embryogenesis (Fig. 3.14B, C).
- The male and female pronuclei approach each other either by cytoplasmic streaming within the yolk system of the egg (e.g., the mayfly *Ephemera japonica* [Tojo and Machida, 1998]) or through action of

astral microtubules contributed by the sperm's centrosome (Fig. 3.14C, D) (Wolf, 1972; Foe et al., 1993).

- The two nuclei then fuse and their chromosomes intermingle (syngamy) on the equator of the first cleavage spindle (Fig. 3.14E–F', F'').

By use of confocal microscopy and various immunocytochemical and computer enhancement techniques, Karr (1991, 1996) and Karr and Pitnick (1996) reconstructed the position and configuration of sperm within newly inseminated eggs of 12 species of *Drosophila*. In species whose males produce sperm less than 20 mm long (*D. bucksii, D. ezoana, D. pachea, D. pseudoobscura, D. melanogaster,* and *D. simulans*), the entire sperm enters the egg and assumes a species-specific, three-dimensional configuration within the yolk, usually toward its anterior end. In *D. hydei* (sperm 23.3 mm long) and *D. bifurca* (58.3 mm long), only 1.31 and 1.6 mm of the sperm, respectively, enters the egg, the remainder extending through the micropyle and over the surface of the chorion. In eggs of both species, remnants of the mitochondrial derivatives persist throughout

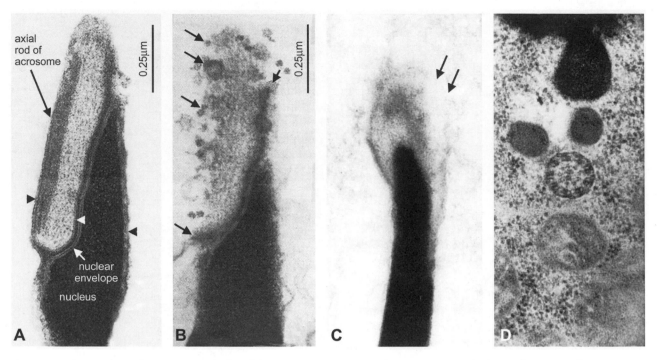

Fig. 3.13. Transmission electron micrographs of the acrosome reaction and sperm entry in the house fly, *Musca domestica* (Diptera: Muscidae). A. Acrosome of unactivated sperm in spermatheca of inseminated female. B. Dehiscense of acrosomal membrane resulting from exposure of sperm to a secretion of the female's accessory gland. C. Acrosome of sperm within micropylar cap substance of fertilization chamber prior to egg entry. Arrows indicate broken acrosomal membrane. D. Transverse section through sperm within the ooplasm directly under the micropyle. The nucleus (black object at top) has separated from the axoneme and mitochondrial derivatives but has yet to swell into a male pronucleus. (A and B reprinted from *Int. J. Insect Morphol. Embryol.*, 14, M. E. Degrugillier, *In vitro* release of housefly, *Musca domestica* L. (Diptera: Muscidae), acrosomal material after treatments with secretion of female accessory gland and micropyle cap substance, pp. 381–391, copyright 1985, with permission from Elsevier Science. C and D reprinted with permission from M. E. Degrugillier and R. A. Leopold. Ultrastructure of sperm penetration of house fly eggs. *J. Ultrastruct. Res.* 56: 312–325. © 1976 Academic Press, Inc.)

embryogenesis and can be found within the midguts of newly hatched larvae and in their feces (Pitnick and Karr, 1998). Cladistic analysis of evolutionary relationships between these species suggests that the existence of short sperm that completely enter the egg is ancestral in the subgenus *Drosophila* and that giant sperm have evolved independently several times, the sperm of sister species sometimes differing from each other in length by an order of magnitude.

Although polyspermy is widespread in insects, only one male pronucleus typically fuses with the female pronucleus (Foe et al., 1993; Longo, 1997; Yu and Omholt, 1999), the other sperm degenerating. For example, in the collembolan *Tetrodontophora bielanensis*, two to four sperm enter the egg, their heads all transforming into male pronuclei, but only one of these fuses with the female pronucleus (Jura and Krgyslofowicz, 1992). The existence of polyspermy in the newly fertilized eggs of some insects suggests that the fertilization response is not effective in preventing entry of other sperm and that competition between sperm of different males to fertilize an egg could occur even after they have entered it.

Successful attempts have even been made to fertilize insect eggs artificially. For example, Takemura and Coworkers (2000) artificially inseminated females of the commercial silkworm *Bombyx mori* with sperm preserved in liquid nitrogen for 356 d at −196°C. Such females deposited an almost normal number of eggs, with normal fertilization and hatch rates. And Hatakeyama and colleagues (2000) transferred previtellogenic ovarioles of the tenthredinid sawfly *Athalia rosae* into the body cavities of females of the related sawfly *A. infumata*. After the transplanted eggs had completed vitellogenesis and choriogenesis, they fertilized them with cryopreserved sperm of *A. rosae*, a few of them eventually developing into fertile female adults.

3.6.1 MOLECULAR ASPECTS

That molecular aspects of fertilization are just beginning to be examined in eggs of *D. melanogaster* is no doubt due to observational difficulties related to their internal fertilization, small size, and tough choria (Foe et al., 1993; Perotti et al., 1996; Page and Orr-Weaver, 1997; Fitch and Wakimoto, 1998). However, results to date

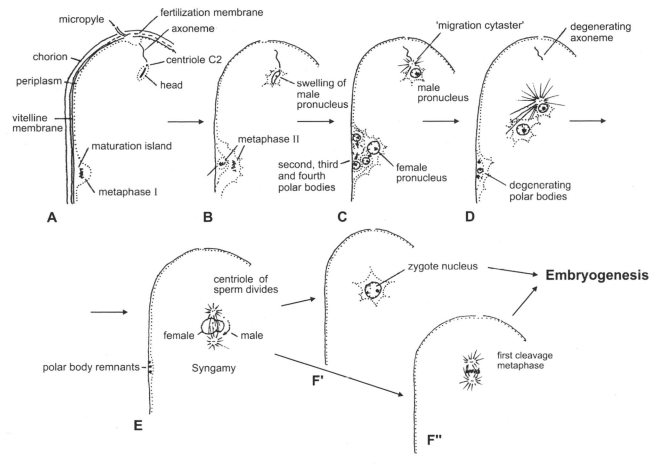

Fig. 3.14. Summary of fertilization, meiosis, and syngamy in an insect egg. All aspects vary in insects of different species.

show this insect to share characteristics demonstrated to occur in sea urchin and mouse eggs (Szabo and O'Day, 1983; Alberts et al., 1994, chapter 20; Longo, 1997).

Specific glycoconjugate residues are restricted to the micropyle of the egg of *D. melanogaster*, while the vitelline membrane within has alpha-mannose (αMan) and beta-*N*-acetylglucosamine (βGlNAc) moieties in its micropylar region that are no longer present following fertilization (Perotti et al., 1996). Binding sites, for both αMan and βGlNAc, are present on the acrosomal and axonemal membranes of wild-type sperm, as are the enzymes beta-*N*-acetylglucosaminidase and alpha-mannosidase. Sperm of the male autosomal sterile mutant *ms(93) HB156* are motile and are transferred to the female but cannot enter the egg. They lack these binding sites and their enzymatic activity is reduced by 50%, suggesting that beta-*N*-acetylglucosaminidase on the sperm plasmalemma and βGlNAc residues on the vitelline membrane play a crucial role in sperm-egg interaction in this fly.

Unlike those of most investigated bisexual insects, the eggs of *D. melanogaster* are activated by passage down the lateral oviduct following ovulation rather than by sperm entry, even though fertilization in this insect occurs a few minutes before oviposition within the female's genital tract (Loppin et al., 2000). Page and Orr-Weaver (1997) developed a technique for activating eggs in vitro in large numbers, in hopes of investigating the mechanics of this process at the molecular level. Using this system, they found the stages and timing of female meiosis to be similar to those occurring in vivo. They also discovered that all proteins required for this process are synthesized prior to metaphase I and that passage of eggs down the lateral oviduct affects the vitelline membrane: before activation, this membrane is permeable to passage of small molecules; after, its components become cross-linked and impermeable.

3.6.2 CONTRIBUTIONS OF THE SPERM TO SUCCESSFUL FERTILIZATION

It has long been recognized that sperm and egg contribute equal haploid genomes to the zygote during fertilization and that most cytoplasm is contributed by the egg (Fig. 3.14). In *D. melanogaster*, screens for mutations affecting male fertility have revealed that over 500 genes are required for normal differentiation of sperm (1.3.7.3). However, few of these mutations are known to interfere with egg penetration and initiation of embryogenesis, suggesting that most steps in fertilization and early embryogenesis are controlled by the female genome through the egg. The maternal effect mutation *sésame* (*ssm*), for example, adversely affects formation of the male pronucleus such that homozygous *ssm* mutant females produce haploid embryos that die before hatch and develop with only the maternal chromosome complement (Loppin et al., 2000). Pronuclear migration and apposition occur normally in mutant eggs (compare Fig. 3.14C, D), but the chromatin of the male pronucleus fails to fully decondense and the male chromosomes fail to enter the first cleavage spindle and, instead, degenerate. Thus, maternally provided Ssm protein may be required for late-stage remodelling of sperm chromatin.

Paternal effect mutations provide a means to identify male-derived factors specifically required by the sperm after it enters the egg (Fitch and Wakimoto, 1998). The paternal effect mutations *paternal loss* (*pal*) and *ms(3)K81* interfere with normal male pronucleus formation and maintenance of the male genomic complement after sperm entry. In *ms(3) sneaky* (*snky*)–fertilized eggs, the chromatin of the sperm fails to decondense, a functional male pronucleus is not formed, female meiosis is not initiated, and the sperm aster, necessary for pronuclear fusion and embryonic cleavage, does not form. These findings suggest that the product of the wild-type gene is required for the initial sperm response to cytoplasmic cues within the egg and for initiating embryogenesis.

Even though all centrioles involved in clone formation within the ovariole of the female end up within the cytoplasm of the oocyte (2.2.2.5), the sperm aster functioning in pronuclear apposition is reconstituted by the centriole contributed by the sperm (Fig. 3.14A–D). This centriole must recruit maternally provided centrosomal proteins if it is to function properly in pronuclear apposition and syngamy, including Centrosomin, CP 190, and gamma-tubulin (Fitch and Wakimoto, 1998). The increasing concentration of gamma-tubulin is believed to nucleate the microtubules to form the sperm aster and to attract additional gamma-tubulin molecules (Schatten, 1994). When the female pronucleus is contacted by lengthening astral microtubules from the sperm's centrosome (Fig. 3.14D), it moves through the yolk toward the aster, using dynein-like motors associated with these microtubules. Interaction between the protein *centrin* in the sperm's centrosome and Ca^{2+} ions released within the egg during activation causes the sperm flagellum to separate from the centrosome and sperm head, and transforms its centriole into a functional centrosome. The axonemal microtubules, mitochondrial derivatives, and other materials (Fig. 3.13D) are thought to disassemble following sperm incorporation into the egg, although, as mentioned earlier, elements of the mitochondrial derivatives of the sperm of *Drosophila bifurca* persist throughout embryogenesis and end up within the midguts of first instar larvae (Pitnick and Karr, 1998).

Embryos of *centrosomin* (*cnn*) null mutants of *D.* *melanogaster* fail to assemble functional mitotic centrosomes and subsequently develop few astral microtubules (Megraw et al., 2001). Nevertheless, such embryos develop into apparently normal larvae and adults, an observation challenging the belief that these organelles are required for normal development!

3.7 Sperm Competition and Sexual Selection

In many insects, the female is *polyandrous*, mating with more than one male following her emergence (Ridley, 1988), and the question arises as to what factors might influence the chances of sperm from one or another of these males to fertilize her eggs. Results of double-mating experiments on a variety of insects, in which one or the other of two different conspecific males has been irradiated or injected with a radioisotope, indicate that sperm contributed by different males compete to fertilize eggs. This competition is intimately related to a special kind of selection operating during reproduction called *sexual selection*. First proposed by Darwin (1874) in *The Descent of Man and Selection in Relation to Sex*, the principal components of this kind of selection are as follows (Parker, 1970; Williams, 1975; Thornhill, 1976; Emlen and Oring, 1977; Smith, 1978; Blum and Blum, 1979; Siwinski, 1980; Thornhill and Alcock, 1983; Anderson, 1994; Birkhead and Møller, 1998; Ben-Ari, 1999; Birkhead, 2000):

- In most sexually reproducing insects, the female invests more in producing offspring (a few, relatively large eggs) than the male (countless small sperm) because she provides most of the resources required for embryogenesis (2.2.1.2). Thus, it is to her and the species' advantage to preferentially accept sperm from the fittest males (i.e., those having the greatest potential to contribute to success of her offspring).
- For the same allocation of resources, a male can produce many more gametes than a female and is more likely to mate because his genetic contribution to the next generation increases with each female he inseminates, whereas a female's entire production of eggs can often be fertilized by sperm from a single male.
- This difference between the sexes results in competition between males for access to females and the evolution, through intrasexual selection, of characteristics contributing to success in this race.
- In females, there is simultaneous selection for characteristics acting as barriers to insemination that provide a base for female "choice" of mate.
- Female choice results in intersexual (epigamic) selection on males for attributes enabling them to overcome these barriers. Only males having these attributes well developed are likely to fertilize eggs (i.e., the success of a male in surmounting such obstacles can provide information to the female about his evolutionary fitness and provide a basis for choosing to mate with him or not).
- Such characteristics can be molecular, behavioral, or

structural; can be present in members of both sexes; and may be influenced by an evolutionary race between male and female as to which sex will control fertilization of eggs (e.g., Clark et al., 1999).

Action of epigamic selection may also account for the external genitalia of male insects being species-specific. These structures are known to function as "courtship devices" (e.g., Eberhard, 1985, 1991, 1996) that may provide an additional basis for female choice, an explanation supported by Arnquist's (1998) discovery that external genitalia are more than twice as diverse in males of polyandrous as in males of monandrous species belonging to sister lineages of Ephemeroptera, Coleoptera, Lepidoptera, and Diptera.

- Finally, female choice of mate may be influenced by male symmetry (Thornhill, 1993). For example, females of the Japanese scorpion fly *Panorpa japonica* (Mecoptera: Panorpidae) seem to prefer the sex attractant pheromone of symmetrical over that of asymmetrical conspecific males. Bodily symmetry is a measure of evolutionary fitness, and an organism stressed during its ontogeny may be asymmetrical when adult, indicating to a female his possible inferiority as a father for her offspring (9.7.7).

Generally, the last male to mate predominates in fertilizing eggs ("last in–first out" or "last male precedence") because sperm are maintained alive within a female's spermatheca by nutritive secretions in the semen or from the female's spermathecal gland (compare Figs. 3.7, 3.11A). Thus, with subsequent insemination of a previously inseminated female, the later-arriving sperm either dilute those already present or displace them farther from the mouth of the spermathecal duct and reduce their chances of fertilizing eggs (Walker, 1980).

However, Price (1997) showed that the situation is far more complex. She examined results of conspecific and heterospecific inseminations between males and females of three sibling species of *Drosophila: D. simulans, D. mauritiana,* and *D. sechellia*. When a female was inseminated by both a conspecific and a heterospecific male, the conspecific sperm fertilized most of her eggs regardless of mating order. In addition, she showed that heterospecific sperm fertilized fewer eggs after double than after single matings. When she inseminated females with spermless (XO) males, she showed the seminal fluid of the conspecific male to confer conspecific sperm precedence. Also, when two conspecific males mated sequentially with a heterospecific female, a highly variable sperm precedence replaced the second-male precedence usually found within insects.

These results indicate that females can recognize the species identity of sperm and can mediate sperm competition within themselves and that second-male-sperm precedence does not automatically result from the mechanics of sperm storage. Also, pairwise mating experiments by Clark and colleagues (1999) on six isogenic lines of *D. melanogaster* showed that the success of a wild-type male's sperm in both "offense" (ability of a second male to displace the sperm placed in the spermatheca by a first, resident "tester" male) and "defense" (ability of a first male to prevent his sperm from being displaced by that of a second tester male) depends on the genotype of the female with which he mates.

To investigate the mechanics of second-male precedence in *D. melanogaster*, Price and coworkers (1999) doubly mated females to males either expressing or not a green fluorescent protein (GFP) on the flagella of their sperm, and showed that the second males both physically displaced and incapacitated sperm stored in females from earlier copulations. Displacement occurred only if the second male transferred sperm (XO second males could not displace first male sperm) and only from her seminal receptacle, not from her two spermathecae. Incapacitation of first-male sperm was caused by factors in the ejaculate of both normal and spermless second males, but required at least 7 d between matings to take effect (this interval would prevent males from inactivating their own sperm). They also found the sperm contributed by sequential males to mix freely within the female's three sperm storage organs rather than stratifying.

In other insects, competition between males for females has selected for innumerable mechanisms enabling either the first male to reduce the success of a second male in fertilizing a female's eggs or for the second male to achieve precedence over sperm supplied by a first male. A number of examples follow:

- Sperm plugs prevent subsequent males from inseminating an already inseminated female (Fig. 3.5: bottom, 3b). In the Australian butterfly *Cressida cressida*, the sperm plug (sphragis) is large and physically prevents copulation (Orr and Rutowski, 1991). In fact it is so large that a male chasing a previously mated female on the wing can see it before contacting her and breaks off the chase. If one progressively reduces the size of the sphragis, chasing males will eventually not see it and will attempt to make contact, land, and copulate with that female. Thus, the sphragis functions as a signal of female mating status.
- An *insemination reaction* (swelling of the anteroventral uterine wall) in response to receipt of male accessory gland substances temporarily prevents subsequent males from inseminating that female (*Drosophila* spp.) (Markow and Ankney, 1988).
- First-male ejaculate may be pressed against the walls of the spermatheca by the inflated internal sac of the intromittent organ of the second male before his ejaculate is released, increasing the chances for sperm in the second ejaculate to fertilize eggs (Ridley, 1989).
- The sperm may be species-specific in size and form and co-adapted to the length of the spermatheca or spermathecal duct, thus preventing entry of second-male sperm (Figs. 3.15A, 3.16) (Dybas and Dybas, 1981; Heming-van Battum and Heming, 1989; Pitnick et al., 1999; Presgraves et al., 1999; Morrow and Gage, 2000).
- The spermatophore may act as a physical deterrent to success of subsequent males and may produce an anti-

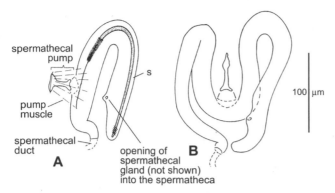

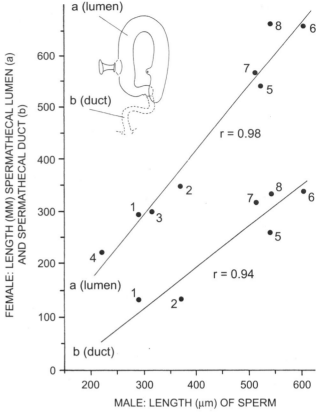

Fig. 3.15. Spermathecae in female ptiliid beetles (Coleoptera). A. *Bambara invisibilis*. A single spermatozoon is depicted showing its normal position within the spermatheca (usually it is packed with sperm so orientated). B. *Bambara* new species 6 (pump muscles not shown) has a different form of spermatheca with a much longer lumen. The spermatozoa of this species are correspondingly longer. (Reprinted with permission from L. K. Dybas and H. S. Dybas. Coadaptation and taxonomic differentiation of sperm and spermathecae in featherwing beetles. *Evolution* 35: 168–174. © 1981 Society for the Study of Evolution.)

Fig. 3.16. Correlation between length of sperm and length of spermathecal lumen and duct in *Bambara* species (Coleoptera: Ptiliidae). Species 1, *B. invisibilis*; species 2–8, undescribed species of the *B. invisibilis* species group in Sri Lanka. (Reprinted with permission from L. K. Dybas and H. S. Dybas. Coadaptation and taxonomic differentiation of sperm and spermathecae in featherwing beetles. *Evolution* 35: 168–174. © 1981 Society for the Study of Evolution.)

aphrodisiac pheromone that discourages additional conspecific mating (Peschke, 1986).

- Prolonged copulation, postcopulatory riding, or non-contact guarding may prevent later-appearing males from mating with an already inseminated female.
- Male defense of territories around good oviposition sites keeps conspecific males, but not ovipositing females, away.
- Male-male fighting or threat displays ensure that the largest males usually inseminate females. The size of male secondary sexual characteristics associated with conspecific fighting over females, such as abdominal forceps in earwigs (Dermaptera), cephalic and pronotal horns in scarabs, enlarged mandibles in lucanids (Coleoptera), and long eyestalks in diopsid flies (Diptera), usually scales with body size either linearly, sigmoidally, or discontinuously, and their degree of development is associated with whether they developed under optimal (large) or suboptimal (small or absent) conditions (Emlen and Nijhout, 2000).
- Removal of sperm placed in the spermatheca by a previous male is common in odonates (Miller, 1990). In these, flexible horns on the "penis" of the accessory genitalia of the male bear recurved microtrichia (Fig. 3.6E). These enter the bursa copulatrix of the female and remove any sperm already there (they seem unable to enter the spermatheca). Their own sperm then replaces them.
- The same occurs in second males of the flour beetles *Tribolium castaneum*, except that such first-male sperm may persist on the aedeagus and be translocated back into the reproductive tract of a new, previously unmated female, where they are still capable of

fertilizing eggs (45/240 crosses) (Haubruge et al., 1999). In this instance, selection on males to remove rival sperm may have resulted in counterselection on spermatozoa to survive removal and to be translocated into new females where they go on to fertilize eggs in significant numbers.

- In the bush cricket *Metaplastes ornatus*, the male, in a first copulation, inserts his subgenital plate into the genital chamber of the female, moves it back and forth several times, and then grasps the wall of the chamber with spines on his plate and turns it inside out, at which time the female cleans it of sperm with her mouthparts. The male then copulates with the female again to place his spermatophore within the vagina of the female (von Hilversen and von Hilversen, 1991). Back-and-forth movements in the first copulation stimulate movement of an egg into the genital chamber and induce the female to release sperm into it from her spermatheca, which she eats when it is everted by the male.

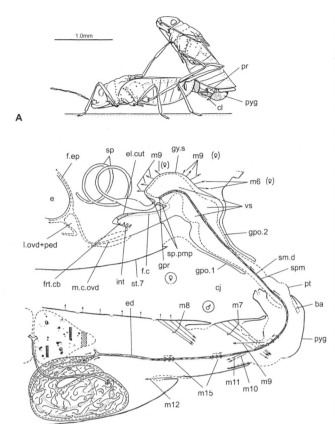

Fig. 3.17. Probable mechanism of sperm transfer in *Hebrus pusillus* (Hemiptera: Hebridae) in which individual spermatozoa are over 2 mm long. A. Position of male (top) and female during sperm transfer. B. Diagrammatic sagittal section through the terminal abdomens and exit systems of female (top) and male during transfer (this position is false male above, the phallus being rotated prior to its eversion up the shaft of the ovipositor). Increased blood pressure causes the endosoma of the phallus to evert through the phallotreme and up the shaft of the female's ovipositor (gpo.1, gpo.2) and its vesica (vs) to evert into her gynatrial sac (gy.s) in such a way that the secondary gonopore (gpr) is adjacent to the opening into the female's spermathecal pump (sp.pmp). Extension of circumferential pleats in the wall of the seminal duct allows it to lengthen as endosomal eversion proceeds. Contraction of the female's gynatrial sac dilator muscles (m6 [female], m9 [female]) and waves of peristalsis in the circular muscles (m15) about the male's ejaculatory duct (ed) probably propel sperm (short, horizontal dotted arrows) out of his seminal vesicles and up the seminal duct (sm.d). Sperm enters the sac only after the male has begun to withdraw his endosoma. Relaxation of the male's abdominal tergosternal muscles and elastic recoil of his tergites (vertical dotted arrows) reduce his internal blood pressure, causing this withdrawal to occur while contraction of muscles m7–m12 begin to withdraw his genital segments. ba, basal articulatory apparatus of phallus; cl, clasper; e, egg in terminal follicle of female's ovariole; el.cut, elastic cuticle at apex of female's fecundation canal (f.c.); f.ep, follicular epithelium; frt.cb, fertilization chamber; int, cuticular intima of ectodermal common oviduct; l.ovd + ped, lateral oviduct and pedicel; m.c.ovd, mesodermal common oviduct; pr, proctiger; pt, phallotheca; pyg, pygophore (abdominal segment 9); sp, spermatheca; spm, sperm moving down ejaculatory and seminal ducts of male; st.7, sternum 7. (Structure, function and evolutionary significance of the reproductive system in males of *Hebrus ruficeps* and *H. pusillus* (Heteroptera, Gerromorpha, Hebridae), K. E. Heming-van Battum and B. S. Heming, *J. Morphol.*, Copyright © 1989 Wiley-Liss, Inc. Reprinted by permission of Wiley-Liss, Inc., a subsidiary of John Wiley & Sons, Inc.)

- In males of the staphylinid beetle *Aleochara curtula*, rupture of the inflated apex of the primary tube of the spermatophore within the spermatheca of the female backflushes previously present sperm out the spermathecal duct (Fig. 3.8C, D) (Gack and Peschke, 1994).

- Males of certain empid flies (Diptera), hangingflies (Mecoptera: Bittacidae), grasshoppers, crickets, and cockroaches present females with a "nuptial gift," the female choosing the male based on its size and quality. The gift can be a previously caught insect, a large spermatophore or associated spermatophylax, accessory secretions of the male, or the male itself, all used by the female to facilitate reproduction. This greater investment by males in offspring production can reverse the roles of males and females such that males become the choosy sex. Gwynne and Simmons (1990) showed that certain katydid species can switch from one behavior to the other based on food availability.

  In empid flies, the ancestral function of the gift insect was probably to placate the predatory instincts of the female until sperm had been fully transferred into her spermatheca. In its absence, the female tended to eat the male, as still happens in members of certain basal lineages within the family.

- A female can regulate the movement of sperm within her exit system in such a way that she, not the male, controls fertilization of the eggs. For example, females of the dung fly *Scatophaga stercoraria* (Diptera: Scathophagidae) can allocate the sperm received sequentially from two males to different spermathecae (Hellriegel and Bernasconi, 2000). And, in the velvet water bug *Hebrus pusillus* (Heteroptera: Hebridae), the female seems to suck sperm out of the seminal duct within the endosoma of the male's phallus (Fig. 3.17) (Heming-van Battum and Heming, 1986, 1989). Such barriers to insemination (there are seven in *H. pusillus*) epigamically select for mechanisms in males to circumvent them, such as their huge sperm (they are longer than the female). In addition, the sperm reverse themselves after entering the long, coiled, tubelike spermatheca of the female so that their heads point toward its mouth.

- In lepidopterans, apyrene sperm of the second male have been suggested to displace eupyrene sperm placed in the spermatheca by a first male (Silberglied et al., 1984). However, Cook and Wedell (1999), in the pierid butterfly *Pieris napi*, showed instead that apyrene sperm protect a male's reproductive investment by delaying remating by that female. Females receiving smaller spermatophores from already mated males remate sooner and more often than do those receiving larger spermatophores from virgin males. In spite of transferring smaller spermatophores, previously mated males introduce more eupyrene sperm (8469 ± 1046) than virgin males (5757 ± 911) but fewer apyrene sperm ($44,954 \pm 5622$ vs. $49,980 \pm 4475$).

- In certain butterflies, Gage (1994) showed testis size to correlate with body size. However, after controlling for body size, he found relative testis size to increase with risk of sperm competition as defined by male mating frequency. Both eupyrene and apyrene sperm length also correlate positively with body size, but after controlling for body size, he found the relative length of eupyrene sperm to be greater in species whose males experience greater risk of sperm competition. This suggests that sperm competition in butterflies selects for increased investment in spermatogenesis, specifically for longer sperm. Because these may swim faster and be more powerful, eupyrene sperm may compete energetically (i.e., they are not selected to be small to maximize numbers, as one might expect). Apyrene sperm length seems not to be affected by risk of encountering rival sperm. Instead, they correlate closely in length with body size, which is consistent with the hypothesis that apyrene sperm function to retard female sexual receptivity by moving within the spermatheca.
- Males of *Drosophila subobscura* produce both long and short sperm (Bressec and Hauschteck-Jungen, 1996). Of some 20,000 sperm transferred to the female in each copulation, 66% are short and 34% long. By 30 h after copulation, approximately 900 have entered the spermatheca and seminal receptacle, with the remaining 19,000 degenerating in the uterus. Females preferentially select long over short sperm from the ejaculate, with 80% in the spermatheca and 84% in the receptacle being long. After 7 d of fertilization and oviposition, only one-third of sperm remain in the storage organs, but with a higher incidence of short sperm, indicating that females preferentially select long sperm to fertilize eggs.

3.7.1 INFLUENCE OF BACTERIAL PARASITES ON FERTILIZATION AND SPERM COMPETITION

Bacteria of the species *Wolbachia pipientis* behave like a sexually selected trait in the flour beetle *Tribolium confusum* (Coleoptera: Tenebrionidae) and enhance male fertility at the expense of female fecundity (Wade and Chang, 1995). Infected females leave fewer offspring than uninfected females, whereas infected males have a major fertility advantage over uninfected males when multiply mated with either infected or uninfected females. This male-fertility effect accelerates the spread of *W. pipientis* through the host population and expands the opportunity for hitchhiking of host nuclear genes. This was the first example of sperm competition in a host being facilitated by an endosymbiont.

3.7.2 COPULATORY INHIBITORS IN *DROSOPHILA MELANOGASTER*

After a male of *D. melanogaster* inseminates a female, that female rejects other males for about 1 wk at 25°C and increases her rate of ovulation and oviposition (Chapman et al., 1995; Soller et al., 1999). This occurs because the male transfers a sex peptide to the female during copulation. This peptide is synthesized in cells of the male's paragenital (accessory) glands and acts in conjunction with a factor contributed by the sperm, as XO males produce the peptide but not viable sperm. When a female copulates with such a male, rejection and heightened ovulation also persist, but only for 2 d. The gene is normally expressed only in males and is committed to do so by action of the sex determination genes in cells of the glands (4.2.1). If a singly mated female is remated to two sterile males, one producing seminal fluid and the other not, the transfer of fluid strongly but temporarily reduces egg hatch (Prout and Clark, 2000). The sex peptide is produced by males of species closely related to *D. melanogaster* but not in more distantly related ones. Frequent receipt of this peptide or of a closely related secretion also shortens the life of multiply mated females and contributes to destruction or removal of sperm from previous males.

3.8 Copulatory Courtship

In his books *Sexual Selection and Animal Genitalia* (1985), and *Female Control: Sexual Selection by Cryptic Female Choice* (1996), Eberhard proposed that the complex, species-specific genitalia of most male animals arose through action of sexual selection, function in courtship, and provide an additional basis for female choice of mate. He and his students tested this hypothesis in a variety of insects and showed the genitalic and secondary sexual characters of males to have a variety of stimulatory functions prior to and during copulation. In addition, genitalic diversity is greater in male members of polyandrous species than of monandrous species in sister lineages (Arnquist, 1988).

3.9 Oviposition

The final responsibility of a female insect, unless she is of a species providing maternal care to her offspring (Zeh et al., 1989), is to deposit her fertilized eggs in a site suitable for their development and for survival of the offspring when they hatch (Hinton, 1981). This function is performed by the ovipositor.

Ovipositor function is similar in females of all species in which this structure is derived from paired ap-

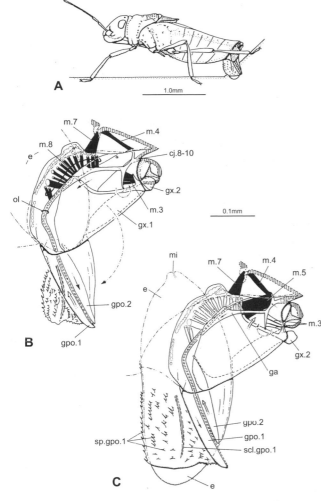

Fig. 3.18. Probable mechanism of ovipositor action during oviposition in a female of *Hebrus pusillus* (Hemiptera: Hebridae). A. Position of the female when depositing an egg. B and C. Alternate contraction of ovipositor muscles during oviposition. Muscles shown in black in each diagram (m.3, m.4, m.5, m.7, m.8) contract alternately to move the ovipositor valves (gpo.1, gpo.2) in opposite directions (arrows) on the ridge and groove system, the olistheter (ol) linking them together, resulting in the relatively large egg (e) moving into the ovipositor, down its shaft, and onto the substrate. In this species, both anterior (gpo.1) and posterior (gpo.2) valves are linked to each other across the midline and function as single units. Passage of the egg forces the mesial cuticle linking gonapophyses 1 ventrally outward between them (B). cj.8–10, conjunctival membrane linking abdominal segment 8 to 10; ga, gonangulum; gx.1, gx.2, gonocoxae 1 and 2; mi, egg micropyle; scl.gpo.1, mesial sclerite of gonapophysis 1; sp.gpo.1, mesial spines of gonapophysis 1. (Structure, function and evolution of the reproductive system in females of *Hebrus pusillus* and *H. ruficeps* (Hemiptera, Gerromorpha, Hebridae), K. E. Heming-van Battum and B. S. Heming, *J. Morphol.*, Copyright © 1986 Wiley-Liss, Inc. Reprinted by permission of Wiley-Liss, Inc., a subsidiary of John Wiley & Sons, Inc.)

pendages of the eighth and ninth abdominal segments (Scudder, 1971). For such an ovipositor to be effective, there must be mechanisms acting to move the egg from the posterior opening (vulva) of the vagina to the apex of the valves and to coordinate their back-and-forth movement (Smith, 1969). These have been achieved through development of the following (Fig. 3.18B, C):

- posteriorly pointing microtrichia or scales (*pectines*) on the inner faces of the valves (Smith, 1969; Austin and Browning, 1981);
- an interlocking system of ridges and grooves, the *olistheters*, between the upper and lower valves of each side; and
- a system of tractor muscles inserting into their basal plates, the first and second gonocoxae.

Alternate contraction and relaxation of these muscles move the upper and lower valves relative to each other, the olistheters allow them to slide longitudinally back and forth on each other without separating, and the pectines clasp the surface of the egg and "walk" it to the tip of the ovipositor and into the substrate (Fig. 3.18) (Smith, 1969; Scudder, 1971; Austin and Browning, 1981; Heming-van Battum and Heming, 1986).

If the valves are to move effectively during oviposition, the *fulcra* (articulation points) of the first and second gonocoxae, respectively on the eighth and ninth terga, must remain fixed in relation to each other (Scudder, 1971). This was accomplished through development of the *gonangula* (singular, gonangulum), a pair of struts cross-connecting between the two segments and effecting the movement of the first and second valves relative to each other (Fig. 3.18B, C). Movement of the first pair against the second, either independently or together (this depends on whether or not the valves are fused on the midline), is the principal result of these modifications: the first producing a sawing action; the second, penetration.

Ovipositor structure in females of a particular species relates, in part, to the nature of her usual oviposition substrate. In insects ovipositing into solid substrates, such as wood, stems, or leaves, or into the bodies of other insects, a sawing action is important and the ovipositor is sawlike or *laciniate*. The gonangula are prominent and sclerotized; the olistheters, strongly developed; the first gonapophyses, separate from each other; and the edges of all four valves, heavily sclerotized and equipped with rasps or teeth on their upper and lower margins. In insects ovipositing on surfaces, a sawing action is not important and the ovipositor is reduced and platelike. The gonangula and olistheters are weakly developed or absent, the first gonapophyses are united, and all four valves are smooth, short (Fig. 3.18B, C), membranous, or even absent. But these are the extremes, and there are unique adaptive changes in ovipositor structure in female members of some lineages, such as the sting of the worker honeybee *Apis mellifera* (Snodgrass, 1956: 154–164), the marvelous telescopic ovipositor of scelionid wasps (Field and

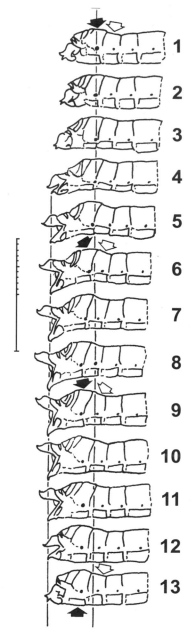

Fig. 3.19. Oviposition in *Locusta migratoria migratorioides* (Orthoptera: Acrididae). Tracings from successive frames (numbered) of a film taken at 18 frames/s, showing the action of the ovipositor valves. The vertical lines provide references through the spiracle of the eighth abdominal segment and the tip of the lower ovipositor valve. On some frames, segment 7 is indicated by an open arrow and segment 8 by a closed arrow. Scale bar is in centimeters and millimeters. (Reproduced with permission from J. F. V. Vincent. How does the female locust dig her oviposition hole? *J. Entomol. (A)* 50: 175–181. © 1975 The Royal Entomological Society of London.)

Austin, 1994), and the diversity of ovipositors in other parasitoid hymenopterans (Quicke et al., 1994).

Sensilla characteristically present on the valves of the ovipositor and in the wall of the genital chamber (Fig.

3.12) function in perceiving substrate cues and during insemination and oviposition (Scudder, 1971; Sugawara, 1993).

Judging by its widespread occurrence in ectognath insects (Fig. 3.20: 1, 1A), the valvular type is ancestral and has been secondarily lost (2A), to be replaced, in certain beetles, caddisflies, lepidopterans, and flies, by extension of the terminal abdominal segments into an extensible oviscapt. It is primitively absent in entognath apterygotes (2) and has been lost and secondarily derived in female snakeflies (Raphidioptera) and in certain female neuropterans (1B) (Mickoleit, 1973).

3.9.1 OVIPOSITOR FUNCTION

Detailed biomechanical studies of oviposition by living female insects are in short supply. The ovipositor in females of the grasshopper *Locusta migratoria* and other species is short and strong and consists of two pairs of stubby valves, the lower, first gonapophyses and the upper gonoplacs, that are adapted for digging the tip of the abdomen into the soil (Fig. 3.19). Vincent (1975) found the thrust of the ovipositor to be at an angle to the axis of the hole, allowing its far side to provide reaction to its movement. The lower valves lever the tip of the abdomen into the soil on each downward movement while the upper valves dig the hole. The maximum rate of vertical penetration into the soil recorded was $2.8\,mm\,s^{-1}$ at 35°C, and maximum displacement per cycle was $2\,mm$. The intersegmental cuticle between abdominal segments 4, 5, 6, and 7 stretch as the tip of the abdomen digs itself more deeply into the soil.

In females of the locust *Schistocerca gregaria*, this alternate contraction and relaxation of ovipositor muscles and movement of valves are controlled by 20 motor neurons in the eighth and by 26 in the ninth neuromere of the composite (8–11), terminal ganglion of the ventral nerve cord (Thompson et al., 1999).

Dambach and Igelmund (1983) provided excellent time-lapse photographs of the crickets *Gryllus bimaculatus*, *Phaeophilacris spectrum*, and *Oecanthus pellucens* ovipositing and of passage of eggs between the first and second gonapophyses into the substrate (soil or vegetation), and time-sequence diagrams of relative valve movement during ovipositor insertion, egg passage, exit of accessory gland secretions, and ovipositor removal. Drilling was shown to involve cyclic anteroposterior movement of the lower valves, which force the upper, tooth-edged valves apart, a movement continuing even after the ventral nerve cord was sectioned anterior to the terminal ganglion.

Following deposition of the egg in or on the substrate, the anteroposterior and dorsoventral axes of the egg usually correspond to those of the ovipositing female unless the egg is rotated or tumbled during oviposition. Downwardly curved ovipositors tumble an egg over during its insertion so that its axes are reversed in its deposited position; straight and upwardly curved ovipositors do not.

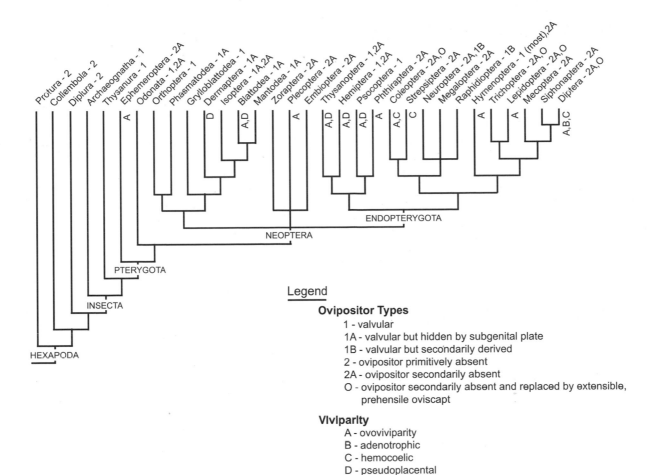

Fig. 3.20. Evolutionary relationships of the hexapod orders as inferred by Gullan and Cranston (1994), showing the distribution of known ovipositor types and of viviparity. (Data from various sources. Cladogram adapted with permission from P. J. Gullan and P. S. Cranston. *The insects. An outline of entomology*, p. 193, Fig. 7.5. 1st ed. Chapman and Hall, London. © 1994 Blackwell Science, Ltd.)

3.9.2 ADAPTIVE SIGNIFICANCE OF THE OVIPOSITOR

Results of sister group comparisons of species diversity and oviposition substrate among members of 14 different pairs of higher arthropod taxa led Zeh and coauthors (1989) to propose that evolution and adaptive radiation of the ovipositor in ectognathous insects may have had a major role in generating insect diversity. Such adaptation drastically increased the number of sites available for egg deposition and enabled these animals to diversify into microhabitats previously closed to them. This in turn resulted in selection for self-sufficient eggs and for embryos having characteristics of the vitelline membrane, chorion, and embryonic membranes (6.1), reducing the possibility of death due to osmotic rupture, desiccation, and drowning in their place of deposition and perhaps contributing, to the low incidence of postzygotic parental investment among insects compared with other terrestrial arthropods. They suggested also that insects would have been unable to exploit the potential of flight and of complete metamorphosis without the

capacity to ensure survival of eggs in the diversity of habitats provided by terrestrial environments. This is because postzygotic parental investment would have restricted the capacity for divergent specialization between adults and immatures, as it may have done in many terrestrial arachnids.

3.10 Viviparity

Most insects are oviparous, depositing eggs that do not begin to develop until after deposition. However, in scattered representatives of many orders (Fig. 3.20), the female deposits free-living, active young, embryogenesis having occurred within the body of the female (Hagan, 1951; Engelmann, 1970; Wigglesworth, 1972; Tremblay and Caltagirone, 1973; Retnakaran and Percy, 1985; Zeh et al., 1989; Meier et al., 1999). Four types have been shown to occur in insects:

- Ovoviviparity: Eggs are normal (i.e., with chorion and

yolk) but are retained within the vagina (uterus) or lateral oviducts until just before or just after hatch. Examples are certain sarcophagid flies.

- Adenotrophic viviparity: Eggs are normal, but after hatch the young insect is retained within the vagina (uterus) where it is nourished by secretions of the female's accessory glands until ready to pupariate, at which time it is deposited, a process Denlinger and Zdárek (1997) recently showed to be induced by secretion of a peptide from cells in the wall of the uterus. The larva is never free-living. Examples are tsetse flies, bat flies, sheep keds, and other Diptera: Pupipara.

- Pseudoplacental viviparity: Eggs lack yolk and chorion, and developing embryos are nourished instead by blood-borne nutrients circulating within the hemocoel of the female. Follicle cells of the oocyte, or cells of one or other of the embryonic envelopes (serosa, amnion; see 6.9) or pleuropodia (6.10.1) may function as a *trophamnion*, transferring nutrients to the developing embryo. Examples are *Hemimerus* earwigs (Dermaptera), aphids, *Diploptera punctata* (Blattodea), and *Hesperoctenes* bugs.

- Hemocoelous viviparity: Eggs lack yolk and chorion. Developing embryos float free in the larval female's blood and are nourished by it; such females are paedo-genetic (5.5.3). These hatch into larvae, which devour most tissues of the mother before escaping through the integument (certain cecidomyiid Diptera) or through special brood canals (strepsipterans).

As can be seen in Fig. 3.20, viviparity is derivative and has evolved independently in members of many orders on numerous occasions. For example, it has been recorded in members of 22 families of Diptera and probably has originated independently at least 61 times, with a single origin in some families and several independent origins in others (Meier et al., 1999). The selective advantages of viviparity in insects are that it shortens the life cycle, reduces the requirement for investment in yolk by females, protects the young during embryonic or larval (adenotrophic viviparity) development, and enables females to produce fewer offspring than those of related, oviparous species because of reduced embryonic and larval (adenotrophic viviparity) mortality (Hagan, 1951; Engelmann, 1970; Meier et al., 1999). The even greater prevalence of ovoviviparity and viviparity in other terrestrial arthropods (scorpions, uropygids, schizomids, amblypygids, and pseudoscorpions) may have inhibited in them the evolution of the resistant, self-sustaining eggs that may have contributed to the proliferation of insect diversity (Zeh et al., 1989).

4/ Sex Determination

Up to this point, we have considered the male and female reproductive systems, gametogenesis (chapters 1 and 2), how sperm are passed from male to female and manipulated to fertilize eggs, fertilization, and oviposition (chapter 3). That there are, in fact, separate males and females is due to expression of certain sex-determining genes during gametogenesis and early embryogenesis, and in this chapter, I summarize how determination of sex is implemented and how it is influenced by various environmental factors.

4.1 Sex Chromosomes

In most insects, determination of sex depends on the distribution of a special class of chromosomes called the *sex chromosomes*, whose behavior during meiosis differs from that of the remaining chromosomes, the *autosomes* (A) (White, 1973). The sex chromosomes usually comprise a pair of similar chromosomes in one sex (usually XX in females) and a dissimilar pair in the other (usually XY in males). The XX-bearing sex is said to be *homogametic* because all gametes produced contain an X-chromosome, whereas the XY-bearing sex is *heterogametic* because half its gametes contain the X-chromosome and half the Y (Jablonka and Lamb, 1990; Charlesworth, 1991).

Sex chromosomes are usually heterochromatic. At periods during the cell cycle (G1, S, G2) when chromatin of the autosomes is uncoiled and diffuse, that of the sex chromosomes is coiled and compact and thus easily recognized in sections and squash preparations (Fig. 4.1). Heterochromatic chromosomes are unable to transcribe RNA while in the condensed state ([^{3}H]uridine autoradiography indicates that no RNA synthesis occurs in them), and it has been suggested that their condensation is a meiotic adaptation to prevent initiation of potentially damaging recombination in nonhomologous regions of the X- and Y-chromosomes (McKee and Handel, 1993).

All insects have heterogametic males except for amphiesmenopterans (Lepidoptera + Trichoptera), in which the females are heterogametic (female: ZW or ZO; male: ZZ) (White, 1973; Kristensen, 1984). In addition, males are normally XO (i.e., they lack a Y-chromosome)

in certain odonates, orthopteroids, hemipterans, coleopterans, neuropterans, and dipterans (White, 1973; Crozier, 1975; Smith and Virkki, 1978; Hewitt, 1979; Ueshima, 1979; Matuszewski, 1982). In those taxa in which the Y-chromosome has been lost, it has been regained secondarily in members of numerous lineages (neo-XY), while those of other lineages have evolved complex mechanisms involving several different X- and Y-chromosomes (multiple sex chromosome systems).

4.1.1 EVOLUTION

In virtually all animal lineages, the sex chromosomes are believed to have originated independently as an ordinary pair of autosomes that happened to carry a major sex-determining locus (Charlesworth, 1991; Morell, 1994; Rice, 1994). As time passed, the Y- (W-) chromosome was selected to stop recombining with the X- (Z-) chromosome during meiosis, perhaps in response to an accumulation of alleles beneficial to the heterogametic but harmful to the homogametic sex. This resulted in the X- (Z-) and Y- (W-) chromosomes becoming increasingly dissimilar during evolution, until in many species the Y (W) became "... a shrunken shadow of its partner carrying a paltry complement of active genes" or actually disappeared to produce an XO system (Morell, 1994). The sequence of events would be as follows:

hermaphrodite → genetic sex determination →
semi-chromosomal sex determination → chromosomal
sex determination (XY) → XO or ZO sex determination

In 1931, R. A. Fisher predicted that a nonrecombining Y-chromosome would eventually degenerate (Morell, 1994), and in 1994, Rice tested this idea in *D. melanogaster* by engineering a large, genetically active "Y"-chromosome, which he prevented from recombining with the X (he made the two chromosomes behave as if they were one Y-chromosome and kept the population small). After 35 generations (he repeated the experiment five times with the same results), this Y began to deteriorate and to lose active genes (once a gene ceases to be expressed, there is no selective pressure to maintain it and deletion of nucleotides can occur). If the chromosome was able to recombine with the X, such deterioration was reduced.

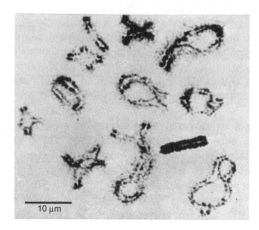

Fig. 4.1. Squash preparation of a diplotene-stage primary spermatocyte of the grasshopper *Schistocerca gregaria*, showing the chiasmata and four chromatids of each autosomal bivalent and the dark-staining rod of the X-chromosome with two chromatids (n = 11 + X). (From *Cytogenetics z/e* by C. P. Swanson, T. Merz, and W. J. Young © 1967. Reprinted by permission of Pearson Education, Inc., Upper Saddle River, N.J.)

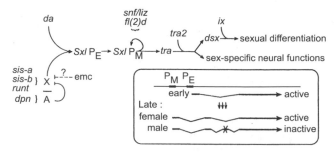

Fig. 4.2. Summary of the genes and their interactions governing somatic sex determination in *D. melanogaster*. See text for details. (Reprinted with permission from S. M. Parkhurst and P. M. Meneely. Sex determination and dosage compensation: lessons from flies and worms. *Science* 264: 924–932. Copyright 1994 American Association for the Advancement of Science.)

4.1.2 DISCOVERY

In 1891, H. Henking first described an X-chromosome in male germ cells of the pyrrhocorid bug *Pyrrhocoris apterus* (2n = 22 + XO). He noted that of four spermatids descending from each primary spermatocyte at meiosis, two received an X-chromosome and two did not (see Fig. 56 in Moore, 1993). In 1901, C. E. McClung proposed that this X-chromosome was associated with the determination of sex, and in 1905, E. B. Wilson and N. M. Stevens, after studying meiosis in a number of insects, found that accessory chromosomes, though differing in number and form in different species, were widespread.

4.2 Sex Determination in *Drosophila melanogaster*

Among insects, the genetics of sex determination is best understood in *D. melanogaster*, in which sex depends on a balance between female-determining loci on the X-chromosome and chromosome 4, and male-determining loci on chromosomes 2 and 3 (in this species, males are 2n = 6A + XY; females are 2n = 6A + XX). The Y-chromosome in males has no direct role in sex determination but is required for normal development of sperm (XO flies are normal males but produce abnormal sperm).

4.2.1 SOMATIC SEX DETERMINATION

All investigated flies seem to use the same basic mechanism to determine the sex of their offspring:
- a primary genetic signal that differs in males and females (the X/A ratio in *Drosophila*);
- a key gene that responds to the primary signal (the *Sex-lethal* [*sxl*] gene in *Drosophila*);

- and a double-switch gene that eventually selects between two alternative sexual programs (the *doublesex* [*dsx*] gene in *Drosophila* [Schütt and Nöthiger, 2000]).

However, these parallels do not extend to the molecular level, as can be seen below.

In *D. melanogaster*, sex is determined by action of a hierarchical genetic cascade that acts independently in each cell of the developing embryo (Fig. 4.2) (Bownes, 1992; Parkhurst and Meneely, 1994; Schütt and Nöthiger, 2000). In zygotes of both sexes, protein of the maternal gene *daughterless* (*da*) is stored in the ooplasm. In female zygotes, the X/A ratio is 1.0, and Da protein, in conjunction with those of the X-linked genes *sisterless-a* (*sis-a*), *sisterless-b* (*sis-b*), and *runt* (*r*) (and others), initiates expression of a key, regulatory switch gene, *Sxl*. *Sxl* encodes an RNA-binding protein (SXL) that regulates sex-specific RNA splicing of its own RNA and that of at least one downstream gene. The *Sxl* gene has two promotors: one initiating *Sxl* activity in the embryo ($Sxl\ P_E$) and the other maintaining it throughout subsequent development ($Sxl\ P_M$). Initiation of *Sxl* activity is transcriptionally controlled through the early embryonic promotor, $Sxl\ P_E$, and is activated only in females (i.e., in animals having an X/A ratio of 1.0 and thus expressing *sis-a*, *sis-b*, and *runt* in diploid amounts) just before blastoderm formation (6.3). Processed transcripts of this promotor encode a full-length, active SXL protein. Maintenance of *Sxl* activity during subsequent development is regulated posttranscriptionally through action of the gene's late promotor, $Sxl\ P_M$. Members of both sexes begin to transcribe *Sxl* RNA from $Sxl\ P_M$ at blastoderm, at about the time that $Sxl\ P_E$ is turned off in female eggs. Transcripts from the late promotor are spliced either to include or to exclude a stop codon. Removal of this codon requires *Sxl* activity. Since only females have active SXL protein from the early promotor, only they can splice the late transcript to make full-length, active SXL protein. Thus, a positive autoregulatory feedback loop that maintains *Sxl* expression in females is established. Since males do not accumulate active SXL protein from the early promotor, they cannot remove the stop codon, and the SXL protein they produce is short and inactive in splicing.

Expression of active *Sxl* activates the downstream *transformer* (*tra*) gene and proteins of *tra* and *transformer2* (*tra2*) together regulate expression of the bifunctional gene *dsx* in the female mode. Female DSX protein, together with Intersex (IX) protein, represses expression of male structural genes but not that of female genes, resulting in development of a female.

In male zygotes, the X/A ratio is 0.5. Although the products of *da*, *sis-a*, *sis-b*, *runt*, and so on, are present, *Sxl* does not produce a functional product, as insufficient *sis-a*, *sis-b*, and *runt* product is made from the single X-chromosome of male eggs to activate the early *Sxl* P_E promotor. As a result, *tra* is not expressed. Expression of *tra2* alone selects a male mode of expression for *dsx*. Male DSX protein represses expression of the female differentiation genes but allows that of male structural genes, resulting in development of a male.

Sxl is also involved in regulating dosage compensation in males, a process equalizing the amount of product derived from genes on the single X-chromosome of males with those on the two X-chromosomes of females. When *Sxl* is active, it represses expression of the genes *male specific lethal* (*msl*) and *male lethal* (*mle*) and probably others. Expression of both these genes is necessary to get hypertranscription of genes on the male's single X-chromosome. With no *msl/mle* expression, the lower female transcription rate per chromosome prevails.

In other investigated flies, particularly the well-known house fly *Musca domestica*, the ortholog of the *Sxl* gene occurs but does not have a sex-determining role (Schütt and Nöthiger, 2000). Recent progress in understanding molecular aspects of sex determination is summarized by Kuroda and Villeneuve (1996) and Schütt and Nöthiger (2000).

4.2.2 SEX DETERMINATION OF GERM CELLS

The sex of germ cells (GCs) in *D. melanogaster* embryos is determined both by cell autonomous (originating within the GCs) and by inductive (originating within gonadal mesoderm surrounding the GCs) signals (Steinmann-Zwicky, 1994). XY (male) GCs will begin spermatogenesis when developing within a female (XX) host, while XX (female) GCs will undergo spermatogenesis when developing in a male host. In female first instars, XX GCs will enter the female or male pathway, depending, respectively, on the presence or absence of *tra* activity in cells of the gonadal mesoderm (see 8.4.1.2). In these, the products of *tra* and *tra2* regulate expression of *dsx* which, as summarized already, can form a male- or a female-specific product. In *dsx* mutant larvae, lacking *dsx* function, both XX and XY GCs develop an intersex phenotype, indicating that female-specific, somatic *dsx* product normally feminizes XX GCs while male-specific *dsx* product masculinizes both XX and XY GCs. These

observations indicate that expression of both *tra* and *dsx* control early inductive signals determining the sex of XX GCs and affecting development of XY GCs. In addition, XX GCs developing in pseudomales lacking the sex-determining function of *Sxl* are spermatogenic. If female-specific *tra* functions are expressed in these larvae, then XX GCs become oogenic. Also, transplanted XX GCs can become oogenic and form eggs in XY animals expressing the female-specific function of *tra*.

Expression of genes additional to those of the sex-determining pathway can also influence GC sex. For example, expression of the *indora* (*idr*) gene of *D. melanogaster* is induced by gonadal mesoderm in GCs enclosed within the gonad but not in those yet to be enclosed (Mukai et al., 1999), and if such expression is reduced, the GCs fail to differentiate into female gametes.

4.3 Sex Determination in Haplodiploid (Arrhenotokous) Insects

In hymenopterans, fertilized eggs usually develop into diploid females and unfertilized eggs into haploid (rarely diploid) males (Fig. 4.3). All sperm are female producing and sex chromosomes are not recognizable (White, 1973). Sex determination seems to depend on a sex locus, X, being heterozygous in diploid females (Xa Xb, Xb Xc, Xa Xc), and *hemizygous* (Xa, Xb, Xc) in haploid or homozygous (Xa Xa, Xb Xb, Xc Xc) in diploid males, this in turn influencing whether female- or male-differentiation genes are turned on (Bownes, 1992).

In scale insects in the families Lecaniidae and Diaspididae of the Coccoidea (Homoptera), the chromosome system is either lecanoid or of the Comstockiella type (Nur, 1971). In both systems, sex chromosomes are lacking and sex is apparently determined by the number of uncondensed (euchromatic) sets of autosomes. The set contributed by the sperm becomes heterochromatic in some eggs but not in others, and those with one euchromatic set, the maternal one, develop into males and those with two, into females.

Haplodiploid sex determination also occurs in whiteflies (Homoptera: Aleurodidae), thrips (Thysanoptera), bark beetles (Coleoptera: Scolytidae), and *Micromalthus debilis* (Coleoptera: Micromalthidae). These are considered in more detail in 5.4.1.

4.4. Chromosome Elimination and Sex Determination in Symphypleon Collembola

Chromosome elimination influences the determination of sex in at least some symphypleon springtails (Dallai et al., 1999). In *Dicyrtomina ornata*, females have a diploid chromosome number of 12 and males 10, while in *Ptenothrix italica*, females have 14 chromosomes and males 12. Embryogenesis is normal in female embryos of both species, but in males, two sex chromosomes are

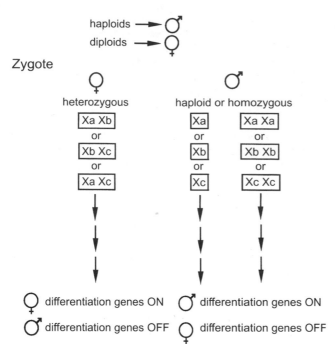

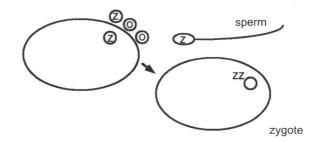

At 30−37°C: Sex ratio = 162 males / 100 females

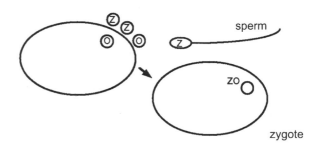

At 3−8°C: Sex ratio = 100 females / 65 males

Fig. 4.3. Sex determination in the haplodiploid (arrhenotokous) insect *Athalia rosae* (Hymenoptera). Under normal circumstances, haploids develop into males and diploids into females (top). This depends on a locus X being heterozygous in females (left column). Any zygotes hemizygous (middle column) or homozygous for an allele at X also develop into males (right column). How such a system could operate, where multiple alleles for a given locus can promote male or female development depending on whether it is homozygous, hemizygous, or heterozygous, is unknown. (Adapted with permission from M. Bownes. Molecular aspects of sex determination in insects. In J. M. Crampton and P. Eggleston (eds.), *Insect molecular science*, p. 86, Fig. 3. 16th Symp. R. Entomol. Soc. London. © 1992 Academic Press, Ltd.)

Fig. 4.4. Temperature effects on sex determination in eggs of *Talaeporia* sp. (Lepidoptera: Psychidae) in which females are the heterogametic sex (male is ZZ; female, ZO). Female meiosis is induced by sperm entry, and temperature influences whether the sex chromosome Z is allocated to a polar body or to the female pronucleus. If to the former, a female develops following fertilization (bottom); if to the latter, a male develops (top). (Drawn from information in Wigglesworth, 1965: 667–668.)

lost during cleavage. Later, at the end of meiosis I in males, a second elimination of chromosomes occurs, resulting in half the secondary spermatocytes produced receiving a full haploid complement of chromosomes and the other half, fewer. The latter spermatocytes degenerate before undergoing meiosis II so that males produce only half their spermatozoa. Oogenesis is normal in females of both species.

4.5 Environmental Effects on Sex Determination

Expression of normal chromosomal sex determination is sometimes influenced by action of various environmental factors (Bergerard, 1972).

4.5.1 TEMPERATURE
The stick insect *Carausius morosus* (Phasmatodea) is normally parthenogenetic, and its eggs develop into females if they are incubated at relatively low temperatures (23°C) during the first 30 d of embryogenesis. However, if they are incubated at 30°C, most embryos develop as males (Bergerard, 1972).

In certain psychid moths (*Talaeporia* spp.), temperatures over 30°C alter the sex ratio of offspring (Fig. 4.4) (Wigglesworth, 1965). If females experience high temperatures (30–37°C) during ovulation and fertilization, males predominate in their offspring; at low temperatures (3–8°C), females predominate. The temperature effect occurs during female meiosis following fertilization. At high temperatures, the single Z-chromosome (females are ZO; males, ZZ) remains in the female pronucleus after meiosis in most eggs, resulting in males (ZZ) at fertilization. At low temperatures, the Z-chromosome is usually allocated to one of the polar bodies, and at fertilization, females (ZO) result.

In the mosquito *Aedes stimulans* and in some other *Aedes* spp. living at high latitudes in Canada and Alaska, the gonadal rudiments of male embryos and larvae have the potential to develop into either testes or ovaries, depending on water temperatures experienced during larval development (Horsfall and Anderson, 1961). If male larvae are reared in water at temperatures above normal (i.e., ≥18°C), the rudiments of the testes become ovaries. If reared throughout larval life at temperatures of 18°C or below or if they are moved into water at these temperatures during the second instar or earlier, their gonads develop normally as testes (Anderson and Horsfall, 1964).

The source primordia for this differentiation are located within the gonadal rudiments. Transplanted testicular rudiments develop into testes in both male and female host larvae if these are reared at 18°C or lower and into ovaries if reared at 27°C. In addition to gonads, the structure of most other sexually dimorphic, imaginal characters (antennae, mouthparts, pretarsal claws, reproductive system and external genitalia, etc.) is changed from male to female when male larvae are reared at high temperatures (Anderson and Horsfall, 1963, 1964). The degree of suppression of male and stimulation of female structures depends entirely on larval water temperatures above 18°C and below 28.4°C, with all intermediates possible; at 28.4°C the male develops into a phenotypic "female" that is normal in every way except for a few minor structural characteristics (e.g., it lacks female reproductive accessory glands) and its inability to take a blood meal and produce eggs (the mouthparts of such females are female-like, but the insect is not "wired" to take a blood meal). If such females are artificially fed blood on a soaked cotton pad, a few vitellogenic oocytes can be induced to develop within their ovaries. Also, egg follicles in ovarioles of thermally induced females are capable of vitellogenesis if they are transplanted into the bodies of genetic females (eggs were not tested for their further ability to develop).

It is now known that male mosquito larvae have both male and female genital primordia but females have only female primordia (Ronquillo and Horsfall, 1969; Horsfall and Ronquillo, 1971; Vorhees and Horsfall, 1971; Horsfall et al., 1972). It is also known that sex in mosquitoes is determined by action of a single M locus on an autosome (Anderson and Horsfall, 1963). Males are heterozygous for the allele (Mm) and females homozygous (mm).

Brust (1968) eventually evaluated 12 other univoltine species of *Aedes* in Canada having an arctic or subarctic distribution, and found all of them to respond to larval-rearing temperature in the same way, the temperature range involved varying with the species and their normal latitudinal distribution. Since thermally induced intersexes and females are infertile, they contribute nothing to the gene pool of the population to which they belong, leading one to doubt if the phenomenon has any evolutionary significance.

4.5.2 NUTRITION AND INHIBITORY PHEROMONES

In colonies of eusocial hymenopterans there may be thousands of females of which only one or a few produce viable eggs (Brian, 1980). In colonies of the domestic honeybee *Apis mellifera*, for example, most genetic females are prevented from developing eggs by an inhibitory pheromone (queen substance: 9-oxo-*trans*-2-decanoic acid) produced by the queen's mandibular glands and spread throughout the colony by *trophallaxis* or food exchange. During metamorphosis of worker larvae, the number of ovarioles per ovary is reduced from 100 to approximately 12, a reduction influenced by the amount of royal jelly fed to female larvae by nurse bees. At a hive temperature of 33–35°C, female larvae fed throughout their larval lives (7 d, five instars) on royal jelly by nurse bees emerge as queens on day 13, whereas potential workers are switched to worker jelly on day 4 and do not emerge until day 18 (nurse bees know what to feed a larva because of the size and shape of the wax cell in which it is developing) (Fig. 12.35). If the queen is removed from the hive, the inhibition is lifted, and some workers are able to develop and deposit a relatively small number of viable eggs, which develop parthenogenetically into haploid drones (males). Other workers modify one worker cell or more, each already containing a young (<4 d; instar 1–4) female larva, into emergency queen cells and feed royal jelly to the larva throughout its life; these develop as one or more replacement queens. Drones develop from unfertilized eggs deposited by the queen in larger drone cells; they are fed the same as worker larvae and emerge on day 22 (Brian, 1980).

Workers in bumblebee (Apidae) and paper wasp (Vespidae) colonies can develop viable eggs but usually don't because of nutritional castration (Brian, 1980); the food they collect is fed to larvae rather than being used as a source of protein for vitellogenesis of their own eggs. They are forced to do so by the dominant queen since colonies of both groups of insects operate in a dominance hierarchy. On removal of the queen, another female will take her place and begin developing and depositing unfertilized, male eggs.

4.5.3 ANDROGENIC HORMONES

Most insects do not produce sex hormones, as countless castration and transplantation experiments (references in Wigglesworth, 1965) have demonstrated:

- Castrated male and female larvae develop into normal but sterile adults.
- Transplanted ovary rudiments develop into ovaries in male hosts and transplanted testicular rudiments, into testes in female hosts.
- Infusion of female blood into male hosts and vice versa has no effect on the development of secondary sexual characteristics.

However, in larvae of the firefly *Lampyris noctiluca* (Coleoptera: Lampyridae), a factor released from neurosecretory cells in the pars intercerebralis of the brain induces the synthesis and release of an androgenic factor from mesodermal cells at the apex of each developing sperm tube in testicular rudiments of male larvae. This factor in turn controls development of the male's primary and secondary sexual characters (Fig. 4.5) (Naisse, 1966a; Raabe, 1982).

Males of these insects have five larval instars, a pupa,

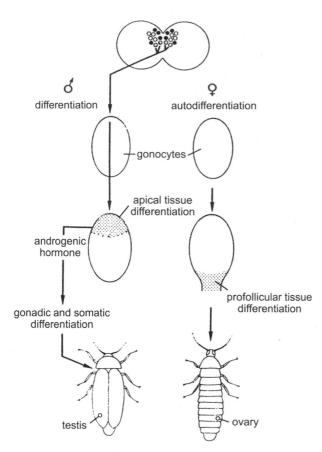

Fig. 4.5. Sex determination in the firefly *Lampyris noctiluca* (Coleoptera: Lampyridae). A neurosecretory factor is released from the brain in fourth instar male larvae. This factor induces the differentiation of apical tissue in the testicular rudiments. Apical cells synthesize and release an androgenic factor that induces male somatic and germinal differentiation. If this sequence does not occur, the gonads differentiate into ovaries and a larviform female results. (Reprinted with permission from M. Raabe. *Insect hormones*, p. 102, Fig. 35. Plenum Press, N.Y. © 1982 Kluwer Academic/Plenum Publishers.)

and a fully winged adult, while female adults are larviform (neotenous) and develop through six larval instars. In instars 1–3, larvae of both sexes are identical, including their gonadal rudiments. After they molt into instar 4, an "androgen" is released from newly proliferated apical cells in the sperm tubes of male larvae and causes them to develop as males (Fig. 4.5). If this factor is not released, a female results (the apical tissue of males degenerates at the onset of metamorphosis). If ovary rudiments are implanted into male larvae after the third molt, but before pupation, they develop into testes (Fig. 4.6A). If testicular rudiments are transplanted into female larvae at this time, the host larvae develop into males (Fig. 4.6B). However, if adult testes, lacking the apical tissue, are implanted into fourth instar females, they degenerate. Finally, if an allatectomized (i.e., corpora allata have been removed), debrained, fourth instar

female is joined in parabiosis to a fourth instar castrated male, the ovary rudiments of the female still become testes, suggesting a direct androgenic effect of the neurosecretory factor from the brain of male larvae (Fig. 4.6C).

To explain these results, Naisse (1966b) proposed that expression of the sex determination loci induces secretory activity in certain neurosecretory cells of the brain of male larvae that in turn stimulates formation of apical tissue in their testicular rudiments. To my knowledge, this fascinating work has not been followed up using modern methods of molecular biology.

4.5.4 MATERNAL HEMOLYMPH

In paedogenetic (larval) reproduction of the fungivorous cecidomyiid midge *Heteropeza pygmaea*, the larval ovaries produce either female- or male-determined eggs, or both, that develop parthenogenetically (Fig. 4.7) (Went and Camenzind, 1977; Went, 1982). Determination of eggs to develop into male or female larvae occurs during oogenesis and seems to be controlled by a factor or factors from the brain of the mother larva. When ovaries are removed and cultured in hemolymph from either male-mother or female-mother larvae, they generate eggs differentiating into either male or female larvae, depending on the hemolymph of the donor. In addition, the quality of this hemolymph varies based on how well the donor larvae were fed.

4.6 Hermaphrodites

Hermaphrodites are individuals having functional reproductive systems of both sexes. In insects the phenomenon is rare; it is known to occur only in three members of the scale insect genus *Icerya* (Hemiptera: Margarodidae): *I. purchasi*, *I. bimaculata*, and *I. zeteki* (Hughes-Schrader, 1927; Nur, 1971). In a typical population of the cottony cushion scale *I. purchasi*, there are usually two kinds of individuals (Hughes-Schrader, 1927; White, 1973). About 99% are diploid hermaphrodites and the rest are haploid males. The hermaphrodites are female-like, have an ovotestis, and may copulate with the rare males. Each ovotestis has a central core of haploid sperm and a peripheral layer of diploid eggs (the testicular portion is haploid as a result of a reduction division that occurred in early embryogenesis in some primordial GCs that eliminated one chromosome set). Most eggs undergo meiosis, are fertilized by sperm from the same individual, and then develop into additional diploid hermaphrodites. The eggs not fertilized undergo meiosis and develop into haploid males.

Nonfunctional or accessory hermaphroditism is known to occur in males of the termite *Neotermes zuluensis* (Geyer, 1951), of the perlid stonefly *Perla marginata* (Junker, 1923), and of the German cockroach *Blattella germanica* (Brooks and Kurtti, 1972). In male juveniles and adults of these insects, both testes and rudimentary but nonfunctional panoistic ovarioles are present, a situ-

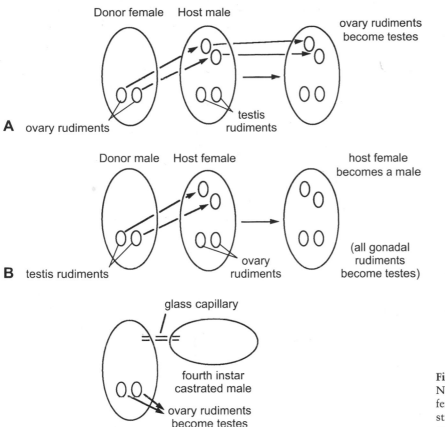

Donor female   Host male

ovary rudiments
become testes

ovary rudiments

testis
rudiments

Donor male   Host female

host female
becomes a male

testis rudiments

ovary
rudiments

(all gonadal
rudiments
become testes)

B

glass capillary

fourth instar
castrated male

ovary rudiments
become testes

fourth instar female
(debrained, allatectomized)

C

Fig. 4.6. Experiments executed by Naisse on fourth instar male and female larvae of *L. noctiluca* demonstrate the role of neurosecretory and apical tissue factors in determining sex. (Drawn from information in Naisse, 1966a, 1966b, 1966c.)

ation probably persisting from the sexually indifferent period of reproductive embryogenesis (Fig. 8.53A).

4.7 Gynandromorphs

Gynandromorphs are individuals in which some body parts are female and others are male, the boundary between such parts being abrupt (Tremblay and Caltagirone, 1973). For example, as mentioned earlier (4.1), *D. melanogaster* has the XX (female): XY (male) sex chromosome system. Elimination of one of the two X-chromosomes, as an unstable ring chromosome, in early cleavage of female embryos eventually results in some imaginal tissues having male (XO) and others having female (XX) cells (Fig. 4.8A).

Most hymenopterans are haplodiploid (4.3), the females of some species sometimes producing binucleate eggs if reared under abnormally high temperatures (Wigglesworth, 1965). If one of the two female pronuclei of such an egg is fertilized by a sperm and the other is not, some parts of the insect developing from that egg will be male (from proliferation of the unfertilized, haploid pronucleus) and other parts, female (from that of the fertilized, diploid zygote pronucleus) (Fig. 4.8B).

And in rare binucleate ova of normal, diplodiploid (bisexual) insects having the XO or XY chromosome system, a gynandromorph can result from double fertilization, one pronucleus fertilized by a Y- or O-bearing sperm to generate male tissue and the other by an X-bearing sperm to generate female tissue (Fig. 4.8C).

4.8 Intersexes

Intersexes are insects having a variable combination of male and female characteristics, with the boundary between them diffuse (Wigglesworth, 1965). They result from a disturbance in the balance between the action of products of male- and female-determining loci due to an extra set of chromosomes, as in triploid *D. melanogaster* and *Solenobia* spp. (Lepidoptera: Psychidae), or by action of various environmental effects such as high temperature, as in our earlier, *A. stimulans* example (4.5.1). A particularly fine study is that by Seiler (1969) on *Solenobia triquetrella* (Lepidoptera: Psychidae) and *Lymantria dispar* (Lymantriidae), which summarized a lifetime's work on this phenomenon. The photomicrograph in Fig. 4.9 illustrates a portion of the gonad of an intersex

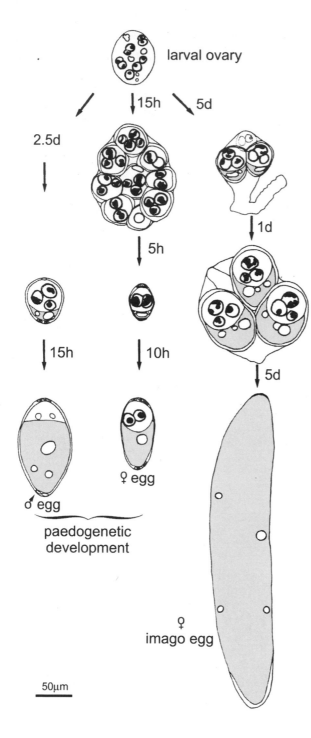

larval ovary

15h   5d

2.5d

1d

5h

5h

15h   10h

5d

♂ egg   ♀ egg

paedogenetic
development

♀
imago egg

50μm

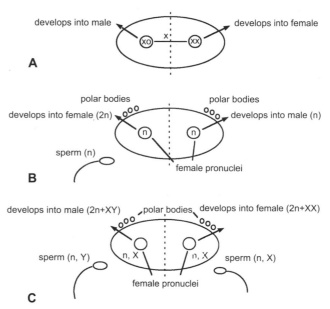

Fig. 4.7. Summary of the different types of oogenesis occurring in females of *Heteropyza pygmaea* (Diptera: Cecidomyiidae). The top drawing is one of the two ovary rudiments in a female larva newly escaped from its mother larva. The pathway on the right summarizes ovarian stages during normal adult development in the pharate pupa (5 d after hatch), in the pupa just after pupation 1 d later, and in the adult stage. When the larva reproduces paedogenetically, female-determined or male-determined follicles are liberated from the larval ovary after 15 h and 2.5 d, respectively (middle and left pathways). The three eggs shown at the bottom are at comparable stages of development just before completion of the meiotic division(s). The small nuclei in male-determined eggs and in the imago eggs are of mesodermal (somatic) origin and fuse with the respective female pronucleus after meiosis is completed. (From D. F. Went. 1982. Egg activation and parthenogenetic reproduction in insects. *Biol. Rev.* 57: 319–344. Reprinted with the permission of Cambridge University Press.)

develops into male   develops into female

A

polar bodies   polar bodies
develops into female (2n)   develops into male (n)

sperm (n)   female pronuclei

B

develops into male (2n+XY)   polar bodies   develops into female (2n+XX)

sperm (n, Y)   sperm (n, X)

female pronuclei

C

Fig. 4.8. Some mechanisms of gynandromorph formation in insects. A. *D. melanogaster*. One of two X-chromosomes is lost during the first cleavage division of a female egg (XX), resulting in half the embryo developing into male (XO) tissue and half into female (XX). B. Binucleate egg of a haplodiploid wasp has one pronucleus fertilized with a sperm and the other not. C. Binucleate egg of a typical, bisexual insect (female is XX; male, XO or XY) is doubly fertilized: one pronucleus by a Y- or O-bearing sperm and the other, by an X-bearing sperm.

S. triquetrella with an ovariole and parts of three sperm tubes.

4.9 Parasite Effects on Sex Determination

Invasion of a developing host larva by a parasite can influence development of its sexual identity (Wigglesworth, 1965). For example, parasitization of late instar female larvae of the chironomid midge *Chirono-*

mus hyperboreus by mermithid nematodes causes them to develop into intersexes having normal female secondary sexual characteristics but fully differentiated male internal and external reproductive systems (Rempel, 1940). The nematode develops at the expense of the host female's reproductive primordia, and loss of these seems to induce development of male reproductive

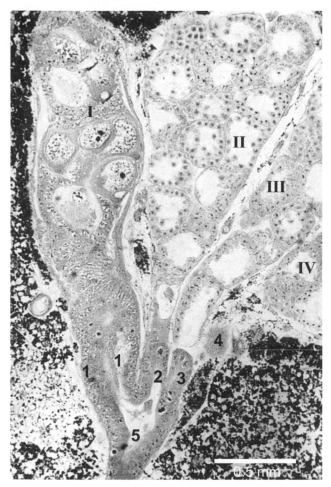

Fig. 4.9. Photomicrograph of a frontal section of the gonad of an intersex *Solenobia triquetrella* (Lepidoptera: Psychidae), showing an ovariole and parts of three sperm tubes. (Reproduced with permission from J. Seiler. Intersexuality in *Solenobia triquetrella* F. R. and *Lymantria dispar* L. (Lepid.). Questions of determination. *Mon. Zool. Ital. (N. S.)* 3: 185–212. © 1969 by *Monitore Zoologico Italiana*.)

primordia, the degree of intersexuality attained in the adult depending on when the larva was parasitized. This is an intriguing difference from the phenotypic females, discussed in 4.5.1, induced by high water temperatures in developing male *Aedes* larvae, but does indicate the presence of both male and female reproductive primordia in female larvae of this fly.

4.10 Why Are There Only Two Sexes?

Most insect species are bisexual and have separate males and females or are parthenogenetic and produce only females (chapter 5). If the purpose of one sex meeting the other is to combine DNA, increase genetic variation, and start the next generation, what could be worse than

having only two sexes? With two sexes, an individual can choose a partner from only 50% of the population, such that valuable time must be wasted in tracking down mates. Wouldn't it be better to have 20 sexes and to be able to mate with any partner except one of your own sex (Anderson, 1992)?

The answer may relate to the fact that all organelles contributed to the egg by a sperm usually degenerate, except for its centriole and nucleus (Fig. 3.14). L. D. Hurst (1992) proposed that the two sexes evolved to manage genomic conflict. He suggested that such management is essential in the organisms reproducing by fusing two cells and so bringing together genetic material of nuclei, mitochondria, and in plants, chloroplasts. Unlike the chromosomes of the male and female pronuclei, both of which contribute to zygotic success, the mitochondria have no "reason" to share ooplasm with mitochondria contributed by the other gamete. Instead, they tend to "look out for themselves" (they are known to actively degrade one another). Hurst suggested the potential damage to the zygote from such conflict to be so great that sexes become essential to avoid conflict between mitochondria. In a two-sex system, the gamete of one sex (the male or – type) surrenders its organelles, while that of the other (the female or + type) maintains them. Thus, the males of insects and other animals usually produce relatively tiny sperm that contribute no mitochondria to the zygote (actually they do, but they usually soon degenerate; see 3.6.2), whereas females generally produce relatively large eggs that do. A nuclear gene that unilaterally destroys its own mitochondria to avoid a costly battle (plus a gene that chooses mates of opposite type, to avoid two disarmers meeting and ending up with no organelles) can thus become evolutionarily fixed (Hurst, 1992).

Hurst searched the literature on reproduction in organisms and found an incredible amount of evidence to support this model (in fact he found only one exception: a slime mold, *Physarum polycephalum*, with 13 or more mating types or sexes). He was able to allocate, to one of two types, all creatures having sex: those fusing cells during mating (*fusion* sex) and those passing only nuclei between them (*conjugatory* sex). Hurst predicted that only organisms of the first type should have evolved two sexes because it is only these that risk warfare between their mitochondria. Organisms of the second group have no need for separate sexes because the mitochondria from separate cells are never mixed. His search of the literature produced only one exception: a hypotrich ciliate protist able to reproduce either way! When reproducing by fusion sex, this ciliate has two sexes; when doing so by conjugatory sex, it has multiple mating types, as do other organisms reproducing this way (e.g., most ciliate protists and ascomycote fungi such as yeasts). In the slime mold with 13 sexes, these are in a hierarchy, each one yielding to the one above.

Why only *two* sexes? If there were more than two, as in the slime mold, "for any particular sex the cytoplasmic genes will sometimes be inherited, and sometimes not . . . depending on mate . . . so that it has an inherent vulnerability to 'cheat' (what happens if one set of mutant mitochondria 'refuses to shut down'?)" (Anderson, 1992). In other words, once you have a system that is not as rigidly fixed as in the two-sex world, there is a greater risk of the "mitochondrial wars" that the sexes evolved to prevent.

5/ Parthenogenesis

As already seen, insects of most species are bisexual (zygogenetic or diplo-diploid) and develop from fertilized eggs. However, a significant number of species throughout the Hexapoda are known to be *parthenogenetic*; that is, they produce offspring in the absence of males and without the stimulus of sperm entry (Suomalainen et al., 1987). A variety of mechanisms are known, including accidental or *tychoparthenogenesis*; impaternate development of males (*arrhenotoky*), females (*thelytoky*), or both (*amphitoky* or *deuterotoky*); and *cyclical parthenogenesis* (the alternation of bisexual and parthenogenetic generations or *heterogony*) (Engelmann, 1970; Wigglesworth, 1972; Retnakaran and Percy, 1985; Suomalainen et al., 1987; Mable and Otto, 1998). The principal selective advantage of such reproduction for a species is that it doubles its innate rate of natural increase, owing to exclusive production of females that can start new colonies immediately without being inseminated (Fig. 5.1) (Cuellar, 1977; Mable and Otto, 1998). Its principal disadvantage is that it reduces the spread of genetic variability, particularly if combined with *paedogenesis* (sexual maturity of juveniles) and *viviparity* (deposition of free-living young instead of eggs), compromising the long-term survival of such species in constantly changing environments.

5.1 Arrhenotoky

In the case of arrhenotoky (male-producing or haplodiploid parthenogenesis), males are haploid in their germ line and develop parthenogenetically from unfertilized eggs, whereas females are diploid in theirs and develop from fertilized eggs (Fig. 5.2). (The parthenogenetic false spider mite *Brevipalpus phoenicis* is exceptional among animals in having haploid females, owing to infection by an undescribed endosymbiotic bacterium [Weeks et al., 2001]. If its eggs are exposed to antibiotics or high temperature, these bacteria are destroyed and the eggs parthenogenetically develop into haploid males. That some other species in the genus are haplodiploid indicates this to be their ancestral mode of reproduction.) Meiosis is normal in eggs of both sexes but continues directly into the first cleavage division of embryogenesis following oviposition in unfertilized eggs without the intervention of a metaphase I block

(2.2.1.5). Without the stimulus of sperm entry (3.6), male eggs are induced to begin embryogenesis in other ways, for example, by passing down the shaft of the ovipositor, a process that can be duplicated by passing eggs through a glass capillary of appropriate caliber (Fig. 5.3) (Went and Krause, 1973; Went, 1982). Since male germ cells are already haploid, their meiosis is functionally mitotic, even though there may still be two divisions, as in the drone honeybee *Apis mellifera* (Hoage and Kessel, 1968), so that sources of genetic variability within the species are reduced by half.

In newly fertilized diploid eggs of the parasitoid pteromalid wasp *Nasonia vitripennis*, the centrioles functioning in cleavage descend, as usual, from that contributed by the sperm (Fig. 3.14) (Tram and Sullivan, 2000). However, centrioles also function during cleavage in unfertilized haploid eggs, and it has remained unclear when and where these organelles originate (the meiotic divisions in eggs occur without them), though in *Drosophila* eggs those of the nurse cells enter the oocyte through the ring canals linking them and the cystocytes together (2.2.2.5). In fact, a large number of maternal centrosomes appear simultaneously in the cortical cytoplasm of both fertilized and unfertilized eggs of this insect immediately after meiosis (Tram and Sullivan, 2000). In unfertilized eggs these begin to migrate toward the female pronucleus, and the first two to arrive organize the spindle for the first cleavage division. The same begins to occur in fertilized eggs, but the paternal centriole contributed by the sperm gets to the female pronucleus first and in some way prevents the two maternal centrosomes from associating with the female pronucleus (centrosomes remaining in the cortex disappear during the first cleavage division). Nothing is known about how cortical centrosome formation is regulated in these eggs nor about how widespread it is in those of other bisexually and parthenogenetically reproducing insects (Karr, 2001), though it occurs also in eggs of another pteromalid, *Muscidifurax uniraptor* (Ripparbelli et al., 1998). In any case, the race between paternal and maternal centrosomes to organize the cleavage divisions represents another example of potential male-female conflict.

Haplodiploid parthenogenesis has arisen independently at least nine times within the Metazoa: twice in Hemiptera (Aleurodidae and Coccoidea: Iceryini), once

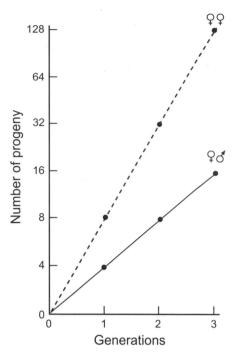

Fig. 5.1. Comparison of the theoretical intrinsic rates of increase between a bisexual (male and female) and a unisexual (female-only) species where clutch size is four and longevity is 1 y. The first generation consists of two individuals, one male and one female for the bisexuals, and two females for the unisexuals. (Adapted with permission from O. Cuellar. Animal parthenogenesis. *Science* 197: 837–843. Copyright 1977 American Association for the Advancement of Science.)

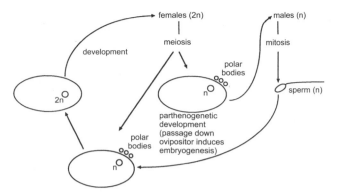

Fig. 5.2. Arrhenotokous (haplodiploid or male-determining) parthenogenesis.

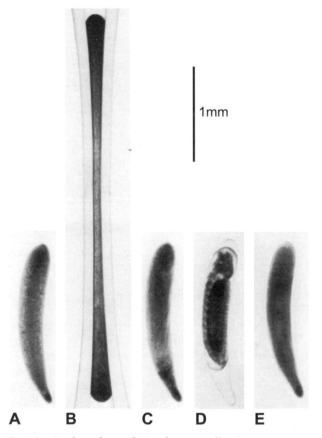

A B C D E

Fig. 5.3. Explanted egg of *Pimpla turionellae* (Hymenoptera: Ichneumonidae) before (A), during (B), and after (C and D) passage through a glass capillary tube to simulate passage down the ovipositor, and an egg (E) after normal oviposition. A, C, D, and E, eggs in Ringer's solution; B, egg in paraffin oil; C and E, eggs 4 min after artificial and normal oviposition, respectively; D, embryo 32 h after treatment. (From D. F. Went. 1982. Egg activation and parthenogenetic reproduction in insects. *Biol. Rev.* 57: 319–344. Reprinted with the permission of Cambridge University Press.)

in Thysanoptera, twice in Coleoptera (Micromalthidae and scolytine weevils), and once in Hymenoptera among insects (Fig. 5.12: 1) and in rotifers (Order Monogonata), nematodes (superfamily Oxyuroidea), and mites (Arachnida: Acarina) (Hartl, 1971; White, 1973; Suomalainen et al., 1987; Heming, 1995).

In some males of *N. vitripennis*, the sperm contain a paternal sex ratio chromosome (PSR). If an egg is fertilized with such a sperm, the paternal autosomes contributed soon degenerate and the egg develops into a haploid male even though it was fertilized (Nur, 1980). Members of some populations of this wasp are also infected by *Wolbachia* bacteria, which cause cytoplasmic incompatibility, also resulting in chromosome loss. Thus, the question arises as to whether PSR transmission or the ability to induce chromosome loss depends on interactions with *Wolbachia*. To answer this question, Dobson and Tanouye (1997) generated a strain of PSR males artificially cleared of *Wolbachia* by treatment with antibiotics, and made test crosses and cytological observations of the eggs of this strain. Their results indicated that the function of PSR in causing

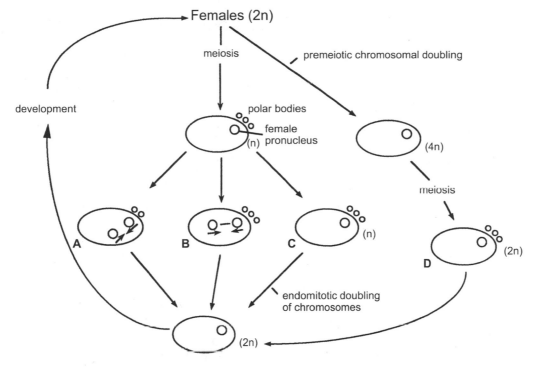

Females (2n)

meiosis

premeiotic chromosomal doubling

development

polar bodies

female
pronucleus
(n)

(4n)

meiosis

A

B

C

(n)

D

(2n)

endomitotic doubling
of chromosomes

(2n)

Legend

A. Fusion of a polar body with the female pronucleus just after meiosis.
B. Haploid cleavage nuclei fuse during cleavage (at any time up until blastoderm
formation in eggs of certain crickets and stick insects).
C. Endomitotic doubling of chromosomes can occur either before or after cleavage begins.
D. Premeiotic doubling of chromosomes.

Fig. 5.4. Automictic (meiotic) thelytoky, showing four ways in which the diploid chromosome number can be reconstituted after meiosis.

chromosome loss does not depend on presence of *Wolbachia*, and in fact, they were unable to demonstrate interactions between the two.

5.2 Thelytoky

In thelytoky (female-producing parthenogenesis), males either are absent or are rare and nonfunctional (*spanandric*). It occurs in representatives of many orders (Fig. 5.12: 2, 3), with perhaps 1000 species known, and is usually one of two types (Suomalainen et al., 1987), *automictic* (mixing with self, meiotic) or *apomictic* (without mixing, ameiotic).

5.2.1 AUTOMICTIC
Meiosis is normal in eggs, and various cytological mechanisms have evolved to restore the diploid chromosome number following reduction. Examples of such mechanisms are as follows:
• fusion of polar body 2 with the female pronucleus

just after meiosis (Fig. 5.4A) (e.g., various scale insects [Hemiptera: Lecaniidae] [Nur, 1971]);
• fusion of haploid cleavage energids following the first cleavage division (Fig. 5.4B) (e.g., *Pulvinaria hydrangeae* [Hemiptera: Lecaniidae] [Nur, 1971]) or in pairs at any time up until blastoderm formation;
• endomitotic doubling of chromosomes either before or after commencement of cleavage (Fig. 5.4C) (e.g., *Bacillus rossius* [Phasmatodea: Phasmida] [Pijnacker, 1969]); and
• premeiotic doubling of chromosomes in primary oocytes to 4n so that eggs are diploid following meiosis (Fig. 5.4D) (e.g., *Ptinus clavipes* form *mobilis* [Coleoptera: Ptinidae] [Smith, 1971] and *Carausius morosus* [Phasmatodea] [Koch et al., 1972]).

5.2.2 APOMICTIC
Meiosis is functionally mitotic with no reduction in chromosome number in the germ line (Fig. 5.5A). This type of parthenogenesis is often combined with viviparity (3.10) and paedogenesis, and there is no source of

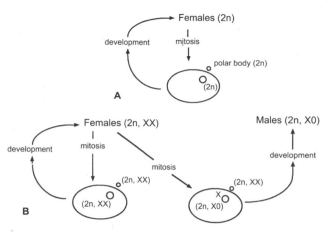

Fig. 5.5. A. Apomictic (ameiotic) thelytoky. B. Amphitoky (male- and female-producing parthenogenesis or deuterotoky) occurring in some insects having the XO sex determination mechanism. In male development, one of the two X-chromosomes is lost from the spindle during mitosis.

genetic variability except mutation and the possible insertion of transposable elements (jumping genes).

5.3 Amphitoky

Amphitoky (male- and female-producing parthenogenesis) resembles apomictic parthenogenesis in that meiosis is functionally mitotic in members of both sexes. Females usually have an XX and males, an XO sex chromosome sytem (4.1). During the single meiotic division of female eggs, both polar body and pronucleus receive two X-chromosomes, whereas in male eggs, one X-chromosome is lost from the spindle, resulting in development of XO (male) individuals (Fig. 5.5B). This type of parthenogenesis is characteristic of many insects having cyclical parthenogenesis.

5.4 Evolutionary Considerations

The kind of parthenogenesis can influence the effects of selection on evolution and diversification.

5.4.1 ARRHENOTOKY
Haplodiploid reproduction is known to be correlated with the following:
- reduced heterozygosity;
- faster rates of evolution;
- rapid loss of recessive lethal mutations (males are hemizygous and their recessive alleles are subject to instant selection);
- exploitation of inbreeding;
- a tendency to develop eusociality (as in ants and in some bees, beetles, and thrips); and

- the advantages but not the disadvantages of genetic recombination (Hartl, 1971; Wrensch, 1993; Crespi et al., 1997).

5.4.2 THELYTOKY
Thelytoky, particularly apomictic thelytoky, is correlated with reduced genetic variability and mating costs and with the ability of single females to establish new colonies (White, 1973; Lokki and Saura, 1980; Suomalainen et al., 1987). Most thelytokous "species" live in open habitats in isolation from closely related bisexual species (Cuellar, 1977). The presence of bisexual parents in a habitat usually prevents newly formed parthenogens from establishing clones, owing to either hybridization or competition, since parthenogens are usually less competitive. Features allowing newly formed parthenogenetic species to invade and occupy open habitats faster than bisexuals might include their doubled intrinsic rate of increase (Fig. 5.1) and the ability of single females to establish new colonies.

With both automictic and apomictic thelytoky, polyploidy of the germ line is common. Such polyploid parthenogenetic species are thought to have a hybrid origin and to result from insemination of a 2n, 4n, and so on, thelytokous female by haploid sperm from males of a related, 2n bisexual species as in certain shortnose weevils (Smith, 1971). In squash preparations of the germ cells of such hybrids, chromosomes contributed by the male can sometimes be recognized because of differences in size, number, shape, and centromere position.

According to Astaurov (1972), such polyploid parthenogenetic races can also sometimes give rise to hybrid, polyploid bisexual species, a common phenomenon in plants but thought to be rare in insects and other animals. He postulated a 4n bisexual species to arise via the following sequence:

2n bisexual species → 2n parthenogenetic race (species)
→ 4n parthenogenetic race through chromosome doubling → production of a 3n parthenogenetic race by hybridization of a 4n parthenogenetic female with a 2n male of an ancestral or related bisexual species → creation of a 4n bisexual species by hybridization of a 3n parthenogenetic female with a diploid male of a related bisexual species

In fact, Astaurov was able to create just such a 4n bisexual species of moth in the laboratory (4n = 112) from two related bisexual silkworm species, *Bombyx mori* (2n = 56) and *B. mandarina* (2n = 58), via an intermediate 3n parthenogen (3n = 84) that, by 1972, had bred true for 14 generations.

Finally, in some thelytokous insects, sperm from males of closely related, sympatric bisexual species are essential for initiating development of the embryo even though the chromosomes they contribute are not used (*pseudogamy* or *gynogenesis*). An excellent example is the triploid pseudogamous female race of the European delphacid leafhopper *Muellerianella fairmairei*, whose embryogenesis is induced by sperm of males of the re-

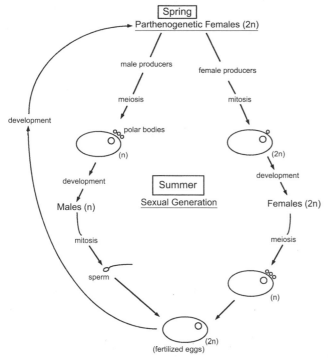

Fig. 5.6. Cyclical parthenogenesis (alternation of generations) in gall wasps (Hymenoptera: Cynipidae).

lated, diploid bisexual form of this species and its sibling species *M. brevipennis* (Drosopoulos, 1976, 1978). Such parthenogenetic leafhoppers are "reproductively parasitic" on males of the bisexual species since they require their sperm to start the next generation but don't use their genes. This may seem a waste of sperm for the male but in fact could facilitate his chances of mating successfully with a conspecific female through a phenomenon called *mate copying*. A female of his own species, witnessing his performance, may preferentially mate with him, as has been demonstrated to occur in sailfin mollies (*Poecilia formosa*—gynogenetic; *P. latipinna*—bisexual) (Schlupp et al., 1994).

The known phylogenetic distribution of thelytoky and amphitoky throughout the hexapods is shown in Fig. 5.12 (2–4).

5.5 Cyclical Parthenogenesis

Species in seven different lineages have independently evolved the advantages of both parthenogenesis (high innate rate of increase) and bisexuality (genetic variability), in which parthenogenetic generations alternate with sexual ones (Hebert, 1987). All such insects either are haplodiploid or have a male XO sex-determining mechanism (White, 1973).

5.5.1 GALL WASPS

Many gall wasp species (Hymenoptera: Cynipidae) have two generations per year, one haplodiploid and the other thelytokous (Fig. 5.6) (Wigglesworth, 1972; Bulmer, 1982; Mable and Otto, 1998). Spring females are thelytokous and either male or female producing. Eggs of the former undergo meiosis and develop into haploid males; those of the latter are apomictic and develop into diploid females of the summer generation; both such males and females reproduce as typical haplodiploids. Members of these two generations can differ in appearance and often form galls on different parts of the same host plant.

5.5.2 APHIDS

Many aphid species (Hemiptera: Aphididae) living at higher latitudes have a male XO sex-determining mechanism and are *holocyclic* (i.e., they have a series of thelytokous generations during the growing season followed by a single sexual generation in early fall) (Fig. 5.7B, C) (Moran, 1992; Hales et al., 1997). Such species generally overwinter as normal fertilized eggs, which then hatch in spring, and develop into wingless, thelytokous, viviparous (pseudoplacental) females, the *fundatrigenia* (singular, fundatrix) (e.g., *Aphis fabae*) (Fig. 5.7C). Fundatrigenia feed on vascular tissues of the primary host plant, usually some tree, and their liveborne offspring rapidly develop apomictically into wingless, thelytokous, viviparous *virginoparae*, whose telotrophic ovarioles are highly adapted for churning out offspring at great speed (low nurse cell ploidy, shortened previtellogenic period, no vitellogenesis or choriogenesis, and up to six embryos at various stages of development within one ovariole at one time). As a result, members of three generations can overlap within the same individual: developing embryos within ovarioles of the female each contain younger embryos (Hardie, 1987; Büning, 1998b). Several overlapping generations of virginoparae follow each other in quick succession, the number varying with species, latitude, food quality, and temperature, so that prodigious numbers of aphids soon infest the primary host.

As their numbers increase, the aphids increasingly contact each other, and this, together with higher temperatures, dietary changes, presence of predators (Weisser et al., 1999), and long days, induces production of *emigrantes*, females similar to virginoparae but with fully developed wings (alate virginoparae). These fly to a secondary host species that is often herbaceous, and apomictically produce additional generations of virginoparae. In late summer and early fall, increasing numbers of aphids induce production of a second, winged generation, the *immigrantes*, which fly back to the primary host. Offspring of these develop parthenogenetically into wingless, viviparous *gynoparae* (sexuparae), which differ from previous generations of females in being amphitokous. The first few oocytes produced by each ovariole in these females develop apomictically into winged or wingless, viviparous, sexually reproduc-

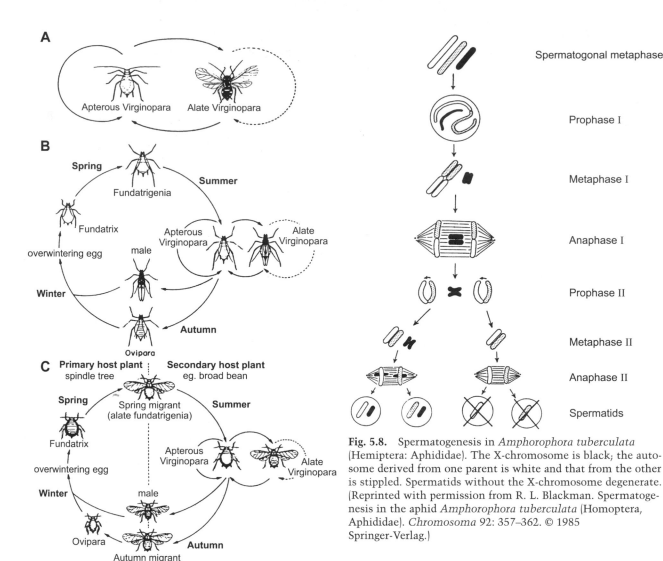

Fig. 5.8. Spermatogenesis in *Amphorophora tuberculata* (Hemiptera: Aphididae). The X-chromosome is black; the autosome derived from one parent is white and that from the other is stippled. Spermatids without the X-chromosome degenerate. (Reprinted with permission from R. L. Blackman. Spermatogenesis in the aphid *Amphorophora tuberculata* (Homoptera, Aphididae). *Chromosoma* 92: 357–362. © 1985 Springer-Verlag.)

Fig. 5.7. Three aphid life cycles (Hemiptera: Aphididae). A. Strict parthenogenesis (e.g., *Myzus persicae*): alate (winged) virginoparae usually produce only apterous offspring but the latter may produce both apterous and alate offspring. B. Alternation between sexual and parthenogenetic generations (e.g., *Megoura viciae*): sexual generations are usually produced in fall, and eggs overwinter in diapause. C. Alternation between sexual and parthenogenetic generations and primary and secondary hosts (e.g., *Aphis fabae*). (Reprinted from J. Hardie and A. D. Lees, Endocrine control of polymorphism and polyphenism. In G. A. Kerkut and L. I. Gilbert (eds.), *Comprehensive insect physiology, biochemistry and pharmacology*, Vol. 8, pp. 441–490, Pergamon, Copyright 1985, with permission from Elsevier Science.)

ing females (*sexuales*), followed by a single oocyte which degenerates. During the single, apomictic meiotic division of subsequently produced oocytes, only the X-chromosomes are reduced, resulting in XO oocytes that subsequently develop into wingless or fully winged males (compare Fig. 5.5B) (Bulmer, 1982; Blackman, 1985b; Blackman and Hales, 1986).

Production of male offspring usually correlates with exposure of gynoparae to shorter photoperiods and lower temperatures in fall while they are still embryos (Searle and Mittler, 1981). Secretion of juvenile hormone (JH) by the corpora allata is reduced below a certain threshold, which in turn results in reduction of the X-chromosomes during the single meiotic division and in production of XO eggs (Fig. 5.5B) (Blackman and Hales, 1986). Application of precocenes (Fig. 12.11I) to virginoparae cause their corpora allata to degenerate, eliminates production of JH, and likewise causes them to produce male offspring (Hales and Mittler, 1983).

In sexual females, meiosis is normal and they produce haploid X-chromosome–containing eggs (Fig. 5.4). In males, the secondary spermatocytes not receiving the X-chromosome at anaphase I are smaller and divide into spermatids, which degenerate so that only X-bearing sperm are produced after meiosis II and spermiogen-

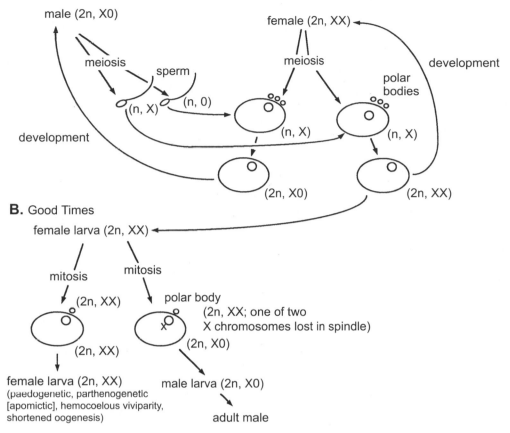

A. Hard Times (Parthenogenetic or Bisexual Adults)

male (2n, X0)

meiosis

sperm

(n, X) (n, 0)

development

female (2n, XX)

meiosis

polar bodies

development

(n, X) (n, X)

(2n, X0) (2n, XX)

B. Good Times

female larva (2n, XX)

mitosis mitosis

(2n, XX) polar body
(2n, XX; one of two
X chromosomes lost in spindle)

(2n, XX)

(2n, X0)

female larva (2n, XX)
(paedogenetic, parthenogenetic
[apomictic], hemocoelous viviparity,
shortened oogenesis)

male larva (2n, X0)

adult male

Fig. 5.9. Cyclical parthenogenesis: various gall midges (Diptera: Cecidomyiidae: *Miastor* and *Heteropeza* spp.).

esis (Fig. 5.8). When males inseminate females, all eggs produced have two X-chromosomes and develop into wingless, thelytokous, oviparous *oviparae* that have typical telotrophic ovarioles and oogenesis (highly polyploid nurse cells, and normal vitellogenesis and choriogenesis) and that deposit normal, fully yolked, and chorionated eggs that overwinter and start the cycle anew the following year (Fig. 5.7C) (Büning, 1998b).

Other aphid species can have life cycles either more or less complicated than this, including those having a single woody host (*autoecy*), a single herbaceous host (autoecy, e.g., *Megoura viciae*) (Fig. 5.7B), or no sexual generation (*anholocycly*, e.g., *Myzus persicae*) (Fig. 5.7A). Moran (1992) and Hales and colleagues (1997) provide convincing discussions of the evolution of such life cycles.

5.5.3 GALL MIDGES

Investigated fungus-feeding cecidomyiids (Diptera) such as *Heteropeza pygmaea* have male XO sex determination and can reproduce bisexually or parthenogenetically

as adults (Figs. 5.9A, 5.10) or parthenogenetically and paedogenetically as larvae (Figs. 5.9B, 5.10), the mode of reproduction depending on food conditions experienced while larvae (Went, 1979). Amphitokous female larvae produce male or female larvae apomictically and paedogenetically, the males resulting from loss of an X-chromosome during the single meiotic division, as in aphids (Fig. 5.9B: right). If conditions are suboptimal, both male and female larvae metamorphose into normal adults (Fig. 5.9A, 5.10: top). Meiosis is normal in both sexes of these adults, the eggs fertilized by O- and X-bearing sperm developing, respectively, into additional males or females, although unfertilized eggs can also develop apomictically into females.

Paedogenetic (larval) females develop when food conditions are ideal. In these, the larval ovaries produce male- or female-determined follicles apomictically that are released into the body cavity (Figs. 5.9B, 5.10) (i.e., they have hemocoelic viviparity—see 3.10). According to sex of progeny, such larvae can be male-mother larvae, female-mother larvae, or male/female-mother lar-

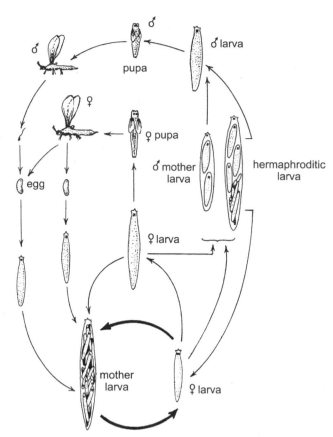

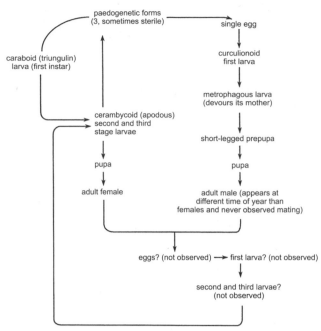

Fig. 5.11. The life cycle of *Micromalthus debilis* (Coleoptera: Micromalthidae). (Adapted with permission from J. A. Pringle. 1938. A contribution to the knowledge of *Micromalthus debilis* LeC. (Coleoptera). *Trans. R. Entomol. Soc. Lond.* 87: 287–290. © 1938 The Royal Entomological Society of London.)

Fig. 5.10. Cyclical parthenogenesis in the gall midge *Heteropyza pygmaea* (Diptera: Cecidomyiidae). The hermaphroditic larva (right) can produce male or female larvae parthenogenetically and paedogenetically (arrows). The male larva pupates and develops into an adult male (top). The female larva can also pupate and develop into an adult (middle of figure). It can also develop into an hermaphroditic larva or into a female mother larva. The latter can produce additional larvae parthenogenetically and paedogenetically. If a female larva pupates, it will deposit fertilized eggs after insemination (left). If it fails to mate, it will deposit eggs parthenogenetically. Both fertilized and unfertilized eggs can develop into female mother larvae. (Reprinted with permission from F. E. Schwalm. *Insect morphogenesis*, p. 83, Fig. 25. Monogr. Dev. Biol. 20. © 1988 S. Karger.)

vae (amphitokous) (Fig. 5.10). Floating in the blood of the mother, the follicles enlarge and their oocytes undergo embryogenesis and larval development, the latter being completed after the mother dies as a hemipupa and her tissues degenerate. Embryogenesis is thus shifted back into oogenesis, greatly increasing the rate of population growth.

The proximate cause of precocious ovarian differentiation in the paedogenetic cecidomyiids *H. pygmaea* and *Mycophila speryeri* was recently shown, by use of tagged monoclonal antibodies to each protein, to be precocious activation of the ecdysone receptor (EcR) and its cofactor

Ultraspiracle (USP) (Hodin and Riddiford, 2000b) (Fig. 12.16). Both proteins are expressed in *Drosophila* ovary rudiments in the late third instar and correlate with the commencement of imaginal ovarian differentiation (Hodin and Riddiford, 1998, 2000a). When these midges are reared under conditions promoting metamorphosis, upregulation of EcR and USP likewise occurs in the final larval instar. However, under optimal conditions, both proteins are upregulated early in the first instar, leading to precocious differentiation of ovaries.

5.5.4 *MICROMALTHUS DEBILIS*
The life cycle of the peculiar haplodiploid beetle *Micromalthus debilis* (Coleoptera: Micromalthidae; its phylogenetic position within the Coleoptera is problematic although it is presently placed within the basal clade Archostemata [Lawrence, 1991]) is more complex than that of any other insect known but has independently evolved close similarities to that of the cecidomyiid flies just discussed (Fig. 5.11) (Pollock and Normark, 2002). The germ line of females is diploid (2n = 20) and that of males haploid (n = 10). Larvae occur in decaying wood of many species of trees infested with brown or white rot fungi (Scott, 1941; Smith, 1971; Künhe, 1972; Künhe and Becker, 1976; Lawrence, 1991). The first instar, *caraboid* larva, is an active, dispersing *triungulin* (i.e., it has well-developed legs, each ending in three apparent claws), which molts into a legless, elongate,

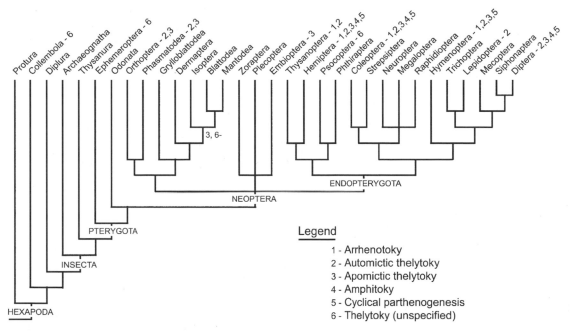

Fig. 5.12. Evolutionary relationships of the hexapod orders as inferred by Gullan and Cranston (1994), showing the known distribution of parthenogenetic species. (Data from various sources. Cladogram adapted with permission from P. J. Gullan and P. S. Cranston. *The insects. An outline of entomology*, p. 193, Fig. 7.5. 1st ed. Chapman and Hall, London. © 1994 Blackwell Science, Ltd.)

feeding second instar, *cerambycoid* larva. This molts once or twice, and then either metamorphoses into an adult diploid female or develops into one of the following three types of paedogenetic (larviform, neotenic) reproductives:

- It can develop into a thelytokous paedogenetic female, producing a number of triungulins viviparously.
- It can develop into an arrhenotokous paedogenetic female, depositing a single large egg that remains attached to the outside of her body. In about 10 d this hatches into a *curculionoid* larva (i.e., it resembles the legless larvae of curculionoids) that inserts its head into her exit system, devours her body contents, and then metamorphoses into a haploid, winged, but apparently nonfunctional adult male. However, if the female is not devoured, she produces a new crop of eggs that develop into females.
- Finally, the paedogenetic larva can develop into an amphitokous paedogenetic female that can produce either form.

Production of these diverse larval types seems to be affected by environmental conditions experienced while burrowing through wood. The last detailed work on this insect was published in 1976 by Künhe and Becker, and I find it amazing that its life cycle has not been investigated in more detail using modern methods. Pollock and Normark (2002) recently discussed the evolution of the life cycle of this beetle in relation to that of other cyclically parthenogenetic insects, and speculate that the species depends on maternally transmitted bacteria for

the ability of its larvae to digest rotten wood. These bacteria are suggested to be senescent in males, causing them to be obligately cannibalistic on their mothers. Male cannibalism increases the cost of males and has strongly selected for development of cyclic thelytoky and for other characteristics of the life cycle minimizing the role of males.

5.6 Microorganisms and Parthenogenesis

Cytoplasmically inherited bacteria of the genus *Wolbachia* occur in perhaps 16% of all insect species and induce female parthenogenesis in thrips and in several dipterans and parasitoid wasps (Stouthamer et al., 1992; Enserink, 1997; Bourtzis and O'Neill, 1998; Hadfield and Axton, 1999; Arabaki et al., 2001). If such females are exposed to tetracycline or high temperatures, these bacteria are killed and normal bisexual or haplodiploid reproduction is restored. Parthenogenesis increases the frequency of such bacteria in a population of susceptible insects by biasing the sex ratio toward females, the transmitting sex, and is thus advantageous for the bacteria (spermatozoa cannot pass them to offspring because they have insufficient cytoplasm). Such bacteria are cytoplasmically inherited through the egg, concentrate in their germ plasm (Hadfield and Axton, 1999), and alter the segregation patterns of host chromosomes during meiosis of unfertilized eggs. For example, unfertilized eggs of parasitoid wasps in the genus *Trichogramma* normally develop parthenogenetically into haploid males (Fig. 5.2). However, if such eggs are infected with *Wolbachia*, chromosomal segregation does not occur during meiosis, resulting in a diploid egg developing parthenogenetically into a female.

The bacteria can apparently jump from one species to another via horizontal transmission, since there is little correlation between *Wolbachia* and host phylogenies, based on comparison of 16S rRNA sequences (Bourtzis and O'Neill, 1998).

The known phylogenetic distribution of parthenogenetic species within the Hexapoda is summarized in Fig. 5.12.

6/ Early Embryogenesis

Once an insect egg is activated, either by sperm entry or by other means, it begins to transform into a fully developed juvenile able to hatch and fend for itself. In this chapter, I summarize the early stages of embryogenesis including cleavage and formation of the blastoderm, germ band, germ layers, segments, and appendages. In Chapter 7, I describe some key experiments executed to reveal how the insect body plan is specified during embryogenesis and recent work on the genes involved, and in Chapter 8, I summarize organogenesis.

Descriptive study of normal embryos is usually based on eggs incubated under a variety of conditions and staged in hours after oviposition (Sander et al., 1985; Schwalm, 1988), though *Drosophila* workers, without exception, use a 17-stage system, with each stage of characteristic length at 22°C or 25°C (Campos-Ortega and Hartenstein, 1997). However, comparison between embryos of different species incubated under differing conditions is facilitated by using percentage of total embryogenesis to stage embryos, with 0% at oviposition and 100% at hatch (Bentley et al., 1979). This approach emphasizes differences in timing in the development of homologous structures and is what I use to stage embryos throughout these chapters.

6.1 Egg Structure

A typical, newly fertilized insect egg is translucent and contains a large amount of yolk, in which the zygote nucleus is suspended (Fig. 6.1A). This nucleus is connected to a thin layer of cytoplasm immediately beneath the oolemma, the *periplasm*, by way of a fine meshwork of cytoplasm, the *cytoplasmic reticulum*, in which the yolk is suspended. The relative proportions of yolk, periplasm, and cytoplasmic reticulum within an egg vary with species and are correlated, very roughly, with type of postembryogenesis. According to Anderson (1972a, 1972b), apterygote and exopterygote insects tend to have relatively large eggs with lots of yolk, a thin periplasm, and a poorly developed cytoplasmic reticulum (Fig. 6.1A, B), whereas eggs of holometabolous insects tend to be relatively small and to have less yolk, a thick periplasm, and a well-developed reticulum (Fig. 6.1C). There are all grades between these extremes and

many exceptions. These differences are also correlated with the type of ovariole in which the eggs developed (2.1.1) and with whether the germ band is short, intermediate, or long at the time of formation (6.7).

Insect egg shells or choria (singular: chorion) usually consist, from inside to out, of a wax layer (generally on the surface of the vitelline membrane but sometimes within the inner chorion), endochorion, trabecular meshwork, and exochorion, the meshwork functioning in gas exchange and the wax layer in reducing desiccation (Fig. 6.1D) (Hinton, 1981; Margaritis and Mazzini, 1998). The details of these layers are species-specific, have been strongly selected to support embryogenesis of eggs deposited in diverse habitats, and have undoubtedly contributed to insect diversification (Legay, 1977; Hinton, 1981; Zeh et al., 1989). The chorion is usually penetrated by three types of perforations: one or more micropyles through which individual spermatozoa pass to fertilize the egg, aeropyles for gas exchange (Fig. 6.1D), and chorionic hydropyles for uptake of water. The number, position, and complexity of these holes vary with the species (2.2.1.6.2).

6.2 Cleavage

Cleavage of the zygote nucleus usually begins at a species-specific site within the egg called the *cleavage center*, usually where the male and female pronuclei fuse during syngamy to form the zygote nucleus (Fig. 3.14). Because of their substantial yolk content, insect eggs are said to be *centrolecithal* (i.e., there is a central yolk mass inside the blastoderm after blastoderm formation) and meroblastic (intralecithal), the cleavage divisions being nuclear not cellular. During cleavage, the egg is transformed into a multinuclear syncytium, its cytoplasmic and yolk components are rearranged by cytoplasmic streaming and contractions of the yolk system, and the ratio of nuclei to cytoplasm is increased (Fig. 6.2). Each daughter nucleus resulting from a division is embedded within an island of cytoplasm continuous with the cytoplasmic reticulum, the cytoplasm of other nuclei, and ultimately, the periplasm and is called a *cleavage energid* (Fig. 6.2: 3.2–4.8).

At first, cleavage divisions are synchronous because all daughter nuclei share a common cytoplasm. In eggs

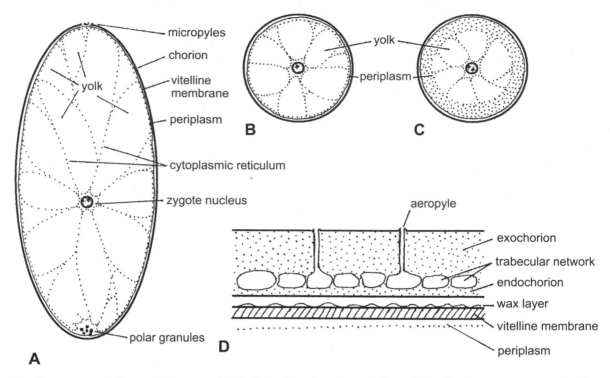

Fig. 6.1. Egg structure in insects (diagrammatic). A. Sagittal section of a typical, newly fertilized exopterygote egg. B. Transverse section of the same, showing cytoplasmic reticulum and thin periplasm. C. Transverse section of the egg of a higher dipteran, showing a thick periplasm. D. Sagittal section through part of the chorion and vitelline membrane of an egg.

of members of advanced lineages (e.g., *D. melanogaster*, Fig. 6.3), this synchrony persists throughout cleavage; in others, it starts to fade (*parasynchrony*) and eventually disappears (*asynchrony*). As they multiply, the energids separate from each other and begin to move apart, eventually reaching the periplasm and entering it to form a syncytial preblastoderm (Figs. 6.2: 5.5; 6.3: 9–14).

The rate of mitosis during cleavage varies with temperature and species and can be very rapid. For example, in *D. melanogaster* eggs, the cell cycle of cleavage divisions 2–9 lasts 8.4–8.8 min at 25°C, with only S (synthesis) and M (mitosis) phases occurring (Foe et al., 1993; Orr-Weaver, 1994; Edgar and Lehner, 1996; see Table IX in Schwalm [1988] for mitotic rates in eggs of other species). This rapid rate of DNA synthesis and chromosome replication is only possible because of the transfer of RNAs, ribosomes, centrosomes, and other materials into the egg from the nurse cells during oogenesis (Fig. 2.13A, B). With each division, the yolk system contracts or twitches, and both yolk and ooplasm are moved about within the egg in ways characteristic for the species (Fig. 6.2). Van der Meer (1988) discussed these movements in relation to changing concentrations of Ca^{2+} ions within the egg.

In known eggs of species of Collembola (Jura, 1972), of viviparous aphids, and of some parasitoid wasps (6.7.2.1) and strepsipterans (6.7.2.2) in which there is little or no yolk, cleavage is total or *holoblastic* and is undoubtedly secondarily derived (Scholtz, 1998).

6.2.1 CONTROL OF CLEAVAGE RATE

Control of the nuclear division cycle during cleavage may be mediated by a mitotic oscillator that is suggested to drive DNA and centrosome replication and chromosome segregation in all eukaryotes (Fig. 6.4A) (Foe et al., 1993; Nasmyth, 1996). In *D. melanogaster* eggs, increasing levels of regulation act on the cleavage divisions as the embryo develops. The first 13 rapid cycles of mitosis (Figs. 6.3: 1–13; 6.4B) rely on gene products contributed by the mother during oogenesis. A maternally derived tyrosine phosphatase is present at sufficient levels to allow progress through the cell cycle (Gerhart and Kirschner, 1997: 66). As the cleavage rate slows in later divisions, decreased levels of this phosphatase limit progress through the cell cycle until differential zygotic phosphatase production begins following blastoderm formation in cleavage 14, causing cell cycles to lengthen to different extents in differing regions of the embryo and resulting in formation of mitotic domains (Fig. 6.10) (Skaer, 1998).

Zygotic regulation is introduced together with a G2 phase in cleavage cycle 14 when the complex of mitotic regulators comes under overall control of the product of the embryo's *string* (*stg*) gene (Fig. 6.5). Expression of zygotic *stg* is necessary for normal subsequent development of the embryo and is one of the first of the embryo's own genes to be expressed after fertilization. Its protein takes over from maternal String (the *Drosophila* homolog of the cell cycle regulator CDC 25),

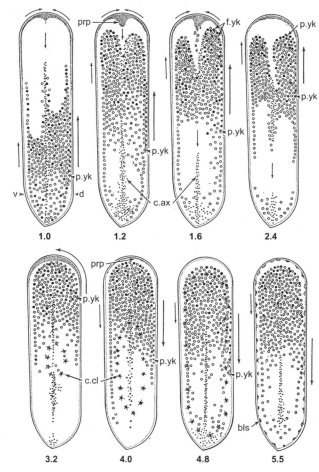

Fig. 6.2. Diagrammatic sagittal sections showing cytoplasmic streaming in the yolk system of a *Lytta viridana* (Coleoptera: Meloidae) egg from 2.5 (1% of embryogenesis) to 14 h (5.5%) after oviposition at 25 ± 0.5°C. Cleavage begins between 2.4% and 3.2% of embryogenesis and the cleavage energids (c.cl) enter the periplasm (prp) to form the preblastoderm (bls) at about 5.5%. c.ax, central axis of egg; d, dorsal; f.yk, lipid yolk; p.yk, protein yolk; v, ventral. (Reprinted with permission from J. G. Rempel and N. S. Church. The embryology of *Lytta viridana* LeConte (Coleoptera: Meloidae). I. Maturation, fertilization and cleavage. *Can. J. Zool.* 43: 915–924. © 1965 NRC Research Press.)

the control of mitotic rate following cell formation at blastoderm, although the former protein is also still involved. The timing of nuclear division begins to vary when the cell membrane invaginates between adjacent nuclei at blastoderm formation (Fig. 6.3: 15). Edgar and Lehner (1996) and Skaer (1998) critically reviewed the subject in *Drosophila* embryos, whereas King and coauthors (1996) emphasized proteolysis; Stillman (1996), control of DNA replication; and Elledge (1996), mitotic checkpoints.

6.2.2 MOVEMENT OF CLEAVAGE ENERGIDS TO THE PERIPHERY

The mechanism by which cleavage energids separate from each other and move to the periphery of an insect

egg during cleavage has been a topic of endless speculation and experimentation (Counce, 1973) and was discussed in detail by Foe and coauthors (1993) for *D. melanogaster* embryos. It is believed to involve elements of the egg's cytoskeleton, particularly the microtubules (Fig. 6.6). At least four kinds of microtubules are associated with the centrosomes of dividing cleavage energids:

- *Kinetochore* microtubules attach the centrosomes to the chromosomes.
- *Spindle* (interpolar) microtubules extend to the equator of the spindle from the centrosomes at either end, surround the chromosomes, and overlap each other at their tips.
- *Astral* microtubules extend into the yolk in all directions from the centrosomes and to the periplasm of the egg.
- A fourth set of microtubules link the centrosomes to the chromosome arms (Sharp et al., 2000).

Kinetochore and spindle microtubules function in segregating the chromosomes at anaphase and telophase and in separating the daughter energids. Most astral microtubules overlap at their tips with microtubules from adjacent energids and are thought to push the energids away from each other by sliding of microtubules over each other. In fact, Raff and Glover (1989) separated the nuclear and cytoplasmic components of cleavage by injecting *aphidicolin*, an inhibitor of DNA synthesis, into cleaving eggs of *D. melanogaster*. This drug inhibits nuclear division but allows centrosome replication and other aspects of the cell cycle to continue. If injected between cleavage cycles 7 and 8 (Fig. 6.3), it completely inhibits the normal migration of nuclei to the periplasm. However, the centrosomes continue to proliferate normally and to migrate in a coordinated way to the periplasm, where they organize elements of the egg's cytoskeleton and overlying plasma membrane. In addition, the centrosomes migrating to the posterior pole of the egg initiate formation of anucleate pole cells (compare Fig. 6.3: 10).

As mentioned earlier (3.6.2), the maternally expressed protein Centrosomin is required for normal assembly and function of centrosomes during cleavage in *D. melanogaster* embryos (Megraw et al., 1999) and hemizygous *male and female-sterile centrosomin* mutant (*cnn^{mfs}*) embryos show multiple spindle anomalies up to cycle 12 and fail to cellularize at the time of blastoderm formation.

In arthropod embryos having holoblastic cleavage, including those of collembolans, viviparous aphids, and parasitoid hymenopterans, this cleavage type is secondarily derived from the meroblastic type; however, the nuclei of the blastomeres, still move out to the periphery of the egg toward the end of cleavage (Scholtz, 1998).

6.2.3 CHROMOSOME ELIMINATION IN EMBRYOS OF LOWER DIPTERA

In eggs of several species of Cecidomyiidae, Chironomidae, and Sciaridae (Diptera), one to many

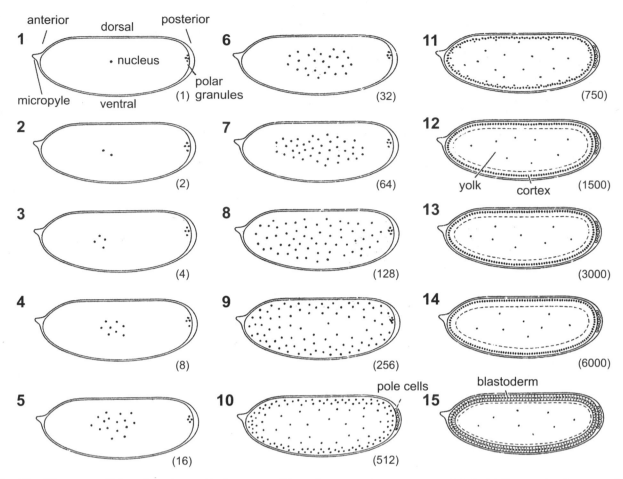

Fig. 6.3. Cleavage and blastoderm formation in an egg of *D. melanogaster* (Diptera: Drosophilidae) following fertilization. The sperm enters the egg through a micropyle in the chorion at its anterior end, and its nucleus unites with the egg nucleus to form the zygote nucleus (1). Nuclear fusion is followed by 13 synchronous, nuclear, cleavage divisions, which double the number of nuclei (cleavage energids) every 8–12 min at 25°C (2–8). After the eighth division (256 nuclei), the energids begin to migrate to the periphery of the egg (9), and after the ninth, when there are 512 nuclei, some of those entering the posterior pole plasm of the egg take up polar granules in their cytoplasm and are surrounded by cell membranes to form the pole cells (10). The other nuclei continue to divide synchronously within the periplasm of the egg (11–14) until there are about 6000, at which time the cell membrane invaginates between adjacent nuclei to form a monolayer of cells, the blastoderm (15). Nuclei remaining in the yolk or re-entering it from the blastoderm become vitellophages, which function in cytoplasmic streaming and mobilize yolk for subsequent use by the developing embryo. (Reprinted with permission from W. J. Gehring. The molecular basis of development. *Sci. Am.* 253(4): 153–162. © 1985 Scientific American.)

chromosomes are lost during cleavage of most nuclei (Haget, 1977; Perondini, 1998). For example, in eggs of the cecidomyiid *Mayetiola destructor* (2n = 40), following the fourth cleavage division, two energids enter the posterior pole plasm, take up polar granules in their cytoplasm, and are budded off as pole cells. During the fifth division (it cannot be switched to a different cleavage division by experimental manipulation), these cells do not divide, whereas the other 14 energids do but in such a way that 31–34 (usually 32) chromosomes remain on the equator of the spindle, fail to enter the daughter nuclei, and degenerate in the yolk (these are called the E [elimination] *chromosomes*; the 8 chromosomes remaining are called S [somatic] *chromosomes*). In the sixth cleavage division, all nuclei, including those of the

two pole cells, divide normally, but only progeny of the latter still contain the original 40 chromosomes.

If the posterior energids are prevented from interacting with polar granules after the fourth cleavage, they too lose their E-chromosomes during the fifth division, and as Bantock (1970) demonstrated, the E-chromosomes retained in the pole cells contain genes essential for normal oogenesis and spermatogenesis.

The somatic cells ultimately descending from most energids can be looked on as altruists within the embryo and the pole cells, as recipients (Darlington, 1980: 149). This altruism is potentially reciprocal, since any cleavage energid can become either a pole or a somatic cell, depending on where it enters the periplasm. All cells profit from the altruism as all contain the same genes,

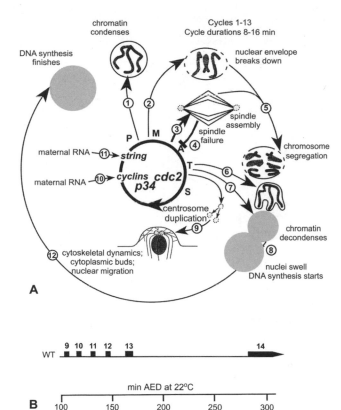

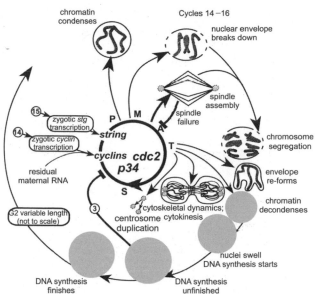

Fig. 6.5. The cell cycle during cellular divisions 14–16. (Reprinted with permission from V. E. Foe, G. M. Odell, and B. A. Edgar. Mitosis and morphogenesis in the *Drosophila* embryo: point and counterpoint. In M. Bate and A. Martinez Arias (eds.), *The development of* Drosophila melanogaster. Vol. 1, p. 280, Fig. 20. © 1993 Cold Spring Harbor Laboratory Press.)

Fig. 6.4. A. Action of a mitotic oscillator during cleavage divisions 1–13 (see Fig. 6.3) of a *D. melanogaster* embryo as postulated by Foe and colleagues (1993). In both this figure and in Fig. 6.5, the central loop represents a biochemical oscillator triggering cyclic changes in the structure of the nucleus and cytoskeleton during each cleavage division. B. Timing of cleavage divsions 9–14 in a wild-type (WT) embryo shown in minutes after egg deposition (AED) at 22°C. The black boxes are mitoses. Notice the large increase in the length of interphase that correlates with formation of cell membranes between adjacent blastoderm nuclei after cleavage division 13. (A and B adapted with permission from V. E. Foe, G. M. Odell, and B. A. Edgar. Mitosis and morphogenesis in the *Drosophila* embryo: point and counterpoint. In M. Bate and A. Martinez Arias (eds.), *The development of* Drosophila melanogaster. Vol. 1, p. 240, Fig. 11, and p. 280, Fig. 20. © 1993 Cold Spring Harbor Laboratory Press.)

though not in the examples mentioned above, and this altruism increases the chances their genes will be represented in the next generation.

6.2.4 VITELLOPHAGES

Not all cleavage energids enter the periplasm to form blastoderm. Those lagging behind in the yolk, in *D. melanogaster* embryos, become primary vitellophages in cleavage cycle 10 and change their cell cycle from S and M to S and G, the "endo cell cycle," in which DNA and chromosomal replication continue but not mitosis. Their nuclei become polyploid and their centrosomes tend to separate from them, perhaps contributing to the

change in their cell cycle (Foe et al., 1993). By division cycle 14, between 175 and 225 vitellophages are present within the yolk, and in eggs of some insects, including those of *D. melanogaster*, additional secondary vitellophages later leave the blastoderm and re-enter the yolk (Fig. 6.3: 15).

Histological evidence suggests that the vitellophages function in mobilizing yolk for the developing embryo (Yamashita and Indrasith, 1988; Fausto et al., 1994, 1997), but in embryos of the German cockroach *Blattella germanica*, yolk vitellins instead are processed, at least partly, by a yolk-borne cysteine proteinase derived from a proenzyme identified immunocytochemically in the fat body, follicle cells, and cortical cytoplasm of egg chambers in the mother (Giorgi et al., 1997).

Vitellophages are known to form the midgut epithelium in embryos of some apterygotes (Mori, 1983) and the large yolk cells during yolk cleavage in some exopterygote and endopterygote embryos (Sander, 1976a).

6.3 Blastoderm Formation

Invagination of oolemma between adjacent nuclei in the periplasm occurs simultaneously in eggs of some insects (e.g., *D. melanogaster*, Fig. 6.3: 15), but as isolated patches that gradually spread in those of others (Johannsen and Butt, 1941; Hagan, 1951; Ando, 1962; Anderson, 1972a, 1972b, 1973; Haget, 1977; Sander

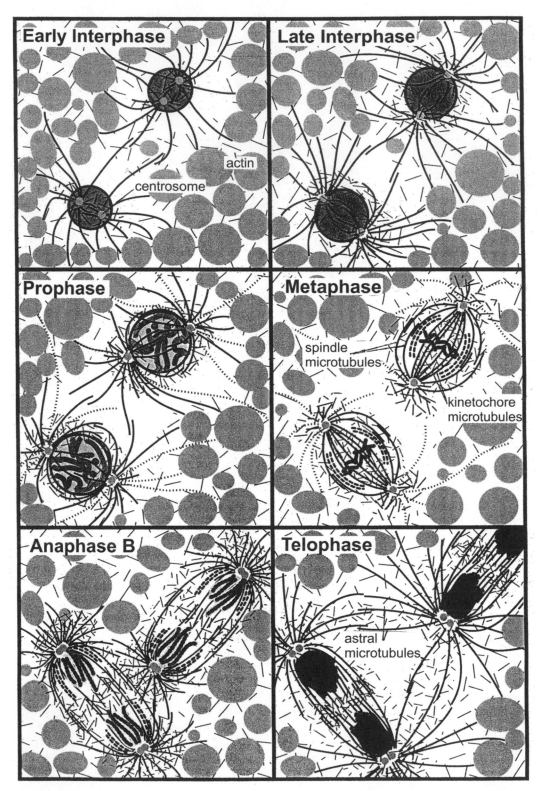

Fig. 6.6. The cell cycle in a pair of dividing cleavage energids within an egg of *D. melanogaster*, showing changes in the number and distribution of centrosomes, of kinetochore, spindle, and astral microtubules, and of actin as mitosis proceeds. Notice that the pattern of overlap between the tips of the microtubules changes and that the centrosomes have divided in anaphase B and telophase in preparation for the next cleavage division. (Reprinted with permission from V. E. Foe, G. M. Odell, and B. A. Edgar. Mitosis and morphogenesis in the *Drosophila* embryo: point and counterpoint. In M. Bate and A. Martinez Arias (eds.), *The development of* Drosophila melanogaster. Vol. 1, p. 172, Fig. 4. © 1993 Cold Spring Harbor Laboratory Press.)

et al., 1985; Ando and Miya, 1985; Ando and Jura, 1987; Schwalm, 1988; Campos-Ortega and Hartenstein, 1997). And in embryos of the locust *Schistocerca gregaria*, individual cleavage energids develop cellular membranes prior to blastoderm formation, as indicated by an end to their uptake of rhodamine dextran from the yolk (Ho et al., 1997).

As might be expected, an incredible amount of new cell membrane is synthesized at this time: a 23-fold increase in 1 h at 25°C around 6000 blastoderm nuclei in *D. melanogaster* eggs, according to Foe and colleagues (1993). This membrane may originate, at least partially, by withdrawal of microvilli from the apices of the nucleus-containing buds, which ultimately form the blastoderm cells; these are present during syncytial blastoderm of many insect eggs. Deepening of furrows between adjacent nuclei involves formation and contraction of rings of actin filaments at the bottom of each furrow (Fig. 6.7).

The cyclic reorganization of filamentous actin and microtubules in embryos of *D. melanogaster* during blastoderm formation was recently examined by Foe's group (2000) using drugs, time-lapse video microscopy, and laser-scanning confocal microscopy, which revealed some spectacular, multiprobe, three-dimensional reconstructions. The completed blastoderm (Fig. 6.3: 15) is the equivalent of the *blastula* (hollow) or *blastosphere* (solid) of other animal embryos.

6.4 DNA, RNA, and Protein Synthesis during Cleavage and Blastoderm Formation

Both DNA and protein synthesis can be detected immediately following fertilization in insect eggs, but that of RNA does not commence at high levels until preblastoderm or later (Fig. 6.8) (Berry, 1982; Andéol, 1994). Zygotic transcripts detected at very low but measurable levels earlier than this in insect (Fig. 6.8) and other animal embryos seem to promote later, zygotic gene expression by degrading specific maternal transcripts interfering with this transition (Andéol, 1994: Table 1).

DNA synthesis commences with fertilization and occurs during the S phase of the cell cycle of dividing cleavage energids, as can be demonstrated by monitoring the uptake of [3H]thymidine autoradiographically (Fig. 6.8). Isotope-labeled amino acids are also taken up by an egg immediately following fertilization (Fig. 6.8), whereas injection of puromycin or cycloheximide or other inhibitors of protein synthesis into a cleaving egg stops development immediately.

However, injection of actinomycin D, alpha-amanitin, tetracycline, or other inhibitors of RNA synthesis into a cleaving egg does not obviously affect development adversely until just before blastoderm formation or later, while extensive uptake of [3H]uridine by an embryo only begins at preblastoderm or later (Fig. 6.8).

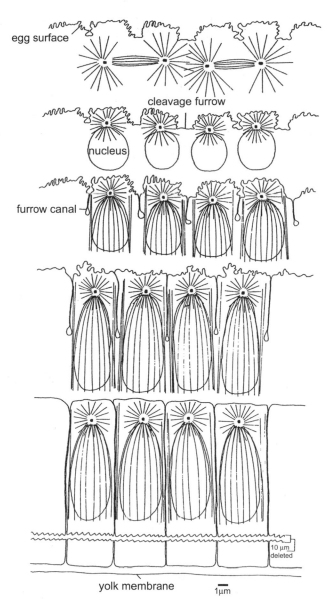

Fig. 6.7. Blastoderm formation in an embryo of *Drosophila montana*. Notice the microvilli in the egg plasmalemma above each nucleus and how these disappear as furrow formation separates adjacent nuclei to form the blastoderm. Notice also how the nuclei greatly enlarge during cell formation. (Reproduced with permission from S. L. Fullilove and A. G. Jacobson. 1971. Nuclear elongation and cytokinesis in *Drosophila montana*. *Dev. Biol.* 26: 560–577. © 1971 Academic Press, Inc.)

That so little RNA synthesis occurs in cleaving eggs suggests that protein synthesis at this time takes place on maternal transcripts and ribosomes synthesized during previtellogenesis (Fig. 2.5; 2.2.1.1.3 and 2.2.2.3.1). Schmidt (1980), in fact, isolated by micromanipulation the periplasm of newly fertilized leafhopper (*Euscelis plebejus*) eggs and found it to contain transcripts labeled with [3H]uridine taken up and contributed by the nurse cells during previtellogenesis (this insect has telotrophic ovarioles).

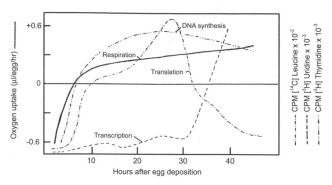

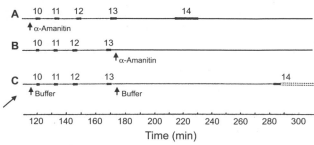

Fig. 6.8. Respiration and DNA, RNA, and protein synthesis during early embryogenesis of tobacco hornworm *Manduca sexta* (Lepidoptera: Sphingidae) eggs. Initial rates of DNA, RNA, and protein synthesis were determined with unfertilized eggs removed from the ovary and were measured by direct injection of labeled precursors into developing eggs. Respiration was measured with a Warburg respirometer. Syncytial blastoderm formation occurs at 20–24 h and gastrulation at 30–35 h after oviposition. CPM, counts per minute. (Adapted with permission from S. J. Berry. RNA synthesis and storage during insect oogenesis. In L. Browder (ed.), *Developmental biology. A comprehensive synthesis.* Vol. 1. *Oogenesis*, p. 352, Fig. 1. Plenum Press, N.Y. © 1985 Kluwer Academic/Plenum Publishers.)

Fig. 6.9. Effects of injecting alpha-amanitin, an inhibitor of RNA synthesis, into cleaving eggs of *D. melanogaster*. Heavy lines with numbers indicate cleavage mitoses. A. When alpha-amanitin is injected early enough to block all zygotic transcription, the maternal division pattern continues, resulting in a premature and defective mitosis 14. B. Alpha-amanitin injected at a time when zygotic *string* RNA levels are very low blocks subsequent mitoses. C. The normal time course of cleavage mitosis. Control injections of buffer early or late in cleavage did not disturb this time course. The dotted line segments to the right indicate the asynchrony of cleavage mitosis 14 that occurs following cellularization of the blastoderm and appearance of mitotic domains (see Fig. 6.10). The time scale indicates time after egg deposition at 22°C. (Reprinted with permission from P. H. O'Farrell, B. A. Edgar, K. Lakich, and C. F. Lehner. Directing cell division during development. *Science* 246: 635–640. Copyright 1989 American Association for the Advancement of Science.)

These observations suggest that embryogenesis up until preblastoderm in insect eggs is under the control of products of the maternal genome and that expression of zygotic genes only commences when cleavage energids enter the periplasm. This change in gene expression is best studied in *D. melanogaster* eggs, where, after fertilization, the embryo undergoes 13 rapid, synchronous mitotic cycles (Fig. 6.3: 1–13) controlled maternally by a mechanism reading the nuclear-cytoplasmic ratio (O'Farrell et al., 1989). This causes the cycles to lengthen progressively after the ninth cleavage division and to lapse into an extended interphase in cycle 14 (Fig. 6.9C). Transcription of zygotic genes is first detected in the brief interphases of cycles 11–13 and increases dramatically in the extended 14th interphase when cellularization occurs. Zygotic transcription can be induced precociously by arresting development in interphase 10 or later with cycloheximide or alpha-amanitin (Fig. 6.9A, B), suggesting that mitosis precludes transcription in early cycles, with zygotic transcriptional activation resulting from the extended interphase of cycle 14.

As mentioned earlier (6.2.1), mitosis in early cleavage is controlled by *maternal* Stg protein (Fig. 6.4A) and after mitosis 14, by *zygotic* Stg (Fig. 6.5). O'Farrell and colleagues (1989) suggested that transition to zygotic control might be initiated by a titration mechanism progressively slowing the later cleavage stages to allow zygotic gene expression to commence. In addition, transition to zygotic control can be provoked by selective destabilization of certain maternal transcripts, in particular that of the *stg* gene.

The stage at which the maternal morphogenetic program is replaced by the zygote's is characteristic for members of each species of animal and is called the *midblastula transition* (O'Farrell et al., 1989). In long germ eggs, as in *D. melanogaster*, it is the preblastoderm stage; in short germ eggs, the late protopod embryo (Fig. 6.22A); in annelids, the trochophore larva; and in vertebrates, the neurula. The subject was reviewed by Andéol (1994).

6.5 Mitotic Domains in the Blastoderm of Flies

As mentioned in 6.2.1, all nuclei within the cleaving egg of *D. melanogaster* divide synchronously during the first 13 cleavage cycles (Fig. 6.3). But with cellularization of the embryo prior to cycle 14, this synchrony ends. At about 1 h into cleavage cycle 14 at 25°C, *mitotic domains* (clusters of cells having locally synchronous mitosis) begin to partition the blastoderm sequentially into a complex, repeatable pattern (Fig. 6.10) (Foe, 1989; Foe and Odell, 1989; Orr-Weaver, 1994; Campos-Ortega and Hartenstein, 1997). These domains were discovered by staining dechorionated embryos with a fluorescently labeled antibody to tubulin in the spindle microtubules of dividing cells. When so labeled and viewed by fluorescence microscopy, the spindles appear bright and enhance the limits of each domain because of their similar appearance in each. Domains were found to be constant in position, order of appearance, and number from one embryo to another, not only in *D. melanogaster* but also in other flies including the sciarid midge *Rhynchosciara americana* (Carvalho et al., 1999), and their cells were found to divide in the same temporal sequence. The length of phase G2 differs in cells of different domains owing to differential expression of zygotic Stg protein (compare Fig. 6.5) (Skaer, 1998). Some domains consist of single cell clusters straddling the dorsal or ventral midline (e.g., Fig. 6.10: 3, 8, 18, 20), others form paired clusters on either side of the midline (1, 5, 6, etc.), while a few comprise a series of paired metameric (segmental) repeats (11, 21, etc.) on each side (domains are numbered in the sequence in which mitosis is initiated within them). Each domain also occupies a specific position along the anteroposterior axis of the blastoderm in relation to expression of the Engrailed protein (in cells of the posterior half of each segment for a total of 14 transverse bands) and along the dorsoventral axis, as determined by cell counts from the ventral midline after gastrulation. In addition, the cells of each domain or domain pair share specific morphogenetic traits (cell shape, spindle orientation during mitosis, participation by all cells of a domain in invagination, etc.) and later give rise to primordia of particular larval structures. Additional mitotic domains appear in older *Drosophila* embryos during gastrulation and the second and third postblastoderm mitoses (Campos-Ortega and Hartenstein, 1997: chapter 14).

Mitotic domains constitute the first visible manifestation of cell commitment other than pole cell specification and resemble the blastodermal fate maps constructed by earlier workers who followed, to time of hatch, the final position of defects caused by experimental injury to the blastoderm (7.1.4).

6.6 Pole Cell Formation

In eggs of many insects, but particularly in those having meroistic ovarioles and complete metamorphosis, the germ-line cells are segregated just before blastoderm formation (Fig. 6.3: 10) (Johannsen and Butt, 1941; Hagan, 1951; Anderson, 1972a, 1972b, 1973; Haget, 1977; Sander et al., 1985; Ando and Miya, 1985; Ando and Jura, 1987; Schwalm, 1988; Campos-Ortega and Hartenstein, 1997). At the posterior pole of the egg are *polar granules* (germ tract determinants, oosome) (Figs. 2.7E; 6.1A; 6.3: 1) consisting, in eggs of *D. melanogaster*, of the transcripts and proteins of various genes (Fig. 6.11) (Mahowald, 1992; Strøme, 1992). These granules differentiate during oogenesis (2.2.1.3) and are taken up by the cytoplasm of the cleavage energids entering this part of the periplasm (Fig. 6.3: 9, 10) (the number of energids that enter varies with the species). Once containing this pole plasm, such cells are determined to form germ cells (GCs) and can be recognized as pole cells, the first cells to form in the embryo, segregating well before blastoderm formation (Fig. 6.3: 10). Eggs having polar granules and pole cells are said to have *preformistic* GCs (Nieuwkoop and Sutasurya, 1981).

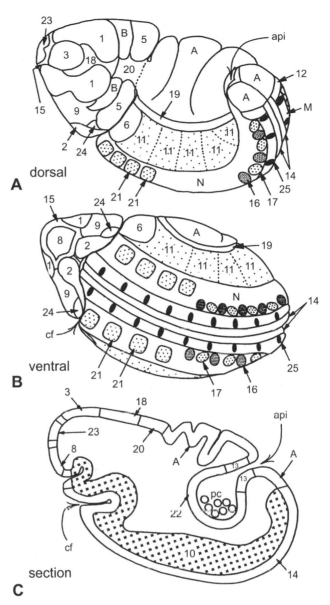

Fig. 6.10. Map of mitotic domains during division cycle 14 at the beginning of gastrulation and germ band elongation (stage 7; 14% of embryogenesis) in an embryo of *D. melanogaster* as seen dorsolaterally (A), ventrolaterally (B), and in midsagittal section (C); head is to the left and dorsal is up. The domains are numbered according to the sequence in which they enter mitosis. However, at this stage of development, only the first seven domains have entered mitosis (later-dividing domains are projected back to the positions their cells had at this stage). N and M denote two domains of asynchronously dividing cells and A and B, of those nondividing cells. Domain 10 (in C) is mesoderm undergoing gastrulation. A, amnioserosa; api, amnioproctodeal invagination; cf, cephalic furrow; pc, pole cells. (Reproduced with permission from V. E. Foe and G. M. Odell. Mitotic domains partition fly embryos, reflecting early cell biological consequences of determination in progress. *Am. Zool.* 29: 617–652. © 1989 J. S. Edwards and the American Zoologist and courtesy of Virginia Foe and Garrett Odell.)

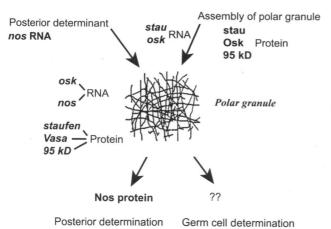

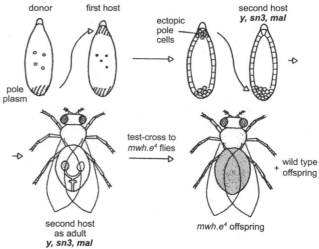

Fig. 6.11. Components of the polar granule in a newly fertilized egg of *D. melanogaster*. (Reprinted with permission from A. P. Mahowald. Germ plasm revisited and illuminated. *Science* 255: 1216–1217. Copyright 1992 American Association for the Advancement of Science.)

Fig. 6.12. The experiment of Illmensee and Mahowald (1974) demonstrating the role of the pole plasm in specifying the pole cells during *D. melanogaster* embryogenesis. Pole plasm from a wild-type donor embryo is injected into the anterior end of a cleaving egg of a different genotype (*mwh. e⁴*). After blastoderm formation, the resulting anterior pole cells are transplanted to a posterior position in a second host egg of different genotype (*y, sn3, mal*). Some of these embryos develop with a chimeric germ cell population and when test crossed to *mwh. e⁴* flies, produce double recessive offspring, proving that some of the nuclei from the first host have become incorporated into the germ line of the second. (From J. M. W. Slack. 1983. *From egg to embryo. Determinative events in early development*, p. 84, Fig. 4.8. Developmental and Cell Biology 13. Reprinted with the permission of Cambridge University Press.)

In the eggs of females of basal and intermediate lineages, polar granules are absent or do not take up conventional stains, or the GCs do not become apparent until later in embryogenesis, often at about the time the germ band forms, or they emerge even later from mesoderm in paired intersegmental clumps on either side of the midline. Such embryos are said to have *intermediate* GCs (Nieuwkoop and Sutasurya, 1981).

If these granules are removed experimentally prior to the arrival of energids, by ligature or microcautery, or if they are rendered ineffective by ultraviolet (UV) irradiation, adults developing from these eggs are sterile (their gonads form but are devoid of sperm or eggs), suggesting that the polar granules or their surrounding cytoplasm contain germ-line determinants. Also, if this pole plasm is removed by a glass microcapillary from a donor egg before it has begun to develop and is injected into the anterior pole of another host egg at the same stage of development, pole cells form at both the anterior and posterior poles of that embryo prior to blastoderm (Fig. 6.12).

In 1986, Technau and Campos-Ortega devised a technique for labeling all cells of developing, donor *D. melanogaster* embryos by injecting horseradish peroxidase (HRP) into them just before blastoderm formation. After cellularization, individual, labeled pole cells of such donor embryos were removed with a glass capillary and injected into unlabeled host eggs of various age among its pole cells. There, the labeled, transplanted pole cells developed together with the unlabeled pole cells of the host. Host eggs were then fixed and stained at various times with tagged anti-HRP to demonstrate the presence of HRP in the progeny of the transplanted cells. Using this technique, they found the gonads to be the only organs to which the transplanted pole cells contributed; the pole cells not entering the gonad degenerated in the yolk. They also found the number of GCs accumulating in each gonadal rudiment to be the same

(7–13) regardless of the number of pole cells originally forming (23–52), suggesting some mechanism to regulate the final number of GCs encapsulated within each gonad. When they injected large numbers of additional, labeled pole cells among the pole cells of host eggs, they still ended up with the same final number of GCs, both labeled and unlabeled, within the gonads. Finally, the GCs were found to divide up to two times between their segregation and encapsulation in the gonadal rudiments but to not do so again until after hatch.

The same occurs in embryos of the exoptergote phlaeothripid thrips *Haplothrips verbasci*, where male GCs (haploid) are much smaller than female GCs (diploid) (Heming, 1979). In addition, each male gonad of this insect, at the time of formation, contains significantly more GCs (a mean of 13) than a female gonad (a mean of 7), suggesting an additional, sex-limited control over GC enclosure. A similar sexual difference in GC number was demonstrated in newly formed gonads of *D. melanogaster* embryos (Poirié et al., 1995).

6.6.1 GENES AND POLE CELL FORMATION IN *DROSOPHILA MELANOGASTER* EMBRYOS

In 1992, Ephrussi and Lehmann showed that the maternal *oskar* (*osk*) gene directs pole cell assembly in

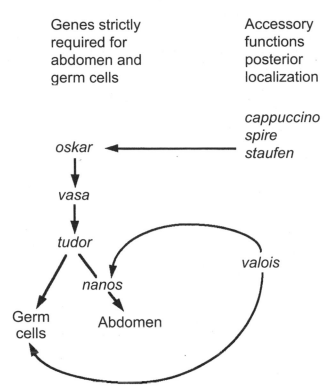

Genes strictly required for abdomen and germ cells

Accessory functions posterior localization

cappuccino
spire
staufen

oskar ←————————

vasa

tudor

valois

nanos

Germ cells

Abdomen

Fig. 6.13. The expression sequence of some genes involved in specifying the germ cells and abdomen during *D. melanogaster* embryogenesis. (Adapted with permission from *Nature* [A. Ephrussi and R. Lehmann. Induction of germ cell formation by *oskar*. 358: 387–392]. © 1992 Macmillan Magazines Limited.)

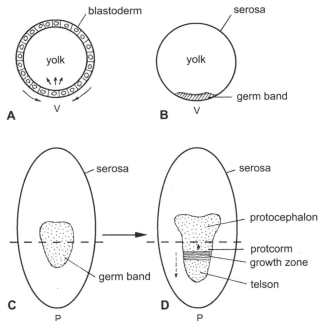

Fig. 6.14. Summary of germ band formation in an insect having a short germ embryo. A and B. Transverse sections showing how the yolk system contracts away from the poster-oventral side of the egg, creating a space into which blastoder-mal cells on either side crowd in to form the germ band. Blastodermal cells remaining around the yolk attenuate to form the serosa. C. The newly formed short germ embryo in ventral aspect. D. Same, a little later, showing the anterior protocephalon, the posterior protocorm, the telson, and the growth zone at the anterior end of the protocorm, which by its mitotic activity generates germ band cells in front of it and pushes itself posteriorly during germ band elongation.

D. melanogaster embryos. By mislocalizing *oskar* (*osk*) mRNA to the anterior pole of the egg, they showed it to be necessary and sufficient to direct formation of the abdomen and of functional pole cells at this ectopic site (Fig. 6.13). They also demonstrated that of the seven other genes (*cappuccino*, *spire*, *staufen*, *vasa*, *valois*, *mago nashi*, and *tudor*) required for pole cell formation at the posterior pole, only *vasa* and *tudor* are essential for *osk*-induced pole cell and abdomen formation. Finally, they showed the amount of *osk* product to determine the number of GC components recruited and the number of pole cells formed: embryos with a wild-type dose produced 10–15 pole cells and those with four copies of *osk*, 40–60.

Expression of the other genes seems to be required to localize the germ tract determinants to the posterior pole of the egg. The products of *osk*, *vasa*, and *tudor* probably provide a scaffold, perhaps in the form of polar granules (Fig. 6.11), for the abdominal signal protein Nanos and for a germ-line signal. The subject was reviewed by Wylie (1999).

6.6.2 SUBSTITUTIVE EMBRYOGENY AND THE ORIGIN OF GERM CELLS IN ORTHOPTEROID EMBRYOS
The GCs, in embryos of species in more basal lineages, do not appear until much later in embryogenesis. For example, in certain stick insects such as *Carausius morosus* (Phasmatodea), the embryo forms in the usual way (6.7) and has paired groups of GCs in some of its abdominal segments (Cavallin, 1971). However, after the embryo enters the yolk at anatrepsis (6.9.1), it degenerates, to be replaced by a second embryo differentiating from *serosal cells* (the layer of blastoderm cells remaining about the yolk after the embryo forms) (Fig. 6.14A, B). This second embryo also develops GCs that, with subsequent development, eventually differentiate into normal viable eggs. The same occurs when an embryo of the cricket *Acheta domesticus* is irradiated during gastrulation (6.8); this embryo degenerates and a new one that is formed from serosa subsequently develops into a normal larva with a normal complement of GCs (Schwalm, 1965).

6.6.3 GERM CELL ORIGIN IN POLYEMBRYONIC INSECTS
Another peculiar example concerns the problematic origin of GCs in polyembryonic insects (6.7.2). Each tiny (25 μm) egg of the parasitoid wasp *Copidosoma floridanum* (Hymenoptera: Encyrtidae) contains polar

granules at its posterior end but has the potential to develop into over 2000 embryos in the body of the host caterpillar *Trichoplusia ni* (Lepidoptera: Noctuidae) (Strand and Grbíc, 1997). Because it lacks yolk, its cleavage is holoblastic, the first division resulting in two equal-sized blastomeres. In the second cleavage division, one blastomere forms two equal daughter cells and the other forms one large and one small blastomere, the latter containing the polar granules. As summarized in 6.7.2.1, subsequent mitosis converts the embryo into an increasingly large mass of totipotent blastomeres surrounded by a multinucleate, enveloping membrane derived from the polar body cast off during meiosis and collectively called the *primary morula* (Fig. 6.16). As the number of blastomeres in the mass increases, the polar granules are continually allocated into progeny of the small blastomere but quickly become too dispersed to visualize. The membrane then intrudes into the mass and divides it into thousands of *secondary morulae*, each developing into a "precocious" or a "reproductive" larva, with each of the latter eventually developing into an adult wasp (Fig. 6.16).

With tagged antibodies to the *Drosophila* Vasa protein (Fig. 6.13), Strand and Grbíc were only able to follow dispersal of polar granules into six blastomeres in an early primary morula. Nevertheless, each of the thousands of reproductive larvae eventually generated by that single egg contains GCs in its gonads and eventually metamorphoses into an adult with fully developed gametes.

6.7 Germ Band Formation

After the blastoderm has formed, the yolk system contracts away from the ventral side of the egg, creating a space into which ventrally and laterally situated blastoderm cells crowd to form a thickened plate, the *germ band* (germ anlage, embryonic primordium), a mode of formation constituting a shared derived character for arthropods (Fig. 6.14A–C) (Johannsen and Butt, 1941; Hagan, 1951; Ando, 1962; Weber, 1966; Anderson, 1972a, 1972b, 1973; Haget, 1977; Sander et al., 1985; Ando and Miya, 1985; Ando and Jura, 1987; Schwalm, 1988; Kobayashi and Ando, 1988; Campos-Ortega and Hartenstein, 1997; Scholtz, 1998). The cells of this band proliferate and differentiate into the embryo while the remaining, extraembryonic serosal cells about the yolk cease dividing, attenuate, and polyploidize their nuclei through endomitosis.

The size of the germ band relative to the egg at the time of formation is characteristic for each species (Krause, 1939). Short (Figs. 6.14C; 6.22A) and semi-long (intermediate) germ eggs are characteristic of ectognath apterygote, exopterygote, and basal endopterygote insects, while long germ eggs occur in collembolans and diplurans and in members of intermediate and advanced endopterygote clades. All three types are known in representatives of other arthropod classes, in Heteroptera, Coleoptera, and Hymenoptera, and perhaps in other insect orders, although the embryos of too few representatives have been examined to know for sure (Johannsen and Butt, 1941; Hagan, 1951; Ando, 1962; Cobben, 1968; Anderson, 1972a, 1972b, 1973; Haget, 1977; Sander et al., 1985; Ando and Miya, 1985; Ando and Jura, 1987; Schwalm, 1988; Kobayashi and Ando, 1988; Patel et al., 1994; Strand and Grbíc, 1997; Ikeda and Machida, 1998; Scholtz, 1998). Long germ eggs are probably ancestral in entognath, apterygote hexapods (members of the four myriapod classes also have long germ embryos) but derived in pterygotes. All intermediates between the two exist (Fig. 6.24).

Differences in germ band size relate to degree of development of gnathal (mouthpart), thoracic, and abdominal portions of the germ band at the time of germ band formation, that portion developing into the anterior head (acron + antennal and intercalary segments) being relatively the same in all three (Fig. 6.22A, C). In long germ eggs, proportions of the germ band for head, thorax, and abdomen are about the same as those of the juvenile at hatch (Fig. 6.22C). In short germ eggs, the germ band is heart- or keyhole-shaped (Figs. 6.14C, D; 6.22A) while semi-long embryos are intermediate in length (Fig. 6.20A). The gnathal, thoracic, and abdominal segments of short germ embryos arise during germ band elongation partly from mitosis within a growth zone in front of the telson near the posterior end of the embryo (Figs. 6.14D, 6.22A) and partly from cell rearrangement. This zone buds off segments anteriorly and is pushed posteriorly by its own mitotic activity, a mode of germ band elongation probably derivative in arthropods (Scholtz, 1998).

6.7.1 FACTORS AFFECTING ORIENTATION OF THE GERM BAND IN PENTATOMID (HETEROPTERA) EGGS
Pentatomid females deposit their eggs in batches of 10–100 placed in two or more rows (Fig. 6.15A) (Lockwood and Story, 1986b). When larvae hatch, they generally stay together until they molt into the second instar, a behavior known to positively influence their survival (Lockwood and Story, 1986a). During their development, the germ bands of eggs at the periphery of a batch originate on the side of each egg facing the center of the batch (Fig. 6.15B), while those within the batch originate on the posterior side of each egg (caption to Fig. 6.15A). By means of a series of simple but elegant experiments, Lockwood and Story (1986b) demonstrated convincingly that an adhesive produced by the female's accessory glands and used to stick the eggs to the substrate during oviposition covers only the inner face of peripheral eggs and both physically (perhaps by reducing gas exchange through the chorion on that side) and chemically induces germ band formation on that side during the first 12 h following oviposition (Fig. 6.15B). Germ bands of eggs within the batch and completely surrounded by glue seem not to be so influenced. These researchers suggested that chemicals within the glue could diffuse into the peripheral eggs and induce an

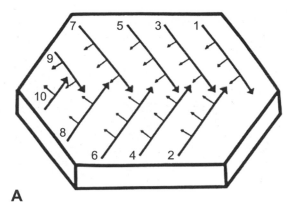

A

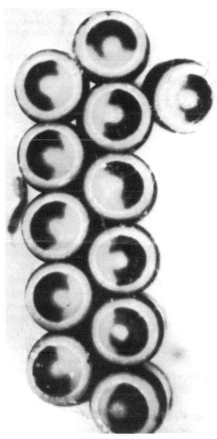

B

Fig. 6.15. The effect of ovipositional secretions on germ band formation in eggs of the pentatomid *Nezara viridula* (Hemiptera). A. Schematic diagram of the ovipositional pattern of the female. Large arrows indicate the direction in which the female proceeds in laying each row of eggs in a batch, and the numbers, the order in which the rows are deposited. The small arrows indicate the direction the female faces during the deposition of each egg in a row and point in the opposite direction to the orientation of inner row embryos. B. An egg batch of the pentatomid *Murgantia histrionica*. Note the inward orientation of the embryos as indicated by the white pattern visible through the chorion. (Reprinted with permission from J. A. Lockwood and R. N. Story. Embryonic orientation in pentatomids: its mechanism and function in southern green stink bug (Hemiptera: Pentatomidae). *Ann. Entomol. Soc. Am.* 79: 963–970. © 1986 Entomological Society of America.)

ooplasmic gradient in them that in turn could influence site of germ band formation, possibly by affecting expression of the dorsoventral patterning genes (7.2.1.1.1).

This adaptation, referred to as *extrinsically mediated embryonic orientation*, seems to ensure that peripheral larvae, on hatch, receive the stimulation from their siblings and the egg mass necessary for their inclusion in the first instar aggregation. First instars do not use visual cues to aggregate and do not possess an aggregation pheromone until 2 d after eclosion (Lockwood and Story, 1985).

6.7.2 POLYEMBRYONY

As mentioned in 6.6.3, the eggs of some insects characteristically generate more than one embryo. *Polyembryony* is a form of asexual reproduction in which a clone of genetically identical embryos develops from a single egg (Craig et al., 1997). It is known to occur in representatives of 18 taxa in seven phyla (Cnidaria, Platyhelminthes, Arthropoda, Annelida, Bryozoa, Echinodermata, and Chordata [some armadillos]) (Craig et al., 1997; Strand and Grbíc, 1997) including some 30 species of parasitoid Hymenoptera in the families Braconidae, Encyrtidae, Platygasteridae, and Dryinidae (Ivanova-Kasas, 1972: Table 1; Strand and Grbíc, 1997: Table 1) and one species of Strepsiptera. Polyembryonic organisms are usually endoparasitic or live in environments whose quality is not predictable by the mother. In some instances, polyembryony may also compensate for constraints on zygote number. For example, a small mother, capable of generating only a small number of eggs, could still produce many offspring (Craig et al., 1997). Though genetically identical, polyembryonic offspring differ genetically from their mother.

According to Craig and colleagues (1997), the occurrence of polyembryony is paradoxical since it clones unproven genotypes at the expense of genetic diversity in a clutch of offspring. They predicted that it is likely to evolve when offspring have more information about resources for their development than their mother has or when the female is constrained to produce fewer eggs than the immediate environment will support.

As an adaptation to their parasitoid lifestyle, the eggs of polyembryonic insects tend to be tiny and yolk-poor and to have reverted to total cleavage during early embryogenesis from the intralecithal cleavage typical of most insects (Tremblay and Calvert, 1971; Strand and Grbíc, 1997).

6.7.2.1 PARASITOID HYMENOPTERA

In parasitoid wasps, nutrition of embryos has evolved in two directions. In one, embryogenesis is reduced and a protopod larva hatches precociously within the host in a highly undeveloped state (Fig. 9.8A) (protopod larvae of some species lack appendages and have no nervous, respiratory, or circulatory systems). In the other, the serosa, or in some species the syncytial progeny of the polar nuclei resulting from meiosis, develops into an envelop-

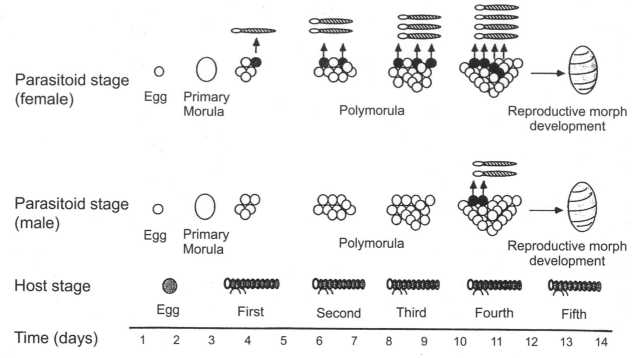

Fig. 6.16. Life cycle of the parasitoid wasp *Copidosoma floridanum* (Hymenoptera: Encyrtidae) in its host *Trichoplusia ni* (Lepidoptera: Noctuidae). The top two lines diagram the developmental stages of the female and male broods; the bottom two, the associated stages of the host and total development time in days. The life cycle begins when the adult wasp oviposits into the host's egg. See text for other details. (Reprinted from *J. Insect Physiol.*, 43, M. Grbíc, D. Rivers, and M. R. Strand, Caste formation in the polyembryonic wasp, *Copidosoma floridanum* (Hymenoptera: Encyrtidae): *in vivo* and *in vitro* analysis, pp. 553–565, Copyright 1997, with permission of Elsevier Science.)

ing membrane or *trophamnion* that sequesters nutrients from the host. It is in this group that polyembryonic species have evolved (Ivanova-Kasas, 1972).

Polyembryony was first described by Marchal (1898) in members of three closely related genera in the tribe Copidosomatini of the Encyrtidae and has since been examined in several other species (Ivanova-Kasas, 1972; Strand and Grbíc, 1997). However, it is best understood in the encyrtid *Copidosoma floridanum*, an egg/larva parasitoid of the cabbage looper *Trichoplusia ni* (Lepidoptera: Noctuidae), owing to the recent remarkable work of Strand and coworkers at the University of Wisconsin (Strand and Grbíc, 1997; Grbíc et al., 1998). The embryogenesis of most larvae in this species takes 264h at 27 ± 2°C, 70% relative humidity, and 16-h-light–8-h-dark cycles (Baehrecke and Strand, 1990). The female wasp inserts one or two tiny (25 μm), yolkless, practically shell-less eggs, either fertilized or not, into a host egg (unfertilized eggs develop into males; fertilized eggs into females; 5.4.1). Their subsequent development is summarized in Table 6.1 and Fig. 6.16 (Baehrecke and Strand, 1990; Strand and Grbíc, 1997).

The sequence summarized in Table 6.1 is for the predominate "reproductive" larvae, which molt three times, pupate late in the host's fifth instar, and kill the host before emerging. However, up to 200 female secondary morulae develop more rapidly than the others during the first four instars of the host, with 1 or 2 in the first instar and 2–10 in each of the second through fourth instars (male embryos only hatch late in the third or in the fourth instar). These eclose as mandibulate "precocious" larvae, which function as soldiers both to manipulate the sex ratio of reproductive larvae (female precocious larvae will kill male reproductive and precocious larvae, and female reproductives in host eggs superparasitized with both male and female eggs), and to defend the brood from interspecific competitors resulting from multiparasitism (Fig. 6.16) (Grbíc et al., 1998; Strand and Grbíc, 1997). Such larvae fail to molt or to metamorphose into adults.

Polyembryony has undoubtedly arisen independently at least four times within the parasitoid Hymenoptera, as the four families in which it occurs are not closely related and contain a majority of monoembryonic species (Strand and Grbíc, 1997: Fig. 7).

6.7.2.2 STREPSIPTERA

Polyembryony in Strepsiptera is, again, associated with endoparasitism and with hemocoelic viviparity (3.10) and so far has been recorded in only one species,

Table 6.1 Polyembryony in *Copidosoma floridanum* at 27 ± 2°C, 70% Relative Humidity, and 16-h Light–8-h Dark Cycle

Hours after oviposition	Developmental stage of wasp	Stage of host	Remarks
0–2	Maturation	Egg	Activation and meiosis of parasitoid egg.
2–15	Cleavage	Egg	Asynchronous, holoblastic cleavage of the zygote generates some 200 blastomeres (the embryonic mass) while polar nuclei and their surrounding cytoplasm proliferate to form a syncytial enveloping membrane about the mass but within the chorion, the whole constituting a primary morula (the membrane functions as a trophamnion and sequesters nutrients from the host egg into the developing morula).
16	Primary morula	Egg	Chorion disappears.
17–191	Secondary polymorulae	Larva 1, day 1– larva 4, day 1	As they continue to proliferate, the enveloping membrane grows into the mass and segregates the blastomeres into cell clusters of varying size, each a secondary morula. Eventually, 1000–2000 secondary morulae are produced, each with the potential to become a larva.
192	Complete polymorula	Larva 4, day 2	Secondary morulae stop proliferating.
216	Early morphogenesis	Larva 5, day 1	All secondary morulae undergo synchronous embryogenesis.
240	Pharate first instars	Larva 5, day 2	1000–2000, completely segmented prolarvae each with a head capsule.
264	Larval hatch	Larva 5, day 3	Simultaneous hatching of first instars.

Source: Baehrecke and Strand, 1990.

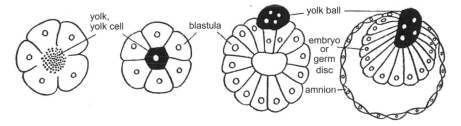

Fig. 6.17. Early embryogenesis in the monoembryonic strepsipteran *Stylops* sp. (Strepsiptera: Stylopidae). (Reprinted with permission from O. M. Ivanova-Kasas. Polyembryony in insects. In S. J. Counce and C. H. Waddington (eds.), *Developmental systems: insects*. Vol. 1, p. 267, Fig. 8. © 1972 Academic Press, Ltd.)

Halictoxenos simplicis (Ivanova-Kasas, 1972). To understand it one must have some knowledge of the events occurring during embryogenesis of less derived species. Eggs of species in the genus *Stylops* have little yolk and cleavage is holoblastic, resulting in formation of an epithelial periblastula surrounding a single yolk cell (Fig. 6.17). The nucleus of this cell divides twice to generate a syncytium of four nuclei within a yolk ball. This passes between the cells of the periblastula to the exterior, where that part of the periblastula adjoining it becomes the germ band and the remainder, the amnion.

In eggs of *H. simplicis*, the central yolk mass is also invaded by nuclei from the periblastula and then dissociates into separate yolk cells that exit the periblastula between its cells and surround it in a continuous, syncytial yolk membrane (Fig. 6.18). Some nuclei of this membrane then differentiate into a thin outer trophamnion; the remainder accumulate into clumps that induce the formation of individual germ bands from the periblastula within. Each germ band then develops into an individual embryo, with 40–50 developing within each egg. The trophamnion then grows between adjacent embryos and isolates each one. The whole eventually

breaks up into many independent vesicles, each containing an embryo. These develop into larvae that eventually escape together from the mother larva through brood canals in her body wall.

6.8 Gastrulation and Germ Layer Formation

Insects are bilaterally symmetrical and *triploblastic*— that is, their various tissues differentiate from cells belonging to three germ layers—ectoderm, mesoderm, and endoderm—that are specified early in embryogenesis. Ectoderm and endoderm are considered to be primary, since they were first to arise phylogenetically in the diploblastic grade of metazoan organization as in helminths, and are specified by expression of maternal genes (Hall, 1998). Mesoderm is secondary, is phylogenetically derived, and is specified by expression of zygotic genes. Although an essential component of metazoan body plans, the germ layers can give rise to nonhomologous structures in members of different lineages.

During germ band elongation in hexapod embryos, longitudinally situated midventral cells in the germ

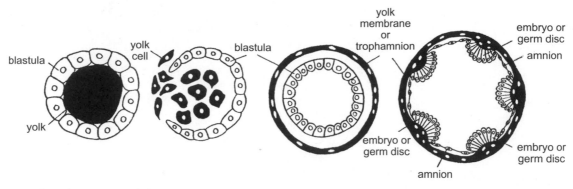

Fig. 6.18. Early embryogenesis of the polyembryonic strepsipteran *Halictoxenos simplicis*. (Reprinted with permission from O. M. Ivanova-Kasas. Polyembryony in insects. In S. J. Counce and C. H. Waddington (eds.), *Developmental systems: insects*. Vol. L, p. 267, Fig. 8. © 1972 Academic Press, Ltd.)

band pass upward into an epineural space between the embryo and yolk to form an inner layer of mesoderm, the lateral cells of the band remaining on the surface as ectoderm. The three known modes of inner layer formation in insect embryos are

- by formation of a longitudinal, midventral furrow that subsequently closes over (Fig. 6.19) (embryos of many orders [Johannsen and Butt, 1941; Hagan, 1951; Ando, 1962; Anderson, 1972a, 1972b, 1973; Haget, 1977; Sander et al., 1985; Ando and Miya, 1985; Ando and Jura, 1987; Schwalm, 1988; Campos-Ortega and Hartenstein, 1997]);
- by lateral overgrowth of a midventral midplate by lateral ectodermal plates (*Apis mellifera* and other Hymenoptera: Apocrita [Nelson, 1915; Schnetter, 1934]); and
- by ingression of individual, ventral midline cells (many lepidopterans [Kobayashi and Ando, 1988]).

6.8.1 GENES AND GASTRULATION IN *DROSOPHILA MELANOGASTER* EMBRYOS

Leptin and Grunewald (1990) have examined gastrulation in great detail in *D. melanogaster* embryos using immunocytochemical methods (Fig. 6.19). In these embryos, the inner layer forms by way of a midventral furrow, takes 20 min at 25°C, and occurs in two phases. During the first phase, the apices (ventral ends) of these cells contract, generating the ventral furrow by a series of cell shape changes unaccompanied by mitosis (Fig. 6.19A, B) (this block to mitosis appears to be due to expression of the *tribbles* gene [Sehar and Leptin, 2000]). In the second phase, the cells of the resulting tube of mesoderm dissociate, divide, and disperse above the ectoderm to form the inner layer (Fig. 6.19C, D).

To examine how ventral furrow formation depends on cell fate in the mesoderm, these workers altered these fates genetically using maternal and zygotic genetic mutations. Some aspects of cell behavior during furrow formation were shown to be cell autonomous—that is, they arose within the cells, the forces driving invagination being generated within the ventral furrow rather than by mechanical or other factors acting elsewhere within the egg.

Two zygotic regulatory genes, *twist* (*twi*) and *snail* (*sna*), necessary for mesoderm formation are expressed in nuclei of the ventral furrow cells (Fig. 6.19; fluorescently labeled antibodies to their proteins are seen only in the nuclei of these cells when whole mounts or sections are examined by fluorescence microsopy), and induction of changes in cell shape within the mesoderm and necessary for gastrulation depends on expression of both genes. However, both genes are expressed also in cells not gastrulating, so expression of other genes is also involved (Reuter and Leptin, 1994). The anterior and posterior limits of the ventral furrow are established by expression of the regulatory gap gene *huckebein* (*hkb*) (Fig. 7.24D). This gene represses expression of *sna* in the anterior and posterior midgut rudiments at the two ends of the inner layer, which are endodermal.

Expression of *twi* and *sna* is activated, in turn, by the maternally expressed transcription factor protein Dorsal (Dl), which forms a gradient along the dorsoventral axis of the egg, with its high point midventrally (2.2.2.6). Expression of *sna* is restricted to presumptive mesoderm, and the sharp lateral limits of this expression define the border between mesoderm and mesectoderm (Fig. 7.20). Costa and coauthors (1993) provided a detailed review of the genetics of gastrulation in *D. melanogaster*.

6.9 Blastokinesis and Embryonic Envelopes

In eggs of most insects, the ventral face of the germ band becomes temporarily covered by an amniotic membrane (amnion) that forms in various ways depending on species (Johannsen and Butt, 1941; Hagan, 1951; Ando, 1962; Sharov, 1966; Cobben, 1968; Anderson, 1972a, 1972b, 1973; Haget, 1977; Sander et al., 1985; Ando and Miya, 1985; Ando and Jura, 1987; Kobayashi and Ando, 1988; Schwalm, 1988; Machida and Ando, 1998). In insects having short and intermediate germ embryos, its

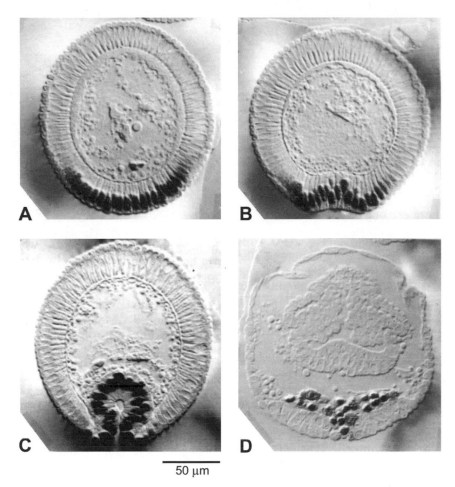

50 µm

Fig. 6.19. Gastrulation in a *D. melanogaster* embryo. Differential interference contrast photomicrographs of transverse sections made at 50% of egg length from the posterior end and stained with tagged anti-Twist antibody at cellular blastoderm (A), beginning of invagination (B), completion of invagination (C), and dispersal and mitosis of mesodermal cells (D) (the intensity of Twist staining differs because the cells are at different stages in the cell cycle). Notice that Twist protein is located within the nuclei of the mesodermal cells. (Reproduced with permission from M. Leptin and B. Grunewald. Cell shape changes during gastrulation in *Drosophila. Development* 110: 73–84. © 1990 Company of Biologists, Ltd.)

formation involves an embryonic movement called *blastokinesis* in which the embryo, posterior end first, is pulled into the yolk while reversing its axes from those of the egg (*anatrepsis*) and then, much later in development, returns to its original position on the ventral side of the egg with a reverse movement called *katatrepsis* (revolution) (Fig. 6.20A–G); the period between entry and exit from the yolk is called *intertrepsis* (Sander et al., 1985). The amount of development completed at the time of anatrepsis and katatrepsis varies with the species, as do the details of both movements, although katatrepsis usually occurs between 40% and 50% of embryogenesis and includes a 90- or 180-degree longitudinal rotation (Fig. 6.24) (Anderson, 1972a, 1973; Cobben, 1968).

6.9.1 ANATREPSIS
In short and intermediate germ embryos, anatrepsis is accompanied by gastrulation and by germ band elongation resulting from mitotic activity within its posterior

growth zone (Fig. 6.14D) and by cellular rearrangement. There are two types:

- In superficial anatrepsis, the posterior end of the germ band moves about the posterior end of the egg just within the serosa, both by extending, attaching (to the inner surface of the serosa), and shortening filopodia from cells at its posterior margin (activities that seem to pull the germ band posteriorly around the end of the egg) and by periodic contractions of the yolk system (e.g., *Acheta domesticus*; Orthoptera: Gryllidae [Vollmar, 1972]).
- In immersion anatrepsis, the germ band is pulled lengthwise into the yolk by periodic contraction of elements within the yolk system collectively referred to as the *anatrepsis center* (Fig. 6.20A–D) (Heming and Huebner, 1994).

In living eggs of the leafhopper *Euscelis plebejus* (Hemiptera: Cicadellidae) (Sander, 1968, 1976a) and of the water strider *Gerris paludum insularis* (Hemiptera: Gerridae) (Mori, 1985), this center is visible by trans-

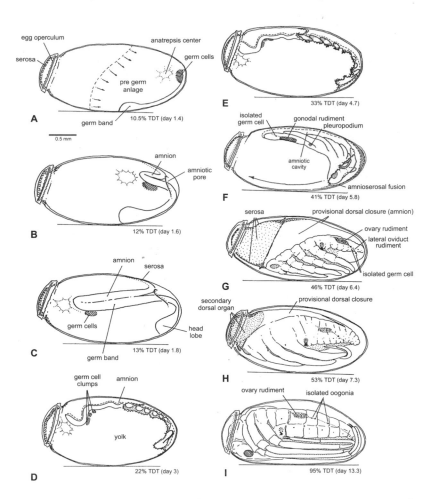

Fig. 6.20. Germ band formation (A), anatrepsis (B and C), intertrepsis (D and E), and katatrepsis (F–I) in an embryo of *Rhodnius prolixus* at 27°C. TDT, total development time. (Reprinted with permission from B. S. Heming and E. Huebner. Development of the germ cells and reproductive primordia in male and female embryos of *Rhodnius prolixus* Stål (Hemiptera: Reduviidae). *Can. J. Zool.* 72: 1100–1119. © 1994 NRC Research Press.)

mitted light as a dark spot. Sander demonstrated its role by transversely squeezing eggs at various sites along their lengths early in anatrepsis and following their subsequent development (Fig. 6.21). If an egg was squeezed within the germ band, only its posterior fragment continued to move forward within the egg (a_1, a_2). If squeezed in front of the germ band, it failed to undergo anatrepsis in 18 of 25 eggs (b_1, b_2). And if anatrepsis was blocked with an incomplete ligature through which a ball of symbiotic bacteria at the rear of the egg could not pass, anatrepsis did not occur and yolk cells within the anterior isolate were pulled back toward the embryo to be replaced by others (c_2). These results suggest the germ band is pulled into the yolk by a system within, which repeatedly contracts, both concentrically toward its longitudinal axis and anteriorly in a longitudinal direction.

Surprisingly, the cytoskeletal elements functioning in anatrepsis have yet to be identified but could be demonstrated by use of drugs disrupting the formation of microtubules, microfilaments, and other cytoskeletal elements or by immunocytochemistry.

In both types of anatrepsis, serosal cells continuous with the periphery of the embryo are carried into the yolk with the embryo to form an amnion, which thins to a veil over its ventral surface and which, on completion of anatrepsis, encloses an amniotic cavity (Fig. 6.20B–D).

And in embryos of some species, anatrepsis is accompanied by cleavage of yolk into mononucleate yolk cells. Following entry into the yolk, the opening to the surface through which the embryo entered, the *amniotic pore* (Fig. 6.20B, C), may (in some ectognath apterygotes) or may not (most exopterygotes) remain open (Fig. 6.24) (Machida and Ando, 1998). In the latter embryos, the lips of the invagination fuse and the amnion separates from the serosa, completely isolating the embryo within the yolk throughout intertrepsis (Fig. 6.20D, E).

6.9.2 KATATREPSIS

Katatrepsis begins when the amnion fuses to the serosa at the head end of the embryo and breaks at this fusion point (amnioserosal fusion) (Fig. 6.20F, G) (Johannsen and Butt, 1941; Hagan, 1951; Ando, 1962; Cobben, 1968; Anderson, 1972a, 1972b, 1973; Haget, 1977; Sander et al., 1985; Ando and Miya, 1985; Ando and Jura, 1987; Schwalm, 1988). The serosa then contracts toward the front of the egg to form a serosal plug or secondary dorsal organ and pulls the embryo, head end first, out of the yolk and back to its original position on the ventral side of the egg (Fig. 6.20F–H). As the embryo re-emerges from the yolk, the amnion rolls back over the surface of the yolk, reversing its method of formation, replacing the serosa, and forming a provisional dorsal closure

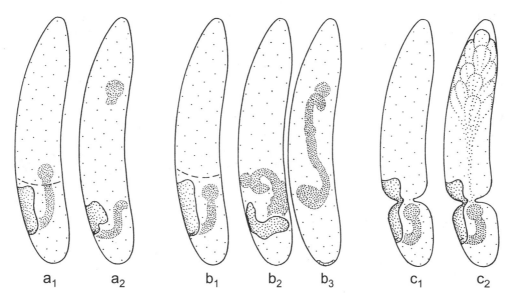

Fig. 6.21. Sander's experiments on embryos of *Euscelis plebejus* (Hemiptera: Cicadellidae) demonstrating that the germ band is pulled into the egg by the yolk system during anatrepsis. See text for details. (Reproduced with permission from K. Sander. Morphogenetic movements in insect embryogenesis. In P. A. Lawrence (ed.), *Insect development*, pp. 35–52. R. Entomol. Soc. Lond. Symp. 8. Blackwell Scientific Publications, Oxford. © 1976 The Royal Entomological Society of London.)

about the yolk and over the back of the embryo (Fig. 6.20G, H). This covering is replaced by the body wall during dorsal closure as the sides of the embryo grow upward around the remaining yolk later in embryogenesis (Fig. 6.20H, I). The contracted serosa is engulfed by yolk behind the head during dorsal closure (Fig. 6.20H) and eventually degenerates within the midgut (Cobben, 1968; Pétavy, 1975, 1976; Tamarelle, 1981).

If katatrepsis is prevented, the amniotic cavity remains intact and later becomes lined with larval cuticle. Appendages and body hairs grow into the cavity while internal organs "dangle" outside in the space originally occupied by yolk because dorsal closure cannot occur (Sander, 1976a). Also, application of actinomycin D, an inhibitor of RNA synthesis, to eggs of the termite *Odontotermes badius* interferes with katatrepsis by preventing the serosa from contracting and pulling the embryo out of the yolk, suggesting that transcription of one or more genes is required for this to occur (Truckenbrodt, 1979).

The possible functional and adaptive significance of blastokinesis is not known, but the movement may facilitate mobilization of yolk to the developing embryo during intertrepsis while katatrepsis accommodates its further growth, change in shape, appendage development (particularly in species having long-legged first instars), and enclosure of the remaining yolk by the body wall and midgut (Cobben, 1968; Anderson, 1972a).

6.9.3 BLASTODERMAL AND SEROSAL CUTICLES
In myriapod, collembolan, and dipluran embryos, cells of the blastoderm deposit a blastodermal cuticle before or after germ band formation (Machida and Ando, 1998)

(this can be partly deposited by embryonic cells in the latter). In collembolans, dorsal cells of the blastoderm then separate from this cuticle and thicken to form a primary dorsal organ, which invaginates into the dorsal yolk. As this occurs, the apices of some of its cells grow out through the opening of this invagination as tendrils and around the blastoderm beneath the blastodermal cuticle and produce a fluid that seems to facilitate blastokinesis, as destruction of the organ prevents this from occurring (Jura, 1972).

In archaeognath and zygentoman eggs and in those of most pterygotes, after the embryo has completed anatrepsis or otherwise becomes surrounded by embryonic membranes, the serosal cells secrete a serosal cuticle under the vitelline membrane which functionally replaces the chorion. In addition, circumstantial evidence suggests that its cells synthesize enzymes that may assist in mobilizing yolk, most of which is due to the action of vitellophages and yolk membrane cells within or surrounding the yolk.

6.9.4 BLASTOKINESIS IN LONG GERM EMBRYOS AND SPECIFICATION OF THE AMNIOSEROSA
Blastokinesis is reduced to sequential elongation (extension) and shortening (retraction) of the germ band in long germ eggs (Fig. 6.23: 8–13), except in embryos of some lepidopterans in which the embryonic movements are exceedingly complex and not homologous to those of short germ insects (Anderson, 1972b, 1973; Kobayashi and Ando, 1988). The embryo remains on the ventral side of the egg throughout embryogenesis and the amnion, if present (it is absent in collembolan and dipluran embryos [Uemiya and Ando, 1987; Ikeda and

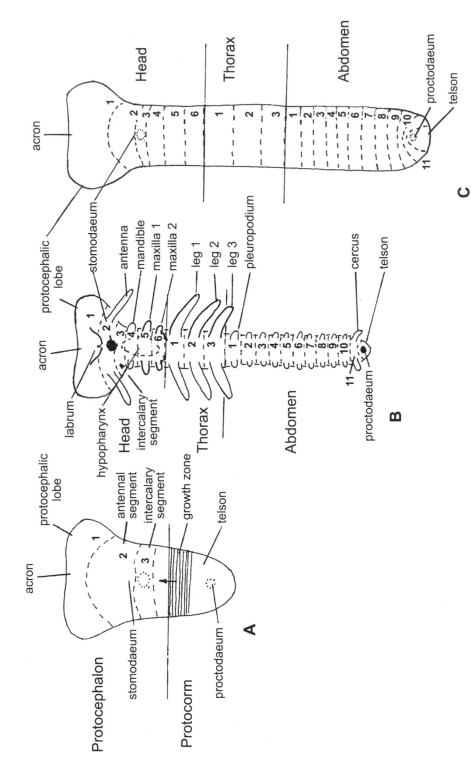

Fig. 6.22. Short (A and B) and long germ (C) embryos in ventral aspect. A. Short germ embryo at the protopod stage, just after its formation, showing the protocephalon or primary head region, the protocorm or primary trunk region, and the growth zone, which will later bud off segments anteriorly. At this time, segments, appendages, stomodaeum, and proctodaeum have yet to appear. B. Short germ embryo at the polypod stage after completion of germ band elongation, gastrulation, and segmentation. The stomodaeum and proctodaeum have invaginated, and all segments except the intercalary segment bear a pair of appendages. C. Long germ embryo just after formation. Proportions of the germ band for head, thorax, abdomen, and each body segment are already about what they will be at the time of hatch. In all three diagrams, segment 1 in the protocephalon (A) or head (B and C) is the ocular or preantennal segment.

Machida, 1998; Machida and Ando, 1998]), originates as an extended amnioserosal fold that arises about the periphery of the germ band and folds inward toward its ventral midline, the folds from each side meeting and fusing midventrally and then separating into amnion and serosa (Handel et al., 2000).

In a variation of this mode of membrane formation, the *fault-type* characteristic of embryos of species in advanced lineages of Lepidoptera, the germ band separates from the serosa, and sinks into the yolk (Kobayashi and Ando, 1988). Serosal cells then proliferate between the germ band and vitelline membrane to close the hole in the serosa, and the amnion proliferates independently from the margins of the germ band to enclose the embryo ventrally.

During germ band elongation in long germ embryos, the posterior end of the abdomen extends temporarily around the posterior pole of the egg onto its dorsal surface and forward to the head (Fig. 6.23: 11). Later, amnioserosal fusion and breakage occur, exactly as in short germ embryos, and the germ band shortens again and returns to the ventral side of the egg, a transient, secondary dorsal organ forming from the contracted serosa that ends up within the developing midgut and degenerates.

In the highly specialized embryos of higher flies, only a vestige of the amnion and serosa, the *amnioserosa*, remains dorsally about the yolk prior to dorsal closure (Fig. 6.23: 11–15) (Campos-Ortega and Hartenstein, 1997). In early blastoderm embryos of this insect, proteins of the dorsoventral regulatory genes *decapentaplegic* (*dpp*) and *screw* (*scw*) function synergistically to subdivide the dorsal ectoderm on either side into somatic epidermis below and amnioserosa above (Rusch and Levine, 1997). Peak *dpp* activity is required dorsally for localized expression of the regulatory gene *zerknüllt* (*zen*) (Fig. 7.20). The *zen* gene encodes a transcription factor that directly activates expression of the downstream gene *race* in cells of the amnioserosa and anterior and posterior midgut rudiments.

During germ band elongation, the embryo of *D. melanogaster* increases over 2.5-fold in length and decreases in width (Fig. 6.23: 11). By examining living embryos during this process by epi-illumination and time-lapse video microscopy, Irvine and Wieschaus (1994) showed germ band cells to intercalate between their dorsal and ventral (lateral) neighbors during extension, increasing the number of cells along the anteroposterior axis at the expense of those across its width and suggesting elongation to result from establishment of adhesive differences between the cells. The onset of extension coincides with peaks of serotonin and 5-ht$_{2Dro}$ expression, the latter, a receptor for serotonin, appearing in alternating segments (Colas et al., 1999). If the genes for either of these proteins is not expressed, germ band extension is abnormal and is no longer synchronized with gastrulation.

Expression of the *u-shaped* group of regulatory genes *u-shaped* (*ush*), *hindsight* (*hnt*), *serpent* (*srp*), and *tail-up*

(*tup*) is required both for germ band shortening and to maintain the amnioserosa (Frank and Rusklow, 1996; Yip et al., 1997; Lamka and Lipshitz, 1999). In *ush*, *hnt*, and *srp* mutants, but not in *tup* mutants, amnioserosal cells are correctly specified and develop normally but degenerate following germ band extension, resulting in the germ band being unable to shorten. This suggests a mechanical role for the amnioserosa in germ band shortening of long germ *Drosophila* embryos, similar to that of the serosa in katatrepsis of short germ and of less specialized, long germ embryos. However, Yip and coauthors (1997) and Lamka and Lipshitz (1999) suggested instead that expression of *hnt* could both maintain the amnioserosa and possibly regulate signaling from it to the embryo to coordinate changes in cell shape and position, shortening the germ band.

Germ band formation and blastokinesis in embryos of a variety of short, semi-long, and long germ species is shown in Fig. 6.24 superimposed on an inferred phylogeny of the orders. Similar diagrams could be provided for representative embryos of all orders, except Protura (embryogenesis uninvestigated), but there was not room to do so here.

6.9.5 YOLK EXTRUSION AND CONSUMPTION IN LONG GERM EMBRYOS OF THE TEPHRITID FLY *ANASTREPHA FRATERCULUS*

Before leaving this topic, I should mention a curious phenomenon that occurs during embryogenesis in eggs of several species in the tephritid fly genus *Anastrepha*. Embryogenesis of the Brazilian fruit fly *A. fraterculus* requires about 42 h at 25°C and is similar to that of other higher dipterans with long germ eggs (compare Fig. 6.23) (Selivon et al., 1996). During gastrulation and germ band extension from 20%–28% of embryogenesis, circumferential epidermis about the posterior end of the embryo contracts into a ring and buds off a ball of yolk containing vitellophages and bacterial symbionts, but *not* before the pole cells have moved out of the ball and into the embryo. Later, during head involution (8.3.7) from 43%–76% of embryogenesis, a similar yolk ball is budded off from the anterior end of the embryo in front of the head lobes. Just before hatch, the larva consumes the anterior ball, reverses itself within the egg, consumes the posterior ball, and then ecloses from the posterior end of the egg (this reversal occurs in larvae of species not producing yolk balls).

In 183 eggs, Selivon and coworkers (1996) noted that 58.6% of embryos produced yolk masses at both ends, as summarized already; 14.8%, only the anterior mass; 13.8%, only the posterior mass; and 12.8%, neither. But all these larvae reversed themselves within the egg and hatched, suggesting the process is not critical to the life of the fly. And in eggs of a closely related cryptic species, the same variants were recorded, but the proportions of each type differed significantly (Selivon et al., 1997). The functional and evolutionary significance of this process is unknown, but Selivon and colleagues (1996) speculated that both extrusion and consumption of yolk

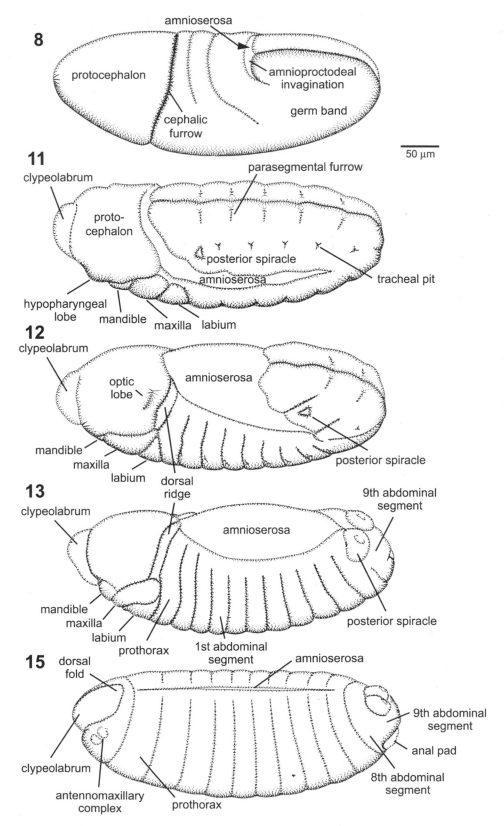

Fig. 6.23. Embryogenesis of *D. melanogaster* in dorsolateral aspect, showing germ band elongation and retraction (the functional equivalent of anatrepsis and katatrepsis in a long germ embryo), dorsal closure, and head involution in a highly derived, long germ embryo. Shown are stages 8 (15%–19% of embryogenesis), 11 (24%–32%), 12 (32%–41%), 13 (41%–46%), and 15 (52%–59%). Note the greatly reduced extent of the amnioserosa. (Adapted with permission from V. Hartenstein. *Atlas of Drosophila development*, pp. 2–4. © 1993 Cold Spring Harbor Laboratory Press.)

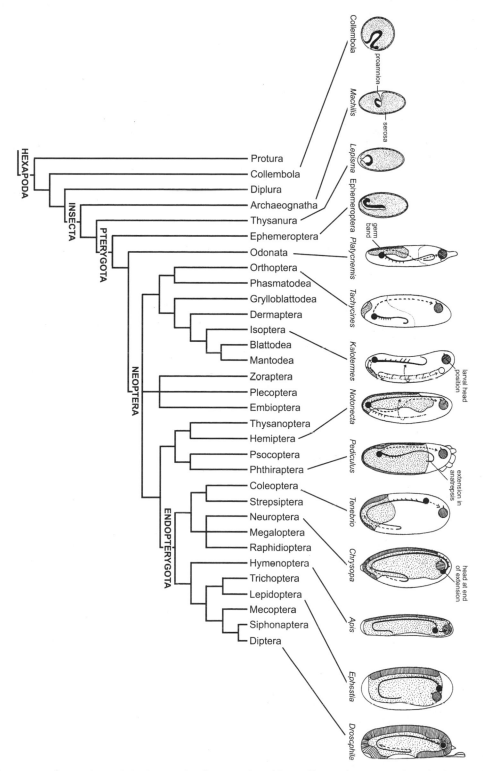

Fig. 6.24. Evolutionary relationships of the hexapod orders as inferred by Gullan and Cranston (1994), showing blastokinesis in selected embryos. In collembolans and diplurans, the germ band arches dorsally into the yolk during elongation and intertrepsis, its ventral face being open to the surface because the amniotic pore does not close. In archaeognaths and zygentomans, the germ band undergoes anatrepsis but the amniotic pore, again, remains open. In mayflies, odonates, and other insects with short or intermediate length embryos, the amniotic pore closes following anatrepsis so that the amnion encloses an amniotic cavity and the embryo is completely isolated from the serosa. The other diagrams indicate the extent of the germ band at the time of formation (stippled and hatched), the position of the embryo during intertrepsis (black), the course taken by the embryo during katatrepsis (dotted arrow), and the position of the head prior to hatch (hatched circle with "nose"). In all diagrams, anterior is to the top and ventral to the left. There is much variation within the embryos of different species within each order. (Embryo images reprinted, with permission, from F. E. Schwalm. *Insect morphogenesis*, p. 185, Fig. 58. Monogr. Dev. Biol. 20. © 1988 S. Karger. Cladogram adapted with permission from P. J. Gullan and P. S. Cranston. *The insects. An outline of entomology*, p. 193, Fig. 7.5. 1st ed. Chapman and Hall, London. © 1994 Blackwell Science, Ltd.)

may facilitate transovarial transmission of bacterial symbionts.

6.10 Segmentation and Appendage Formation

After formation of the inner layer, the embryo becomes divided by transverse furrows into segments, each of which, in members of basal and intermediate clades, subsequently develops a pair of appendages (Johannsen and Butt, 1941; Hagan, 1951; Ando, 1962; Cobben, 1968; Anderson, 1972a, 1972b, 1973; Haget, 1977; Sander et al., 1985; Ando and Miya, 1985; Ando and Jura, 1987; Schwalm, 1988; Campos-Ortega and Hartenstein, 1987).

6.10.1 SHORT GERM EMBRYOS

When first formed, a short germ embryo is oval-, heart- or keyhole-shaped; consists of a *protocephalon* (primary head region) and *protocorm* (primary trunk region), and may occupy less than 10% of the egg's surface (Figs. 6.14C, 6.22A). Until recent molecular studies, the protocephalon had been thought to consist of a non-segmented acron and three pregnathal segments, which later develop, respectively, a pair of labral, antennal, and (rarely) premandibular (intercalary, tritocerebral) appendages. However, expression of the segment polarity gene *engrailed* in embryos of several species supports the existence of an ocular (preantennal) segment in front of the antennal segment but not that of a labral segment. Instead, for good reason, the labrum recently was suggested (Haas et al., 2001a) to derive from the intercalary segment (7.2.1.5). The protocephalic lobes give rise to the protocerebrum and optic lobes of the brain and to the eyes (stemmata or ommatidia) while the *stomodaeum* or foregut invaginates on the midline of the intercalary segment (Haas et al., 2001a), although it appears to do so on the posterior venter of the antennal segment (Fig. 6.22A).

At first, the protocorm consists of an anterior growth zone and a posterior, nonsegmented telson, from which later invaginates the *proctodaeum* or hindgut (its roof often develops from posterior amnion) (Fig. 6.22A). Shortly after germ band formation and continuing throughout anatrepsis and early intertrepsis, the growth zone buds off segments anteriorly. As additional segments are added, the embryo elongates and the growth zone is pushed posteriorly by its own mitotic activity. The first three segments budded off are the gnathal segments: mandibular, and first and second maxillary (Fig. 6.22B). Later each of these develops a pair of appendages—the mandibles and the first and second maxillae—and then consolidates anteriorly with the protocephalon to form a six-segmented head. Both pairs of maxillae subsequently divide longitudinally from apex to base into inner and outer lobes, with the inner lobes dividing again into laciniae and galeae (first maxillae) or glossae and paraglossae (second maxillae) and the outer lobes becoming the stipites (submentum) and maxillary (labial) palpi (Machida, 2000). The second maxillae then approach each other and fuse on the mid-

line to form the labium while the hypopharynx arises as a thickening in the venters of the mandibular and of both maxillary segments. Cephalic segments anterior to the so-called dorsal ridge (Fig. 6.23: 12), of mixed maxillary and labial composition, produce only ventral components (Rogers and Kaufman, 1996).

The proliferation zone then buds off the prothoracic, mesothoracic, and metathoracic segments followed by 11 abdominal segments and then disappears (Fig. 6.22B). In semi-long germ embryos, such as those of grasshoppers, all head and thoracic segments appear shortly after germ band formation.

All these segments, in embryos of some apterygote and exopterygote insects and of basal endopterygotes, later come to bear a pair of appendages: a pair of legs on each thoracic segment, and a pair of small, evanescent abdominal appendages on each of segments 1–11 (Fig. 6.22B). Those of segment 11 persist as the cerci in insects whose juveniles have them, while those of segment 1 usually develop into a pair of glandular pleuropodia having various functions depending on species (Heming, 1993a). In collembolan, dipluran, and myriapod embryos, if appendages appear at all on segment 1, they develop, respectively, into the collophore, eversible vesicles, or walking legs (Uemiya and Ando, 1987; Ikeda and Machida, 1998).

As can be seen, segmentation in short and semi-long germ embryos occurs in two phases: a gradual, antero-posterior partitioning of the germ band into segments, followed by the differentiation of their appendages (Cohen and Jürgens, 1991).

Novák and Zambre (1974) showed the pleuropodia, in embryos of the grasshopper *Schistocerca gregaria*, to synthesize an ecdysteroid (molting hormone)-like factor that seems to induce production of enzymes by serosal cells that function to digest serosal cuticle prior to hatch. Other functions postulated for these appendages, based on when they reach their maximum size and then degenerate (usually either just after katatrepsis or hatch), are in katatrepsis, osmoregulation, secretion, or excretion (Haget, 1977; Heming, 1993a).

All abdominal appendages except the pleuropodia and cerci have usually disappeared by the time of hatch except in certain apterygote and endopterygote larvae, where they persist and differentiate into short, functional abdominal styli, eversible vesicles, or prolegs (e.g., Machida, 1981; Uemiya and Ando, 1987; Ikeda and Machida, 1998), and in orthopterans, where those of the genital segments, 8–10, differentiate directly into components of the external genitalia (Matsuda, 1976)..

In embryos of *Sphodromantis centralis* and other mantids (Mantodea), the cellular rudiment of a hatching thread grows out from the apex of each cercal appendage between 36% and 43% of embryogenesis and eventually attaches, by five or six tip cells, to the inner surface of the serosal cuticle dorsolateral to segment 9 (Kenchington, 1969). From 46%–50% of embryogenesis, each thread is transformed into a helical cellular strand by loss of adhesion between adjacent cells. The filaments then become syncytial and from 54%–57% of embryogenesis, coil proximally at the end of the abdomen between the two cerci. Second embryonic cuticle is deposited about the threads beginning at 50% of embryogenesis, and at hatch they unravel, suspending the larva from the ootheca while it molts to the first instar (9.2).

The temporary appearance of abdominal appendages during embryogenesis supports the old idea, now rejuvenated owing to results of comparative nucleotide sequencing of several mitochondrial and nuclear genes, that collembolans, diplurans, ectognath insects, crustaceans, and myriapods (centipedes, millipedes, pauropods, symphylans) are monophyletic, constituting the taxon Mandibulata, and share a common, multilegged ancestor, an example of "ontogeny recapitulating phylogeny" (13.3.1). And in some juvenile trilobites (Fortey, 2000), myriapods, proturans, and crustaceans, the growth zone remains active even after hatch and adds additional segments to the body during the first few molts, a phenomenon referred to as *anamorphosis* (Anderson, 1973).

Most short germ embryos thus pass through three stages during ontogeny:

- a protopod stage in which both segments and appendages are absent (Fig. 6.22A);
- a polypod (extended germ band) stage in which segmentation is complete, each segment bears a pair of appendages, and the body plan characteristic of the class is mapped out (the phylotypic stage) (Fig. 6.22B); and
- an oligopod stage in which the abdominal appendages are withdrawn in those hexapod juveniles lacking functional abdominal appendages (Fig. 6.20I).

6.10.2 LONG GERM EMBRYOS

In long germ embryos of collembolans (Uemiya and Ando, 1987) and diplurans (Ikeda and Machida, 1998) and of intermediate and higher endopterygotes, the germ band can occupy up to 90% of the egg surface at the time of formation (e.g., Figs. 6.23: 8; 6.24), the protopod stage does not occur, a growth zone is absent, and all three tagmata (head, thorax, abdomen) have close to their final larval proportions at the time of germ band formation (Fig. 6.22C). In these, appendage formation often begins in the prothoracic segment and proceeds anteriorly and posteriorly, and in embryos of higher flies, the last three abdominal segments, 8–10, are fused from their inception as in *D. melanogaster* embryos (Fig. 6.23) (Hartenstein, 1993; Campos-Ortega and Hartenstein, 1997).

Long germ eggs seem to constitute an adaptation for rapid embryogenesis, as such insects organize their entire segmented body pattern at once, within an existing field of cells, rather than first establishing a primary subfield, the protocephalon, then subsequently adding cells to it from a growth zone in the protocorm, as occurs in short and intermediate germ insects (Cohen and Jürgens, 1991; Jürgens and Hartenstein, 1993). However, an echo of these two phases of segmentation persists in long germ embryos, a secondary anatomy according to Gerhart and Kirschner (1997: 307), in the sequential expression pattern of the regulatory genes specifying segments and segmental identity, as discussed in chapter 7 (7.2.1.2, 7.2.1.3).

7/ Specification of the Body Plan in Insect Embryos

In chapter 6, I summarized insect embryogenesis up until the polypod or extended germ band stage at which time the body plan of the juvenile has been roughed out into tagmata, body segments, and appendages. Here, I summarize, in chronological order, some experiments performed to reveal how this pattern is established and more recent discoveries on the role of regulatory genes in controlling this process in embryos of *Drosophila melanogaster* and other arthropods.

7.1 Early Experiments

At the close of the 19th century, the descriptive knowledge of insect embryogenesis was practically the equal of ours, owing to contemporary improvements in microscopy and microtechnique and to availability of motivating concepts such as the germ plasm hypothesis (Weismann, 1892), natural selection and evolution (Darwin, 1859), and Müller-Haeckel's "biogenetic law" (Haeckel, 1866): "ontogeny recapitulates phylogeny." Since phylogeny was accepted by many as the "cause" of ontogeny (13.3.1), there was little stimulus to investigate it experimentally and most work published at that time was descriptive and comparative.

Emphasis on the biogenetic law ended with the rediscovery of Mendel's (1866) "hereditary factors" in 1900 by de Vries and Correns, the success of Roux's "Entwicklungsmechanik" (analytical embryology), and the commencement in 1909 of Morgan's breeding experiments on *D. melanogaster* at Columbia University. These developments stimulated a burgeoning band of ingenious experimentalists to forget about evolution and to focus instead on uncovering and explaining the role of proximate factors in controlling insect embryogenesis.

7.1.1 DETERMINANTS AND MOSAIC DEVELOPMENT

Three hypotheses guided early attempts to explain how the body pattern of insects is established during embryogenesis (Sander, 1997). In his germ plasm hypothesis, Weismann (1892) proposed that development of an egg into a complex organism results from the differential allocation of a host of invisible determinants among daughter cells during cleavage and subsequent cell divisions. Eventually, each cell in the embryo has only one kind present and differentiates in the direction specified.

The first experimental evidence for the existence of such determinants in insect eggs was provided by Hegner (1908), who destroyed posterior polar granules in newly fertilized eggs of the Colorado potato beetle *Leptinotarsa decimlineata*, which then developed into sterile adults, implying that these determinants specify the germ cells (6.6). Also, microcautery of particular portions of newly deposited house fly *Musca domestica* eggs was found by Reith (1925) to eliminate only those parts of the resulting larva that would have formed from these regions, again implying the presence of organ-specific determinants. However, the subsequent production of normal, duplicate larvae resulting from splitting experiments on damselfly (*Platycnemis pennipes*), stone cricket (*Tachycines asynomorus*), and fly (*D. melanogaster*, *Protophormia terrae novae*) embryos soon showed such determinants for other cell types not to be present (Sander, 1997).

7.1.2 PHYSIOLOGICAL CENTERS

Between 1926 and 1929, Friedrich Seidel published four classic papers on embryonic determination in semilong germ eggs of the damselfly *P. pennipes* (Odonata: Zygoptera) that seemed to suggest embryo formation and differentiation to be under control of two physiological centers: an *activation* (formative) *center* (AC) at the posterior pole of the egg and a *differentiation center* (DC) in the vicinity of the presumptive maxillary to mesothoracic segments (Seidel, 1929; Richards and Miller, 1937; Kühn, 1971; Counce, 1973). Evidence for action of such centers during embryogenesis of other insects was soon forthcoming, as was that for a third, earlier acting *cleavage center* (CC) (Counce, 1973).

The CC is usually located within the egg at the site where fusion of sperm and egg pronuclei occurs (Fig. 3.14E), appears to be a focus for the autonomous formation of cytoplasmic islands, and appears to be required for normal blastoderm formation (Counce, 1973). It was thought to be activated by sperm entry, ovulation, or oviposition, depending on the species, and proved difficult to examine because experimental interference in this area usually destroys the zygote nucleus or the first cleavage energids.

In the house cricket *Acheta domesticus* (Orthoptera: Gryllidae), 5% of unfertilized eggs undergo pseudocleavage in which cytoplasmic islands, devoid of nuclei and

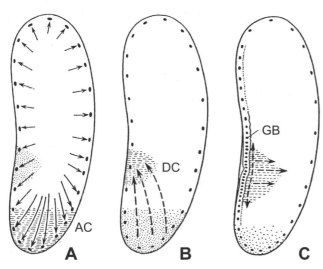

Fig. 7.1. Summary of the function of the activation (AC) and differentiation (DC) centers in germ band (GB) formation as proposed by Seidel, based on the results of experiments on embryos of the damselfly *Platycnemis pennipes* (Odonata). Cleavage energids entering the posterior pole of the egg (A) stimulate production of an agent that diffuses forward, liquefying the yolk (B) and inducing contraction in the ventral yolk system (C), creating a space into which blastoderm cells can move to form the germ band. (Reprinted with permission from A. Kühn. *Lectures on developmental physiology*, p. 322, Fig. 408. © 1971 Springer-Verlag.)

smaller than normal cleavage energids, proliferate to a maximum of 512, as in normal cleavage, and move to the periphery to join the periplasm (compare Fig. 6.3) (Mahr, 1960, in Counce, 1973). Pseudocleavage can also occur in enucleate egg isolates produced by ligation, where it begins within the yolk system in the area of the cleavage center.

The AC is usually located at the posterior pole of the egg and appears to be required for germ band formation. It seems to be activated by entry of at least one cleavage energid into it and to act by producing an "agent" from this interaction that spreads forward, causing liquefaction of the yolk (Fig. 7.1A, B). Liquefaction enables the yolk system (yolk and cytoplasmic reticulum) to contract more strongly and to cause cell displacement (Fig. 7.1C).

If the posterior pole of the egg is eliminated before or during early cleavage, a blastoderm forms but soon becomes serosal, and no germ band forms. If the posterior pole is loosely ligatured with a hair loop such that cleavage energids cannot pass through its neck, again, a temporarily normal blastoderm forms but no embryo, implying that the AC must interact with one or more cleavage energids to be activated (Fig. 7.1A, B). After the cleavage energid/AC interaction has occurred, progressively more of the posterior end of the egg can be destroyed without preventing formation of a normal germ band, suggesting that some agent diffuses forward after the interaction (Fig. 7.1B). With similar methods,

ACs have since been demonstrated in eggs of representative species of Orthoptera, Hemiptera, Coleoptera, Hymenoptera, and Diptera, but not in all beetles and flies studied (Counce, 1973).

A third DC is "activated" when the agent from the AC reaches the area of the presumptive prothorax, and is the point from which subsequent segmentation, organogenesis, and differentiation spread forward and backward in the developing body, perhaps due to instructions spreading fore and aft from the DC (Fig. 7.1C). Its exact position varies between long and short germ eggs and it can be shifted fore or aft experimentally (Counce, 1973). Also, if the germ bands in the eggs of the honeybee *Apis mellifera* are transected into fragments at various points along their lengths, only those portions containing the DC will differentiate; those lacking it eventually degenerate (Schnetter, 1934).

The first effect of DC activity is to induce localized contractions of the yolk system ventrally (Fig. 7.1C). This causes the yolk to retract dorsally away from the ventral blastoderm and creates a space into which blastodermal cells from either side can enter to form the germ band.

According to Sander (1976b) and Lawrence (1992: 205), Seidel was opposed to the concept of gradients. Sander (1976b) suggested that Seidel's results could be better explained as resulting from interference in the production or diffusion of one or more gradient factors from either end of the egg (Fig. 7.6), as he had earlier demonstrated experimentally in embryos of the leafhopper *Euscelis plebejus* (Hemiptera: Cicadellidae) (Sander, 1960). Thus, Seidel's DC is more likely to be a *commitment center* where blastoderm cells first commit to specific developmental fates owing to their position in relation to these gradients (Sander, 1997). Results of more recent experimental genetic and molecular work on pattern formation in *Drosophila* embryos shows this to be so, implying that the concept of physiological centers has outlived its usefulness.

7.1.3 GRADIENTS AND PATTERN FORMATION
The basic body pattern of an insect embryo is first recognizable in the extended germ band or polypod stage of embryogenesis when all segments and appendages are present (Fig. 6.22B) (Sander, 1976b). This phylotypic stage (Cohen and Massey, 1983) is also the stage of maximum similarity between embryos of different species in different orders.

The pattern consists of two linear patterns superimposed on each other: one bilateral or dorsoventral (D-V), and the other longitudinal or anteroposterior (A-P). Both arise progressively during early embryogenesis, beginning just before blastoderm formation in long germ eggs and during gastrulation, germ band elongation, and segmentation in short and intermediate germ eggs.

7.1.3.1 DORSOVENTRAL PATTERN FORMATION
In short germ embryos of the stone cricket *Tachycines asynomorus* (Orthoptera: Tettigoniidae), Krause (1953)

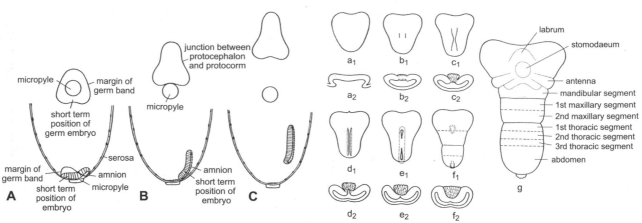

Fig. 7.2. Position of the short germ embryo of the stone cricket *Tachycines asynomorus* (Orthoptera: Tettigoniidae) in ventral aspect as viewed from the posterior end of the egg (above) and in sagittal section (below) at 3 (A), 9 (B), and 16 (C) d after the beginning of anatrepsis at 26°C. Note how the germ band begins to lengthen during this movement. (Reprinted with permission from A. Kühn. *Lectures on developmental physiology*, p. 330, Fig. 421. © 1971 Springer-Verlag.)

Fig. 7.3. Germ band of *T. asynomorus* in ventral aspect (a_1–f_1, g) and in transverse section (a_2–f_2) during gastrulation and elongation at 3 (a), 6 (b), 9 (c), 12 (d), 15 (e), 18 (f), and 26 (g) d after germ band formation at 26°C. a, amniotic folds; b, middle plate; c, beginning of gastrulation; d, gastrular groove; e and f, groove closure; g, fate map of segments midway through germ band elongation. (Reprinted with permission from A. Kühn. *Lectures on developmental physiology*, p. 330, Fig. 422. © 1971 Springer-Verlag.)

demonstrated that a bilateral body pattern can redevelop in any longitudinal strip of early germ band if it is wide enough, a new midline for a bilateral pattern becoming established between the lateral margins of the strip.

Krause devised this model based on results of experiments on germ bands split longitudinally just before, during, or after gastrulation. Early in anatrepsis of embryos of this insect, the germ band approaches the posterior pole of the egg next to the micropyle (Fig. 7.2A). If a small hole is made in the chorion at this point, and if the egg is then placed on a dab of plasticine and gently squeezed, the germ band emerges in a drop of ooplasm and can be cut in various ways with fine glass needles. If the plasticine is then spread, pressure is released from the egg, the operated germ band is sucked back inside, and the hole in the chorion is sealed with paraffin. Later, the egg can be opened, following various periods of embryogenesis, and the results of these manipulations examined, while other eggs are allowed to complete their embryogenesis. Results are shown in Figs. 7.3–7.5.

Normal early embryogenesis in this insect is summarized from both the ventral aspect and in transverse section in Fig. 7.3. Depending on stage of embryogenesis at the time of longitudinal splitting, subsequent bilateral regulation could be achieved in either of two ways (Fig. 7.4). In experiments a_1–c_1 represented in Fig. 7.4, both embryo and amnion were split, whereas in experiments d_1–f_1, only the embryo was sectioned. Notice that in all experiments, the amount of lateral regulation toward normal development occurring after section is reduced as the embryo ages. When splitting occurred before gastrulation, each half formed a new midline, regenerated the missing side, gastrulated, and developed into a nor-

mal embryo (Fig. 7.4: a_1–a_4). With later splitting (Fig. 7.4: b_1–b_4, c_1–c_4, e_1–e_4, f_1–f_4), the halves of the germ band more or less regulated toward normal development but not completely, except in f_1–f_4 where wound healing occurred.

Bisection during or after gastrulation (Fig. 7.4: b_1–b_4, c_1–c_4) also resulted in a new midline becoming established in each half, with bilaterality possibly being regained by assimilatory (intradermal) induction of neighboring amniotic cells (Fig. 7.5D–F). The capacity to replace a missing lateral half was first lost in the prothoracic region and later in more anterior and posterior regions.

If entire cleaving eggs of the leafhopper *E. plebejus* were bisected longitudinally in either the sagittal or the frontal plane, a normal embryo developed from both the left and the right or dorsal and ventral halves (Sander, 1971).

7.1.3.2 ANTEROPOSTERIOR PATTERN FORMATION

Results of exquisitely designed and executed ligaturing and centrifugation experiments in the late 1950s and early 1960s on recently deposited insect eggs by workers such as Sander and Yajima suggested temporary production of two agents, one at either end of the egg, to create two mutually permeating gradients within it (Fig. 7.6). Each agent was believed to originate from a high point or *source* at one end of the egg and to be eliminated at a low point or *sink* at the other. Depending on the relative concentrations of the two agents at particular positions along the length of the egg, it was suggested that a cell could "recognize" its position within it and differentiate accordingly.

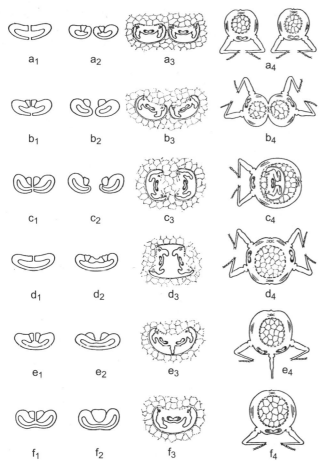

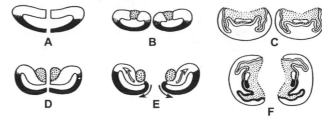

Fig. 7.5. Bilateral embryonic regulation depicted in transverse sections through the prothorax of *T. asynomorus* embryos, showing the source of cells forming the germ band when both the germ band and amnion are split longitudinally before (A–C) and just after (D–F) gastrulation. When the embryo is longitudinally bisected after gastrulation, half of each embryo is derived from amniotic cells (shown in black in D–F) by intradermal induction. (Reprinted with permission from A. Kühn. *Lectures on developmental physiology*, p. 333, Fig. 426 after Krause, 1953. © 1971 Springer-Verlag.)

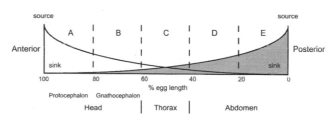

Fig. 7.6. Graph showing how two permeating gradients within an egg in preblastoderm, and each with a source at one end and a sink at the other, could provide positional information to nuclei distributed along its anteroposterior axis.

Fig. 7.4. Bilateral embryonic regulation, as shown in transverse sections through the prothorax, in embryos of *T. asynomorus* split lengthwise before (a_1, d_1), during (b_1, e_1), and after (c_1, f_1) gastrulation. a_1–f_1, before healing; a_2–f_2, after healing; a_3–f_3, midway through intertrepsis; a_4–f_4, after dorsal closure. a–c, median longitudinal section of both germ band and amnion; d–f, same, through germ band only. (Reprinted with permission from A. Kühn. *Lectures on developmental physiology*, p. 332, Fig. 425 after Krause, 1953. © 1971 Springer-Verlag.)

The concept of morphogenetic gradients in animal development first arose from T. H. Morgan's experiments in 1897 on regeneration in annelid worms (Lawrence, 1992: 204). He found that the speed at which a worm was able to regenerate its head following decapitation depended on where the cut was made—neck cells regenerating heads quickly and trunk cells, more slowly—and concluded that factors varying quantitatively along the length of the body controlled this response.

The possible role of gradients in specifying positional information was more rigorously presented in a model called the *French flag* devised by Lewis Wolpert (1968, 1978) (Fig. 7.7), in which pattern formation was envisaged to be a two-step process: the setting up of positional information within the egg, followed by its interpretation by cells according to both their position and their genetic program. Wolpert suggested that this positional information could be provided by a gradient in concentration of a diffusible morphogen along the length of the egg, the cells using its absolute concentration as a measure of their position and their threshold of response to it depending on their previous developmental history. Increasing experimental evidence is now available for the existence and function of just such gradients.

7.1.3.2.1 *EUSCELIS PLEBEJUS*

Leafhoppers of *E. plebejus* have intermediate germ eggs. Sander (1959) transversely ligatured eggs with a hair loop at distances 30%–70% from their posterior ends and at various times (0–40h) following oviposition and monitored the development of embryos in the posterior isolate. When eggs in late cleavage were ligatured at more than 60% from the posterior end, a complete embryo usually formed in the posterior compartment. When they were ligatured farther back, progressively more anterior segments were missing from the embryo that formed. Also, as time passed, the minimum distances required for total development and for development of

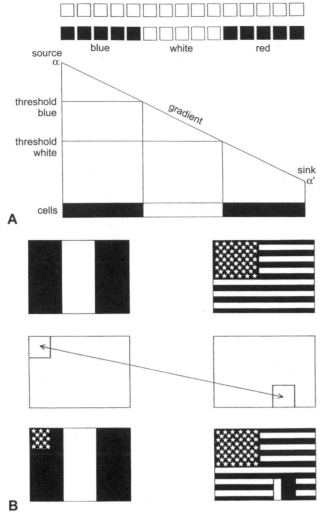

Fig. 7.7. The French flag model. A. This diagram shows how a line of cells (top) that can differentiate as blue, white, or red (second line) can be organized to form the French flag. Positional information is delivered by a gradient of diffusible morphogen and the cells have appropriate thresholds. A source keeps the gradient concentration fixed at α at one end, and a sink keeps the concentration at α′ at the other. Thresholds are cellular properties that enable the gradient in positional information to be interpreted. For example, above threshold blue, a cell becomes blue. B. Grafted flag material develops on the basis of position and "genetic program." Undifferentiated American flag material grafted into the upper left corner of the French flag develops as stars; similarly, a piece of French flag grafted into the American flag develops in accordance with its own genetic program, which, near bottom right, specifies vertical white and red patches. (Reprinted with permission from L. Wolpert. Pattern formation in biological development. *Sci. Am.* 239(4): 154–164. © 1978 Scientific American.)

particular anterior segments of embryos in the posterior isolate decreased. These results suggested that some factor produced in the anterior end of the egg and necessary for embryogenesis of anterior segments diffuses posteriorly as development proceeds.

When eggs were ligatured near the posterior end, complete embryos developed in the anterior isolate regardless of when the ligature was placed, but when ligatured at over 50% of egg length, no germ band developed there. Also, when eggs were ligatured between 23% and 45% of egg length, complementary partial embryos developed in both fragments, those in anterior isolates always beginning with anterior head segments and those in posterior ones with posterior abdominal segments. And when eggs were ligatured during cleavage, one or more middle segments were missing from the embryos that formed (the *gap phenomenon*), suggesting that factors diffusing in either direction from both poles are required if all segments are to form. These results stimulated Sander (1960) to investigate these factors further in the following way (Fig. 7.8).

In *E. plebejus*, the posterior 10% of egg length contains a ball of symbiotic bacteria that can be used as a marker to indicate the position of posterior ooplasm (Fig. 7.8) (see Müller [1940] for a detailed and beautifully illustrated analysis of the development of these bacteria in fulgoroid embryos). Both this ooplasm and the ball can be pushed anteriorly by poking in the posterior end of the egg with a blunt needle and then ligaturing the egg with a hair loop. Results of such experiments proved that diffusible morphogens, produced in the periplasm at both poles, are involved in specifying longitudinal body pattern in embryos of this insect (Fig. 7.8).

Normal embryogenesis is summarized in Fig. 7.8A. Anterior fragments of eggs, ligatured during cleavage, produced only head structures or nothing, indicating the presence of an anterior morphogen (Fig. 7.8B, D). However, if posterior pole material was pushed anteriorly before ligaturing, such anterior fragments became capable of forming partial or complete embryos, presumably owing to the anterior presence of a posterior morphogen (Fig. 7.8C). Partial or complete polarity reversal of structures within posterior fragments could be produced by pushing posterior pole plasm to just behind the ligature (Fig. 7.8E, F), while complete embryos sometimes formed in the anterior fragment if the posterior material, after having been present from late cleavage to early germ band stages, was subsequently removed from the fragment (Fig. 7.8F).

Results varied depending on where and when the ligature was placed. For example, the posterior pole material, when provided to an anterior fragment, sometimes resulted in only a partial germ band being formed (Fig. 7.8C, left egg). In addition, the length of the germ band produced could vary by a factor of two or more.

Sander concluded that each pole initiates a gradient of one morphogen, the local concentrations of the two morphogens at a particular distance along the egg telling the blastoderm cells at that site to differentiate into those characteristic of a particular body segment (Fig. 7.6). Similar results have since been obtained in eggs of *Bruchidius* sp. (Coleoptera: Curculionidae), *Pimpla turionellae* (Hymenoptera: Ichneumonidae), *Protophormia terrae novae* (Diptera: Calliphoridae), and *D.*

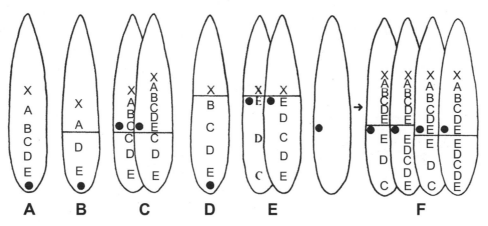

Fig. 7.8. Results of experiments demonstrating the properties of the posterior pole plasm in *Euscelis plebejus* eggs. In each diagram the filled circle represents a symbiont ball in the posterior cytoplasm of the egg that serves as a marker to indicate the position of posterior cytoplasm displaced in the experiments. A. Normal sequence of tagmata in the germ band: x, serosa; A, protocephalon; B, gnathocephalon; C, thorax; D and E, abdomen. B. Following ligation during early cleavage, a gap appears in the pattern (B and C are missing). C. A more complete set of structures is formed by the anterior isolate if posterior cytoplasm is moved forward before ligation. D. A different gap results if the ligature is placed farther forward in the egg early in cleavage. E. Partial or complete polarity reversal of tagmata within the posterior isolate can be produced by displacement of posterior cytoplasm. F. If a time interval is left between displacement and tying the ligature, then both effects can be produced simultaneously. (Adapted from K. Sander. 1984. Embryonic pattern formation in insects: basic concepts and their experimental foundations. In G. M. Malacinski (ed.), *Pattern formation: a primer in developmental biology*, p. 257, Fig. 10, after Sander, 1960. © 1983 Macmillan, N.Y.)

melanogaster (Herth and Sander, 1973; Counce 1973) using similar techniques.

7.1.3.2.2 LOWER DIPTERA

When cleaving eggs of the chironomid midges *Chironomus dorsalis* and *C. samoensis* were placed within a fragment of fine glass capillary tube and spun in a centrifuge toward either of their poles, their contents were stratified according to specific gravity and their embryos developed as "double abdomens" or "double cephalons" depending on the direction of centrifugation (Fig. 7.9B, C) (Yajima, 1960, 1964, 1970, 1983). When spun with the centrifugal force acting toward either end, the polarity of the A-P gradients was reversed and normal pattern elements were replaced by elements usually forming in the other half of the egg. The larvae resulting were normal except for the absence of labial, thoracic, and abdominal segments in double cephalons and of cephalic, thoracic, and first abdominal segments in double abdomens and except for duplication and reversed polarity of the remaining segments (Fig. 7.9B, C, F). However, gonads occurred in only the anterior abdomen of double abdomens (Fig. 7.9F: right) and germ cells (pole cells) only in the posterior cephalon of double cephalons (Fig. 7.9F: left); germ cells in the latter instance became tetranucleate and did not lodge within the gonadal rudiments, presumably because appropriate mesoderm was not there for the gonad to form.

Missing middle segments can also be induced by transversely squeezing the long germ eggs of higher dipterans (*P. t. novae*, *D. melanogaster*) across their length with a blunt razor blade at any time during cleavage, the number of middle segments missing from the larva being larger or smaller depending on whether the squeeze was made early or late in cleavage (Herth and Sander, 1973).

Finally, mutant expression of the gene *bicaudal* (*bic*) early in oogenesis of *D. melanogaster* females can cause the production of double-abdomen embryos (Bull, 1966), and mutant expression of the gene *dicephalic* (*dic*) results in double-cephalon embryos (Lohs-Schardin, 1982) (the latter develop from egg follicles having nurse cells at both ends), suggesting the anterior and posterior diffusion factors to be gene products.

All these findings can be construed as resulting from disruption in the diffusion of factor(s) emanating from the two ends of the egg after fertilization, factors that are necessary to generate a complete body pattern (Fig. 7.6).

In later experiments, Kalthoff and colleagues (1982, 1983) showed that double abdomen could be induced in embryos of the chironomid *Smittia parthenogenetica* by irradiating the anterior pole region with ultraviolet (UV) light, by centrifuging the eggs toward either pole, or by puncturing the anterior pole before or during cleavage (Fig. 7.10). From his results, Kalthoff concluded

- that factors specifying formation of "head" and "abdomen" exist in both anterior (determinants *a*, *p'*) and posterior (determinants *a'*, *p*) egg halves;
- that determinant *p'* is not expressed as long as an anterior determinant *a* is present and active at the proper site and time; and

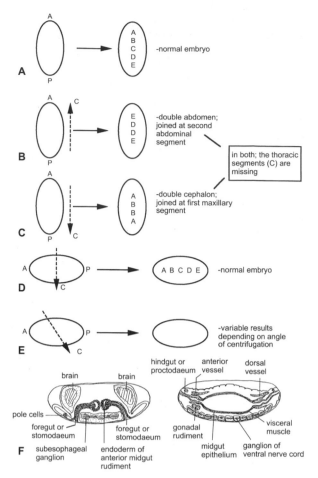

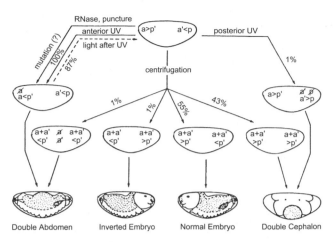

Fig. 7.9. A–E. Effects of centrifugation on pattern formation in cleaving eggs of *Chironomus dorsalis* (Diptera: Chironomidae). A. Normal development. Letters refer to same tagmata as in Fig. 7.8. Changes in pattern formation are shown for when the egg was centrifuged (C) in the direction indicated. A, anterior pole; P, posterior pole. F. Sagittal sections of double-cephalon (left) and double-abdomen (right) prolarvae just before hatch. (F reproduced with permission from H. Yajima. Study of the development of the internal organs of the double malformations of *Chironomus dorsalis* by fixed sectioned materials. *J. Embryol. Exp. Morphol.* 24: 287–303.© 1970 Company of Biologists, Ltd.)

Fig. 7.10. Kalthoff's model of anteroposterior specification in *Smittia parthenogenetica* (Diptera: Chironomidae) embryos. Dorsal is up; anterior, to the left. a and a', anterior determinants; p and p', posterior determinants. Slashed letters indicate partial inactivation or loss of determinant. Determination for anterior versus posterior development was thought to occur independently in each egg half according to the relative strength of anterior and posterior determinants (symbolized by < or >). Application of ultraviolet (UV) irradiation or RNase to the anterior pole causes double-abdomen formation with high yield and was thought to inactivate a, rendering p' dominant. Exposure to visible light after UV appeared to reactivate a. Posterior UV irradiation caused double-cephalon formation, but with low yield, and was thought to inactivate a' and p to a similar extent, rendering a' dominant only in exceptional instances. Four body patterns were found after centrifugation. They were thought to result from a variable redistribution of a and a' after centrifugation. Percentages indicate the frequency of body patterns obtained under the experimental conditions indicated. (Cytoplasmic determinants in dipteran eggs, K. Kalthoff, In W. R. Jeffrey and R. A. Raff (eds.), *Time, space and pattern in embryonic development*, pp. 313–348, Copyright © 1983 Wiley-Liss, Inc. Reprinted by permission of Wiley-Liss, Inc., a subsidiary of John Wiley & Sons, Inc.)

methods applied to eggs immediately following oviposition.

- Double abdomen resulted whether or not nuclei were present at the time of treatment; thus, the determinant is passed into the egg during oogenesis.
- Injection of the enzyme RNase into the anterior pole region removed this agent and caused double abdomen, suggesting the agent to be maternal mRNA.
- Repair, by exposure to light, of UV damage to mRNA could be delayed until blastoderm formation. Thus, cooperation of a and p' does not occur before this time.

In summary, results of both transverse and longitudinal disruption experiments and the phenotypic expression of certain mutant genes strongly imply that patterning mechanisms act successively and differently along the two axes of the embryo and at right angles to each other rather than by simultaneous, two-dimensional patterning. The results also suggest that

- that removal of *a* does not necessarily remove or destroy *p'*.

Kandler-Singer and Kalthoff (1976) then demonstrated the anterior determinant in *S. parthenogenetica* eggs to be, at least partly, maternal mRNA bound to a protein and provided by the nurse cells to the egg's periplasm during oogenesis. This transcript was suggested to be inactive until fertilization, at which time it was removed from its protein and translated, the A-P polarity of the egg then being established by production of an A-P gradient of its protein (compare Fig. 7.6). This sequence was suggested by the following results:

- Double abdomen could be induced by any of these

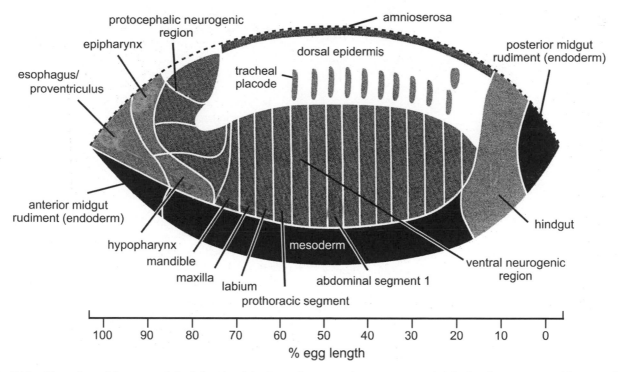

Fig. 7.11. Blastodermal fate map of the left side of the *D. melanogaster* (Diptera: Drosophilidae) embryo as it would appear if flattened to show the true distances between positions. The dashed line at the top represents the dorsal midline, and the ventral margin, the ventral midline. The scale shows distances in percentage of body length from the posterior pole. (Reprinted with permission from V. Hartenstein. *Atlas of* Drosophila *development*, p. 5. © 1993 Cold Spring Harbor Laboratory Press.)

these mechanisms are under the control of various genes (Sander, 1984).

7.1.4 FATE MAPS

When partial determination of an embryo's cells has occurred, it is possible to recognize organ-forming areas or fields on the blastoderm giving rise to each body part. For example, Foe's (1989) "mitotic domains" (Fig. 6.10) constitute a naturally existing fate map. The easiest way to construct a fate map, the histological method, is simply to follow the embryonic development of each body part of an insect from blastoderm to hatch in whole mounts and serial sections and then to indicate, on an outline diagram of the egg, which blastoderm cells you believe gave rise to it. This was the method used by Poulson (1950: Fig. 39) to construct his much-used fate map of the *D. melanogaster* preblastoderm and by Anderson (1973) for the fate maps so fundamental to his phylogenetic conclusions in *Embryology and Phylogeny in Annelids and Arthropods*.

A more rigorous method is to label or destroy one or more cells in the blastoderm and follow them, their progeny, or the lesion resulting from their destruction to the position they eventually occupy in the fully formed juvenile. One then indicates its extent on the blastoderm by outlining it on a diagram of the egg. Other blastodermal cells elsewhere in the embryo are similarly marked and followed and the procedure repeated until one has constructed an entire fate map. Figure 7.11 is a fate map of the *D. melanogaster* blastoderm based on data of Hartenstein and colleagues (1985) and Jürgens and coworkers (1996) as presented in Hartenstein (1993).

Over the years, many methods have been devised for marking embryonic cells so that they and their progeny can be followed during their subsequent development, including the following:

- Natural markers (lipid vacuoles, yolk or pigment granules, etc.) may be present in certain blastoderm cells that are allocated to particular body parts or organs during organogenesis, and can serve to identify them (e.g., the polar granules of pole cells).
- Localized injury to an embryo caused by microcautery with a hot wire, UV microbeam irradiation, ligaturing, puncturing, centrifuging, and so on, can be followed during subsequent ontogeny.
- Vital staining of blastoderm cells with harmless, long-lasting dyes can be used to mark them. For example, Technau (1987) used horseradish peroxidase (HRP) to mark cells that were transplanted from donor to host embryos by employing suction, injection, and wound-sealing capillaries (Fig. 7.12; see 6.6.1). The number and distribution of the progeny of such labeled transplanted cells can then be observed later in develop-

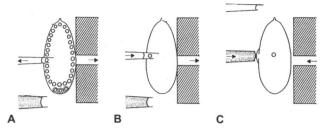

A   **B**   **C**

Fig. 7.12. Illmensee's procedure (1972) for transplanting embryonic nuclei between eggs of *D. melanogaster*. A. Removal of nucleus from donor egg. B. Its injection into an unfertilized recipient egg. C. Sealing wound with sulfosalicylic acid. In all three diagrams, the injection capillary is above and the wound-sealing capillary below, while the larger suction capillary, used to hold an egg during nuclear removal or injection, is hatched. (Reprinted with permission from K. Illmensee. Developmental potencies of nuclei from cleavage, preblastoderm, and syncytial blastoderm transplanted into unfertilized eggs of *Drosophila melanogaster*. *W. Roux's Arch. Entwicklungsmech. Organ.* 170: 267–298. © 1972 Springer-Verlag.)

ment in whole mounts of host embryos examined by differential interference contrast (DIC) or fluorescence microscopy or in serial sections (only the transplanted cell and its progeny are labeled with HRP). Both *homotypic* (from the same area of the donor into the same area of the host) and *heterotypic* (from one area of the donor into another area of the host) transplants can be made, the latter providing information about the state of cellular commitment at the time of transfer and the roles of cellular interaction, positional formation, and cell lineage on this process.

- Gynandromorphs (4.7) induced in *Drosophila* embryos can be used to identify the position, at the blastoderm stage, of different organ primordia. For example, Sturtevant (1929) calculated the frequency with which various pairs of epidermal structures were of different genotype in gynandromorphs of *D. simulans* and devised the first fate map for embryos of this genus of flies.
- Monoclonal antibodies (Mabs) can be raised in mice or rabbits against specific embryonic antigens such as the membrane-bound protein characterizing a particular cell type. Such antibodies can then be labeled with a fluorescent marker injected into the embryo, and the specimen examined by fluorescence microscopy. The labeled antibody binds to the antigen wherever it occurs and thus functions as a probe specifically identifying all cells in the embryo having that antigen. Hartenstein and colleagues (1995) used this technique with great success to identify the time and site of origin and proliferation, and the eventual number and location of the 87 cell types known to comprise a ready-to-hatch *D. melanogaster* larva.

7.1.5 DEVELOPMENTAL GENETICS
In 1909, T. H. Morgan and colleagues commenced their classic breeding experiments at Columbia University on the fruit fly *D. melanogaster* and by 1915 had discovered more than 100 mutant genes in four linkage groups that equaled the number of chromosome pairs in the cells of this insect (Morgan et al., 1915).

Detecting a gene mutation adversely affecting a particular developmental process implies that the normal (wild-type) allele of that gene participates in that process. The nature of the mutant phenotype can sometimes suggest which developmental step is affected. Thus, the sequence of gene-controlled steps in development of a particular body part can be reconstructed by carefully analyzing a series of mutant genes affecting its development, a method of analysis called *genetic dissection*. Many genes now known to function in *D. melanogaster* development have been identified in such screens, with the often whimsical names applied to them being based on the mutant phenotypes induced.

In the early 1940s, Poulson published a series of papers first demonstrating the importance of genes in controlling the development of *D. melanogaster* embryos (Poulson, 1945). With 12 mutants, all deletions, he showed closely coordinated but separate steps in embryogenesis to be under the control of different genes.

Since then, expression of four classes of regulatory, pattern-control genes has been shown to specify characteristics of the developing *Drosophila* embryo:
- those specifying the spatial (A-P, D-V, and terminal) coordinates of the egg and future embryo;
- those specifying cell identity along the D-V axis;
- those specifying the number and polarity of body segments; and
- those specifying the identity and sequence of each segment.

A model greatly facilitating our perception of how such genes might work was proposed by García-Bellido and colleagues in 1973 based on their discovery of the compartment boundary in the wing discs of *D. melanogaster* larvae (García-Bellido et al., 1973, 1979) and was further perfected by Kauffman (1981) and Meinhardt (1982). Changing views on the roles of genes in insect embryogenesis from 1886 to 1983 were perceptively reviewed by Sander (1986).

7.1.6 THE COMPARTMENT HYPOTHESIS
Beginning just prior to blastoderm formation in *D. melanogaster* eggs and continuing progressively throughout their subsequent embryonic and postembryonic development, there appears to be sequential commitment of neighboring groups of cells into progressively smaller and alternate developmental pathways (compare Fig. 7.17A) (García-Bellido et al., 1979; Kauffman, 1981). These commitments are invisible in normal eggs but can be demonstrated, by use of appropriate methods, to involve the subdivision of the blastoderm, the body of the embryo, the segments, and after hatch, the imaginal discs of the larva into progressively smaller morphogenetic fields or compartments, each bounded by lines of clonal restriction called *compartment boundaries*.

The first experimental demonstration of compartments depended on the discovery by T. S. Painter, in 1934, that somatic cells of *D. melanogaster* exhibit pairing of homologous chromosomes similar to that occurring in primary spermatocytes and oocytes during meiotic prophase (Fig. 1.10D). In 1936, C. Stern observed that recombination (crossing-over) between these chromosomes can occur during mitotic prophase just as it does in prophase I in male and female germ cells (Figs. 1.10E, F; 4.1) in a process called *mitotic recombination*. Although the frequency of such recombination is low in nature, it can be increased by irradiating embryos and larvae with x-rays or by treating them with chemical mutagens such as ethylmethyl sulfate.

In developmental studies, stocks of *Drosophila* are used that are heterozygous for a recessive, mutant marker gene. If such a heterozygous cell is irradiated in prophase in a developing embryo or larva, crossing-over may be induced such that one or both of its daughter cells become homozygous with respect to the mutant gene and subsequently exhibit the mutant phenotype following metamorphosis (Fig. 7.13: top). The clone of cells arising from such initial daughter cells will appear in the adult as a patch of mutant cuticle surrounded by wild-type cuticle (Fig. 7.13: middle, bottom).

Such clones of homozygous mutant cells can be induced in flies at any stage of development from blastoderm to third (last) instar larva: the earlier they are induced, the fewer there are and the larger the patch of mutant cuticle resulting in the adult (Fig. 7.14). This is because individual mutant cells induced early in development comprise a larger proportion of the cells present and divide more times before adult emergence than do those induced later.

It was soon noticed that such clones of mutant tissue have certain limits to their growth (Fig. 7.15). In this example, cells of a clone were unable to cross an invisible line dividing the wing into anterior and posterior halves, this line being a compartment boundary. It is now known that as development proceeds, the blastoderm and, later, embryonic segments and the larva's imaginal discs are subdivided into smaller and smaller compartments by formation of more and more compartment boundaries, as shown in Fig. 7.16.

Based on such results, García-Bellido, Ripoll, and Morata, between 1973 and 1976, developed the *compartment hypothesis* (García-Bellido et al., 1979; Lawrence, 1990). They proposed that whenever a group of cells is subdivided into two compartments, a regulatory selector gene is expressed (1) in the cells in one compartment but remains switched off (0) in those of the other, thereby controlling the developmental distinction between them (Fig. 7.17A). Additional selector genes were thought to be activated as compartmentalization continues, the cells in the last-formed terminal compartments each containing a record of these determinative steps (for example, as 111 in terminal compartment A in the bottom line of Fig. 7.17A).

García-Bellido and Kauffman then suggested that this

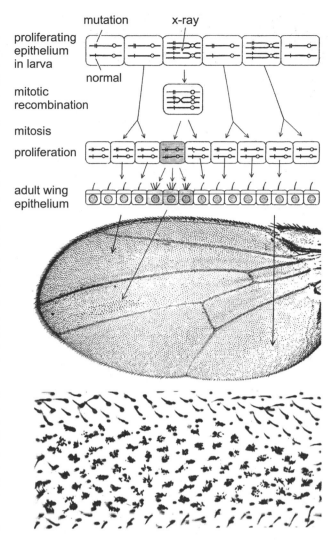

Fig. 7.13. Cell lineage is examined in *D. melanogaster* flies heterozygous for a recessive "marker" mutation (*multiple wing hair* [*mwh*]) affecting the deposition of adult cuticle. In this example, a mutation followed by mitotic recombination results in development of a clump of microtrichia by that cell instead of just one. Usually, one chromosome of the homologous pair carries the mutation; the other does not, and the recessive mutation is not expressed. The fly embryo or larva is exposed to x-rays. Irradiation may break the chromosomes in such a way that during mitosis, crossing-over occurs and a daughter cell becomes homozygous for the mutation. That cell and its descendants (shaded) express the mutation and produce multiple microtrichia that appear as a patch in the adult wing following metamorphosis, as shown in the photomicrograph of the wing (middle) and the enlargement (bottom). (Reprinted with permission from A. García-Bellido, P. A. Lawrence, and G. Morata. Compartments in animal development. *Sci. Am.* 241(1): 102–110. © 1979 Scientific American.)

progressive compartmentalization might involve production of a sequence of overlapping and progressively smaller protein patterns, with each pattern the product of a single, expressed selector gene and each accounting for formation of an additional compartment boundary.

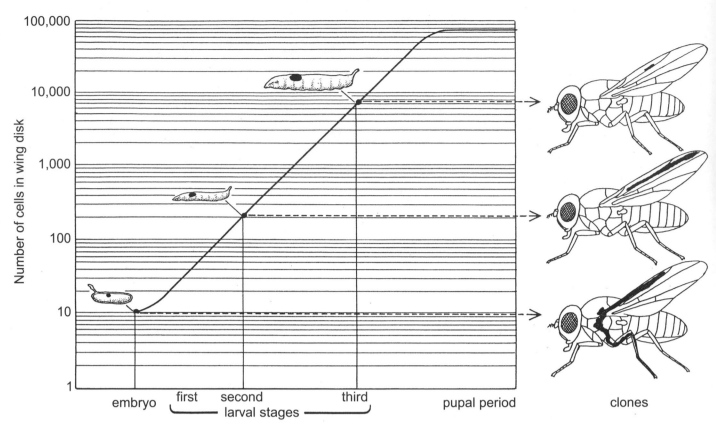

Fig. 7.14. Clues to the development of the wing and leg in *D. melanogaster* derive from studies of clones induced at successive stages of development. If cells in the embryo that will develop into a wing or leg are irradiated, the induced clones are few (because the target is small and has relatively few cells) but can be large, extending from the wing to the leg of the adult after metamorphosis. In the larva, the cells that will form each adult part proliferate within discrete imaginal discs. As the number of cells in each wing disc increases, irradiation induces clones with higher frequency, but each clone is smaller and populates a more limited region of the adult because a cell irradiated later leaves fewer descendants. (Reprinted with permission from A. García-Bellido, P. A. Lawrence, and G. Morata. Compartments in animal development. *Sci. Am.* 241(1): 102–110. © 1979 Scientific American.)

When compartmentalization is complete in the late larva, each terminal compartment in an imaginal disc would be specified by products of a unique combination of active selector genes (Fig. 7.17A: bottom row) collectively constituting its "genetic address."

This hierarchy of selector gene proteins, in turn, was thought to regulate expression of blocks of downstream structural genes in each compartment so that appropriate differentiation could occur within it. The hypothesis thus envisaged the animal as developing in a series of modules, each module, once specified, growing and shaping itself relatively independently of the others and each subject to somewhat independent genetic control (Lawrence, 1990). Finally, it was suggested that cell determination during development of *D. melanogaster* might have a three-level hierarchy of control:

- The gradients of one or more morphogens, resulting from expression of maternal genes during oogenesis, were proposed to be created in the periplasm of the egg following fertilization and to influence which selector genes were expressed in blastoderm cells and which

were not based on their threshold of response to the morphogen (Fig. 7.17B).
- The differing combination of selector gene proteins translated in each compartment were then thought to influence which structural genes were transcribed in that compartment (Fig. 7.17B).
- Finally, expression of the structural genes in each cell of each compartment was suggested to produce a particular mix of proteins and the cells, to differentiate as instructed.

Fig. 7.17C is a diagram of the blastoderm of a *D. melanogaster* egg with four hypothetical compartment boundaries indicated on it and with the genetic address of each compartment specified.

7.2 Molecular Genetics of Pattern Formation in *Drosophila melanogaster* Embryos

Many genetic mutations known to affect embryogenesis in *D. melanogaster*, though eventually lethal, allow the

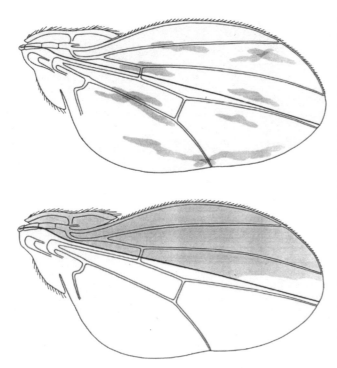

Fig. 7.15. Plotting typical wing clones induced in different flies on a diagram of the adult wing shows that they overlap (top); descendants of each marked cell make different parts of it. The lineage is not completely determinant, however. Even when marked cells are made to grow relatively quickly by use of the *Minute* technique (the marked cells divide more rapidly than the wild-type cells), they never fill more than half the wing (bottom), either the anterior (as here) or the posterior half. Note that the small clones (top) also respect this boundary. This shows that development is compartmentalized. (Reprinted with permission from A. García-Bellido, P. A. Lawrence, and G. Morata. Compartments in animal development. *Sci. Am.* 241(1): 102–110. © 1979 Scientific American.)

embryo to develop to a stage at which it can deposit its first instar cuticle. Analysis of disruptions in the pattern of sensilla and ventral denticles in such cuticles (these are segment specific) enabled E. B. Lewis and colleagues to search the chromosomes for genes specifying segment identity. Since 1978, this method has been used with great success, first by C. Nüsslein-Volhard and E. Wieschaus (1980) and more recently by hosts of other developmental geneticists, to systematically search the genome for regulatory genes specifying the two ends (termini), the D-V and the A-P axes of *D. melanogaster* embryos, and their segmental body pattern (Akam, 1987; Ingham, 1988; Lawrence, 1992; Driever, 1993; St. Johnston, 1993; Sprenger and Nüsslein-Volhard, 1993; Chasan and Anderson, 1993; Pankratz and Jäckle, 1993; Jürgens and Hartenstein, 1993; Campos-Ortega and Hartenstein, 1997).

The first clue as to how such genes might regulate pattern formation came over 85 years ago with the dis-

covery of mutations in *D. melanogaster* that disrupted development of the adult body pattern (Alberts et al., 1994: 1093). The mutation *Antennapedia* (*Antp*), for example, causes legs to replace antennae on the head of the fly. Such mutations that transform structures characteristic of one segment into those of another are said to be *homeotic* (Bateson, 1894; then spelled "homoeotic") and in their normal, wild-type expression, are referred to as *homeotic selector* or *HOM* genes (now usually called *Hox* [*Homeobox*] genes; see 7.2.1.3.5). Their discovery stimulated many workers to perform genetic dissection experiments that gradually revealed the body of a fly to consist of a series of discrete modules, each expressing a different set of one or more *HOM* genes, *exactly as envisaged in the compartment hypothesis.*

7.2.1 PATTERN-CONTROL GENES

The *HOM* genes are now known to belong to a larger group of *pattern-control genes* whose coordinated expression specifies development of the normal body pattern (Alberts et al., 1994; Rivera-Pomar and Jäckle, 1996; Gerhart and Kirschner, 1997). As mentioned earlier, the system consists of four classes of regulatory genes: (1) Products of the mother's egg-polarity (maternal effect) genes act first to establish the spatial (terminal, D-V, and A-P) coordinates of the embryo by generating a series of morphogen gradients within the periplasm of the egg. (2) The zygote's D-V genes then interpret this D-V positional information, their products specifying cellular identity along the D-V axis. (3) Simultaneously, the zygote's segmentation genes interpret the A-P positional information, their products both dividing the body of the embryo into a series of identical segments and, along with products of the egg polarity genes, activating the (4) homeotic genes, which specify segment identity and position. Combined expression of the D-V, segmentation, and *HOM* genes provides the cells in each segment with a precise indication of their position and functions to guide their subsequent determination and differentiation.

In milk bottle cultures in the laboratory, development of *D. melanogaster* from newly fertilized egg to adult takes about 10 d at 25°C, with embryogenesis requiring 24 h and postembryogenesis the remainder (Fig. 7.18) (see Sgro and Partridge, 2000, for significant differences in the life history of this species between members of wild-caught populations and those cultured in the lab in bottles or cages). Adults have a six-segmented head, a three-segmented thorax (T1–T3), and an eight-segmented abdomen (A1–A8) (Fig. 7.19: top).

However, based on the changing pattern of selector gene expression in embryos as they develop, developmental geneticists now recognize a "secondary anatomy" (Gerhart and Kirschner, 1997) of 14 parasegments, P1–P14, from the mandibular segment to abdominal segment 8, each one-half segment out of register posteriorly with respect to traditionally recognized segments (Fig. 7.19: middle). Parasegmental rather than segmental posi-

clones　　　　　　　　compartments　　　　　　polyclones

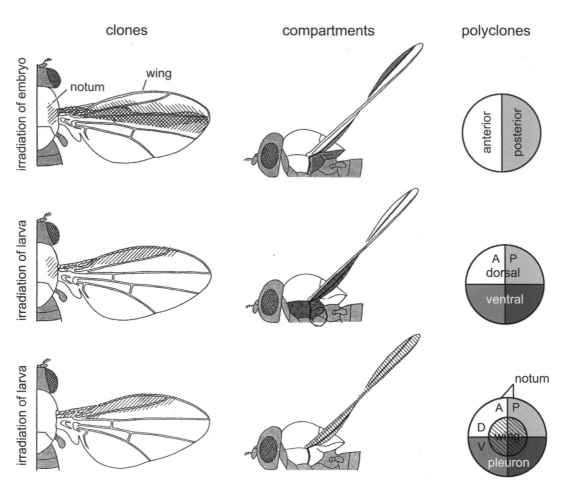

Fig. 7.16. Polyclones in the wing disc build compartments in the wing, notum, and pleuron of an adult *D. melanogaster*. Irradiation of the embryo (top) induces a large clone (hatched, left) that can extend throughout either the anterior (white) or posterior (shaded) compartment (center), but not into both compartments because the cells are already divided into anterior or posterior polyclones (right). Clones induced in the larva (middle) are further confined to either the dorsal (white) or ventral (dark gray) region of the anterior or posterior compartment (center); the wing disc has become divided into four polyclones (right). At about the same time, the clones become still further confined (bottom) to either notum and pleuron (white) or wing (hatched). The preexisting polyclones have become subdivided again; each of the eight polyclones (right) has a different combination of three properties: anterior (A) or posterior (P), dorsal (D) or ventral (V), and wing or notum. (Reprinted with permission from A. García-Bellido, P. A. Lawrence, and G. Morata. Compartments in animal development. *Sci. Am.* 241(1): 102–110. © 1979 Scientific American and Ikuyo Tagawa Garber, executrix of the estate of the artist Bunji Tagawa.)

tion is routinely used to specify where a particular gene is active in the A-P axis. The two ends (termini) of the animal, the acron and telson, are not segmentally derived and are under control of a separate set of terminal group genes.

7.2.1.1 MATERNAL EFFECT (EGG-POLARITY) GENES
As indicated by results of early experiments summarized in 7.1.3, D-V and A-P coordinates appear to define cellular position within the blastoderm, and results of countless genetic experiments on embryos of *D. melanogaster* have revealed that the specification of body pattern within the embryonic body depends on

the function of three, relatively independent, patterning mechanisms: one for D-V, one for terminal, and one for A-P pattern:

- Sequential expression of at least 18 regulatory genes establishes the D-V axis of the egg, and a mutation in any one of them generates embryos that are either *dorsalized* (lack ventral structures) or *ventralized* (lack dorsal structures).

- Expression of some 24 terminal group genes specifies the two ends of the body, the acron and telson, and mutation in any of them generates embryos lacking these body parts.

- Expression of some 50 additional regulatory genes,

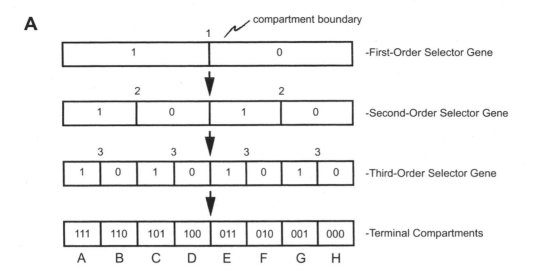

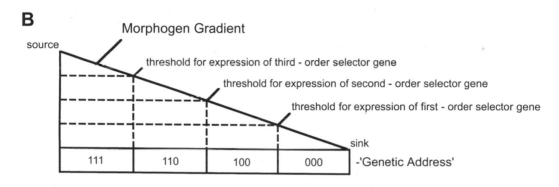

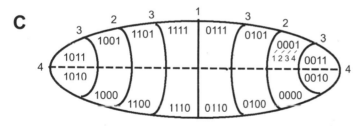

Fig. 7.17. A. How sequential expression of first-, second-, and third-order selector genes could specify eight terminal compartments in the embryonic blastoderm, each containing the products of a unique combination of active (1) and inactive (0) genes to give each one a specific genetic address. B. How the gradient of a diffusible morphogen with its source at the anterior pole of the egg and its sink at the posterior pole could influence which selector genes are expressed. C. The blastoderm of an egg with 16 hypothetical compartments induced by sequential expression of the four selector genes 1–4. (C adapted with permission from S. A. Kauffman, R. Shymko, and K. Trabert. Control of sequential compartment formation in *Drosophila*. *Science* 199: 259–270. Copyright 1978 American Association for the Advancement of Science.)

including the segmentation and *HOM* selector genes, establishes the A-P axis. Mutation in these genes generates embryos having defects or duplications in their A-P pattern.

As described in 2.2.2.6, factors inducing the formation of D-V and terminal morphogen gradients within the egg are supplied to its surface during oogenesis by genes expressed in the follicle cells about each egg while it is still within the ovary, whereas the nurse cells of each follicle contribute those defining A-P polarity.

7.2.1.1.1 DORSOVENTRAL AXIS
As summarized in Fig. 2.20, establishing the D-V axis requires the production of signals during oogenesis from

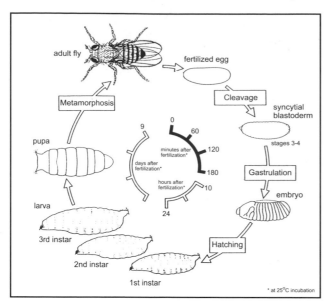

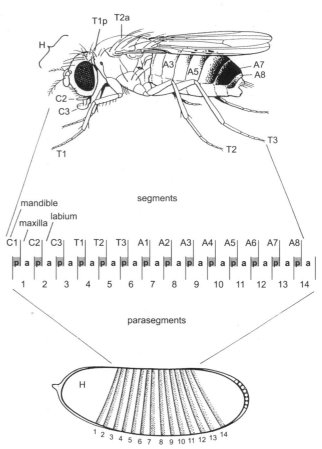

Fig. 7.18. The life history of *D. melanogaster*. (Reprinted by permission of Oxford University Press from L. Wolpert. *Principles of development*, p. 45, Fig. 2.29. Current Biology Ltd. © 1998 Oxford University Press.)

the oocyte nucleus to the follicle cells and back to the oocyte that ultimately result in a gradient of Spätzle (Spz) protein becoming established in the vitelline membrane and inner chorion of the egg, with its high point on the ventral midline (Govind and Steward, 1991; Lawrence, 1992; Chasan and Anderson, 1993; Morisato and Anderson, 1995; Gerhard and Kirschner, 1997; Wolpert, 1998; Klingler and Tautz, 1999). Release of Spz following fertilization results in a similar gradient of Dorsal (Dl) protein becoming established within ventral preblastoderm nuclei that peters out laterally at 90 degrees on either side. Dl is a transcription factor that activates expression of *twist* (*twi*) and *snail* (*sna*) at high concentrations and of *rhomboid* (*rho*) at intermediate concentrations; represses that of *zerknüllt* (*zen*), *decapentaplegic* (*dpp*), and *tolloid* (*tld*), even at very low concentrations; and has no effect on other genes such as *short gastrulation* (*sog*) (Fig. 7.20) (the complex role of activators, co-activators, repressors, and co-repressors in these downstream responses to Dl is reviewed in Mannervik et al., 1999). Thus, *twi* and *sna* are expressed ventrally where their transcription factors specify mesoderm (6.8.1), and *zen*, *dpp*, and *tld* are only expressed dorsally (*dpp* is upstream of *zen*).

A gradient of the secreted protein Decapentaplegic (Dpp) is generated dorsally with a middorsal high point because the Short gastrulation (Sog) and Twisted gastrulation (Tsg) proteins, secreted ventrally, inhibit Dpp activity by complexing with it and preventing it from binding to its receptor (Gerhart and Kirschner, 1997; Harland, 2001). The proteinase Tolloid breaks down Sog

Fig. 7.19. Lateral aspect of an adult female of *D. melanogaster* (above), showing the relationship between body segments (C1–A8) and parasegments (1–14) (middle). Parasegments represent units of gene expression that are roughly a half segment out of register posteriorly with respect to segments (they do not occur in the head anterior to the mandibular segment or in the acron or telson). Shading in the lower figure indicates the expression pattern of the segment-polarity gene *engrailed* (*en*) in the posterior compartment of each prospective segment at the time of blastoderm formation in lateral aspect and, in the middle figure, in relation to segmental and parasegmental boundaries. a, anterior compartment; H, anterior head; p, posterior compartment. (Reprinted with permission from P. A. Lawrence. *The making of a fly: the genetics of animal design*, p. 4, Fig. B1.1. © 1992 Blackwell Science, Ltd.)

and releases Dpp in its active form, with Tsg accelerating this process (Harland, 2001). Since Sog concentrations are high ventrally, degradation of Sog will have little effect ventrally because there is lots around to complex with Tsg and Dpp. But more dorsally, where Sog is less available, Dpp will be released.

Due to the presence of the Dpp gradient dorsally and the Dl gradient ventrally, five longitudinal compartments of differential gene expression are established in the blastoderm; from top to bottom these are the amnioserosa (unpaired), dorsal ectoderm (paired), neuro-

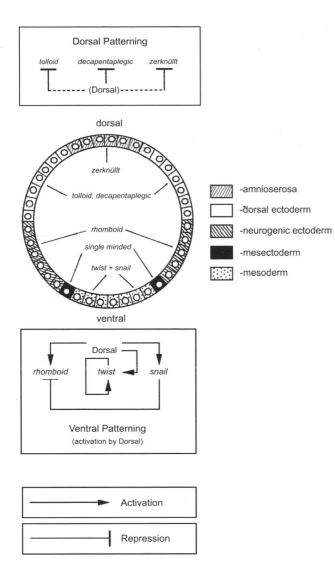

Fig. 7.20. Summary of dorsoventral compartment specification in a *D. melanogaster* embryo at blastoderm. (Adapted by permission of Oxford University Press from L. Wolpert. *Principles of development*, p. 139, Fig. 5.13. Current Biology Ltd. © 1998 Oxford University Press.)

genic ectoderm (paired), mesectoderm (paired), and mesoderm (unpaired) (Fig. 7.20).

Neurogenic ectoderm is specified because Dl activates the *rho* gene and Snail (Sna) represses it (Fig. 7.20). Because *twi* is activated by Dl in a slightly wider blastoderm domain than *sna*, and since Sna inhibits expression of the *single-minded* (*sim*) gene, a narrow row of cells on either side between mesoderm and neurogenic ectoderm can express Single-minded (Sim) protein, and these cells are specified as mesectoderm (Fig. 7.20) (Gerhart and Kirschner, 1997: 459).

Is Ventral in Insects Dorsal in Vertebrates?
The idea that the vertebrate body plan is an inverted arthropod body plan was first proposed by Étienne Geoffroy St.-Hilaire in 1822. His concept proposed that the ventral nerve cord of insects and the dorsal nerve cord of vertebrates are homologous and derive from a ventral nerve cord present in a common ancestor of these two lineages. Early acceptance of this idea led to the classification of bilateral invertebrates as Gastroneuralia and of vertebrates as Notoneuralia (Arendt and Nübler-Jung, 1994; Nübler-Jung and Arendt, 1994).

Recently, it was discovered that orthologs of some of the same genes function during development of their differently located nervous systems (Arendt and Nübler-Jung, 1994; Nübler-Jung and Arendt, 1994; De Robertis and Sasai, 1996; Gould, 1997; Holley et al., 1995; Holley and Ferguson, 1997; Gerhart and Kirschner, 1997). For example, the *achaete-scute* gene complex of *D. melanogaster* is expressed in neural precursor cells in embryos of both insects and mice. Likewise, *dpp*-related genes are involved in D-V patterning in members of both taxa. However, in *D. melanogaster*, *dpp* is expressed dorsally and controls differentiation of dorsal structures, as summarized earlier (Fig. 7.20), whereas the vertebrate ortholog *Bone morphogenetic protein-4* (*BMP-4*) is expressed ventrally and does the same for ventral structures. Similarly, the *sog* gene encodes a ventralizing factor in *D. melanogaster* while its vertebrate ortholog *chordin* has a dorsalizing effect (Arendt and Nübler-Jung, 1996).

These observations suggest that the A-P axis of the metazoan body plan was established before evolutionary divergence of gastroneuralians and notoneuralians and that D-V axis inversion occurred early in vertebrate evolution, as the deuterostome gastrulation of chordates can be conceptually derived from the protostome gastrulation seen in the presumably older protostome gastroneuralians (in deuterostomes, only the anus derives from the blastopore; in protostomes, including insects, both mouth and anus do; see Lacalli, 1995, 1996, for a full discussion). Arendt and Nübler-Jung (1994) thus concluded that the nerve cords of the two are homologous even though in different positions.

Gerhart (2000), however, suggested that this so-called inversion can be better explained by envisaging both protostome and deuterostome lineages having diverged from a common, hemichordate-like ancestor with a diffuse D-V organization followed by selection for oppositely directed condensation of the central nervous system and relocation of the heart. D-V patterning mechanisms were reviewed by Holley and Ferguson (1997).

7.2.1.1.2 TERMINAL SYSTEM
Blastodermal cells at the two ends (termini) of the embryo in *D. melanogaster* eventually differentiate as acron and telson. As summarized in Fig. 2.20 and Table 2.1, induction of termini depends on the activation of a transcription factor cascade, beginning with a transmembrane receptor tyrosine kinase (RTK) in the oolemma at each end of the egg encoded by the *torso* (*tor*) gene.

This binds an inducer protein encoded by the *trunk* (*trk*) gene whose release from the vitelline membrane is activated indirectly by release of Torso like protein from the inner chorion and vitelline membrane at each end of the egg follicle just after fertilization (Lawrence, 1992; Sprenger and Nüsslein-Volhard, 1993; Jürgens and Hartenstein, 1993; Gerhart and Kirschner, 1997; Klingler and Tautz, 1999). The Torso receptor is homogeneously distributed throughout the oolemma of the egg but only enough Trunk ligand is released to occupy receptors at the two ends.

Activated Torso induces expression of the terminal, zygotic gap genes *tailless* (*tll*) and *huckebein* (*hkb*) at either end of the egg (Fig. 7.24D) but, to do so, must inactivate co-repressors of these genes encoded by the *groucho* (*gro*) and *capicua* (*cic*) genes (Jiménez et al., 2000) (both these repressors also act in specifying the D-V axis). Expression of *bicoid* (*bcd*), *tll*, and *hkb* at the anterior end of the egg activates expression of *forkhead* (*fkh*) and *orthodenticle* (*otd*), which specify acron, foregut, and anterior midgut, while expression of *tll* and *hkb* at the posterior end of the egg activates *fkh* and *caudal* (*cad*) expression, which specifies telson, hindgut, posterior midgut, and Malpighian tubules (Fig. 7.24B, D) (details in Jürgens and Hartenstein, 1993).

In hexapods and in other arthropods with short germ embryos (6.10.1), the termini are specified before and independently of the body segments. In the nauplius larva of certain crustaceans, they establish a continuous functioning gut before any body segments are added between them through activity of the posterior growth zone (Gerhart and Kirschner, 1997).

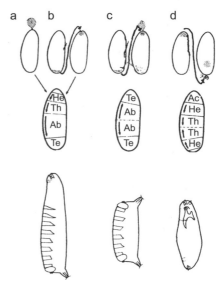

Fig. 7.21. Experimental manipulation of wild-type *D. melanogaster* embryos demonstrating the presence and action of anterior and posterior pole plasm in specifying anteroposterior body pattern. The upper row indicates the experimental procedure; the second row, the fate maps of the operated embryos; and the lower row, drawings of the resulting larvae. Ab, abdomen; Ac, acron; He, head; Te, telson; Th, thorax. See text for details. (Reprinted with permission from C. Nüsslein-Volhard, H. G. Frohnhofer, and R. Lehmann. Determination of anteroposterior polarity in *Drosophila*. *Science* 238: 1675–1681. Copyright 1987 American Association for the Advancement of Science.)

7.2.1.1.3 ANTEROPOSTERIOR AXIS

L. von Ubisch was probably the first to suggest that gradients might influence gene expression during development (Lawrence, 1992: 206), whereas beginning in 1940, H. Piepho, V. B. Wigglesworth, M. Locke, H. Stumpf, P. Lawrence, and others demonstrated the relationship between segmental polarity and gradients in the larvae of several butterflies, moths, and bugs by removing, rotating, and transposing pieces of integument (cuticle and epidermis) or portions of appendages or trachea and analyzing pattern disruption in the cuticle of subsequent instars (11.2).

When a newly deposited *D. melanogaster* egg is carefully punctured at its anterior end so that a small drop of ooplasm leaks out, the embryo developing in that egg lacks the acron (Fig. 7.21: a) (Nüsslein-Volhard et al., 1987; Nüsslein-Volhard, 1996). However, when ooplasm from the posterior end of another egg is injected into the site from which the anterior ooplasm escaped, a second set of abdominal segments, with mirror-image symmetry, develops in the anterior half of the recipient egg, replacing the head and thorax and giving it a double-abdomen phenotype) (Fig. 7.21: c) (7.1.3.2). And when ooplasm from the anterior end of a donor egg is injected into the posterior end of a posteriorly punctured egg,

that egg develops a double-cephalon phenotype (Fig. 7.21: d).

Many mutations in *D. melanogaster* have since been discovered to cause similar disturbances in body pattern in the anterior or posterior half of the embryo or in its two ends. At least 15 egg-polarity genes have been so identified:

- anterior: *bcd*, *exuperantia* (*exu*), *swallow* (*swa*), and *staufen* (*stau*)
- posterior: *cappuccino* (*capu*), *cad*, *spire* (*spir*), *mago nashi* (*mago*), *staufen* (*stau*), *oskar* (*osk*), *vasa* (*vas*), *valois* (*val*), *tudor* (*tud*), *nanos* (*nos*), and *pumilio* (*pum*)

These genes are the first to be expressed in a hierarchical system of genes encoding transcription factors that function to progressively specify the A-P axis of the body (Nüsslein-Volhard, 1991; Driever, 1993; St Johnston, 1993). They are transcribed from the maternal genome in nurse or follicle cell nuclei during oogenesis (2.2.2.3.1, 2.2.2.6) and are translated and begin to act immediately following fertilization.

For example, mothers homozygous for the egg-polarity mutation *bcd* produce embryos without acron, head, and thorax and with an abnormally large abdomen

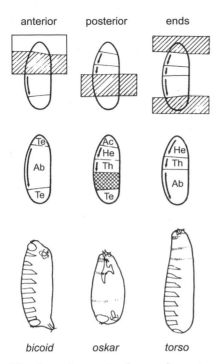

| anterior | posterior | ends |

Fig. 7.22. The three phenotypic classes of genetic mutations disrupting development of anteroposterior body pattern in *D. melanogaster* embryos as exemplified by *bicoid*, *oskar*, and *torso*. In the upper row, the wild-type origin of the regions deleted in mutant embryos is indicated by hatching (abbreviations as in Fig. 7.21). The second row shows the corresponding changes in the fate maps of the remaining areas of the embryo, while the lower row presents drawings of the larval phenotypes resulting. (Reprinted with permission from C. Nüsslein-Volhard, H. G. Frohnhofer, and R. Lehmann. Determination of anteroposterior polarity in *Drosophila*. *Science* 238: 1675–1681. Copyright 1987 American Association for the Advancement of Science.)

with a telson at either end (Fig. 7.22: left); those homozygous for the *osk* mutation generate embryos with most of the body consisting of head and thorax, with a normal acron and telson, but with the abdomen missing (Fig. 7.22: middle); while those mutant in *tor* produce embryos lacking acron and telson (Fig. 7.22: right). All three classes of mutations can be rescued by injecting ooplasm from the appropriate end of a wild-type embryo (Nüsslein-Volhard et al., 1987). The wild-type gene transcript, in each instance, appears to be localized at one end of the egg and to be translated into a factor controlling specification of A-P positional values (Alberts et al., 1994).

By in situ hybridization with a *bcd* cDNA probe, it has been shown that *bcd* mRNA is first concentrated at the anterior tip of the oocyte, after being synthesized by nurse cells connected to the oocyte, in response to release of a transcription factor encoded by the *Serendipity* (*Sry*) gene (Fig. 2.20A) (Lawrence, 1992; Alberts et al.,

1994; Stauber et al., 1999; Klingler and Tautz, 1999). As *bcd* mRNA passes by way of the ring canals into the anterior end of the oocyte from the nurse cells (Fig. 2.13B, C), it anchors to a component of the cytoskeleton (Figs. 2.20A, 7.24A). Proteins encoded by the maternal genes *exu*, *swa*, and *stau* assist in *bcd* mRNA localization, as its distribution is abnormal when any of these genes are mutant (Lawrence, 1992). Translation of *bcd* mRNA following ovulation and fertilization creates a concentration gradient of Bicoid (Bcd) protein within the periplasm, with its high point at the anterior end of the egg (Figs. 7.23: middle column; 7.24B, C).

At the same time that Bcd protein is diffusing posteriorly, a second protein, Nanos (Nos), translated from mRNA of the *nos* gene, begins to diffuse anteriorly from the posterior pole of the egg as a second gradient, with its high point at the posterior pole of the egg (Fig. 7.24A–C). The *nos* mRNA is localized to the posterior end of the oocyte during oogenesis in the same way that *bcd* mRNA is to the anterior pole, and is likewise translated into protein at fertilization (Fig. 2.20A, B). Nos protein selectively binds to the mRNAs of certain other pattern-control genes and blocks their translation, resulting in regional differences in availability of their proteins (Gerhart and Kirschner, 1997: 447). For example, mRNA of the maternal gap gene *hunchback* (*hb*) is uniformly distributed throughout the periplasm of the egg during oogenesis and is translated into protein following fertilization, except in the posterior third of the egg where Nos prevents it (Fig. 7.24C). The result is a gradient of maternal Hb protein that is lowest in the posterior end of the egg.

However, Caudal (Cad), not Nos, is the primary posterior gradient factor in *Drosophila* embryos (Rivera-Pomar et al., 1996). Prior to fertilization, *cad* mRNA is also evenly distributed throughout the periplasm of the egg. However, Bcd protein inhibits its translation into protein in the anterior third of the egg, leading to formation of a gradient of Cad, with its high point posteriorly just as for Nos (compare Fig. 7.24B, C).

To discover how widely distributed the *bcd* gene is in insects, Schröder and Sander (1993) compared Bcd activity in the eggs of six *Drosophila* species, of *Musca domestica*, of three blowfly species, of a phorid fly (all Diptera), and of the honeybee *Apis mellifera* (Hymenoptera: Apidae), by injecting anterior egg ooplasm from cleaving eggs of these insects into *bcd* mutant embryos of *D. melanogaster*. Mutant host embryos were rescued by injection of anterior ooplasm from eggs of the other drosophilids, the house fly, and two of the three blowflies but not from the other blowfly, the phorid, or *A. mellifera*, suggesting (1) that *bcd* functions in A-P pattern specification only in eggs of flies closely related to *Drosophila*; (2) that its protein is sufficiently different in eggs of more distantly related species that it fails to work in the foreign environment; or (3) that binding sites in the regulatory sequences of its target genes differ.

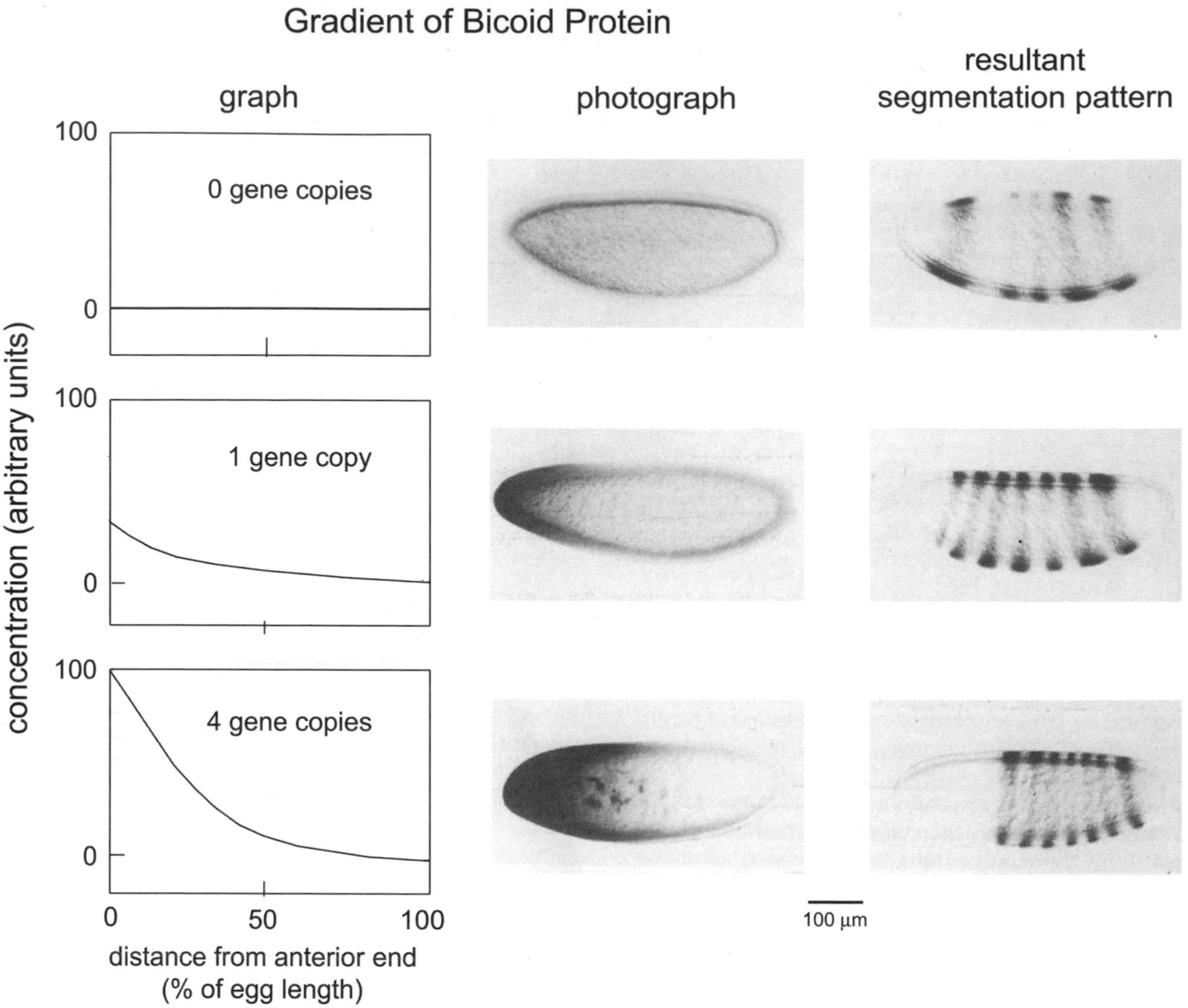

Gradient of Bicoid Protein

Fig. 7.23. The gradient of Bicoid (Bcd) protein in the *D. melanogaster* egg and its effects on anteroposterior pattern formation. The gradient is revealed by staining with a labeled antibody against Bcd protein, and the presumptive segmental pattern, by a labeled antibody against the protein of the pair-rule gene *even-skipped* (*eve*). Three embryos are compared containing zero (top row), one (center row), and four (bottom row) copies of the wild-type *bicoid* (*bcd*) gene. With zero dosage of *bcd*, segments with an anterior character do not form; with increasing gene dosage, they form progressively farther from the anterior pole of the egg, as expected if their position is determined by the local concentration of Bcd protein (measurements of this concentration, as indicated by intensity of staining [middle column], are shown in the graphs at the left). Despite the differences in *eve* expression pattern (right), both upper and lower embryos will develop into normally proportioned larvae and adults. (© 1994 From *Molecular biology of the cell*, 4th ed., by B. Alberts et al. Reproduced by permission of Routledge, Inc., part of The Taylor & Francis Group.)

Intrigued by these results, Stauber and coworkers (1999) cloned the ortholog of the *bcd* gene, *Ma-bcd*, in the same phorid fly, *Megaselia abdita*, examined in vain by Schröder and Sander (1993), and by in situ hybridization to its cDNA, found it to be expressed during oogenesis and embryogenesis exactly as *bcd* is in *Drosophila*. When they compared amino acid sequences in the DNA-binding motif of the Ma-Bcd protein, the homeodomain (7.2.1.4), with those of other homeotic genes (7.2.1.3) in both *Drosophila* and other animals, they found it to have 70% sequence similarity with *Drosophila* Bcd and 48.3% similarity with *Drosophila* Zerknüllt (Zen) protein (the D-V protein mentioned in 7.2.1.1.1) but with less than 45% similarity to homeodomains of the other genes. This suggested that the genes *bcd* and *zen* might be related (in *D. melanogaster*, *bcd* is situated on the

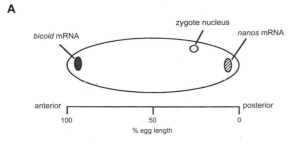

A

bicoid mRNA   zygote nucleus   nanos mRNA

anterior   posterior

100   50   0
% egg length

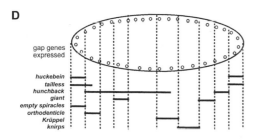

B

Bicoid   cleavage energids
protein   entering periplasm
gradient

Nanos
protein
gradient

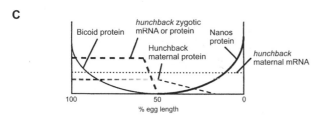

C

Bicoid protein   hunchback zygotic
                        mRNA or protein   Nanos
                        Hunchback   protein
                        maternal protein
                                                 hunchback
                                                 maternal mRNA

100   50   0
% egg length

D

gap genes
expressed

huckebein
tailless
hunchback
giant
empty spiracles
orthodenticle
Krüppel
knirps

Fig. 7.24. The organization of gap gene expression during early embryogenesis of *D. melanogaster*. In all diagrams the mRNAs and proteins are located in the periplasm of the egg. A. Egg just after fertilization, showing the distribution of *bicoid* and *nanos* mRNA. B. Translation and diffusion of Bicoid and Nanos proteins from anterior and posterior ends during cleavage establish two mutually permeating concentration gradients. C. Differential concentration of Bicoid and Nanos proteins along the length of the egg influences the distribution of maternal Hunchback protein. The *hunchback* mRNA from the nurse cells is distributed evenly throughout the periplasm during oogenesis. Its translation into protein is blocked in the posterior end of the egg by Nanos, creating a gradient of maternal Hunchback in the anterior end. Bicoid protein activates zygotic *hunchback* expression so that the anterior periplasm is filled with *hunchback* mRNA and protein. D. At syncytial blastoderm, the cleavage nuclei are exposed to different concentrations of Bicoid and Hunchback protein along the length of the egg, which differentially activates, derepresses, or represses eight gap genes in different regions along the length of the egg. (Adapted with permission from J. Gerhart and M. Kirschner. *Cells, embryos, and evolution. Toward a cellular and developmental understanding of phenotypic variation and evolutionary adaptability*, p. 449, Fig. 9-1. © 1997 Blackwell Science, Ltd.)

right arm of chromosome 3 between *zen* and *Deformed* [*Dfd*] of the homeotic gene complex; 7.2.1.3). They cloned the ortholog of the *zen* gene, *Ma-zen*, in *M. abdita* and found it to be expressed dorsally in prospective amnioserosal cells (6.9.4) of embryos before, during, and just after gastrulation just as *zen* is in *Drosophila* embryos (Fig. 7.20), showing the two genes to have separate and similar functions in embryos of both insects. Finally, phylogenetic analysis of homeodomain sequences of the proteins of *Ma-bcd*, *Ma-zen*, and the *Drosophila HOM* genes showed the two *Megaselia* genes to be sisters. They thus concluded that

- the *bcd* gene of higher flies originated relatively recently by duplication of the *zen* gene;
- it first functioned along with *zen* in specifying amnioserosa; and
- it was subsequently recruited to function in A-P pattern specification in embryos of the Diptera-Schizophora.

This suggestion is supported by the discovery that maternal Hunchback (Hb) protein can control zygotic Hb expression in the absence of Bcd protein (Winner et al., 2000).

An ortholog of the posterior *cad* gene, *Tc-cad*, is expressed in short germ embryos of the flour beetle *Tribolium castaneum* (Coleoptera: Tenebrionidae) as was shown by in situ hybridization with *Tc-cad* cDNA and by staining with tagged antibodies to its protein (Schulz et al., 1998). At first, *Tc-cad* mRNA was expressed homogeneously throughout the periplasm, but by the time of blastoderm formation, its protein assumed a posterior to anterior gradient as in *Drosophila*, though the basis for this is unknown. Because *T. castaneum* has short germ embryos, Tc-Cad is also expressed in the prospective head and thorax, not just in the abdomen as in the long germ embryo of *D. melanogaster*. During germ band elongation, its expression becomes limited to the proliferation zone in the protocorm (compare Fig. 6.21A) and, at its end, to a terminal stripe similar to that in the late blastoderm of *D. melanogaster*.

If *Tc-cad* mRNA is transferred into *Drosophila* embryos under a maternal promotor, it is translationally repressed anteriorly by *Drosophila* Bcd protein, even though a *bcd* gene has yet to be identified outside the Diptera (Wolff et al., 1998). The topic of axis specification in animals was reviewed by Goldstein and Freeman (1997).

7.2.1.2 SEGMENTATION GENES

In eggs of *D. melanogaster*, normal expression of the egg-polarity genes induces formation of body segments by selectively activating at least 28 segmentation genes (Gerhart and Kirschner, 1997). Mutations in these genes alter either the number of segments formed or their internal organization, the wild-type genes beginning to be expressed in preblastoderm while the egg is still a syncytium. Since the phenotype produced by expression of these genes is determined by the genotype of the embryo, they are referred to as *zygotic-effect genes*.

Mutations in most segmentation genes are recessive (those whose names are not capitalized) and cause homozygous mutant larvae to die before hatch so that their effects cannot be evaluated in adults. However, heterozygous mutant stocks are viable and can be maintained indefinitely. When a pair of heterozygous parents breed, one in four offspring are homozygous mutant. These die prematurely as late embryos or young first instars, but reveal their mutant phenotype in their altered larval cuticular patterns (Fig. 7.29: bottom) (Nüsslein-Volhard and Wieschaus, 1980).

By use of these methods, three classes of segmentation genes have been revealed. First to be expressed are eight gap genes—*huckebein* (*hkb*), *tailless* (*tll*), *hunchback* (*hb*), *giant* (*gt*), *empty spiracles* (*ems*), *orthodenticle* (*otd*), *Krüppel* (*Kr*), and *knirps* (*kni*)—whose proteins, all transcription factors, function to divide the preblastoderm into eight broad regions within a 2-h period just before blastoderm formation (Figs. 7.24D, 7.29: middle) (Pankratz and Jäckle, 1993). Mutation in a gap gene eliminates a large block of contiguous segments in the larva developing from that egg, with those in other gap genes causing different but partially overlapping defects (Fig. 7.25: right). In the mutant *Kr*, for example, the larva lacks segments T1–A5 (parasegments 3–10).

The differential expression of these genes is due to differences in their regulatory regions (7.2.1.4). *Kr* and *kni*, for example, are repressed posteriorly by absence of Bcd protein and by binding of Hb protein to their regulatory sites, while *otd* and *gt* are activated anteriorly by binding of both Bcd and Hb (Fig. 7.24C, D).

The pattern of gap protein distribution ultimately produced (Figs. 7.24D, 7.29: middle) is transient but serves to activate three primary pair-rule genes: *runt* (*run*), *hairy* (*h*), and *even-skipped* (*eve*) (Fig. 7.26). The transcription factors of these genes then activate five secondary pair-rule genes: *fushi tarazu* (*ftz*), *odd-paired* (*opa*), *odd-skipped* (*odd*), *sloppy paired* (*slp*), and *paired* (*prd*) (Fig. 7.26).

Mutation in any of these genes causes a series of deletions affecting alternate segments and results in a larva with only half its usual number of body segments (Fig. 7.25: middle). Although all pair-rule mutants display this two-segment periodicity, they differ in the location of their deletions along the length of the embryo. For example, embryos homozygous for the pair-rule mutant *eve* lack all even-numbered segments, while those homozygous for *ftz*, *odd*, or *opa* lack all odd-numbered segments (Fig. 7.25: middle). Both the transcripts and proteins of each pair-rule gene are expressed circumferentially about the egg within the periplasm in seven, evenly spaced stripes along its length (Fig. 7.23: right column), the stripe for each gene being slightly displaced relative to the stripes of the others and differing characteristically in length (Fig. 7.26).

Finally, there are about 12 segment-polarity genes, of which the best known are *engrailed* (*en*), *wingless* (*wg*), *patched* (*ptc*), *naked* (*nk*), and *hedgehog* (*hh*), whose mutant expression causes part of each segment to be lost and to be replaced by a mirror-image duplicate of all or part of the rest of that segment. In *gooseberry* (*gsb*) mutants, for example, the posterior half of each segment is replaced by a mirror image of the adjacent anterior half-segment (Fig. 7.25: left).

Examination of the phenotypes of segmentation mutants belonging to all three classes (gap, pair-rule, segment-polarity) suggests that sequential expression of the wild-type genes progressively subdivides the embryo into smaller and smaller domains distinguished by differing patterns of gene expression (Alberts et al., 1994).

All known segmentation genes have been cloned, and tagged cDNA probes derived from them have been used to locate their transcripts in whole mounts of wild-type embryos by in situ hybridization (compare Fig. 7.23: right column). All have been shown to generate positional signals influencing pattern formation in sequentially more localized neighborhoods. Mutants defective in the gap gene *Kr*, for example, show abnormalities extending throughout the region (T1–A5) where that gene's transcripts are detected in a normal embryo and for several segments beyond (Fig. 7.27A) (Alberts et al., 1989). This gene, like many other segmentation genes, is known to encode a DNA-binding transcription factor capable of influencing whether a competent downstream gene is activated or inhibited (7.2.1.4).

The segment-polarity genes influence spatial patterning within a parasegment, some exerting effects on cells neighboring the regions in which they are transcribed (e.g., the secreted proteins Wingless [Wg] and Hedgehog [Hh]) and others affecting development only in the same regions (Alberts et al., 1994).

If one selects an embryo mutant for the gap gene *Kr* and evaluates it for expression of the wild-type, pair-rule gene *ftz* by in situ hybridization, the *ftz* stripes fail to develop in that region of the blastoderm corresponding to the defect in the *Kr* mutant, indicating that Kr protein, directly or indirectly, regulates *ftz* gene expression (Alberts et al., 1994). However, in a *ftz* mutant embryo, the distribution of wild-type Kr protein is not disturbed, indicating that Ftz protein does not regulate *Kr* expression.

Interactions also occur between genes in the same tier (Fig. 7.28: left). The gap genes *Kr* and *hb*, for example, mutually inhibit each other such that preblastoderm nuclei cannot express both genes at once. Instead, they are normally expressed in adjacent regions with a sharp boundary between *hb* expression anteriorly and *Kr* expression posteriorly (Fig. 7.29: middle). And when the protein of either gene is absent, the domain of expression of the other extends beyond its usual boundary. This mutual inhibition functions to gradually generate sharply defined, nonoverlapping territories in response to a smooth gradient of morphogen.

The regular, periodic expression pattern of the pair-rule genes (Fig. 7.26) is similarly generated (Alberts et al., 1994). Thus, different circumferential bands on the

Segment polarity   Pair-rule   Gap

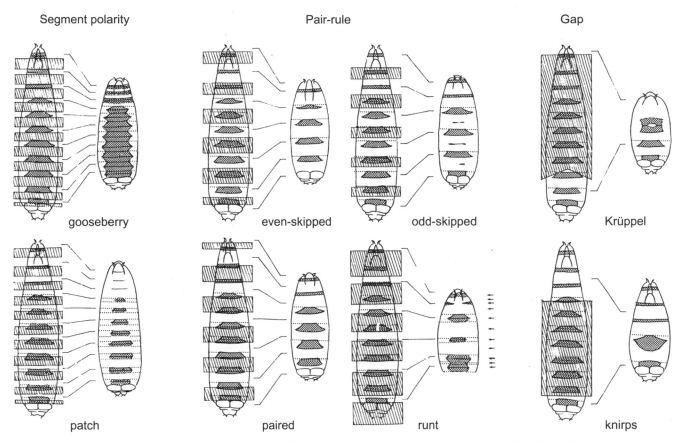

gooseberry   even-skipped   odd-skipped   Krüppel

patch   paired   runt   knirps

Fig. 7.25. The regions deleted from the wild-type body pattern in *D. melanogaster* larvae mutant for selected segmentation genes. Shaded areas indicate denticle belts and dotted lines, the segmental boundaries. The regions missing in mutant larvae are indicated by hatched bars. The transverse lines connect corresponding regions in mutant and wild-type larvae. Planes of polarity reversal in *runt* and *Krüppel* mutants are indicated by arrows. (Reprinted with permission from *Nature* (C. Nüsslein-Volhard and E. Wieschaus. Mutations affecting segment number and polarity in *Drosophila*. 287: 795–801). © 1980 Macmillan Magazines Limited.)

preblastoderm can be distinguished by different combinations of pair-rule gene expression, as can be demonstrated by use of double and triple labeling experiments using differently tagged antibodies to the proteins of each gene.

The segment-polarity genes are similarly expressed in part of every parasegment. Transcripts of *en*, for example, can be demonstrated by in situ hybridization to form a series of 14 transverse circumferential stripes in the cellular blastoderm, each stripe being one cell wide and corresponding to the anteriormost portion of each parasegment, or to the posteriormost portion of each segment, and in a fixed relationship to protein bands of the pair-rule genes (compare Fig. 7.32F of a stretched-out, extended germ band embryo of *D. melanogaster*). And from the altered pattern of *en* bands in various pair-rule mutants, one can infer the pair-rule expression requirements for *en* transcription. This shows that one or other of two specific combinations of pair-rule gene products is required (compare Fig. 7.26). In even-numbered parasegments, *en* is switched on in those cells expressing a combination of *ftz*, *odd*, and *naked*. The gene *odd* represses *ftz* expression, a known activator of *en*, whereas *naked* inhibits the activation of *en* by *ftz* without affecting *ftz* expression (Mullen and DiNardo, 1995). The result is that *en* expression is limited to a narrow strip of cells at the anterior boundary of each even-numbered parasegment (Fig. 7.30).

Expression of *wg* gene is similarly regulated by expression of *odd*, *prd*, and *slp*. The gene *odd* represses *wg* expression, whereas *prd* restricts the domain of expression of *odd*. Expression of *wg* is thus allowed only in a narrow strip of cells at the posterior boundary of each parasegment (Fig. 7.30) (or near the anterior boundary of each segment). Fig. 7.28 illustrates the hierarchy of inhibition and activation in three tiers of pattern-control genes leading to expression of the segment-polarity genes (Petit and Scott, 1992).

The end result of this transcription factor cascade is that by the time of blastoderm formation, each future

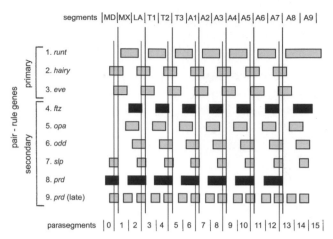

Fig. 7.26. Seven stripes of expression, in relation to segmental (top) and parasegmental (bottom) borders, of each of eight pair-rule genes in a *D. melanogaster* embryo at the syncytial blastoderm stage. Gap proteins (Fig. 7.24D) activate or repress transcription of the primary pair-rule genes, and the proteins encoded by these genes (all transcription factors) activate or repress the secondary pair-rule genes. Pair-rule proteins also affect the expression of other pair-rule genes so that a dense network of interactions is established. Pair-rule transcripts and proteins are transient (present from hour 2–3 after fertilization) and disappear as the segment-polarity genes and their encoded proteins become autoactivating with cellularization of the blastoderm. The thicker vertical lines represent segmental, and more delicate lines, the parasegmental boundaries. Genes are *runt*, *hairy*, *even-skipped* (*eve*), *fushi tarazu* (*ftz*), *odd-paired* (*opa*), *odd-skipped* (*odd*), *sloppy-paired* (*slp*), and *paired* (*prd*). Paired protein resolves into 14 stripes later in embryogenesis. (Adapted with permission from J. Gerhart and M. Kirschner. *Cells, embryos, and evolution. Toward a cellular and developmental understanding of phenotypic variation and evolutionary adaptability*, p. 451, Fig. 9-2. © 1997 Blackwell Science, Ltd.)

parasegment is already subdivided into four distinct regions that persist, long after the pair-rule proteins have disappeared, by continued transcription of at least some segment-polarity genes. And *wg*, *en*, and *hh* continue to be expressed throughout subsequent embryonic and postembryonic development, the products of *wg* and *hh* acting as local signaling molecules in a feedback loop between each parasegment and functioning to stabilize and maintain the compartment boundary (Fig. 7.30).

In fact, recent work on *D. melanogaster* embryos by Liang and Biggen (1998) indicates that by the time of blastoderm formation, 20%–50% of all genes whose transcription can be monitored are regulated by both Eve and Ftz proteins and that by late embryogenesis, up to 87% of all genes are directly or indirectly controlled by proteins of the pattern-control genes!

Proteins of the pair-rule genes not only regulate expression of the segment-polarity genes but also collaborate with products of the egg-polarity, gap, and segment-polarity genes to differentially activate the final tier of genes in the hierarchy, the *Homeotic selector genes* (Fig. 7.28), which function to permanently distinguish one parasegment from another (7.2.1.3). But first a few words about the distribution of segmentation genes in other animals.

7.2.1.2.1 SEGMENTATION GENES IN OTHER ANIMALS

Now that knowledge of the pattern-control genes in *D. melanogaster* embryos is extensive, increasing numbers of developmental biologists are searching for their orthologs in embryos of other animals using tagged antibody probes prepared from proteins of the *Drosophila* genes (Schmidt-Ott et al., 1994a; Abzhanov and Kaufman, 1999a, 1999b; Abzhanov et al., 1999; Damen et al., 2000) (see computer simulation of von Dassow et

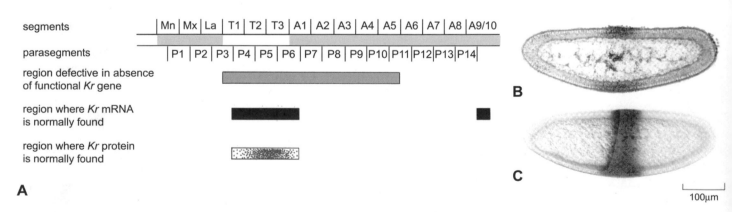

Fig. 7.27. Spatial domains of action of the gap gene *Krüppel* (*Kr*) mapped on the blastoderm of *D. melanogaster*. A. How the defect caused by the absence of functional Kr product extends far beyond the region where Kr transcripts and protein are normally found. B. Wild-type distribution of *Kr* mRNA as seen by in situ hybridization. C. Same, of Kr protein as seen by antibody binding. The phenotype of a Kr mutant is shown in Fig. 7.25: upper right. (© 1989 From *Molecular biology of the cell*, 3rd ed. by B. Alberts et al. Reproduced by permission of Routledge, Inc., part of The Taylor & Francis Group.)

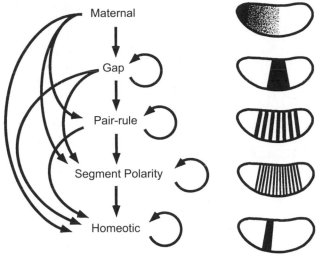

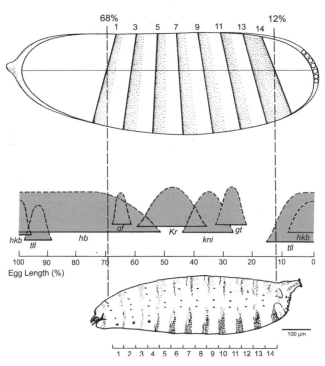

Fig. 7.28. Summary of the gene regulatory hierarchy controlling segmentation in *D. melanogaster* embryos. Left: a simplified schematic representation of the gene classes specifying the larval anteroposterior body pattern and the regulatory relationships between them. Earlier-acting genes regulate the spatial expression pattern of later-acting genes. In many instances, genes within a given class regulate other members of the class (depicted as arrows beginning and ending at the same class). Other genes, not shown in the diagram, also play a role in segmentation. Not shown are the putative downstream targets of regulation by segmentation and homeotic genes. Right: representative expression patterns of each gene class are shown on the outline of the preblastoderm or blastoderm embryo. (Reprinted with permission from M. G. Petit and M. P. Scott. Gene control systems affecting insect development. In J. M. Crampton and P. Eggleston (eds.), *Insect molecular science*, p. 112, Fig. 6. © 1992 Academic Press, Inc.)

Fig. 7.29. The approximate pattern of gap gene expression in the blastoderm stage of *D. melanogaster* embryos. The stripes in the embryo at top indicate the anterior borders of odd-numbered parasegments specified by expression of *even-skipped* (*eve*). Compare with Fig. 7.24D. Genes are *gt*, *giant*; *hb*, *hunchback*; *hkb*, *huckebein*; *kni*, *knirps*; *Kr*, *Krüppel*; and *tll*, *tailless*. The bottom drawing is of a newly hatched first instar larva in lateral aspect, showing position of the parasegments. (Reprinted with permission from P. A. Lawrence. *The making of a fly: the genetics of animal design*, p. 58, Fig. 3.4. © 1992 Blackwell Science, Ltd.)

al., 2000). For example, Patel and others (1994) compared expression of the pair-rule gene *eve* in embryos of the tenebrionid *Tribolium castaneum* (short germ), the dermestid *Dermestes frischi* (semi-long germ), the chrysomelid *Callosobruchus maculatus* (long germ) (Coleoptera), and the locust *Schistocerca americana* (short germ). In the beetle embryos, Eve protein is expressed in pair-rule spatial patterns identical to those of *D. melanogaster* embryos, except that eight rather than seven Eve stripes are established from an initial, posterior domain of *eve* expression through elimination of Eve protein in interstripe regions as each new segment is budded off from the growth zone. Thus, even though each beetle has a different germ band type, beetles seem to use identical mechanisms of pair-rule expression.

The relationship of *eve* expression to that of the segment-polarity gene *en* is also similar to that in *D. melanogaster* embryos. However, Eve and En stripes are gradually established from front to back in beetle embryos rather than almost simultaneously as in *D. melanogaster* embryos, suggesting slight differences in

pair-rule gene interaction. Also, the pattern of expression of both *eve* and *en* is established by the onset of gastrulation in *D. melanogaster* embryos but has only proceeded as far back as the gnathal segments in embryos of *T. castaneum*, the anterior abdomen in those of *D. frischi*, and the posterior abdomen in those of *C. maculatus* at time of gastrulation. And in *S. americana* embryos, striped *eve* expression does not begin until well after gastrulation. Eve is expressed in a similar posterior domain, but no pattern of pair-rule stripes is established and there is no overlap between *eve* and *en* expression. Thus, establishment of segments in embryos of this grasshopper embryo seems not to involve pair-rule patterning.

The same labeled antibody also labels Eve protein in embryos of a dragonfly (Odonata), of the earwig *Anisolabis annulipes* (Dermaptera), of the cricket *Acheta domesticus* (Orthoptera), of the honeybee *Apis mellifera*, of the carpenter ant *Camponotus laevigatus* (Hymenoptera), of the mosquito *Anopheles gambiae*, of the house fly *Musca domestica*, of the blowfly *Lucilia*

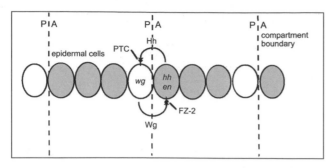

Fig. 7.30. Interactions between the segment-polarity genes *hedgehog* (*hh*), *wingless* (*wg*), and *engrailed* (*en*) and their proteins at the compartment boundary in an embryo of *D. melanogaster*. The gene *en* is expressed in the nuclei of cells along the anterior margin (A) of the parasegment. Its protein is a transcription factor that induces expression of *hh* and secretion of its protein, Hh. By way of its receptor protein Patched (PTC), Hh activates and maintains expression of *wg* in adjacent posterior cells (P) across the compartment boundary. Wg protein is secreted and feeds back on *en*-expressing cells by way of its receptor protein Frizzled-2 (FZ-2) to maintain expression of *en* and *hh*. These interactions stabilize and maintain the compartment boundary throughout embryonic and postembryonic development. (Adapted by permission of Oxford University Press from L. Wolpert. 1998. *Principles of development*, p. 156, Fig. 5.29. Current Biology Ltd. © 1998 Oxford University Press.)

cuprina (Diptera), and of the spider *Cupiennius salei*, suggesting that Eve protein is produced by all arthropod embryos (Damen et al., 2000).

And what about the expression of these genes in the incredibly diverse embryos of parasitoid Hymenoptera? Grbić and coworkers (1996) and Strand and Grbić (1997) examined expression of Eve and En in embryos of three quite different parasitoids (Fig. 7.31): the braconids *Bracon hebetor* (a monoembryonic ectoparasitoid of lepidopteran caterpillars with a normal, fully yolked and chorionated egg, superficial cleavage, and a long germ embryo) and *Aphidius ervi* (a monoembryonic endoparasitoid of aphids with a tiny, yolkless egg, a thin chorion, holoblastic cleavage, a trophamnion of polar body origin, and a short germ embryo) and the polyembryonic encyrtid *Copidosoma floridanum* (6.7.2.1). In *B. hebetor* embryos, Eve is first expressed in a broad posterior domain in the preblastoderm, followed by a split into seven pair-rule stripes, followed by sequential A-P resolution into 14 segmental stripes (Fig. 7.31: a–c). In *C. floridanum* embryos, Eve expression at first is similar to that in *B. hebetor* but arises in a cellularized rather than a syncytial environment. Also, at the onset of gastrulation, it resolves directly and sequentially, from front to back, into 15 segmental stripes (i.e., there is no pair-rule expression) (Fig. 7.31: d–f). Finally, in *A. ervi* embryos, Eve is not even expressed until well after gastrulation and germ band elongation, when it first appears in dor-

solateral mesoderm and in neuroblasts (nerve mother cells) of the developing central nervous system (Fig. 7.31: g–i).

Fifteen stripes of En expression are established rapidly from front to back in *B. hebetor* and *C. floridanum* embryos during germ band elongation, and more slowly and sequentially in those of *A. ervi*, as is typical of short germ embryos (Fig. 7.31: j–o). In all three species, Ubx/abd-A is expressed in the posterior thorax and in the abdomen in the retracted germ band stage (Fig. 7.31: p–r).

Expression of *en* has since been observed in embryos of a leech (Annelida), of *Thermobia domestica* (Zygentoma), of the milkweed bug *Oncopeltus fasciatus* (Hemiptera), of the tobacco hornworm *Manduca sexta*, of the butterfly *Precis coenia* and the silkworm *Bombyx mori* (Lepidoptera), of the honeybee *Apis mellifera* (Hymenoptera), of the brine shrimp *Artemia franciscana* and a crayfish, *Procambarus* sp. (Crustacea), of *Branchiostoma* sp. (*Amphioxus*; Cephalochordata), of the zebrafish *Brachydanio rerio* (Vertebrata: Pisces), of the chicken *Gallus gallus* (Vertebrata: Aves), and of the house mouse *Mus musculus* (Vertebrata: Mammalia) (Fig. 7.32) (Patel et al., 1989; Patel, 1994; De Robertis, 1997; Peterson et al., 1998, 1999). However, though very similar in annelid and arthropod embryos, *en* expression in *Branchiostoma* and vertebrate embryos occurs relatively much later, after body segmentation is complete, and principally in the brain and mesodermal somites (Fig. 7.32B).

7.2.1.2.2 WAS THE COMMON ANCESTOR OF ARTHROPODS AND VERTEBRATES SEGMENTED?

In 1822, Étienne Geoffroy St.-Hilaire homologized the body segments of arthropods with those of vertebrates, and recent work on gene expression in *Branchiostoma* (Cephalochordata) and zebrafish embryos provides evidence for his scenario (Kimmel, 1996). The *Drosophila* pair-rule gene *h* is the ortholog of the zebrafish gene *her-1*. In zebrafish embryos, the first *her-1* band of expression forms in mesodermal somite pair 5 and the second in 7, so it too is pair-rule in its expression pattern. Altogether, more than 11 bands of *her-1* expression form sequentially in the posterior germ ring (the equivalent of the growth zone of short germ insect embryos [Fig. 6.21A], but only in developing mesodermal somites) of these embryos, though in not more than three at one time because expression fades in older, more anterior bands as new ones develop from the ring in the somite-forming (paraxial) mesoderm. That *her-1* is expressed in only three developing somite pairs at a time is because its expression is downregulated 4 h before a somite actually forms. *her-1* is not expressed in head segment primordia of zebrafish embryos, nor is its ortholog *h* expressed in pregnathal cephalic segments of *Drosophila* embryos (7.2.1.5).

In the lancelet *Branchiostoma*, the ortholog of the segment-polarity gene *en* is only expressed bilaterally in

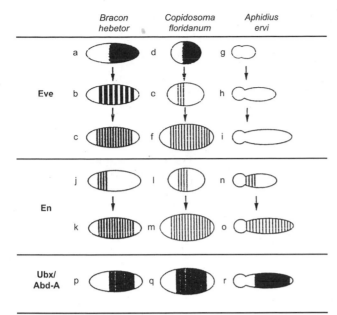

	Bracon hebetor	Copidosoma floridanum	Aphidius ervi

Fig. 7.31. Expression pattern of the pair-rule protein Even-skipped (Eve), the segment-polarity protein Engrailed (En), and the homeotic protein Ultrabithorax/abdominal-A (Ubx/abd-A) in embryos of the parasitoid wasps *Bracon hebetor* (Hymenoptera: Braconidae), *Copidosoma floridanum* (Encyrtidae), and *Aphidius ervi* (Aphidiidae). See text for details. Note that this figure is highly schematic, since periods of expression for each protein are dynamic. (Reproduced with permission from M. R. Strand and M. Grbíc. 1997. The development and evolution of polyembryonic insects. In *Current topics in developmental biology.* Vol. 35, pp. 121–159. © 1997 Academic Press, Inc.)

the posterior compartment of that segmental mesoderm, giving rise to the first eight somites, and only before these somites form. (Embryos of this cephalochordate eventually have ≥50 somite pairs that originate by outpocketing from the primitive gut [*enterocoely*] [De Robertis, 1997]). In vertebrate (zebrafish, chick, and mouse) embryos, *en* orthologs are expressed in equivalent cells of the somites but only after they form by hollowing out of mesenchymal blocks (*schizocoely*). Also, *Branchiostoma* has only one *en* gene whereas the zebrafish has three.

That *en* is expressed in both insect and chordate metameres implies that segmentation was present in the common ancestor from which their lineages diverged over 500 million years ago, the Urbilateria (from *ur,* "primeval," and *bilateria,* "bilateral animal") (De Robertis, 1997). Also, in both *Drosophila* and chordate embryos, expression of the segmentation genes exhibits *resegmentation* (i.e., it is parasegmental not segmental). In vertebrates, each vertebra results from cells of the posterior half of one sclerotome pair fusing to those of the anterior half of the next, suggesting that resegmentation has originated only once.

Expression of other orthologous regulatory genes suggests that this segmented common ancestor also had

- D-V patterning (*sog/chd; dpp/BMP-4*) (7.2.1.1.1);
- A-P patterning (*Hox* genes) (7.2.2);
- orthologs of the anterior head genes *otd* and *ems* (7.2.1.5);
- a primitive photoreceptor (*Pax 6/eyeless*); and
- a contractile blood vessel (Tinman; Nkx2.5 and DMEF2).

All these expression patterns suggest that the common ancestor of vertebrates and arthropods (Urbilataria) may have been segmented (Kimmel, 1996; De Robertis, 1997).

7.2.1.3 HOMEOTIC (*HOM*) SELECTOR GENES

Many mutations in *D. melanogaster* cause one body part to substitute for another during embryogenesis and are said to be homeotic (Lewis, 1978; Gehring, 1998). The wild-type alleles of most of these genes occur in one or the other of two tight gene clusters on the right arm of chromosome 3 known as the *Antennapedia* (*ANT-C*) and *Bithorax* (*BX-C*) complexes (Fig. 7.38: *Drosophila*), or collectively as the *Homeotic* (*HOM-C*) or *Homeobox* (*Hox-C*) gene complex because they each encode a transcription factor containing a homeodomain (see 7.2.1.4). The two together contain 11 genes occupying about 600 kb of DNA: the three genes in *BX-C* (*Ultrabithorax* [*Ubx*], *abdominal-A* [*abd-A*], and *Abdominal-B* [*Abd-B*]) specifying differences between thoracic and abdominal segments from the posterior mesothorax (T2) to abdominal segment 8 (A8), and the five in *ANT-C* (*labial* [*lab*], *proboscipedia* [*pb*], *Deformed* [*Dfd*], *Sex combs reduced* [*Scr*], and *Antennapedia* [*Antp*]) specifying differences between segments from the intercalary to anterior T3. Three other regulatory genes also occur in *ANT-C* but are not homeotic: the maternal effect gene *bcd*, the pair-rule gene *ftz*, and the D-V gene *zen*. Like the segmentation genes, each *HOM* gene has a characteristic domain of expression, indicated by that region of the body transformed by a mutation in that gene, and has sharp boundaries of expression corresponding to those of the parasegments (Fig. 7.36: Hexapoda).

Though the *BX-C* genes were first thoroughly investigated by E. B. Lewis and colleagues at Cal Tech beginning in the 1940s (Lewis, 1978), the first homeotic mutation, *bithorax* (*bx*), was discovered by Bridges in 1915, and the phenotypes of some of their mutations were first described by Bateson in insects and crabs in 1894 (see Lawrence, 1992: 211–215, and Gehring, 1998, for a history of *HOM* gene research). Like many of the segmentation genes, some mutations in this complex have a recessive lethal phenotype enabling homozygous mutant embryos to survive only to the time of hatch. Thus, their effects are rarely seen in adults but can be observed in embryos or in the cuticle of early first instars.

First instar larvae deficient in all *BX-C* genes have the head and anterior thorax normal as far back as paraseg-

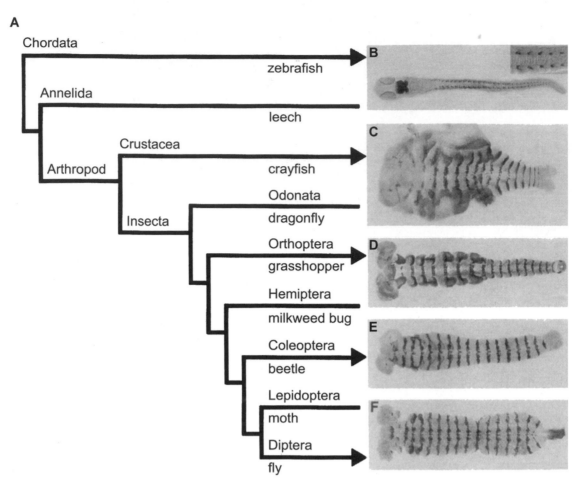

Fig. 7.32. A. The inferred phylogenetic relationships of several insect orders and their relationship to Crustacea, Annelida, and Chordata. B–F. The expression pattern of the segment-polarity gene *engrailed* (*en*) at the extended germ band (polypod) stage of embryogenesis. The pattern is conserved in embryos of *Procambarus* sp. (Crustacea) (C), *Schistocerca gregaria* (Orthoptera) (D), *Tribolium castaneum* (Coleoptera) (E), and *D. melanogaster* (F). In the zebrafish *Brachydanio rerio* (B), *en* orthologs are expressed in the midbrain-hindbrain region of the central nervous system and in a segmentally repeated pattern in the mesodermal somites (shown at higher magnification in the inset). The segmentally repeated *en* patterns in vertebrate embryos occur only after morphological segmentation is already present. (Reprinted with permission from N. Patel. Developmental evolution: insights from studies of insect segmentation. *Science* 266: 581–590. Copyright 1994 American Association for the Advancement of Science.)

ment P4 (anterior compartment of T2), but the remaining 10 parasegments converted to P4 (T2; Fig. 7.33: *BX-C⁻*), while partial deletions of *BX-C* cause less extensive transformation. These observations and similar findings for the *ANT-C* genes (Fig. 7.33: *Antp⁻*, *ANT-C⁻*) begun by T. C. Kaufman and colleagues (Kaufman et al., 1990) established the critical role of these genes in defining differences among parasegments. Thus, if all *BX-C* and *ANT-C* are deleted, all parasegments from P1 (posterior compartment of mandibular segment) back differentiate as P4 (T1; Fig. 7.33: *ANT-C⁻ BX-C⁻*) and, in embryos of the flour beetle *Tribolium castaneum*, a similar deletion results in all appendages developing as antennae (Brown et al., 2000).

Like the segment-polarity genes, the *HOM* genes of *D. melanogaster* are first expressed at blastoderm. Since all DNA in both complexes has been cloned, cDNA probes are available to map the transcription pattern of each gene by in situ hybridization and of their protein products by tagged antibodies to each of their proteins. Results of such study show each gene to be normally expressed in those regions developing abnormally when that gene is mutant or deleted. The proteins of these genes thus constitute the compartmental molecular address labels mentioned earlier, the cells of each parasegment having its unique set (Fig. 7.17).

Because expression of the *HOM* genes is controlled by products of the egg-polarity, gap, and pair-rule genes (Fig. 7.28), their pattern of expression is in exact register with the parasegmental boundaries defined by the distribu-

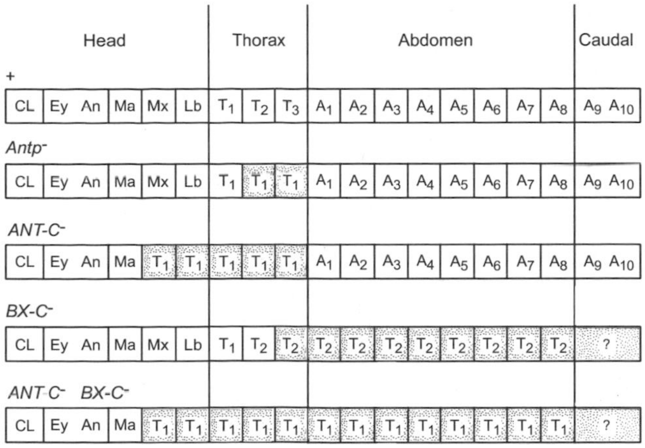

	Head						Thorax			Abdomen								Caudal	
+	CL	Ey	An	Ma	Mx	Lb	T_1	T_2	T_3	A_1	A_2	A_3	A_4	A_5	A_6	A_7	A_8	A_9	A_{10}
Antp⁻	CL	Ey	An	Ma	Mx	Lb	T_1	*T_1*	*T_1*	A_1	A_2	A_3	A_4	A_5	A_6	A_7	A_8	A_9	A_{10}
ANT-C⁻	CL	Ey	An	Ma	*T_1*	*T_1*	*T_1*	*T_1*	*T_1*	A_1	A_2	A_3	A_4	A_5	A_6	A_7	A_8	A_9	A_{10}
BX-C⁻	CL	Ey	An	Ma	Mx	Lb	T_1	T_2	*T_2*	*T_2*	*T_2*	*T_2*	*T_2*	*T_2*	*T_2*	*T_2*	*T_2*	?	
ANT-C⁻ BX-C⁻	CL	Ey	An	Ma	*T_1*	*T_1*	*T_1*	*T_1*	*T_1*	*T_1*	*T_1*	*T_1*	*T_1*	*T_1*	*T_1*	*T_1*	*T_1*	?	

Fig. 7.33. The segmental transformations observed in the body pattern of *D. melanogaster* after deletion of portions or all of *ANT-C* and *BX-C*. The segments transformed are stippled and their new identity indicated. The genes or gene complexes deleted are shown at the left above each box diagram. The transformations indicated are a simplification and compilation of effects on adults, larvae, and embryos. CL, clypeolabrum; Ey, ocular segment; An, antennal segment; Lb, labial segment; Ma, mandibular segment; Mx, maxillary segment; T_1, prothoracic segment; T_2, mesothoracic segment; T_3, metathoracic segment; A_1–A_{10}, abdominal segments 1–10. (Adapted from R. A. Raff and T. C. Kaufman. 1983. *Embryos, genes and evolution*, p. 249, Fig. 8-9. © 1983 Macmillan, N.Y.)

tions of the protein products of these genes. Likewise, this pattern undergoes a period of adjustment following the first appearance of transcripts in the blastoderm and becomes nonuniform within each parasegment, the definitive pattern of their proteins not becoming established until after gastrulation and germ band elongation. Fig. 7.36 (Hexapoda) shows the expression pattern of all of these genes at the extended germ band stage of *D. melanogaster* (28% of embryogenesis) 5 h after fertilization (Martinez Arias, 1993; Alberts et al., 1994).

Expression patterns of the *HOM-C* genes are similarly modulated by interactions between their products. And if the genes are arranged in sequence from front to back in order of their expression (Fig. 7.36: Hexapoda), each gene appears to be repressed farther back by proteins of subsequently expressed members of the series (posterior dominance or prevalence) (Alberts et al., 1994; Gerhart and Kirschner, 1997; Gehring, 1998). Therefore, if one "posterior" gene product or more are missing, a given gene in the series will be strongly expressed both in its usual domain and more posteriorly. This longitudinal ordering of genes according to their pattern of expression and control is identical to that in which they are arranged along the chromosome in each of the two complexes (Fig. 7.38: Drosophila).

7.2.1.3.1 HOM GENE EXPRESSION IN IMAGINAL DISCS
Although *HOM* gene expression in *D. melanogaster* begins at blastoderm, it determines the phenotypes of both larva and adult. As illustrated in Fig. 7.13, one can create a patch of marked homozygous cells mutant for a particular *HOM* gene in an imaginal disc by mitotic recombination and examine their behavior against a heterozygous background. Such cells transform only if they lie within the normal expression domain of that gene (Fig. 7.36: top) regardless of whether the recombination event was provoked early or late in development (Alberts et al., 1994). This is because each cell's "knowledge" of its position depends on continued autonomous expression of the normal *HOM* gene.

Cellular distinctions specified by *HOM* gene expression are discrete with an abrupt change in gene expression between cells in adjacent parasegments (Fig. 7.36: top). Thus, through differential expression of *BX-C* and *ANT-C* genes, the body is subdivided into a series of modules, each containing cells in different states of determination—exactly as envisaged by the compartment hypothesis of García-Bellido and Kauffman (Fig. 7.17).

As shown in Fig. 7.15, a compartment boundary is a frontier between two populations of cells in different states of determination (Alberts et al., 1994). Because this state is normally not reversible, each compartment is autonomous: it cannot recruit cells from adjacent compartments or transfer cells into them (but see Blair and Ralston, 1997). However, each compartment has a characteristic internal pattern within its boundaries that is generated through cell-cell interactions during larval development and that governs both the pattern and differentiation of structures developing from cells within that compartment.

It is generally assumed that the presence or absence of selector gene product autonomously drives the expression of compartment-specific adhesion or recognition molecules that inhibit the intermixing of cells between compartments (Blair and Ralston, 1997). However, during pattern specification in wing discs (11.3.2.2), an early step is for cells in the posterior compartment to express the secreted segment-polarity protein Hh (Fig. 7.30). This passes to cells in the anterior compartment where it binds to a receptor encoded by the *patched* (*ptc*) gene. Anterior cells lacking the *ptc* gene, and thus unable to receive the Hh signal, no longer obey the A-P compartment boundary and are able to move into the posterior compartment, suggesting that compartmentalization involves intercompartmental signaling, as shown in Fig. 7.30 (Blair and Ralston, 1997).

7.2.1.3.2 CAN *HOM* GENES REGULATE THEMSELVES?
HOM genes can regulate themselves by positive feedback in which the product of a gene stimulates its own transcription (Alberts et al., 1994). This can be demonstrated by transfecting cultured cells with engineered bacterial plasmids containing either the protein-coding sequence of a *HOM* gene coupled to a foreign promotor, or the *HOM* gene's promotor coupled to some other gene sequence encoding a readily assayed enzyme such as beta-galactosidase (β-GAL). The first combination lets

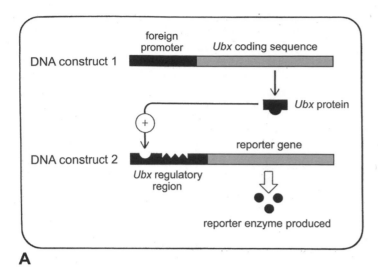

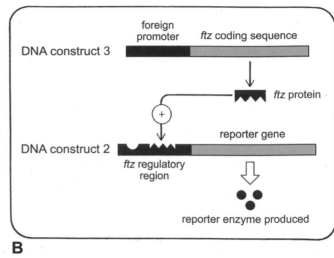

Fig. 7.34. Experiments with cultured cells transfected with artificially constructed bacterial plasmids carrying portions of the *Ultrabithorax* (*Ubx*) and *fushi tarazu* (*ftz*) genes show that both the Ftz protein (B) and the Ubx protein (A) can bind to the *Ubx* regulatory region to stimulate gene expression. The self-stimulation of *Ubx* (A) comprises a feedback loop that would tend to make *Ubx* expression self-sustaining. In a real embryo of *D. melanogaster*, there are many regulatory proteins that combine or compete to control expression of these genes. (© 1989 From *Molecular biology of the cell*, 3rd ed. by B. Alberts et al. Reproduced by permission of Routledge, Inc., part of The Taylor & Francis Group.)

one generate the protein encoded by the *HOM* gene at will; the second allows one to evaluate the control of that *HOM* gene's protein synthesis by producing a labeled reporter enzyme. Results of such experiments indicate, for example, that the Ultrabithorax (Ubx) protein can bind both to the *Ubx* regulatory sequence to stimulate additional *Ubx* expression (Fig. 7.34A) and to the *Antp* control sequence to inhibit *Antp* expression, while the Fushi tarazu (Ftz) protein can stimulate expression of both *Ubx* (Fig. 7.34B) and *Antp*.

7.2.1.3.3 REGULATION OF *HOM* GENE EXPRESSION
The *HOM* genes are incredibly complex, and interactions between their proteins and the genes themselves seem insufficient to stabilize their initial pattern of expression. Also, although expression of the *BX-C* genes defines the identity of nine parasegments (Fig. 7.35) (nine loss-of-function mutations are known [Lewis, 1978]), the complex contains only three genes: *Ubx*, *abd-A*, and *Abd-B* (Fig. 7.35). All *HOM* genes are unusually large, have large complicated regulatory regions, and are capable of producing a variety of differently spliced transcripts (Gehring, 1998). The *Ubx* gene, for example, is about 75 kb long and generates five mRNAs that are spliced differently according to stage of development and parasegmental location within the embryo (Lopez and Hogness, 1991). Also, the various mutations known in this gene map to different sites within it and affect different parasegments or parts of parasegments. Because such mutations occur in slightly different positions on

the genetic map and affect distinct body regions, Lewis (1978) thought that they corresponded to two different genes: *bithorax* (*bx*) and *postbithorax* (*pbx*) (Fig. 7.35). Thus, many *HOM* gene mutations affecting distinct parasegments or parts of parasegments are alterations of different parts of the same gene (Alberts et al., 1994).

The five, alternatively spliced transcripts of *Ubx* are known to encode five protein isoforms differing in internal amino acid sequence immediately adjacent to their DNA-binding sequence. A series of monoclonal antibodies specific to each of these proteins has been raised and labeled, and each isoform has been shown to be expressed in different stage- and tissue-specific patterns.

All *HOM* genes are also subject to global control by products of an additional set of 10 regulatory genes called the *Polycomb* (*Pc*) group that *maintain* transcriptional repression of *HOM* genes where they are initially turned off (Martinez Arias, 1993; Gellon and McGinnis, 1998). If any *Pc* genes are inactivated by mutation, the *HOM* genes, though initially switched on in the correct pattern, later become activated all over the embryo, creating an open competition between their proteins for the regulatory sites of downstream genes. This competition is usually won by the proteins of *abd-A* and *Abd-B* because of posterior dominance, inducing a partial transformation toward posterior abdominal identity.

Products of additional regulatory genes of the *trithorax* (*trx*) group maintain *HOM* gene expression in the cells in which they are active (Martinez Arias, 1993;

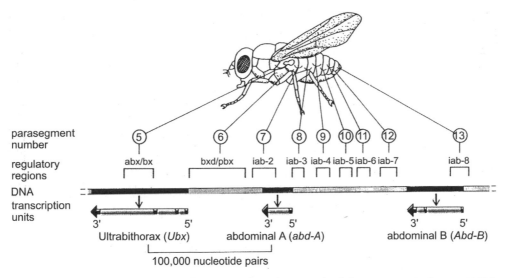

parasegment number

regulatory regions

DNA transcription units

Ultrabithorax (*Ubx*)   abdominal A (*abd-A*)   abdominal B (*Abd-B*)

100,000 nucleotide pairs

Fig. 7.35. Organization of the *D. melanogaster* bithorax complex. This stretch of chromosome 3 of some 300 kb contains three genes—*Ultrabithorax* (*Ubx*), *abdominal-A* (*abd-A*), and *Abdominal-B* (*Abd-B*)—encoding master regulatory proteins specifying development of posterior thoracic and of abdominal segments in the fly (see Fig. 7.36: top). Homeotic mutations have further defined nine groups of regulatory DNA sequences. Each is required for development of the parasegment indicated and more posterior parasegments. These regulatory DNA sequences act as enhancers to control the expression of one of the nearby genes, and their order on the DNA corresponds to the order of the body segments they affect. (© 1989 From *Molecular biology of the cell*, 3rd ed. by B. Alberts et al. Reproduced by permission of Routledge, Inc., part of The Taylor & Francis Group.)

Gerhart and Kirschner, 1997; Gellon and McGinnis, 1998).

7.2.1.3.4 DIFFERENCES IN *HOM* GENE FUNCTION BETWEEN *DROSOPHILA MELANOGASTER* EMBRYOS AND THOSE OF OTHER ARTHROPODS

Drosophila melanogaster has a highly specialized legless larva with an invaginated head and mouthparts and a poorly differentiated thorax (Fig. 7.29: bottom), whereas lepidopteran caterpillars have a well-developed head capsule with external biting and chewing mouthparts, functional thoracic legs, and a species-specific number of paired abdominal prolegs developing directly from the embryonic abdominal appendages (Fig. 9.9A). Warren and coworkers (1994) showed that the presence of abdominal limbs in embryos of the nymphalid butterfly *Precis coenia* result from striking changes in *BX-C* gene regulation compared with that occurring in *D. melanogaster* embryos.

At 20% of embryogenesis, expression of *Ubx* (posterior T3–anterior A8) and *abd-A* (posterior A1–anterior A8) is identical in *P. coenia* and *D. melanogaster* embryos (Fig. 7.36: Hexapoda), indicating that early specification of abdominal segments is the same. However, just after 20% of embryogenesis, paired larval limb buds begin to arise in *P. coenia* embryos, coincident with expression of the downstream regulatory gene *Distal-less* (*Dll*) in primordia of the antennae, first and second maxillae (but not mandibles), thoracic legs, pleuropodia

(A1), and prolegs on each of A3–A6 and A10. *Dll* expression is required for distal outgrowth of appendages but is repressed by products of *Ubx* and *abd-A* in the abdomen of *Drosophila* embryos. To release the abdominal limb program in *P. coenia* embryos, expression of *Ubx* and *abd-A* is repressed in paired, circular patches of cells on either side of the ventral midline, later developing into pleuropodia and prolegs. (*Ubx* and *abd-A* are activated again in the abdomen prior to metamorphosis so that the adult abdomen again lacks appendages, except for the external genitalia in males.) The change in expression of these two genes with respect to abdominal limb formation suggests that changes in number, size, and pattern of serially homologous structures such as appendages involve changes in the timing and spatial regulation of expression in the *HOM* genes themselves or of their downstream target genes (Gellon and McGinnis, 1998).

However, such repression of abdominal appendage formation by *Ubx/abd-A* is not universal in arthropod embryos, since their proteins do *not* repress expression of *Dll* or formation of appendages in myriapod, crustacean, and onychophoran embryos (Fig. 7.36) (Grenier et al., 1997; Popadíc et al., 1998a; Grenier and Carroll, 2000) or of abdominal prolegs in embryos of the tobacco hornworm *Manduca sexta* (Zheng et al., 1999). And in embryos of the flour beetle *Tribolium castaneum*, abd-A but not Ubx protein represses early expression of *Dll* in the abdomen whereas Ubx modifies development of A1 appendages so that they become pleuropodia instead of

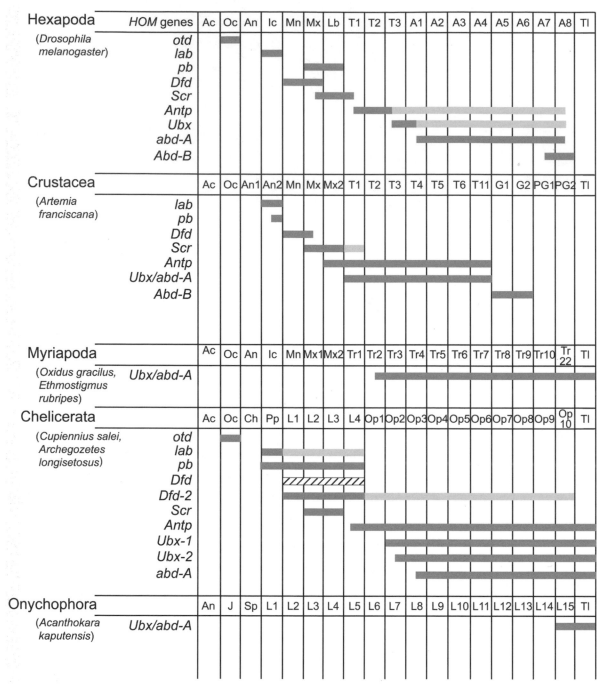

Fig. 7.36. Segmental expression of the homeotic (*HOM*) and an anterior gap gene (*orthodenticle* [*otd*]) in embryos of *D. melanogaster* (top) compared with that in embryos of other arthropods (see text for references). Subtle variations in the patterns of different species within a taxon, of the same species at different stages of development, and dorsally and ventrally within each embryo are not shown. Dark shading indicates strong expression; light shading, weak expression; and hatching, expression only at the tips of the legs. For ease of comparison, orthologs of the *Drosophila* genes are given the same names: *Dfd*, *Deformed*; *lab*, *labial*; *otd*, *orthodenticle*; *pb*, *proboscipedia*; *Scr*, *Sex combs reduced*; *Antp*, *Antennapedia*; *Ubx*, *Ultrabithorax*; *Ubx/abd-A*, *Ultrabithorax/abdominal-A*; *abd-A*, *abdominal-A*; *Abd-B*, *Abdominal-B*. Body segments are Ac, acron; An, antennal; A1–A8, abdominal; Ch, cheliceral; G1–G2, genital; Ic, intercalary; J, jaw; L1–L15, leg; Mn, mandibular; Mx, maxillary; Oc, ocular; Op, opisthosomal; Pp, pedipalpal; Sp, slime papilla; T1–T11, thoracic; Tl, telson; Tr1–Tr22, trunk.

legs (Lewis et al., 2000). The same could be true in embryos of all insects having abdominal appendages.

This suggestion is supported further by extensive, recent comparative work on *HOM* gene expression along the A-P axis in representative hexapod, chelicerate, myriapod, crustacean, and onychophoran embryos by Carroll (1995), Zrzavy and Stys (1995), Rogers and Kaufman (1996, 1997), Rogers and colleagues (1997), Gerhart and Kirschner (1997), Grenier and coworkers (1997), Averof (1998), Damen's group (1998), Telford and Thomas (1998b), Popadíc and colleagues (1998a), Peterson and others (1998, 1999), Abzhanov and K aufman (1999a, 1999b, 2000a), Abzhanov and coworkers (1999), Grenier and Carroll (2000), Hughes and Kaufman (2000), and Carroll and colleagues (2001). Some of these segment- and taxon-specific expression patterns are illustrated in Fig. 7.36 and listed below for Hexapoda (based on examination of embryos of *Thermobia domestica* [Zygentoma], *Acheta domesticus* [Orthoptera], *Oncopeltus fasciatus* [Hemiptera], *Tribolium castaneum* [Coleoptera], *Ctenophilades felis* [Siphonaptera], and *D. melanogaster*):

- *orthodenticle* (*otd*): preantennal (ocular) segment;
- *labial* (*lab*): intercalary;
- *proboscipedia* (*pb*): maxillary and labial (in maxillary and labial appendages in *T. domestica* embryos but only in labial appendages in *O. fasciatus*);
- *Deformed* (*Dfd*): mandibular and first maxillary;
- *Sex combs reduced* (*Scr*): posterior first maxillary to anterior T1 (expression required for fusion of second maxillae to form the labium);
- *Antennapedia* (*Antp*): posterior T1 to anterior T3 (strong) to anterior A8 (weak; to A10 in *T. domestica*);
- *Ultrabithorax* (*Ubx*): posterior T3 to anterior A1 (strong) to anterior A8 (weak);
- *abdominal-A* (*abd-A*): posterior A1 to anterior A8 (to A10 in *T. domestica*); and
- *Abdominal-B* (*Abd-B*): posterior A7 and A8 (to A10 in *T. domestica*).

However, even though the segmental expression of these genes is relatively similar in all investigated hexapod embryos, there are differences in their mutant phenotypes. For example, in larvae of *D. melanogaster* mutant in *Dfd* and *Scr*, the mouthparts are abnormal but not homeotically transformed. However, loss of function of the *T. castaneum* ortholog *Tc-Dfd* causes the mandibles to develop as antennae and the maxillae to lose the mala (galeal-lacinial lobe) (Brown et al., 2000). Loss of *Tc-Scr* and *Tc-Antp* function causes the labial and thoracic appendages to develop as antennae and loss of all *HOM-C* function, the development of antennae on all body segments. And loss of *Dfd/pb/Scr* function in *O. fasciatus* embryos causes the mandibles to develop as distal antennal segments and the maxillae and labium, as normal antennae (Hughes and Kaufman, 2000).

Based on these observations, Brown and coauthors (2000) suggested the ancestral insect *HOM-C* to have been considerably modified during the evolution of the highly specialized maggot of *D. melanogaster*. An ances-tral *Hox* gene that originated close to the root of the Eumetazoa, and that originally served to repress anterior development and to confer a trunk-specific identity, appears to have given rise to the *Dfd*, *Scr*, and *Antp* homeobox class of genes through tandem duplication and divergence.

HOM gene expression is summarized below for other selected arthropod embryos.

Crustacea (*Artemia franciscana* [Branchiopoda] and *Porcellio scaber* [Malacostraca: Isopoda]; expression differs in embryos of these two crustaceans—that of *A. franciscana* is shown in Fig. 7.36; orthologs of *Drosophila HOM* genes are given the same names)
- *lab*: second antennal segment;
- *pb*: posterior second antennal;
- *Dfd*: mandibular and anterior first maxillary;
- *Scr*: first and second maxillary (strong) and T1 (weak);
- *Antp*: second maxillary to back of thorax;
- *Ubx/abd-A*: all thoracic (but not abdominal) segments in members of primitive taxa, from T2, T3, or T4 to back of thorax in those of advanced taxa; and
- *Abd-B*: genital segments.

Myriapoda (*Oxidus gracilis, Ethmostigmus rubripes*)
- *Ubx/abd-A*: posterior trunk segment 2 to end of telson (also in first trunk segment in early development before differentiation of the maxillopeds).

Chelicerata (*Limulus polyphemus, Steatoda triangulosa, Archaearanea tepidariorum, Cupiennius salei, Archegozetes longisetosus*; orthologs of *Drosophila HOM* genes are given the same names)
- *otd*: ocular segment;
- *lab*: pedipalp (strong), leg 1 to leg 4 (weak);
- *pb*: pedipalp to leg 4;
- *Dfd*: tips only of leg 1 to leg 4;
- *Dfd2*: leg 1 to leg 4 (strong), to end of opisthosoma (weak);
- *Scr*: leg 2 to leg 3 (*C. salei, A. longisetosus*); leg 2 (weak), leg 3, leg 4 (complex; *S. triangulosa, A. tepidariorum*);
- *Antp*: posterior leg 4 to end of telson;
- *Ubx 1*: opisthosomal 2 to end of telson;
- *Ubx 2*: posterior opisthosomal 2 to end of telson; and
- *abd-A*: posterior opisthosomal segment 3 to end of telson.

(*Note*: this expression pattern suggests that the chelicerate prosoma could be homologous to the head of mandibulate arthropods.)

Onychophora (*Acanthokara kaputensis*)
- *Ubx/abd-A*: only in terminal trunk segment with reduced lobopods and telson.

Grenier and Carroll (2000) have expressed the *A. kaputensis Ubx* ortholog, *OUbx*, in *D. melanogaster* embryos mutant for *Ubx* and shown that it generated many of the same gain-of-function tissue transformations and activated and repressed many of the same target genes as does wild-type *Drosophila Ubx*. However, the OUbx protein did not transform the segmental identity of *Drosophila* embryonic ectoderm or repress the *Ubx* target gene *Dll*. Sequence differences between the two Ubx proteins were principally outside the homeodomain. Thus, the authors suggested that the *Drosophila* Ubx protein may have acquired new cofactors or activity modifiers since the divergence of the two lineages. Notice that in all these arthropods, the anterior border of *Ubx/abd-A* expression correlates with a change in appendage form, function, or presence/absence (Fig. 7.36).

To understand more fully the developmental and molecular bases for differences in body plan in members of lower arthropod taxa, Carroll and associates (2001) and many others suggest that we will require comparative expression studies on embryonic representatives of many more taxa and will have to compare the complex regulatory regions of both the *HOM* genes and the downstream genes they regulate. Only by this means will we be able to reconstruct the extensive shifts in timing and location of gene expression that might have contributed to the altered segment and appendage development in members of different lineages.

7.2.1.3.5 EVOLUTION OF THE *HOM* GENES

The striking sequence homology of DNA-binding homeodomains in proteins of the *HOM* genes (see 7.2.1.4) suggests that they evolved by repeated tandem gene duplication followed by divergence. Such an origin is suggested also by the close clustering of many of these genes in the *D. melanogaster* genome (Fig. 7.38). ANT-C, for example, includes the egg-polarity gene *bcd*, the pairrule gene *ftz*, and the D-V gene *zen* in addition to its five homeotic selector genes (*bcd* and *ftz* are omitted in Fig. 7.38: *Drosophila*; see 7.2.1.1.3). This clustering probably reflects their evolutionary history rather than functional necessity, since flies having genetic rearrangements separating the *Ubx* or other genes from their companions in BX-C show no abnormality in body plan (Alberts et al., 1994). Thus, throughout metazoan evolution, a series of altered, homeobox-containing genes were added to the genome that contributed to the evolution of increasingly complex body plans in representatives of the different phyla (Gellon and McGinnis, 1998).

The evidence? Similar homeotic gene complexes (called *Hox* genes in nonarthropod animals) are now known from embryos of *Thermobia domestica* (Zygentoma), *Schistocerca gregaria*, *S. americana* and *Acheta domesticus* (Orthoptera), *Blattella germanica* (Blattodea), *Oncopeltus fasciatus* (Heteroptera), *Tribolium confusum* (Coleoptera: where *ANT-C* and *BX-C* are combined into a single complex), *Precis coenia*, *Bombyx mori* and *Manduca sexta* (Lepidoptera), and *Apis mellifera* (Hymenoptera) among insects, and in embryos of representative onychophorans (*Acanthokara kaputensis*), crustaceans (*Artemia franciscana*), myriapods (*Ethmostigmus rubripes*, *Oxidus gracilis*), chelicerates (*Limulus polyphemus* [two sets], *Cupiennius salei*, *Archegozetes longisetosus*), brachiopods, priapulids, annelids, cnidarians (*Hydra* has orthologs of *lab*, *Dfd*, and *Antp*; Fig. 7.38), nematodes (*Caenorhabditis elegans*), sea urchins, acorn worms and pterobranchs (Hemichordata), tunicates (Urochordata), cephalochordates (*Branchiostoma*), hagfish, lampreys, teleost fishes, chickens, mice, and humans (Gehring, 1998; de Rosa et al., 1999; Carroll et al., 2001; Davidson, 2001). Therefore, the progenitors of the most widespread of these regulatory genes seem to date from the origin of the Metazoa in the Late Precambrian and could be construed as a shared, derived character defining the Metazoa: the *zootype* (Fig. 7.37) (Miller and Miles, 1993; Slack et al., 1993).

The *HOM* gene cluster in chromosome 3 of *D. melanogaster* (Fig. 7.38) and containing eight (plus three other) genes probably derives from splitting of a single ancestral gene cluster like that in the beetle *T. confusum* (Akam, 1991) and occurs in all investigated arthropods including onychophorans (Fig. 7.36) (Grenier et al., 1997; Popadíc et al., 1998a). Organization of the orthologous vertebrate *Hox* gene clusters resembles that inferred for the ancestral insect but appears to have been replicated twice in the lineage leading to vertebrates, so that each genome contains four clusters (*Hox A* to *Hox D*), each of 9–13 genes (for a total of 39), located on four chromosomes (Fig. 7.38) (see McGinnis and Kuziora, 1994, and Slack et al., 1993; Kenyon, 1994; Holland and Garcia-Fernandez, 1996; Valentine et al., 1996; and Erwin et al., 1997). And the zebrafish *Brachydanio rerio* has seven *Hox* clusters on seven different chromosomes, two clusters resembling each of *Hox* clusters *A*, *B*, and *C* and one of *D*, indicating that an additional replication of the cluster with the second copy of *Hox D* has been lost (Vogel, 1998). Many genes in each of these clusters are at corresponding (*paralogous*) positions to those of the *HOM* genes in *Drosophila* (*Hox* classes 1–13) and show extensive sequence similarity.

In addition, the linear order of genes within the clusters has been conserved between vertebrates and arthropods, and the genes of each cluster are likewise activated in a series along the A-P axis of the body that parallels the order of genes on the chromosomes (Fig. 7.38). Finally, the proteins of several vertebrate *Hox* genes expressed in *Drosophila* embryos mutant for their orthologs have resulted in their normal embryogenesis, emphasizing their close functional similarity (Rogers and Kaufman, 1997; Gellon and McGinnis, 1998; Leuzinger et al., 1998). These observations suggest that the first *Hox* gene clusters arose early in animal evolution, perhaps over 700 million years ago (Fig. 7.38)!

Based on known *Hox* gene sequence similarity, the metazoan ancestor probably had three to seven such genes (Fig. 7.38) (Carroll, 1995; Valentine et al., 1996; Erwin et al., 1997). Cnidarian *Hox* genes most closely resemble those on the 3' or anteriorly expressed ends of the arthropod and mammalian clusters (Galliot and

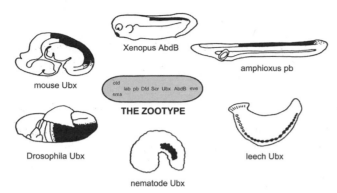

Fig. 7.37. The zootype (center), showing the spatial order of anterior expression limits of the *HOM* gene cluster and some other regulatory genes as revealed by study of *D. melanogaster* embryos. Around the zootype are displayed the phylotypic stages of various animal embryos, showing the expression of individual *Hox* genes as seen in whole mounts. *Drosophila* nomenclature is used for the gene orthologs: otd, *orthodenticle*; ems, *empty spiracles*; lab, *labial*; pb, *proboscipedia*; Dfd, *Deformed*; Scr, *Sex combs reduced*; Ubx, *Ultrabithorax*; AbdB, *Abdominal-B*; eve, *even-skipped*. The specific nomenclature for each animal and gene illustrated is Mouse, Hox A7; *Xenopus*, XlHbox 6; *Amphioxus*, AmphiHox 3; *Drosophila*, Ubx; *Caenorhabditis elegans*, mab-5; *Helobdella*, Lox-2. (Reprinted with permission from *Nature* (J. M. W. Slack, P. W. H. Holland, and C. F. Graham. The zootype and the phylotypic stage. 361: 490–492) © 1993 Macmillan Magazines Limited.)

Miller, 2000). Thus, it has been suggested that these original genes were probably involved in specifying the anteriormost structures of the metazoan body plan. The genes required for the formation of more posterior structures apparently evolved later, presumably by duplication and divergence of preexisting genes, to yield a cluster with about eight genes in the stem species of arthropods and chordates (Fig. 7.38) (Valentine et al., 1996; Grenier et al., 1997). Continued evolution in vertebrate complexity probably involved the addition of more posterior genes. In fact, some mouse and human homeotic genes located at the 5′ end of the *Hox A, C*, and *D* clusters (10–13) and expressed in the post-anal tail of the mouse embryo have no close relatives in *D. melanogaster*, suggesting that they arose after divergence of vertebrate and arthropod clades (Fig. 7.38). Finally, during vertebrate evolution, and following divergence of the Cephalochordata, the whole cluster duplicated at least twice, possibly as a result of chromosome duplication, as the body plans of species became more complex, ultimately producing the four clusters presently seen in mammals (Fig. 7.38) (see Holland and Garcia-Fernandez, 1996, for a discussion of how this might have occurred).

A final question concerns the evolutionary significance of the homeobox-containing genes no longer functioning in A-P specification. In embryos of *Schisto-*

cerca gregaria (Orthoptera), *Tribolium castaneum* (Coleoptera) and *Drosophila*, the *zen* gene, located in *Drosophila* on chromosome 3 between *pb* and *Dfd* (Fig. 7.38), functions to specify the D-V axis and amnioserosa (Fig. 7.20). In embryos of the spider *Cupiennius salei* (Damen and Tautz, 1998) and of the oribatid mite *Archegozetes longisetosus* (Telford and Thomas, 1998a), its orthologs, respectively, *Cs-Hox3* and *Al-Hox3*, are expressed in a continuous, longitudinal domain including the pedipalp- and four leg–bearing segments, suggesting a role in A-P specification in chelicerates that has been lost in insects, perhaps owing to functional redundancy with *pb* (Telford and Thomas, 1998b).

7.2.1.4 DNA-BINDING PROTEINS AND CONTROL OF EXPRESSION OF PATTERN-CONTROL GENES

When the pattern-control genes of *D. melanogaster* began to be cloned in the early 1980s, a sequence of 180 nucleotide base pairs called the *homeobox* was discovered, by Scott and Weiner (1984) and McGinnis and coworkers (1984), to occur in many of them (Gehring and Hiromi, 1986; Beardsley, 1991; Krumlauf and Gould, 1992; Alberts et al., 1994: 410; Gehring et al., 1994; Lawrence and Morata, 1994; Duboule, 1994; Gehring, 1998). The pair-rule gene *ftz*, for example, contains a homeobox whose base pair sequence is 77% identical to that of the *HOM* gene *Antp*, and the match between the 60–amino acid homeodomain encoded by these sequences is 83%. Over 100 genes in *D. melanogaster* contain a homeobox including the egg-polarity gene *bcd*; the D-V gene *zen*, the pair-rule genes *ftz*, *eve*, and *prd*; the segment-polarity genes *gsb* and *en*; and all *HOM* genes (see Alberts et al., 1994: 408, and gene tables by various authorities in Bate and Martinez Arias, 1993).

Genes containing a homeobox encode proteins localized in the cell nucleus, suggesting a role in the control of gene expression (Alberts et al., 1994). Also, the homeodomain enables proteins containing it to bind to specific regulatory (enhancer) DNA sequences of other genes and to regulate their expression. (Outside the homeodomain, the amino acid sequences of different homeobox genes differ greatly, perhaps indicating extensive diversification in their interaction with other, already bound transcription factors [Gerhart and Kirschner, 1997: 322]). However, it should be mentioned that the proteins of many other regulatory genes in *D. melanogaster* have other DNA-binding motifs, including zinc finger, basic helix-loop-helix, and leucine zipper (see Alberts et al., 1994: 408–413 and gene tables in Bate and Martinez Arias, 1993).

The pair-rule gene *eve*, for example, has a gigantic control sequence of 20 kb containing a series of regulatory modules, each with multiple regulatory sequences and each responsible for specifying one of the seven stripes of *eve* expression along the length of the embryo (Stanojevic et al., 1991; Small et al., 1992; Alberts et al., 1994: 426–429; Gerhart and Kirschner, 1997: 135–138). This was demonstrated by deleting successive portions

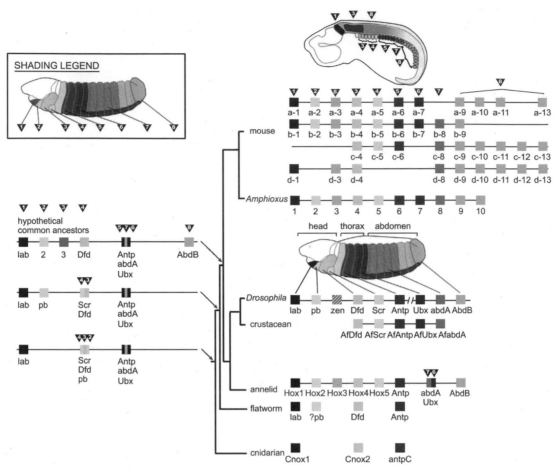

Fig. 7.38. Modifications in metazoan body plans correlate with modifications in the expression pattern of control genes—for example, through duplications of genes and subsequent sequence changes—such as those found in the *Hox* gene clusters. Mapping *Hox* genes on a phylogenetic tree of selected metazoans suggests that the common ancestor of *D. melanogaster* and mice had one gene that presumably controlled the development of the animal's midsection. In *Drosophila*, this gene has split into three separate genes. These genes then underwent further expansion, duplication, and sequence modifications in evolution so that in the mouse there are a larger number of related genes to control and direct the formation of an abdominal region that is much more complex than the abdominal region of representives of more basal lineages. Similar expansions and modifications can be seen in other *Hox* genes as one goes from more primitive to more advanced animals and body plans. Numbered triangles in the boxed shading legend for a *Drosophila* embryo are used to indicate the expression distribution of orthologs of these genes in the embryos of other animals. (Adapted with permission from D. H. Erwin, J. Valentine, and D. Jablonski. 1997. The origin of animal body plans. *Am. Sci.* 85: 126–137, drawing by Linda Huff, as adapted from Carroll, 1995.)

of the DNA from the regulatory region; the number of sequences removed reduced the seven stripes of *eve* expression to six, five, four, and so on. The regulatory module for stripe 2, of 480 bp, contains multiple recognition sequences for binding four different transcription factors: five for Bicoid (egg-polarity protein) and one for Hunchback (gap protein) that activate *eve* transcription and three for Krüppel and two for Giant (both gap proteins) that repress it (Fig. 7.39C). The relative concentration of these four proteins along the length of the egg (Fig. 7.39A) determines whether *eve* is transcribed in stripe 2 (Fig. 7.39B). Both Bicoid and Hunchback must bind to maximally activate the module, while binding of either Giant or Krüppel inhibits it. Thus, only preblastoderm nuclei located at that length of the egg where Bicoid and Hunchback concentrations are high and where proteins of Krüppel and Giant are absent will express stripe 2 *eve* (Fig. 7.39A, B).

Six additional combinations of transcription factors activate *eve* expression in the other six stripes, with one for each stripe, while other combinations prevent stripes from forming in the interstripe regions. Altogether there are binding sequences for 20 different transcription factors in the 20 kb of the entire *eve* control region (Alberts et al., 1994). Reinitz and Sharp (1995) did a computer analysis of gene circuits to evaluate *eve* stripe forma-

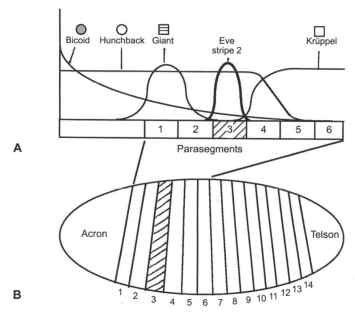

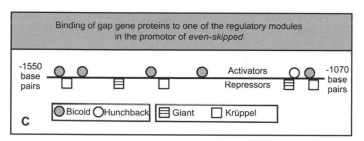

Fig. 7.39. Specification of *even-skipped* (*eve*) stripe 2 expression in a preblastoderm embryo of *D. melanogaster* by the binding of gap gene proteins. A and B. The different concentrations of transcription factors encoded by the maternal effect gene *bicoid* and by the gap genes *hunchback*, *giant*, and *Krüppel* localize *eve* expression in parasegment 3. Bicoid and Hunchback proteins activate the gene in a broad domain, and its anterior and posterior borders are specified through repression by Giant and Krüppel proteins, respectively. C. A promotor region of about 500 bp, located between 1070 and 1550 bp upstream of the transcription start site, directs formation of the second *eve* stripe. Expression of *eve* occurs when the Bicoid and Hunchback transcription factors are present above a threshold concentration, with the Giant and Krüppel proteins acting as repressors where they are above threshold levels. The repressors may act by preventing binding of activators. (Adapted by permission of Oxford University Press from L. Wolpert. *Principles of development*, p. 149, Figs. 5.22, 5.23. Current Biology Ltd. © 1998 Oxford University Press.)

tion. They used gene expression data to infer how concentrations of products of a given gene change with time and how these changes are influenced by the activating or repressing effects of the products of other genes.

Other regulatory genes such as the terminal (7.2.1.1.2), segment-polarity (7.2.1.2), HOM (7.2.1.3), and proneural genes have equally complex control regions likely functioning in the same way (Gerhart and Kirschner, 1997; Rudolph et al., 1997; Gehring, 1998). However, there is a paradox. Although the homeodomain proteins select and regulate diverse developmental processes in a host of organisms (7.2.1.2.1, 7.2.1.3.4, 7.2.1.3.5), they bind in vitro to similar DNA sequences. So, how is target gene specificity achieved in vivo? As mentioned, the protein of the pair-rule gene *ftz* contains a homeodomain and must bind appropriately to regulatory DNA of target genes to specify alternate segments. Copeland and colleagues (1996) created a *Drosophila* homeodomain-deleted Ftz protein (FtzΔHD) that was incapable of binding to DNA in vitro. However, such a protein could still directly regulate *ftz*-dependent segmentation, suggesting that it can control target gene expression not only by binding to DNA in regulatory sequences in the regulatory module but also by interact-

ing with other transcription factors (often called cofactors) that are already bound. In this example, a likely candidate is the pair-rule protein Paired (Prd). FtzΔHD bound directly to Prd in vitro and required Prd to repress *wg* expression in vivo.

In a further experiment, Nasiadka and coworkers (2000) produced a chimeric Ftz protein by fusing it to a strong activation domain of the viral VP16 protein. Genes previously thought not to be targets of Ftz remained unaffected by FtzVP16. However, previously identified target genes were regulated by FtzVP16 at times and in regions that Ftz alone cannot, and two genes that are normally repressed by Ftz were activated.

Finally, in an attempt to understand how rapidly the pattern-control genes can interact, Nasiadka and Krause (1999) analyzed the response kinetics of 11 known and possible target genes of Ftz in *Drosophila* embryos, including *wg*, *en*, *prd*, *odd*, and *slp*, by providing a brief pulse of Ftz expression and then measuring the time required for each downstream gene to respond (the time required for Ftz to bind to and regulate its own enhancer was used as a standard). Both positively and negatively regulated downstream genes were found to respond to Ftz with the same kinetics as autoregulation, the rate-limiting step between successive interactions (<10 min) being the time required for Ftz protein to either enter or be cleared from target cell nuclei. This matching of response seems to be critical for rapid synchronous progression from one class of segmentation genes to the next during embryogenesis.

Such protein-protein interactions are now known to influence the affinity, specificity, and function of the products of other pattern-control genes such as *bcd* (Ma et al., 1996), *lab* and *Dfd* (Gellon and McGinnis, 1998), *Ubx* (Passner et al., 1999), and *Abd-B* (Castelli-Gair, 1998). For example, if Ubx complexes with Extradenticle (Exd) when binding to the DNA, it increases its affinity and specificity for the binding sites of target genes as shown by x-ray crystallography of the Ubx-Exd-DNA complex. And combinations of binding sites for Scalloped (Sd; a transcription factor expressed in the ventral wing pouch) and for the transcription effectors of various signaling molecules are necessary and sufficient to specify wing-specific responses to different signaling pathways (Guss et al., 2001).

7.2.1.5 MOLECULAR GENETICS OF HEAD SEGMENTATION IN *DROSOPHILA* EMBRYOS

As summarized above, the larval body pattern in embryos of *D. melanogaster* is specified by an elaborate cascade of hierarchical and cross-regulatory gene interactions (Gerhart and Kirschner's [1997] intermediate processes) that function to subdivide the trunk of the embryo into segments from the intercalary segment back and to specify their identity. However, an additional mechanism must define segmentation in front of the intercalary segment, since products of the *HOM* genes, except for *forkhead* (*fkh*) in the anterior midgut (AMG), are not expressed in these segments (Fig. 7.41) (Finkelstein and Perrimon, 1991; Jürgens and Hartenstein, 1993; Finkelstein and Boncinelli, 1994; Grossniklauf et al., 1994; Rogers and Kaufman, 1996, 1997; Schöck et al., 2000).

After decades of frequently vociferous debate (Rempel, 1975), most insect embryologists and morphologists agree that the insect head consists of a nonsegmented procephalon (acron) and six segments—three pregnathal segments (labral, antennal, intercalary) and three specialized gnathal segments (mandibular, maxillary, labial)—the so-called *linear model* (Fig. 6.22B; see 6.10.1). However, by carefully monitoring expression of the segment-polarity genes *en* and *wg*, Schmidt-Ott and Technau (1992) provided solid evidence indicating that there are actually seven head segments, at least in *D. melanogaster*, an additional *ocular* (preantennal) segment occurring between the labral and antennal segments (Fig. 7.40B: note expression of both genes in the protocephalon). Such also occurs in crustacean embryos (Scholtz, 1998). In addition, the distribution of cephalic sense organs in *D. melanogaster* and of the nerves of both the peripheral and stomatogastric nervous systems also support a seven-segment head (Schmidt-Ott and

Technau, 1994; Schmidt-Ott et al., 1994b, 1995; Schöck et al., 2000).

More recent work, using the expression pattern of *en* only, in embryos of the silverfish *Thermobia domestica*, the cat flea *Ctenophalides felis*, the milkweed bug *Oncopeltus fasciatus*, the house cricket *Acheta domesticus* (Rogers and Kaufman, 1996, 1997; Gallitano-Mendel and Finkelstein, 1997; Peterson et al., 1998), the spiders *Cupiennius salei* (Damen et al., 1998), *Steatoda triangulosa*, and *Archaearanea tepidariorum* (Abzhanov et al., 1999), the oribatid mite *Archegozetes longisetosus* (Telford and Thomas, 1998b), and several crustaceans (Scholtz, 1998; Abzhanov and Kaufman, 1999a, 1999b) does not support the presence of a labral segment in any except *Drosophila* embryos, since *en* is not expressed in its cells. In fact, Scholtz (1998) suggested instead that the labrum probably represents the anterior tip of the acron, and Schöck and colleagues (2000) refer to the acron as the acron/labrum.

The issue finally seems to have been resolved by discovery of a homeotic mutation *Antennagalea* (Ag^5), in embryos of the flour beetle *Tribolium castaneum*, that transforms both antennal and labral structures into those resembling the gnathal appendages (Haas et al., 2001b) and by a thorough and critical review of both histological and molecular evidence (Haas et al., 2001a). Based on this evidence Haas and coauthors (2001a) proposed a new *L-/bent Y- model* of head segmentation in insects in which both the labrum and stomodaeum derive from the intercalary segment, the labrum arising through median fusion of the proximal endites (galeae, laciniae) of a pair of ancestral intercalary appendages.

In the three gnathal segments of *D. melanogaster* embryos, *en* is expressed under control of the pair-rule genes as in the thorax and abdomen (Figs. 7.26, 7.41). However, in segments in front of the mandibular segment, *en* expression arises much later in development and without instructions from the pair-rule genes (Cohen and Jürgens, 1991; Jürgens and Hartenstein, 1993; Rogers and Kaufman, 1996, 1997). Pregnathal *en* expression is identical in extended germ band–stage embryos of other flies and of bugs, beetles, locusts, spiders, mites, and crustaceans, whether they be short or long germ, suggesting a common mechanism for all arthropods, with the head segments homologous throughout the phylum including the prosoma of chelicerates (Fig. 7.36) (Schmidt-Ott et al., 1994a; Patel, 1994; Rogers and Kaufman, 1996, 1997; Gerhart and Kirschner, 1997; Rogers et al., 1997; Damen et al., 1998; Telford and Thomas, 1998a, 1998b; Scholtz, 1998; Abzhanov and Kaufman, 1999a, 1999b; Abzhanov et al., 1999; Brown et al., 1999; Peterson et al., 1999; Knoll and Carroll, 1999).

In *D. melanogaster* embryos, expression of the maternal egg-polarity genes *tor* and *bcd* of the terminal (7.2.1.1.2) and A-P (7.2.1.1.3) pattern-forming systems specifies the spatial domains of gap gene expression, which in turn subdivides the pregnathal head region into

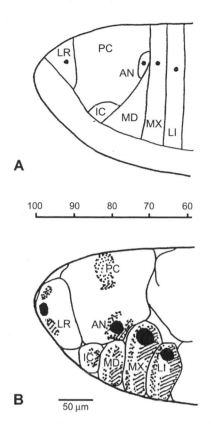

A

```
100    90    80    70    60
```

B

50 μm

Fig. 7.40. Fate map of the head segments in an embryo of *D. melanogaster* based on UV ablation studies. A. At the blastoderm stage just before gastrulation. B. At the extended germ band stage (about 28% of embryogenesis). The three anterior head segments—labral (LR), antennal (AN), and intercalary (IC)—are cephalic; the three posterior segments—mandibular (MD), maxillary (MX), and labial (Ll)—are gnathal. The protocephalic region (PC) includes the nonsegmented terminal acron and an additional ocular (preantennal) segment. The spots in LR, AN, MX, and Ll indicate the relative position of primordia of the cephalic sense organs, and at the extended germ band stage, these express the limb-specific gene *Distalless* (*Dll*) required for distal outgrowth of these appendages. The scale bar indicates percentage of egg length from the posterior end. Hatched and stippled areas in B, respectively, indicate the expression patterns of the segment-polarity genes *engrailed* (*en*) in the posterior compartment and of *wingless* (*wg*) in the anterior compartment of each segment. (Reprinted from *Trends Genet.*, 7, S. Cohen and G. Jürgens, *Drosophila* headlines, pp. 267–272, Copyright 1991, with permission from Elsevier Science.)

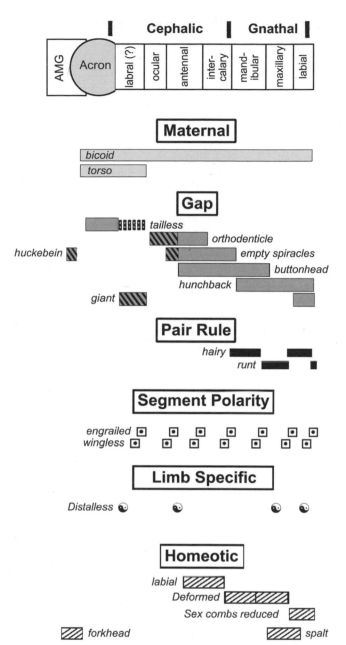

Fig. 7.41. Expression pattern of maternal egg polarity and of zygotic gap, pair-rule, segment-polarity, homeotic, and limb-specific (*Distal-less*) genes that specify segment identity in the head of the *D. melanogaster* embryo. Expression of *wingless* and *engrailed* in the labral (?) segment occurs in *Drosophila* (Fig. 7.40B) but not in other insect embryos examined. AMG, anterior midgut. (Copyright 1993 from Phylogenetic pattern and developmental process in *Drosophila.* by R. DeSalle and D. Grimaldi. 1993. *Syst. Biol.* 42: 458–475. Adapted by permission of Taylor & Francis, Inc., http://www.routledge-ny.com.)

particular segments (Fig. 7.41). The A-P gradient of Bcd protein (Figs. 7.24B, C; 7.39A) provides positional information along the length of the anterior head by transcriptionally regulating expression of seven gap genes (Fig. 7.41). The affinity of Bcd protein for the specific array of binding sites in the regulatory region of each gap gene determines the threshold concentration at which expression of that gene is activated or repressed, as explained more fully for the pair-rule gene *eve* in 7.2.1.4. Expression of *hb*, for example, is directly activated by binding of Bcd protein to its promotor, and changing the distribution of Bcd in the embryo artificially shifts the posterior border of *hb* expression forward or posteriorly along the A-P axis.

Three additional gaplike genes—*orthodenticle* (*otd*), *empty spiracles* (*ems*), and *buttonhead* (*btd*)—function anterior to *hb* in the head (Fig. 7.41), and mutations in these genes cause deletions of groups of adjacent, pregnathal segment primordia. Domains in which the three genes are required overlap but are out of register by one segment at their posterior borders: *btd* mutants lack the antennal, intercalary, and mandibular segments and also the cephalic furrow that forms between the intercalary and mandibular segments from 15% to 19% of embryogenesis (Fig. 6.23: 8 [Vincent et al., 1997]); *ems* mutants, the antennal and intercalary segments and the posterior compartment of the ocular segment; and *otd* mutants, the ocular and antennal segments (Fig. 7.41). Both *ems* and *otd* genes begin to be expressed in preblastoderm and depend on *bcd* expression: differences in their threshold concentrations for Bcd protein establish the posterior borders of their expression. Anterior limits of *hb* and *otd* expression are determined by inhibition by products of the terminal genes *tor* and *tll* that specify acron (Fig. 7.41; 7.2.1.1.2).

In *D. melanogaster*, the labral segment (clypeolabrum of Rogers and Kaufman, 1997) is said to be determined partly by action of *tor* and *tll* and of *cap'n'collar* (*cnc*), a selector gene also contributing to specification of the mandibular segment (Mohler et al., 1995) (however, recall that Haas et al. [2001a] provided strong evidence that the labrum is intercalary). Finally, expression of a newly discovered gene, *knot* (*kn*), seems to be required for development of the hypopharyngeal lobe and its derivatives, the ventral arms and lateral grate of the cephalopharyngeal skeleton (Fig. 8.45), which at least partly derive from the mandibular segment (Seecoomar et al., 2000).

Following determination of segment primordia, expression of the limb-specific gene *Dll* (Panganiban, 2000) is required for distal outgrowth of the labral, antennal, intercalary (second antennal; not in *Drosophila*), maxillary, and second maxillary (labial) appendages or their homologs in chelicerate, myriapod, insect, and crustacean embryos but not for outgrowth of the mandibular appendages in mandibulate embryos (Figs 7.40, 7.41) (Cohen et al., 1989; Rogers and Kaufman, 1996, 1997; Popadíc et al., 1998b; Scholtz, 1998; Scholtz et al., 1998). However, *Dll* is expressed in the jaws of onychophoran embryos (Grenier et al., 1997). That *Dll* activity is not required for mandibular outgrowth is perhaps because such appendages are gnathobasic and reduced from an ancestral, whole-limb mandible by loss of the distal exopod segments, as suggested by certain fossils and by comparative functional and embryonic studies (Popadíc et al., 1998b; Scholtz et al., 1998). However, *Dll* is expressed early in mandibular primordia in crustacean and myriapod embryos and in the mandibular telopodite (palp) in the nauplius and early larval stages of indirect-developing crustacean embryos.

That the mechanism functioning to specify segment identity within the pregnathal head of mandibulate arthropods differs from that in the rest of the body led Scholtz (1998) to propose that division of the head into pregnathal and gnathal components is a shared derived character (apomorphy) for members of the subphylum Mandibulata. Rogers and Kaufman (1996, 1997) discussed this entire topic in much more detail with respect to development of the head and mouthparts in silverfish, cricket, milkweed bug, flea, and *Drosophila* embryos.

7.2.2 EVOLUTIONARY CONSIDERATIONS

Existence of two segmentation systems establishing segmental identity in the head and trunk of *D. melanogaster* embryos reflects the mechanism of segmentation occurring in the short germ embryos of members of basal and intermediate clades of insects and of other arthropods (Cohen and Jürgens, 1991; Scholtz, 1998). Recall that in short germ insects, the embryo first appears posteroventrally in the egg as a small, keyhole-shaped germ band consisting of protocephalon and protocorm, with gnathal, thoracic, and abdominal segments budding off anteriorly from a growth zone in front of the telson during germ band elongation (Figs. 6.14D, 6.22A). Establishing the body pattern of such insects occurs in two phases: the budding off of segments from the front of the growth zone followed by sequential specification of segment identity. In long germ embryos of *D. melanogaster*, however, the entire segmented body pattern is specified by the time of blastoderm formation (6.10.2), the two phases unfolding concurrently.

All investigated arthropod embryos use *en* expression to define the posterior compartment of each segment and expression of the cephalic gap genes both to control partition of the anterior head into segments and to specify their identity (compare Fig. 7.32) (Patel, 1994; Scholtz, 1998). Semi-long germ insects could result from recruitment of additional gap genes to specify segment identity in the gnathal and thoracic segments. In the transition from semi-long to long germ insects, the overlapping patterns of gap gene proteins in the thorax might have spread posteriorly to include abdominal segment primordia. To extract more resolution from existing gap gene overlap, pair-rule gene functions might have been interposed between those of gap and segment-polarity genes. This view predicts that pair-rule genes will be found unnecessary for segmentation in short and semi-long germ insects, and Patel and coworkers (1992) and Strand and Grbíc (1997) have evidence to indicate that this is so, respectively, in locust and *Aphidius ervi* (Hymenoptera: Braconidae) embryos (Fig. 7.31) (but see Sommer and Tautz, 1993). That pair-rule genes such as *eve* are known to function identically in development of the ventral nerve cord in both locust and *Drosophila* embryos may indicate that this is the original function of the pair-rule genes, their having been co-opted subsequently to function in A-P axis specification in the long germ embryos of *Drosophila* (Gerhart and Kirschner, 1997: 466–467; Rogers and Kaufman, 1997).

In fact, many recent students of evolutionary developmental biology have proposed that mutations changing spatial, temporal, or cellular asymmetry in activity of the pattern-control genes might have contributed to evolution of the diverse body forms in all metazoan lineages. Syntheses of recent comparative work support this suggestion (e.g., Carroll et al., 2001), but West-Eberhard (1989) and Budd (1999) are not convinced and provide interesting alternative explanations (Figs. 13.10, 13.11).

8/ Organogenesis

In chapter 6, I summarized early embryogenesis up until the polypod or extended germ band stage when the primordia of all segments and appendages have formed (Fig. 6.22B), and in chapter 7, I discussed the role of pattern-control genes in specifying development of this body pattern. Here, I describe how cells differentiate within the various organ systems of the juvenile body and comment on the role of genes and other factors in directing their ontogeny.

Fig. 8.1, by the late insect functional morphologist Hermann Weber, illustrates the external structure and principal internal organ systems of a generalized adult male and female insect and the spatial arrangement of organs within the female's fourth abdominal system in transverse section. During organogenesis, previously determined cells in the embryo segregate into organ primordia and differentiate so that at its end, a functioning young insect hatches from the egg. In this chapter, the formation of each organ system is discussed according to its germ layer of origin: mesoderm, endoderm, and ectoderm.

8.1 Mesoderm

Except at its anterior and posterior ends, all cells of the inner layer originating through gastrulation (Fig. 6.19) are specified as mesoderm (Bate, 1993). After their ingression, these cells dissociate and move to each side of the embryo to form two longitudinal bands; a few remain on the midline as the middle layer (Fig. 8.2A, B, D). Commencing in the presumptive prothorax and progressing forward and backward, these bands segment transversely into a pair of somites in each body segment, which then hollow out into a pair of coelomic sacs for a total of 19 or 20 pairs: 6 in the head, 3 in the thorax, and 10 or 11 in the abdomen (Fig. 8.2C, E). Based on classic descriptive study of their subsequent history in embryos of many insects (summarized in Johannsen and Butt, 1941; Hagan, 1951; Ando, 1962; Anderson, 1972a, 1972b, 1973; Haget, 1977; Sander et al., 1985; Ando and Miya, 1985; Ando and Jura, 1987; Kobayashi and Ando, 1988; and Schwalm, 1988), cells of the inner, splanchnic walls of the sacs facing the yolk differentiate into visceral (gut) muscle, gonadal sheath cells, primary exit system, and fat body, and those of the outer, somatic walls, into

skeletal muscles of the body wall and appendages, dorsal and ventral diaphragms, and pericardial cells (Fig. 8.2F, G). Cardioblasts in a dorsal row between the somatic and splanchnic walls on each side come together during dorsal closure to form the aorta and heart, while cells of the middle layer differentiate into hemocytes (Fig. 8.2C, F, G).

In smaller embryos of the members of the more advanced hexapod lineages, coelomic sacs either fail to form (e.g., *D. melanogaster* and other higher flies) or form only in the thoracic segments, or the somites fragment without becoming hollow. Nevertheless, the mesoderm is segmentally arranged in all investigated insect embryos, and the fates of its cells are similar to that just summarized, as recently confirmed in embryos of *D. melanogaster* by use of molecular and genetic techniques (Hartenstein and Jan, 1992; Borkowski et al., 1995; Campos-Ortega and Hartenstein, 1997; Riechmann et al., 1997, 1998; Baylies et al., 1998; Kusch and Reuter, 1999). In embryos of *Drosophila*, precursor cells for each mesodermal organ originate in specific areas on each side of each parasegment: visceral muscle anterodorsally, cardioblasts and pericardial cells posterodorsally, fat body anterolaterally, and somatic muscle posterolaterally (Figs. 8.3B, 8.4) (Borkowski et al., 1995; Riechmann et al., 1997).

These observations suggest that insect embryos are both *pseudocoelomate* (body cavity is incompletely lined by tissues derived from mesoderm) and *schizocoelomate* (coelomic cavities arise through separation of mesodermal cells within each somite).

8.1.1 GENES AND MESODERMAL FATE IN *DROSOPHILA* EMBRYOS

Borkowski and colleagues (1995) followed the fate of recently gastrulated mesoderm in *D. melanogaster* embryos using a cell marker in which the *twist* (*twi*) promotor (6.8.1) directed the synthesis of a cell-surface protein CD2 (twi-CD2) (Fig. 8.3A). They found early mesoderm in either side of each parasegment to be divided into an anterior region with relatively low levels of *twi* and twi-CD2 expression and a posterior region where expression of both *twi* and twi-CD2 remains high. This subdivision coincides with regional assignment of cells to form different progenitors within each parasegment (Figs. 8.3B, 8.4):

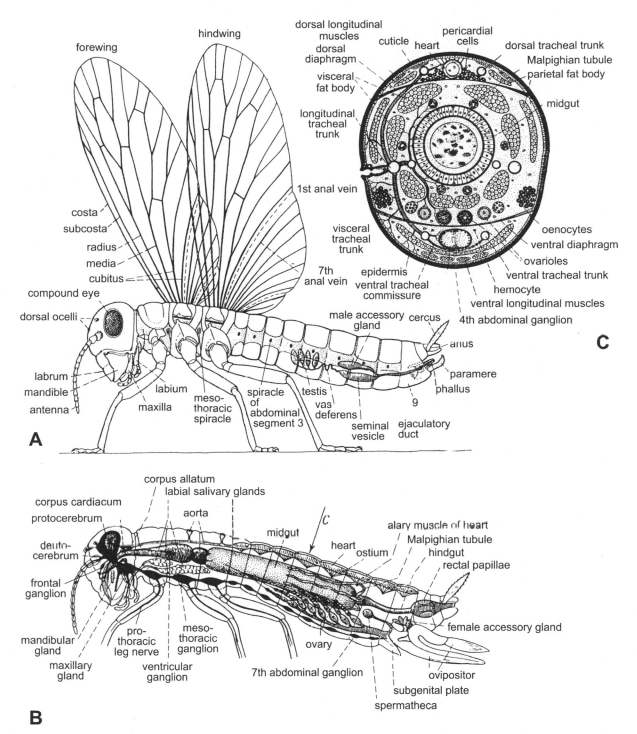

Fig. 8.1. External and internal structure of a generalized male (A) and female (B) insect in lateral aspect, illustrating the principal organ systems. C is a transverse section through abdominal segment 4 of the female (note arrow in B). (Adapted with permission from H. Weber. *Grundriss der Insektenkunde*, p. 36, Fig. 22. 2 ed. © 1966 G. Fischer Verlag.)

- Dorsal anterior cells invaginate to form an interior layer from which the visceral mesoderm and fat body derives.
- Ventral anterior cells form progenitors of mesodermal glial cells about the central nervous system (CNS).
- Dorsal and ventral posterior cells, respectively, form dorsal vessel and somatic muscles.

Correct specification of these cells in each side of each parasegment depends on where they are located in relation to ectodermal expression of the pair-rule and segment-polarity genes (7.2.1.2) (Riechmann et al., 1997; Baylies et al., 1998) and results in progenitors of the various mesodermal primordia

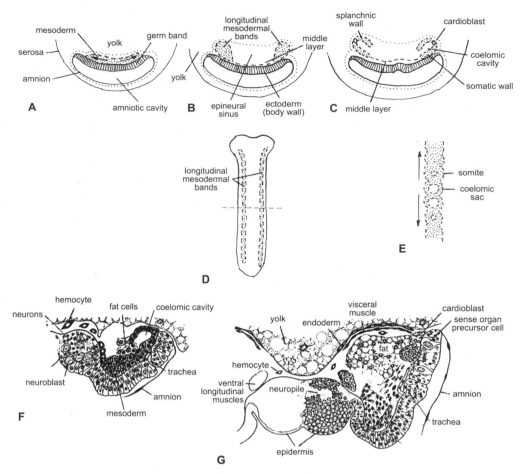

Fig. 8.2. Formation and differentiation of mesodermal coelomic sacs in an unspecialized insect embryo. A. Transverse section, showing spread of mesoderm to either side of germ band following gastrulation. B. Transverse section at the point shown by the dotted line in D, showing formation of two longitudinal mesodermal bands along either side of the germ band. C. Hollowing out of somites in each segment to form a pair of coelomic sacs. D. Dorsal aspect of embryo, showing longitudinal mesodermal bands. E. Dorsal aspect of the right mesodermal band, showing each somite hollowing out to form a coelomic sac. In embryos of species in more basal lineages, this often begins in the presumptive prothorax and proceeds anteriorly and posteriorly. F. Transverse section through the abdomen of an embryo of *Chrysopa perla* (Neuroptera: Chrysopidae), showing cells of a coelomic sac beginning to dissociate to form the various mesodermal tissues. G. Transverse section later in organogenesis. (F and G reprinted with permission from H. Weber. *Grundriss der Insektenkunde*, p. 19, Fig. 10i, m. 2 ed. © 1966 G. Fischer Verlag.)

in each parasegment having the distribution shown in Fig. 8.4.

This distribution was revealed by observing the mesodermal phenotypes of embryos mutant in one or another of these pattern-control genes (Riechmann et al., 1997). For example, somatic muscle primordia were lost in *sloppy-paired* (*slp*) mutants while those of visceral muscle were expanded. Both visceral muscle and fat body primordia require expression of *even-skipped* (*eve*), and if this gene was mutant, the mesoderm lost its segmental organization. The domain of *eve* function was found to be regulated by *slp*.

Expression of the *bagpipe* (*bap*) gene specifies anlagen of visceral muscle and expression of *serpent* (*srp*), those of fat body (Azpiazu et al., 1996). Both these genes are expressed in dorsal anterior mesodermal cells of each parasegment, while expression of *tinman* (*tin*) marks

precursors of cardiac (dorsal vessel) and somatic muscles in posterior positions within each parasegment. The gene *eve* is expressed autonomously within the mesoderm, whereas *hedgehog* (*hh*) and *wingless* (*wg*) mediate pair-rule functions within the mesoderm partly by acting within the mesoderm and partly by means of inductive signaling from the ectoderm (8.6.2) (proteins of both genes are secreted). Expression of *hh* is required to activate expression of *bap* and *srp* in anterior mesoderm of each parasegment, whereas *wg* suppresses expression of these genes in posterior mesoderm.

Following the specification of mesoderm as larval muscle progenitors, the amount of Twist (Twi) protein remains high in nuclei of prospective somatic muscle cells and is essential for their differentiation but rapidly falls in nuclei of other mesodermal cells. If high levels of Twi persist in the nuclei of these cells, they fail to dif-

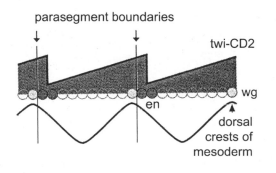

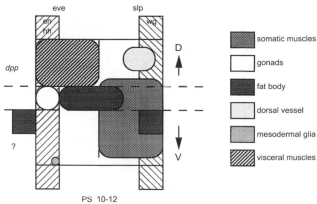

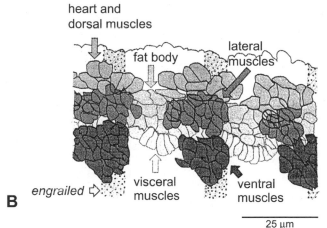

Fig. 8.4. Fate map, in lateral aspect, of mesodermal cells in each of parasegments 10–12 (abdominal segments 4–7) of a *D. melanogaster* embryo at 24%–32% of embryogenesis, showing the expression pattern of the genes specifying their fates. *dpp*, *decapentaplegic*; en, *engrailed*; eve, *even-skipped*; hh, *hedgehog*; slp, *sloppy-paired*; wg, *wingless*. The map is the same in parasegments 4–9 (mesothoracic to abdominal segment 4) except that the cells specified as gonadal develop into additional fat body. (Adapted with permission from V. Reichmann, K.-P. Rehorn, R. Reuter, and M. Leptin. The genetic control of the distinction between fat body and gonadal mesoderm in *Drosophila. Development* 125: 713–723. © 1998 Company of Biologists, Ltd.)

Fig. 8.3. Gene expression and the distribution of mesodermal primordia in three parasegments of a *D. melanogaster* embryo at 24%–32% of embryogenesis. A. The relative positions of parasegmental borders, the sharp anteroposterior boundary for twi-CD2 expression, and the distribution of mesodermal crests. en, *engrailed* expression; wg, *wingless* expression. B. Lateral aspect of an embryo at the same stage of development, showing the distribution of mesodermal primordia in three parasegments. The more heavily shaded cells are just under the epidermis; the more lightly shaded ones are more deeply situated. The vertical stippled area indicates the anterior parasegmental compartment in the epidermis in which *en* is expressed. (Reproduced with permission from O. M. D. Borkowski, N. H. Brown, and M. Bate. Anterior-posterior subdivision and the diversification of the mesoderm in *Drosophila. Development* 121: 4183–4193. © 1995 Company of Biologists, Ltd.)

ferentiate into their appropriate, mesodermal derivatives, whereas ectopic expression of Twi, in the nuclei of ectodermal cells, causes these to differentiate as somatic muscle (Baylies and Bate, 1996). Therefore, *twi* not only functions in gastrulation and to specify mesoderm (6.8.1) but also directs a specific subset of mesodermal cells in each segment to differentiate into skeletal muscle. Finally, recent mRNA transcription profiles, obtained by microarray analysis in stage 9 (19% of embryogenesis) to stage 12 (41%) embryos, of *twi* loss-of-function embryos, of embryos with ubiquitous *twi* expression (*Toll* mutants), and of wild-type embryos show hundreds of genes downstream

of *twi* to be involved in mesodermal ontogeny (Furlong et al., 2001).

8.1.2 MUSCLE PIONEERS AND MUSCLE FORMATION

As first shown in locust embryos (*Schistocerca americana*), paired, segment-specific, muscle founder cells or *pioneers* originate through dissociation of coelomic sacs, with one founder cell for each future muscle (Fig. 8.5) (Ho et al., 1983). Each cell acts as a scaffold on which subsequent, "fusion competent cells" or myoblasts fuse to form the muscle precursors, within both the body segments (Fig. 8.5) and appendages (Fig. 8.6) (Ball and Goodman, 1985a, 1985b; Ball et al., 1985). The resulting muscle syncytia then extend growth cones to specific, epidermal insertion sites while in embryos of *D. melanogaster*, simultaneously expressing a specific protein of the spectrin superfamily at their tips that seems to function in muscle-ectoderm and muscle-muscle attachment (Volk, 1992).

Ectodermal, segmental border cells also have specific molecular cues on their surfaces (Fig. 8.32) that enable these growth cones to recognize their attachment sites (Volk and VijayRaghavan, 1994), and in embryos mutant for the segment-polarity genes *wg* and *naked*, the muscles resulting are severely disorganized. In addition, the muscle founder cells are known to provide guidance cues in the form of specific, membrane-bound glycoproteins, which later guide the growth cones of developing motor neurons as they grow out from the CNS to innervate developing muscles (8.3.2.1) (Ball et al., 1985).

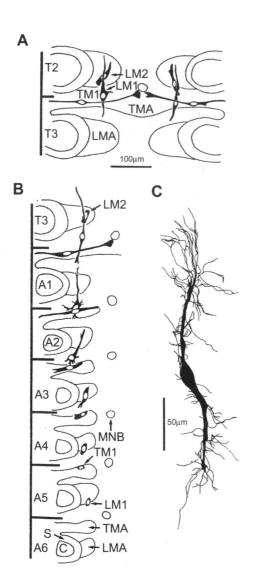

Fig. 8.5. Muscle pioneers in the body wall of an embryo of the grasshopper *Schistocerca americana* (Orthoptera: Acrididae) in dorsal aspect, as revealed by immunocytochemical staining with a monoclonal antibody (I-5 Mab; A and B) and intracellular injection of horseradish peroxidase (C). Camera lucida drawings of segments T2–A6 (in A, both sides of T2 and T3 are shown; in B, only the left sides of T3–A6) in the same embryo. Each successive anterior segment is older owing to action of a rostral to caudal developmental gradient. In each segment, the somite (s) surrounding the coelomic sac (c) spreads ventromedially to form the extrinsic leg muscles, ventral longitudinal muscles, and ventral transverse muscles (B, segment A6). On either side of the midline of each segment, the mesoderm spreads mesially as two distinct masses of small cells that remain separate at their anterior and posterior boundaries and do not fuse: the anterior (TMA) and posterior longitudinal muscle anlagen (LMA) (B, segment A6). The TMA spreads toward the midline and fuses with the equivalent mass on the other side. The LMA spreads anteriorly and posteriorly and does not meet or fuse with its homolog on the other side. Within the LMA, a single cell (LM1) near the posterior margin enlarges and begins to stain with I-5 Mab and to extend processes anteriorly and posteriorly (B, segments A5–T3). Soon after, a single cell (TM1) appears in TMA, begins to stain with I-5 Mab, and extends processes laterally and medially (B, segments A5–T3). A third, large cell (LM2) then appears (A, segment T2; B, segment T3) and later, others. C. Intracellular injection of horseradish peroxidase into a LM1 cell at the stage shown in T3 (B). Note the filopodia on the growth cone at each end. Reprinted with permission from *Nature* (R. K. Ho, E. E. Ball, and C. S. Goodman. Muscle pioneers: large mesodermal cells that erect a scaffold for developing muscles and motoneurons in grasshopper embryos. 301: 66–69) © 1983 Macmillan Magazines Limited.)

of muscle, it is maintained only in EMACs connected to muscle, suggesting that reciprocal signaling occurs between the two cell types:

$$EMAC \rightarrow myotube = myotube\ attraction + adhesion$$

$$myotube\ attached\ to\ EMAC = its\ differentiation$$
$$into\ a\ tendon\text{-}like\ cell$$

8.1.2.1 GENES AND MUSCLE SPECIFICATION IN *DROSOPHILA MELANOGASTER* EMBRYOS

In *D. melanogaster* embryos, the muscle founder cells (see Fig. 4.4 in Campos-Ortega and Hartenstein, 1997) and myoblasts are mitotic progeny of the somatic muscle progenitor cells (Fig. 8.4) (Gómez and Bate, 1997). Genes expressed in a progenitor cell are maintained in one daughter cell (remains as a stem cell) but are repressed in the other (becomes a muscle pioneer or myoblast), and just as in neuroblasts of the CNS (Fig. 8.14; 8.3.1.4) and in sense organ precursor cells of the peripheral nervous system (8.3.2.4.1), the differences depend on asymmetrical segregation of Numb, Prospero, and associated proteins (Baylies et al., 1998).

Each founder cell has a specific gene expression pattern that probably influences characteristics of the muscle it gives rise to, such as size, shape, insertion, and

In *D. melanogaster* embryos, each half of abdominal segments 2–7 develops a stereotyped pattern of 30 muscle fibers by 59% of embryogenesis (Bate, 1993; Campos-Ortega and Hartenstein, 1997: Fig. 4.6), these originating from muscle founder cells that first appear at the onset of germ band shortening at 33% of embryogenesis (Campos-Ortega and Hartenstein, 1997: Fig. 4.4). These cells develop fine filopodia (compare Fig. 8.5C), which probably assist in the fusion of additional myoblasts and in their attachment to specific epidermal muscle attachment cells (EMACs)—all before innervation from the CNS commences.

Autonomous expression of the *stripe* (*sr*) gene in an epidermal cell is necessary and sufficient to induce it to differentiate into an EMAC, and in *sr* mutant embryos, such cells fail to differentiate correctly (Becker et al., 1997). Also, ectopic expression of *sr* in other epidermal cells induces them to also differentiate into EMACs that can attract differentiating myotubes to them, providing the latter have not been influenced earlier by wild-type EMACs. Although initial expression of *sr* is independent

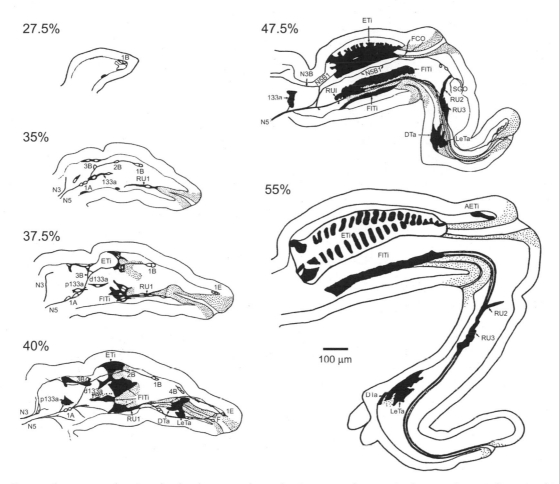

Fig. 8.6. Temporal sequence showing the development of muscle pioneers and nerves in the metathoracic (jumping hind) leg of an embryo of the grasshopper *Schistocerca americana* from 27.5% to 55% of embryogenesis. The muscle pioneers are shown in black. AETi, accessory extensor tibiae; DTa, depressor tarsi; ETi, extensor tibiae; FCO, femoral chordotonal organ; FlTi, flexor tibiae; LeTa, levator tarsi; RU1, RU2, RU3, retractor unguis (flexes claws against substrate); SGO, subgenual organ. The fine lines (N3, N3B, N5, N5B1) represent developing nerves; the lettered numbers (1A, 1B, 2B, 3B, 4B, 1E), various sense organs; and others (133a, d133a, p133a), various muscle precursors. The stippled areas are developing apodemes. (Reproduced with permission from E. E. Ball, R. K. Ho, and C. S. Goodman. Muscle development in the grasshopper embryo. I. Muscles, nerves, and apodemes in the metathoracic leg. *Dev. Biol*. 111: 383–398. © 1985 Academic Press, Inc.)

innervation (Ruiz-Gómez et al., 1997). For example, the gap gene *Krüppel* (*Kr*) is expressed in a subset of founder cells and muscles, regulates specific patterns of downstream gene expression in them, and is required for muscles to assume their correct identity. If *Kr* expression is lost or gained in sibling founder cells, it is sufficient to switch these cells and the muscles they become to other identities.

Expression of the gene *nautilus* (*nau*) is correlated with formation of muscle founders in *D. melanogaster* embryos and coincides with reduction in mesodermal expression of *twi*. However, *twi* expression persists throughout embryogenesis and postembryogenesis in a small number of mesodermal cells in each segment associated with the imaginal discs (wing, haltere, and leg primordia) and with portions of the peripheral nervous system (the *persistent twist cells* or *PTCs*) that are precursors of adult muscles (Campos-Ortega and

Hartenstein, 1997: Fig. 4.12; Farrell and Keshishian, 1999). Thus, by 33% of embryogenesis, three categories of cells have segregated from mesoderm: larval skeletal muscle precursors, larval precursors of other mesodermal structures, and imaginal skeletal muscle precursors characterized by persistent *twi* expression and by their capacity for further division (Bate, 1993).

The *dumbfounded* (*duf*) gene of *D. melanogaster* is expressed by muscle founder cells and encodes an attractant protein that attracts myoblasts to them (Ruiz-Gómez et al., 2000). The *sticks-and-stones* (*sns*) gene encodes cell adhesion molecules that must be expressed in myoblasts if they are to fuse to founder cells during embryogenesis (Bour et al., 2000). If either of these genes are mutant, myoblasts are fusion incompetent and no muscles differentiate.

Translation of *Drosophila* myocyte enhancer binding factor-2 (D-MEF-2) is required if somatic, cardiac, and visceral muscles are to differentiate, since in mutant loss-of-function embryos, none of these muscles do so even though their progenitor myoblasts are normally positioned and specified (Lilly et al., 1995). This protein binds to an adenine/thymidine-rich DNA sequence in the nuclei of these cells and directs expression of muscle-specific structural genes.

Correct attachment of the growth cones of muscle pioneers to specifically positioned epidermal cells in embryos of *D. melanogaster* depends on correct expression of the *derailed* (*drl*) gene (Callahan et al., 1996). This encodes a receptor tyrosine kinase essential for pathfinding by certain motor neurons but is also expressed by a small subset of developing embryonic muscle and neighboring epidermal cells during selection of muscle attachment sites. In *drl* mutants, these muscle progenitors attach to the epidermis at inappropriate locations.

PS integrins are cell matrix receptor proteins functioning in cell–extracellular matrix (ECM) adhesion and in transmitting signals from the ECM into cells to regulate gene expression (Martin-Bermudo, 2000). They are required at muscle attachment sites in *D. melanogaster* embryos to regulate tendon differentiation, and if absent, the expression of tendon-specific genes such as *stripe* and β*1 tubulin* is not maintained and tendons do not develop.

Visceral muscles surrounding the gut originate from cells in the anterodorsal muscle rudiment in each parasegment (Fig. 8.4) (Bilder and Scott, 1998). These dissociate, merge to form a continuous band running longitudinally along either side of the developing midgut (Fig. 8.2G), and express the protein connectin (con) in 11 metameric patches reflecting the anteroposterior expression pattern of the segment-polarity gene *en* in the ectoderm (7.2.1.2). The patches form in response to receipt of secreted signals encoded by the segment-polarity genes *hh* and *wg* in an inductive cascade from ectoderm to mesoderm to endoderm as follows:

| | | P A P | endoderm |
| P A P | P A P | | mesoderm |

hh wg hh   hh wg hh   hh wg hh   hh wg hh

| P A P | P A P | P A P | P A P | ectoderm |

14%   15%–18%   21%–23%   24%–32%   %embryogenesis

Visceral muscles of the larval midgut are lattice like and consist of inner circular and outer longitudinal fibers (Kusch and Reuter, 1999). The longitudinal fibers originate from cells specified at the posterior end of the mesoderm: the *caudal visceral mesoderm* (CVM). Expression of the genes *forkhead* (*fkh*) and *brachyenteron* (*byn*) specifies these cells, and Brachyenteron (Byn) protein establishes the surface properties required for their orderly forward migration along the trunk-derived

visceral mesoderm (TVM). The molecular genetics of muscle specification in *D. melanogaster* embryos was reviewed by Bate (1993), Abmayr and coauthors (1995) and Baylies and colleagues (1998).

8.1.3 HEMOCYTES
Based on histological study of the embryos of many species, insect blood cells or hemocytes are thought to arise throughout the length of the embryo from the middle layer of mesodermal cells stretching across the midline of each segment between the coelomic sacs on each side (Fig. 8.2B, C, F, G) (Mori, 1979). However, in *D. melanogaster* embryos, as shown by use of hemocyte-specific markers, the hemocyte precursors arise exclusively from mesoderm within the head and are specified by expression of the regulatory gene *serpent* (*srp*) (Fig. 8.7A) (Tepass et al., 1994; Lebestky et al., 2000). In fact, in *bicaudal* (*bic*) mutant embryos lacking a head and thorax and having a double-abdomen phenotype (Fig. 7.9F: right), no blood cells form. These precursors then give rise to two classes of hemocytes: plasmatocytes and crystal cells (Fig. 8.7B, C). Plasmatocytes express the gene *glial cells missing* (*gcm*), spread throughout the body cavity, and differentiate into macrophages that subsequently function to engulf degenerating cells. Crystal cells express the gene *lozenge* (*lz*) and, after hatch, function to melanize pathogenic microorganisms in the hemolymph. After hatch, all hemocytes are synthesized in mesodermal lymph glands associated with the dorsal vessel (Fig. 8.7C) (Lebestky et al., 2000). At least in grasshopper (*Locusta migratoria*) embryos, some fully differentiated hemocytes, in addition, later function to deposit the basal lamina subtending the epidermis and the cells of various organs (Ball et al., 1987).

8.1.4 BODY CAVITY (HEMOCOEL)
The body cavity originates from an epineural sinus, above the developing germ band and below the yolk (Fig. 8.2B, C, F, G). As cells of the coelomic sacs dissociate, their cavities become confluent with the epineural sinus (Fig. 8.2C, F, G). When the embryo enlarges and grows upward, on either side, around the diminishing yolk, the body cavity enlarges as well (Fig. 8.9A: i), and when the upper margins of the body wall on either side come together at dorsal closure, formation of the body cavity is complete. The cavity continues to enlarge as yolk remaining within the midgut is used up by the developing embryo.

In *D. melanogaster* embryos, activation of the so-called c-Jun amino-terminal kinase (JNK) cascade in the leading edge of migrating dorsal epidermal cells (Fig. 6.23: 13, 15) induces expression of *decapentaplegic* (*dpp*) and the patterning and movement of adjacent lateral epidermal cells in dorsal closure (Noselli, 1998). During closure, dorsal, marginal cells of the body wall epidermis change from polygonal to columnar. Cytoplasmic myosin colocalizes with actin immediately below the plasma membrane at the apices of cells at the leading edge of the advancing dorsal epithelial sheet on each side

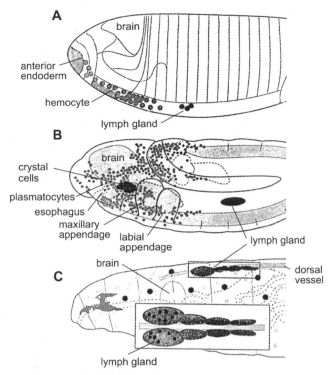

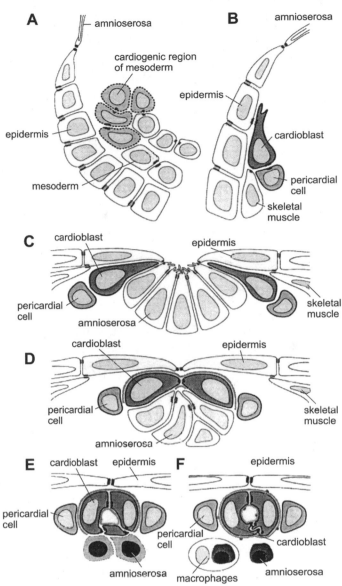

Fig. 8.7. Origin and migration of hemocytes in embryos of *D. melanogaster*. A. Hemocyte and lymph gland precursors originate, respectively, from head and thoracic mesoderm prior to gastrulation at 9%–13% of embryogenesis. B. During germ band elongation (24%–32%), the precursors are specified as crystal cells or plasmatocytes. C. In third instar larvae, crystal cells (black) and plasmatocytes (gray) circulate freely in the hemolymph throughout the body cavity, with all subsequent hemocytes being produced in the lymph glands. (Reprinted with permission from T. Lebestky, T. Chang, V. Hartenstein, and U. Banerjee. Specification of *Drosophila* hematopoietic lineage by conserved transcription factors. *Science* 288: 146–149. Copyright 2000 American Association for the Advancement of Science.)

Fig. 8.8. Diagrammatic transverse sections showing the origin of the dorsal vessel in an embryo of *D. melanogaster*. A. At 30%–38% of embryogenesis. B. At 38%–42%. C. At 46%–49%. D. At 50%–51% (dorsal closure). E. At 59%. F. After 59% (cells of the amnioserosa degenerate and are phagocytosed by macrophages). (Reprinted with permission from A. Rugendorff, A. Younossi-Hartenstein, and V. Hartenstein. Embryonic origin and differentiation of the *Drosophila* heart. *W. Roux's Arch. Dev. Biol.* 203: 266–280. © 1994 Springer-Verlag.)

of the body. This constricts the apex of the cells longitudinally and extends the sheet over the amnioserosa (the small remnant of extraembryonic membrane in this embryo persisting from that blastoderm not becoming part of the germ band), which is overgrown and subsequently degenerates (Figs. 6.23: 13, 15; 8.8C–F).

Mutations in other genes such as *Src42A*, *hemipterous* (*hep*), and *basket* (*bsk*) interfere with dorsal closure in embryos and, much later in development, these same genes function in dorsal fusion of the wing and haltere imaginal discs, to form the mesothoracic and metathoracic nota during metamorphosis (10.2.4.3.1) (Tateno et al., 2000).

8.1.5 DORSAL VESSEL
The principal blood-circulating organ of insects, the *dorsal vessel*, usually extends along the dorsal midline of the body from the posterior end of the abdomen into the head, where it empties into the body cavity below the brain (Fig. 8.1B, C: heart, aorta) (Snodgrass, 1935; Weber,

1966). It arises during dorsal closure through medial fusion of two longitudinal rows of *cardioblasts* (heart progenitor cells) located dorsally between splanchnic and somatic walls of each coelomic sac (Figs. 8.2C, G; 8.8). In *D. melanogaster* embryos, precursors of both pericardial cells and cardioblasts split off from lateral cardiogenic mesoderm in the mesothoracic (T2) to seventh abdominal segment as two longitudinal rows at about 32% of embryogenesis, with each of segments T2–A6 giving rise to six cardioblasts

(Fig. 8.8A, B) (Rugendorff et al., 1994; Campos-Ortega and Hartenstein, 1997). During provisional dorsal closure, at 41%–59% of embryogenesis (Fig. 6.23: 13, 15), these precursors become flattened, polarized, and aligned in a regular longitudinal row on each side. At dorsal closure, the leading edges of the cardioblasts meet their counterparts on the other side (Fig. 8.8D), the lumen of the vessel forming when the trailing edges of the two rows of cells flex medially and contact each other on the midline (Fig. 8.8D–F). The cellular details of dorsal vessel development in insect embryos are practically identical to those of capillary development in vertebrate embryos.

The regulatory gene *tin* is expressed in newly formed mesoderm of *D. melanogaster* embryos under control of the Twi protein (Bodmer, 1993). When this mesoderm dissociates, expression of *tin* becomes restricted to visceral mesoderm surrounding the midgut (Fig. 8.2G) and to cardioblasts. In *tin* mutant embryos, both dorsal vessel and visceral muscles fail to develop, fusion of the anterior and posterior midgut rudiments (8.2.1) is impaired, and somatic muscles develop abnormally. Tin protein, in turn, influences expression of two downstream regulatory genes, *ladybird early* (*lbe*) and *ladybird late* (*lbl*), which are expressed in only the posterior two of the six cardioblasts in each parasegment (Jagla et al., 1997).

8.1.6 FAT BODY

In embryos of *D. melanogaster*, precursors of the fat body cells originate between 24% and 40% of embryogenesis from inner, anterodorsal mesodermal cells within each parasegment (Figs. 8.3B, 8.4) and express the *srp* gene (Azpiazu et al., 1996). Between 40% and 46% of embryogenesis, they form an elongate sheet of cells on each side sandwiched between the developing visceral muscle of the gut and the somatic muscle of the body wall (Figs. 8.2G, 8.3B, 8.4) (Hartenstein and Jan, 1992; Hartenstein, 1993; Campos-Ortega and Hartenstein, 1997). After 68%, large holes and clefts begin to appear within the sheet at places where other organs (trachea, gonads, etc.) push into it.

8.1.7 SUBESOPHAGEAL BODY

Segmentation, at least in silverfish, orthopteroid, megalopteran, beetle, trichopteran, and lepidopteran embryos, is accompanied by formation of the *subesophageal body*, a mass of large, sometimes binucleate cells floating within the developing head below the stomodaeum (Johannsen and Butt, 1941; Anderson, 1972a, 1972b, 1973). Because of their ultrastructural resemblance to nephrocytes and pericardial cells (Kessel, 1961) (there is, as yet, no experimental evidence as to their function), the binucleate cells have been implicated in the regulation of embryonic hemolymph composition (Rempel and Church, 1969; Harrat et al., 1999). In embryos of the blister beetle *Lytta viridana*, they appear to invaginate from midventral ectoderm between the intercalary and mandibular segments beginning at 23%

of embryogenesis (Rempel and Church, 1969), whereas in those of the grasshopper *Locusta migratoria*, they are said to descend from the intercalary (Anderson, 1973) or mandibular (Harrat et al., 1999) coelomic sacs. They begin to degenerate after katatrepsis and dorsal closure at about the time that the pericardial cells and fat body appear fully differentiated and presumably functional. In crickets (Orthoptera: Gryllidae) and lepidopterans, they persist into early larval instars and in certain termites, they persist into the adult (Rempel and Church, 1969). The function of these intriguing cells should be investigated using modern methods.

8.2 Endoderm and Ectoderm

8.2.1 MIDGUT

In insects and many other animals, the only structure differentiating from endoderm is the midgut epithelium and its derivatives (Fig. 8.1B, C). The endoderm originates from the portions of the inner layer above the presumptive foregut (the *anterior midgut rudiment*) and hindgut (the *posterior midgut rudiment*) (Fig. 8.9A: b–f, h, i) (Johannsen and Butt, 1941; Hagan, 1951; Ando, 1962; Anderson, 1972a, 1972b, 1973; Haget, 1977; Sander et al., 1985; Ando and Miya, 1985; Ando and Jura, 1987; Schwalm, 1988). If subtended by visceral mesoderm, these primordia proliferate posteriorly and anteriorly as paired ventral strands that meet below the yolk and later spread around it to form the midgut epithelium at the time of dorsal closure (Fig. 8.9A: c, f, i) (Tepass and Hartenstein, 1994; Campos-Ortega and Hartenstein, 1997). In the absence of visceral mesoderm (*twi* and *snail* double mutants), midgut epithelium of *D. melanogaster* embryos does not form, and in *shotgun* (*shg*) mutants, though proper contact between visceral mesoderm and endoderm is established, the endodermal cells do not form a columnar epithelium, suggesting that *shg* controls adhesion between midgut epithelial cells prior to their differentiation (Tepass and Hartenstein, 1994).

Expression of the regulatory proneural and neurogenic genes (8.3.1.4) is required in *D. melanogaster* embryos if the three principal types of midgut cell— larval midgut epithelial cells (PMECs), imaginal midgut precursors (AMPs), and interstitial precursors (LBCs)— are to be specified (Tepass and Hartenstein, 1995). In embryos mutant in one or more of these genes, the PMECs fail to arise and the AMPs and LBCs develop to excess. And in *achaete-scute* (*as*) and *daughterless* (*da*) mutants (wild-type alleles of both are proneural genes), the LBCs fail to develop and the number of AMPs is greatly reduced. Expression of the homeotic gene *Ultrabithorax*, the dorsoventral regulatory gene *dpp*, the segment-polarity gene *wg*, and the *D-fos* gene in parasegments 3–10 is required for expression of the *HOM* gene *labial* and for induction of endoderm (Riese et al., 1997). And in embryos mutant for the gene *srp*, the entire midgut is missing because there is no endoderm (Reuter,

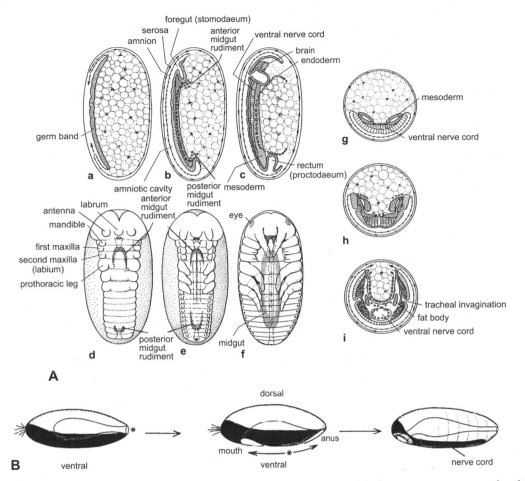

Fig. 8.9. A. Development of embryonic envelopes, gut, central nervous system, and body cavity in an unspecialized insect embryo in sagittal (a–c), ventral (d–f), and transverse (g–i) aspects. B. Possible derivation of the insect mouth and anus through ventral clongation of the blastopore of an ancestral embryo into a slit. (A reprinted, with permission, from A. Kühn. *Lectures on developmental physiology*, p. 318, Fig. 403, after Weber, 1966. © 1971 Springer-Verlag.) (B from Dorsoventral axis inversion: a phylogenetic perspective, T. Lacalli, *Bioessays*, Copyright © 1996 Wiley-Liss, Inc. Reprinted by permission of Wiley-Liss, Inc., a subsidiary of John Wiley & Sons, Inc.)

1994). Such embryos gastrulate normally and form normal stomodaeal and proctodaeal invaginations, but prospective anterior midgut cells act like stomodaeal cells, and posterior midgut cells form a second hindgut contiguous with the normal one. Thus, the mutation transforms anterior midgut to foregut and posterior midgut to hindgut, suggesting its wild-type allele functions in determining endoderm. In apterygote and odonate embryos, the midgut epithelium more often descends from persisting vitellophages and the yolk membrane (Ando, 1962; Mori, 1983; see 6.2.4).

8.2.2 FOREGUT AND HINDGUT
The foregut arises by invagination of the stomodaeum from ectoderm on the ventral midline of the intercalary segment (Haas et al., 2001b), and the hindgut arises a little later by invagination of the proctodaeum from that of the telson and posterior amnion (Fig. 6.22A, B); both subsequently communicate with the lumen of the midgut by dissolution of cells at their apices and eventually

secrete a cuticular intima (Fig. 8.9A: b–f). This type of gut development is thought to have evolved by the single posterior blastopore of the hexapod ancestor, having stretched midventrally in the longitudinal direction to generate a slit blastopore giving rise to both anus and mouth (Fig. 8.9B) (Lacalli, 1996; Gerhart and Kirschner, 1997: 363). The stomodaeum, proctodacum, and endoderm are thus remnants of the archentcron of other protostome embryos developing by invagination of a single blastopore. The origin and differentiation of the hindgut in *D. melanogaster* embryos are influenced by expression of a posterior-acting cascade of nine genes including *wg, hh, dpp,* and *en,* with the maternal effect gene *caudal* (*cad*) playing a pivotal role (7.2.1.1.3) (Wu and Lengyel, 1998; Takashima and Murakami, 2001).

8.2.3 MALPIGHIAN TUBULES
The osmoregulatory and excretory Malpighian (renal) tubules insert into the gut between the midgut and hindgut. They were long thought to be endodermal and

part of the midgut because of their histological similarity: brush border of microvilli and lack of a cuticular intima (Fig. 8.1B) (Snodgrass, 1935; Weber, 1966). However, they are now known to be ectodermal, based on their widespread origin by evagination from cells at the apex of the proctodaeum and on results of recent molecular genetic studies of their embryogenesis in *D. melanogaster* (Skaer, 1993; Campos-Ortega and Hartenstein, 1997). Skaer (1989, 1993) showed, by immunocytochemical localization of the substituted nucleotide 5-bromodeoxyuridine (BrdU) (a marker for DNA synthesis and cell division) in embryos of *D. melanogaster* and *Rhodnius prolixus* (Heteroptera: Reduviidae), that a single cell at the tip of each Malpighian tubule primordium, the *tip cell*, does not enter the cell cycle during tubule development but must be present if proliferation of more proximal cells is to occur; if it is destroyed, other cells in the tube primordium cease dividing. This cell arises by division of a tip mother cell selected from a cluster of equivalent cells in each tubule primordium (Hoch et al., 1994; Wan et al., 2000). Each cluster is marked by expression of the proneural genes, and selection of the tip cell involves lateral inhibition mediated by the neurogenic genes, exactly as occurs during neuroblast selection in neuroectoderm (Fig. 8.30) and sense organ precursor cell selection in the epidermis (8.3.2.4.1). Expression of the segment-polarity gene *wg* is required to maintain the tip cell and, in *wg* mutants, tip cells do not develop (Wan et al., 2000). Other aspects of the molecular genetics of gut formation in *D. melanogaster* were summarized by Skaer (1993; see particularly her Table 4 and Fig. 11).

8.3 Ectoderm

Cells of ectodermal origin differentiate into body wall and appendage epidermis; central, peripheral, and stomatogastric nervous systems; apodemes; foregut and hindgut; Malpighian tubules; tracheal system; salivary glands; oenocytes; and the secondary exit system and external genitalia of both males and females (Johannsen and Butt, 1941; Hagan, 1951; Ando, 1962; Anderson, 1972a, 1972b, 1973; Haget, 1977; Sander et al., 1985; Ando and Miya, 1985; Ando and Jura, 1987; Schwalm, 1988; Campos-Ortega and Hartenstein, 1987). Those derivatives having an epithelial character at some time during their development can be revealed vividly in whole mount preparations of small insect embryos by use of labeled, anti-Crumbs antibody (see numerous photomicrographs of *D. melanogaster* embryos in Campos-Ortega and Hartenstein, 1997). Crumbs protein, one of two key regulators of epithelial polarity in *Drosophila* (the other is DE-cadherin), is bound to the apical surfaces of epidermal cells as they invaginate, and influences morphogenetic movements, patterning, and cell-type determination (Tepass, 1997).

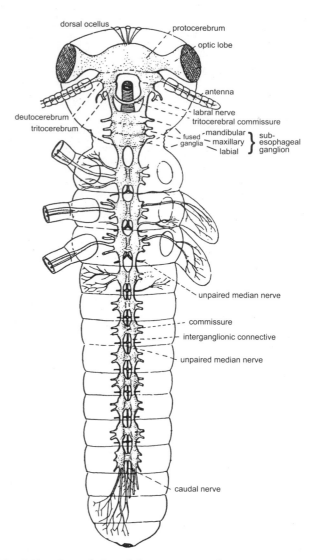

Fig. 8.10. Ground plan of the insect central nervous system in ventral aspect. (Reprinted with permission from H. Weber. *Grundriss der Insektenkunde*, p. 99, Fig. 66. 2 ed. © 1966 G. Fischer Verlag.)

8.3.1 CENTRAL NERVOUS SYSTEM

Beginning in 1976 with Bate's description of pioneer neurons (1976b) and identified nerve mother cells (neuroblasts) (1976a) in locust embryos and continuing to the present, there has been a revolution in our understanding of how insect nervous systems originate and differentiate, principally because of the development and use of genetic and molecular techniques on embryos of various grasshoppers and of *D. melanogaster*.

The principal components of the CNS in an insect are diagrammed in Fig. 8.10. Each segment of the body from the mandibular segment back contains a single ganglion of two neuromeres, one on either side of the midline. Each pair of neuromeres is linked fore and aft to neuromeres of adjacent ganglia by interganglionic connectives, and across the midline by anterior and posterior

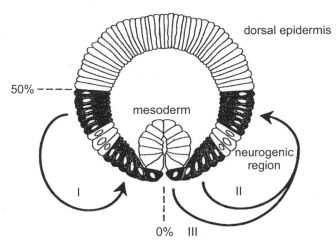

dorsal epidermis

50% ----

mesoderm

neurogenic region

I

II

0% III

Fig. 8.11. Heterotopic transplantations performed by Udolph and coworkers (1995) along the dorsoventral axis of a *D. melanogaster* embryo just after gastrulation (shown in transverse section of the abdomen). The neurogenic region includes the ventral half and the dorsal epidermis, the dorsal half of the ectoderm. Cells were removed from horseradish peroxidase–labeled donor embryos and transplanted into unlabeled hosts (experiments I, II, and III). Black indicates the midline mesepidermal cells; shading, ventral (5%–20% V-D) and dorsal area (30%–50% V-D) of the neurogenic region from or to which cells were transplanted. See text for details. (Reprinted with permission from G. Udolph, K. Lüer, T. Bossing, and G. M. Technau. Commitment of CNS progenitors along the dorsoventral axis of *Drosophila* neuroectoderm. *Science* 269: 1278–1281. Copyright 1995 American Association for the Advancement of Science.)

cross commissures, collectively forming the ventral nerve cord (VNC). Ganglia of the gnathal segments eventually fuse longitudinally to form the subesophageal ganglion, while those beside and in front of the stomodaeum contribute to formation of the brain (8.3.1.2).

8.3.1.1 VENTRAL NERVE CORD
Ganglia of the VNC originate from ventrolaterally situated neuroectodermal cells of the embryo, as summarized in Fig. 8.12, beginning in the prothorax and proceeding anteriorly and posteriorly. At first, the neuroectoderm is more laterally situated on each side of the blastoderm (Fig. 7.20), but then moves to the midline with gastrulation (Fig. 8.11).

The extent of determination of VNC progenitor cells along the dorsoventral axis of just gastrulated *D. melanogaster* embryos was assayed by Udolph and coworkers (1995), who transplanted dorsal neuroectodermal cells into ventral neuroectoderm or among mesectodermal cells on the ventral midline (Fig. 8.11: I) and found them to generate neuronal lineages consistent with their new position rather than their site of origin. However, when ventral neuroectodermal and mesectodermal cells were transplanted into dorsal neuroectoderm (Fig. 8.11: II, III), they migrated ventrally after transfer and produced neuronal lineages consistent with

their site of origin. This suggests that inductive signals from the ventral midline and neuroectoderm confer ventral identity to VNC progenitors and the ability to assume and maintain characteristic positions within the developing VNC.

Each nerve mother cell or *neuroblast* (NB) within a neuromere arises in specific order from neuroectoderm, enlarges, moves inward away from the surface, and commences to divide vertically, unequally, and teloblastically (i.e., as a stem cell) an NB-specific number of times, depending on its row and number in each lateral plate, to yield a characteristic number of ganglion mother cells (GMCs) (Figs. 8.12; 8.14F, B–E) (Doe and Goodman, 1985a, 1985b; Bossing et al., 1996; Schmidt et al., 1997; Lin and Schagot, 1997; Campos-Ortega and Hartenstein, 1997). Each GMC then divides once to yield two ganglion cells, each of which later differentiates into a neuron. Collectively, the approximately 60 NBs in each presumptive thoracic ganglion (there are 46–56 in each abdominal ganglion) generate about 1000 pairs of neurons, each NB contributing 6–100 neurons depending on its identity. These neurons then differentiate and project their axons to specific targets within the VNC (local and interganglionic interneurons) or to effector organs elsewhere in the body (motor neurons) (Figs. 8.12, 8.16). The entire process has been described in exquisite detail and depicted in three dimensions in living embryos of *D. melanogaster* by Schmid and coauthors (1999), who traced the development and fate of the progeny of every known NB, including their axon projections and muscle targets (movies can be viewed at the Web site www.uoneuro.uoregon.edu/doelab/lineages).

In embryos of the locust *Schistocerca gregaria*, the NBs start to divide at about 28% of embryonic development, beginning in the thorax, with division continuing until 90%. In abdominal ganglia, division begins at about 30% and continues until 70%, the fewer divisions correlating with the smaller number of neurons present in each abdominal ganglion (Shepherd and Bate, 1990). Each NB has its own fixed period and number of divisions and eventually degenerates at a time characteristic for its segment and for its row and number within the lateral plate (Fig. 8.13) (Goodman and Spitzer, 1979; Goodman et al., 1984). Abdominal ganglia 2–7 have identical patterns of NB fate, but these differ from those of the thoracic ganglia.

Embryos of the apterygote silverfish *Ctenolepisma longicaudata* (Zygentoma: Lepismatidae) have the same number and position of NBs as *S. gregaria* and the same relative order of NB segregation (Truman and Ball, 1998). However, NB 6-3 (i.e., that neuroblast in row 6, third from the midline) has a much longer proliferative period than this NB in *Schistocerca*, and of the remaining 30 NBs in each neuromere, 14 have shorter proliferative phases and 16 have longer ones. Also, some NBs in the terminal abdominal ganglion continue to divide up until the time of hatch, possibly in relation to extensive sensory input from the well-developed cerci and

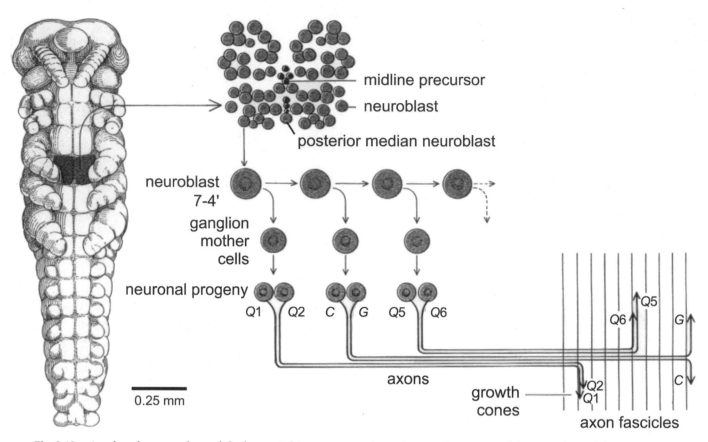

Fig. 8.12. A polypod stage embryo of the locust *Schistocerca americana* in ventral aspect. At this stage, the embryo contains 17 paired neuromeres, each in a different segment and each giving rise to a single ganglion of the ventral nerve cord. Each neuromere has a virtually identical pattern of three kinds of neuronal precursor cells or neuroblasts (NBs): 30 arranged in a lateral plate on either side of the midline in seven rows of two to six cells each, one posterior median NB, and seven midline precursor cells. Each NB divides repeatedly to generate ganglion mother cells (GMCs), each of which in turn divides to yield two ganglion cells, which then differentiate into neurons. NB 7-4, for example, divides about 50 times to yield 50 GMCs and 100 neurons, the first six of which (Q1, Q2, C, G, Q5, Q6) are illustrated (the subsequent differentiation of neuron G is illustrated in Figs. 8.17 and 8.18). The primary growth cone of each of these neurons then follows a specific pathway as it recognizes a previously established fascicle of axons. (Reprinted with permission from C. S. Goodman and M. J. Bastiani. How embryonic nerve cells find one another. *Sci. Am.* 251(6): 58–66. © 1984 Scientific American.)

median filament of this insect (compare Fig. 9.3: Archaeognatha).

Surprisingly, the shift from winglessness to flight in insects was not accompanied by addition of new neuronal lineages (Truman and Ball, 1998). Instead, certain thoracic NBs in the mesothorax and metathorax underwent a proliferative expansion, presumably to supply the additional neurons needed for flight. This expansion was accompanied by a reduction in terminal abdominal lineages, perhaps in response to flying insects deemphasizing their cercal sensory system.

In *D. melanogaster* embryos and probably in those of other holometabolous insects, many NBs do not degenerate after completing their embryonic divisions but instead, are temporarily arrested until long after hatch, owing to production of a glial cell–derived, inhibitory glycoprotein encoded by the *anachronism* gene (Prokop and Technau, 1991; Ebens et al., 1993). Toward the end of larval development, this inhibition is lifted and these NBs enlarge and become mitotically

active again to generate a large number of quite different, imaginal neurons (10.2.4.1.1).

Fig. 8.13 summarizes the distribution of NBs in each thoracic and abdominal segment of the embryonic body of a locust and emphasizes its general similarity in all of them (Doe and Goodman, 1985a).

Vertical asymmetrical division of NBs in the VNC to generate GMCs in *D. melanogaster* embryos is associated with asymmetrical segregation of the homeodomain protein Prospero (Hirata et al., 1995; Lin and Schagot, 1997). This is synthesized in the NB, is retained in its dorsal cytoplasm, and at mitosis is exclusively allocated to the GMC in which it is translocated into the nucleus (Fig. 8.14B–E) (the same protein is differentially segregated in the stem cell divisions of midgut endodermal, of muscle, and of sense organ precursor cells). A 13–amino acid sequence in the Prospero protein is also present in Numb protein, which is likewise segregated asymmetrically but into the cortex, not the nucleus, of the GMC (Fig. 8.14C–E: hatched).

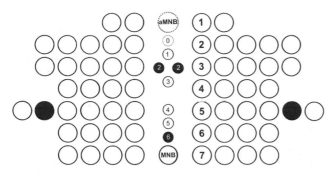

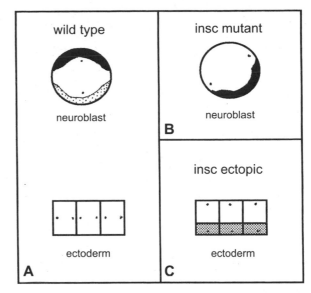

Fig. 8.13. The distribution of neuroblasts in other body segments of the *S. americana* embryo compared to that of the mesothorax. The clear cells are in all segments; those drawn in dotted outline are absent from the mesothorax but are present in some other segments; the black cells are in the mesothorax but absent in some other segments. The pattern is very different in the anterior head segments (Fig. 8.25). aMNB, anterior median neuroblast; MNB, median neuroblast. (Reproduced with permission from C. L. Doe and C. S. Goodman. Early events in insect neurogenesis. I. Development and segmental differences in the pattern of neuronal precursor cells. *Dev. Biol.* 111: 193–205. © 1985 Academic Press, Inc.)

Fig. 8.15. Inscuteable protein distribution (stippled or shaded) is correlated with spindle orientation and is necessary and sufficient for vertical spindle alignment. A. Wild-type neuroblast showing basal localization of Inscuteable protein and apical localization of the cell fate determinants Prospero and Numb (black) at mitosis. The location of the centrosomes (·) in both diagrams indicates a vertical spindle in neuroblasts and a horizontal spindle in the ectoderm. B. In the absence of Inscuteable (insc), neuroblasts show randomized spindle orientation and Prospero/Numb crescents. C. Misexpression of Inscuteable in the ectoderm results in its ventral localization and triggers a vertical spindle orientation. (Adapted with permission from *Cell*, 86, C. L. Doe, Spindle orientation and asymmetric localization in *Drosophila*: both *inscuteable?*, pp. 695–697. Copyright 1996, with permission from Elsevier Science.)

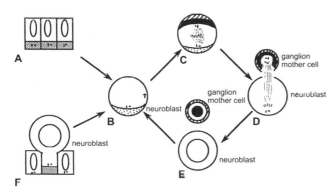

Fig. 8.14. Spindle orientation and localized protein distribution during neurogenesis. A. Protocephalic neuroectoderm: all cells show ventrally located Inscuteable protein (shaded) prior to entering mitosis and dividing asymmetrically and vertically. Following division (B–E), the fate of each sibling is unknown. The large ventral cell may produce epidermis; the smaller, apical cell, neurons. F. Ventral neuroectoderm: single cells show basal Inscuteable protein as they delaminate from the epidermis to form neuroblasts. B–E. The neuroblast cell cycle (clockwise). B. Prophase. C. Metaphase with apical crescents of Numb (hatched) and Prospero (black) protein. D. Anaphase/telophase. E. Interphase. Prospero translocates into the nucleus of the ganglion mother cell, whereas Numb remains associated with its membrane. (Adapted with permission from *Cell*, 86, C. L. Doe, Spindle orientation and asymmetric localization in *Drosophila*: both *inscuteable?*, pp. 695–697, Copyright 1996, with permission from Elsevier Science.)

The protein of a third gene, *miranda*, colocalizes with Prospero, tethers it to the dorsal cortex of dividing NBs, directs it into the GMC at anaphase, and releases it from the cortex of the GMC so that it can enter the nucleus (Ikeshima-Kataoka et al., 1997). The protein of *partner-of-numb* (*pon*) does the same for Numb protein except that it remains in the cortex of the GMC (Lu et al., 2001).

During NB mitosis, both the vertical orientation of the spindle and the asymmetrical localization of Numb and Prospero proteins to the GMC are influenced by the product of the *inscuteable* (*insc*) gene (Doe, 1996; Lin and Schagot, 1997; Jan and Jan, 2000). During NB mitosis, one centrosome remains at the ventral end of the cell while the other migrates dorsally, resulting in a vertical division (Fig. 8.15A). The Inscuteable (Insc) protein is asymmetrically localized to the ventral (apical) cytoplasm of an NB assisted by expression of the genes *bazooka* (*baz*) (Schober et al., 1999) and *partner of inscuteable* (*pins*) (Yu et al., 2000). However, Insc is degraded or delocalized by anaphase and is not segregated into either daughter cell (Fig. 8.14C, D). If Insc is not produced, the spindle that forms is horizontal as in surrounding, nonneural epidermal cells (Fig 8.15A, B), whereas in *insc* mutants, Prospero and Numb proteins are uniformly distributed in the cortex of the NB as crescents not aligned with the spindle (Fig. 8.15B). Thus, *insc* is necessary and sufficient for dorsal localization of these two proteins during NB mitosis.

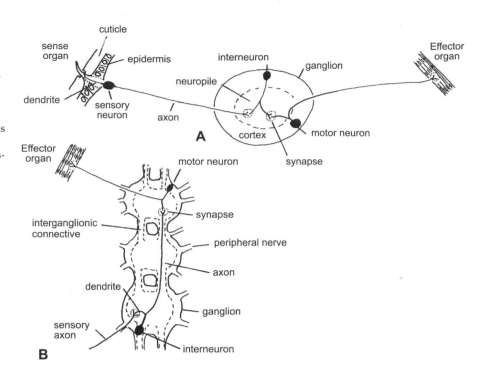

Fig. 8.16. Greatly simplified summary of the positions of somata and the projection patterns and synaptic contacts between the axons and dendrites of interneurons and sensory and motor neurons in an insect nervous system. A. Transverse section. B. Dorsal aspect of three ganglia in the ventral nerve cord, showing how interneurons carry information up and down its length. Notice that synaptic contacts between the axons and dendrites of all three neuron types occur in the neuropile of each ganglion and that the somata of the sensory neurons of surface sensilla are associated with the epidermis, while those of interneurons and motor neurons are located in the cortices of ganglia in the ventral nerve cord.

Two sets of polarity cues influence spindle position in epidermal cells: a planar cue mediated by the adherens junctions between adjacent epithelial cells, and an apical-basal cue regulated by Bazooka (Baz) protein (Lu et al., 2001). In wild-type cells, the planar cue is dominant but can be overwhelmed by the presence of Insc.

Establishing the specific gene expression of each NB according to its row and number within a neuromere is under control of the segmentation, homeotic, and dorsoventral genes (7.2.1) (Skeath, 1999). The gap gene *huckebein* (*hkb*) is expressed in a subset of NBs and contributes to neuronal and glial cell determination (McDonald and Doe, 1997). Its product is a nuclear protein that is first detected in a subset of neuroectodermal cells and then in the NBs derived from them. The secreted, segment-polarity proteins Wingless (Wg) and Hedgehog (Hh) activate expression of *hkb* in distinct but overlapping clusters of neuroectodermal cells and NBs, whereas the proteins of *en* and *gooseberry* (*gsb*) repress it. Integration of these effects is required to establish the precise neuroectodermal expression pattern of *hkb* subsequently required for development of specific NB lineages.

8.3.1.1.1 NEURONS

Neurons are the functioning nerve cells of the CNS (Fig. 8.16) (Strausfeld and Meinertzhagen, 1998). The first to arise, the *central pioneer neurons*, use the underside of the ectoderm's basal lamina or the surfaces of glial cells as a substrate for their growth; later-appearing neurons use the surfaces of these pioneers. The details of these processes are complex and have been worked out in great detail in locust and *D. melanogaster* embryos (Goodman and Doe, 1993; Schmid et al., 1999). When complete, the combined axons, dendrites, and synapses of the neurons within each ganglion, together with afferent axons projecting to the CNS from peripheral sense organs, comprise its neuropile (Fig. 8.16A). Somata (nuclei) of the CNS neurons (all motor or local and interganglionic interneurons) come to occupy the periphery (*cortex, rind*) of each ganglion.

Figs. 8.17 and 8.18 illustrate the developmental history and projection pattern of a single neuron: the mesothoracic G neuron and the projection of its segmental homologs in the metathoracic and first abdominal ganglia of a locust (Pearson et al., 1985). Figs. 8.19 and 8.20 summarize the results of some classic experiments on NBs and GMCs from locust (*Schistocerca nitens*) embryos conducted by members of Goodman's group (Goodman et al., 1984). These results and those of many other experiments suggest that the specific identity of a neuron depends on its mitotic history or *lineage* (i.e., its NB and GMC of origin) and that the growth of its axon to its proper synaptic site within the developing VNC is guided by recognition molecules on the surfaces of previously formed pioneer neurons (Goodman, 1984; Goodman and Bastiani, 1984; Ganfornina et al., 1996a; Tear, 1999).

Six of these recognition markers are illustrated in Fig. 8.21 (Goodman and Doe, 1993). Neuroglian is expressed on most axons and glial cells, and Fasciclin I, II, III, and IV (Semaphorin I) and connectin, on particular subsets of axon pathways within the VNC. These molecules are recognized by constantly "searching" filopodia borne by growth cones at the tips of growing axons.

The molecules in Fig. 8.21 constitute one of four types of signals known to guide axons to appropriate targets within the CNS (Marx, 1995): short-range repulsive

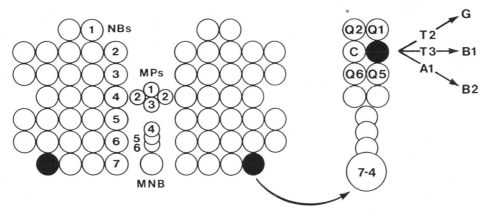

Fig. 8.17. Lineage of the mesothoracic G neuron and its segmental homologs in the metathoracic and first abdominal ganglia of *Locusta migratoria* (Orthoptera: Acrididae). Left: Neuron precursors consist of 60 neuroblasts (NBs) in two plates of 30 arranged in seven rows on either side of the midline, a single midline neuroblast (MNB), and 7 midline precursor cells (MPs). The G neuron arises from NB 7-4 (black). Right: NB 7-4 divides repeatedly to generate a chain of ganglion mother cells, each of which divides to give a pair of neurons. In the mesothorax (T2), the first three pairs of neurons arising from NB 7-4 are the Q1 and Q2 cells, the G (filled) and C cells, and the Q5 and Q6 cells. Homologs of the mesothoracic G cell in the metathoracic (T3) and first abdominal segment (A1) develop into the B1 and B2 neurons, respectively, as shown in Fig. 8.18. (Heterogeneous properties of segmentally homologous interneurons in the ventral nerve cord of locusts, K. G. Pearson, G. S. Boyan, M. Bastiani, and C. S. Goodman, *J. Comp. Neurol.*, Copyright © 1985 Wiley-Liss, Inc. Reprinted by permission of Wiley-Liss, Inc., a subsidiary of John Wiley & Sons, Inc.)

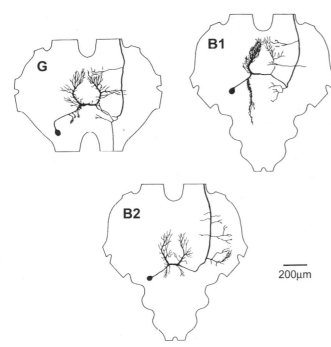

Fig. 8.18. Drawings of the fully differentiated G, B1, and B2 neurons in the adult *L. migratoria* ventral nerve cord as seen in whole mount preparations viewed from above. The soma of the G neuron is located in the mesothoracic; that of the B1 neuron, in the metathoracic; and that of the B2 neuron, in the first abdominal ganglia. The first three abdominal ganglia are fused with the metathoracic ganglion in adults of this insect. (Heterogeneous properties of segmentally homologous interneurons in the ventral nerve cord of locusts, K. G. Pearson, G. S. Boyan, M. Bastiani, and C. S. Goodman, *J. Comp. Neurol.*, Copyright © 1985 Wiley-Liss, Inc. Reprinted by permission of Wiley-Liss, Inc., a subsidiary of John Wiley & Sons, Inc.)

and attractive cues provided by molecules on the nerve cell surface, and diffusible chemoattractants and repellants. The protein Semaphorin I, demonstrated in grasshopper and *D. melanogaster* embryos, regulates sensory axon guidance and branching in the limb buds (see Fig. 8.32), while Semaphorin II inhibits synapse formation of a subset of motor axons.

8.3.1.1.2 CROSSING THE MIDLINE

It is critical, during VNC embryogenesis, that commissural interneurons within each ganglion project across the midline so that the two sides of the body can "talk to each other." The growth cones of such neurons have receptors for the secreted proteins Netrin, Slit, and Commissureless (Comm) (Fig. 8.22A) and are attracted to the midline by secretion of Netrin from ventral midline cells (Fig. 8.22B) (Harris and Holt, 1999). Robo protein, encoded by the gene *roundabout* (*robo*), is expressed at high levels on growth cone filopodia of most neurons in the VNC (Fig. 8.22A). Robo is the receptor for the repellant molecule Slit, encoded by the *slit* gene secreted from ventral midline glial cells (Fig. 8.22A). If a growth cone has a low concentration of Robo, it can cross the midline. Certain midline glial cells produce a second protein known as Comm, encoded by the gene *commissureless* (*comm*) (Harris and Holt, 1999). Comm binds to receptors on the growth cones of commissural axons when they reach the midline and functions to downregulate Robo (Fig. 8.22C). With little Robo receptor available to bind to Slit, such axons cannot sense the midline so they can cross (Fig. 8.22C, D). However, as they cross, they upregulate Robo to its orginal high level and so cannot cross back (Fig. 8.22D).

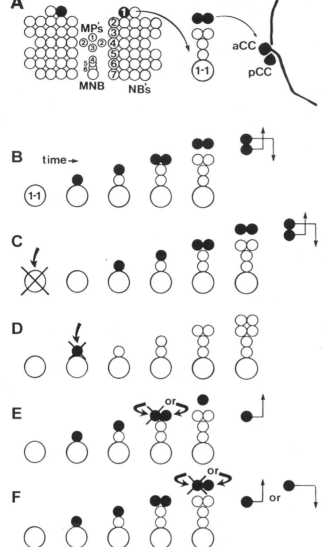

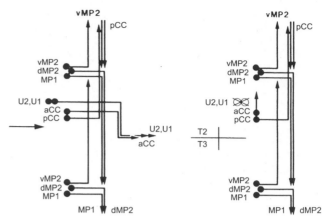

Fig. 8.20. Experiment demonstrating the selective affinity of the growth cone of the posterior corner (pCC) neuron for the MP1-dMP2 fascicle and that of the anterior corner (aCC) neuron for the U2, U1 fascicle. These schematic diagrams show the first three longitudinal axon fascicles (vMP2, MP1-dMP2, and U2-U1) in each hemisegment of the grasshopper (*Schistocerca nitens*) ventral nerve cord and the seven identified neurons whose axons fasciculate to form these three bundles. Left: Normal. Right: Selectively ablating the U1 and U2 neurons prevents the posterior extension of aCC. (Reprinted with permission from C. S. Goodman, M. J. Bastiani, C. Q. Doe, S. Du Lac, S. Helfand, J. Y. Kurada, and J. B. Thomas. 1984. Cell recognition during neuronal development. *Science* 225: 1271–1279. Copyright 1984 American Association for the Advancement of Science.)

Fig. 8.19. Experiments demonstrating that both cell lineage and cell interactions contribute to formation of the anterior (aCC) and posterior (pCC) corner neurons (abbreviations as in Fig. 8.17). A. Pattern of neuroblasts and the cell lineage of aCC and pCC neurons from NB 1-1. The first ganglion mother cell (GMC) from NB 1-1 generates the aCC and pCC neurons, which then migrate anteriorly and differentiate. B. Normal lineage of aCC and pCC neurons. C. Ablation of NB 1-1. A new NB 1-1 appears and produces the aCC and pCC neurons. D. Ablation of the first GMC from NB 1-1. The second GMC does not produce the aCC and pCC neurons. E. Ablation of either of the progeny of the first GMC within 5 h of their birth. The remaining cell differentiates into the pCC neuron. F. Ablation of either of the progeny of the first GMC 5–10 h after their birth: the remaining cell differentiates into either the aCC or the pCC neurons with equal probability. (Reprinted with permission from C. S. Goodman, M. J. Bastiani, C. Q. Doe, S. Du Lac, S. Helfand, J. Y. Kurada, and J. B. Thomas. Cell recognition during neuronal development. *Science* 225: 1271–1279. Copyright 1984 American Association for the Advancement of Science.)

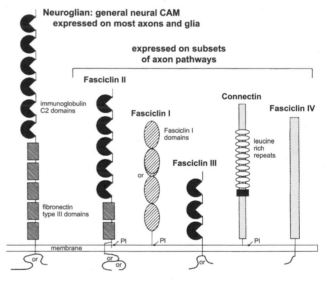

Fig. 8.21. Structure of six axonal glycoproteins functioning in neuronal pathfinding in *D. melanogaster* embryos. See text for details. As shown schematically, several come in multiple forms either with different membrane linkages (phosphoinositol [PI] lipid of PI linkages versus a transmembrane domain; Fasciclin II), with different cytoplasmic domains (Neuroglian, Fasciclin II and III), or with alternately spliced microexons in the extracellular domain (Fasciclin II). (Reprinted with permission from C. S. Goodman and C. Q Doe. Embryonic development of the *Drosophila* central nervous system. In M. Bate and A. Martinez Arias (eds.), *The development of* Drosophila melanogaster. Vol. 1, p. 1168, Fig. 12 . © 1993 Cold Spring Harbor Laboratory Press.)

Midline

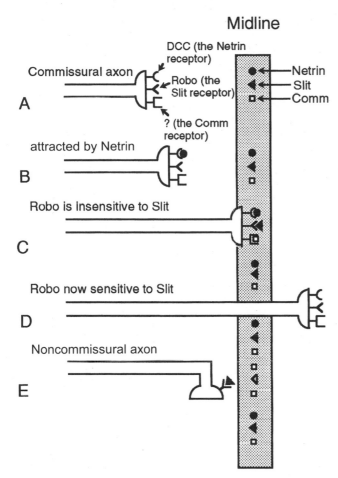

Fig. 8.22. The function of the Slit protein in axon guidance across the midline in embryos of *D. melanogaster*. A. Commissural axons express receptors for Netrin, Slit, and Commissureless (Comm). B. The commissural axon is attracted to the midline by binding Netrin. C. Comm downregulates the Slit receptor, Roundabout (Robo), so the axon is insensitive to Slit and can cross the midline. D. Having crossed the midline, the axon re-expresses Robo, making it sensitive to Slit and preventing it from recrossing the midline. E. Noncommissural axons are repelled by Slit, so cannot cross the midline. (Adapted with permission from *Nature* (W. A. Harris and C. E. Holt. Slit, the midline repellant. 398: 462–463) © 1999 Macmillan Magazines, Limited.)

Two additional Slit receptor proteins, Robo 2 and Robo 3, also bind Slit protein and influence how far laterally the growth cone of a commissural neuron can extend, after crossing the midline, before it fasciculates with neuron fascicles extending longitudinally. Axons of these neurons express the surface glycoprotein Fasciclin II and direct the commissural axon to grow anteriorly or posteriorly (Vogel, 2000).

Axons of *robo* and *slit* mutant embryos meander back and forth across the midline, ignoring the barrier normally separating the animal into two halves, while axons in *comm* mutants run up and down each side of the midline so that no cross commissures form.

Recent comparative work on embryos of the nema-

tode *Caenorhabditis elegans* and on those of zebrafish, birds, and mice indicates that orthologous genes are involved in directing axonal growth across the midline in other animals, suggesting these genes are evolutionarily conserved (Harris and Holt, 1999; Vogel, 2000).

8.3.1.1.3 MIDLINE PRECURSORS

Special median NBs arising from two longitudinal rows of mesectodermal cells (Fig. 7.20) between the lateral plates of NBs, the *midline precursor cells* (MPs), give rise to certain VNC pioneer interneurons, as summarized in Fig. 8.23 (Bate and Grunewald, 1981). Together, these interneurons lay down an axon scaffold within each segment that specifies the growth pattern of subsequently formed neurons. Commitment of these cells requires the expression, in *D. melanogaster* embryos, of some of the dorsoventral patterning genes (7.2.1.1.1) (Campos-Ortega and Hartenstein, 1997).

In each body segment, each of the two MP2 precursor NBs divides asymmetrically into ventral (vMP2) and dorsal (dMP2) interneurons, with dMP2 projecting its axon anteriorly and vMP2 posteriorly to the adjacent segmental borders where their growth cones meet and fasciculate with those projecting forward or backward from equivalent cells in adjacent segments (Fig. 8.23B–D). During MP2 mitosis, Numb protein is localized in dMP2 but is excluded from vMP2, and ectopic segregation of Numb into vMP2 transforms it into dMP2 (Spane et al., 1995).

As shown by Nambu and colleagues (1993), the *single-minded* (*sim*) gene encodes a transcription factor that acts as a master regulator for the midline lineage, but other proteins and transcription factors must be present for the midline neurons and glial cells to differentiate.

8.3.1.1.4 PHYLOGENETIC ASPECTS

Recent comparative work on embryos of *D. melanogaster*, the moth *Manduca sexta*, the apterygote silverfish *Ctenolepisma longicaudata* (Zygentoma), and several crustaceans (e.g., *Cherax destructor* [Decapoda] and *Porcellio scaber* [Isopoda]) indicate that their pattern of lateral and median NBs is similar to that in locust embryos, leading to the suggestion that developmental processes generating the VNC of all arthropods might have been conserved during evolution (Thomas et al., 1984; Patel, 1994; Whitington et al., 1996; Truman and Ball, 1998; Duman-Scheel and Patel, 1999). Unfortunately for this idea, embryos of the centipede *Ethmostigmus rubripes* do not even have NBs (Whitington et al., 1991; Whitington and Bacon, 1998) in spite of the formerly widely accepted notion that myriapods constitute the sister group of the hexapods (e.g., Fortey et al., 1997). In this centipede, the earliest VNC pathways do not arise from segmentally repeating pioneer neurons, as in insects, but rather by posteriorly directed growth of neurons whose cell bodies are located within the brain. Only later does axonogenesis by segmental

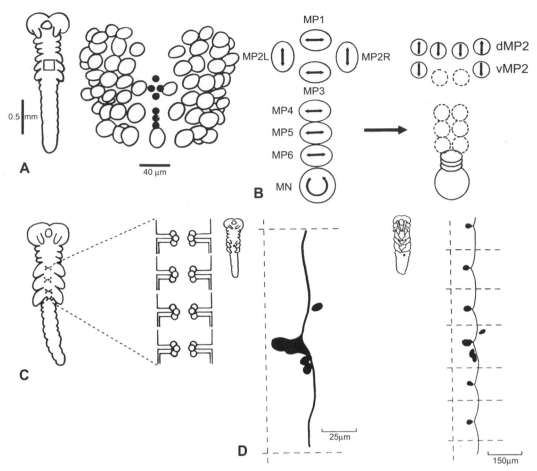

Fig. 8.23. Origin and differentiation of the midline precursor cells in embryos of the migratory locust *Locusta migratoria*. A. Ventral aspect of neuronal precursor cells in the mesothoracic ganglion. Open circles represent neuroblasts; filled circles, midline precursors (MPs). B. Diagram of MPs and their division pattern. Left: Seven MPs and the single posterior median neuroblast (MN). Arrows indicate axis of mitosis of the MPs and the repetitive divisions of the MN. Right: Products of the MP and MN divisions. Progeny of MP3–MP6 are dotted circles; those of MP1 and paired MP2, closed circles. Arrows indicate direction of axon growth of MP1 and MP2 progeny. C. Pattern of axon pathways formed by the cell trios on the midline of segments T1–A1 (MP2s and MP6 are absent in some abdominal segments). D. Tracings of Lucifer yellow–filled interneurons at 96 h (left) and 132 h (right). Cells are coupled both intrasegmentally (left) and intersegmentally (right). Dotted lines represent median and segmental borders. (Reproduced with permission from C. M. Bate and E. B. Grunewald. Embryogenesis of an insect nervous system II: a second class of neuron precursor cells and the origin of intersegmental connectives. *J. Embryol. Exp. Morphol.* 61: 317–330. © 1981 Company of Biologists, Ltd.)

neurons begin, but with a pattern not obviously homologous to the conservative set of central pioneer neurons occurring in the embryos of insects and crustaceans. Recent comparison of homologous gene nucleotide sequences in various genes suggests that the hexapods constitute the sister group of the crustaceans (Fig. 13.7).

The little that is known about neurogenesis in chelicerate embryos (Anderson, 1973) likewise suggests little similarity to that occurring in insects and crustaceans (Whitington and Bacon, 1998). Most surprising is that in spite of these differences within the Arthropoda, there are striking similarities in nerve cord neurogenesis in insect and vertebrate embryos (Arendt and Nübler-Jung, 1999):

• Neurons differentiate toward the body cavity in embryos of both lineages.

• In early neurogenesis, neural progenitor cells (NBs) are arranged in three longitudinal columns on either side of the midline, and orthologs of the vertebrate genes *NK-2/NK-2-2*, *ind/Gsh*, and *msh/Msx* specify identity, respectively, of the medial, intermediate, and lateral columns.

• Within any neurogenic column, some emerging cell types are similar and thus may be phylogenetically old.

• Lateral column NBs produce nerve root ganglia and peripheral glia.

• Midline precursors give rise to glial cells that enwrap outgrowing commissural axons.

• Midline glia express Netrin orthologs to attract commissural axons from a distance (8.3.1.1.2).

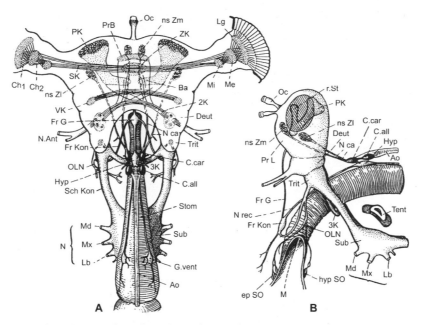

Fig. 8.24. General structure of the brain, subesophageal ganglion, and stomatogastric nervous system in an adult insect. A. Dorsal aspect with brain in frontal section. B. Lateral aspect. Ao, aorta; Ba, beta lobe of corpus pedunculatum; C.all, corpus allatum; C.car, corpus cardiacum; Ch$_1$, outer and Ch$_2$, inner chiasmata of optic lobe; Deut, deutocerebrum; ep SO, epipharyngeal sense organ; Fr G, frontal ganglion; Fr Kon, frontal connective; G.vent, ventricular ganglion; Hyp, hypocerebral ganglion; hyp SO, hypopharyngeal sense organ; Lg, lamina ganglionaris; Me, medulla; Mi, lobula; N (Md, Mx, Lb), mandibular, maxillary, and labial nerves; N.Ant, antennal nerve; N ca, nervi corpora allata I and II; N rec, recurrent nerve; ns Zm, median neurosecretory cells; ns Zl, lateral neurosecretory cells; M, mouth; Oc, median dorsal ocellus; OLN, labral nerve; PK, Kenyon cells in calyx of corpus pedunculatum; PrB, protocerebral bridge; Pr L, protocerebral lobe; r. St, alpha lobe of corpus pedunculatum; Sch Kon, circumesophageal connective; SK, projection tract of optic interneurons; Stom, foregut; Sub, subesophageal ganglion; Tent, tentorial bridge; Trit, tritocerebrum; VK, accessory lobe; ZK, central body; 2K, 3K, commissures. (Adapted with permission from Weber 1966, p. 100, Fig. 67. © 1966 G. Fischer Verlag.)

8.3.1.2 BRAIN

The insect brain encompasses the supraesophageal and subesophageal ganglia (Fig. 8.24) (Weber, 1966; Strausfeld, 1976). The latter is part of the VNC and consists of the fused mandibular, maxillary, and labial ganglia. The former comprises the paired, protocerebral, deutocerebral, and tritocerebral lobes and contains various synaptic centers or glomeruli including the paired optic lobes (each containing a lamina ganglionaris, medulla, and lobula); corpora pedunculata (mushroom bodies); the median pars intercerebralis, protocerebral bridge, and central body of the protocerebrum; the paired accessory and antennal lobes, respectively, of the protocerebrum and deutocerebrum; and all their interconnecting tracts and commissures.

Until recently, little detailed work had been done on brain development in any insect embryo using modern methods, so knowledge lagged far behind that available for the more accessible VNC. However, in 1993, Zacharias and coauthors described early brain development in embryos of the locust *Schistocerca gregaria* using a variety of techniques (toluidine blue staining, BrdU incorporation [for monitoring DNA synthesis and mitosis], and immunocytochemistry in both whole mounts and sections), and found the first NBs to originate before 25% of embryogenesis and to generate GMCs and neurons in the

same way as do NBs of the VNC (8.3.1.1). By 30%–45%, about 260 identifiable, mitotically active NBs, each characterized by cell-specific molecular labels (tagged antibodies to the Engrailed, Fasciclin I, and TERM-I proteins), are present in positions appropriate for generating the major brain centers (Fig. 8.25).

Between 26% and 28% of embryogenesis, these NBs and their progeny become segregated into six pairs of clustered aggregates, the *proliferation clusters* (PC), by growth of glial cells around and between them, with one cluster for each brain center on each side of the head (Fig. 8.26) (Boyan et al., 1995b). In each cluster, 2–5 individually identifiable pioneer neurons arise and, from 29% to 40%, collectively construct a *primary axon* scaffold within the brain resembling that generated by the midline precursors in the VNC (Fig. 8.23). The axon of each pioneer navigates over and between the glial borders separating adjacent clusters (Fig. 8.26). Expression of the genes *extradenticle* (*exd*) and *homothorax* (*hth*) is essential for these scaffolds to develop normally (Nagoo et al., 2000).

During this process, and simultaneously in each brain hemisphere, a pioneer axon outgrowth cascade establishes a pathway from each optic ganglion to the brain midline (1 in Fig. 8.26) and between the various centers. A primary preoral commissure is pioneered by a pair of

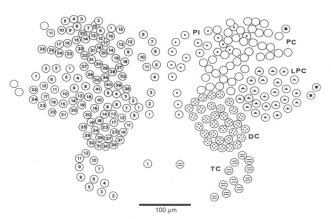

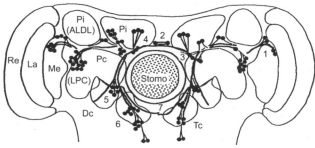

Fig. 8.25. An idealized map of brain neuroblasts (NBs) in an embryo of the locust *Schistocerca gregaria* at 45% of embryogenesis in ventral aspect. On the right, groups of NBs belonging to each brain region are indicated by the same symbol (pars intercerebralis [PI], single dot; lobe 1 of the protocerebrum [PC], blank; lateral protocerebrum [LPC], filled triangles; deutocerebrum [DC], stippling; tritocerebrum [TC], stripes; NBs of the optic lobe, single square). On the left, the NBs in each region are numbered except for the three optic lobe NBs. (Reproduced with permission from D. Zacharias, J. L. D. Williams, T. Meier, and H. Reichert. Neurogenesis in the insect brain: cellular identification and molecular characterization of brain neuroblasts in the grasshopper embryo. *Development* 118: 941–955. © 1993 Company of Biologists, Ltd.)

Fig. 8.26. Semischematic summary of neurons establishing the primary axon scaffold in the developing brain of the migratory locust *S. gregaria*, based on anti–horseradish peroxidase immunocytochemistry at 29%–34% of embryogenesis. Dc, deutocerebrum; La, lamina ganglionaris; Me, medulla; Pc, protocerebrum; Pc (LPC), lateral protocerebrum; Pi, pars intercerebralis; Pi (ALDL), anterolateral dorsal lobe of pars intercerebralis; Re, retina; Stomo, stomodaeum; Tc, tritocerebrum; 1–7, groups of interneurons whose outgrowth pioneers early pathways in the developing brain. (Adapted with permission from G. Boyan, S. Therainos, J. L. D. Williams, and H. Reichert. Axogenesis in the embryonic brain of the grasshopper *Schistocerca gregaria*: an identified cell analysis of early brain development. *Development* 121: 75–86. © 1995 Company of Biologists, Ltd.)

midline-derived pioneer neurons (2) and a second preoral commissure, by two pairs of pioneers in the pars intercerebralis (3). Descending tracts connecting the brain to the VNC are pioneered by progeny of identified NBs in the pars intercerebralis (4), deutocerebrum (5), and tritocerebrum (6) on each side of the brain, and the postoral commissure, by a pair of tritocerebral pioneers (7). During this process, some neurons seem to function as "guidepost" cells, similar to those functioning in developing appendages (8.3.2.1) (guidepost cells bear specific surface molecules that are contacted by searching, axon-borne filopodia), and all pioneers and guidepost cells express the cell adhesion molecule Fasciclin I during initial axon outgrowth and fasciculation (8.1.1).

Once established, this axon scaffold guides the growth of subsequently differentiating neurons and by 40% of embryogenesis, axonal projections characterizing the fully developed brain become evident. Later in embryogenesis, axons from sensilla on the head capsule and its appendages grow into this scaffold and depend on it to find their appropriate synaptic targets within the brain and subesophageal ganglion.

These events are accompanied by an upward 90-degree flexion in the frontal axis of the brain relative to that of the VNC, by the movement of clusters of proliferating NBs and their progeny in relation to each other and to the stomodaeum, and within each cluster, by extensive displacement of individual NBs and their progeny. All of these actions function to change the brain

from a simple sheet of NBs at 25% of embryogenesis (compare Fig. 8.25) to a complex, three-dimensional structure by 70% (Boyan et al., 1995a).

The 125 glomeruli (antennal synaptic centers) in each antennal lobe of larval and adult deutocerebra of the cockroach *Periplaneta americana* (Rospars, 1988) differentiate asynchronously between 63% and 77% of embryogenesis some time after entry and first synaptic contact of antennal afferents at 33%–43% (Salecker and Boeckh, 1995). The latter establish an adultlike projection pattern between 73% and hatch.

From this summary, it is obvious that the early embryonic brain of *S. gregaria* is not a miniature of the adult brain and that the primary scaffold of pioneer axons is established in the absence of certain centers that are later prominent (central body, protocerebral bridge, and mushroom bodies). These centers only begin to appear in a second phase of brain embryogenesis after 40% (Fig. 8.24).

Brain development is presently under intense scrutiny in *D. melanogaster* embryos, because of the well-understood genetics of this fly, and a host of different methods are being used, including computer-assisted three-dimensional reconstruction from confocal image stacks made from living embryos (Therianos et al., 1995; Younossi-Hartenstein et al., 1996, 1997; Campos-Ortega and Hartenstein, 1997; Reichert and Boyan, 1997; Meinertzhagen et al., 1998; Nassif et al., 1998a; Hartenstein et al., 1998; Schmid et al., 1999). A Web site (Flybrain: www.flybrain.org) shows many of the resulting images. Though this fly is only distantly related to grasshoppers (Fig. 0.2: Orthoptera, Diptera) and has complete rather than incomplete metamorphosis, brain embryogenesis in it proceeds in much the same way as in *S. gregaria*, not surprising considering the general similarity in brain organization throughout the hexapods (Bullock and Horridge, 1965).

NBs generating the brain in *D. melanogaster* embryos segregate in specific order from protocephalic neuroectoderm at 24%–32% of embryogenesis to form the protocerebrum, deutocerebrum, and tritocerebrum (Younossi-Hartenstein et al., 1996, 1997; see Fig. 11.8 in Campos-Ortega and Hartenstein, 1997). Just as in the VNC, expression of proneural genes of the *achaete-scute* complex (*as-c*) is required for neuroectodermal cells to acquire competence to form NBs. For example, expression of the *lethal of scute* (*l'sc*) gene is required for development of most protocephalic NBs and is controlled by proteins of the cephalic gap genes *tailless* (*tll*), *orthodenticle* (*otd*), *buttonhead* (*btd*), and *empty spiracles* (*ems*), which are expressed in partly overlapping domains in the head neuroectoderm (Fig. 7.41). Loss-of-function mutations of a given gap gene result in absence of *l'sc* expression in its domain of expression followed by absence of the NBs normally segregating from it. If *tll* is mutant, no protocerebral NBs arise (Rudolph et al., 1997).

Mutations in the *comm* gene affect growth cone guidance of midline pioneers toward the midline (Fig. 8.26: 2, 3, 7) and cause a marked reduction in the development of brain commissures, while those in the *sim* gene prevent the differentiation of CNS midline cells (Hirth et al., 1996). These mesectodermal cells ultimately give rise to the optic lobes (8.3.1.2.2), larval stemmata (8.3.2.4.3), ventromedial and dorsomedial parts of the brain, and the stomatogastric nervous system (SNS) (8.3.3) (Dumstrei et al., 1998). Localized activation of the *spitz* (*spi*) gene in mesectodermal cells is induced by receipt of Rhomboid (Rho) and Star (S) proteins early in embryogenesis (Fig. 7.20). Spi protein binds to the regulatory site of the *epidermal growth factor receptor* (*egfr*) gene, activates it, and triggers the Ras pathway required for normal determination, differentiation, and maintenance of these tissues. Loss of EGFR signaling results in the total failure of the optic lobes and stemmata to develop and perturbs development of the medial brain and SNS, whereas its overexpression results in hyperplasia and deformity of these structures.

8.3.1.2.1 CORPORA PEDUNCULATA

Until 1997, little was known about the embryogenesis of corpora pedunculata in insect brains other than the classic descriptive study of Malzacher (1968) on those of the stick insect *Carausius morosus* and the cockroach *Periplaneta americana*. However, their development is now being intensely investigated in embryos of *D. melanogaster*, in which it has been shown that the mushroom body on each side originates from the progeny of only four NBs (Tettamanti et al., 1997).

8.3.1.2.2 OPTIC LOBES

Each eye in a juvenile insect is innervated by an optic lobe associated with the brain (Fig. 8.27G). The anlagen of these lobes arise by separate, ectodermal delamination (apterygotes, exopterygotes) or invagination (endopterygotes) of protocephalic ectoderm in the ocular segment between protocerebral lobes 1 and 2 on each side of the developing head (Heming, 1982; Green et al., 1993; Schmidt-Ott et al., 1995; Campos-Ortega and Hartenstein, 1997). In *D. melanogaster* embryos, each optic lobe contains about 85 cells; invaginates from the side of the head, between 32% and 46% of embryogenesis, in contact with the developing brain; separates from head epidermis to form a vesicle; and at hatch, gives rise to inner and outer optic anlagen that will later generate the imaginal optic lobe of the compound eye during metamorphosis (10.2.4.1.2; see Fig. 9.17 in Campos-Ortega and Hartenstein, 1997).

Events are practically identical in embryos of the blister beetle *Lytta viridana* (Coleoptera: Meloidae), except that the lobes invaginate relatively earlier in embryogenesis (26%–38%) (Fig. 8.27A–C) (Heming, 1982). In early locust (*S. gregaria*) embryos, each optic lobe originates from only three NBs (those clear or marked with squares in Fig. 8.25) (Zacharias et al., 1993).

8.3.1.2.3 PARS INTERCEREBRALIS AND CENTRAL COMPLEX

The pars intercerebralis of the brain is situated anterodorsally on either side of the midline and houses somata of medially situated neurosecretory cells (Fig. 8.26; 10.2.4.1.2). Its development was followed by Boyan and Williams (1997) in embryos of the grasshopper *S. gregaria*, by labeling NBs and their progeny with tagged antibodies to the cell-surface protein Lachesin and by analyzing uptake of 5-BrdU to monitor cell division.

The first NBs of this structure begin to separate from neuroectoderm at 26% of embryogenesis, and by 28%, a total of 40 NBs, in two sheets of 20, are present on either side of the midline (compare Fig. 8.25). The NBs on each side then redistribute themselves in both the horizontal and the vertical plane into discrete subsets, each generating a cluster of GMCs that extends in a stereotyped direction into the developing brain. The four NBs of one subset (NB 5, 6, 7, and 8 in Fig. 8.25) were found to generate four clusters of embryonic neurons, W, X, Y, and Z, which begin to project their axons into the central complex, at 55% of embryogenesis, in four discrete fascicles (w, x, y, z) and to influence its differentiation (this complex is thought to function in multimodal information processing and, in adults, to influence head movements, flight, walking, jumping, and stridulation).

Between the two lobes of the pars intercerebralis, in embryos of this grasshopper, the brain midline is divided into dorsal and ventral domains (Ludwig et al., 1999). At 25% of embryogenesis, a single large midline precursor NB differentiates in the dorsal domain that buds off six neurons before degenerating. If the dye Lucifer yellow is injected into this NB, only this cell and its progeny are stained, so they are not coupled to adjacent neurons. The primary preoral commissure pioneer pair (Fig. 8.26: 2) and other nearby NBs remain unlabeled.

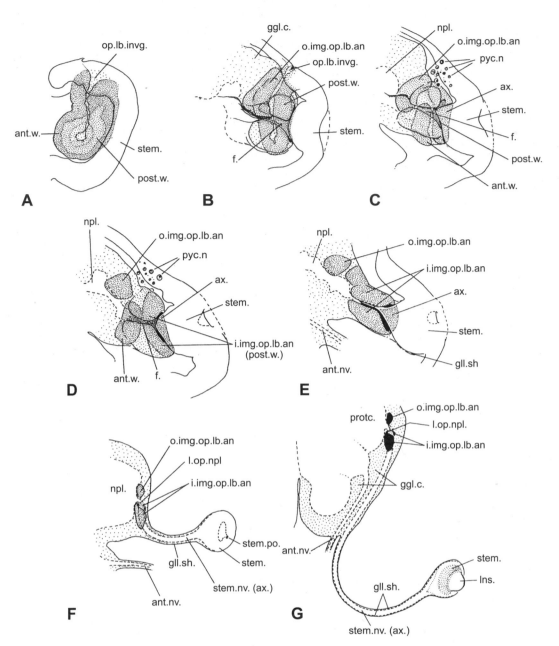

Fig. 8.27. Development of the larval visual system (right side) in an embryo of the blister beetle *Lytta viridana* (Coleoptera: Meloidae). The drawings are semidiagrammatic and are reconstructed from transverse serial sections. In each drawing, the observer is looking forward within the head from the level of the stemma (stem.). A. At 29% of embryogenesis. B. At 35%. C. At 38%. D. At 41%. E. At 48%. F. At 52.5%. G. At hatch. ant.nv., antennal nerve; ant.w., anterior wall of optic lobe invagination; ax., photoreceptor axons; f., fold in posterior wall of optic lobe invagination; ggl.c., ganglion cells (protocerebral neurons); gll.sh., glial sheath; i.img.op.lb.an., inner anlage of imaginal optic lobe; lns., stemmatal lens; l.op.npl., larval optic neuropile; npl., protocerebral neuropile; o.img.op.lb.an., outer anlage of imaginal optic lobe; op.lb.invg., optic lobe invagination; post.w., posterior wall of optic lobe invagination; protc., protocerebrum; pyc.n, pycnotic nuclei remaining from degeneration of base of optic lobe invagination; stem., optic placode of stemma; stem.nv (ax.)., stemmatal nerve (photoreceptor axons); stem.po., stemmatal pore. (Structure and development of the larval visual system in embryos of *Lytta viridana* Leconte (Coleoptera, Meloidae), B. S. Heming, *J. Morphol.*, Copyright © 1982 Wiley-Liss, Inc. Reprinted by permission of Wiley-Liss, Inc., a subsidiary of John Wiley & Sons, Inc.)

8.3.1.2.4 NEUROSECRETORY CELLS

Neurosecretory cells in the brains of insects are important sources of hormones influencing reproduction (1.3.7.2, 2.3.2.2), molting, and metamorphosis (chapter 12). In larvae of the tobacco hornworm *Manduca sexta*, the group III lateral neurosecretory cells (L-NSC III) synthesize "big" prothoracicotropic hormone (PTTH), which induces the prothoracic glands to synthesize and release the ecdysteroids causing a molt (Fig. 12.10A, C). A monoclonal antibody raised against one variant of this molecule has been labeled and used to follow the embryogenesis of these neurosecretory cells by immunocytochemistry (Westbrook and Bollenbacher, 1990). PTTH is first detected in L-NSC III somata between 25% and 30% of embryonic development, shortly after these cells arise. By 48%, their neurites have grown to the contralateral (opposite) brain lobe, where they fasciculate with axons of the L-NSC IIIs of this side. The axons of both then exit the brain at about 60% and grow via the nervi corpora cardiaci I+II through the corpora cardiaca into the developing corpora allata at about 65% of embryogenesis, where they begin to arborize. By 98%, the system is fully differentiated.

8.3.1.2.5 ARE INSECT AND VERTEBRATE BRAINS HOMOLOGOUS?

Building on a recent resuscitation of interest in É. Geoffroy St.-Hilaire's old (1822) idea of dorsoventral axis inversion between insects and vertebrates (7.2.1.1.1) and on recent progress in understanding the molecular genetics of brain development in representative insect and vertebrate embryos, Salzberg and Bellen (1996), Arendt and Nübler-Jung (1996), Reichert and Boyan (1997), and Sharma and Brand (1998) have marshalled structural, functional, and molecular genetic evidence to suggest that insect and vertebrate brains share a ground plan inherited from a common ancestor. They demonstrated that such brains form in comparable body regions; contain comparable regions of similar function expressing similar, evolutionarily conserved regulatory genes during development; and have similar early axonal scaffolds developing in similar ways and similarly guiding growth of afferent sensory axons. Brain areas homologized are as follows:

Insect	Vertebrate
syncerebrum (protocerebrum + deutocerebrum + tritocerebrum)	forebrain (prosencephalon + mesencephalon)
gnathocerebrum (subesophageal ganglion)	hindbrain (rhombomeres 1, 2, 3, etc.)
corpora pedunculata	pyriform cortex, hippocampus, neocortex
central body complex	cerebellum
a gap between the brain and VNC in development and in expression of regulatory genes	a gap between the forebrain and hindbrain in development and in expression of regulatory genes
CNS midline cells	CNS midline cells
stomodaeum	ventral diencephalic region (infundibulum + hypothalamus)
SNS	hypothalamus
central and peripheral brain areas separated by an	central and peripheral brain areas separated by an axonal circle around the infundibulum

Insect	Vertebrate
axonal circle around the stomodaeum, the circumesophageal connectives	

Among the similarities in development and embryonic gene expression are the following:

Insect	Vertebrate
NBs of anterior syncerebrum express tll, with eyes and optic lobes arising in the area of expression.	NBs of anterior prosencephalon express tll ortholog, with eyes and optic lobes arising in the area of expression.
The gene eyeless (ey) plays an early and fundamental role in eye development.	Orthologous Pax-6 gene plays an early and fundamental role in eye development.
Deutocerebral lobes express ems and otd genes, and otd mutant embryos are rescued by expression of vertebrate orthologs Otx1 and Otx2.	Olfactory lobes express orthologs of the ems (Emx1, Emx2) and otd (Otx1, Otx2) genes, and Otx mutants are rescued by expression of insect ortholog otd.
Tritocerebral NBs segregate early from those of deutocerebrum.	NBs of mesencephalon segregate early from those of prosencephalon.

Memory functions and molecular mechanisms of learning are similar in insect protocerebrum and in the vertebrate prosencephalon.
Metameric subunits of the insect gnathocerebrum and vertebrate hindbrain are similar.
Midline NBs share expression of three sets of orthologous genes: fkh, short gastrulation (sog), and netrin.

Insect	Vertebrate
The gene dpp is expressed in the roof of the stomodaeum	The dpp ortholog bone morphogenetic protein (BMP-4) is expressed in the future infundibulum.
SNS evaginates from the roof of the stomodeaum.	The hypothalamus evaginates from the infundibulum.

As mentioned earlier (Fig. 8.9B), these researchers proposed embryos of the putative common ancestor of insects and vertebrates, the Urbilataria, to form a mouth and anus, respectively, at the anterior and posterior ends of an elongate, ventral, slit blastopore, the CNS differentiating anterior to and along the lateral margins of the blastopore, with the midline region (mesectoderm) of the cord marking the ancestral line of fusion of the lateral blastopore lips in both insects and vertebrates (Fig. 8.28) (Lacalli, 1996). With dorsoventral inversion of the vertebrate ancestor after divergence (7.2.1.1.1), the old mouth in the middle of the brain was obliterated (the

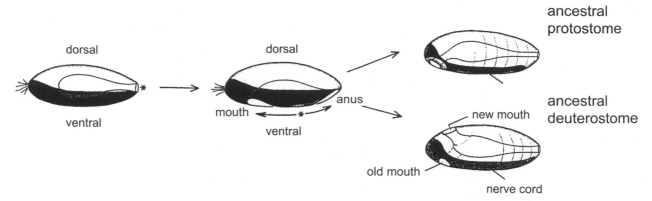

dorsal

ventral

dorsal

mouth

ventral

anus

ancestral protostome

ancestral deuterostome

new mouth

old mouth

nerve cord

Fig. 8.28. Derivation of the ventral (in protostomes, the lineage leading to insects) and dorsal (in deuterostomes, the lineage leading to vertebrates) central nervous system in animals. See text for details. The asterisk represents the blastopore. (Dorsoventral axis inversion: a phylogenetic perspective, by T. Lacalli, *Bioessays*, Copyright © 1996 Wiley-Liss, Inc. Reprinted by permission of Wiley-Liss, Inc., a subsidiary of John Wiley & Sons, Inc.)

stomodaeum of insects) and replaced by the infundibulum in the venter of the brain, a new mouth invaginating at the new ventral (formerly dorsal) side (Fig. 8.28).

8.3.1.3 GLIAL CELLS

Neurons are not the only cell type to be found within the CNS. *Glial cells* support and surround the neurons and contribute to their migration, differentiation, nutrition, and function (Edwards and Tolbert, 1998). In *D. melanogaster* embryos, 54–64 glial cells of three types originate within each VNC ganglion during embryogenesis and are of neural origin (Fig. 8.29) (Ito et al., 1995; Bossing et al., 1996; Schmidt et al., 1997): surface associated (16–18 subperineural, 6–8 channel), cortex associated (6–8 cell bodies), and neuropile associated (8–10 nerve root, 14–16 interface, and 3–4 midline).

Expression of the gene *glial cell deficient* (*glide*) is required for glial cell specification; if this gene is mutant, their progenitor cells become neurons (Vincent et al., 1996). Expression of *pointed* (*pnt*) is required for neuronal-glial interaction, as loss-of-function mutations of *pnt* lead to changes in migration of midline glial cells and to fusion of anterior and posterior cross commissures within each ganglion (Klämbt, 1993).

When the growth cone of a searching neuron contacts the appropriate glial cell or axon fascicle, it immediately changes its adhesion molecule(s) to match those of the guidepost cell. Then a complex interaction seems to occur between the cells and can involve changes in motility, adhesion, signal transduction, cytoskeletal changes, or gene expression.

Udolph (1993), Bossing (1996), and Schmidt (1997) and their respective colleagues reconstructed the complete lineage progeny of many NBs in the VNC of *D. melanogaster* embryos and showed some to generate only neurons, some only glial cells, and others both cell

types. The gene *glial cells missing* (*gcm*) encodes a nuclear transcription factor transiently expressed in nearly all glial cells (Jones et al., 1995). In loss-of-function mutant embryos, most glial cells fail to differentiate and instead transform into neurons, whereas in gain-of-function mutants, many presumptive neurons differentiate as glia. Therefore, *gcm* seems to function as a binary switch gene: with Gcm protein, presumptive neurons differentiate as glial cells; without it, presumptive glial cells become neurons. Glial cell development and function were reviewed by Klämbt et al. (1996).

8.3.1.4 GENES AND CENTRAL NERVOUS SYSTEM DEVELOPMENT IN EMBRYOS OF *DROSOPHILA MELANOGASTER*

The neuroectoderm of insects contains an initially indifferent population of cells which, during subsequent embryogenesis, give rise to various progenitor cells of the CNS and epidermis (Campos-Ortega, 1993, 1994; Goodman and Doe, 1993; Burrows, 1996; Campos-Ortega and Hartenstein, 1997; Schmid et al., 1999). A hierarchy of five classes of regulatory genes is known to influence cell fate in the developing CNS of *D. melanogaster* embryos (Fig. 8.30) (Goodman and Doe, 1993). Sequential expression of such genes during development

- provides positional cues to cells in the neuroectoderm (segmentation, dorsoventral, and homeotic genes; 7.2.1);
- specifies NB identity (neuroblast identity genes);
- induces NB differentiation (proneural genes);
- inhibits other neuroectoderm cells in a proneural cluster from becoming NBs through lateral inhibition (neurogenic genes); and
- specifies the identity of GMCs and of individual neurons (GMC and neural identity genes).

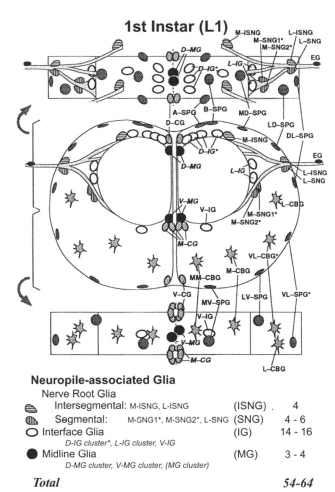

1st Instar (L1)

Neuropile-associated Glia

Nerve Root Glia

⊜	Intersegmental: M-ISNG, L-ISNG	(ISNG)	4
⬟	Segmental: M-SNG1*, M-SNG2*, L-SNG	(SNG)	4 - 6
○	Interface Glia	(IG)	14 - 16
	D-IG cluster*, L-IG cluster, V-IG		
●	Midline Glia	(MG)	3 - 4
	D-MG cluster, V-MG cluster, (MG cluster)		

Total   ***54-64***

Fig. 8.29. Type and distribution of glial cells occurring on one abdominal ganglion of the ventral nerve cord of a first instar larva of *D. melanogaster*. Top and bottom panels show distributions of dorsally (top) and ventrally (bottom) situated cells in frontal aspect; the middle panel, a transverse section. (Reprinted with permission from K. Ito, J. Urban, and G. M. Technau. Distribution, classification, and development of *Drosophila* glial cells in the late embryonic and early larval ventral nerve cord. *W. Roux's Arch. Dev. Biol.* 204: 284–307. © 1995 Springer-Verlag.)

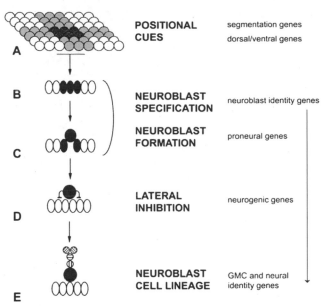

Fig. 8.30. Classes of genes specifying neuroblasts (NBs), ganglion mother cells (GMCs), and neurons in the developing CNS of *D. melanogaster*. A. Positional cues (shaded) in the neuroectoderm that are provided by expression of the antero-posterior and dorsoventral pattern-control genes provide unique positional values to clusters of four to six proneural cells (black). B and C. Cells of each cluster express one or more NB identity and proneural genes at a specific time, depending on their position in the neuroectoderm, that render them competent to differentiate into NBs. One cell of the cluster is singled out to enlarge and delaminate as a NB and its identity is specified (NB identity genes). D. The NB inhibits other cells of the cluster from developing into NBs by expressing the neurogenic genes, and they lose expression of proneural and NB identity genes. E. The newly formed NB initiates a cell lineage to produce a series of GMCs, which divide into pairs of neurons. GMCs and neurons are specified by expression of a unique combination of GMC and neural identity genes. (Reprinted with permission from C. S. Goodman and C. Q. Doe. Embryonic development of the *Drosophila* central nervous system. In M. Bate and A. Martinez Arias (eds.), *The development of* Drosophila melanogaster. Vol. 1, p. 1140, Fig. 5. © 1993 Cold Spring Harbor Laboratory Press.)

The principal elements of this incredibly complex network are two groups of DNA-binding proteins encoded by the *achaete* (*ac*), *scute* (*sc*), and *lethal of scute* (*l'sc*) genes (the *proneural* genes), which activate the neural developmental pathway, and others (the *antineural* genes *extramacrochaetae* [*emc*], *hairless* [*h*], *Suppressor of Hairless* [*Su(H)*], *mastermind* [*mam*], and *Enhancer-of-split* [*E(spl)-C*]), which suppress it (Gerhart and Kirschner, 1997: 273–285; Artavanis-Tsakonas et al., 1999). Segregation of a single NB from each proneural cluster in *D. melanogaster* relies on Notch-mediated lateral signaling (Muskavitch, 1994; Artavanis-Tsakonas et al., 1995; Seugnet et al., 1997). Activation of the Notch (N) receptor protein by binding of the Delta (Dl) ligand inhibits expression of the *ac-sc* genes in other proneural cells, which is required for their specification as NBs. Commitment to a neural or nonneural fate results when one of the two groups of regulatory proteins predominates in a given cell. N and Dl also function in mesoderm formation, germ-line and ovarial follicle cell development, formation of larval Malpighian tubules, sense organ differentiation, eye development, and wing blade formation during *D. melanogaster* ontogeny. Their orthologs have diverse functions in other developing organisms including nematodes (*Caenorhabditis elegans*) and vertebrates (zebrafish, mice, rats, and humans) (Artavanis-Tsakonas et al., 1995, 1999; Fleming et al., 1997).

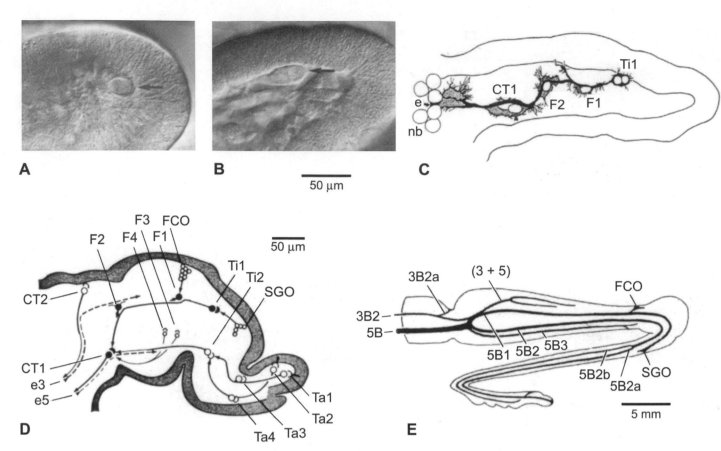

Fig. 8.31. Development of the peripheral pioneer pathway in the metathoracic limb bud of a grasshopper (*Schistocerca nitens*) embryo. A. Interference contrast optical section of the stem cell (arrow) of the tibial (Ti1) pioneer neuron pair at 29%–30% of embryogenesis (at 30°C) just after it has moved into the lumen of the limb from the tibial epidermis. B. Same, at 34%. The stem cell has divided and growth cones of the axon pair have reached the CNS proximally by growing between ectoderm and mesoderm. C. Camera lucida drawing of a limb bud at 34%, showing projection of the Ti1 pioneers to neuroblasts (nb) in the CNS by way of the femoral (F1), trochanteral (F2 [=Tr1]), and coxal (CT1) guidepost cells. D. Diagram of limb bud at 45%, showing those cells establishing the principal peripheral nerve branches. Solid lines indicate afferent and dashed lines, efferent axons. Filled somata are those of the first pioneer neurons and of guidepost cells along their route; unfilled somata, those of later-appearing sensory neurons. E. Principal nerve branches in the adult leg. (A and B reproduced with permission from H. Keshishian. The origin and morphogenesis of pioneer neurons in the grasshopper metathoracic leg. *Dev. Biol.* 80: 388–397. © 1980 Academic Press, Inc.) (C reprinted with permission from *Nature* (D. Bentley and M. Caudy. Pioneer axons lose directed growth after selective killing of guide post cells. 304: 62–65) © 1983 Macmillan Magazines Limited.) (D and E reprinted with permisson from D. Bentley and H. Keshishian. Pathfinding by peripheral pioneer neurons in grasshoppers. *Science* 218: 1082–1088. Copyright 1982 American Association for the Advancement of Science.)

Detailed information on the molecular genetics of CNS development in *D. melanogaster* embryos can be found in publications by Campos-Ortega (1993, 1994), Goodman and Doe (1993), and Haasen and Vaessin (1996). Salzberg and Bellen (1996) compared the genetics of neurogenesis between *D. melanogaster* and vertebrates.

8.3.2 PERIPHERAL NERVOUS SYSTEM
The peripheral nervous system (PNS) consists of thousands (about 5000 in adult *D. melanogaster*) of sense organs (sensilla) distributed outside and within the body and of peripheral nerves containing afferent sensory axons projecting to ganglia of the CNS from the sensilla

and efferent motor axons projecting to various effector organs (Fig. 8.16) (Snodgrass, 1935).

8.3.2.1 PERIPHERAL PIONEER NEURONS IN GRASSHOPPER EMBRYOS
Innervation of the muscles within each developing appendage by efferent motor neuron axons projecting from ganglia of the CNS seems to involve the function of a separate class of peripheral pioneer neurons, first observed by Bate (1976) in antennal and leg primordia of *Locusta migratoria* (Orthoptera: Acrididae) embryos. In leg primordia, these originate from ectoderm at the tip (actually the presumptive tibia, hence Ti1 pioneers) at 28%–30% of embryogenesis (Fig. 8.31A,

202   *Organogenesis*

B) (Bentley and Keshishian, 1982a, 1982b). The growth cones of these pioneers grow proximally within the limb bud via a convoluted zigzag path into the appropriate thoracic ganglion by 34% (Fig. 8.31C) and forward within the *ipsilateral* (on the same side) connective, and pioneer a path by which the growth cones of motor neuron axons can find their appropriate target muscles within each leg. They also guide the growth cones of afferent axons from later-appearing appendage sensilla to appropriate targets within the CNS and die after these connections are established (Fig. 8.31D) (Klose and Bentley, 1989).

Two Ti1 pioneers are normally essential for development of nerve 5B1 (Fig. 8.31E) (5B1 innervates the tibial subgenual organ) in each hind leg in intact *Schistocerca nitens* and *S. americana* embryos. However, if these neurons are destroyed by dye injection and photoinactivation, they can be replaced by growth of the first pair of femoral (F1) neurons (Bentley and Keshishian, 1982). The growth cones of these pioneers traverse the femoral-tibial and trochanteral-femoral segment boundaries before these border cells assume their molecular identities (expression of the transmembrane proteins annulin and Semaphorin I; Fig. 8.32). If they fail to cross these boundaries, later differentiating afferents are unable to reach the CNS because they are unable to cross these borders after these identities are established. *Guidepost cells* (immature, later-appearing pioneer neurons yet to begin axonogenesis) in the femur (F1, F2) and trochanter (CT1) are normally contacted by growth cones of the Ti1 pair and constitute part of the path by which these pioneers normally find their way to the CNS; they, too, later project to the CNS (Lefcort and Bentley, 1987). However, and in spite of early evidence to the contrary (Bentley and Caudy, 1983), these cells can be removed experimentally without stopping effective growth of Ti1 axons to their correct targets (Condic and Bentley, 1989).

When they reach the coxal-trochanteral boundary, the Ti1 growth cones grow circumferentially from top to bottom parallel to this boundary, owing to expression of Semaphorin I (Sema I) by its cells (Fig. 8.32A) (Wong et al., 1997). If these boundary cells are exposed to Sema I monoclonal antibodies, the Ti1 growth cones fail to change direction, cross this boundary at many points, and never reach their appropriate targets within the CNS. Eventually they reach a ventral pair of coxal guidepost cells (Cx1) and project proximally into the appropriate CNS ganglion (Lefcort and Bentley, 1987).

The proteins annulin and Sema I are expressed at all limb segment boundaries between coxa, trochanter, femur, tibia, and tarsus before these segments are recognizable morphologically (Fig. 8.32), and influence filopodial searching and growth both by Ti1 pioneers (Singer et al., 1992; Bastiani et al., 1992; Ganfornina et al., 1996a) and by axons of the later-appearing subgenual organ (SGO, a chordotonal sense organ) of the tibia (Wong et al., 1997). Growth cones of the SGO axons normally fasciculate with Ti1 axons (Fig. 8.32A–C). If the Ti1 pathway is destroyed, the SGO growth cones enter but are unable to cross cells of the Sema I–expressing band between tibia and tarsus (Fig. 8.32D). When Sema I expression is blocked by exposure to Sema I monoclonal antibodies, axon outgrowth from the SGO does not occur (Fig. 8.32E), indicating that Sema I in some way attracts SGO growth cones to cells expressing it (Wong et al., 1997).

During proximal growth of Ti1 axons, there are places where their growth cone filopodia are not in contact with guidepost cells, for example, when they first start to project their neurites proximally (Fig. 8.31A, B). Possible cues directing such growth in these areas include extrinsic cues, contact guidance by limb contour, an axial electrical field, a diffusion gradient generated by a localized source, internal polarity of the ectoderm, and mesoderm or a proximodistal adhesive gradient in the ectoderm or its basal lamina. All have been eliminated by use of ingenious experiments (Berlot and Goodman, 1984; Lefcort and Bentley, 1987) except the last (Norbeck et al., 1992).

Cell-surface glycoproteins anchored to cell membranes via covalently attached glycosylphosphatidylinositol (GPI) (Fig. 8.21) have been implicated in neuronal adhesion, promotion of neurite outgrowth, and directed neuronal migration. Chang and colleagues (1992) treated grasshopper embryos (*S. americana*) with bacterial phosphatidylinositol-specific phospholipase C (PI-PLC), an enzyme cleaving this GPI anchor, and disrupted the highly stereotyped migrations of Ti1 growth cones and those of later-appearing sensory neurons (compare Fig. 8.31D). If distal limb regions were treated at early stages of pioneer axon outgrowth, their growth cones lost their normal proximal orientation toward the CNS and turned distally.

Nervous connections with the CNS are first established while the limb bud is very short, and with subsequent development, the cell bodies of these sensory neurons are displaced to relatively great distances by distal growth of the appendages.

8.3.2.2 INTERSEGMENTAL NERVE FORMATION IN *DROSOPHILA* EMBRYOS

Extended germ band embryos of *D. melanogaster* have reduced gnathal appendages and no thoracic or abdominal ones (Fig. 6.23: 11); thus, it is not surprising that peripheral pioneers seem to be lacking (Ghysen et al., 1986). Nevertheless, the lateral intersegmental nerves in larvae follow reproducible paths as they extend laterally and dorsally from the VNC on each side of the body near the anterior border of each segment (Giniger et al., 1993; see Fig. 11.11 in Campos-Ortega and Hartenstein, 1997), the growth of their axons being guided partly by expression of the transmembrane proteins Dl and N. Presence of the N receptor on the principal, lateral, dorsoventral tracheal branch on each side of each segment provides a pathway for growth of the nerves through the lateral part of each body segment to developing muscle fibers, the growth cones of their axons using the N ligand on their surfaces to recognize this path. Disruption of the trachea

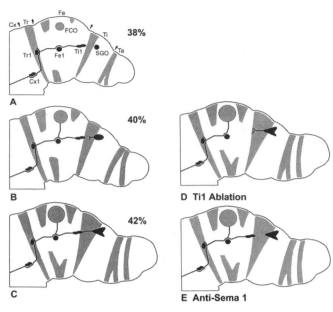

Fig. 8.32. Experiments demonstrating the influence of the epidermal transmembrane protein Semaphorin I (Sema I) on axon outgrowth from the subgenual organ (SGO) in metathoracic limb buds of embryos of the locust *Schistocerca gregaria.* A–C. Normal; D and E. Experimental. A. Location of the SGO and femoral chordotonal organs (FCO) in relation to Sema I expression (patterned stripes) at 38% of embryogenesis. The Ti1 pioneers are located on the proximal margin of a band of Sema I near the femorotibial segment boundary (segmental boundaries marked by arrowheads). The cells of the SGO and FCO are just beginning to differentiate and those of the FCO also express Sema I. B. By 40%, axons of the SGO extend proximally into the femorotibial band of Sema I expression, and those of the FCO have contacted the femoral guidepost cells (Fe1) and fasciculated with the Ti1 pathway. C. At 42%, the SGO axons have crossed the band of Sema I expression and have fasciculated with the Ti1 pathway. D. At 42%, when the Ti1 pathway is eliminated by heat shock, axons of the SGO are blocked at the proximal margin of femorotibial Sema I expression. E. In limb bud fillets cultured in media containing an antibody to Sema I, SGO axons fail to extend whether Ti1 neurons are present or absent. Cx, coxa; Cx1, coxal guidepost neurons; Fe, femur; Fe1, femoral guidepost neurons; Ta, tarsus; Ti, tibia; Tr, trochanter; Tr1, trochanteral guidepost neurons. (Reproduced with permission from J. T. W. Wong, W. T. C. Wu, and T. P. O'Conner. Transmembrane Semaphorin I promotes axon outgrowth in vivo. *Development* 124: 3597–3607. © 1997 Company of Biologists, Ltd.)

abolishes the ability of these nerves to establish the correct innervation of somatic muscles.

8.3.2.3 SYNAPSE FORMATION

In insect embryos, motor neurons from the CNS form stereotyped synapses on identified muscle fibers throughout the body. As summarized in 8.1.2.1, each half of a larval abdominal segment contains 30 muscle fibers defined by location, attachment, and sometimes,

molecular expression pattern (see Fig. 4.6 in Campos-Ortega and Hartenstein, 1997). These are innervated by 36 glutamatergic motor neurons in each segment.

Broadie and Bate (1993) followed development of the innervation of these muscles in cultured fillets made by flattening a dorsally incised embryo against a glass slide, using whole-cell voltage clamping to monitor neuronal activity. They used a mutant gene (*prospero* [*pro*]) that prevents or delays innervation to assay the role of the presynaptic neuron in developing the receptive field of the postsynaptic muscle fiber (*pro* is not expressed in muscles or their precursors) and found the muscle itself to define the correct synaptic zone by restricting the distribution of putative guidance molecules (Fasciclin III, connectin) to this membrane region. The muscle also expresses functional transmitter receptors at the correct developmental time without innervation but does not localize them to the synapse without instruction from the motor neuron. Neither does a second, much larger synthesis of receptors occur in muscles deprived of innervation.

In muscles receiving delayed innervation, or in those innervated at *ectopic* (other than normal) synaptic sites, both receptor clustering and synthesis are delayed or are redirected, consistent with the new pattern of innervation. Thus, the muscle seems to define the synaptic site autonomously, whereas the motor neuron directs development of the muscle's receptive field by stimulating the synthesis and localization of transmitter receptors.

Actual formation of synapses between adjacent neurons and between motor neurons and their effector organs seems to be stimulated by release of the protein Agrin (named because it causes transmitter receptor proteins to aggregate at the synapse in the postsynaptic effector) from each axon terminal (Wallace, 1996). This protein binds to a muscle-specific receptor tyrosine kinase to trigger postsynaptic differentiation. Agrin is synthesized in the cell bodies of presynaptic neurons and is transported to the terminals, where it is released at the appropriate time.

8.3.2.4 SENSE ORGANS

Most external sense organs in insects develop from individual epidermal sense organ precursor cells (SOPs) (stem cells, sense organ mother cells) through a characteristic number of asymmetrical, differentiative mitoses, as shown in Fig. 8.33 for a singly innervated, four-celled mechanosensory hair. The number of divisions varies with the complexity of the sensillum and the number of accessory and sensory cells it contains (Bate, 1978). The sensory axons of the sensilla then project to appropriate target interneurons within a ganglion of the CNS on the surfaces of the pioneer neurons mentioned already (Fig. 8.31D) or on the axons of earlier-formed sensilla, as shown in Fig. 8.34 for the cercus of a locust embryo, *S. americana* (Shankland and Bentley, 1983). Fig. 8.35 illustrates the relationship between the

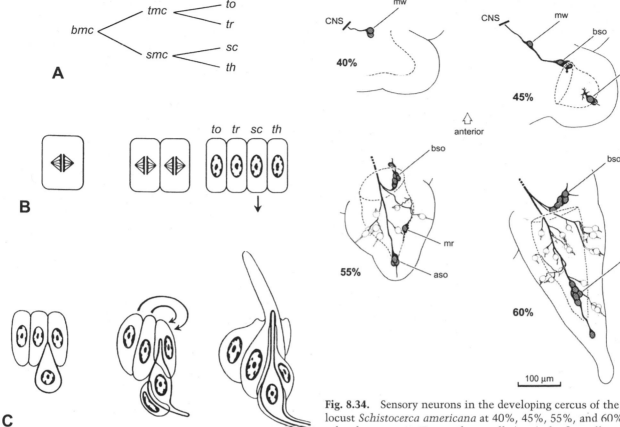

Fig. 8.33. Early development of a mechanoreceptor in a *D. melanogaster* embryo. A. Differential mitosis of bristle mother cell (bmc). smc, mother cell of sensory neuron (sc) and thecogen (th) cell; tmc, mother cell of trichogen (tr) and tormogen (th) cells. B. Creation of the early, side-by-side arrangement of cells. C. The prospective neuron sinks beneath the plane of the epidermis and develops a primary dendrite, which becomes enveloped by a glial cell and then grows through the thecogen cell. The trichogen cell then develops an apical sprout around which larval cuticle is subsequently deposited to form the hair. (Reprinted from *Int. J. Insect Morphol. Embryol.*, 26, T. A. Keil, Comparative morphogenesis of sensilla: A review, pp. 151–160, Copyright 1997, with permission from Elsevier Science.)

Fig. 8.34. Sensory neurons in the developing cercus of the locust *Schistocerca americana* at 40%, 45%, 55%, and 60% of embryogenesis. Two midway cells (mw), the first afferent neurons to appear, originate halfway between the cercus and the developing terminal ganglion of the ventral nerve cord and may serve as guidepost cells for the growth cones of proximally growing axons of the basal (bso) and apical (aso) scolopidial organs. The apical scolopidial organ establishes a nerve down the lumen of the appendage, to which the axons of later-appearing epidermal sensilla (white) fasciculate. Still later-appearing afferents form a lateral nerve branch, which fuses to that of the basal scolopidial organ proximal to the base of the appendage. In some embryos, this branch contains the axon(s) of one or more luminally situated mural cell(s) (mr). (Reproduced with permission from M. Shankland and D. Bentley. Sensory receptor differentiation and axonal pathfinding in the cercus of the grasshopper embryo. *Dev. Biol.* 97: 468–482. © 1983 Academic Press, Inc.)

primary dendrite of a cercal mechanosensory neuron and the second embryonic (C$_2$) and first instar (C$_3$) cuticles of the same insect (note that although this dendrite is attached to C$_2$, no sensillum is present).

In addition to these obvious, surface-borne sensilla, all juvenile insects contain a homologous, subepidermal peripheral nerve plexus of multidendritic sensory neurons, each projecting an axon to interneurons in ganglia of the VNC. In larvae of *Manduca sexta* about 350 of these in each half-segment are sensitive to nitric oxide and originate in three waves during embryonic and larval development (Grueber and Truman, 1999). The

12–16 neurons of the primary plexus are uniquely identifiable and originate at 35%–45% of embryogenesis. Those of the secondary plexus arise in two waves at 70%–80% and during the larva I–larva II molt. Each neuron of the secondary plexus arises from a five-cell cluster, the other four cells forming a mechanoreceptor, as in Fig. 8.33.

8.3.2.4.1 GENES AND SENSE ORGAN SPECIFICATION IN EMBRYOS OF *DROSOPHILA MELANOGASTER*

The molecular genetics of sense organ specification and differentiation in *D. melanogaster* embryos is

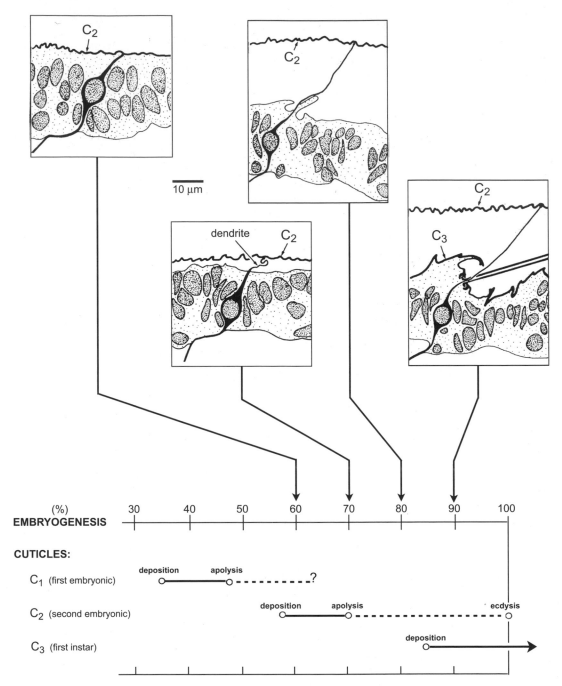

Fig. 8.35. Formation of a mechanoreceptive hair on the cercus in relation to deposition of the second embryonic (C_2) and first instar (C_3) cuticles during embryogenesis of the locust *S. americana*. The dendritic sheath surrounding the primary dendrite of its single sensory neuron is attached to the underside of C_2 at time of its deposition. With its apolysis at 70% of embryogenesis, both sheath and dendrite elongate to bridge the ecdysial space. This connection breaks with eclosion of the first instar and ecdysis of cuticle C_2. The time lines depict the history of each cuticle: the solid line represents the period when the cuticle adheres to the epidermis; the dashed line, the period when the cuticle has separated from it but still surrounds the embryo, the pharate stage. (Reproduced with permission from M. Shankland and D. Bentley. Sensory receptor differentiation and axonal pathfinding in the cercus of the grasshopper embryo. *Dev. Biol. 97*: 468–482. © 1983 Academic Press, Inc.)

presently under intense scrutiny, and many of the same genes shown to specify NBs, GMCs, and neurons in the CNS (Fig. 8.30) seem to be involved (Jan and Jan, 1993: Table 2; Campos-Ortega, 1994; Posakony, 1994; Brewster and Bodmer, 1996; Artavanis-Tsakonas et al., 1999; Ghysen and Dambly-Chaudière, 2000). Sequential expression of these genes

- provides positional information to the epidermis (anteroposterior and dorsoventral prepattern genes);
- endows groups of cells with the potential to become neuronal precursors (proneural genes);
- selects a cell from each proneural cluster to become an SOP and inhibits its neighbors from doing so (neurogenic genes);
- confers neuronal properties on that cell (neuronal precursor genes);
- specifies type of SOP (neuronal precursor type selector genes); and
- controls the asymmetrical mitosis of the SOP and its progeny and assigns a fate to each cell in the resulting cluster (cell cycle and sense organ cell fate genes) (Jan and Jan, 1993: 1224, 1230–1232).

Regulatory interactions between the neuronal precursor type selector genes *pox-neural* (*poxn*) and *cut* determine sense organ identity (Vervoort et al., 1995). If *cut* is expressed in an SOP, an external sense organ differentiates; if not, an internal (e.g., chordotonal) sense organ does. If *poxn* is expressed in an SOP, a multiply innervated sense organ results; if not, a singly innervated one results. Also, *poxn* can induce expression of *cut* but not vice versa.

Two mechanisms acting alone or in conjunction have been proposed to specify cell identity within a cluster (Posakony, 1994). According to the intrinsic mechanism, each stem cell is polarized with respect to plane of division and produces daughter cells that differ from each other, owing to segregation of a specific molecule to one side of the stem cell that only one daughter cell receives. The membrane-associated protein Numb and the nuclear protein Prospero mentioned earlier (Fig. 8.14) are such molecules (Rhyu et al., 1994; Spane and Doe, 1995). Numb is localized as a crescent to one side of each stem cell and is allocated preferentially to the nucleus of one daughter cell during mitosis. Loss of wild-type *numb* function results in the neuron and *thecogen* (dendritic sheath–forming) cell of a four-celled mechanoreceptor differentiating both as additional *trichogen* (hair-forming) and as *tormogen* (socket-forming) cells. The protein thus functions to determine the fates of the two secondary precursors resulting from division I of the SOP. Prospero is a transcription factor necessary for proper expression of downstream genes in the daughter cell that contains it.

A second, extrinsic mechanism proposes that action of external cues asymmetrically influences the fates of sister cells initially having identical potential. An obvious source of such cues is cell-cell interaction within the sensillar cluster, and mounting genetic evidence is available for just such activity in *D. melanogaster* embryos, particularly the *Notch/Delta* system (Fig. 8.36) (Posakony, 1994; Artavanis-Tsakonas et al., 1999).

8.3.2.4.2 EVOLUTIONARY CONSIDERATIONS
Is the pattern of larval SOPs evolutionarily conserved? This seems to be so, at least in embryos of the grasshopper *S. gregaria* and the fly *D. melanogaster*. Using a

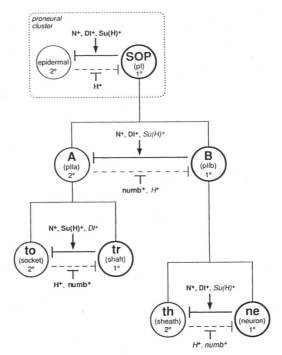

Fig. 8.36. A model of the genetic control of sense organ cell fates in *D. melanogaster* as proposed by Posakony (1994). Genes whose role in a particular cell fate decision is supported by published evidence are shown in bold; italics indicates that the function of the gene is hypothetical or based on unpublished data. The interaction of the presumptive sense organ precursor cell (SOP) with a single presumptive epidermal cell is illustrated (multiple SOP-epidermal and epidermal-epidermal interactions are known to take place within the proneural cluster). Inhibitory interactions are represented as occurring only between sister cells in each of the three asymmetrical cell divisions of the lineage, but more complex interactions are likely. Only three genes in the neurogenic group are represented in the figure, but it is likely that other identified neurogenic genes (see Fig. 8.30) have similar roles in sense organ development. to, tormogen cell; tr, trichogen cell; th, thecogen cell; ne, sensory neuron; N, Notch, Dl, Delta; H, Hairless; Su(H), Suppressor of Hairless. (Reprinted from *Cell*, 76, J. W. Posakony, Nature versus nurture: asymmetric cell divisions in *Drosophila* bristle development, pp. 415–418, Copyright 1994, with permission from Elsevier Science.)

labeled neuron-specific antibody, Meier and coworkers (1991) followed the origin and differentiation of SOPs in embryos of both insects. In both, a simple peripheral nerve scaffold is first established within each body segment. Then, identifiable sets of sensory neurons differentiate in a stereotyped, spatiotemporal pattern in dorsal, lateral, and ventral clusters on each side of each segment and project their axons onto these nerves. Although segment-specific differences do exist between the two species, serial homologs of the developing nerves and sensory neurons have been identified to have virtually the same pattern in both species. Since these two insects belong to lineages that diverged over 300 million years ago (Fig. 13.1: Orthoptera, Diptera), this developmental mechanism must have been conserved for at least that length of time and probably occurs in embryos of insects in orders situated phylogenetically between Orthoptera and Diptera.

The specific time sequence for development of the sensory nervous system in relation to embryogenesis in *D. melanogaster* was provided by Jan and Jan (1993: Fig. 7) and for sensilla on the larval antennae of the blister beetle *Lytta viridana*, by Heming (1996: Figs. 31 and 33).

8.3.2.4.3 LARVAL STEMMATA IN HOLOMETABOLOUS INSECTS

If not blind, the juveniles of apterygote and exopterygote insects have compound eyes similar to but smaller than those of their adults (Paulus, 1979), the cells of each visual unit or *ommatidium* originating through proliferation, invagination, and differentiation of preommatidial rudiments within an optic placode on either side of the embryonic head (Such, 1978). However, except for larval scorpion flies (Mecoptera), the larvae of endopterygote insects have one to several pairs of specialized larval eyes or *stemmata* (singular, stemma), probably derived from compound eyes in the juveniles of an exopterygote ancestor of these insects. These stemmata degenerate or are withdrawn into the brain during metamorphosis, to be replaced by compound eyes that develop from imaginal eye discs during metamorphosis (10.2.4.2.2) (Paulus, 1979; Gilbert, 1994).

In embryos of *L. viridana* (Coleoptera: Meloidae), a single stemma arises at 28% of embryogenesis from an optic placode below and behind the optic lobe invagination on each side of the head. The optic placode subsequently becomes cup shaped, closes over, and differentiates into a planoconvex corneal lens, a corneagenous layer (secretes the lens), and a retina of numerous pigmented photoreceptor cells, each with a terminal rhabdomere (Fig. 8.27) (Heming, 1982). Between 38% and 41%, proximal ends of the photoreceptor axons grow posteromedially into a horizontal fold in the posterior wall of each optic lobe invagination and along its length to the protocerebral neuropile, which they contact by 44% of embryogenesis (Fig. 8.27C–E). As the brain withdraws posteriorly within the head, these axons are surrounded by glial cells and elongate to form the stemmatal nerve (Fig. 8.27F, G).

Details are practically identical in embryos of *D. melanogaster*, except that the two stemmata (*Bolwig's organs* [BOs]) are much smaller (each contains only 12 photoreceptor cells) and are carried into the anterior end of the embryo during head involution (8.3.7), to nestle against the sides of the cephalopharyngeal skeleton in larvae (Green et al., 1993). Both BOs and optic lobes originate from a few cells in the center of mitotic domain 20 at the time of gastrulation (15%–18% of embryogenesis or stage 8; Fig. 6.10A) (Namba and Minden, 1999). Expression of *Notch* (*N*) is essential between 24% and 46% of embryogenesis (stages 11–13) if the correct number of cells is to be incorporated into each BO and for retaining the epithelial character of the optic lobes after they invaginate (Green et al., 1993). At this time, expression of the gap gene *tll* (Fig. 7.24D) is confined to cells of the optic lobe primordium and specifies them as optic lobe, while *atonal* (*ato*) is first expressed in a subset of photoreceptor cells, the Bolwig organ founders (BOFs), and specifies them as photoreceptors (Daniel et al., 1999). Expression of *tll* is thought to restrict the ability of optic lobe cells to respond to Spitz (Spi) signaling from the BOFs, which specifies and maintains additional secondary BO cells as photoreceptors. Secretion of the Beaten path protein, an anticohesion molecule, by cells of the optic lobes as they invaginate on each side, allows the Fasciclin II–expressing photoreceptor cells to detach from the optic lobes as a cohesive cluster (Holmes and Heilig, 2000). Projection of the larval optic nerves into the brain later in embryogenesis (see Fig. 9.11 in Campos-Ortega and Hartenstein, 1997) was genetically analyzed by Schmucker and colleagues (1997), who isolated 13 mutations disrupting this process. They found such growth to proceed in three phases, during which time the nerves change direction twice at two intermediate targets, P1 and P2.

8.3.2.4.4 THE MOLECULAR EVOLUTION OF EYES AND EYE DEVELOPMENT

Although the evolution of animal eyes might seem outside the topic being addressed, recent discoveries on the genetic control of eye development in a variety of animals have shed new light on such evolution. The simplest visual organs in animals are eye spots on the surface of the head or body or in a shallow pit. These result from aggregation of a few photoreceptor cells, are sensitive to differences in light intensity, and have evolved independently 40–65 times within the Metazoa (Salvini-Plawen and Mayr, 1977). In fact, members of only 5 of 35 phyla have failed to evolve such simple eyes (Raff, 1996).

From this common substrate, members of six phyla (Cnidaria, Annelida, Onychophora, Mollusca, Arthropoda, and Chordata) have evolved eyes capable of forming images (Gould, 1994). In each instance, such eyes probably arose by deepening the eye pit into an optic cup, narrowing of its aperture to produce a pinhole camera, and adding structures to refract or reflect an image on a retina of photoreceptor cells; all these steps are recapitulated during stemmatal ontogeny in *L. viridana* embryos (Fig. 8.27) (Heming, 1982). Most aquatic animals use a cuticular lens as the primary image-forming device, whereas members of many terrestrial taxa have evolved an outer cornea and use the underlying lens primarily for adjusting focus. *Compound eyes* employ multiple lenses in which each ommatidium forms one part of a total mosaic image integrated by the brain. Compound eyes are typical of trilobites (Fortey, 2000), insects, and crustaceans but occur also on the tentacles of some tube-dwelling annelids and on the mantle edge of some clams (Gould, 1994).

Ever since Darwin, such diversity in eye structure has been thought to serve as one of the best examples of convergent evolution within the Metazoa (Fryer, 1996). However, Quiring and colleagues (1994) and Halder and coworkers (1995) discovered a basic homology in the genetic pathway for building eyes throughout the Metazoa. The *Pax* ("paired box") genes owe their name to initial discovery of their key DNA-binding nucleotide sequence within the segmentation gene *paired* (*prd*) of *D. melanogaster* (7.2.1.2). Orthologs of *Pax* have since been found in all vertebrates investigated, including zebrafish, mice, and humans, with nine separate loci identified in mammals (Raff, 1996). Of these, *small eye* (*Pax-6*) is best known and is critical to eye development, as mutations at this locus disrupt eye development (e.g., *Sey*, the *small eye* mutation of *Pax-6* in mice, results in loss of eyes in homozygotes; and mutation of its ortholog, *Aniridia*, in humans interferes with normal eye development).

The *Drosophila eyeless* (*ey*) is a large regulatory gene encoding a transcription factor having both a paired domain and a homeodomain that bind to DNA (7.2.1.4). The gene *ey* is the ortholog of the mouse *Pax-6* gene and of the human *Aniridia* gene and shares extensive sequence identity with them in the paired (94%) and homeodomains (90%), in the position of three intron splice sites, and in their sites of expression within the developing nervous system and eye. Loss-of-function mutations in both the *Drosophila* and mammalian genes likewise lead to reduction or absence of eyes, suggesting the wild-type *ey* locus functions in eye morphogenesis.

Halder and associates (1995) devised a method for expressing *ey* cDNA in various imaginal discs of *D. melanogaster* larvae and were able to induce development of ectopic compound eyes in the wings, legs, and antennae of adults that were identical to wild-type compound eyes (they didn't investigate whether these eyes were functional or whether their photoreceptor axons projected to normal synaptic sites within the optic lobes of the brain). They also expressed the mouse *Sey* gene in *Drosophila* imaginal discs and, again, were able to induce development of normal but ectopic compound eyes in antennae, wings, and legs.

These results suggest that *ey* and its mammalian orthologs are at the top of the regulatory cascade for eye development, since mutations of other genes functioning in eye development in *D. melanogaster* such as *eyes absent*, *sine oculis*, *eye gone*, and *eyelisch* do not affect expression of *ey*. Halder's group (1995) estimated that more than 2500 genes are involved in eye morphogenesis in *Drosophila*, with *ey* at the top, and suggested that *ey* may function as a transcriptional regulator at successive steps of eye morphogenesis.

Orthologs of *Pax-6* have been found since in the flatworm *Dugesia trigrina*, in the nemertean worm *Lineus sanguineus*, and in ascidians and squids, suggesting the gene is fundamental to eye development in all metazoans (Quiring et al., 1994). However, N. Patel (in Barinaga, 1995) observed that such evidence of common ancestry for this gene in no way contradicts the prior view that image-forming eyes evolved independently on numerous evolutionary occasions. It merely means they diverged from a common ancestral eye, probably in the late Precambrian, and that development of that eye was likely governed by the ancestor of the *ey* gene. But as descendants of this animal diversified into the lineages we see today, they continued to use the *ey* gene to control the ever more divergent and complicated processes of eye development.

Shen and Mordon (1997) recently showed that the *Drosophila* regulatory gene *dachshund* (*dac*) and *ey* mutually induce expression of each other and that *dac* is required for the ectopic eye development driven by *ey* misexpression. In *dac* mutants, developing eyes either abort or are reduced in size, just as in *ey* mutants, suggesting that complex interactions of multiple genes are required for eye development. An additional *Drosophila* homeobox gene, *optix*, is capable of inducing ectopic eyes in wing and haltere imaginal discs independently of the *ey* gene (Seimiya and Gehring, 2000).

8.3.3 STOMATOGASTRIC (ENTERIC) NERVOUS SYSTEM

The stomatogastric nervous system (SNS) in insects varies with species (Ganfornina et al., 1996b) but generally consists of the frontal ganglion (FG), recurrent nerve (RN), hypocerebral ganglion (HG), ventricular ganglion, corpora cardiaca (CC), and corpora allata (CA) and their connecting nerves, all resting on the roof of the foregut (Figs. 8.24, 8.37B). Some of its components synthesize (CA) or release (CC) developmental hormones into the hemolymph (chapter 12), whereas others innervate muscles of the foregut and midgut (Fig. 8.32A) and, in the embryo, originate partly by evagination from the roof of the stomodaeum and partly by invagination from the bases of the maxillary appendages (CA).

Copenhaver and Taghert (1989a, 1989b, 1990, 1991) investigated the origin and differentiation of this system in great detail in embryos of the tobacco hornworm *Manduca sexta*, using cell-specific markers and immunocytochemical techniques. In this insect, the SNS (Fig. 8.37A) contains about 600 neurons and consists of two parts: an anterior domain containing the FG, HG, and RN on the roof of the foregut (Fig. 8.27B) and a posterior domain forming a branching, enteric nerve plexus (EP) of some 360 neurons spanning the foregut-midgut boundary, extending over most of the midgut, and innervating the visceral musculature (Fig. 8.37A).

Neurons of the anterior domain arise from three pouches or *neurogenic zones* (Z_1–Z_3) in the roof of the stomodaeum beginning at 24% of embryogenesis shortly after it has begun to invaginate (Fig. 8.38A).

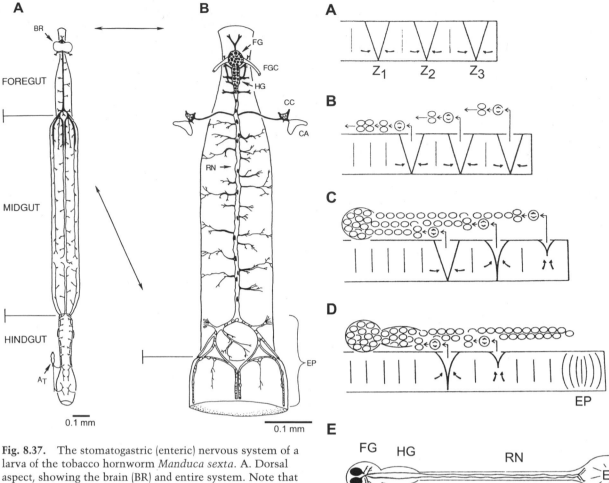

Fig. 8.37. The stomatogastric (enteric) nervous system of a larva of the tobacco hornworm *Manduca sexta*. A. Dorsal aspect, showing the brain (BR) and entire system. Note that the posterior midgut and hindgut are innervated by nerves from the terminal ganglion of the ventral nerve cord (A_T). B. Dorsal aspect of the foregut and anterior midgut at higher magnification, showing the frontal ganglion (FG), frontal connectives (FGC; connect to tritocerebra of brain), hypocerebral ganglion (HG), corpora cardiaca (CC), corpora allata (CA), recurrent nerve (RN), and anterior enteric plexus (EP). (Reproduced with permission from P. F. Copenhaver and P. H. Taghert. Origins of the insect enteric nervous system: differentiation of the enteric ganglia from a neurogenic epithelium. *Development* 115: 1115–1132. © 1991 Company of Biologists, Ltd.)

Fig. 8.38. Summary of the development of the stomatogastric ganglia and recurrent nerve of *M. sexta* from three neurogenic zones (Z_1, Z_2, Z_3) in the roof of the stomodaeum at 24% (A), 27% (B), 30% (C), 36% (D), and 100% (E) of embryogenesis. A. Zones 1–3 appear at 24% of embryogenesis. B–D. During the proliferation phase, from 26%–39%, individual cells are recruited into the zones from surrounding stomodaeal epithelium, pass through them individually onto the roof of the stomodaeum, divide once or twice, and differentiate into two to four neurons. Neurons from all three zones migrate anteriorly, intermingle, and aggregate into a bulbous cluster at the anterior end of the foregut. The zones disappear sequentially from back to front at 30% (Z_3), 36% (Z_2), and 39% (Z_1). E. Postmitotic neurons coalesce to form the frontal (FG) and hypocerebral (HG) ganglia and recurrent nerve (RN) while stomodaeal cells originally including Z_3 are incorporated into the enteric placode (EP) giving rise to the enteric plexus. (Reproduced with permission from P. F. Copenhaver and P. H. Taghert. Origins of the insect enteric nervous system: differentiation of the enteric ganglia from a neurogenic epithelium. *Development* 115: 1115–1132. © 1991 Company of Biologists, Ltd.)

These are not sites of active mitosis but serve instead as sites through which epithelial cells of the stomodaeum are recruited into a program of neuronal differentiation (Fig. 8.38B–E). As they emerge on the dorsal surface of the stomodaeum, the cells divide once or twice and their progeny move forward over its surface and gradually assemble into the ganglia and nerves of the anterior SNS. The system is essentially complete by 65% of embryogenesis, at which time each FG contains about 70 neurons and each HG about 40.

Neurons of the posterior domain are generated from a large enteric placode (EP) invaginating from neurogenic zone 3 at the posterodorsal lip of the stomodaeum beginning at 24% of embryogenesis (Fig. 8.38D, E). The cells of this placode divide prior to leaving it, each cell generating one to four neurons. These neurons subsequently disperse, first slowly and circumferentially around the posterior end of the stomodaeum (Fig. 8.39: top), then more rapidly and posteriorly in eight longitudinal rows over the surface of the midgut to form the plexus (Fig. 8.39: bottom). Expression of their specific, peptidergic transmitters begins at 65% only after this migration is complete. Prior to completion of SNS neurogenesis, an additional class of precursor cells is generated from all three neurogenic zones. These precursors divide and give rise to glial cells, which surround and service the neurons of the enteric system (Copenhaver, 1993).

Neurons of the EP disperse posteriorly over the midgut along eight visceral muscle bands, L1–L4 and R1–R4, formed by coalescence of longitudinal muscle fibers apparently in response to regional cues from underlying midgut epithelium (Fig. 8.39: bottom) (Copenhaver et al., 1996). Prior to migrating, individual EP cells extend filopodia onto both band and interband regions of the midgut, but as the muscle bands coalesce, the filopodia of each EP cell become confined to a specific band onto which each cell subsequently migrates. EP cells prevented from contacting the bands do not migrate.

Development of the SNS is similar in embryos of the grasshopper *S. americana* (Ganfornina et al., 1996b) and of *D. melanogaster* but is complicated in the latter embryo by anterior involution of the head between 46% and 68% of embryogenesis (8.3.7) (Hartenstein et al., 1994; Forjanic et al., 1997; Campos-Ortega and Hartenstein, 1997). In *Drosophila* embryos, precursor cells of the SNS (stomatogastric nervous system precursor cells, SNSPs) originate, between 15% and 18.5% of embryogenesis, from blastodermal mitotic domain 23 within that (domain 15) from which the stomodaeum developed (Fig. 6.10), and undergo four rounds of mitosis, respectively, at 15%–19%, 21%–24%, 24%–32%, and 32%–41% of embryogenesis. Most SNSPs emerge from three stomodaeal neurogenic zones homologous to those in *M. sexta* and other insect embryos. Later, the cells bud off as vesicles from the roof of the stomodaeum, dissociate, and migrate to various sites on the foregut, where they differentiate as neurons. Cells of zones 1 and 2 contribute to formation of the paired FG; those of zone 2, to the HG; cells of zone 1, to the paraesophageal ganglion; and those of zone 3, to the ventricular ganglion. Three additional clusters of SNSPs delaminate anterior to zone 1 (dSNPs), and a single additional SNSP emerges from the apex of each pouch (tSNP); all these cells contribute to formation of the FG. Beginning at 46%, these ganglia interconnect by formation of the recurrent and esophageal nerves, and the FGs connect to the tritocerebrum of the brain by projecting paired frontal connectives, both nerves being pioneered by a small subpopulation of earlier differentiating stomatogastric neurons. On hatch, the FGs of *D. melanogaster* together

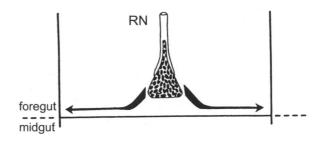

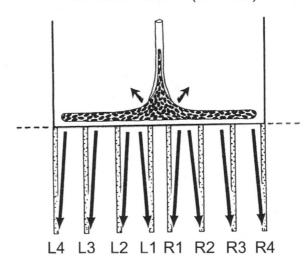

Fig. 8.39. The patterns of cell migration seen during formation of the enteric plexus in embryos of *M. sexta*. Arrows indicate the prospective direction of migration. RN, recurrent nerve; L_1–L_4 and R_1–R_4, longitudinal midgut bands over the left and right surfaces of the midgut. (Reproduced with permission from P. F. Copenhaver and P. H. Taghert. Development of the enteric nervous system in the moth. II. Stereotyped cell migration precedes the differentiation of embryonic neurons. *Dev. Biol.* 131: 85–101. © 1989 Academic Press, Inc.)

contain 25–39 neurons; the PG and HG, 10–12 each; and the ventricular ganglion, about 10. All steps in SNS embryogenesis in *D. melanogaster* are under control of the same proneural and neurogenic genes involved in the CNS (Fig. 8.30) and sense organ (8.3.2.4.1) formation (Hartenstein et al., 1996).

The CA (source of juvenile hormone) and prothoracic glands (PGs) (source of ecdysteroids) have long been known to invaginate from the anterior bases of the first and from the posterior bases of the second maxillary appendages, respectively (Sander et al., 1985), but the site of origin of the CC, traditionally thought to be

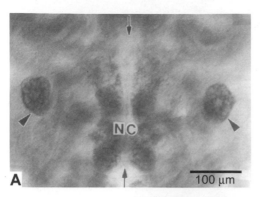

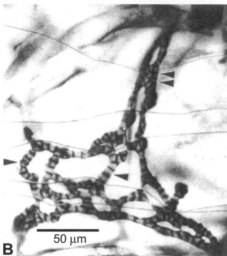

Fig. 8.40. Origin and differentiation of the prothoracic glands in embryos of *M. sexta*. A. Whole mount of labial segment in dorsal aspect at 32% of embryogenesis immunolabeled with Mab 3B11 at a time when invagination of presumptive gland cells is complete (arrowheads). B. Same, at hatch (100%). Cells of the gland have rearranged to increase their secretory surface area. Some cells (double arrowheads) are arranged in two-dimensional sheets; others (singe arrowheads), in one-dimensional filaments. Neuroglian is expressed only at sites of cell-cell contact. NC, ventral nerve cord. (Reproduced with permission from C.-L. Chen, D. J. Lampe, H. M. Robertson, and J. B. Nardi. Neuroglian is expressed in cells destined to form the prothoracic glands of *Manduca* embryos as they segregate from surrounding cells and rearrange during morphogenesis. *Dev. Biol.* 181: 1–13. © 1997 Academic Press, Inc.)

neurogenic zone 2, has yet to be determined experimentally. In *D. melanogaster* embryos, the CC and CA of the ring gland appear to originate in "deep layers of the dorsal part of the gnathal segments" (Hartenstein et al., 1994; Campos-Ortega and Hartenstein, 1997: Fig. 10.4), but their exact embryonic origin in this insect remains problematic because of the reversal of cephalic spatial relationships resulting from head involution (8.3.7).

Embryogenesis of the PGs in *M. sexta* embryos has been studied by Chen and coworkers (1997), who found

the adhesion/recognition molecule Neuroglian (Fig. 8.21) to begin to be expressed on the surfaces of two circular PG placodes, one on either side of the labial segment lateral to the salivary gland placode and medial to the labial appendages, at 25% of embryogenesis (Fig. 8.40A). Between 27% and 35% these cells invaginate without dividing and by 80% of embryogenesis have differentiated into long filaments of cells associated with the spiracle on each side of the prothorax behind the head (Figs. 8.40B, 12.2).

According to Campos-Ortega and Hartenstein (1997: Fig. 10.4), precursors of the PGs in *D. melanogaster* embryos originate from dorsal ectoderm of the prothoracic segment, migrate dorsally, and form bilaterally symmetrical plates in the developing ring gland on either side of the aorta. After hatch, the PG cells enter the endo cell cycle (S and G) (Reed and Orr-Weaver, 1997) and become polyploid. As mentioned earlier in relation to nurse cell endopolyploidy (2.2.2.4), this is caused by expression of the *morula* (*mr*) gene, and in larvae with strong lethal alleles of *mr*, the PG cells regress into mitosis and form spindles.

8.3.4 TRACHEAL SYSTEM

In insects, the respiratory system consists of a complex of ramifying tracheal tubes opening to the exterior through a maximum of 10 pairs of spiracles (Figs. 8.1C, 8.41: 17), the finest branches, the *tracheoles*, extending to every attached cell in every tissue of the body. This system originates in the embryo from 10 pairs of ectodermal, tracheal placodes: one on each side of each of the mesothoracic to eighth abdominal segments (Fig. 8.41: 11) (in *Drosophila* embryos there is an additional pair of placodes for the posterior spiracles) (Manning and Krasnow, 1993; Hu and Castelli-Gair, 1999). Prospective tracheal cells invaginate from these placodes and branch; some of these branches eventually fuse to those of the same side of adjacent segments to form longitudinal trunks or of the opposite side of each segment to form transverse commissures at fusion points called *tracheal nodes* (Fig. 8.41: 12, 17). In embryos of this insect, each placode contains about 80 cells and generates six buds, which collectively invaginate and differentiate through cell migration and changes in shape, but surprisingly without mitosis, into about 500 branches, eventually to form a larval tracheal system of some 10,000 branches (Fig. 8.41: 17) (Metzger and Krasnow, 1999).

Three levels of branching occur, each involving different mechanisms of tube formation and the expression of at least 50 genes (Samakovilis et al., 1996a, 1996b; Metzger and Krasnow, 1999):

- primary branches: multicellular tubes arising by cell migration and intercalation;
- secondary branches: unicellular tubes formed by individual tracheal cells at the tips of the branches rolling up; and

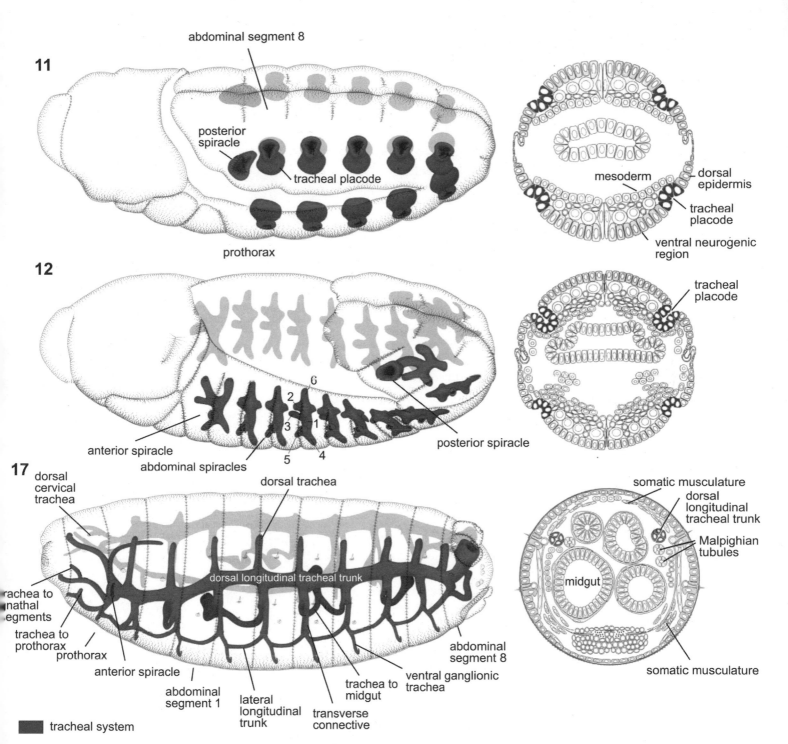

Fig. 8.41. Development of the tracheal system in an embryo of *D. melanogaster* in dorsolateral aspect (left) and in transverse section (right) at stages 11 (24%–32% of embryogenesis), 12 (32%–41%), and 17 (68%–hatch). (Adapted with permission from V. Hartenstein. *Atlas of* Drosophila *development*, pp. 18, 20. © 1993 Cold Spring Harbor Laboratory Press.)

- terminal branches: subcellular tracheoles forming within long cellular extensions of individual tracheoblasts at the tip of each branch.

This sequential branching is known to occur during the embryogenesis of all investigated hexapods, even in those lacking or having a reduced number of spiracles. The system is absent or greatly reduced in proturans and collembolans and lacks longitudinal tracheal trunks in diplurans. In embryos of fully aquatic insects, the spiracles close after tracheal invagination and, after hatch, periodically open at each ecdysis, to allow shedding of the cuticular lining of the previous instar, good evidence that a reduced number of spiracles is secondary and that insects were primitively terrestrial (Pritchard et al., 1993).

Based on her interpretation of the structure of Paleozoic insect fossils, Kukalová-Peck (1987) believed the spiracles originally to have been ecdysial only, respiration at first occurring by means of thoracic and abdominal gills. She thus believed insects to be primitively aquatic. Recent comprehensive reviews marshalling evidence of all sorts (Labandeira and Beall, 1990; Pritchard et al., 1993) provide overwhelming support for a terrestrial origin for hexapods, though they undoubtedly descended from a marine, probably crustacean-like ancestor, the freshwater realm having been entered secondarily and independently on at least 10 evolutionary occasions.

Some of the genes known to function in tracheogenesis in *D. melanogaster* embryos (Metzger and Krasnow, 1999) are as follows:

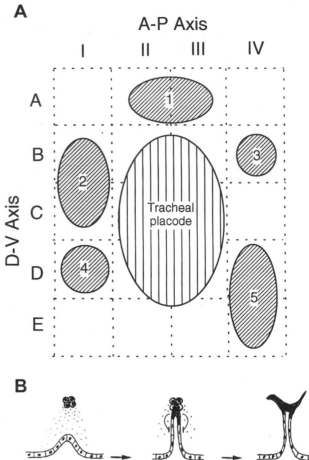

A

A-P Axis

I   II   III   IV

Tracheal placode

D-V Axis

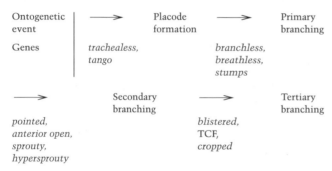

Ontogenetic event	→	Placode formation	→	Primary branching
Genes	*trachealess, tango*		*branchless, breathless, stumps*	

→		Secondary branching	→	Tertiary branching
pointed, anterior open, sprouty, hypersprouty			*blistered, TCF, cropped*	

Epidermal cells expressing *trachealess* (*trh*) are specified as placodal (Fig. 8.42A). Trachealess protein binds with Tango protein to form a heterodimer presumably regulating expression of target genes encoding the structural proteins responsible for placode formation and branching. Just before this begins, the *branchless* (*bnl*) gene is expressed in five clusters of epidermal cells surrounding each placode (Fig. 8.42A). The secreted Branchless protein binds to a receptor protein encoded by *breathless* (*btl*) in adjacent placodal cells and, by way of a signaling cascade, guides the outgrowth of primary branch cells (Fig. 8.42B: left). However, as each branch approaches a secreting cluster, the *bnl* gene is turned off and the branch stops growing (Fig. 8.42B: center) unless an adjacent patch of cells farther away begins to express this gene, in which case the branch grows toward this new site. Misexpresson of *bnl* in abnormal sites within the embryo causes ectopic branch outgrowth.

Secondary branching is also caused by expression of Branchless and Breathless but by a different mechanism (Fig. 8.42B: center). As primary branches extend toward Branchless-secreting sites, cells at their tips are exposed to higher levels of this signal, which induces expression

Fig. 8.42. Gene expression during tracheogenesis in an embryo of *D. melanogaster*. A. Five domains of *branchless* mRNA expression (diagonal hatching) surrounding a tracheal placode whose cells are expressing *trachealess* (vertical hatching) at about 6h of embryogenesis. Primary branches bud at these five positions and at a sixth position (not shown) below the plane of the diagram. Expression domains are shown within the gridlike pattern of positional values established by expression of the anteroposterior (A-P axis) and dorsoventral (D-V axis) patterning hierarchies. B. Model of Branchless patterning of tracheal branching. Secreted Branchless protein (dots) guides the migration of tracheal cells as they form primary branches. High levels of Branchless induce expression of secondary branch genes such as *pointed* in cells at the tips of the primary branches (gray), which reprogram these cells to form secondary branches. Another induced gene, *sprouty*, encodes an inhibitor that limits the range of Branchless signaling (inhibitory arrows) and restricts secondary branch formation to cells closest to the Branchless signaling center. (Adapted with permission from R. J. Metzger and M. A. Krasnow. Genetic control of branching morphogenesis. *Science* 284: 1635–1639. Copyright 1999 American Association for the Advancement of Science.)

of secondary branch genes such as *pointed* that stimulate the outgrowth of secondary branches and of Sprouty protein, which blocks Branchless signaling. Terminal branching begins several hours after secondary branching, continues throughout larval life, and differs from primary and secondary branching in that it is regulated by tissue oxygen requirements (Mill, 1998; Jarecki et al., 1999). Additional genes that are expressed at this time such as *blistered* change the expression pattern of *bnl* so that it is oxygen-sensitive.

The position and shape of each tracheal placode appear to be established by differing combinations of anteroposterior and dorsoventral patterning gene products acting on *trh* expression (Fig. 8.42A), as does *bnl* expression by epidermal cells about each placode (Metzger and Krasnow, 1999); different domains of *bnl* expression depend on different genes in the anteroposterior and dorsoventral patterning hierarchy.

A similar though more complex pattern of gene expression is known to specify lung branching in mouse and human embryos, suggesting the respiratory systems of insects and mammals to be *homologous*. A similar though more simple pathway perhaps served to specify an ancestral branched structure that was then co-opted to pattern a variety of branched organs in descendants (Metzger and Krasnow, 1999).

In 1990, Nardi used a monoclonal antibody to label a membrane-bound protein characteristic of tracheal node cells in embryos and larvae of *M. sexta*. This antibody first labeled a single presumptive nodal cell at the tip of each branch immediately prior to linking of branches from successive and opposite spiracles, and then an increasing number of nodal cells as embryogenesis and postembryogenesis proceeded. He suggested the protein facilitates coupling of tracheal branches during tracheogenesis and influences the deposition of the special cuticle that breaks at the nodes prior to each larval ecdysis (10.2.4.3.4).

Similar nodal cells at developing fusion points in *D. melanogaster* embryos express a sequence of specific markers as each branch grows out and contacts a similar cell from another branch. These first adhere to each other, then form an intercellular junction, and finally become doughnut-shaped, the lumina of the two tubes becoming confluent (Samakovilis et al., 1996a).

Expression of the gene *hindsight* (*hnt*) is required for correct tracheal morphogenesis, to maintain epithelial integrity, and for the assembly of *taenidia* (the spiral cuticular thickening of its intima) in differentiating tracheal tubes (Wilk et al., 2000). In *hnt* mutant embryos, the tracheal placodes form, invaginate, and undergo primary branching, but at midembryogenesis, the tracheal epithelium either collapses or expands to form sacs of tissue.

8.3.5 SALIVARY GLANDS

Mandibular, maxillary, hypopharyngeal, and labial glands are more or less diverse and widely distributed in the heads of insects (Snodgrass, 1935: 153–154), particularly in adult honeybees (*Apis mellifera*) (Snodgrass, 1956: 51–55) and ants (Hölldobler and Wilson, 1990: 229–244), in which their products influence the behavior of other colony members. In trichopteran, lepidopteran, and black fly larvae, the labial glands produce silk and the mandibular glands are either salivary or poorly developed (Kobayashi and Ando, 1988). When present, each gland invaginates from an ectodermal placode behind the base of the mandibular appendage shortly after it appears.

In most insect embryos, the salivary glands have a paired origin on the venter of the labial segment, the two openings later fusing to form a common salivary duct (Anderson, 1972a, 1972b). In *D. melanogaster* embryos, they originate between 24% and 32% of embryogenesis from a single, median salivary placode in the venter of this segment (Panzer et al., 1992; Campos-Ortega and Hartenstein, 1997: Fig. 7.1). The anteroposterior extent of this placode is established by positive expression of the homeotic gene *Sex combs reduced* (*Scr*) (Fig. 7.36), and in *Scr⁻* embryos no placode develops. If *Scr* is expressed ectopically, then salivary glands develop in segments in which they do not normally occur. Dorsal and ventral extent of the placode is negatively specified by products of the dorsoventral gene *decapentaplegic* (*dpp*) (dorsal) and the ventral genes *dorsal* (*dl*) and *spitz* (*spi*) (7.2.1.1.1).

After they are determined, the cells of the placode divide into dorsally situated, pregland cells, specified by expression of the Forkhead (Fkh) transcription factor, and into more ventrally situated preduct cells, by expression of the *Drosophila* epidermal growth factor receptor (EGFR) signaling pathway (Kuo et al., 1996). EGFR signaling blocks *forkhead* expression in preduct cells, whereas Fkh protein blocks expression of duct-specific genes in pregland cells. Expression of the gap gene *huckebein* (*hkb*) (Fig. 7.24D) dictates the initial site of salivary gland invagination, the order in which invagination progresses, and the final shape of the glands (Myat and Andrew, 2000).

The Trachealess (Trh) transcription factor mentioned earlier acts downstream or independently of Fkh and EGFR: in *trh* mutants, the ventral cells remain undetermined, so that the dorsal ones develop as glands. The *Pax* gene *eye gone* (*eyg*) is required for development of each individual gland duct and for specification of the imaginal ring cells that later generate the imaginal salivary gland during metamorphosis (10.2.4.3.5) (Jones et al., 1998).

8.3.6 EPIDERMIS AND IMAGINAL DISCS

The epidermis and, in some holometabolous insects, the imaginal discs and histoblasts develop from the ectodermal cells that do not contribute to formation of the nervous, tracheal, salivary, alimentary, and reproductive systems (Martinez Arias, 1993; Cohen, 1993). In *D. melanogaster* embryos, imaginal disc cells are specified shortly after blastoderm formation (Fig. 8.43: 5) and can be detected soon after germ band shortening at 40% of embryogenesis as clusters of cells of characteristic number and with characteristic shapes, sizes, and

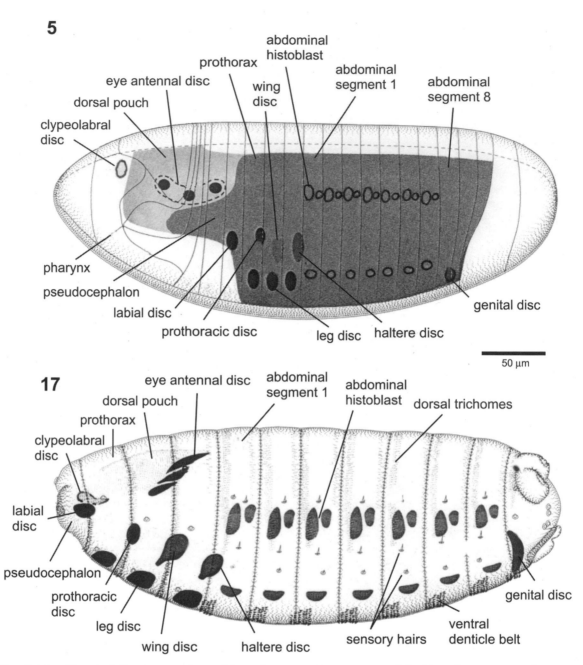

Fig. 8.43. Origin of imaginal discs in an embryo of *D. melanogaster*. Top: Fate map of the blastoderm at stage 5 (9%–13% of embryogenesis), showing those cells giving rise to larval epidermis (dark gray), to the dorsal pouch (light gray), and to the various imaginal discs and histoblasts. Bottom: Fate map in a ready-to-hatch prolarva. Notice how most head segments in front of the prothorax in the fate map invaginate into the front of the body during head involution. (Reprinted with permission from V. Hartenstein. *Atlas of* Drosophila *development*, p. 24. © 1993 Cold Spring Harbor Laboratory Press.)

behaviors, their development seeming to be tightly linked to that of the larval epidermis (Fig. 8.43: 17) (Bate and Martinez Arias, 1991; Martinez Arias, 1993; Campos-Ortega and Hartenstein, 1997). In older larvae of higher flies, cells of the imaginal discs and histoblasts can be readily distinguished from other epidermal cells in histological sections because they are small and diploid, while larval cells are large and polyploid (9.7.1, 10.2.4.3.1). Transition of presumptive larval tissues to polyteny begins late in embryogenesis and results from a change in the cell cycle of these cells from the standard cycle with G1, S, G2, and M phases to an endo cell cycle having only G1 and S phases (Smith and Orr-Weaver, 1991).

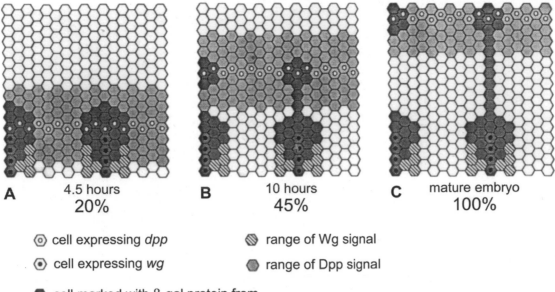

◎ cell expressing *dpp*　　　　◈ range of Wg signal

◉ cell expressing *wg*　　　　⬡ range of Dpp signal

⬢ cell marked with β-gal protein from
　 expression of the early *Dll* enhancer

Fig. 8.44. A model for the allocation and separation of the thoracic imaginal primordia in an embryo of *D. melanogaster* in lateral aspect. A. At 4.5 h (20% of embryogenesis): Lines of cells expressing *wingless* (*wg*) and *decapentaplegic* (*dpp*) form a ladder-like pattern on the ventral half of the embryo. Both Wg and Dpp proteins are secreted; the former can be detected over a range of two or three cells and the latter over a range of three cells. Groups of cells situated near the intersections between the sources of Wg and Dpp signals become imaginal progenitor cells. Expression of the early *Distal-less* (*Dll*) enhancer is thought to be an independent response to these signals. B. At 10h (45%): As the germ band shortens, the lines of cells expressing *wg* and *dpp* separate. Cells that expressed *Dll* enhancer are displaced dorsally in register with the line of cells expressing *dpp*. C. Ready-to-hatch prolarva (100%): The dorsalmost cells of this cluster ultimately contribute to formation of the dorsal discs (wings in mesothorax; halteres in metathorax). (Reproduced with permission from B. Cohen, A. Simcox, and S. M. Cohen. Allocation of the thoracic imaginal primordia in the *Drosophila* embryo. *Development* 117: 597–608. © 1993 Company of Biologists, Ltd.)

In *escargot* (*esg*) mutant larvae of *D. melanogaster*, diploid cells of the imaginal discs and abdominal histoblasts also switch to the endo cell cycle and likewise become polyploid (Hayashi et al., 1993). When such larvae metamorphose to adults, they lack patches of imaginal abdominal cuticle and have various defects in their appendages, suggesting that one function of this gene is to maintain diploidy in imaginal disc cells (*esg* RNA is expressed in imaginal disc and abdominal histoblast primordia in older embryos and in larvae).

Genetic and molecular events that lead to determination and segregation of wing, haltere, and leg disc cells in embryos of this insect are summarized in Fig. 8.44. The first cells of these discs are specified in response to signals from the segment-polarity gene *wingless* (*wg*) and the dorsoventral gene *decapentaplegic* (*dpp*), both of which encode secreted proteins (Fig. 8.44A) (Cohen et al., 1993). Vertical rows of cells expressing Wingless (Wg) protein in each thoracic parasegment intersect with a single longitudinal row of cells expressing Decapentaplegic (Dpp) protein on the parasegmental boundary on each side of the embryo, to form a ladder-like pattern in the ectoderm of the extended germ band embryo, when viewed from below. The two thoracic imaginal discs in each

of segments T1–T3 originate as groups of cells near these intersection points at 20% of embryogenesis (Fig. 8.44A). Wg but not Dpp protein initially specifies limb primordia, the first cells to be specified having a distal identity (Goto and Hayashi, 1997). Loss of Dpp function in older embryos causes deletion of proximal leg structures in contrast to deletion of distal structures in larval imaginal discs (11.3.2.1). That these limb primordia are restricted to a specific lateral position in relation to the ventral midline is due to negative control by the early function of Dpp in determining the dorsoventral pattern (Fig. 7.20).

By following beta-galactosidase (β-GAL) protein expression induced by an early enhancer of the *Distal-less* (*Dll*) gene (see Fig. 7.34 for the method), Cohen and coworkers (1993) showed these cell populations, in the mesothoracic and metathoracic segments, to contain progenitor cells for both dorsal (wing and haltere) and ventral (leg) discs that are induced by the presence of Wg, but not Dpp, to synthesize Distal-less (Dll) protein. Some cells of each of these populations later shift dorsally, along with dorsal migration in cellular expression of Dpp protein, to form those of the wing and haltere discs (Fig. 8.44B, C). Thus, both wing and haltere cells derive from the leg discs, a finding providing support for Kukalová-Peck's (1991) hypothesis of a leg origin for wings based on her study of Paleozoic fossils. The same is known to occur in other extant insects, including exopterygotes where nymphal wing anlagen move dorsally away from the leg with increasing age (references in Kukalová-Peck, 1978).

Specification of differences between wing and haltere discs in *Drosophila* embryos and larvae is associated with differences in expression of the homeotic gene *Ultrabithorax* (*Ubx*) (Warren et al., 1994). *Ubx* is expressed in posterior T3 to the anterior compartment of A8 (Fig. 7.36: Hexapoda) and thus in cells of the haltere disc, in which it represses wing development. However, it is also expressed in cells of the metathoracic wing disc in fifth instar caterpillars of the butterfly *Precis coenia* and in beetle larvae, where it does not repress wing development. Warren and coauthors (1994) suggested that these differences probably result from divergence in downstream, wing-patterning genes in flies, butterflies, and beetles that are, nevertheless, all under control of *Ubx*.

8.3.7 HEAD INVOLUTION IN EMBRYOS OF HIGHER FLIES

The larvae of higher flies, including those of *D. melanogaster*, are *apodous* (legless) and apparently *acephalous* (headless) (Fig. 9.9D), are commonly referred to as maggots, and are undoubtedly the most specialized juveniles in the Insecta, except for the protopod larvae of certain parasitoid Hymenoptera (Fig. 9.8A) and apodous larvae of sting-bearing Hymenoptera (9.6.1). The heads of these larvae are greatly reduced and are withdrawn almost completely into the thorax to form a complex, sclerotized cephalopharyngeal skeleton, with the mouthparts reduced to mouth hooks (Fig. 8.45).

In spite of these differences, these structures differentiate during embryogenesis from cephalic segments similar to those of other extended germ band stage embryos (21%–24% of embryogenesis or stage 10 in *Drosophila* embryos) but with the mandibular, maxillary, and labial appendages reduced to small lobes (Figs. 6.23: 11; 8.46A). From 32% to 68% of embryogenesis, almost all cells of the head segments roll posteriorly into the front of the body in a complicated morphogenetic movement known as *head involution* (Figs. 6.23: 11–15; 8.46) (Schoeller, 1964; Younossi-Hartenstein et al., 1993; Hartenstein, 1993; Jürgens and Hartenstein, 1993; Campos-Ortega and Hartenstein, 1997).

Involution comprises three different but interrelated processes (Fig. 8.46) that were reconstructed by following the movement of labeled anlagen of specific, cephalic (labral, antennomaxillary, hypopharyngeal, and labial) sense organs and of the salivary and tentorial invaginations (all formed before involution) in relation to expression of the segment-polarity gene *engrailed* (Fig. 8.46; see chapter 9 in Campos-Ortega and Hartenstein, 1997):

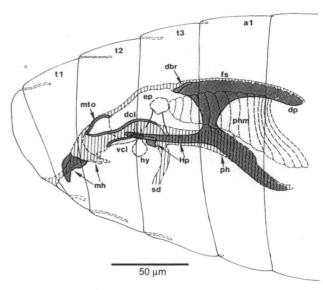

Fig. 8.45. Lateral aspect of the invaginated head of a late prolarva of *D. melanogaster*. The salivary duct (sd) and frontal sac (fs) are hatched, and sclerotized parts of the cephalopharyngeal skeleton, mouth hooks (mh), and median tooth (mto) are shaded. t1–a1, thoracic and abdominal segments; dbr, dorsal bridge; dcl, dorsal clasps; dp, dorsal process; ep, epiphysis; Hp, H-piece; hy, hypophysis; ph, pharynx; phm, pharyngeal musculature; vcl, ventral clasps. (Reprinted with permission from J. A. Campos-Ortega and V. Hartenstein. *The embryonic development of* Drosophila melanogaster, p. 195, Fig. 6.4. 1st ed. © 1985 Springer-Verlag.)

- formation of the atriopharyngeal cavity from 32% to 68% of embryogenesis (ventral regions of the clypeolabrum and of the intercalary, and gnathal segments invaginate through the stomodaeal opening to form the pharynx and atrium);
- formation of the dorsal pouch from 46% to 68% (dorsal part of the head, including the acron, the dorsal ridge, and dorsal parts of the maxillary and labial segments, rolls in and folds over to create a vaulted pouch covering the pharynx); and
- retraction of the clypeolabrum and formation of an anterior lobe from 59% to 68% (the clypeolabrum at the anterior end of the head withdraws posteriorly into the thorax, creating two lateral folds whose medial cells later secrete the cuticle of the cephalopharyngeal skeleton [Fig. 8.45]).

Simultaneously with the last process, the fused lateral regions of the antennal, mandibular, and maxillary segments shift to the extreme anterior tip of the embryo, forming an anterior lobe bearing the antennomaxillary sense organs (Figs. 6.23: 15; 8.46E).

The acron and dorsal regions of the head segments give rise to the protocerebrum, dorsal pouch, and Bolwig's organs (larval stemmata); the acron, to the roof of the atriopharyngeal cavity; and the intercalary and gnathal segments, to its floor. The anteroposterior order

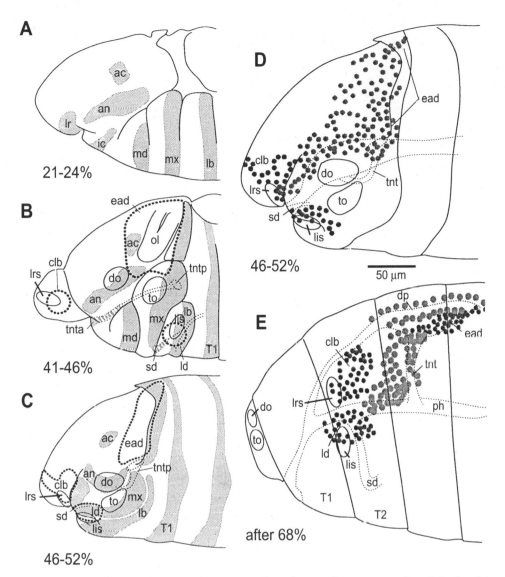

Fig. 8.46. Development of the larval head in *D. melanogaster* in lateral aspect from 21% to after 68% of embryogenesis, showing the relationship between head segments and primordia of the imaginal head. In A–C, the expression pattern of the segment-polarity gene *engrailed* is indicated in the posterior compartment of each segment (shaded). B and C show the positions of various larval sensory complexes (lis, labial complex [hypophysis]; to, terminal [maxillary] organ; do, dorsal [antennal] organ; lrs, labral complex [epiphysis]) and of certain presumptive cephalic imaginal discs (ld, labial; ead, eye-antennal; clb, clypeal labral). D and E show the expression pattern of the P *lac* Z insertion 1(2)4B7. Each cell expressing the insertion is shown as a gray circle, with most corresponding to the three imaginal discs. The more lightly shaded cells in E form part of the dorsal pouch (dp) and tentorium (tnt). Abbreviations not previously mentioned are ac, acron; an, antennal segment; ic, intercalary segment; lb, labial segment; lr, labrum; md, mandibular segment; mx, maxillary segment; ol, optic lobe invagination; ph, pharynx; sd, salivary duct; tnta, anterior opening of tentorium in intercalary segment; tntp, posterior opening of tentorium in maxillary segment; T1, T2, first and second thoracic segments. (Reprinted with permission from A. Younossi-Hartenstein, U. Tepass, and V. Hartenstein. Embryonic origin of the imaginal discs of the head of *Drosophila melanogaster*. *W. Roux's Arch. Dev. Biol.* 203: 60–73. © 1993 Springer-Verlag.)

of all these segmental components is reversed from that before involution. All segments also contribute narrow longitudinal strips of cells to the dorsal pouch and to the lateral walls of the atriopharyngeal cavity. The mouth hooks are maxillary.

The cephalic complex also includes the three pairs of imaginal discs that later generate the imaginal head during metamorphosis: the clypeolabral, eye-antennal, and labial discs (Figs. 8.43: 17; 8.46B–E) (10.2.1, 10.2.4.3.1); the eye-antennal discs contain cells from the acron, antennal, intercalary, and gnathal segments. None of these discs are recognizable morphologically until after hatch but can be demonstrated by use of the enhancer trap technique from 46% of embryogenesis on (Younossi-

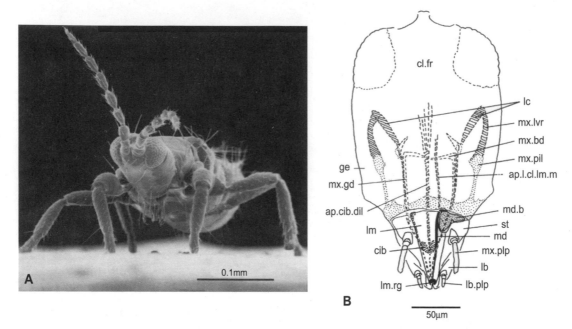

Fig. 8.47. Mouthparts of the thrips *Haplothrips verbasci* (Thysanoptera: Phlaeothripidae). A. Scanning electron micrograph of an adult male showing the mouthcone attached to the ventral side of the head between the forelegs. B. The head in ventral aspect, showing the principal structures used in feeding. ap.cib.dil., apodeme of cibarial dilator muscle; ap.l.cl.lm.m., apodeme of left lateral clypeolabral muscle; cib, cibarium; cl.fr, frontoclypeus; ge, gena; lb, labium; lb.plp, labial palpus; lc, lacinia (maxillary stylet); lm, labrum; lm.rg, labral ring; md, left mandibular stylet; md.b, left mandibular base; mx.bd, maxillary bridge; mx.gd, maxillary guide; mx.lvr, maxillary lever; mx.pil, maxillary pillar; mx.plp, maxillary palpus; st, maxillary stipes. (Reprinted with permission from B. S. Heming. Structure, function, ontogeny and evolution of feeding in thrips (Thysanoptera). In C. W. Schaefer and R. A. B. Leschen (eds.), *Functional morphology of insect feeding*, pp. 3–41. Thomas Say Publications in Entomology. Entomological Society of America. Lanham. © 1993 Entomological Society of America.)

Hartenstein et al., 1993; Jürgens and Hartenstein, 1993; Campos-Ortega and Hartenstein, 1997).

All morphogenetic movements involved in head involution are marked by extensive areas of cell death and by expression of the cell death effector gene *reaper* (*rpr*) (8.6.1) (Nassif et al., 1998b). In *rpr* deficient embryos, cell death does not occur and head involution is disrupted, except for earlier events such as invagination of the stomodaeum and formation of the dorsal ridge.

8.3.8 EMBRYOGENESIS OF PIERCING AND SUCKING MOUTHPARTS

Immature thrips (Thysanoptera), bugs (Hemiptera), and some lice (Phthiraptera) have piercing and sucking mouthparts that differentiate during later embryogenesis from gnathal appendages, similar to those of the polypod-stage embryos of insects having biting/chewing mouthparts (Fig. 6.22B). The development of these is briefly summarized below.

8.3.8.1 THYSANOPTERA

Larval and adult thrips feed on the hyphae, spores, or digestive products of fungi; on algal cells; on the leaves,

flower parts, stems, fruits, seeds, nectar, or pollen of higher plants; or on other small arthropods by use of asymmetrical, punch-and-suck mouthparts arranged into a cone below the head (Fig. 8.47A). This consists of the labrum, maxillary stipites, and labium; bears paired maxillary and labial palpi; and contains the left mandibular stylet, two maxillary stylets (laciniae), and cibarial and salivary pumps (Fig. 8.47B) (Heming, 1978, 1993b). The two laciniae are co-adapted to each other along their lengths and, on their protraction through a sclerotized labral ring, interlock to enclose a single canal functioning both to deliver saliva into the host and to take up food. Co-adaptation (co-aptation) is the mechanical adjusting that occurs between two or more independent structures during ontogeny such that they can later function together.

In embryos of *Haplothrips verbasci* (Phlaeothripidae), these mouthparts originate during intertrepsis from paired appendages of the gnathal segments similar to those of other polypod stage embryos (Fig. 8.48A) (Heming, 1980). With subsequent development, the labral appendage flexes posteroventrally over the stomodaeum, the right mandibular appendage degenerates, the maxillary appendages divide into inner (lacinial) and outer (stipital) lobes, and the hypopharynx arises from

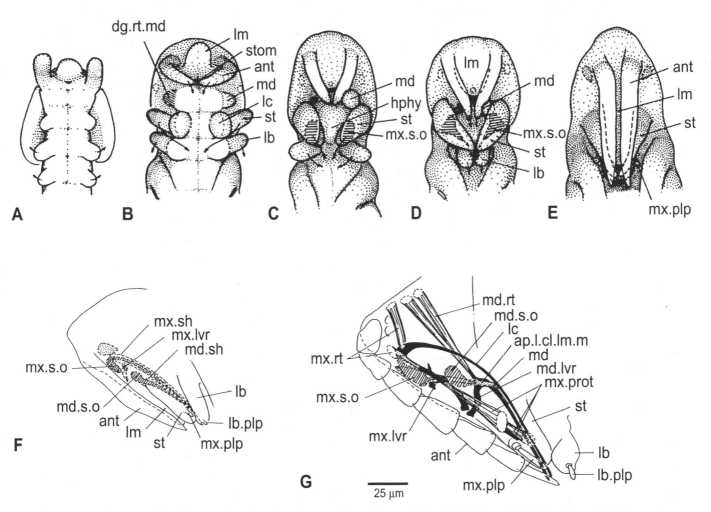

Fig. 8.48. Embryogenesis of the punch-and-suck mouthparts in embryos of *H. verbasci*. Ventral (A–E) and lateral (F, G) aspects of the head showing progressive development of the mouthparts. Note ventral outpocketing of the maxillary sheath (mx.sh) in F, in which the cuticle of the maxillary lever (mx.lvr, G.) will be deposited. The stylet-secreting organs (md.s.o, mx.s.o) persist throughout postembryogenesis and deposit the cuticle of the second instar and adult stylets (lc, md) and the membranous cuticle of their nonfunctional rudiments in the three quiescent stages. A. At 22% of embryogenesis. B. At 27%. C. At 38%. D. At 43%. E. At 53%. F. At 78%. G. At 98%. ant, antenna; ap.l.cl.lm.m, apodeme of left lateral clypeolabral muscle; dg.rt.md, degenerating right mandibular appendage; hphy, hypopharynx; lc, lacinial lobe; lb, labial appendage, labium; lb.plp, labial palpus; lm, labrum; md, mandibular appendage; md.lvr, mandibular lever; md.rt, mandibular retractor muscle; md.sh, mandibular sheath; mx.plp, maxillary palpus; mx.prot, maxillary protractor muscles; mx.rt, maxillary retractor muscles; st, stipital lobe of maxilla; stom, stomodaeum. (Reprinted with permission from B. S. Heming. Structure, function, ontogeny and evolution of feeding in thrips (Thysanoptera). In C. W. Schaefer and R. A. B. Leschen (eds.), *Functional morphology of insect feeding*, pp. 3–41. Thomas Say Publications in Entomology. Entomological Society of America. Lanham. © 1993, Entomological Society of America.)

the venters of the mandibular and maxillary segments (Fig. 8.48B, C). Just before katatrepsis, all cephalic segments consolidate anteriorly, their appendages flex ventrally, and the labial appendages fuse medially to form the labium and anlagen of the salivarium and salivary glands (Fig. 8.48C–E). During katatrepsis, the bases of the left mandible and of the lacinial lobes of the maxillae begin to invaginate into the head to form their respective stylet-secreting organs (Fig. 8.48D–F), about which cuticle is later deposited to form the stylets (Fig. 8.48G). Cuticle of the mandibular lever is deposited by labral cells at the apex of the mandibular sheath during and after hatch, and that of each maxillary lever is deposited simultaneously into the lumen of a ventrally directed pouch originating from stipital cells at the apex of each maxillary sheath (Fig. 8.48F, G). Maxillary and labial palpi differentiate, respectively, from cells in the outer wall of each stipital lobe and at the apex of the labium shortly after katatrepsis, while muscles of the mouthparts arise simultaneously from cephalic mesoderm and differentiate before cuticle of the mandibular and maxillary levers has been deposited (Fig. 8.48E, G).

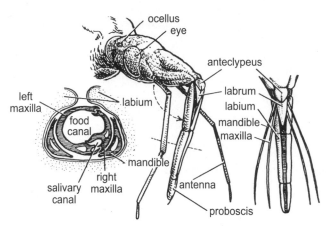

Fig. 8.49. Piercing and sucking mouthparts of an adult blood-feeding bug, *Triatoma protracta* (Heteroptera: Reduviidae). (From D. P. Furman and E. P. Catts. 1970. *Manual of medical entomology*, p. 11. 3rd ed. National Press Books, Palo Alto. Reprinted with the permission of Cambridge University Press.)

8.3.8.2 HEMIPTERA

In true bugs the mouthparts are adapted to pierce and suck, to imbibe either fluids from plant parenchymal or vascular tissues or the blood of other animals. Their maxillary and labial palpi are reduced to a few sensilla, the labium is greatly extended as a three- or four-segmented beak, and the mandibles and maxillary laciniae are transformed into four long stylets co-adapted to each other along their lengths to form a fascicle enclosed within the labium when not in use (Fig. 8.49). The maxillary stylets usually enclose between them separate food and salivary canals and are sheathed laterally by the mandibular stylets (Cobben, 1978). Lateral to the bases of the mandibular stylets are a pair of mandibular (loral) plates, missing in thrips, that are believed to be of hypopharyngeal origin (Parsons, 1974). Both cibarial and salivary pumps are relatively larger and more efficient than equivalent structures in thrips.

Embryogenesis of these mouthparts was fully described for the milkweed bug *Oncopeltus fasciatus* (Hemiptera: Lygaeidae) by Newcomer (1948) and at the molecular level by Rogers and Kaufman (1996, 1997), Rogers and others (1997), Popadíc and coauthors (1998b), and Hughes and Kaufman (2000). Pesson (1944) described the ontogeny of the mouthparts of three scale insects (Hemiptera: Coccoidea).

8.3.8.2.1 ONCOPELTUS FASCIATUS

Mouthpart development in *O. fasciatus* embryos is similar to that in thrips except that both mandibles persist (Fig. 8.50A, B). Following invagination of the mandibular appendages and of the inner (lacinial) lobes of the maxillae to form the stylet-secreting organs (Fig. 8.50A–E), they each coil 2.5 times within the head, the cuticle of each stylet being secreted by cells of an elongate hyaline area extending the length of each coiled appendage (Fig. 8.50F). At hatch, the stylets and sheaths uncoil and the

stylets are protruded into the labial groove (Fig. 8.49: right), where they are said, *incorrectly*, to acquire their co-adaptations (rather, these are formed sequentially as the cuticle of each stylet is deposited). The mandibular and maxillary levers do not develop until after hatch (one reason why first instars do not feed) and arise separately from their stylets, just as in thrips embryos. In larvae of *O. fasciatus*, each mandibular lever is in two parts at an angle to each other. The distal portion, attached to the stylet, is deposited by cells of the mandibular sheath, just as is the entire lever of thrips embryos. Proximally, it articulates with an apodemal invagination from the surface of the head near the antenna. Each maxillary lever arises as a lateral pouch from each maxillary sheath, which grows laterally and attaches to the side of the head ventral to the eye, its cuticle not being deposited within it until after hatch.

Gene expression during head development in *O. fasciatus* embryos is practically identical to that in *D. melanogaster* embryos (Figs. 7.36, 7.41), the differences relating to development of the mandibular and maxillary stylets (Rogers and Kaufman, 1996, 1997; Hughes and Kaufman, 2000). During segmentation, the segment-polarity gene *engrailed* (*en*) is expressed in cells of the posterior compartments of the ocular, antennal, intercalary, mandibular, maxillary, and labial segments, indicating a typical six-segmented head (compare Fig. 7.32). During mouthpart formation, *proboscipedia* (*pb*) is expressed only in the labial appendages (Fig. 13.8). In embryos of *Thermobia domestica* (Zygentoma) (Fig. 13.8), *Tenebrio molitor* (Coleoptera), and *Drosophila*, *pb* is also expressed in cells of the maxillary segment (Figs. 7.36, 7.41, 13.7, 13.8). *Deformed* (*Dfd*) is expressed in the mandibular and maxillary stylets and maxillary plates, and *Sex combs reduced* (*Scr*), in the labial appendages. The zygotic gene *cap'n'collar* (*cnc*) and the *Dll* gene, required for distal outgrowth of appendages, are expressed in the labial appendages, but only *cnc* is expressed in the mandibular stylets. *Dll* is not expressed in the mandibular stylets, just as in embryos of biting and chewing insects (Popadíc et al., 1998b).

8.3.8.2.2 SCALE INSECTS

Development is similar in scale embryos. However, because of the relatively much greater length of the stylets after hatch, the stylet-forming organs coil much more during their invagination and are located within the thorax, and the entire maxillary appendages apparently contribute to stylet formation rather than just their lacinial lobes (Fig. 8.51) (Pesson, 1944).

8.3.8.3 PHTHIRAPTERA

Juveniles and adults of the sucking lice (Anoplura) feed on blood of various mammals and birds. Their mouthparts consist of a soft proboscis (*haustellum*) lacking palpi and bearing small internal teeth, which evert to grip the host during feeding, and of three stylets within a trophic sac (salivarium) opening ventrally into the cibarium.

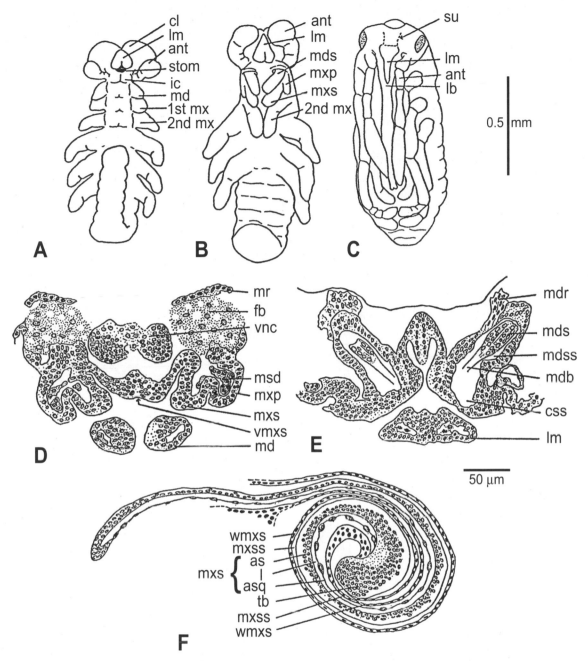

Fig. 8.50. Embryogenesis of the piercing and sucking mouthparts of the milkweed bug *Oncopeltus fasciatus* (Hemiptera: Lygaeidae). A. Ventral aspect of embryo at 47% of embryogenesis showing mouthpart appendages. B. Same view at 53%. The bases of the mandibular (mds) and maxillary (mxs) stylets are beginning to invaginate. C. Same view just before hatch, at 100%. D. Transverse section through the maxillary segment of an embryo at 50%, showing maxillary plates (mxp), mesoderm (msd), and inner lobes (mxs) of the maxillae that will form the maxillary stylets (laciniae). E. Same view through the mandibular segment at 83%, showing the mandibular stylet-secreting organ (mds). F. Sagittal section of a maxillary stylet-secreting organ at 73%, showing how it coils within the head. ant, antenna; as, cuticle-secreting cells of maxillary appendage; asq, attenuated epithelium of maxillary appendage; cl, protocephalic lobe; css, common stylet chamber; fb, fat body; ic, intercalary segment; lm, labrum; l, lumen of stylet sheath; md, mandible; mdb, mandibular stylet; mdr, mandibular retractor muscle; mdss, mandibular stylet sheath; mr, anterior midgut rudiment; stom, stomodaeum; su, suture; tb, terminal bulb of stylet-secreting organ; vmxs, venter of maxillary segment (hypopharynx); vnc, maxillary ganglion; wmxs, wall of maxillary sheath; 1st mx, first maxilla; 2nd mx, second maxilla. (Embryological development of the mouthparts and related structures of the milkweed bug, *Oncopeltus fasciatus* (Dallas), W. S. Newcomer, *J. Morphol.*, Copyright © 1948 Wiley-Liss, Inc. Reprinted by permission of Wiley-Liss, Inc., a subsidiary of John Wiley & Sons, Inc.)

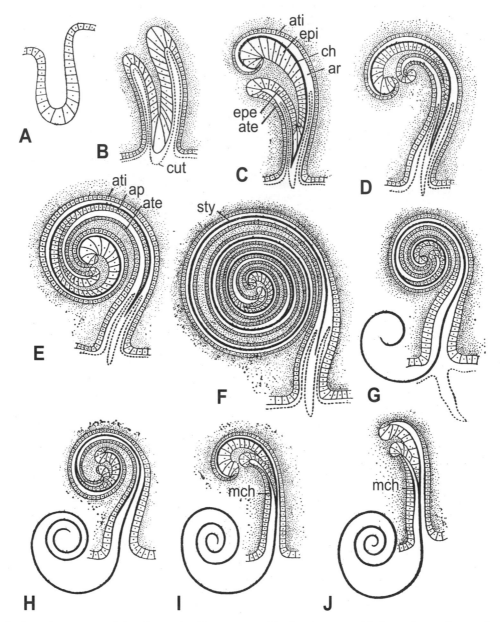

Fig. 8.51. Sagittal sections summarizing the mode of formation of the four long mandibular and maxillary stylets of a scale insect embryo (Hemiptera: Coccoidea). A. Mandibular or maxillary appendage before invagination. B. Invagination of the base of the appendage into the head pulls in surrounding epidermis to form a stylet sheath. C–E. Three stages in the elongation of the stylet appendage to form a retortiform (stylet-secreting) organ. Its apex remains fixed in relation to the surface of the body while it and its sheath elongate into a spiral. Simultaneously, anterodorsal appendage cells progressively deposit stylet cuticle. F. Maximum invagination. As they coil, the stylet cells deposit a long filament of cuticle on their anterior faces. G–J. On completion of invagination, the stylet-secreting cells are reduced to a small mass at the apex of the invagination and the coil unwinds, protracting the stylet out of the sheath where it coils (occurs simultaneously in all four stylets). During protraction, the co-adaptations of the stylets interlock to form the stylet fascicle. ap, appendage epidermis depositing stylet cuticle; ar, lumen of stylet sheath; ate, posterior wall of stylet sheath; ati, anterior wall of stylet sheath; ch, stylet cuticle; cut, embryonic cuticle; epe, posterior epidermis of stylet appendage; epi, anterior epidermal cells depositing the stylet cuticle; mch, thickened cuticle in the atrium enclosing the stylet; sty, stylet. (Adapted from P. Pesson. 1944. Contribution à l'étude morphologique et fonctionelle de la tête, de l'appareil buccal, et du tube digestif des femelles de coccides. In *Monographies publiées par les Stations et Laboratoires de Recherches Agronomiques*, p. 83, Fig. 51. Paris.)

In embryos of the pig louse *Haematopinus suis*, these mouthparts arise during intertrepsis in the same way as do those of *H. verbasci* and *O. fasciatus* (Young, 1953). Unlike those of thrips and bug embryos, however, the mandibular appendages give rise to mandibular vestiges and a food channel, and the maxillary appendages, to paired structures functioning as guides for the dorsal stylet. After it forms through expansion of the salivarium during katatrepsis, the trophic sac opens distally into the floor of the cibarium and encloses three stylets—dorsal and median stylets of hypopharyngeal origin and a ventral one of labial origin—totally unrelated to those of thrips or bugs.

In all these insects, the stylets are more or less co-adapted to each other along their lengths to form a sucking tube. Nuclei of cells producing cuticle of the stylets, at least in thrips embryos, are located far proximal to the region in which it is being deposited and in which it will subsequently interlock, and the question arises as to whether formation of these co-adaptations is mechanically or chemically induced in the apices of these cells by their close approximation or whether the nuclei of the cells producing the cuticle are specified earlier in development to produce their characteristic patterns. By transplanting the stylet-secreting organs of larval *Rhodnius prolixus* (Hemiptera: Reduviidae) into the body cavities of other larvae, Pinet (1969) showed that the stylet-secreting cells are programmed beforehand to form their characteristic co-adaptations.

Gene expression in thrips embryos will no doubt be found to be similar to that of *O. fasciatus* during stylet ontogeny, but will differ in that of sucking lice, because their stylets are of hypopharyngeal and labial rather than of mandibular and maxillary origin.

8.3.9 SKELETAL APODEMES

The heads of most juvenile and adult insects have a well-developed internal skeleton or *tentorium* and a system of sclerotized apodemes in the thorax, abdomen, and appendages on which skeletal muscles originate or insert (Snodgrass, 1935; Ando, 1962; Weber, 1966). All these apodemes arise by invagination of surface ectoderm during embryogenesis and are subsequently lined with cuticle (Rempel, 1975). Based on their development in embryonic heads of the blister beetle *Lytta viridana* (Coleoptera: Meloidae) (Rempel and Church, 1971) (Fig. 8.52), Rempel (1975) concluded that those of the labral segment generate labral apodemes; of the antennal segment, the mandibular extensor (abductor) apodemes; of the intercalary segment, the anterior tentorial arms; of the mandibular segment, the mandibular flexor (abductor) apodemes; of the maxillary segment, the posterior tentorial arms; and of the labial segment, the salivary glands. The anterior and posterior tentorial invaginations later fuse to form the tentorial bridge, and the point of fusion is where the cuticle of the tentorium breaks prior to ecdysis in each molt after hatch (Sharplin, 1965).

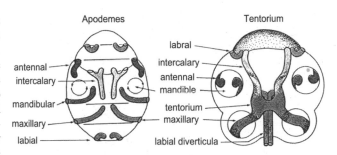

Fig. 8.52. Left: Arrangement of apodemal invaginations in the head of a *Lytta viridana* (Coleoptera: Meloidae) embryo at 53% of embryogenesis as viewed from above. Right: Tentorium and other apodemes at time of hatch. (Reproduced with permission from J. G. Rempel. 1975. The evolution of the insect head: the endless dispute. *Quaest. Entomol.* 11: 7–25.)

Paired apodemes, coelomic sacs, appendages, and neuromeres, were the principal criteria used for recognizing segments (Rempel, 1975; Heming, 1978) until the expression pattern of some segment-polarity genes such as *wg* and *en* (Fig. 8.46A–C) was discovered. Development of apodemes in the legs of locust embryos were described in great detail in relation to muscle origin, differentiation, and innervation by Ball and Goodman (1985a, 1985b) and Ball and others (1985) (Fig. 8.6).

8.4 Primordia of Mixed Germ Layer Origin

Some aspects of insect embryogenesis do not lend themselves well to consideration according to germ layer. Organ systems descending from more than one layer, such as the reproductive system, are a case in point.

8.4.1 REPRODUCTIVE SYSTEM

At the time of hatch, the reproductive system in larvae of both sexes consists of a pair of germ cell (GC)–containing gonadal rudiments, each linked to the back of the seventh (female) or ninth (male) abdominal sternum by the rudiment of a primary exit duct (Fig. 8.53A: left and right). The origin of the pole cells and their descendants, the GCs, were described in 2.2.1.3 and 6.6. Here I summarize how they migrate to a species-specific site within the developing abdomen and become surrounded by mesodermal cells to form the gonadal rudiments. I also consider the origin of the primary exit system, ovarioles, and sperm tubes.

8.4.1.1 GERM CELL MIGRATION WITHIN THE EMBRYO

In short and semi-long germ embryos, the GCs are carried into the yolk on the posterodorsal surface of the embryo at anatrepsis (Fig. 6.20A–D) (Johannsen and Butt, 1941; Heming, 1979; Heming and Huebner, 1994; Huebner and Diehl-Jones, 1998). With gastrulation they are lifted above the ectoderm by ingressing mesoderm and, upon formation of the coelomic sacs, divide into two groups on either side of the embryo. They lodge in their splanchnic walls (genital ridges) in one or more abdominal segments, the number of segments being characteristic for the species. GC-specific mesodermal cells, the *somatic gonadal precursors* (SGPs), then proliferate about the GCs to form a primary epithelial sheath from which the various support tissues of the testis or ovary will later descend (1.1.1, 2.1.1). Events are similar in long germ embryos but are usually simplified by a drastic reduction in blastokinetic movements and, in more specialized species, by an absence of coelomic sacs (Richard-Mercier, 1982).

Until recently, it was not known how GCs reach their final encapsulation sites within the embryo. However, Nardi (1993) was able to immunolabel the GCs in the embryos of *M. sexta* with a monoclonal antibody as soon as they emerged from the posterior midline of the germ band, and to follow their migration in great detail in whole mounts of embryos removed from the yolk. His observations suggested the presence of adhesion/recognition molecules on the surfaces of both GCs and SGPs that enable the GCs to find their final locations within the embryo. They suggested also that anterior migration of GCs results both from elongation of the embryo and from the GCs' active movement due to the temporary extension, adhesion, and withdrawal of filopodia from their surfaces.

A similar analysis of GC movement in *D. melanogaster* embryos by others (Warrior, 1994; Jaglarz and Howard, 1994; Zhang et al., 1997; Broihier et al., 1998; Forbes and Lehmann, 1998; Moore et al., 1998; van Doren et al., 1998) showed migration to have seven phases, each influenced by expression of different genes:

- passive displacement during gastrulation;
- amoeboid migration (maternal *nanos* [*nos*]; GCs in *nos* mutant embryos fail to migrate and become transcriptionally active prematurely [Deshpande et al., 1999]);
- migration on the surface of endodermal midgut cells;
- migration through the endoderm (*serpent* [*srp*] and *huckebein* [*hkb*]);
- attachment to mesoderm (*zfh-1*, *columbus* [*clb*], *heartless* [*htl*], *wunen*);
- alignment with SGPs (*abdominal-A* [*abd-A*], *Abdominal-B* [*Abd-B*], *trithorax* [*trx*], *trithoraxgleid* [*trg*], *tinman* [*tin*], and *columbus* [*clb*]); and
- gonadal coalescence (*fear of intimacy* [*foi*]).

By use of embryonic pattern mutants, the researchers were able to demonstrate that such migration behavior was specific to GCs but was controlled by somatic tissues. Those SGPs competent to react to the presence of GCs by forming a sheath around them were found to express the transposable element 412 and to be capable of forming a sheath even in the absence of GCs.

Proteins of the *wunen* and *zfh-1* genes function to guide the GCs when they migrate from the lumen of the hindgut to the SGPs (Zhang et al., 1997; Broihier et al., 1998). Wunen protein can transform a permissive cellular environment into a repulsive one and is expressed in the hindgut in a pattern that guides the GCs toward the SGPs; in *wunen* mutant embryos the GCs remain dispersed and fail to enter the gonadal rudiments. In *zfh-1* mutants, development of caudal visceral and gonadal mesoderm is disrupted, resulting in failure of the GCs to migrate properly and to associate with SGPs (Broihier et al., 1998), and ectopic expression of ZFH-1 protein induces production of SGPs in abnormal sites. Finally, *3-hydroxy-3-methylglutaryl coenzyme A reductase^columbus* (*hmcgr^clb*) is normally expressed in SGPs and attracts the GCs to them; if misexpressed in epidermal and ventral nerve cord cells, it attracts the GCs to these ectopic sites (van Doren et al., 1998). The subject was reviewed by Wylie (1999).

8.4.1.2 GONADAL MESODERM

When gonad-specific mesoderm is destroyed in embryos of the Colorado potato beetle *Leptinotarsa decimlineata* (Coleoptera: Chrysomelidae), the GCs cannot induce mesoderm elsewhere to form gonadal tissue, presumably because it has already been committed to other fates; such "homeless" GCs degenerate before hatch (Haget, 1969). This regional capacity of mesoderm to generate gonadal tissue is probably induced by underlying ectoderm just after gastrulation (Counce, 1973), as demonstrated by Cumberledge and colleagues (1992), Boyle and DiNardo (1995), and Reichmann and coworkers (1998) in *D. melanogaster* embryos.

Dorsoventral (7.2.1.1.1), segmentation (7.2.1.2), and homeotic (7.2.1.3) genes are all involved in specifying SGPs in the mesoderm of parasegments 10–12 (abdominal segments 4–7) (Fig. 8.4). These cells develop from the same mesodermal primordia as do fat body cells in more anterior segments (Reichmann et al., 1998). Expression of both dorsoventral and segmentation genes defines the three regions in each parasegment in which fat body precursors can develop (Fig. 8.4). Two anterior regions require expression of *en* and *hh* for their development and a posterior region, expression of *wg*. Expression of *dpp* and one or two additional genes determines the dorsoventral extent of these regions.

At least five genes control the mesodermal switch from fat body to SGPs in parasegments 10–12 (Fig. 8.4) (Reichmann et al., 1998; Broihier et al., 1998). Expression of *tin*, *zfh-1*, *en*, and *wg* permits mesoderm to develop into SGPs, while that of *srp* counteracts the action of these genes and promotes fat body development.

Expression of these genes, in turn, is influenced by expression of *abd-A* and *Abd-B* of the bithorax complex (Fig. 7.36) (Cumberledge et al., 1992; Boyle and DiNardo, 1995). Expression of *abd-A* specifies determination of anterior SGPs, and expression of both *abd-A* and *Abd-B* specifies determination of posterior SGPs. Expression of *abd-A* also limits expression of *srp* and is required if these SGP precursors are to move to abdominal segment 5, the site of gonad formation, and to ensheathe the GCs. In embryos mutant for *abd-A*, the SGPs are not specified as gonadal and no gonads form.

Mutations in the gene *clift* (*cli*) likewise abolish gonad formation (Boyle et al., 1997). In wild-type embryos of *D. melanogaster*, *cli* is expressed within SGPs as soon as they form, with 9–12 cells being selected in each of parasegments 12–14 (abdominal segments 6–8). These cells are specified as SGPs in the absence of *cli* expression but fail to maintain their identity, so that the GCs fail to be surrounded by gonadal sheath cells.

8.4.1.3 DEVELOPMENT OF SPERM TUBES AND OVARIOLES

The ovaries and testes of adult insects are characteristically divided into sperm tubes (Fig. 1.1) or ovarioles (Fig. 2.1). In most exopterygote and in some endopterygote embryos, partition of ovary rudiments into ovarioles and of testicular rudiments into sperm tubes proceeds by invagination of dorsally or anteriorly situated primary epithelial sheath cells and begins well before hatch. For example, in *Rhodnius prolixus* (Hemiptera: Reduviidae) embryos, it occurs from 37%–46% of embryonic development (Heming and Huebner, 1994; Huebner and Diehl-Jones, 1998). In certain apterygote and basal exopterygote embryos, the gamete tubes appear to be primary: paired, intersegmental groups of GCs arising from mesoderm on either side in each of seven abdominal segments and each group subsequently becoming the germarium of a single ovariole or sperm tubule (Figs. 2.23, 2.24). Finally, in holometabolous insects, division of ovary and testicular rudiments into ovarioles or sperm tubes also proceeds by distal/proximal splitting but does not commence until metamorphosis in the late larval and pupal stages (10.2.3.1).

8.4.1.4 DEVELOPMENT OF PRIMARY EXIT DUCTS

Paired primary exit ducts of the male (Fig. 1.1) and female (Fig. 2.1) reproductive systems originate in the embryo from strands of splanchnic mesoderm within the abdomen behind the gonadal rudiments and end, respectively, at the back of sterna 9 and 7 (Fig. 8.53A: left and right) (Johannsen and Butt, 1941). The ontogeny of these primordia has been followed in great detail by Nardi and Cattani (1995) in embryos of *M. sexta* using labeled antibodies to cell-surface molecules related to *D. melanogaster* Fasciclin II and Neuroglian (Fig. 8.21). The strands arise simultaneously in abdominal segments 5–9 between 25% and 30% of embryogenesis, and from 30% to 36%, components of both male and female ducts are present in embryos of both sexes, the male primordia ending on a nerve at the posterior margin of sternum 9 and the female primordia on ectoderm at the posterior margin of 7 (Fig. 8.53A: center). Following this sexually indifferent period, the duct components of the opposite sex gradually retract (the branches to 7 and 8 in male embryos and to 8 and 9 in females) (Fig. 8.53A: left and right). Later, in embryos of both sexes, each fully formed primordium ends in a swollen genital ampulla of mesodermal origin (Fig. 8.53B).

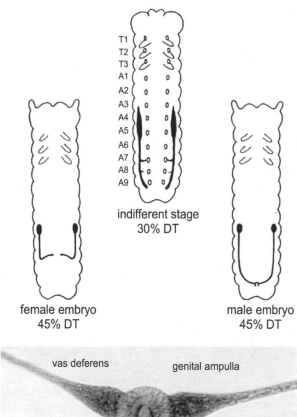

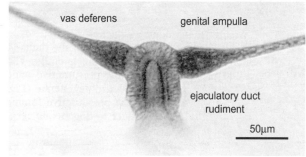

Fig. 8.53. A. Dorsal aspect of three embryos of *Manduca sexta*, summarizing the transformation of primary genital duct primordia from a sexually indifferent period at 30% of embryogenesis (center) to the differentiated female (left) and male (right) states at 45%. Paired, ectodermal cell clusters from the prothoracic to ninth abdominal segments are indicated in the middle diagram as ellipses. B. Photomicrograph, in dorsal aspect, of the genital ampullae and ejaculatory duct rudiment of a third instar larva immunolabeled with a monoclonal antibody to surface protein 2F5. The protein is present on both mesodermal cells of the vasa deferentia primordia and their genital ampullae and on ectodermal cells of the ejaculatory duct rudiment. (Reprinted with permission from J. B. Nardi and E. G. Cattani. Expression of a cell surface protein during morphogenesis of the reproductive system in *Manduca sexta* embryos. Both moths and mammals have an indifferent stage of genital differentiation. *W. Roux's Arch. Dev. Biol.* 205: 21–30. © 1995 Springer-Verlag.)

The posterior end of each duct in the seventh segment ends on a median, ectodermal cell cluster expressing the same antibody as the duct, and a pair of these labeled epidermal clusters is present at corresponding ventral positions at the base of the appendages in all body segments from T1 to A9 (Fig. 8.53A: center). In crustaceans and millipedes, the genital ducts open on modified ambulatory appendages (Sharov, 1966: 228–233). Nardi and Cattini (1995) suggested that these clusters may represent a relict developmental feature in insect embryos.

Beginning at 40% of embryogenesis, the ends of the ducts in male embryos move toward the ventral midline, and ectodermal cells in contact with them begin to express the same antibody. These paired groups of cells then invaginate and fuse to form the rudiment of the ejaculatory duct (Fig. 8.53B), perhaps induced to do so by the presence of the ducts. A similar median invagination does not arise in females until the beginning of metamorphosis (Fig. 8.53A: left). Aspects of this development have been observed in embryos of many insects for over a century (Johannsen and Butt, 1941; Matsuda, 1976), except in those of higher flies, in which the primary exit system is mostly of ectodermal origin (10.2.3.1) (Matsuda, 1976).

8.5 Other Tissues

Hartenstein and Jan (1992) developed a number of markers specific for particular cell types using P-lac Z enhancer lines. With these markers they were able to follow, in great detail and for the first time, the embryonic origin and early development of little-known tissue types such as oenocytes, imaginal discs, abdominal histoblasts, and fat body, dorsal vessel, and perineurial cells and to generate a fate map of the mesoderm at 24%–32% of embryogenesis in *D. melanogaster* embryos. In fact, because of the efforts of Hartenstein and colleagues we now know the source and identity of practically every one of the 60,000 cells and 87 cell types in a newly hatched larva (Hartenstein et al., 1995; Campos-Ortega and Hartenstein, 1997).

8.6 Miscellaneous Events

In addition to those described so far here and in chapters 6 and 7, certain other interactions also occur between cells during embryogenesis.

8.6.1 THE ROLE OF APOPTOSIS IN INSECT EMBRYOGENESIS

Genetically controlled suicide (*apoptosis*) of particular embryonic cells at particular sites and times during development is critical in shaping the organs of the embryo, larva, and adult as these differentiate (Whitten, 1969a; Abrams et al., 1993; Steller, 1995; Jacobson et al., 1997; Meier and Evan, 1998; Meier et al., 2000; Bango and White, 2000). In *D. melanogaster* embryos, most cell death normally occurring is blocked in embryos homozygous for a small deletion on chromosome 3 including the *reaper* (*rpr*) gene. Mutant embryos contain many additional cells and fail to hatch, and deletions including *rpr*

protect such embryos from cell death caused by x-irradiation and developmental defects. Results of other experiments on these embryos suggest the basic cell death program to be intact but not activated.

The *rpr* gene encodes a 65–amino acid peptide. Its mRNA is expressed in cells destined to degenerate, and its peptide, along with those of the cell death genes *grim* and *hid*, serves as a regulator of the suicide-inducing program in response to various death-inducing signals. Other genes involved in apoptosis encode proteins functioning as receptors, modulators, effectors, protectors, adaptors, and inhibitors influencing cell death (Meier and Evan, 1998; Meier et al., 2000), with some of these conserved in the nematode *Caenorhabditis elegans* and in mammals (Steller, 1995; Jacobson et al., 1997).

A prominent example of apoptosis in another insect embryo is the degeneration of the right mandibular appendage during embryogenesis of the mullein thrips *Haplothrips verbasci*, to generate the asymmetrical punch-and-suck mouthparts characteristic of larvae and adults (Fig. 8.48B) (Heming, 1980).

8.6.2 INDUCTION

Induction is a process whereby one group of cells, brought close to another by a morphogenetic movement, causes a change in the way the second group develops. Induction was first demonstrated in insects in 1941 by E. Bock, in embryos of the lace wing *Chrysopa perla* (Neuroptera: Chrysopidae) (Fig. 8.54). By destroying portions of the middle (mesoderm) or lateral (ectoderm) plates of the germ band prior to gastrulation and following subsequent development, he showed the ectoderm to be autonomous or self-differentiating but to induce the normal spread and differentiation of mesoderm.

The egg of this insect is deposited on the tip of a long stalk and has a transparent chorion, so that each cell of the blastoderm is visible. Thus, it is possible to damage or remove small parts of the embryo with great accuracy by microcautery. In Fig. 8.54 (A_1–A_3), the normal embryogenesis of *C. perla* is summarized in transverse section. If the midplate or parts of it are destroyed before gastrulation, the lateral plates differentiate into normal ectodermal structures except with less than normal extension (Fig. 8.54: B_1, B_2). However, ectoderm induces differentiation of mesoderm. When a portion of the midplate is destroyed, the remaining mesodermal cells gastrulate, spread over the inner surface of the ectoderm, and differentiate into as many structures as their number allow (Fig. 8.54: C_1, C_2, D_1, D_2). However, if parts of the lateral plates are destroyed, mesodermal differentiation occurs only in contact with remaining ectoderm (Fig. 8.54: E_1, E_2, F_1–F_3, G_1–G_3), that mesoderm lacking such contact moving into ectoderm-free space and degenerating. Thus, movement of mesoderm is independent of ectoderm, but mesoderm requires the presence of ectoderm if it is to differentiate properly.

This inductive relationship has since been shown to occur in embryos of *Platycnemis pennipes* (Odonata), *Tachycines asynomorus* (Orthoptera), *Leptinotarsa*

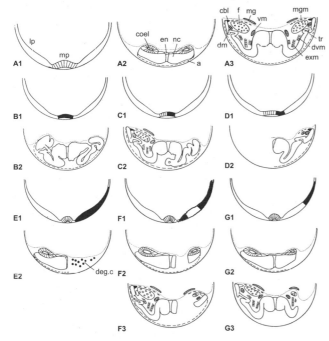

Fig. 8.54. Evidence for ectodermal induction of mesodermal determination and differentiation in embryos of *Chrysopa perla* (Neuroptera: Chrysopidae): diagrammatic transverse sections through the ventral side of a thoracic segment. A_1–A_3, normal development; B–G (read vertically), results of treatment at beginning of gastrulation. Areas eliminated are shown in black in B1–G1; B–D, elimination of mesoderm; E–G, elimination of ectoderm. a, amnion; cbl, cardioblasts; coel, coelom; deg.c, degenerating cells; dm, dorsal muscles; dvm, dorsoventral muscles; en, endoderm; exm, extrinsic muscles of appendages; f, fat body; lp, lateral plate (ectoderm); mg, midgut rudiment; mgm, midgut musculature; mp, midplate prior to gastrulation; nc, ventral nerve cord; tr, tracheal invagination; vm, ventral muscles. (Reprinted with permission from A. Kühn. *Lectures on developmental physiology*, p. 334, Fig. 427, after Seidel, Boch and Krause, 1940. © 1971 Springer-Verlag.)

decimlineata and *Dermestes lardarius* (Coleoptera), *Apis mellifera* (Hymenoptera), and *Calliphora erythrocephala* and *D. melanogaster* (Diptera) by means of similar techniques (references in Counce, 1973).

Mesodermal induction is now being reinvestigated at the molecular level in *D. melanogaster* embryos, with intriguing results. For example, most mesodermal cells in mutant *gastrulation-arrested* (*ga*) embryos do not gastrulate and differentiate but instead express latent germ layer–specific genes appropriate for their changed positions. Baker and Schubiger (1995) showed that ventral ectoderm induces gastrulated mesoderm to express *nautilus* (*nau*) and to differentiate as somatic muscles, while dorsal ectoderm induces such mesoderm to express visceral and cardiac muscle–specific genes by secreting the dorsoventral protein Dpp (Fig. 8.4). Thus, as in Bock's (1941) earlier experiments, muscle determination in these embryos seems to be induced in mesoderm by segment- and dorsoventral-specific ectoderm during and following gastrulation.

In addition, Lawrence and colleagues (1995) demonstrated that segmentally arranged clusters of mesodermal cells in *D. melanogaster* embryos express the pair-rule gene *even-skipped* (*eve*) upon receipt of Wg protein from the ectoderm (since mesodermally expressed Wg protein also induces this *eve* expression, ectodermal induction is not essential). However, when patches of *wg* mutant mesoderm were overlaid with wild-type ectoderm, their cells began to express *eve*, suggesting that both ectoderm and mesoderm assist in patterning mesoderm.

Induction of mesoderm by ectoderm in insect embryos is exactly opposite to what occurs in various vertebrate embryos, in which mesoderm of the notochord and adjacent cells induces the ectodermal neural plate above them to roll up into the neural tube, eventually to differentiate as CNS (Gerhart and Kirschner, 1997). This difference between insects and vertebrates may relate to the usually small size of insect embryos and to their relatively large surface-volume relationship, where ectoderm might be expected to dominate mesoderm. In the usually larger embryos of vertebrates, surface area is relatively less and mesoderm might be expected to dominate. Other examples of induction already mentioned in this chapter include induction of midgut epithelium by visceral mesoderm (8.2.1) and of gonadal mesoderm by abdominal ectoderm (8.4.1.2).

8.6.3 HORMONES AND EMBRYOGENESIS
Embryos of members of basal and intermediate clades are known to deposit two to four cuticles prior to hatch, with production of both molting (20-OH ecdysone [20E]) and juvenile hormones (JH) being required for each of these molts, just as in juvenile molts after hatch (12.5) (Hoffman and Lagueux, 1985; Truman and Riddiford, 1999).

8.6.3.1 20-OH ECDYSONE
First instars of the grasshopper *Locusta migratoria* hatch at 264h at 33°C (Lagueux et al., 1979). During embryogenesis, four peaks of ecdysteroid concentration correlate with the deposition of four embryonic cuticles: serosal (at 48h [18% of embryogenesis]; see 6.9.3), first embryonic (76h [29%]), second embryonic (pronymphal; 120h [45%]), and larval (200h [76%]) (Fig. 8.55).

Most endocrine organs (neurosecretory cells, corpora cardiaca, corpora allata, prothoracic glands) originate some time before katatrepsis but show little ultrastructural or hormonal evidence of function until afterward. The peaks in hormonal concentration occurring before this movement result from mobilization of hormonal conjugates supplied to the yolk by the mother during oogenesis (2.2.1.4) (Dorn, 1983; Hoffman and Lagueux, 1985; Dorn et al., 1987; Westbrook and Bollenbacher, 1990).

In a series of carefully timed and executed ligaturing experiments on eggs of the locusts *Locustana pardalina* and *Locusta migratoria*, Jones (1956a, 1956b) showed

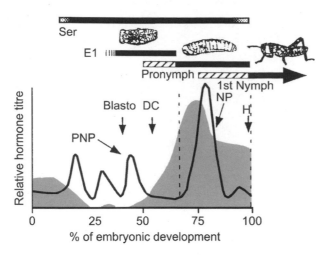

Fig. 8.55. Titers of ecdysteroid (black line) and juvenile hormone (shaded) in relation to deposition of serosal (Ser), embryonic (E1), pronymphal, and first instar nymphal cuticles during embryogenesis of the locust *Locusta migratoria* (Orthoptera: Acrididae). Bars over the hormone titers show the times when the four cuticles are present. Cross hatching represents the pharate stages; the vertical dashed lines, the time of ecdysis of the embryonic and pronymphal cuticles. Blasto, katatrepsis; DC, dorsal closure; H, hatch; PNP, pronymphal peak of ecdysteroid release. (Adapted with permission from *Nature* (J. W. Truman and L. M. Riddiford. The origins of insect metamorphosis. 401: 447–452) © 1999 Macmillan Magazines Limited; based on Lagueux et al. [1979] and Temin et al. [1986].)

that a factor produced by ventral (prothoracic) gland cells in the head of the embryo stimulates secretion of a chitinase by cells of the pleuropodia on abdominal segment 1. This chitinase functions to digest the inner white layer of serosal cuticle before hatch (6.10.1) (Slifer, 1937). Hagedorn (1985: 255) suggested that ecdysteroids for the second embryonic molt might be produced by pleuropodial cells, particularly in eggs not containing a large supply of maternally contributed ecdysteroids.

Experimental evidence for Hagedorn's (1985) suggestion had been provided earlier by Novák and Zambre (1974) for embryos of another locust, *Schistocerca gregaria*. Injection of either live pleuropodia from these embryos or centrifuged homogenates of them into posterior isolates of third instar *D. melanogaster* larvae that had been ligatured before the end of the critical period for release of their own ecdysteroids (12.2) induced pupariation (10.2.1), as did injection of dihydroxyecdysone extracted from the fern *Polypodium vulgare*. The researchers concluded that pleuropodia in embryos of this insect secrete an ecdysteroid into the amniotic fluid, which secondarily induces serosal cells to produce the chitinase necessary for hatch. To my knowledge, this work has yet to be followed up using modern methods.

8.6.3.2 JUVENILE HORMONE

JH III (Fig. 12.11E) is present in relatively low concentrations in newly deposited *L. migratoria* eggs. The concentration drops to zero by 25% of embryogenesis during and just after the serosal peak of ecdysteroid release; has a small peak coincident with the second peak of ecdysteroid release (associated with secretion of the first embryonic cuticle, E1); rises to a large peak at 75%, well after dorsal closure and just before the fourth peak of ecdysteroid release (associated with the secretion of larval cuticle); and stays relatively high until hatch (Fig. 8.55) (Temin et al., 1986).

8.6.3.3 INSULIN-LIKE PEPTIDES WITH PROTHORACICOTROPIC ACTIVITY

Sevala and Loughton (1992) monitored the levels of insulin-like peptides (ILPs) during embryogenesis of the grasshopper *L. migratoria* by use of radioimmune assay for human anti-insulin, and observed three peaks closely correlating with peaks of ecdysteroid concentration (Fig. 8.55). They postulated that ILPs stimulate the synthesis and release of ecdysteroids. ILPs with prothoracicotropic activity have also been isolated and characterized from brains of *Bombyx mori* (bombyxin).

8.6.3.4 BURSICON

As discussed in 12.4.10, bursicon is a neuropeptide hormone that induces cuticular sclerotization in newly ecdysed insects at all stages of postembryogenesis. A factor having bursicon-like activity can first be detected in locust embryos (*S. gregaria*) at about 75% of embryogenesis by use of the ligated fly assay (Honegger et al., 1992). Its concentration rises rapidly until 95% of embryogenesis and remains high until hatch. In fact, these late embryos contain as much bursicon-like activity as the entire CNS of adults!

8.6.3.5 FEMALE ACCESSORY GLAND SECRETIONS PROMOTING EMBRYOGENESIS

Females of the chironomid midge *Chironomus dorsalis* deposit a cylindrical egg mass 1.5–3.0 cm long into water (Fig. 8.56A) (Takeda and Ohishi, 1976). This mass contains up to 540 eggs, in up to 27 transverse rows of 20 eggs each, embedded within a jelly coat attached proximally to an adhesive cap. Embryogenesis and hatching of eggs occur sequentially, the processes being complete in about 48 h after oviposition at 20°C. When Takeda and Ohishi (1976) followed egg hatch in proximal, middle, and distal thirds of the mass, they found eggs in the proximal third to develop and hatch more rapidly than those in the middle and distal thirds, as if influenced by a gradient of physiologically active substance emanating from the adhesive cap (Fig. 8.56B). The researchers removed the adhesive cap from the egg mass (Fig. 8.56C); isolated proximal, middle, and distal thirds of the mass both from each other and from the adhesive cap; and juxtaposed middle and distal thirds individually with an adhesive cap. They then monitored the effect

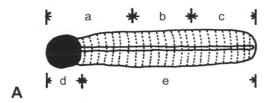

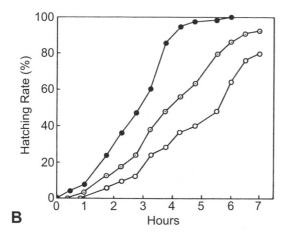

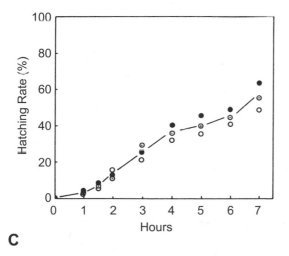

Fig. 8.56. Effect of a development-stimulating pheromone on embryos of *Chironomus dorsalis* (Diptera: Chironomidae). A. The egg mass. a, proximal part; b, middle part; c, distal part; d, adhesive cap; e, egg portion. B. Hatching rate in the proximal (filled circle), middle (dotted circle), and distal (unfilled circle) parts of an intact egg mass. C. Effect of removing the adhesive cap on the hatching rate within each part of the egg mass. Symbols represent the same as in B. (Reprinted from *J. Insect Physiol.*, 22, N. Takeda and T. Ohishi, Effect of a development-stimulating pheromone on the embryo of *Chironomus dorsalis*, pp. 1327–1330, Copyright 1976, with permission from Elsevier Science.)

of these manipulations on embryogenesis and hatching. Their results showed that the cap gradually releases a pheromone, synthesized in the female's accessory glands, that percolates distally into the mass, where it

stimulates embryogenesis and hatching. They did not speculate on what the adaptive function of such a component might be.

8.6.4 ENVIRONMENTAL FACTORS

8.6.4.1 TEMPERATURE

Because most insects, including their eggs, are ectothermic, embryogenesis tends to accelerate at higher temperatures and to slow at lower ones (Howe, 1967; Bursell, 1974; Jones and Heming, 1979), the embryos of each species having an optimum temperature for development and upper and lower threshold temperatures beyond which development ceases.

8.6.4.2 RELATIVE HUMIDITY

Relative humidity so strongly influences temperature effects that the two must be studied together (Bursell, 1974; Jones and Heming, 1979). In habitats having a dry season and in ponds drying up in late summer, the eggs of some insects can become *cryptobiotic* (*anhydrobiotic*)—that is, they can lose much of their water and become highly resistant to a variety of environmental insults. This state usually can be ended quickly by adding water.

8.6.4.3 PHOTOPERIOD

Diapause is a genetically determined state of suppressed development and low metabolic rate that can be induced by environmental factors. It occurs in insects at higher latitudes or in areas with a pronounced dry season and often in a species-specific stage of the life cycle, and enables them to overwinter or to survive extended periods of drought (Danks, 1987).

Exposing female embryos of the silkworm *Bombyx mori* to long days and a high temperature induces them to develop into adults that deposit diapausing eggs, whereas exposure to short days and a low temperature results in the development of females that deposit nondiapause eggs (Fig. 8.57) (Denlinger, 1985). Neurosecretory cells in the subesophageal ganglion of diapause-producing females synthesize and secrete a diapause hormone (DH). This 24–amino acid peptide is known to induce an increase in glycogen content, triglycerides, and 3-hydroxykynurenine in individual follicles in her ovarioles, which collectively function to arrest development of the embryo following germ band formation (Nagasawa, 1992). Female pupae injected with DH develop into adults whose eggs always enter diapause, whether or not they would have without injection.

Results of more recent anatomical and pharmacological experiments by Shimizu and coworkers (1997) reveal additional complexity. They monitored the distribution of diapause follicles in the ovarioles of ovipositing females by staining with anti-3-hydroxykynurenine. When they removed the subesophageal ganglion of nondiapause producers 2 d after the larval/pupal molt,

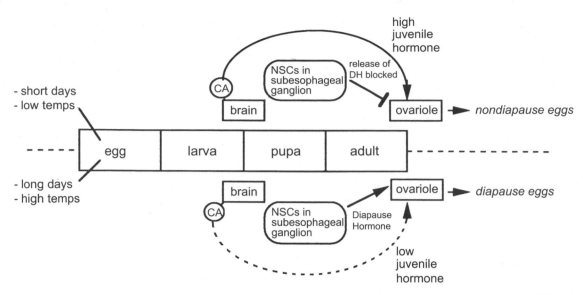

Fig. 8.57. Function of the diapause hormone on development of diapause and nondiapause eggs in *Bombyx mori* (Lepidoptera: Bombycidae). CA, corpora allata; DH, diapause hormone; NSCs, neurosecretory cells. (Adapted from D. L. Denlinger. 1985. Hormonal control of diapause. In G. A. Kerkut and L. I. Gilbert (eds.), *Comprehensive insect physiology, biochemistry and pharmacology*, Vol. 8, pp. 353–412, Pergamon Press, Copyright 1985, with permission from Elsevier Science.)

these females were able to produce diapausing eggs, whereas if the brain–corpora cardiaca–corpora allata axis was also removed, such eggs did not appear. When they transected the protocerebrum of nondiapause producers, diapause eggs, again, appeared. Finally, they found that injection of gamma-aminobutyric acid (GABA) or picrotoxin into diapause producers inhibited production of diapause eggs. Based on these results, they concluded that after DH is produced by the subesophageal ganglion,

- it is transferred to the corpora cardiaca–corpora allata complex;
- its release from this complex is suppressed in nondiapause producers;
- there is an inhibitory center in the protocerebrum that governs its release; and
- this center may function by releasing GABA.

In diapausing eggs of other species, developmental arrest occurs at characteristic embryonic stages, usually at blastoderm or intertrepsis, or in the prolarva just prior to hatch, but proximate factors inducing such arrest have yet to be examined.

8.6.4.4 PHYSIOLOGICAL CONDITION OF PARENT

In the grasshopper *Aulocara ellioti*, viability of eggs and rate of embryogenesis are influenced by age and physiological state of the mother, the embryos of young females varying more in developmental rate and tending to emerge later (Visscher, 1971). Also, females reared under conditions of stress (high population density, high temperature, too many males, etc.) deposit eggs that develop more slowly than do those of normally reared females.

8.6.4.5 HOST HORMONES

As summarized in Table 6.1 and Fig. 6.16, a single egg of the parasitoid encyrtid wasp *Copidosoma floridanum*, deposited in the egg of its host, the cabbage looper moth *Trichoplusia ni*, may proliferate into 2000 or more "secondary morulae" (embryos) during the first four larval instars of the host caterpillar (Grbíc et al., 1997; Strand and Grbíc, 1997). Development of individual embryos into the majority "reproductive" larvae following their proliferation is synchronized with onset of metamorphosis in the host, the embryos having to develop for 9 d before acquiring competence to do so. Such embryos differentiate when injected into host larvae or pupae having high concentrations of ecdysteroids but not when transplanted into adults lacking these hormones. In addition, ontogeny is arrested in embryos if the host's prothoracic glands are destroyed but can be restarted by injecting 20E, suggesting that the trigger for parasitoid embryogenesis is host ecdysteroid.

The less numerous, "precocious" larvae function as soldiers within the host and manipulate the sex ratio of "reproductive" larvae by killing siblings of the opposite sex and by defending those of the same sex from interspecific competitor larvae resulting from multiparasitism. Their embryogenesis from secondary morulae is synchronized with specific stages of the host larvae in any of the first four instars and can begin following ecdysis in any of them (Fig. 6.16) (Grbíc et al., 1997). By transplanting both precocious and reproductive secondary morulae into various host instars, Grbíc's group determined that fluctuations in host JH concentration did not influence embryogenesis, whereas a pulse of host ecdysteroid was required for embryogenesis and eclosion

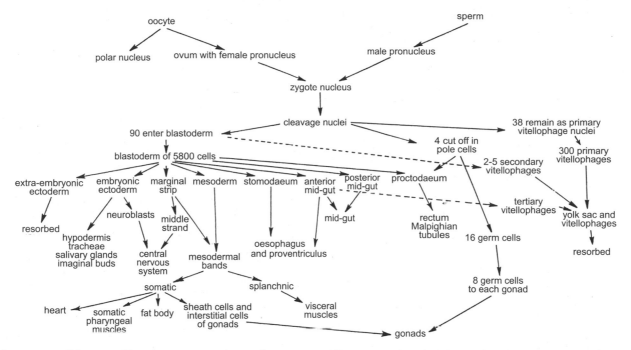

Fig. 8.58. Cell lineage of larval organs in embryos of *Dacus tryoni* (Diptera: Tephritidae). (Reproduced with permission from D. T. Anderson. The embryology of *Dacus tryoni* (Frogg.) [Diptera, Trypetidae], the Queensland fruit-fly. *J. Embryol. Exp. Morphol.* 10: 248–292. © 1962 Company of Biologists, Ltd.)

of parasitoid larvae of both types. Just as for reproductive larvae, embryogenesis of precocious larvae could be arrested by removing host prothoracic glands and rescued by injecting 20E.

8.7 Overview

Fig. 8.58 summarizes the embryogenesis of a tephritid fruit fly, *Dacus tryoni* (Anderson, 1962), having a long germ embryo similar to that of *D. melanogaster*, and indicates the source of cells constituting the various organ systems that make up the body. Development of any organ within the body of an embryo characteristically occurs in three phases:

- Proliferation. Much differential mitotic activity is visible. The cell cycle of these divisions is modified during this proliferative period, depending on the tissue primordium to which the cells are allocated (Foe et al., 1993).
- Morphogenesis. Changes in cell adhesion, polarity (Müller, 2000), position, and shape, and often programmed cell death (apoptosis; 8.6.1), cause changes in the shape and position of organ rudiments as they differentiate. In *D. melanogaster* embryos, the position-specific (PS) integrins, a family of cell-surface heterodimers functioning as receptors for extracellular matrix molecules, seem to function in cell movement, since embryos lacking expression of the maternal and zygotic *PS integrin* genes show abnormal morphogenesis (Roote and Zusman, 1995). The genetics of cell

migration in *D. melanogaster* was reviewed by Montell (1999).

- Differentiation. The organs assume their final shapes, and their cells take on the structure and function characteristic of them in the free-living juvenile.

8.8 Computer Simulation of Gene Expression during Embryogenesis of *Drosophila melanogaster*

Hartenstein and associates (1995) constructed a graphic digital database of *D. melanogaster* embryogenesis by assembling a complete series of transverse, sagittal, and frontal histological sections and optical sections of whole, cleared embryos that had been treated with various labeled antibodies specific for particular cell types and examined by confocal microscopy. They imported the digitized sections into the Macintosh drawing program Adobe, where they served as templates for defining the contours of the embryo and of organ primordia, and of the positions of individual cells. From these they generated surface- and point-closed models of all embryonic stages.

Gene expression data can be entered into this database by translating the expression domain of a given gene into the three-dimensional coordinate (Cartesian) system of the database. Once the data are entered, the embryo can then be displayed on the monitor at any stage of development and from any angle in three dimensions, and one can pull out sections from any position or angle to illustrate the expression pattern of a particular gene in a particular cell type at a particular stage of embryogenesis. Similar models are available for embryogenesis of the nematode *Caenorhabditis elegans* and for those of the zebrafish and house mouse (Meinertzhagen et al., 1998).

Computer simulation and modeling is currently being used to good effect in analyzing gene expression in early embryos (Bodnar, 1997; von Dassow et al., 2000) and in the developing CNS of *Drosophila* larvae (Meinertzhagen et al., 1998; Schmid et al., 1999) and is being augmented by incredible new developments in light microscopy such as the multiphoton imaging and automated interactive microscopes (Thomas et al., 1996; Gura, 1997). The latter fuses several different microscopes (differential interference contrast, multiphoton, confocal) into one and, by use of powerful computer techniques, allows a researcher to interact in three dimensions directly with a living embryo as it develops, the data being collected, processed, and displayed as embryogenesis proceeds. The article by Schmid and coauthors (1999) on CNS and PNS development in *D. melanogaster* embryos shows what can be accomplished.

9/ Postembryonic Development and Life History

When they have completed their embryogenesis, young insects are ready to hatch, feed, and grow and to differentiate into adult males and females able to reproduce. In this chapter, I summarize information on hatching, life histories, juvenile stages, the origin and function of the pupal stage in holometabolous insects, and quantitative aspects of growth. But first a few words about life history trade-offs and the effects they are thought to have on life history evolution.

9.1 Life History Theory

According to Stearns (1992), consideration of life history trade-offs has successfully explained why adults of particular species are large or small, why they mature early or late, why they have few or many offspring, why they have a short or long life, and why they must grow old and die. Trade-offs are thought to occur because resources for members of a given species are limiting, resulting in a trade-off between their allocation to reproduction and survival (Roff, 1992; Stearns, 1992; Nylin and Gotthard, 1998; Fox and Czesak, 2000). The optimum allocation to each of these functions depends on the shape of that species' survivorship curve: diminishing returns for allocation to defense result in higher allocation to reproduction and vice versa.

Such trade-offs are intuitively satisfying, are analytically tractable, and occur in a variety of animals, including many insects (Roff, 1992; Stearns, 1992). However, advances in theory have substantially outpaced empirical data (Fox and Czesak, 2000), evidence for their underlying assumptions are known to be absent in some insects (e.g., in water striders [Klingenberg and Spence, 1997]), and their effects are constrained by phylogeny (Stearns, 1992; Klingenberg, 1996b) and by resistance to genetic reorganization (Raff, 1996). These constraints result from past adaptations becoming progressively irreversible as they became more deeply embedded within the developmental pathways of descendant species (Scheiner, 1992: 1820). Finally, the recent discovery of mutant genes in *Drosophila* such as *methuselah* (*mth*) (Lin et al., 1998) and *Indy* (Rogina et al., 2000) that drastically increase adult life span (35% for the first and almost 100% for the second) without exacting other costs suggests that some life history theory is naive.

Also, it is significant that the number (Table 9.4) and form (Fig. 9.3) of life history stages in insects are often characteristic for members of a given higher taxon regardless of such trade-offs.

9.2 Hatching

A young insect usually hatches when it has fully differentiated and is ready to fend for itself. Hatching is initiated by the young insect swallowing amniotic fluid and air, and then repeatedly contracting its abdominal and other muscles in characteristic ways. These contractions rhythmically force blood into the head, neck, and thorax; repeatedly expand and collapse them; and eventually rupture the chorion or serosal cuticle by applying focused pressure on their inner surfaces in ways characteristic for the species (Müller, 1951; Cobben, 1968). Known mechanisms include the use of an egg burster or hatching spines, usually borne by the second embryonic cuticle; of eversible brustia, or cervical bladders; of extensible, apical serosal cuticle; or of the larval mandibles (Müller, 1951; Cobben, 1968). A line of weakness in the chorion or an *operculum* (Fig. 6.20A) that facilitates hatching may be present. The insect then escapes and generally, though not always, begins to feed.

In an ingenious and carefully executed quantitative study of hatching of first instars of the grasshopper *Schistocerca gregaria*, Bernays (1971, 1972a, 1972b, 1972c) revealed both primary and secondary accessory muscles in the neck, thorax, and abdomen, and permanent muscles throughout the body to function in hatching, in digging to the soil surface from the egg pod, and in shedding the second embryonic cuticle, the so-called intermediate molt of the vermiform larva. The first group of 5 paired cervical fibers degenerates following the molt into the first instar, the second group of 34 paired fibers degenerates in the newly emerged adult, while all permanent muscles continue to function throughout postembryogenesis and in the adult. Similar muscles are no doubt present in first instars of other insects.

In ready-to-hatch prolarvae of the mantid *Sphodromantis centralis* and of other mantids, a coiled hatching thread of second embryonic cuticle extends from the apex of each cercus to the inner surface of the serosal

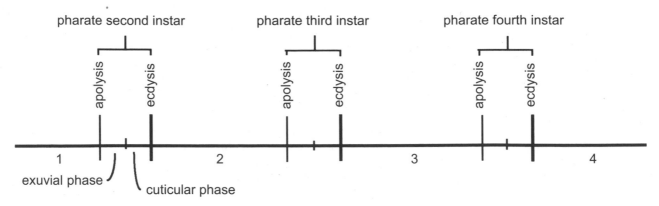

Fig. 9.1. The relationship between apolysis, ecdysis, and the exuvial and cuticular pharate phases of three molts in an immature insect.

cuticle at the posterior end of the egg (6.10.1). When the insect ecloses, these threads uncoil and temporarily suspend it from the ootheca as it molts to the first instar (Kenchington, 1969). Although older mantid nymphs are voracious predators capable of eating each other, the significance of this adaptation is obscure since newly hatched first instars apparently do not feed (Daly et al., 1998: 369).

9.3 Stage and Instar

Growth in arthropods occurs in bursts. As a young hexapod grows, it periodically sheds the remains of its previous, small cuticle (*ecdysis*) and replaces it with a new, larger one, the whole process being called a *molt*. In addition, most insects assume several different forms during postembryogenesis, for example, in snakeflies (Raphidioptera), the larva, pupa, and adult (Fig. 9.3). Each form is called a *stage*, and each stage can have one or more instars, each instar separated from other instars by molts; in snakeflies, for example, the larval stage has eight instars. In addition, each molt has hidden stages not visible externally (Fig. 9.1).

Separation of the epidermis from the old cuticle is called *apolysis*, and between it and ecdysis, there is usually a characteristic time interval during which the insect is a *pharate* member of the next instar. This pharate period can be divided further into *exuvial* and *cuticular* phases. In the former, the epidermis has yet to begin depositing new cuticle, and it is at this time that differential proliferation, growth, movement, and death of epidermal cells can cause changes in shape between one instar and the next. Deposition of the cuticle of the next instar only begins in the cuticular phase after such change is largely complete. As one might expect, the temporal extent of both exuvial and cuticular pharate phases varies in relation to instar length and correlates with the amount of change in form between any two instars, particularly in the molts from larva to pupa and from pupa or last nymphal instar to adult. Apolyses, not

ecdyses, separate instars, although field biologists, for obvious practical reasons, use the term *ecdysis*.

9.4 Metamorphosis

If one compares a newly hatched pterygote (winged) insect with its adult (Fig. 9.3), the following differences are more or less obvious depending on species:

Newly hatched juvenile	*Adult*
relatively small	relatively large
wingless	often fully winged
sexually immature	sexually mature
of characteristic form	similar to or more or less differing in form from juvenile

Because of these differences, a young insect must change its form (metamorphose) to that of the adult after it hatches from the egg. The degree of change occurring during postembryogenesis depends on how far the juveniles and adults have diverged in structure and habit from each other (Fig. 9.2). Although there is a continuum from essentially none to extreme change within insects, all can usually be allocated to one of three categories: Ametabola (slight metamorphosis), Hemimetabola (incomplete metamorphosis), or Holometabola (complete metamorphosis) (Fig. 9.3).

Immatures of representatives of known North American hexapod orders and families (and of Coleoptera, for the world) are described, illustrated, and discussed by the numerous authorities in Stehr's edited volumes (1987b, 1991), probably the single most important sources of descriptive information on immature insects available (but see also Joly's [1977] and Sehnal's [1985b] wonderful overviews).

9.4.1 AMETABOLA
If one includes the *entognath* (mouthparts primitively withdrawn into the head capsule) hexapod taxa Collembola + Protura (Parainsecta) and Diplura with

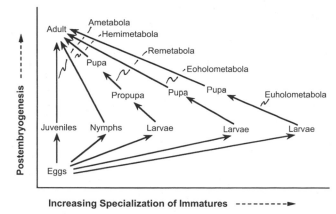

Fig. 9.2. The relationship between increased specialization of immatures (cenogenesis) and type of postembryonic development.

Archaeognatha and Zygentoma within the Insecta, about 8000 species are known to develop by slight metamorphosis (Figs. 0.2; 9.3: Archaeognatha; 9.4). These hexapods are characterized by

- primitively wingless adults (hence, Apterygota, "without wing-bearers");
- young resembling adults in microhabitat, body shape, food, mouthparts, and some behaviors;
- compound eyes (Archaeognatha, most Zygentoma, some Collembola) or not (other Collembola, Protura, Diplura, and a few Zygentoma);
- little or no metamorphosis (hence, Ametabola, "without metamorphosis"); and
- adults continuing to molt and to copulate after each molt.

Life cycle* JO-JO-JO-JO-JO-JO-JO-JO-
AO-AO-AO-AO- etc.

9.4.2 HEMIMETABOLA (*SENSU LATU*) (EXOPTERYGOTA)

The approximately 110,200 known species of insect having incomplete metamorphosis are classified into 16 orders (Figs. 0.2; 9.3: Orthoptera; 9.4) and are characterized by

- a dimorphic life history (nymphs and adult);
- incomplete metamorphosis (hence, Hemimetabola, "some metamorphosis");
- microhabitats, behavior, food, and mouthparts that are usually similar in juveniles and adults;
- compound eyes that are usually present in both juveniles and adults, with additional visual units (ommatidia) being added to the anterior margin of each eye during postembryogenesis;

*In the life cycles summarized in this and subsequent sections (Sehnal et al., 1996), the following abbreviations are used: J, juvenile; A, adult; O, wings absent; w, nonfunctional wing pads present; d, nonfunctional wings present as invaginated wing discs; W, with functional wings; Q, nonfeeding and quiescent; "–," molt; and "=," more or less extensive metamorphosis.

- dorsal ocelli present in juveniles of some taxa (e.g., some Ephemeroptera, Plecoptera);
- adults usually *macropterous* (fully winged), with wings developing gradually as external wing pads in older juveniles (hence, Exopterygota, "outside wing-bearers") (adults of many species may be secondarily wingless due to close adaptation to specific habitats in which wings are disadvantageous [e.g., the ectoparasitic Phthiraptera]); and
- adults that do not molt.

However, there is considerable diversity in the degree of metamorphosis extant among members of this paraphyletic lineage. This diversity is recognized by allocating them to the following subcategories (Fig. 9.4) (Weber, 1966; Nüesch, 1987; Sehnal et al., 1996):

- Paurometabola (Hemimetabola *sensu latu*): These insects have all the characteristics noted above and include the Orthoptera, Phasmatodea, Grylloblattodea, Dermaptera, Isoptera, Blattodea, Mantodea, Zoraptera, Embioptera, Psocoptera, and most Hemiptera (Figs. 9.3: Orthoptera; 9.4).

Life cycle (e.g., a grasshopper): JO-JO-Jw-Jw-Jw-
Jw=AW (death) (number of molts characteristic
for the species)

- Pseudometabola (secondary ametabola): This category includes the hemimetabolous insects that have secondarily lost their wings by adopting an ectoparasitic lifestyle: Phthiraptera, Hemiptera: Cimicidae. Their postembryogenesis is convergently similar to that of ametabolous hexapods, except that they usually have fewer instars and do not molt as adults.

Life cycle (e.g., a louse): JO-JO-JO-AO (death)

- Hemimetabola (*sensu stricto*): The juveniles of Ephemeroptera, Odonata, and Plecoptera are aquatic and have special adaptations (e.g., streamlined bodies, gnathal, thoracic, and/or abdominal gills, prehensile labium, respiratory filaments, swimming legs, jet propulsion) enabling them to live in freshwater that they lose upon adult emergence (Fig. 9.3: Ephemeroptera). They thus change more at the ultimate or penultimate molt than do most terrestrial exopterygotes. In addition, most mayflies are unique among insects in having a flying juvenile stage called the *subimago*, which molts into the adult (Fig. 9.3) (Edmunds and McCafferty, 1988). Sehnal and colleagues (1996), with good reason, placed mayflies in a developmental category of their own called the Prometabola (Figs. 9.3, 9.4; Table 9.2).

Life cycle (e.g., mayfly): JO-JO-Jw-Jw-Jw-Jw-
Jw-Jw-Jw=JW-AW (death) (number of molts
characteristic for the species)

- Remetabola: The propupal and pupal stages of thysanopterans are quiescent, lack functional mouthparts, and have a level of internal reorganization equivalent to or greater than that during the single pupal stage of members of basal, holometabolous

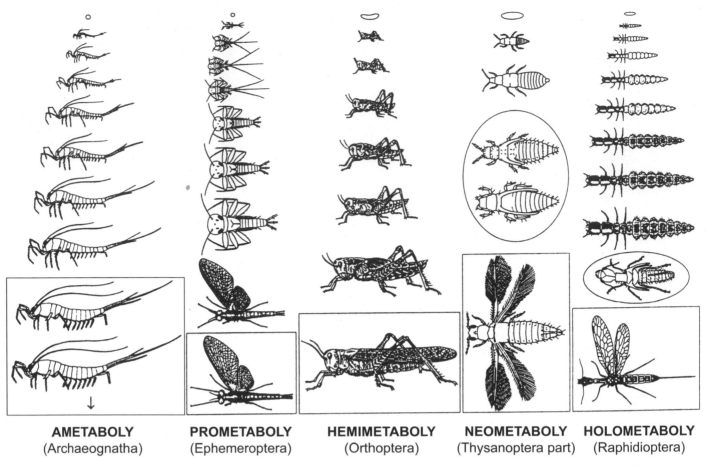

Fig. 9.3. The five principal types of postembryonic development in extant hexapods. Quiescent metamorphic stages are shown in circles while adults are shown in boxes. The primitively wingless hexapods continue to molt as adults and do not have a pre-determined final instar. (Reprinted with permission from F. Sehnal, P. Svácha, and J. Zrzavy. Evolution of insect metamorphosis. In L. I. Gilbert, J. R. Tata, and B. G. Atkinson (eds.), *Metamorphosis. Postembryonic reprogramming of gene expression in amphibian and insect cells*, p. 11, Fig. 1. © 1996 Academic Press, Ltd.)

lineages (Figs. 9.2, 9.3: Thysanoptera) (Heming, 1991; Moritz, 1997). However, these stages do have external wing pads if their adults are winged. The term *propu-pa* is preferred to the often used *prepupa* for the first of these quiescent instars, since this stage is a separate instar; the *prepupa* of the holometabolous insect is the pharate pupa (9.6.2).

Life cycle (suborder Terebrantia):
JO-JO=JwQ=JwQ=AW (death)
(suborder Tubulifera): JO-JO=JwQ (wing pads folded transversely under propupal cuticle between midlegs and hindlegs)=JwQ=JwQ=AW (death)

• Allometabola: In whiteflies (Hemiptera: Aleurodidae) the first instars are dispersing "crawlers" with well-developed functional legs; the second and third instars are sessile, feeding stages; and the fourth has all the characteristics of a true pupal stage (Fig. 9.5) (Weber, 1934). They are unique among insects in that external wings are present only in adults.

Life cycle   JO=JO-JO-JOQ=AW (death)

• Parametabola: In scale insects (Hemiptera: Coccoidea), females have two or three immature instars and males usually four (Fig. 9.6). First instars of both sexes are dispersing "crawlers" with well-developed functional legs. Second instar (and third in some species) and adult females and second instar males, in most families, are sessile, feeding stages usually without functional legs. The third (propupal) and fourth (pupal) instars of males are quiescent, are contained within a cocoon spun by the second instar in some species, undergo extensive metamorphosis, and lack functional mouthparts (Mäkel, 1942). Adult males also lack mouthparts and have peculiar and, in some species, numerous eyes resembling the stemmata of holometabolous larvae (8.3.2.4.3) rather than the compound eyes of other immature and adult insects (Pflugfelder, 1936: plate 21).

Life cycle   JO=JO-(JO)-AO (death) (females)
JO=JO=JwQ=JwQ=AW (death) (males)

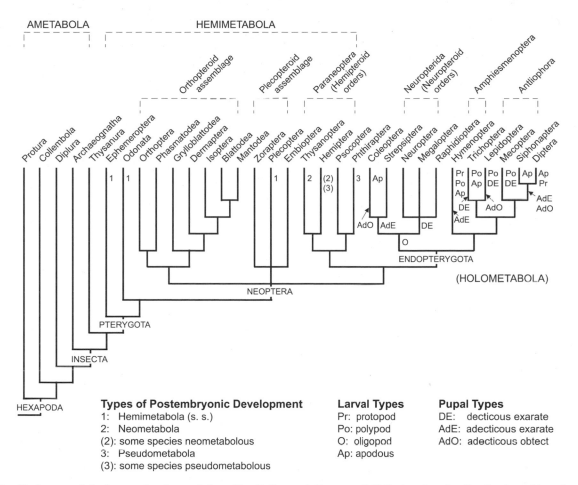

Types of Postembryonic Development
1: Hemimetabola (s. s.)
2: Neometabola
(2): some species neometabolous
3: Pseudometabola
(3): some species pseudometabolous

Larval Types
Pr: protopod
Po: polypod
O: oligopod
Ap: apodous

Pupal Types
DE: decticous exarate
AdE: adecticous exarate
AdO: adecticous obtect

Fig. 9.4. Phylogeny of the hexapod orders as inferred by Gullan and Cranston (1994), showing the distribution of larval and pupal types and the amount of metamorphosis occurring during postembryogenesis. (Cladogram adapted with permission from P. J. Gullan and P. S. Cranston. *The insects. An outline of entomology*, p. 193, Fig. 7.5. 1st ed. Chapman and Hall, London. © 1994 Blackwell Science, Ltd.)

Thrips, whiteflies, and scale insects constitute the aberrant Hemimetabola and, in fact, are holometabolous exopterygotes. The somewhat similar life histories of members of these three taxa have evolved independently through selection for sessile or cryptic juvenile feeding. This has resulted in a level of structural divergence between young and adults requiring the presence of two or three transitional pupal instars to bridge the structural gap between the two (Figs. 9.2–9.6). Sehnal and colleagues (1996) placed all three taxa in the Neometabola (Figs. 9.3, 9.4; Table 9.2).

9.4.3 HOLOMETABOLA
The approximately 765,900 known species in 11 orders of this most successful, monophyletic assemblage of insects (Figs. 0.2; 9.3: Raphidioptera; 9.4) are characterized by Kristensen (1999) as having
• a trimorphic life history (larva, pupa, adult).
• larvae with a structure, behavior, microenvironment, mouthparts, and food more or less differing from those of adults.

• compound eyes usually absent in juveniles and replaced by one or more pairs of specialized larval eyes or stemmata. These can be either a few, isolated ommatidia (Lepidoptera, Mecoptera), highly specialized compound eyes having only one cuticular lens for all ommatidia (sawfly larvae), or a retina of separate photoreceptor cells beneath a single cuticular lens (e.g., Fig. 8.27) (Paulus, 1979; Gilbert, 1994). During metamorphosis, the stemmata either degenerate or are withdrawn into the brain by shortening of photoreceptor axons where they remain photosensitive, and are replaced by eversion and differentiation of the imaginal compound eyes from eye imaginal discs set aside earlier in development (10.2.4.2.2). Dorsal ocelli are absent except for the median one in certain scorpionfly larvae (Sehnal et al., 1996).
• a transitional, preadult, pupal stage in which specifically larval structures transform into or are replaced by quite different imaginal ones (see Svácha, 1992, for a critical discussion).
• wings which develop as internal wing discs that evert

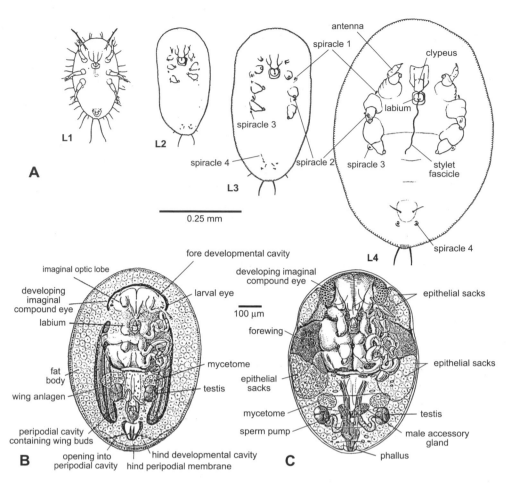

Fig. 9.5. Allometabola. A. The four juvenile instars (L1–L4) of the whitefly *Trialeruodes vaporariorum* (Hemiptera: Aleyrodidae) in ventral aspect. B. Ventral aspect of a whole mount of an early pharate adult male of *T. vaporariorum* within the L4 cuticle, showing an intermediate stage in the development of the compound eyes, legs, wings, and some internal organ systems. C. Same view of a ready-to-emerge adult male. (Reprinted with permission from H. Weber. Die postembryonale Entwicklung der Aleurodinen (Hemiptera-Homoptera). Ein Beitrag zur Kenntnis der Metamorphosen der Insekten. *Z. Morphol. Okol. Tiere* 29: 268–305. © 1934 Springer-Verlag.)

onto the surface in the exuvial pharate pupa (hence, Endopterygota, "inside wing-bearers"). However, in members of the basal lineages of this clade (Megaloptera, Raphidioptera, Neuroptera, some Coleoptera, Mecoptera, and Hymenoptera: Symphyta), wing primordia do not invaginate into the body until the last larval instar and never develop in larvae of some Coleoptera and nematocerous Diptera (Svácha, 1992; Sehnal et al., 1996; Kristensen, 1999). Endopterygotes can be further classified into those having functional, segmented thoracic legs in their larvae (Eoholometabola) and those lacking them (Euholometabola) (Figs. 9.2, 9.8, 9.9) (Weber, 1966).

Life cycle (e.g., Raphidioptera):   JO-JO-JO-JO-JO-JO-JO-Jd=JwQ=AW (death)
(e.g., *D. melanogaster*): Jd-Jd-Jd=JwQ=AW (death)

The phylogenetic distribution of these life history types is indicated in Fig. 9.4.

9.4.4 RECENT CONCEPTS
There recently has been a renewed interest in the classification of hexapod life histories and I consider two new concepts here.

9.4.4.1 NÜESCH
In 1987, Nüesch published a scheme for classifying insects based on their degree of structural change during metamorphosis (Table 9.1):

- Ametamorph hexapods have direct development, with no juvenile structures degenerating or being cast off upon adult emergence.
- Paurometamorph insects have a few specifically juvenile structures (e.g., certain muscles or sensilla) that degenerate or are lost at adult emergence.
- Hemimetamorph insects undergo extensive replacement of juvenile by imaginal structures, particularly in the last molt, but no transitional pupal instar in which this change occurs.

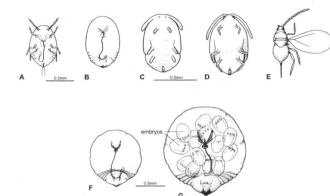

Fig. 9.6. Parametabola. Life history of the San José scale *Quadraspidiotus perniciousus* (Hemiptera: Diaspididae). A–E. Male instars in ventral aspect except E (adult male in dorsal aspect). A. First instar "crawler." B. Second instar. C. Third instar ("propupa"). D. Fourth instar ("pupa"). E. Fifth instar (adult). F and G. Females in ventral aspect. F. Second instar. G. Third instar adult with embryos. (Reprinted with permission from T. E. Woodward, J. W. Evans, and V. F. Eastop. Hemiptera (bugs, leafhoppers, etc.). In I. M. Mackerras (ed.), *The insects of Australia. A textbook for students and research workers*, p. 426, Fig. 26.40. 1st ed. Melbourne University Press, Carlton, Victoria. © 1970 CSIRO Australia.)

- Holometamorph insects also undergo extensive transformation of juvenile into imaginal structures, with this transformation occurring within one, two, or (rarely) three quiescent, pupal instars.

Nüesch recommended the term *juvenile* for the young instars of ametamorph and paurometamorph insects and *larva* for those of the remainder.

9.4.4.2 SEHNAL, SVÁCHA, AND ZRZAVY

In 1996, the whole subject of insect metamorphosis, including that of fossil taxa, but not of entognath apterygotes, was critically reviewed within a phylogenetic framework by Sehnal and colleagues (1996), who recognized six categories:

- apterous ametaboly (Archaeognatha and Zygentoma);
- alate ametaboly (probable in representatives of some extinct, Paleozoic ephemeropterans, protodonates, paleodictyopterans, plecopteroids, polyneopterans [Orthopteromorpha], and paraneopterans [hemipteroid insects] in which older juveniles as well as adults may have had functional wings and were capable of flight) (Fig. 9.14B; see Kukalová-Peck [1991] and 9.6.2.5.9);
- prometaboly (Ephemeroptera);

Table 9.1 Nüesch's (1987) Classification of Hexapod Metamorphosis

Taxon	Ametamorph	Paurometamorph	Hemimetamorph	Holometamorph
Protura	+			
Collembola	+			
Diplura	+			
Archaeognatha	+			
Zygentoma	+			
Ephemeroptera			+	
Odonata			+	
Orthoptera		+		
Phasmatodea	+			
Grylloblattodea				
Dermaptera	+	+		
Isoptera	+			
Blattodea	+			
Mantodea	+			
Zoraptera	+			
Plecoptera			+	
Embioptera	+			
Thysanoptera				+
Hemiptera				
Psylloidea		+		
Aleyroidea			+	
Aphidoidea	+			
Coccoidea—female	+			
Coccoidea—male				+
Cicadomorpha	+	+		
Fulogoromorpha	+	+		
Coleorrhyncha		+		
Heteroptera		+	+	
Psocoptera	+	+		
Phthiraptera	+			
Endopterygota				+

Table 9.2 An Overview of Insect Developmental Types and Their Presumed Evolutionary Relationships

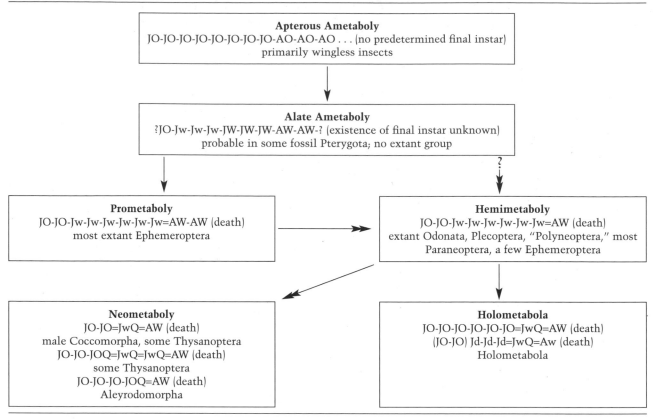

Apterous Ametaboly
JO-JO-JO-JO-JO-JO-JO-JO-AO-AO-AO . . . (no predetermined final instar)
primarily wingless insects

Alate Ametaboly
?JO-Jw-Jw-Jw-JW-JW-JW-AW-AW-? (existence of final instar unknown)
probable in some fossil Pterygota; no extant group

Prometaboly
JO-JO-Jw-Jw-Jw-Jw-Jw-Jw=AW-AW (death)
most extant Ephemeroptera

Hemimetaboly
JO-JO-Jw-Jw-Jw-Jw-Jw-Jw=AW (death)
extant Odonata, Plecoptera, "Polyneoptera," most
Paraneoptera, a few Ephemeroptera

Neometaboly
JO-JO=JwQ=AW (death)
male Coccomorpha, some Thysanoptera
JO-JO-JOQ=JwQ=JwQ=AW (death)
some Thysanoptera
JO-JO-JO-JOQ=AW (death)
Aleyrodomorpha

Holometabola
JO-JO-JO-JO-JO-JO=JwQ=AW (death)
(JO-JO) Jd-Jd-Jd=JwQ=Aw (death)
Holometabola

Source: Sehnal et al. 1996: Fig. 4.

Note: J, juvenile; A, adult; O, wings absent; w, nonfunctional wing pads present; d, nonfunctional wings present as invaginated wing discs; W, with functional wings; Q, nonfeeding and quiescent; -, molt; =, more or less extensive metamorphosis. Arrowhead, single origin; double arrowheads, multiple origin.

- hemimetaboly (Odonata, Plecoptera, a few Ephemeroptera, most Orthopteromorpha, most Paraneoptera);
- neometaboly (Thysanoptera, Hemiptera: Aleyroidea, male Coccoidea); and
- holometaboly (Endopterygota).

They used the term *larva* for the juvenile instars of all pterygotes except the subimago of mayflies, the pupa of endopterygotes, and the quiescent instars of neometabolans. Their ideas on the evolution of these types are summarized in Table 9.2 and Fig. 9.7.

9.5 The Concept of Stase

In 1938, the French acarologist F. Grandjean introduced the concept of stase to cope with the highly diverse postembryogenesis of mites (Acarina). A *stase* is one of a succession of forms in the life history of an arthropod. It differs from adjacent ones by one or more meristic (countable) characters. A stase differs from an instar in the following ways (André, 1989):

Instar	*Stase*
An instar is defined in relation to two molts.	A stase is not necessarily defined in relation to two molts (it may contain many instars).
Two successive instars may be quite similar or even identical.	Two stases may appear quite similar but differ from each other by one or more meristic characters.
Allometric characters are used to discriminate between instars.	Only meristic characters may be used to discriminate between stases.
An instar is related to growth.	A stase is a level of development.
Instars are not necessarily homologous from one species to another.	Stases of different species may be homologous.
The number of instars is constant or variable depending on species.	The number of stases is constant.

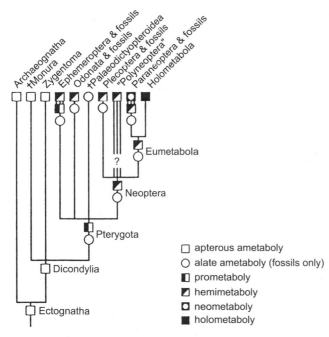

Fig. 9.7. Phylogenetic relationships of the major lineages of Ectognatha (Insecta *sensu stricto*), both fossil and recent, showing the distribution of postembryonic developmental types (details of these are illustrated in Fig. 9.3 and Table 9.2). Triple lines indicate multiple, parallel evolution of a particular type (the "Polyneoptera," containing the orthopteroid orders, are probably polyphyletic). If only recent insects are considered, the evolution of metamorphosis seems straightforward (only once in the stem species of Pterygota or, at most, three times independently in Ephemeroptera, Odonata, and Neoptera). Inclusion of "alate ametaboly" based on the Paleozoic fossil record as interpreted by Kukalová-Peck (1991) suggests that metamorphosis may have originated at least seven times and at least twice in Polyneoptera. (Reprinted with permission from F. Sehnal, P. Svácha, and J. Zrzavy. Evolution of insect metamorphosis. In L. I. Gilbert, J. R. Tata, and B. G. Atkinson (eds.), *Metamorphosis. Postembryonic reprogramming of gene expression in amphibian and insect cells*, p. 14, Fig. 4. © 1996 Academic Press, Ltd.)

| If variable, the number is environmentally determined. | The number is genetically fixed. |

Although the number of instars may equal the number of stases in a particular life history, for example, in thrips (Fig. 9.3: Thysanoptera), the number of instars is often greater. I provide two examples here.

The isotomid springtail *Folsomia quadriculata* usually has about 50 instars, only 5 of which are juvenile (9.4.1). Each juvenile instar differs from the others by differences in the number of setae. However, all 45 imaginal instars are identical except in size, although they differ in setation from all juvenile instars. Thus, this hexapod has 50 instars and 6 stases, 5 immature and 1 imaginal.

The lepidopteran *Eldana saccharina* (Pyralidae) can have 5–10 free-living larval instars, depending on environmental conditions experienced while it is a juvenile. This insect molts once within the egg prior to hatch, an embryonic molt, and hatches as a first instar. This instar molts to a second instar, differing from the first by the presence of additional setae. All subsequent larval instars are identical to the second except in size, whether there be three, four, five, six, seven, or eight. Thus, this insect can have 7–12 instars (including the pupa and adult) but only 5 stases: calyptostase (prolarva or pronymph; the nonfeeding, immotile form within the egg having an embryonic cuticle), larva 1, larval instars 2 to 5–10, pupa, and adult. A stase differs from a stage (e.g., larva, pupa, adult) in that adjacent stases may be practically identical.

9.6 Larvae and Pupae in Holometabola

As mentioned in 9.4.3, holometabolous insects are trimorphic and generally change form twice after they hatch: from the larva into a pupa and from the pupa into an adult.

9.6.1 LARVAE

The great Italian entomologist Antonio Berlese (1913) classified endopterygote larvae according to their superficial resemblance to the three phases of embryogenesis mentioned in 6.10.1—protopod, polypod, and oligopod—because he believed them to be protracted free-living embryos arrested at various stages of embryogenesis at the time of hatch (Fig. 9.13A). This theory, though probably wrong (Sehnal et al., 1996; but see Truman and Riddiford, 1999), provides the basis for our present system of classifying holometabolous larvae:

- Protopod larvae: The highly specialized, precocious first instars of many endoparasitoid Hymenoptera (6.7.2.1) and Diptera develop from eggs having little yolk and are little more than germ bands when they hatch (Fig. 9.8A). They can survive only because they are immersed in their food. Later instars are legless grubs.
- Polypod larvae: Such larvae are said to be eruciform and are characteristic of members of the Hymenoptera: Symphyta, Mecoptera, and Lepidoptera (Fig. 9.9A). They have a species-specific number of well-developed, paired abdominal prolegs and small, functional antennae and thoracic legs. Most are phytophagous or are general feeders and, if they change the nature of their food during larval development, can have correlated change in the structure of their mouthparts (e.g., Dewhurst, 1999).
- Oligopod larvae: Such larvae lack abdominal prolegs but have well-developed thoracic legs. There are two basic types: (1) *Campodeiform* larvae resemble adults of members of the entognath, dipluran family Cam-

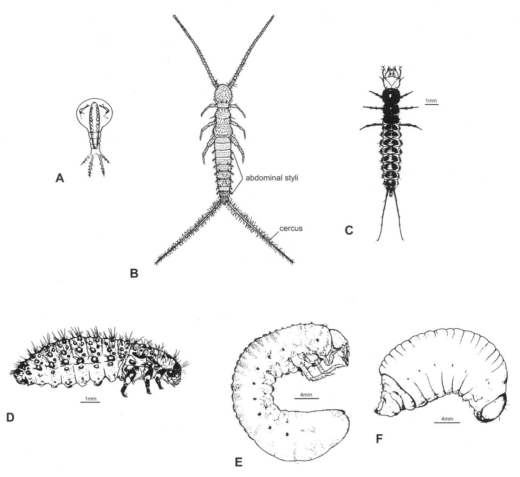

Fig. 9.8. Larval types in Endopterygota. A. Protopod cyclopoid larva (Hymenoptera: Platygasteridae) (from Riek, 1970: 881, Fig. 37.15C). B. Adult of *Campodea montis* (Diplura: Campodeidae) after which form campodeiform larvae are named. C. Campodeiform larva of *Eudalia macleayi* (Coleoptera: Carabidae) (from Britton, 1970: 509, Fig. 30.12A). D. Chrysomeliform larva of *Paropsisterna beata* (Coleoptera: Chrysomelidae) (from Britton, 1970: 509, Fig. 30.12B) . E. Scarabaeiform larva of *Anoplognathus pindarus* (Coleoptera: Scarabaeidae) (from Britton, 1970: 509, Fig. 30.12C). F. Apodous larva of *Trigonotarsus rugosus* (Coleoptera: Curculionidae) (from Britton, 1970: 509, Fig. 30.12D). (A and C–F reprinted with permission from I. M. Mackerras (ed.), *The insects of Australia. A textbook for students and research workers.* 1st ed. Melbourne University Press, Carlton, Victoria. © 1970 CSIRO Australia. B reprinted from H. V. Daly, J. T. Doyen, and A. H. Purcell III. *Introduction to insect biology and diversity*, p. 339, Fig. 18.1. 2d ed. © 1998 Oxford University Press.)

podeidae (Fig. 9.8B). The body is long and more or less sclerotized; the mouthparts are *prognathous* (directed forward; Fig. 9.8C) or *hypognathous* (directed downward; Fig. 9.8D); and the larvae are predacious, phytophagous, or *phoretic* on the bodies of other animals (hitch rides to disperse). They differ from exopterygote nymphs in having stemmata rather than compound eyes and in lacking external wing pads (Megaloptera, Raphidioptera [Fig. 9.3], Neuroptera, many Coleoptera, Strepsiptera, some Trichoptera). (2) *Scarabaeiform* larvae are C-shaped, stout, and cylindrical and have short thoracic legs (Fig. 9.8E) (Coleoptera: Scarabaeidae, Anobiidae, Scolytidae). Many other larval types—all variations on these themes—are illustrated by the numerous authorities in Stehr's publications (1987b, 1991).

• Apodous larvae: These lack functional thoracic legs and are of three types: (1) *Eucephalous* larvae have a well-developed, sclerotized head capsule and normal biting and chewing mouthparts moving in the horizontal plane (Figs. 9.8F, 9.9B) (some Diptera: Nematocera, Coleoptera: Curculionidae, some aculeate Hymenoptera). (2) *Hemicephalous* larvae have the head capsule reduced and more or less retracted into the prothorax, with mandibles moving in the vertical plane (Fig. 9.9C, C") (Diptera: some Nematocera and Brachycera). (3) *Acephalous* larvae have the head capsule greatly reduced and withdrawn into the body to form a cephalopharyngeal skeleton, with the mouthparts reduced to mouth hooks (Fig. 8.45) (Diptera: Cyclorrhapha, including larvae of the model species *D. melanogaster* [Fig. 9.9D]).

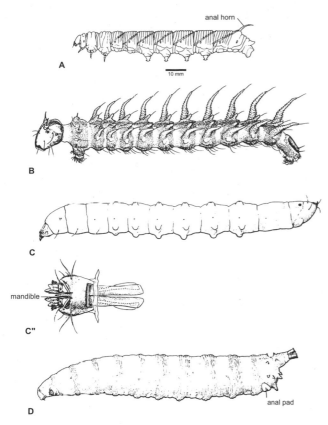

Fig. 9.9. Larval types in Endopterygota. A. Eruciform larva of *Manduca sexta* (Lepidoptera: Sphingidae) B–D. Apodous larvae of various Diptera. B. Eucephalous larva of *Atrichopogon polydactylus* (Ceratopogonidae) (from Downes and Wirth, 1981: 419, Fig. 28.131). C. Hemicephalous larva of *Laphria* sp. (Asilidae) (from Wood, 1981: 567, Fig. 42.77). C″. Head capsule of same partly pulled into front of body (from Wood, 1981: 567, Fig. 42.79). D. Acephalous larva of *Drosophila melanogaster* (Drosophilidae) (from Teskey, H. J. 1981: 125–147). A from *Immature insects* Volume I by Frederick W. Stehr. Copyright 1987 by Kendall/Hunt Publishing Company. Used with permission. B–D reprinted with permission from J. F. McAlpine, B. V. Peterson, G. E. Shewell, H. J. Teskey, J. R. Vockeroth, and D. M. Wood (eds.), *Manual of Nearctic Diptera*, Vol. 1. Monograph No. 27. Research Branch, Agriculture Canada. For the Department of Agriculture and Agri-Food, Government of Canada © Minister of Public Works and Government Services Canada, 1981.)

9.6.1.1 HYPERMETAMORPHOSIS (HETEROMORPHOSIS)
In some insects, two or more different larval forms occur within a single life cycle (Sehnal et al., 1996). Such forms are characteristic of some leaf-mining, predatory, and parasitoid insects in which a change in larval habit or food occurs during postembryogenesis. In members of some taxa the egg is laid in the open or in the soil, and on hatch, the first instar, an active planidium, must search for a host (Fig. 9.10). Such larvae are usually campodeiform and heavily sclerotized, and some appear to have three claws at the end of each leg (hence, the name

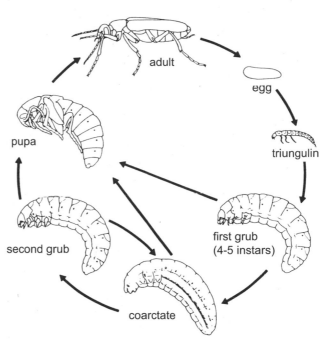

Fig. 9.10. Hypermetamorphosis in Meloidae (Coleoptera). The life cycle of blister beetles of the subfamily Meloinae. The larval stage includes the triungulin, first grub, coarctate, and second grub phases. The coarctate phase, second grub phase, or both may be omitted during development, while the second grub phase may be followed by either the pupal stage or a return to the coarctate phase (note arrows). (From *Immature insects* Volume II by Frederick W. Stehr. Copyright 1991 by Kendall/Hunt Publishing Company. Used with permission.)

triungulin; there is actually only one claw [*unguis*], the other two being large recurved setae).

An example is a blister beetle (Coleoptera: Meloidae) (Fig. 9.10). The planidium climbs into a flower and attaches to a foraging, ground-nesting bee visiting the flower for pollen or nectar (Selander, 1991). The bee carries it back to its nest. The planidium leaves the bee, enters a cell, and kills the host egg or larva that might already be present with its sickle-shaped mandibles. It then molts to a legless, first grub stage of several instars that feeds on the pollen and nectar with which the cell is provisioned (larvae of other species prey on the egg pods of locusts). Additional larval forms follow in some species (Fig. 9.10). The *coarctate larva* is a heavily sclerotized, diapausing, and immobile stage (cuticle in all joints and intersegmental membranes is sclerotized), and most of its internal tissues including its muscles degenerate, only to regenerate prior to its molting into a second grub or pupa (Berríos-Ortiz and Selander, 1979). The second grub stage molts into a normal pupa and then into the adult, but there is much variation in the numbers of each larval form in different species (Fig. 9.10).

Planidia also occur in certain species of Neuroptera: Mantispidae; Coleoptera: some Staphylinidae and

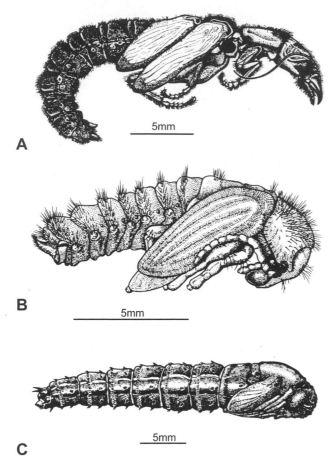

Fig. 9.11. The three types of pupae occurring in Endopterygota. A. Decticous exarate, *Archichauliodes guttiferus* (Megaloptera: Corydalidae). B. Adecticous exarate, *Paropsis atomaria* (Coleoptera: Chrysomelidae). C. Adecticous obtect, *Pelecorhynchus fulvus* (Diptera: Pelecorhynchidae). (Reprinted with permission from H. E. Hinton and I. M. MacKerras. Reproduction and metamorphosis. In I. M. Mackerras (ed.), *The insects of Australia. A textbook for students and research workers*, p. 105, Fig. 4.18. 1st ed. Melbourne University Press, Carlton, Victoria. © 1970 CSIRO Australia.)

Rhipiphoridae; Diptera: Acroceridae, Bombyliidae, Nemestrinidae; Strepsiptera; Lepidoptera: Epipyropidae, Gracillariidae; and Hymenoptera: Perilampidae, Eucharitidae (see various authorities in Stehr, 1987b, 1991, and Wagner et al., 2000).

When eggs are deposited within a host egg or larva, the newly hatched parasitoid larva is often protopod (Fig. 9.8A) (but see Stehr, 1987b, and Sehnal et al., 1996) and subsequently becomes apodous. The large mandibles present in some protopod larvae are used for killing other parasitoid larvae already present in the host due to superparasitism or multiparasitism (6.7.2.1; Hymenoptera: Parasitica).

9.6.2 PUPAE

The pupa is the ultimate quiescent (usually), nonfeeding, transitional juvenile instar of holometabolous insects in which larval structures are replaced by or transform into imaginal structures (Daly et al., 1998).

9.6.2.1 PUPAL TYPES

In 1948, Howard Hinton of Bristol University devised a scheme for classifying pupae based on how the adult escapes from the pupal cuticle and cocoon (Fig. 9.11). *Decticous pupae* have large functional mandibles moved by muscles of the pharate adult (Figs. 9.11A, 9.12A) and use them to cut a hole in the cocoon or pupal cell prior to eclosion. All such pupae are *exarate*, with their appendages free except where they attach to the body. This type is believed to be ancestral in Endopterygota and is known to occur in Megaloptera (Fig. 9.11A), Raphidioptera (Fig. 9.3), Neuroptera and Trichoptera, basal mandibulate Lepidoptera, and Mecoptera.

Adecticous pupae have reduced, nonfunctional mandibles or lack them entirely and are of two types:
- *Exarate adecticous pupae* have their appendages free from the body surface as in decticous pupae and are known to occur in male strepsipterans, certain coleopterans (Figs. 9.10, 9.11B), hymenopterans, siphonapterans, and some dipterans. Higher flies (Diptera: Cyclorrhapha) have pupae of this type (Fig. 10.8G, H), but they are enclosed within the sclerotized, third instar cuticle, the *puparium* (Fig. 10.8A); it is referred to as a "coarctate" pupa in the older literature.
- *Obtect adecticous pupae* have their appendages glued to the body with their exposed surfaces heavily sclerotozed and are characteristic of certain coleopterans, lepidopterans, and nematocerous dipterans (Fig. 9.11C) (the motile pupa of mosquitoes is of this type [Fig. 10.37E]).

The prepupa (pronymph, eonymph) of the life history literature is not a separate instar but rather a pharate pupa within the last larval cuticle. It is sometimes considered a separate instar by field biologists because it deforms the shape of the last larval cuticle and is usually shorter and more robust than previous larval instars. An additional confusing factor is that in some sawflies, the last larval instar can differ in color from previous ones. The phylogenetic distribution of pupal types is shown in Fig. 9.4.

9.6.2.2 PUPAL PROTECTION

Insect pupae are usually quiescent because of the structural reorganization occurring within their bodies associated with changing from larva to adult and hence are vulnerable to predators and parasitoids. Pupae and pharate adults of Megaloptera, Raphidioptera, some Neuroptera, Mecoptera, and mosquitoes are more or less motile. Larvae, therefore, usually enter protected situations (under bark, in leaf litter or soil, etc.) before they pupate, often within a protective "cell" or cocoon spun by the last instar larva of silk produced by the salivary glands or Malpighian tubules (Neuroptera). Insects pupating in the open, as do most butterflies (the *chrysalis*), often have cryptically colored cuticle in the pupal stage.

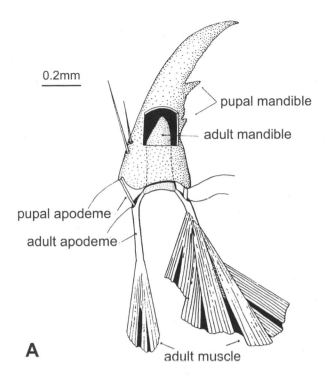

0.2mm

pupal mandible

adult mandible

pupal apodeme

adult apodeme

adult muscle

A

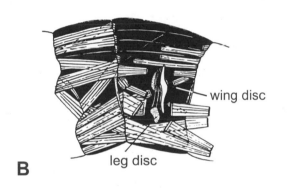

wing disc

leg disc

B

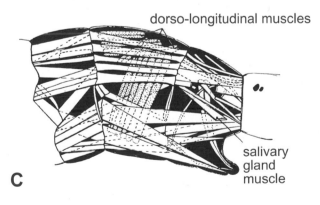

dorso-longitudinal muscles

salivary
gland
muscle

C

Fig. 9.12. A. Pupal mandible of the decticous exarate pupa of *Rhyacophila dorsalis* (Trichoptera: Rhyacophilidae), enclosing and manipulated by the mandible of the pharate adult. B and C. Mesothorax and metathorax of *Simulium ornatum* (Diptera: Simuliidae). B. Penultimate larval instar, showing the rudiments of the indirect flight muscles and tergal depressor of the trochanter associated with leg and wing discs. C. Pharate pupa, showing how these muscles have developed. (A reprinted with permission from H. E. Hinton and I. M. MacKerras. Reproduction and metamorphosis. In I. M. Mackerras (ed.), *The insects of Australia. A textbook for students and research workers*, p. 94, Fig. 4.12C. 1st ed. Melbourne University Press, Carlton, Victoria. © 1970 CSIRO Australia. B and C reprinted with permission from H. E. Hinton. The origin and function of the pupal stage. *Proc. R. Entomol. Soc. Lond.* 38: 77–85. © 1963 The Royal Entomological Society of London.)

staltic contractions in their abdomens. Pupal-adult ecdysis occurs when the surface is reached. Adults of species having adecticous pupae (Fig. 9.11B, C) either shed the pupal cuticle inside the cocoon and use their adult mandibles to escape or have sharp cocoon cutters of sclerotized cuticle on the anterodorsal pupal cuticle, which can be pressed or rubbed against the cocoon to cut a hole through it just like the egg burster of a ready-to-hatch prolarva (9.2). Adults of higher flies, which pupate within the sclerotized last larval cuticle (puparium), pop off its apex via a circular split in its cuticle (Fig. 10.8A; hence, the taxon name Cyclorrhapha for these insects) using rhythmic, hemostatic expansion and collapse of an eversible spinous sac of cuticle in the head above the antennae: the *ptilinum* (Fig. 10.8F–H) (Fraenkel and Bhaskaran, 1973).

In simuliids, deuterophlebiids, blepharocerids, and some other aquatic dipterans that pupate underwater, the adult pops to the surface within a bubble of air and takes flight immediately. In pharate adults of such insects, the wings are fully developed within the pupal cuticle but are characteristically folded, enabling the wings to straighten and function as soon as the insect breaks through the surface film.

On emergence, most adult endopterygotes void a large volume of material from the anus called the *meconium*, which is the remains of larval midgut cells and waste material from the hindgut and Malpighian tubules that accumulated during metamorphosis (10.2.3.2). In neuropterans and some hymenopterans, the midgut and hindgut do not become confluent until the pharate pupal stage or after the molt into the pupal stage (Snodgrass, 1956; Daly et al., 1998).

9.6.2.3 ADULT ECLOSION

Adults of species having decticous pupae (Fig. 9.11A) cut their way out of the cocoon using their pupal mandibles (Fig. 9.12A) and worm their way to the surface, still surrounded by pupal cuticle, by posterior to anterior peri-

9.6.2.4 FUNCTIONAL SIGNIFICANCE OF THE PUPAL STAGE

Structural divergence between young and adult of more than a certain, relatively small amount necessitates the presence of a transitional pupal instar in which the larva can change to an adult (Fig. 9.2). Thus, as juvenile members of intermediate and advanced clades of Holometabola adapted to particular specialized microhabitats, they gradually diverged in structure from their adults; that is, they evolved specifically larval or cenogenetic characters (De Beer, 1962). As this divergence increased, the last juvenile instar gradually became a transitional, quiescent form, allowing transformation of the now quite different, penultimate juvenile into its adult (Fig. 9.13D). Such a transition stage allowed the juveniles to specialize still farther, so that many are now able to exploit habitats closed to their adults and to the adults and young of apterygote and exopterygote insects. For example, because the wing primordia of many holometabolous larvae develop internally as wing discs, these larvae can bore or mine in soil (the only exopterygote juveniles able to do so are those of mole crickets and cicadas), leaves, or wood. Perhaps because of their backwardly directed wing pads, most exopterygote nymphs cannot bore, as they might experience difficulty backing up in their tunnels owing to their pads "hanging up" on the walls. (To my knowledge this has not been observed, and it may be significant that nymphs and winged adults of the exopterygote psocid *Psilopsocus mimulus* were recently shown to bore in wood [Smithers, 1995].) In addition, at least in members of the higher lineages of Endopterygota, there is no longer competition between larvae and adults of the same species for food and habitat because they are so different in structure and behavior, often have different mouthparts, and usually live in different microhabitats. Finally, both larvae and adults are more efficient in using their resources because they are highly adapted to them and are unencumbered with equipment they don't use. That the pupa is a common overwintering stage is probably secondary in representatives of many species (Sehnal et al., 1996).

9.6.2.5 MODELS EXPLAINING THE EVOLUTIONARY ORIGIN OF THE PUPAL STAGE

Since the time of William Harvey (1651), countless authors have speculated on how the pupal stage of holometabolous insects might have evolved from a motile last juvenile instar of a common, hemimetabolous ancestor (Fig. 9.7) (Novák, 1975). All recent and fossil evidence suggests that holometabolous insects are monophyletic and descended from a common ancestor that was hemimetabolous (Table 9.2; Figs. 9.4, 9.7) (Kristensen, 1991, 1999; Kukalová-Peck, 1991; Sehnal et al., 1996; Whiting et al., 1997; Truman and Riddiford, 1999; Wheeler et al., 2001). Of the many theories advanced to explain the origin of the pupa (Novák, 1975; Sehnal et al., 1996), few have been tested and, except for those of Kulalová-Peck (1991), Sehnal and colleagues (1996), and Truman and Riddiford (1999), are weakened by a failure to consider endocrinology, phylogeny, and the fossil record. I summarize 11 of these below.

9.6.2.5.1 BERLESE (1913) AND JESCHIKOV (1941)

According to the theory of Berlese (1913) and Jeschikov (1941), the eggs of exopterygote insects, because of their relatively greater amount of yolk (Fig. 6.1A, B), yield embryos that hatch at a postoligopod stage of development, whereas those of endopterygotes, with relatively less yolk (Fig. 6.1C), yield embryos hatching at protopod, polypod, or early oligopod stages of embryogenesis depending on species (Fig. 9.13A). The abdominal appendages of polypod embryos (Fig. 6.22B) are said to be homologous to the abdominal prolegs of eruciform larvae (Fig. 9.9A) and to be serially homologous with the thoracic legs, an idea now well supported by genetic and molecular data (Warren et al., 1994; Shubin et al., 1997). In addition, the endopterygote larval instars are considered to be free-living embryos, and the pupa is considered to result from compression of all postoligopod instars of the exopterygote ancestor into a single, quiescent transitional instar (Fig. 9.13A).

However, although there is a trend, the eggs of many holometabolous insects do not have relatively less yolk than those of hemimetabolous ones (Hinton, 1981). Hinton (1955) also claimed that larval prolegs, in the various orders in which they occur, have evolved independently and are not serially homologous with thoracic legs (but see Warren et al., 1994). Sehnal and coauthors (1996) provided additional criticisms.

9.6.2.5.2 POYARKOFF (1914) AND HINTON (1948)

According to the theory of Poyarkoff (1914) and Hinton (1948), the embryos of exopterygote and endopterygote insects hatch at an equivalent stage of development, so that nymphs and larvae are homologous (Fig. 9.13B). The adult stage of an exopterygote ancestor with functional wings is suggested to have become divided into two instars—the pupa and adult—the first providing a mold with the proper spatial relationships for development of adult wing muscles and the second allowing these muscles to attach to adult cuticle (the subimago and adult of most extant Ephemeroptera might resemble such instars) (Fig. 9.3). In support of this idea is the fact that the pupa more closely resembles the adult than the larva in members of advanced clades of endopterygotes. Absence of an imaginal molt in most exopterygotes (e.g., Fig. 9.3: Orthoptera) is explained by the fact that the thoracic muscles of the last instar nymph and adult are so similar that no mold is required. Hinton further suggested that the pupa arose before larval wing rudiments had been internalized, the latter resulting from selection for subsequent, additional larval specialization.

This model has been criticized for being teleological (a "need" for a mold), because its structural basis is untrue (some pupae resemble larvae almost as much as the adults, e.g., Fig. 9.3: Raphidioptera), and because the existence of two or three independently evolved pupal stages in aberrant exopterygotes (Neometabola) tends to refute it (Figs. 9.3: Thysanoptera; 9.5B, C; 9.6C, D). In fact, Hinton (1963) himself later demonstrated that primordia of the imaginal flight muscles are present in

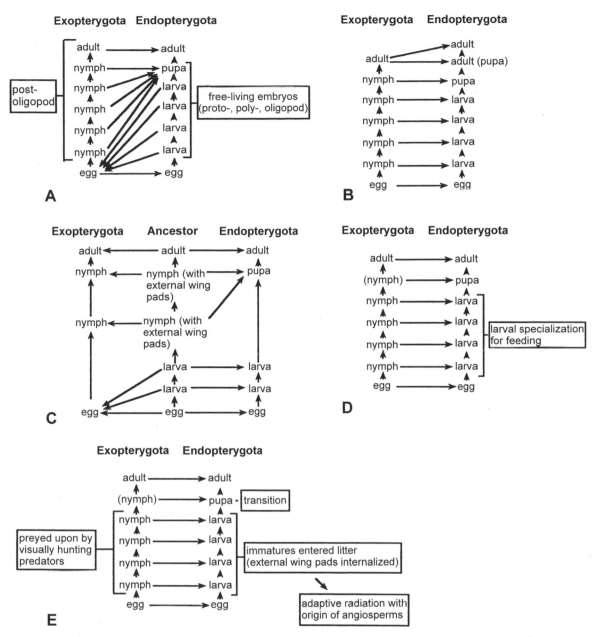

Fig. 9.13. Theories concerning the origin of the endopterygote pupa. A. Berlese (1913) and Jeschikov (1941). B. Poyarkoff (1914) and Hinton (1948). C. Heslop-Harrison (1958). D. Hinton (1963). E. Downes (1987).

younger black fly (Simuliidae) larvae, so that no mold is required (Fig. 9.12B, C).

9.6.2.5.3 HESLOP-HARRISON (1958)

Heslop-Harrison (1958) suggested that both exopterygote and endopterygote insects evolved from a common ancestor having a life history similar to that of extant thrips (neometabolous) and to the reproductive castes of termites (Isoptera) (Fig. 9.13C). In exopterygotes, the earlier larval instars were suggested to be compressed into embryogenesis so that the insect hatched at a postoligopod stage of development, the nymphal stages, corre-

sponding to the nymphal stages of the ancestor, persisting. In endopterygotes, the nymphal instars were suggested to have been compressed into a single pupal instar (one or more molts were lost), while younger larval instars were retained.

This theory was based on knowledge of the life histories of thrips and some termites but lacks supporting evidence. The earliest, Upper Carboniferous, winged fossils are exopterygote, while neometabolous insects do not appear in the fossil record until the Permian or later (Fig. 13.1) (Kukalová-Peck, 1991; Carpenter, 1992a, 1992b). In addition, their presently accepted phylogenetic relationships (Fig. 9.4) would suggest otherwise.

9.6.2.5.4 HINTON (1963)

After he discovered that primordia of adult flight muscles of *Simulium* (black fly) larvae start developing well before the larva-pupa molt (Fig. 9.12B, C), Hinton (1963) modified his own previous theory (Fig. 9.13D). In this model, the pupa is said to be homologous to the last juvenile instar of an exopterygote ancestor, while holometabolous larvae and the remaining exopterygote nymphs are homologous. Hinton proposed that during the evolution of endopterygotes, the last nymphal stage with external wing pads was retained to bridge the structural gap between the increasingly specialized earlier instars and the adult—hence, the similar appearance of pupae and adults in modern endopterygotes. At first, this pupa would also have resembled older nymphs, as in present-day exopterygotes (Fig. 9.3: Orthoptera) and basal endopterygotes (Fig. 9.3: Raphidioptera). These earlier juvenile stages then became progressively adapted for feeding in diverse habitats and increasingly diverged in form from the last nymphal instar (the pupa) and adult, the reproductive and dispersal stage. As larval divergence increased, the pupa assumed the function of a transitional instar between larva and adult, eventually to the extent that it ceased to feed and move because of the magnitude of the structural reorganization occurring within it. The larval-pupal molt then assumed the function of allowing evagination of wings (Fig. 9.3: Raphidioptera). The internalization of wing primordia as imaginal discs, an additional larval specialization, arose later and allowed continuous development of wings but in such a way as to no longer interfere with larval function. The pupal-adult molt ensured successful release of wings from the pupal cuticle after axillary sclerites formed in the wing bases. Once such an instar was available for metamorphic change, the way was open for additional larval specialization and for the gradual loss of larva-adult competition for food and habitat.

The propupae and pupae of Neometabola (Figs. 9.3: Thysanoptera; 9.5B, C; 9.6C, D) are thought to have evolved independently in thrips, whiteflies, and male scale insects, but for similar reasons—adaptation of juveniles to a sessile or cryptic feeding lifestyle—while the single pupal instar of endopterygotes arose only once in the ancestor of the Holometabola (but see Svácha, 1992). As indicated in section 9.6.2.4, this theory is generally accepted but has had additional details added to it by Sehnal and colleagues (1996).

9.6.2.5.5 WIGGLESWORTH (1964)

In his wonderful book *The Life of Insects*, Wigglesworth (1964) suggested that complete metamorphosis was an example of sequential polymorphism and that its origin resulted from the independent, gradual evolution of three relatively independent sets of genes by gene triplication, divergence, and co-option, with one set eventually controlling larval, one set pupal, and one set adult differentiation (Fig. 9.14A). Regulatory genes were suggested to function in switching development from one group of genes to the next, their expression being mediated by molting (20-OH ecdysone [20E]) and juvenile (JH) hormones whose release was evoked by receipt of appropriate environmental cues (12.1, 12.2).

In Fig. 9.15A, this model is compared to a filing cabinet, with a drawer for the gene set of each stage and a drawer at the bottom for the thousands of genes common to all stages ("housekeeping" genes encoding enzymes, structural proteins, etc.) (Willis, 1986). Release of 20E was said to permit a particular drawer to be opened, and release of JH, to regulate which drawer. Fig. 9.15B illustrates how such separate gene sets might have arisen sequentially during evolution.

Willis (1986) suggested that Wigglesworth's model, more a genetic explanation than a theory of origin, was simplistic and refutable by more recent evidence. She and her students systematically compared cuticular proteins in larvae, pupae, and adults of the mealworm *Tenebrio molitor* (Coleoptera: Tenebrionidae) and of the saturniid moth *Hyalophora cecropia* (Lepidoptera) and found many of them to be the same in larvae and pupae but to be somewhat more different in adults (of 152 protein bands in *H. cecropia* cuticle, 105 [69%] were found to occur in two or more stages). In other words, many of the same proteins are used to make the different cuticles of larvae and pupae. They also discovered that when different proteins are used in different regions of the same or of different stages, they perform different functions, not because of a requirement for use of stage-specific genes (Table 9.3). The few proteins restricted to a single stage occurred in only one or two anatomical regions. Willis (1986) predicted that when the cuticular proteins of representative ametabolous and hemimetabolous insects are surveyed, the same diversity of proteins will be found as in endopterygotes (to my knowledge, such comparison has yet to be made).

Willis (1986) also examined promotor utilization by the gene encoding the enzyme alcohol dehydrogenase (ADH) in fat body and Malpighian tubule cells of *D. melanogaster* and found the cells of both tissues to change promotors in mid third (last) instar (larvae use a proximal promotor; adults, a distal one), suggesting that modification of the regulatory elements of genes might have occurred to permit the evolution of metamorphosis. (The Malpighian tubule cells are carried through metamorphosis, whereas a new adult generation of fat cells develops; 10.2.2.4, 10.2.3.2.) There is only one gene and one protein for ADH but its transcripts are in two sizes: one larval, the other adult. Willis thought that such "promotor shifting" resulted from having different metamorphic stages and suggested that different promotors in a gene may be necessary to enable it to respond to the different hormonal concentrations acting in each stage (12.2, 12.5).

Based on these findings, Willis devised a new filing cabinet model with only three drawers: one for flexible, one for sclerotized, and one for specialized cuticles, with

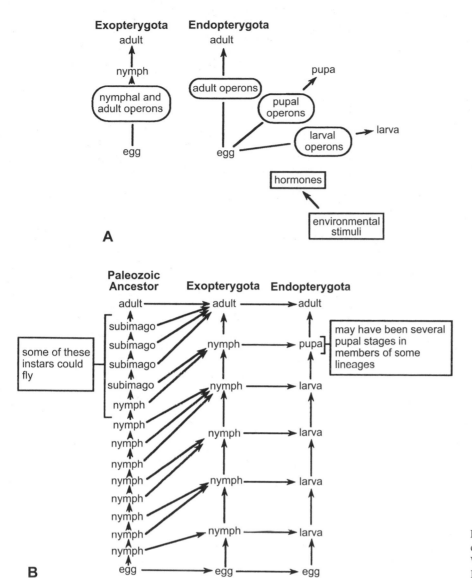

Fig. 9.14. Theories concerning the origin of the endopterygote pupa. A. Wigglesworth (1964). B. Kukalová-Peck (1991).

20E opening all three drawers at each stage (Fig. 9.15C). A single tissue within a single stage can thus use files from all three drawers. JH acting in its antimetamorphic role (12.2, 12.4.5) prevents a cell from using a file not previously used.

9.6.2.5.6 BRYANT (1969)
Bryant (1969) devised a simulation model favoring the evolution of complete metamorphosis based on contrasting modes of habitat selection by hypothetical insects in a spatially heterogeneous environment: either a priori habitat selection by ovipositing females or a posteriori selection by migrating immatures. His model predicted that under these conditions, selection will favor increased habitat discrimination by ovipositing females and loss of migratory ability in juveniles, resulting in a partition of postembryogenesis into that characteristic of endopterygotes: sedentary, feeding juveniles; vagile adults specializing in reproduction and dispersal. Once such partition has occurred, selection will facilitate additional divergence between larvae and adults and development of the last juvenile instar into a transitional, pupal stage, as postulated by Hinton (1963) (Fig. 9.13D). This proposal makes sense and complements the ideas of Hinton (1963) and Sehnal and colleagues (1996).

9.6.2.5.7 BERNAYS (1986B)
In a unique quantitative study, Bernays (1986b) compared feeding efficiency, growth, and respiration between an exopterygote, *Melanoplus sanguinipes* (Orthoptera: Acrididae), and an endopterygote, *Pseudaletia unipuncta* (Lepidoptera: Noctuidae), having similar food requirements, instar number, temperature preferences, and adult body weight. Under identical rearing conditions (light-dark cycle 12:12h; 25–35°C [light], 19–25°C [dark]), the caterpillars grew significantly faster

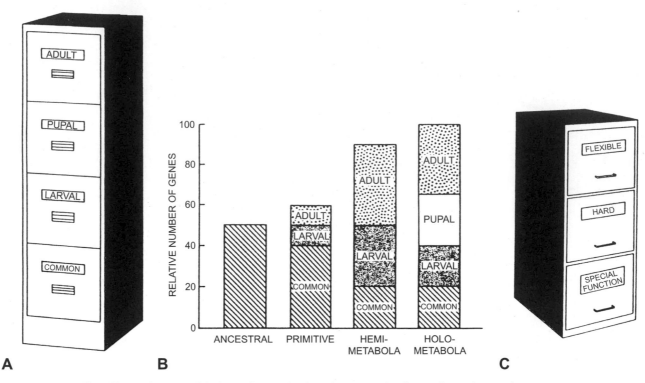

Fig. 9.15. A. Willis's filing cabinet model of Wigglesworth's (1964) sequential polymorphism theory, showing organization of genes into stage-specific sets. B. The evolution of metamorphosis according to the stage-specific gene set paradigm of Wigglesworth (1964). C. Willis's filing cabinet model for gene organization in accord with data obtained from analyses of cuticular proteins in different life history stages. (Reprinted with permission from J. H. Willis. The paradigm of stage-specific gene sets in insect metamorphosis: time for revision. *Arch. Insect Biochem. Physiol. Suppl.* 1: 47–57. © 1986 Entomological Society of America.)

Table 9.3 Distribution of Protein 12

Region	Larval	Pupal	Adult
Sclerite	+	–	+
Intersegmental membrane	+	+	+
Tubercle	–		
Dorsal wing cuticle		–	–
Ventral wing cuticle		+	–

Source: Adapted from Willis, 1986: 52.

than the grasshopper larvae (0.40 vs. 0.21 mg/mg body weight/d) and were more efficient in using energy (no differences in respiration rate in spite of increased growth rate) and in converting food to biomass (8.5 vs. 2.4 mg/d; due to the doubled gut capacity of caterpillars). She postulated the principal cause of this difference in net growth efficiency to be the requirement of the grasshopper to produce more cuticle (49% vs. 4.2% of estimated dry body weight, excluding gut contents for all of postembryogenesis)—considered by Bernays to "represent the great burden of hemimetabolous larval life"—and suggested this greater nutritional efficiency of endopterygotes, if it is widespread, to be an important evolutionary advantage of holometabolous over

hemimetabolous development. To my knowledge, comparable studies are still unavailable. The hard data in this empirical study are a refreshing change from the speculation rampant in some other theories considered in this section.

9.6.2.5.8 DOWNES (1987)

Downes (1987) considered the different levels of metamorphosis occurring in insects to have arisen through selection by terrestrial vertebrate predators (Fig. 9.13E). He suggested that the end-of-Permian extinction of certain Paleozoic insect lineages (Monura, Paleodictyoptera, Megasecoptera, Diaphanopterodea, Permothistida, Caloneurodea, Blattinopsodea; see Fig. 13.1) resulted from increasingly effective predation by visually hunting tetrapod predators, the insects living on the ground's surface being the first to go (vertebrate tetrapods began to diversify in the Carboniferous between 380 and 360 million years ago [Carroll, 1988]). He proposed that such predation stimulated divergence between juvenile and adult modes of life in the ancestors of holometabolous insects by selecting for life in leaf litter and other protected places by immatures (adaptation to such lifestyles would eventually include internal wing development and other larval specializations). Finally, he suggested the origin of wing flexing in neopterans (13.2.1), of piercing and sucking mouthparts in hemipteroids (8.3.8), and of small size in collembolans, proturans, diplurans, psocids and thrips, and so on, and the incredible radiation of holometabolous insects (Fig. 0.2) to have resulted from selection by tetrapod predators.

Though this model relates to life history theory (Stearns, 1992) more than the others, I've seen no published experimental support for or against it, and find it to be naive. Seventeen of the 32 extant hexapod orders contain a significant number or a majority of litter-dwelling species, of which only four are holometabolous (Fig. 13.1). There is also fossil evidence to indicate that wing flexing and complete metamorphosis had arisen by the time of the Upper Carboniferous before effective terrestrial tetrapod predators had evolved (Fig. 13.1) (Kukalová-Peck, 1991; Carroll, 1988; Kristensen, 1999) and that the end-of-Permian extinctions had other causes (13.2.2.1).

9.6.2.5.9 KUKALOVÁ-PECK (1978, 1991)

Kukalová-Peck (1978, 1991) marshalled extensive fossil evidence from the Upper Carboniferous and Permian to suggest that all known insects living at that time were either wingless (apterous ametaboly) or had articulated functional wings, not only in adults but also in older juveniles (alate ametaboly; Table 9.2; Figs. 9.7, 9.14B). In addition, the older, preadult (subimago) instars of some of these insects appear to have had functional, though smaller wings not only on the mesothorax and metathorax but also on the prothorax and on one or more abdominal segments (Fig. 9.16) (the abdominal gills and gill flaps of extant mayfly nymphs are considered to be lineal descendants of these abdominal wings).

Kukalová-Peck (1978, 1991) defined metamorphosis as that period in postembryogenesis when the wings increase dramatically in size and straighten and when their final functional articulations form, and referred to the instar(s) in which this transformation occurs as *metamorphic instar(s)*. Though metamorphosis appears to have been absent in members of some taxa living at that time (Fig. 9.7), it seems to have evolved independently, in parallel, and at different rates in representatives of those lineages giving rise to extant ephemeropterans, odonates, plecopterans, orthopteroids, blattoids, hemipteroids, and endopterygotes.

Most insects living in the Upper Paleozoic appear to have had numerous juvenile and subimaginal instars (Fig. 9.14B). Kukalová-Peck (1991) proposed that several subimaginal instars were condensed into the adult stage of extant descendant insects, the remaining subimaginal and older nymphal instars were condensed into one or more metamorphic instars, and the younger nymphal instars into a smaller number of composite nymphal instars (Fig. 9.14B), the pattern of condensation varying from one lineage to another and correlating with gradual, divergent, juvenile and imaginal specialization. The reduced number of molts in most extant insects (Table

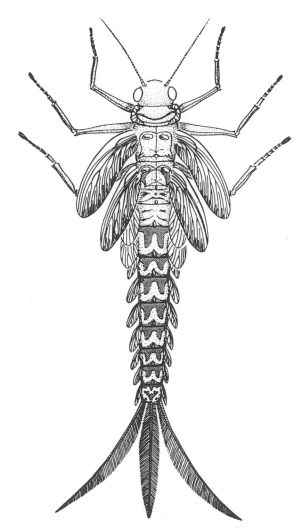

Fig. 9.16. Reconstruction of an older nymph of a Lower Permian mayfly (Ephemeroptera; Protereismatidae). Its mesothoracic and metathoracic wings were curved backward, appear to have been articulated, and were probably used for underwater rowing. Its prothoracic winglets were fused to the pronotum while its abdomen was equipped with nine pairs of veined wings. (Origin and evolution of insect wings and their relation to metamorphosis, as documented by the fossil record, J. Kukalová-Peck, *J. Morphol.*, copyright © 1978 Wiley-Liss, Inc. Reprinted by permission of Wiley-Liss, Inc., a subsidiary of John Wiley & Sons, Inc.)

9.4) was suggested to have arisen because of the vulnerability of insects to predators and other adverse environmental factors during molts. She believed the presently accepted Hemimetabola to be polyphyletic but accepted the Holometabola as holophyletic, as shown in Fig. 9.4, and to be defined by internal wing development, the pupa being a consequence of such development. Her thinking in other respects is identical to that of Hinton (1963), as summarized earlier (Fig. 9.13D).

This model seems to be well supported by the Paleozoic fossil record and by recent developmental genetic and molecular evidence (Raff, 1996; Shubin et al., 1997), but much of the fossil evidence is based on analysis of concentrated assemblages of compression fossils of disarticulated wings and of variously sized nymphs, and there is no way of establishing whether these belong to the same species and actually represent an ontogenetic series. Some researchers (e.g., Fryer, 1996) suggest that Kukalová-Peck reconstructs in her mind an idealized ancestor and then "finds" its characters in the fossils, whether these are present or not.

9.6.2.5.10 SEHNAL, SVÁCHA, AND ZRZAVY (1996)

In many respects, the theory of Sehnal, Svácha, and Zrzavy (1996) resembles that of Hinton (1963) (Fig. 9.13D), and they accept most of its criteria. However, in addition, it includes information from comparative endocrinology and is unique in being related to phylogeny. Their principal conclusions follow:

- All insects, with a few, highly adapted exceptions (e.g., certain endoparasitoid hymenopterans; Fig. 9.8A), hatch at a similar stage of development.
- Metamorphosis may have originated independently several times within the pterygotes, is associated with the evolution of wings, and results from divergent evolution between juveniles and adults (Table 9.2, Fig. 9.7).
- Endopterygotes are characterized by a delay in the appearance of external wing buds until the last, preadult instar (pupa), which is considered to be a quiescent juvenile.
- Amount of delay correlates with degree of larval specialization (Fig. 9.2) and prevents the presence of wing primordia from interfering with larval function.
- That wings usually develop within the bodies of preceding larval instars is secondary and allows wing development to continue during larval development in spite of larval specialization.
- Originally, there was probably no tight correlation, as there is today, between structural metamorphosis, by which the insect acquires the ability to fly, sexual maturation, and obligate termination of molting.
- Suppression of adult molting during evolution may have been due to development of wings and to changes in thoracic sclerotization associated with ability to fly that rendered molting impossible. (But what about the subimago of many extant mayflies and the fossil record indicating the existence of alate ametaboly?)
- In extant pterygotes, JH and 20E coordinate structural metamorphosis with attainment of sexual maturity and an end to molting (12.2). Endocrine changes allowing evolution of metamorphosis included
  —acquisition of a metamorphic function for JH in juveniles (12.4.5) in addition to its original, gonadotrophic one (1.3.7.2, 2.3);
  —degeneration of the prothoracic glands (source of 20E) in response to the presence of ecdysteroids in the absence of JH (12.4.1.3); and
  —diversification in the hormonal control of gametogenesis (1.3.7.2, 2.3).
- Cladistic analysis of comparative information on endocrine function throughout the hexapods could

reveal details relevant to understanding the evolution of metamorphosis, but such knowledge is limited to a few common model species. Study of representative apterygote, mayfly, stonefly, odonate, neometabolan, basal endopterygote (Megalopera, Raphidioptera, Neuroptera, Mecoptera), and hypermetamorphic species (Fig. 9.10) is required.

The critical synthesis of Sehnal and colleagues (1996) is presently the best overview of this topic and the quickest way to access its literature.

9.6.2.5.11 TRUMAN AND RIDDIFORD (1999)

According to Truman and Riddiford (1999), ancestral and extant ametabolous insects had three developmental stages—pronymph, nymph (N1–N4), and adult (Fig. 9.17A)—directly comparable respectively to the larval, pupal, and adult stages of extant holometabolous insects (Fig. 9.17E, F). *Pronymph* is the term used for the specialized hatching stage in ametabolous and basal hemimetabolous insects whose proportions differ from those of the first instar, which doesn't feed, which is mobile or not, and which is still surrounded by a pronymphal (embryonic) cuticle. This cuticle differs from nymphal cuticle in its ultrastructure, in its reduction or lack of functional sensilla, and in the presence of hatching spines or an egg burster used during hatching (Cobben, 1968: 315–320); it is shed during the molt to the first instar (Fig. 8.35).

Truman and Riddiford (1999) proposed that a key proximate cause for the gradual evolutionary change from ametabolous to hemimetabolous and holometabolous development was a subtle shift in the timing of JH secretion during embryonic and postembryonic development, as summarized in Fig. 9.17. Possible steps in this evolutionary sequence are listed below:

- In known members of extant ametabolous and hemimetabolous orders, the pronymph is nonfeeding (Fig. 9.17A).
- The ability of the pronymph to feed is essential if it is to evolve into the larva. A possible scenario for this transformation can be found in the embryogenesis of extant lepidopterans. During dorsal closure, only part of the yolk is enclosed, the remaining extraembryonic part being eaten by the developing caterpillar just before it hatches. If embryos of the ancestor of Holometabola did the same, there would be selection on the pronymph to develop functional mouthparts, while still within the egg, resulting in development of a functioning protolarva (Fig. 9.17B).
- The females of ancestral Holometabola, as do those of many extant exopterygotes, probably had well-developed valvular ovipositors and deposited their eggs in diverse protected substrates (3.9.2). This would result in selection for preadaptations in the protolarva, allowing it to escape from these sites (Cobben, 1968: 316–326). A feeding protolarva might also have encountered digestible items in the substrate as it escaped that are not available to its nymphs (N1–N4) or adults (Fig. 9.17B).

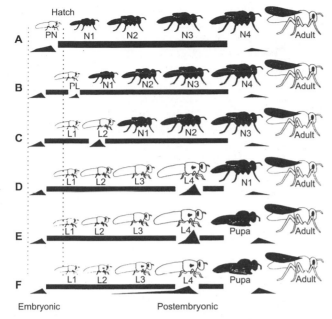

Fig. 9.17. Possible steps in the transition from an ancestral hemimetabolous to a highly derived, holometabolous life history, and their relationships to the timing of juvenile hormone (JH) production (horizontal bars) and to growth directed toward the adult form (black triangles in the absence of JH). The pronymphal (PN), protolarval (PL), and larval (L) stages are shown in white, and the nymphal (N) and pupal (P) stages in dark gray. Wings, wing pads, and wing imaginal discs are shown in black. (Reprinted with permission from *Nature* (J. W. Truman and L. M. Riddiford. The origins of insect metamorphosis. 401: 447–452). © 1999 Macmillan Magazines Limited.)

- If these resources were abundant, there would have been selection to maintain this form in a second larval instar (L2), resulting in the division of postembryogenesis into larval (L1, L2) and nymphal (N1–N3) stages (Fig. 9.17C).
- As nymphs potentially compete with adults for food, as in extant exopterygotes (9.4.2), selection for undergoing all juvenile growth in the larval form would tend to separate the resources used for growth from those used for reproduction, reducing the nymphal form to a single instar (N1) that is no longer required to feed but that could serve as a transitional form between larva and adult; this became the pupa (Fig. 9.17D, E).
- Truman and Riddiford (1999) proposed that the shift from a transitional pronymphal stage (Fig. 9.17A) to a functioning protolarval stage (Fig. 9.17B) is due to a shift in JH secretion in the embryo to an earlier stage, resulting in some aspects of nymphal embryogenesis being suppressed, while the presence of JH during formation of the pronymph would have caused the precocious maturation of embryonic tissues and the transformation of the pronymph into a protolarva with functioning mouthparts.

- Continued presence of JH then would maintain the form of the protolarva and larva (L1, L2) (Fig. 9.17B, C).
- For the protolarva/larva to molt to a nymphal (N1) stage, JH would have had to disappear to allow development of nymphal structures (second triangle in Fig. 9.17B–D).
- JH would then reappear to inhibit their differentiation into adult structures until a second JH-free period prior to the molt into the adult (Fig. 9.17A–D: small third triangle).
- Once the nymphal stage was achieved, it would also be stable as long as JH was present (Fig. 9.17B, C: second horizontal bar).
- As the time spent as a protolarva/larva was extended, the second JH-free period was progressively delayed until it was finally pushed to the end of the growth period (note shift of second triangle to the right in Fig. 9.17C and D and associated shortening of the second JH bar). Thus, the anomalous reappearance of JH during the L4-pupa transition (Fig. 9.17E, F) is directly related to the ancestral "need" for JH during the pronymphal-nymphal (N1) transition (Fig. 9.17A).
- As the larval and nymphal stages diverged in structure and behavior, the amount of tissue and organ reorganization required after the JH decline became progressively greater.
- Finally, the ability of selected tissues to become JH-independent allowed development of imaginal discs to continue during the larval stage and contributed to the marked reduction in the duration of life cycles (note appearance of wing discs first in the L4 stage [Fig. 9.17D, E] and then in L2 and L3 [Fig. 9.17F]).

The ancestral "task" of embryogenesis was to produce an individual, the first instar nymph, N1, that, except for wings and functioning reproductive organs, was a miniature of its adult (Fig. 9.17A). During the evolution of metamorphosis, a shift in the time of secretion of JH to an earlier stage in embryogenesis may have interrupted this ancestral growth trajectory, particularly the genetic cascades patterning and specifying the body and limbs (Fig. 9.17B–D; see 7.2.1, 11.3). In appendage primordia, early JH secretion might have interrupted their progressive patterning into their adult form and stabilized them at an intermediate state. This intermediate level of patterning might then have been the basis for constructing the different and simplified larval appendages of a holometabolous ancestor. With the disappearance of JH in the final larval instar (Fig. 9.17D: triangle below L4), this intermediate condition could no longer be maintained, and the resumption of patterning would provoke the development of imaginal primordia in the pupa, eventually resulting in the differentiation of adult appendages (Fig. 9.17D).

Exogenous application of JH strongly affects embryogenesis in the apterygote *Ctenolepisma longicaudata* (Zygentoma: Lepismatidae) but has minimal effects on its postembryogenesis (Truman and Riddiford, 1999). In members of hemimetabolous and holometabolous taxa, the effect of JH on embryogenesis is reduced while its postembryonic effects are intensified. This trend suggests that the ancestral morphogenetic role of JH was in embryogenesis, with emergence of JH as a major postembryonic hormone accompanying the shift of some phases of embryonic growth into postembryonic life.

This original, highly detailed, and ingenious proposal is supported by much comparative and experimental research on the role of JH and ecdysteroids in embryonic and postembryonic development. However, it is based on detailed study of only four species (the locusts *Schistocerca gregaria* and *Locusta migratoria*, the tobacco hornworm *Manduca sexta*, and *D. melanogaster*), it has some errors in fact (e.g., pronymphal cuticle does not bear functioning external sense organs except in members of the most basal lineages, Archaeognatha, Zygentoma, and Ephemeroptera), some of its generalizations have exceptions (e.g., larval body cuticle is soft and unsclerotized only in some representatives of the Holometabola), and it is weakened by being presented in a linear rather than in a phylogenetic sequence (Fig. 9.17) (Truman and Riddiford do consider phylogeny but only in relation to first appearance of imaginal discs in holometabolous insects). It is certain to stimulate additional research on the centuries' old question of the origin of complete metamorphosis and to induce insect endocrinologists to investigate representatives of Neometabola and of basal endopterygote lineages.

9.7 Growth

Regardless of type, postembryogenesis in insects is a time of growth, of an increase in the mass of living substance. Thus, weight is usually considered an index of growth even though it is accompanied by an increase in linear dimensions.

9.7.1 PATTERNS OF GROWTH
If a growing insect is weighed at regular intervals and these values are plotted against time or instar, a growth curve or trajectory results that is exponential, the fastest weight increase occurring in the older feeding instars (Nijhout, 1994b). There are at least three ways in which this increase can occur:

- With *auxetic growth*, weight increases due to growth of individual cells, not to an increase in their number. As they grow, their chromosomes replicate endomitotically within their nuclei, so that the cells become highly polyploid, as in the larval epidermis of higher flies (Fig. 9.18A). The process begins late in embryogenesis, with only S and G phases occurring in the cell cycle of larval cells (8.3.6).
- When growth results from an increase in cell number with little or no increase in size, *multiplicative growth* results, such as in the epidermis of most apterygote, exopterygote, and basal endopterygote juveniles.
- When growth results from special groups of cells retaining their ability to divide mitotically when

other, differentiated cells have stopped, *accretionary growth* occurs, as in the growth zone of short germ embryos (Fig. 6.22A), the imaginal midgut precursors of the larval midgut (Fig. 10.15), the imaginal discs and histoblasts of higher flies (Fig. 8.43: 17), and the cephalic and thoracic horns of certain scarab beetles (12.8).

In multicellular organisms such as insects, each cell type has a size characteristic for the species that is related to the body size of each juvenile instar and the adult. The control of cell size and proliferation thus influences the proportions and sizes of organs, limbs, and the entire body (Montagne et al., 1999; Emlen and Nijhout, 2000). Within a particular developing tissue a cell must reach a minimum size before it enters the M phase and divides. However, mutations altering progression through the cell cycle do not necessarily inhibit cell growth. Thus, mechanisms must exist to integrate cell growth and proliferation so that the cells maintain a size appropriate to the body size of the species.

In *D. melanogaster*, a family of mutant genes, the *Minutes*, that are defective in ribosomal protein production and induce a delay in rate of cell growth and mitosis has been discovered. Also, the *S6 kinase* gene (*dS6K*), when mutant, slows the rate of the cell cycle in wing discs from 12.5 ± 1 to 24 ± 4 h at 25°C, retards development of the adult by 5 d, and reduces both cell and body size of the adult by about 30% without influencing either cell number (evaluated in wing discs, wings, and ommatidia of the imaginal compound eye) or adult proportions (Montagne et al., 1999). These observations and others suggest that this gene is cell autonomous and that cell size participates in the control of compartment size. The gene is part of a signaling pathway contributing to the regulation of growth.

9.7.2 GROWTH AND MOLTING
Because much of their exoskeleton is sclerotized or calcified, most arthropods exhibit discontinuous growth, their weight usually increasing by stepwise increments punctuated by molts. In terrestrial insects, weight typically increases steadily throughout an instar, then falls slightly at ecdysis owing to the loss of exuviae and water (Fig. 9.18B). In aquatic insects, weight increases steadily throughout an instar, with a sharp increase at each molt due to an uptake of water (Fig. 9.18C). And in immatures of hemimetabolous, blood-feeding insects, the stimulus for each molt is a blood meal, which results in an instantaneous and large increase in weight (Fig. 9.18D). During the nonfeeding part of each instar, this meal is digested and much of its water secreted (as in ticks) or excreted, resulting in a steady weight loss.

No matter what lifestyle, the total amount of growth occurring during postembryogenesis can be substantial. For example, caterpillars of the carpenter worm *Cossus cossus* (Lepidoptera: Cossidae) enlarge 72,000-fold between hatching and pupation!

Final adult weight depends on the conditions under which the juveniles developed as well as on the genome of the species (Nijhout, 1994b; Emlen, 2000; Emlen and

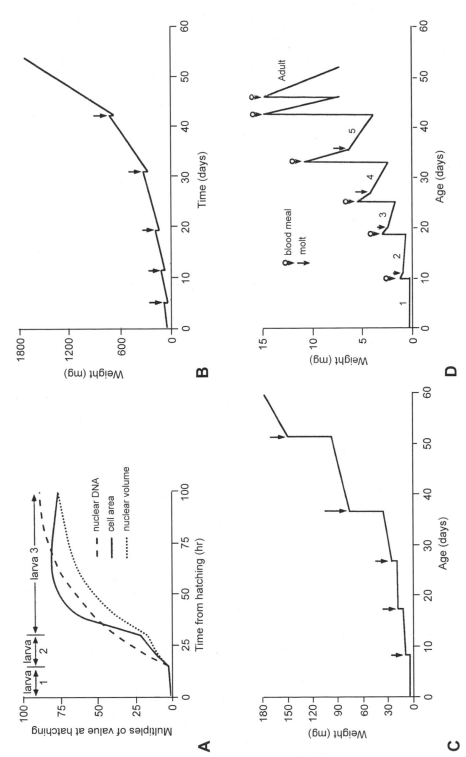

Fig. 9.18. A. Increase in epidermal cell-surface area, nuclear volume, and DNA content during postembryonic development of the larva of *Calliphora erythrocephala* (Diptera: Calliphoridae). Increase in each parameter is expressed in multiples of the initial value at the time of hatch. B. Weight change during the postembryogenesis of a female of *Locusta migratoria* (Orthoptera: Acrididae). C. Same of *Notonecta glauca* (Hemiptera: Notonectidae). D. Same of the human bed bug *Cimex lectularius* (Hemiptera: Cimicidae). (A adapted with permission from M. J. Pearson. The abdominal epidermis of *Calliphora erythrocephala* (Diptera). I. Polyteny and growth in the larval cells. *J. Cell Sci.* 16: 113–131. © 1974 Company of Biologists, Ltd. B adapted with permission from K. U. Clarke. 1957. On the increase in linear size during growth in *Locusta migratoria* L. *Proc. R. Entomol. Soc. Lond. A* 32: 37, Fig. 1. © 1957 The Royal Entomological Society of London. C adapted from *The principles of insect physiology*, 6th ed. 1965, p. 55, V. B. Wigglesworth, Fig. 36B, © 1965 Kluwer Academic Publishers, with kind permission of Kluwer Academic Publishers. D adapted from Titschack, 1930.)

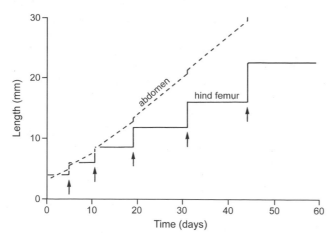

Fig. 9.19. Increase in length of the hind femur and abdomen in a female of *Locusta migratoria* (Orthoptera: Acrididae) during postembryogenesis. The abdomen appears to grow continuously because of the stretching out and unfolding of the arthrodial membrane between segments. (Adapted with permission from K. U. Clarke. On the increase in linear size during growth in *Locusta migratoria* L. *Proc. R. Entomol. Soc. Lond. A* 32: 35–39. © 1957 The Royal Entomological Society of London.)

Nijhout, 2000). For example, rapid development at high temperature and humidity usually results in smaller, lighter adults, compared with those developing under optimal temperature conditions (Bursell, 1974), and adult *Drosophila* flies developing from starved larvae are smaller and have fewer and smaller cells than do those developing from well-fed larvae (Held, 1979). Food quality, relative humidity, population density, stress, and other factors can also influence growth rate and final adult weight.

9.7.3 EFFECTS OF SCLEROTIZATION

Once sclerotized, cuticle cannot expand. Thus, growth of sclerotized body parts can only occur when an insect molts, and its new, larger, flexible cuticle is expanded just after ecdysis. Growth of such parts is thus in steps, as in the hind femur illustrated in Fig. 9.19. Several models have been developed to describe the growth in linear measurements of sclerotized parts in successive instars (Klingenberg and Zimmermann, 1992b):

- The linear progression model, $y = a + bx$, is used if there is a straight-line relationship between untransformed size measurements and instar number, that is, if the absolute growth increment is the same in each molt (y is the measure of size, x is instar number, and a and b are constants).
- The geometric progression model, $y = ab^x$, termed *Dyar's law* and used more often in its log-transformed version, $\log y = \log a + x(\log b)$, assumes a geometric progression in size measures, where growth ratios of succeeding instars (i.e., postmolt size/premolt size) or percentage increments, rather than absolute incre-

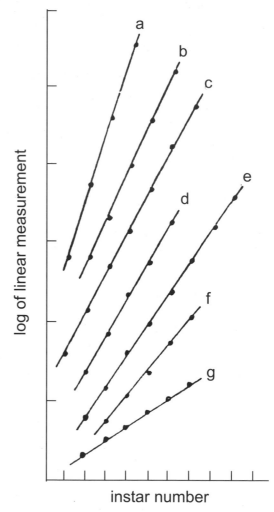

Fig. 9.20. Chart illustrating Dyar's law (head capsule width against instar number). a. *Bombyx mori*, ratio (r) = 1.91 (Lepidoptera: Bombycidae). b. *Philosamia ricini*, r = 1.52 (Lepidoptera: Saturniidae). c. *Crambus mutabilis*, r = 1.44 (Lepidoptera: Pyralidae). d. *Chaetopletus vestitus*, r = 1.40. e. *Sphodromantis bioculata*, r = 1.30 (Mantodea: Mantidae). f. *Cladius isomerus*, r = 1.50 (Hymenoptera: Tenthridinidae). g. *Tenebrio molitor*, r = 1.12 (Coleoptera: Tenebrionidae). (Adapted from *The principles of insect physiology*, 6th ed. 1965, p. 54, V. B. Wigglesworth, Fig. 35, © 1965 Kluwer Academic Publishers, with kind permission from Kluwer Academic Publishers.)

ments, are constant. Therefore, a plot of log-transformed size measures, for example, head capsule width, against instar number, reveals a straight-line relationship (Fig. 9.20).

According to Dyar's law, it is possible to calculate the number of instars from an incomplete series of exuviae, since differences amounting to twice the normal value occurring between any pair of head capsules would indicate that an instar is missing. However, in some insects, head capsule width can increase slightly within stadia, and the number of instars can vary with no change in

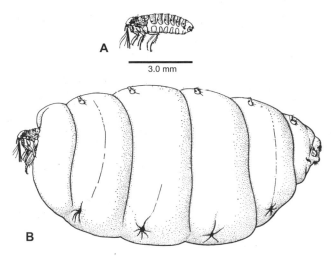

Fig. 9.21. Neosomy in a female of the tunga flea *Dorcadia ioffi* (Siphonaptera: Vermipsyllidae). A. Unfed. B. Fed. (Reprinted with permission from J. R. Audy, F. J. Rodovsky, and P. H. Vercammen-Grandjean. Neosomy: radical intrastadial metamorphosis associated with arthropod symbiosis. *J. Med. Entomol.* 9: 487–494. © 1972 Entomological Society of America.)

final head capsule width (references in Klingenberg and Zimmermann, 1992a).

Since membranous cuticle such as intersegmental (arthrodial) membrane can expand between molts, both by pulling out of folds and by stretching, such parts can grow continuously, as does the abdomen of a grasshopper (Fig. 9.19) or the entire membranous cuticle of a caterpillar, bee larva, or fly maggot. And in certain parasitic, symbiotic, and eusocial arthropods capable of physogastry or neosomy, additional membranous cuticle can be produced after adult emergence and the form of the body transformed, often with monstrous results (Fig. 9.21) (Audy et al., 1972). Bordereau (1982) examined this process by electron microscopy and histochemistry in physogastric queens of the termites *Cubitermes fungifaber* and *Macrotermes bellicosus*.

9.7.4 WHY DON'T CATERPILLARS GROW SHORT AND FAT?

How is it that as a caterpillar grows, it maintains its long cylindrical shape throughout the larval period? Between molts the cuticle stretches in such a way that the body's proportions are the same at the end of an instar as at the beginning (Carter and Locke, 1993). This introduces a paradox, since under uniform internal blood (hemostatic) pressure, the *hoop* (circumferential) stress is twice the *axial* (longitudinal) stress. Thus, if one were to inflate a caterpillar by artificially injecting additional hemolymph, the increased stress would be twice as great in the circumferential as in the longitudinal axes of the caterpillar, causing it to become progressively fatter. This does not occur because circumferential pleats in its cuticle allow it to extend more easily in its longitudinal than in its circumferential axis, thus allowing uniform

growth in spite of the greater hoop stress arising from internal blood pressure.

9.7.5 DISPROPORTIONATE (ALLOMETRIC) GROWTH

Following hatch, most pterygote insects change shape as they grow (e.g., Fig. 9.3). For example, the relative proportions of head, thorax, abdomen, and appendages usually change during postembryonic development in a typical, exopterygote insect (Huxley, 1932). If different structures of a juvenile insect are measured during development, they usually will be found to be growing at different rates. This disproportionate growth is the reason why the juvenile changes shape as it develops and is known as *allometry*, *heterogony*, or *scaling* (Huxley, 1932; Zeger and Harlow, 1987; Klingenberg, 1998; Emlen and Nijhout, 2000; see Gayon, 2000, for a short history). If the organ in question grows relatively faster than some other part taken as the standard, typically body length or head capsule width, it is said to exhibit *positive* allometry; if it grows more slowly, *negative* allometry; and if at the same rate, *isometry*.

Disproportionate growth is expressed in the regression equation $y = bx^k$ but is usually used in its logarithmic form, $\log y = \log b + k \log x$, where y is the dimension of the part being studied, x is the dimension of the standard (body length, etc.), k is the slope in log-log plots of x and y (allometric coefficient), and b is the initial growth index of y (y-intercept) (Fig. 9.22). If $k = 1$, growth of y is isometric; if $k > 1$, y is growing by positive allometry; and if $k < 1$, y is growing by negative allometry. If the logarithms of the growth values for two organs, x and y, growing at different rates are proportional to each other throughout development, a straight-line regression results when these measurements are plotted on log-log paper (Fig. 9.23).

The regression equation can be used to analyze the growth of body parts in groups of specimens of a species whose ages are not known or in the sequentially shed exuviae of single developing individuals. Organs for x and y can be chosen at will, and x and y can be linear measurements as already indicated, or weights, volumes, or concentrations of particular chemical compounds.

Allometric growth characterized by a straight-line relationship will occur only if the growth ratios of the parts being compared are constant throughout development. Since they often change, the slope of the regression line may change as development proceeds. In addition, the excess capacity for growth in a particular body part is often unequally distributed and falls off in either direction from a center of maximum growth (CMG) (Fig. 9.24A) (Katz, 1980). This center, called the *germinal center*, is usually an area of higher mitotic activity or greater cell growth than elsewhere in the structure. As growth of a part proceeds, this center can move around within it so that the shape of the part changes (Fig. 9.24A), it can divide into two or more centers (Fig. 9.24B, C), the mitotic rate of its cells can decrease or increase (Fig. 9.24D), or the longitudinal axes

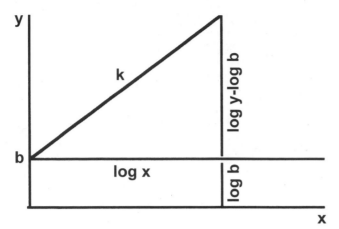

Fig. 9.22. Allometric growth as expressed in the regression equation $y = bx^k$ or, in its logarithmic form, $\log y = \log b + k \log x$, where y is the dimension of the body part being examined; x, the dimension of the standard (body length, etc.); k, the growth coefficient for y (slope); and b, the initial growth index of y (y intercept).

of cell divisions can change, influencing the shape of the CMG.

9.7.5.1 DIET AND ALLOMETRY

Allometric growth in the head capsules of juvenile insects can be influenced by the type of food the insect eats (Bernays, 1986a). For example, caterpillars of the grass-feeding noctuid moth *Pseudaletia unipuncta*, reared from the time of hatch on coarse leaves of various grasses, had head masses twice as great as those of caterpillars fed a soft artificial diet, even though the mature larvae reached the same final body mass (Fig. 9.25A). The muscular effort required in securing and processing the normal food of this insect increases the mass of the mandibular adductor muscles, which in turn has a dramatic morphogenetic effect on head capsule size, degree of sclerotization, and shape. Such allometry is known to be widespread among both moth caterpillars and grasshoppers where grass-feeding species, whose food contains high concentrations of silica, have relatively larger, more sclerotized heads than those feeding on softer, herbaceous vegetation (Fig. 9.25B) (Bernays and Hamai, 1987).

Also, in the sexually dimorphic males of certain species of Dermaptera, Coleoptera (Scarabaeidae, Lucanidae), and Diptera (Diopsidae), the size of secondary sexual structures involved in defense of resources or females from conspecific males scales with body size, large males having the largest structures and small males either having relatively smaller structures or lacking them entirely (Emlen, 2000; Emlen and Nijhout, 2000). The size and position of secondary sexual structures depend on whether larvae develop under ideal or suboptimal conditions and how close they are situated to other structures such as antennae, eyes,

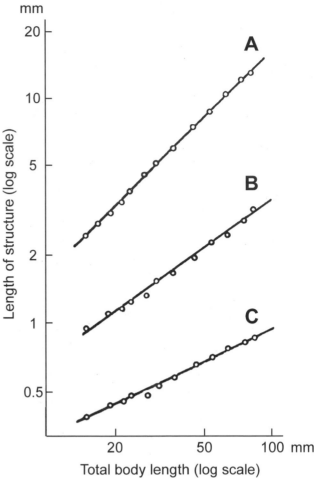

Fig. 9.23. Allometric growth in *Carausius morosus* (Phasmatodea: Phasmatidae) plotted on log-log paper. A. Length of posterior segment of pronotum. B. Width of head capsule. C. Diameter of compound eye. Circles equal different instars. (Adapted from *The principles of insect physiology*, 6th ed., 1965, p. 56, V. B. Wigglesworth, Fig. 37, © 1965 Kluwer Academic Publishers, with kind permission from Kluwer Academic Publishers.)

and wings important in the species' way of life (Emlen, 2001).

9.7.5.2 SOURCES OF ALLOMETRIC VARIATION

Allometry in animals can have three sources (Fig. 9.26) (Klingenberg, 1996a, 1998; Klingenberg and Zimmermann, 1992b; Stern and Emlen, 1999; Emlen and Nijhout, 2000):

- *Static allometry* refers to the pattern of variation occurring within a single instar of a particular species and includes the scaling relationship between organ and body size.
- *Ontogenetic allometry* refers to the growth trajectory of an organ relative to body size during the entire development of a single individual.

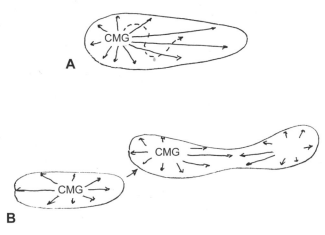

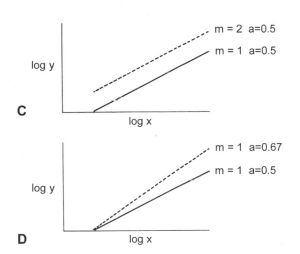

Fig. 9.24. Germinal centers (centers of maximum growth, or CMG) and shape change in a growing structure (a germinal center is an area of higher mitotic activity or greater cell growth than elsewhere within the structure). A. The germinal center can move around within the structure as it develops, causing it to change shape continuously. B. It can divide into two or more centers as development proceeds. C. Change in the number of germinal centers can produce a change in the y intercept of k. D. Change in the relative rate of mitosis in a germinal center can produce a change in the slope of k. (C and D adapted from M. J. Katz. 1980. Allometry formula: a cellular model. *Growth* 44: 93, Fig. 2A, B.)

- *Evolutionary (phylogenetic) allometry* refers to the relationship between organ and body size among homologous instars of the member species of a single clade.

Stern and Emlen (1999) and Emlen and Nijhout (2000) lamented the dearth of experimental study of proximate mechanisms generating static allometry in holometabolous insects and divided the possibilities into two processes: (1) those governing the autonomous specification of organ identity and pattern, perhaps including approximate size due to action of proximate intrinsic factors (hormones, transcription factors of the segmentation and homeotic selector genes, receptor pro-

teins, signaling pathways, etc.); and (2) those determining the final size of organs relative to total body size. They proposed three models to explain static allometry:

- All organs could autonomously absorb nutrients and grow at organ-specific rates.
- A centralized system could translate nutrition or body size into a growth signal and distribute this signal to all growing organs.
- Feedback communication between specific imaginal tissues and the rest of the body could modulate attainment of final sizes.

Their perusal of the literature provided support for the second and third models and for the suggestion that hormones or growth factors are the messengers of size information.

Results of their survey also suggest that future study of the proximate developmental basis of allometry should focus on size differences among individuals of a species, since it is intraspecific variation in size and shape resulting from development under varying conditions that results in selection for allometric differences among members of natural populations. This means that future investigation of allometry should break from standardized laboratory conditions and concentrate on the interface between population biology, developmental biology, and endocrinology (Stern and Emlen, 1999: 1099; Emlen, 2000; Emlen and Nijhout, 2000).

9.7.5.3 QUANTITATIVE ANALYSIS OF ALLOMETRIC VARIATION

Allometric growth can be quantified by use of various multivariate statistical methods and computer programs (Klingenberg, 1996a, and references). Such analysis relies on the use of morphometric methods to assess the spatial relationship between *landmarks*, particular points selected to characterize the size and shape of the structure under investigation. Such points can be located unambiguously on each specimen and are assumed to be homologous between forms (e.g., the longitudinal and cross veins of bee and fly wings). Several methods can be used:

- In *conventional morphometrics*, form is quantified as vectors of distance between these landmarks (variables). Relationships between forms can then be characterized in the multidimensional space spanned by these distances. Variation recorded by this approach is an "integral" of the local variation at the two landmarks.
- *Procrustes analysis* finds an optimal superimposition of shapes after adjusting for overall size, and variation can be characterized by the scatter of individual shapes around a "mean shape" for each landmark.
- The *thin plate spline* method characterizes differences in form as deformations consisting of changes in size and uniform and nonuniform changes in shape and is a modern version of D'Arcy Thompson's (1917) method of transformed Cartesian coordinates.

Fig. 9.27 illustrates allometric change in shape by means of growth contours (intervals of $k = 0.25$) of the

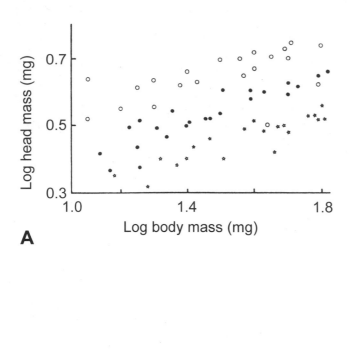

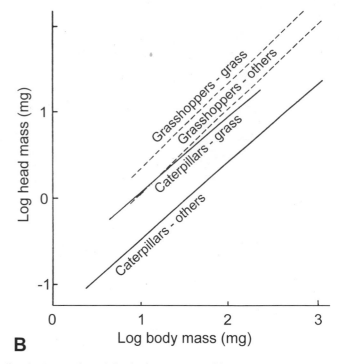

Fig. 9.25. Influence of food consistency on the relation between dry head mass and total dry body mass. A. Fifth instar caterpillars of *Pseudaletia unipuncta* (Lepidoptera: Noctuidae) reared from hatch on three different diets. The equation of the line for *Triticum* (open circles) is $y = -0.001(\pm0.0005) + 0.365(\pm0.035)x$, $r^2 = 0.82$; for *Cynodon* (closed circles), $y = 0.058(\pm0.009) + 0.389(\pm0.065)x$, $r^2 = 0.62$; and for artificial diet (asterisks), $y = -0.019(\pm0.001) + 0.306(\pm0.344)x$, $r^2 = 0.8$. B. Grass feeders and other foliage feeders among grasshoppers and caterpillars. The equation of the line for grasshopper grass specialists is $y = -0.65(\pm0.09) + 0.97(\pm0.15)x$, $r^2 = 0.95$, $n = 56$; for grasshoppers with other diets, $y = -0.92(\pm0.11) + 0.97(\pm0.14)x$, $r^2 = 0.96$, $n = 57$; for caterpillar grass specialists is $y = -0.83(\pm0.23) + 0.86(\pm0.16)x$, $r^2 = 0.64$, $n = 17$; and for caterpillars feeding on other herbaceous plants, $y = -1.37(\pm0.09) + 0.84(\pm0.06)x$, $r^2 = 0.76$, $n = 56$. (A and B reprinted with permission from E. A. Bernays. Diet-induced head allometry among foliage-chewing insects and its importance for graminivores. *Science* 231: 495–497. Copyright 1986 American Association for the Advancement of Science.)

head and trunk segments of the five nymphal instars of the pyrrhocorid bug *Dysdercus fasciatus* (Hemiptera) as measured along the dorsal midline (Blackith et al., 1963). Notice how the values change for each segment from one instar to the next.

9.7.6 DIET AND DEVELOPMENTAL POLYMORPHISM
Diet quality is also known to influence development of polymorphism in the larvae of certain insects (Nijhout, 1994b). An interesting example is the geometrid moth *Nemoria arizonaria* (Greene, 1989). This species occurs in the American Southwest, where it is bivoltine, the caterpillars feeding on several species of oak. Caterpillars of the spring generation develop into mimics of the catkins (inflorescences) on which they feed, whereas those of the summer generation emerge after the catkins have fallen, feed on leaves, and develop into twig mimics. This developmental polymorphism seems to be triggered by the concentration of defensive secondary compounds in the larval diet: all caterpillars raised on catkins, which are low in tannins, develop into catkin morphs, and those raised on leaves high in tannins develop into twig morphs. Time to pupation (Fig. 9.28A), pupal weight (Fig. 9.28B), and female fecundity (Fig.

9.28C) are among the fitness components in this species that are influenced by larval diet.

9.7.7 FLUCTUATING AND DIRECTIONAL ASYMMETRY
Small random differences between sides in otherwise bilaterally symmetrical organisms are referred to as *fluctuating asymmetry* (Fig. 9.29: top) (Palmer, 1996a). If grown under optimal conditions, insects and other animals are usually bilaterally symmetrical, with their left and right sides mirror images of each other. However, structures involved in mating, sperm transfer, and securing of food are often fixed as either "left-" or "right-handed" (*directional asymmetry*) (Fig. 9.29: middle) or are randomly so (*antisymmetry*) (Fig. 9.29: bottom) (Palmer, 1996b). If such animals develop under suboptimal conditions (low quantity or quality of food, too low or too high temperature or humidity, excessive population density, presence of endoparasitoids, pollutants, etc.), they are subject to increasing levels of stress. These stresses may, though not inevitably (Leung and Forbes, 1996), adversely affect *developmental stability* (i.e., the ability to produce identical mirror-image forms on both sides), the degree of asymmetry tending to be correlated with degree of stress experienced. Active feeding instars

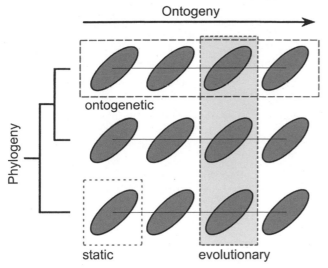

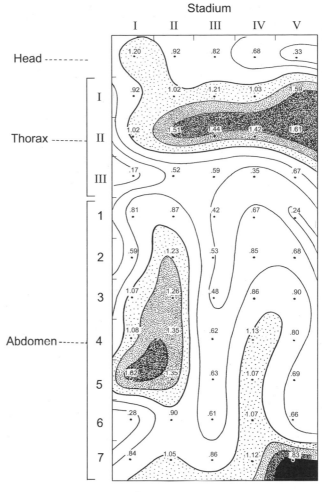

Fig. 9.26. The three levels of allometry. The diagram shows three species, each with four different ontogenetic stages, that are considered to be homologous. Rectangles enclose the species and stage groups included in an analysis of allometry at each of the three levels. Ontogenetic allometry can be analyzed separately for all three species; evolutionary allometry, for each of the four stages; and static allometry, for each of the 12 species and stage groups. (Reprinted with permission from C. P. Klingenberg. Multivariate allometry. In L. F. Marcus et al. (eds.), *Advances in morphometrics*, p. 25, Fig. 1. Plenum Press, N.Y. © 1996 Kluwer Academic/Plenum Publishers.)

in the life cycle are more sensitive to stress than are sessile, nonfeeding stages such as eggs or pupae (Leary and Allendorf, 1989; Graham et al., 1993).

Analysis of fluctuating asymmetry has been used to measure the quality of laboratory cultures of insects long maintained without infusions of wild germ plasm and in assessing environmental quality (e.g., the effects of insecticides on nontarget organisms, or of pollutants on aquatic insects). It has become a fertile field for modeling the role of selection on evolution (e.g., Møller and Swaddle, 1997; Klingenberg and Nijhout, 1999; Stern and Emlen, 1999).

Directional asymmetry (Fig. 9.29: middle) is also widespread in animals (Palmer, 1996b). Its presence suggests the existence of a left-right body axis, in addition to the anteroposterior and dorsoventral axes (2.2.2.6, 7.2.1.1.1, 7.2.1.1.3), that conveys positional identities to developing structures on either side of the midline (Klingenberg et al., 1998). Developmental mechanisms and genes establishing this axis in nematode and vertebrate embryos have been identified (references in Klingenberg et al., 1998), and mutations in the genes *dicephalic* and *wunen* recently were shown to disrupt development of the asymmetry characteristic of the proventriculus in the foreguts of *Drosophila* (Ligoxygakis et al., 2001). By examining left and right wings of three fly species (*D. melanogaster*, *Musca domestica*, and *Glossina palpalis gambiensis*) by geometric mor-

Fig. 9.27. Growth contours, at intervals of $k = 0.25$, of the head and of the thoracic and abdominal segments of the pyrrhocorid bug *Dysdercus fasciatus* during the five nymphal instars (pooled for both sexes) as measured along the dorsal midline. Growth coefficients were calculated for each measurement for each instar, and the resulting values were plotted on coordinates depicting the distribution of coefficients over the body of the insect throughout its postembryogenesis. Numbers above 1 indicate positive allometry and those below 1, negative allometry. Notice the center of positive allometry developing in the mesothorax in instar 2 that later extends to the prothorax; the region of even higher positive allometry in abdominal segment 7 (based principally on males and less prominent in females); and a third region of positive allometry in the mid abdomen in instars 1 and 2. Between these centers of positive allometry are regions of negative allometry, particularly noticeable in the metathorax and first abdominal segment. Growth of the mesothorax is associated with its role in imaginal flight, and that of the posterior abdomen is associated with development of the male and female external genitalia. (Reprinted from R. E. Blackith, R. G. Davies, and E. A. Moy. A biometric analysis of development in *Dysdercus fasciatus* Sign. (Hemiptera: Pyrrhocoridae). *Growth* 27: 317–334. © 1963.)

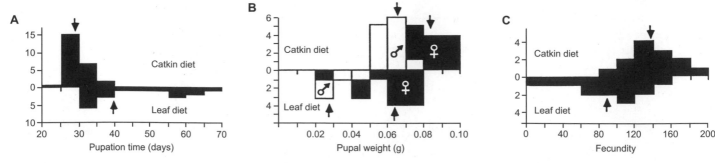

Fig. 9.28. Components of fitness in the moth *Nemoria arizonaria* (Lepidoptera: Geometridae) are influenced by diet, as shown for caterpillars raised at 25°C and 14 h of light. The upper half of each graph shows the frequency distribution for catkin morphs; the lower half, for twig morphs. Arrows indicate the median of the distributions. A. Catkin morphs pupated more quickly than leaf morphs (Mann-Whitney $U = 298$, $P < .001$). There was no significant difference in pupation times between sexes within a diet treatment. B. Catkin morphs pupated at larger sizes than the twig morphs (for females, Mann-Whitney $U = 101$, $P < 0.001$; for males, $U = 36$, $P < 0.001$). The sexes are shown separately since, within a diet treatment, females tend to attain a greater pupal weight than males. C. Female moths raised on catkin diets produced more offspring than those raised on twig diets (Mann-Whitney $U = 90$, $P = 0.0005$). Fecundity measures the total number of eggs produced by a female (eggs laid while alive plus eggs dissected from ovaries). (Reprinted with permission from E. Greene. A diet-induced developmental polymorphism in a caterpillar. *Science* 243: 643–646. Copyright 1989 American Association for the Advancement of Science.)

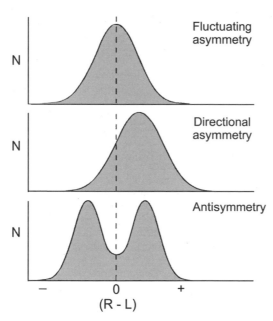

Fig. 9.29. Frequency distributions of the signed difference between sides, illustrating three commonly observed patterns of subtle deviations from bilateral symmetry. Top: Fluctuating asymmetry is defined as a distribution whose mean is zero and whose shape is statistically normal (bell-shaped). Middle: Departures of the mean from zero indicate directional asymmetry. Bottom: Bimodal departures of the shape of the distributions from normal reveal antisymmetry. (Reprinted with permission from A. R. Palmer. Waltzing with asymmetry. Is fluctuating asymmetry a powerful new tool for biologists or just an alluring new dance step? *Bioscience* 46: 518–532. © 1996 American Institute of Biological Sciences.)

Table 9.4 Number of Molts in Hexapods

Order	Number of juvenile molts
Apterygota	
Protura	5 or 6 (not known if adults continue to molt)
Collembola	5–11 (up to 50 as adults)
Diplura	? (up to 30 including adult molts)
Archaeognatha	8–10
Zygentoma	9–13 (dozens as adults)
Exopterygota	
Ephemeroptera	12–52 (15–30 most common)
Odonata	11–16 (vary even within species)
Orthoptera	5–10 (fewer in males)
Phasmatodea	4–12 (fewer in males)
Grylloblattodea	8
Dermaptera	5 or 6
Isoptera	5–11
Blattodea	6–12 (vary even within species)
Mantodea	5–9 (fewer in males)
Zoraptera	5
Plecoptera	13–34
Embioptera	4
Thysanoptera	4 or 5
Hemiptera	3–8 (usually 5)
Psocoptera	6
Phthiraptera	3
Endopterygota	
Coleoptera	Differs widely between species
Strepsiptera	5
Neuroptera	3
Megaloptera	11–13
Raphidioptera	11–16
Hymenoptera	3–6
Trichoptera	6–8
Lepidoptera	3–11 (usually 5 or 6)
Mecoptera	5
Siphonaptera	4
Diptera	4 or 11 (usually 4 or 5)

Sources: Stehr 1987b, 1991; CSIRO, 1991; and others.
Note: Number includes molt into adult.

phometrics, Klingenberg and colleagues (1998) found subtle but similar and statistically significant differences between left and right wings, even though their anlagen seem to develop relatively independent of each other (10.2.4.3.1), suggesting the existence of factors differentially affecting the differentiation of left and right sides.

To identify the proximate sources of such asymmetries, Klingenberg and Nijhout (1998) removed hind wing imaginal discs from one or both sides of caterpillars of the butterfly *Precis coenia* (Lepidoptera: Nymphalidae), which were about to molt from the fourth to the fifth (final larval) instar. When adults emerged, those lacking one or both hind wings were found to have heavier fore wings and mid and hind legs than controls. When only one hind wing was removed, the fore wing and hind leg on the treated side were heavier than those on the untreated side. In addition, the asymmetry and weight increase resulting from removal diminished with increasing distance of the responding tissue from the disc removed. These results suggest that adjacent developing imaginal discs compete for a hemolymph-borne nutrient or growth factor and that perturbation of such competition, as might occur under stress, could influence the development of asymmetry.

Nijhout and Emlen (1998; reviewed in Emlen, 2000) also demonstrated competition between growing structures in beetles of the species *Onthophagus taurus*. In this scarab, males with a pronotal width of 4.9 mm or greater normally bear a pair of long, slender, recurved horns on the back of the head that are used to repel competing males prior to copulation in brood tunnels under dung, whereas smaller males and all females lack them. They found an inverse relationship to exist between horn size and compound eye size in adult males and tested the hypothesis that they compete for resources during their development by applying JH to developing male larvae. Artificial presence of JH delayed metamorphosis, postponed the onset of horn growth, and resulted in emergence of adults with shorter horns than in untreated males but with larger compound eyes. They also selected for long horns in members of a laboratory population for seven generations and found that those adults having the longest horns had the smallest compound eyes with the fewest ommatidia and the most restricted field of vision. Similar trade-offs in size have been found between prothoracic horns and wings in certain dynastine and orthophagine scarabs and between anterior cephalic horns and antennae (Emlen, 2000, 2001).

9.7.8 NUMBER OF MOLTS

The number of molts is characteristic for the species in most pterygote insects (Table 9.4), and the process ends when the prothoracic glands degenerate before or after adult emergence (12.4.1.3). Apterygote adults, however, continue to molt, and their prothoracic glands persist throughout adult life (9.4.1). Members of more basal lineages of hexapods tend to molt more often than those of

intermediate or higher clades and to vary more (Table 9.4).

In members of some taxa, the number of molts and growth rate can be influenced by conspecific social interactions (Nijhout, 1994b). For example, grouped nymphs of the house cricket *Acheta domesticus* (Orthoptera: Gryllidae) had fewer instars and matured faster than did isolated ones (Watler, 1982). They also consumed more food per milligram of body weight and converted it to body tissue more efficiently than did isolated nymphs during the fifth and sixth instars. And solitary male first instars of the cockroach *Diploptera punctata* took significantly longer to mature than did male nymphs paired with either a male or female nymph or grouped with four other nymphs since birth (this species is viviparous) (Holbrook and Schal, 1998). Only 15.8% of solitary male nymphs had three instars before adult eclosion, while 60% of paired males did, the rest requiring four instars. Female first instars, however, were not affected by social factors.

In diapausing strains of the noctuid corn stalk borer *Sesamia nonagrioides*, last instar larvae enter a facultative diapause when exposed to short photoperiods (Gadenne et al., 1997). During diapause development, the larvae continue to molt up to seven times before pupating. They also continue to increase in size with each molt, as indicated by increases in weight and head capsule width, eventually to produce larger pupae and adults than nondiapausing strains produce.

An inadequate diet is known to prolong the larval period and to increase the number of molts enormously in certain insects living on dry foodstuffs. Rearing larvae of the webbing clothes moth *Tineola bisselliella* (Lepidoptera: Tineidae) on a rich or a poor diet varied the larval period from 26 d and 4 molts to 900 d and 40 molts (Titschack, 1926). In the latter instance, the insect actually decreased in size. The larval period can only increase for a certain amount of time; beyond this point the insect dies, probably because of an upper limit to the number of times its cells can divide (e.g., a human infant's skin cells in culture divide about 100 times before dying, while those from a 60-y old only divide about 20 times [Oliwenstein, 1993]). The clock? Telomeres at the ends of each chromosome prevent loss of nucleotides and degradation and fusion of chromosomes during mitosis. In each telomere, the nucleotide sequence TTGGG is repeated over a thousandfold. Each time a cell divides, its telomeres get a little shorter (they lose up to 50 Ts and Gs at each division). However, in a minority of cells, the enzyme telomerase adds nucleotides to the telomeres each time a cell divides, replacing the nucleotides lost. As a result, the telomeres don't get shorter, the command to "stop dividing" never gets sent, and the cells can continue to proliferate (Marx, 1994; de Lange, 1998). In most multicellular organisms, including insects, most cells do not produce telomerase. Telomere shortening is part of programmed cell death (apoptosis; 8.6.1).

10/ Molting and Metamorphosis

In chapter 9, I treated postembryogenesis in broad terms. Here, I address the structural changes occurring during molting and metamorphosis, with an emphasis on endopterygotes. I also include results of recent theoretical and experimental work not discussed elsewhere. The specification of adult body pattern is considered in chapter 11 and the endocrine control of postembryogenesis, in chapter 12.

10.1 Molting

The epidermis and its secretion product, the cuticle, establish the form of the insect and contribute to its ability to move, sense its environment, feed, flee, and in the adult, reproduce. Because its skeleton is on the outside, a juvenile insect, to grow, must periodically shed the remains of its old, smaller cuticle and replace it with a new, larger one. Fig. 10.1A shows the principal layers of the cuticle of an immature insect during the intermolt—the *epicuticle*, *exocuticle*, and *endocuticle*, the latter two constituting the *procuticle* (Chapman, 1982).

With the binding of 20-OH ecdysone (20E) molecules to receptor proteins in the nuclei of epidermal cells, a genetic cascade and signaling pathway is activated that eventually stimulates the epidermal cells to divide (Fig. 10.1A, B) (12.4.1.5). Their numbers increase, they become closely packed and columnar (Fig. 10.1B), and the surface area of the epithelium increases in preparation for depositing the larger cuticle of the next instar. Tension is generated between the epidermis and endocuticle, and the former separates from the latter (*apolysis*), leaving a subcuticular (ecdysial, molting) space (Fig. 10.1B). Molting fluid that contains chitinases and proteases is secreted into this space from the epidermis (Fig. 10.1C). Though active at the time of secretion, the chitinases, at first, are unable to access chitin in the old endocuticle, as this requires protease activity and the proteases are mostly secreted as inactive proenzymes (Reynolds and Samuels, 1996). Before this enzymatic activity begins, the apex of each epidermal cell extends microvilli and begins to deposit a cuticulin envelope as plaques at their tips, which broaden and fuse to adjacent plaques to form a continuous outer layer (Locke, 2001) (Fig. 10.6B, C). The microvilli persist throughout the deposition of cuticle, although they tend to shorten, widen, and become less numerous toward the end of an instar (Locke, 1985, 1990). At first it is smooth, but as it increases in area, the outer epicuticle is thrown into folds and a complex surface pattern results that is characteristic for the species, varies from one area to another, and is determined by the shape of the secreting epidermal cells (Bennet-Clark, 1963). Next, the epidermis begins to deposit an inner protein epicuticle under the cuticulin envelope, stabilizes its proteins by releasing polyphenols and phenoloxidase, and then begins to secrete new procuticle (Figs. 10.1D, 10.2). At the same time the inner epicuticle is being deposited, the polyphenols and phenoloxidase act to sclerotize some inner lamellae of the old endocuticle to form a thin *ecdysial membrane* that is shed along with the remains of the old cuticle at ecdysis (Fig. 10.1D–F) (Chapman, 1982).

When the new epicuticle is complete, the molting enzymes are activated, probably indirectly by a drop in 20E concentration toward the end of its molting pulse. These enzymes digest perhaps 90% of the old endocuticle, including the fusion points between internally fused components of the endoskeleton (8.3.9) and tracheal system (8.3.4), and recycle their components to the epidermis for reuse (Fig. 10.1D, E) (Chapman, 1982) (Reynolds and Samuels [1996] suggested that specific, zymogen-activating proteinases could be released into the molting fluid at this time or that the activity of specific inhibitors could be lifted). These enzymes do not affect the old exocuticle or ecdysial membrane (Fig. 10.1E), muscle attachment fibers (Fig. 10.7), or nervous connections to external sensilla (Fig. 10.5) because all either are of sclerotized cuticle or are surrounded by it.

Just before ecdysis (Fig. 10.4), a layer of wax is deposited on the surface of the new epicuticle via epidermal, epicuticular filaments extending through vertical pore canals in the new procuticle (Fig. 10.2). At ecdysis, this wax layer reduces desiccation in the newly molted insect (even so, in larvae of the mealworm beetle *Tenebrio molitor*, water loss is fourfold to sixfold above normal for the first 24 h after ecdysis [Chapman, 1982]).

As soon as digestion of old endocuticle is complete, molecules of eclosion hormone (EH) are released into the blood, and at least in larvae and pupae of the tobacco hornworm *Manduca sexta*, are carried to nine

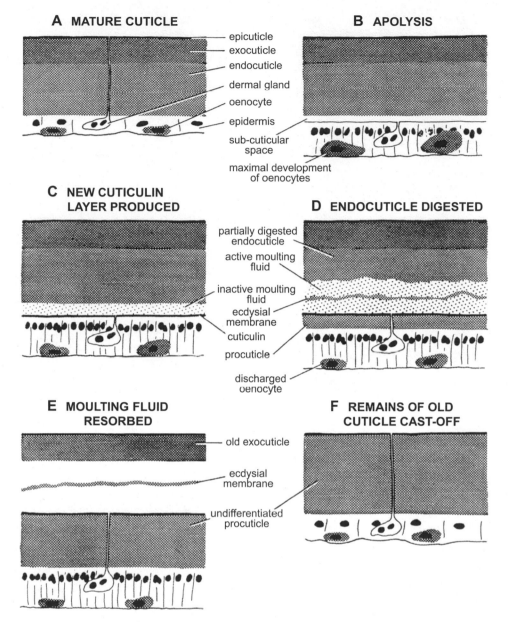

A MATURE CUTICLE

— epicuticle
— exocuticle
— endocuticle
— dermal gland
— oenocyte
— epidermis
— sub-cuticular space
— maximal development of oenocytes

B APOLYSIS

C NEW CUTICULIN LAYER PRODUCED

partially digested endocuticle
active moulting fluid
inactive moulting fluid
ecdysial membrane
cuticulin
procuticle
discharged oenocyte

D ENDOCUTICLE DIGESTED

E MOULTING FLUID RESORBED

— old exocuticle
— ecdysial membrane
— undifferentiated procuticle

F REMAINS OF OLD CUTICLE CAST-OFF

Fig. 10.1. Diagrammatic cross section through the cuticle of an immature insect, showing the principal stages of molting. (Adapted by permission of the publisher from R. F. Chapman. *The insects. Structure and function*, p. 519, Fig. 331. 3rd ed. Harvard University Press, Cambridge, Mass. Copyright © 1969, 1971, 1982 by R. F. Chapman.)

pairs of Inka cells, each attached to the ventral tracheal trunk below each spiracle (Zitnan et al., 1996). These cells are induced by EH to secrete pre-ecdysis– and ecdysis-triggering hormones. These three hormones mutually stimulate each other's secretion, thereby inducing pre-ecdysial and ecdysial behavior (Ewer et al., 1997) (12.4.9). As at hatch (9.2), the insect swallows air and water, and rhythmic, peristaltic contractions of its body muscles from back to front force blood into its head and thorax, causing them to expand and collapse. The old exocuticle and epicuticle split along a preformed line of weakness in the head capsule and mid-dorsal

thoracic cuticle, the *ecdysial line*, in which exocuticle is absent, and the insect extracts itself from the remains of its old cuticle by means of a complex and species-specific behavioral pattern, usually assisted by gravity (Fig. 10.1F). Undigested, sclerotized cuticular parts remaining from the previous instar are shed as the exuviae, including the linings of the tracheal system, foregut and hindgut, and endoskeletal elements but not those of the finest tracheal branches, the *tracheoles*. In some insects, such as the blood-feeding reduviid *Rhodnius prolixus*, special molting muscles function during each molt. Then their contractile elements regress, only

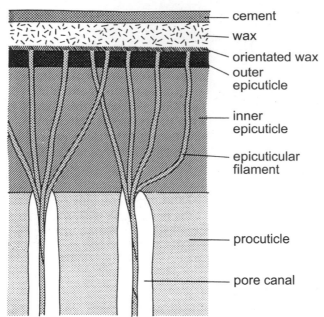

Fig. 10.2. The principal layers of the insect epicuticle highly magnified. (Reprinted by permission of the publisher from R. F. Chapman. *The insects. Structure and function*, p. 509, Fig. 322. 3rd ed. Harvard University Press, Cambridge, Mass. Copyright © 1969, 1971, 1982 by R. F. Chapman.)

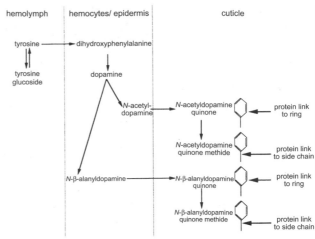

Fig. 10.3. Summary of the synthesis of catecholamines from tyrosine and their role in cuticle sclerotization. (From R. F. Chapman. 1998. *The insects. Structure and function*, p. 434, Fig. 16.20. 4th ed. Reprinted with the permission of Cambridge University Press.)

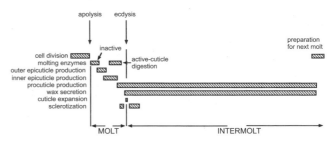

Fig. 10.4. Summary of the sequence of events in molting and cuticle deposition in an insect. The timing of apolysis relative to cell division and the degree of overlap between procuticle production and cell division vary in different insects. The short period of sclerotization before ecdysis involves the prehardening of cuticle at articulation points and in sensilla and their sockets. (Adapted by permission of the publisher from R. F. Chapman. *The insects. Structure and function*, p. 516, Fig. 329. 3rd ed. Harvard University Press, Cambridge, Mass. © 1969, 1971, 1982 by R. F. Chapman.)

to regenerate again prior to the next molt when stimulated by a blood meal (Wigglesworth, 1956).

After ecdysis, the new cuticle is soft, colorless, and unexpanded, and the insect is supported principally by a hemostatic skeleton. Molting behavior continues for some time after ecdysis and functions to expand the new cuticle by stretching out its folds.

Expansion of new cuticle is ended by sclerotization (tanning) (Figs. 10.1F, A; 10.4) evoked by release of molecules of the hormone bursicon (12.4.10). These molecules act to stabilize the chitin and protein chains in the outer procuticle as new exocuticle by inducing formation of cross-linkages between them from *N*-acetyldopamine quinone (Fig. 10.3) (Andersen, 1990; Hopkins and Kramer, 1992). This quinone is derived from the amino acid tyrosine by the pathway shown in Fig. 10.3, with bursicon inducing the response by activating the enzyme tyrosinase that catalyzes the transformation of tyrosine to dihydroxyphenylalanine (DOPA).

In some insects, a layer of cement elaborated by glandular cells in the epidermis (Fig. 10.1A) is then passed through vertical ductules in the new cuticle onto the surface of the wax layer (Fig. 10.2), where it hardens and protects it from abrasion (Chapman, 1982). After sclerotization is complete, the deposition of new procuticle continues, often in daily increments, for most of the intermolt (Fig. 10.4).

New cuticle of the pretarsal claws, of articulation points between segments in the appendages, of external sensilla and their sockets, and of internal apodemes are prehardened (presclerotized) before ecdysis because the ability of an insect to move and to sense its environment immediately after ecdysis depends on these structures being functional (Fig. 10.4). Unlike that of the general body surface, the cuticle of these is deposited in its fully expanded configuration. Release of 20E seems to induce prehardening (Hopkins and Kramer, 1992). The functional significance of both prehardened and yet-to-be-sclerotized cuticle after ecdysis is obvious when one observes a blowfly or fleshfly forced to burrow through sawdust in a test tube immediately following its emergence from the puparium.

Complex ultrastructural changes occurring within epidermal cells during a molt cycle were exhaustively

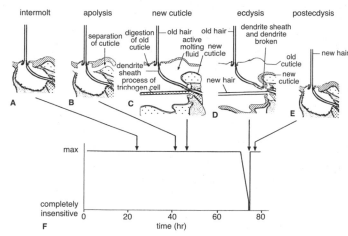

Fig. 10.5. Sensitivity of a mechanoreceptive hair in relation to molting in a caterpillar of the moth *Barathra brassicae* (Lepidoptera: Noctuidae). Diagrams show the structural changes occurring throughout one ecdysis (time 0) to the next. The lower panel shows the sensitivity of the hair to stimulation at different times. Sensitivity is completely lost for only about 30 min at ecdysis. (From R. F. Chapman. 1998. *The insects. Structure and function*, p. 612, Fig. 322. 4th ed. Reprinted with the permission of Cambridge University Press.)

reviewed by Locke (1985, 1990, 1998). Kunkel (1975) showed that molting is delayed in cockroach (*Blattella germanica*) larvae regenerating lost appendages. The sequence is summarized in Fig. 10.4 (Chapman, 1982).

Each molt is under precise neural and hormonal control (12.1, 12.5) and involves the functioning of complex motor patterns operating only at this time (Hughes, 1980, and references; Denlinger and Zdárek, 1994). And although molting is usually a solitary exercise, in aggregations of the collembolans *Hypogastrura lapponica* and *H. socialis*, it is synchronized, probably by chemical communication between individuals, since it can be entrained by placing formerly isolated individuals of any age into groups (Leinaas, 1983).

10.1.1 SENSORY CONTINUITY DURING A MOLT

It is critical that a young insect remains sensitive to its environment and be able to move while it is pharate if it is to escape from predators or parasitoids. Fig. 10.5 illustrates how sensory continuity is achieved during this period for a trichoid mechanosensillum (Gnatzy and Tautz, 1977). In each mechanoreceptive hair on the thorax of *Barathra brassicae* (Lepidoptera: Noctuidae) caterpillars, the distal dendrite of its single sensory neuron connects to its base by an ecdysial canal containing an impulse-transducing, tubular body (Fig. 10.5A) (Gnatzy and Tautz, 1977). At apolysis (Fig. 10.5B, C), the dendrite and its sheath elongate to maintain this connection across the ecdysial space. When the new, hair-forming (trichogen) cell of the sensillum grows out and begins to

deposit the cuticle of the next instar, its dendrite passes through its base and functionally connects to it by forming a new tubular body, even though it is still functionally connected to the base of the previous hair (Fig. 10.5C). With ecdysis, the old dendrite and sheath break above the base of the new hair (Fig. 10.5D) and are shed with the exuviae (Fig. 10.5E), the dendritic exit point in the new hair closing over to form a new ecdysial scar. Response of intermolt fourth instar and pharate fifth instar caterpillars to a 300-Hz tone showed these mechanoreceptors to remain functional throughout the 35-h (at 24°C) pharate period, except for 30–60 min during ecdysis, and the new sensilla to become functional immediately after ecdysis, as soon as their hairs were erect (Fig. 10.5F) (Gnatzy and Tautz, 1977).

Other external mechanoreceptors maintain their connection to the mechanoreceptors in the previous cuticle in similar ways (Zacharuk, 1985), as in a campaniform sensillum, where the ecdysial scar is at the apex of the dome (Moran et al., 1976). Internal, chordotonal organs presumably do so by reattaching their attachment cells and the proximal ends of their ligaments to the underside of new cuticle in the same way as do muscle fibers (10.1.2).

In uniporous, contact, gustatory (taste) sensilla having two or more sensory neurons, dendritic elongation at apolysis occurs in a similar way but through the pore at the tip of each new hair (Zacharuk, 1985). Multiporous, multiply innervated, olfactory (smell) receptors seem to maintain this contact via similar basal or apical dendritic extensions (Zacharuk, 1985).

In pharate adults of the silphid beetle *Necrophorus vespilloides*, differentiation of a new, imaginal, antennal, olfactory, multiporous sensillum not present in pupal cuticle requires 9–12 d at 22–24°C and begins with extension of its ensheathed primary dendrites above the surface of the epidermis exactly as it does after apolysis (Fig. 10.6A [Ernst, 1972]). The trichogen cell then grows out about the dendrites proximally but parallel to them distally (Fig. 10.6B). When the cuticulin envelope begins to be deposited about the surface of the new hair, the primary dendrites degenerate to its base, grow into it, and a few hours after adult eclosion, begin to branch (Fig. 10.6C–G). Pore tubules linking individual dendritic branches to pores in the hair wall begin to form at the same time as the envelope and are complete by the time the sensillar cuticle is 30% its final thickness.

If sensilla or epidermal glands that were missing from the previous instar are added to an instar, each sensillum or gland develops by asymmetrical division of a sense organ or glandular precursor cell during the exuvial pharate period exactly as in embryos (Fig. 8.33) (Sreng and Quennedey, 1976; Johnson and Berry, 1977; Bate, 1978; Keil and Steiner, 1990, 1991; Quennedey, 1998).

10.1.2 MUSCULAR CONTINUITY DURING A MOLT

Muscle fibers attach to the basal surface of epidermal cells by desmosomes (Fig. 10.7) (Caveney, 1969). Within

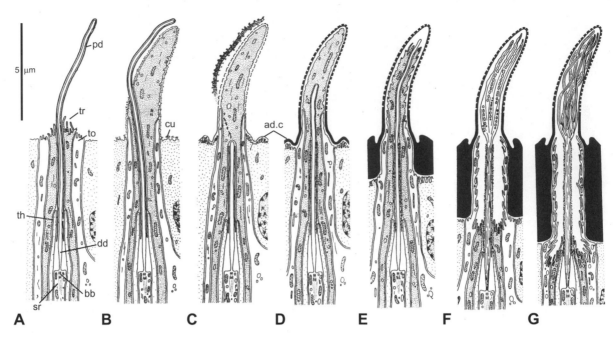

Fig. 10.6. Sagittal sections illustrating the differentiation of an olfactory, multiporous chemosensillum on the antenna of *Necrophorus vespilloides* (Coleoptera: Silphidae) during during the exuvial (A) and cuticular pharate (B–D) and early adult (E–G) stages. A. Outgrowth of primary dendrite (pd) surrounded by dendritic sheath. B. Outgrowth of trichogen cell and deposition of adult cuticulin layer. C. Degeneration of primary dendrite distal to trichogen cell. D. Growth of dendrite into cytoplasm of trichogen cell. E. Withdrawal of trichogen cytoplasm from the newly formed olfactory hair. F and G. Branching of sensory dendrite. ad.c, adult cuticle; bb, basal bodies (centrioles) of cilium; cu, cuticulin of new adult epicuticle; dd, distal dendrite; sr, ciliary root of cilium; th, apex of thecogen (dendritic sheath) cell; to, tormogen cell; tr, trichogen cell. (Reprinted with permission from K.-D. Ernst. Die Ontogenie der basiconischen Riechsensillen auf der Antenne von *Necrophorus* (Coleoptera). *Z. Zellforsch.* 129: 217–236. © 1972 Springer-Verlag.)

each epidermal cell, microtubules extend from the desmosomes to apical hemidesmosomes between the cell apex and the inner surface of the endocuticle and from there to the epicuticle as attachment fibers that pass through the procuticle within pore canals; the microtubules and fibers together constitute the *tonofibrillae* of the classic histological literature. These fibers are sclerotized, are unaffected by molting enzymes, and attach the muscle to the old epicuticle and exocuticle until the connection is broken at ecdysis. Therefore, apolysis and ecdysis are simultaneous at muscle attachment sites. New attachment fibers develop within the new procuticle as it is deposited, are sclerotized at the same time as the exocuticle, and begin to function at ecdysis.

10.2 Structural Change during Metamorphosis

As mentioned earlier (Fig. 9.2), the amount of structural change occurring during metamorphosis varies between species, depending on how far the juveniles have diverged in structure and behavior from the adults (De Beer, 1962: 135). In ametabolous, paurometabolous, hemimetabolous, neometabolous, and basal holo-

metabolous insects, most juvenile structures are carried through to the adult with greater or lesser amounts of reorganization (Snodgrass, 1954, 1956; Whitten, 1968; Sehnal, 1985a, 1985b; Nüesch, 1987; Svácha, 1992; Sehnal et al., 1996). However, in higher flies (Diptera: Cyclorrhapha) and in bees and wasps (Hymenoptera: Apocrita), many larval structures progressively degenerate and are replaced by differentiation of quite different, imaginal structures from epidermal imaginal discs and histoblasts. Below, I first summarize metamorphosis in a higher fly, *D. melanogaster*, in which it is best understood and is most extensive, and then describe that of each organ system in more detail. I also refer to events occurring in exopterygotes and in basal endopterygotes. Fig. 10.8 summarizes the events occurring in pupae of the apple maggot *Rhagoletis pomonella* (Diptera: Tephritidae) (Snodgrass, 1924); these events are almost identical to those of *D. melanogaster*, for which suitable figures are not available.

10.2.1 METAMORPHOSIS IN HIGHER FLIES
A striking characteristic of more advanced, holometabolous insects is the dramatic change in form from larva to pupa occurring at the larval-pupal ecdysis.

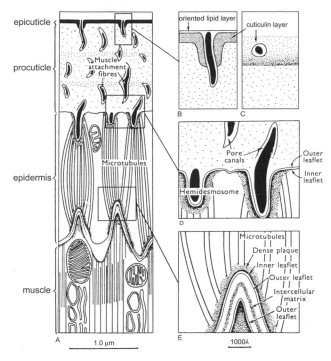

epicuticle
procuticle
epidermis
muscle

Muscle attachment fibres

Microtubules

oriented lipid layer
cuticulin layer

B   C

Pore canals

Hemidesmosome

Outer leaflet
Inner leaflet

D

Microtubules
Dense plaque
Inner leaflet
Outer leaflet
Intercellular matrix
Outer leaflet

A   1.0 μm   E   1000Å

Fig. 10.7. A muscle attachment to the cuticle in an aptery-gote insect. A. General view. B–E. Ultrastructural details of regions indicated in A. The muscle attachment fiber traverses the procuticle within a pore canal into a pitlike depression in the cuticulin layer of the epicuticle (B, C). D. Junction between cuticle and epidermis as shown in A. Muscle attachment fibers originate in conical hemidesmosomes, and microtubules insert into the dense material of these junctions. E. Junction between epidermis and muscle fiber as shown in A. Epidermal microtubules originate in the dense plaque associated with interdigitations in this region. A dense intercellular matrix is present between the two hemidesmosomal components of the desmosome. (Reprinted with permission from S. Caveney. Muscle attachment related to cuticle architecture in Apterygota. *J. Cell Sci.* 4: 541–559. © 1969 Company of Biologists, Ltd.)

More remarkable still is that the pupa has the shape of the future adult, even though at the time of its formation few adult structures are developed (Fig. 10.8G, H) (Snodgrass, 1924; Whitten, 1968).

At 25°C, metamorphosis in *D. melanogaster* requires 96 h (Fig. 10.9) (Fristrom and Fristrom, 1993; Denlinger and Zdárek, 1994). It begins when the third instar larva (Fig. 9.9D) leaves its food during a wandering stage to find a site suitable for pupariation (Denlinger and Zdárek, 1994). In response to a short release of 20E from prothoracic gland cells within a ring gland about the dorsal vessel behind the brain, the larva shortens itself both by muscular contraction and by formation of longitudinally oriented epidermal "feet" (*filopodia*) that contract and shorten each segment (Locke and Huie, 1981), rounds up into a football-shaped white puparium, and sclerotizes its cuticle to form the puparium (this is the

only instance known in insects where sclerotization occurs near the end of an instar rather than near its beginning) (Fig. 10.8A).

When pupariation is complete, the epidermis separates from larval cuticle from front to back, resulting in an apparently headless, wingless, legless, exuvial pharate pupa (Fig. 10.8B). Four hours after puparium formation (4 h APF), the insect draws back from the anterior end of the puparium by muscular contraction and everts its wing and leg primordia onto the surface of the body through their peripodial cords (Fig. 10.8C–E), an event caused by changes in shape and position of imaginal disc cells and by contraction of peripodial membranes, not by hemostatic pressure (Fristrom and Fristrom, 1993). At first, the head is still invaginated; the wings and legs, though everted, are flaccid and unexpanded; and the abdomen is still larval in shape and size (Fig. 10.8C–E). This stage is referred to as the "hidden head" or *cryptocephalic pupa*.

Orchestrated contraction of larval abdominal muscles at 12 h APF reduces the interior volume of the abdomen and forces blood into the thorax and appendages, assisted by a gas bubble that forms within the midgut in the cryptocephalic pupa and moves forward at this time. In about 10 min, the imaginal head everts (Fig. 10.8F), reversing the involution of the larval head that occurred during embryogenesis (8.3.7), the larval cephalopharyngeal skeleton is ejected against the front of the puparium, the legs and wings expand, and the larval tracheal lining is shed. Following eversion, the head is very small at first (Fig. 10.8F) (this stage is called the "small head" or *microcephalic pupa*), but abrupt, additional muscular contraction, increasing the hemostatic pressure up to fourfold, results in formation of a typical adecticous exarate pupa (Fig. 10.8G, H) referred to as the "visible head" or *phanerocephalic pupa*. Head eversion is considered to be the beginning of pupation by drosophilists and the period from pupariation to head eversion, as the prepupal stage (Fristrom and Fristrom, 1993).

About 2 h before disc eversion, a thin, membranous pupal cuticle begins to be deposited about the body and the still invaginated appendages. The cuticle of the head and thorax is deposited by imaginal cells, and about 2 h later, that of the abdomen, by both larval cells and abdominal histoblasts (Fig. 10.8B). Four hours after the pupal cuticle is complete, at 14 h APF, pupal-adult apolysis is induced by a second, small release of 20E, resulting in an exuvial pharate adult enclosed within both larval and pupal cuticles (Fig. 10.9).

The larval epidermis begins to degenerate before disc eversion is complete, first in the head and thorax and later in the abdomen. Its cells are sloughed into the hemocoel by spread of cells from the imaginal discs and histoblasts underneath them and are engulfed by multinucleate spherule cells or *macrophages* (Fig. 10.11). Imaginal cuticle deposition takes place from 36 to 70 h APF, the insect becoming a cuticular pharate adult.

Ninety-six hours after pupariation, and at a species-

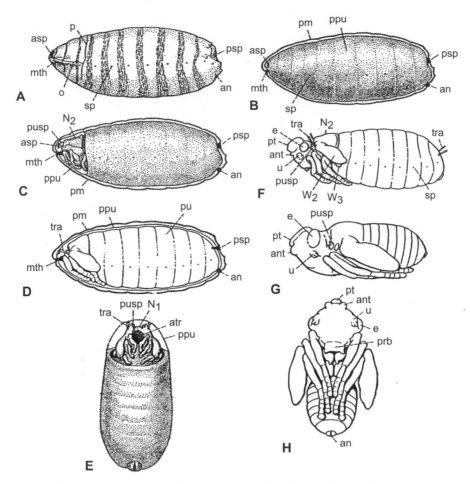

Fig. 10.8. Metamorphosis of the apple maggot fly *Rhagoletis pomonella* (Diptera: Tephritidae). A. Puparium (shortened and sclerotized third instar cuticle) in lateral aspect, showing ptilinal suture (o, p). B. Exuvial pharate pupa 24 h after puparium formation just after larval-pupal apolysis, lateral aspect. C. "Cryptocephalic" pupa after eversion of the appendages but before that of the head. The so-called prepupal larval cuticle (ppu) is probably ecdysial membrane from the third instar larval endocuticle. D. Same, after apolysis of the ecdysial membrane but before the cuticular linings of foregut and hindgut are shed. E. "Cryptocephalic" pupa removed from puparium and in ventral aspect. F. "Microcephalic" pupa just after head eversion but before shedding of tracheal cuticular linings. G. "Phanerocephalic" pupa in lateral aspect after abdominal contraction and high blood pressure has expanded the head and appendages. H. Same, in ventral aspect (this fly has a typical adecticus exarate pupa). an, anus; ant, antenna; asp, anterior larval spiracle; atr, atrium; e, compound eye; mth, larval mouth; N_1, pronotum; N_2, mesonotum; o, horizontal component of ptilinal suture; pm, puparium; ppu, "prepupal larva" (pharate pupa) or "prepupal larval cuticle" (larval ecdysial membrane); prb, proboscis; psp, posterior larval spiracle; pt, ptilinum; pu, pupa; pusp, pupal prothoracic spiracle; sp, lateral thoracic and abdominal spiracles functioning only in ecdysis of tracheal cuticular linings; tra, third instar tracheal lining; u, subocular lobe of pupal head; W_2, wing; W_3, haltere. (Adapted from R. E. Snodgrass. 1924. Anatomy and metamorphosis of the apple maggot, *Rhagoletis pomonella* Walsh. *J. Agric. Res.* 28 (1): 1–36; plates 1–6.)

specific time of day, extrication behavior is induced by release of EH and perhaps pre-ecdysis– and ecdysis-triggering hormones, although this has yet to be demonstrated in *D. melanogaster*. The *ptilinum*, a soft flexible sac of spinous cuticle above the antennae and continuous with the head capsule (Fig. 10.8F–H), is repeatedly expanded and retracted by alternate contraction of thoracic/abdominal and ptilinal muscles. This pops the apex of the puparium off along a preformed line of weakness (Fig. 10.8A) (Segal and Sprey, 1984), allowing the fly to escape. After emergence, the ptilinum continues to be used when the insect bumps into an obstacle in its path to the surface (program for obstacle removal), but its use is interspersed with extended periods of peristaltic contractions from back to front (program for forward movement) that function to move the insect forward through the substrate (Denlinger and Zdárek, 1994).

The final shape of the fly is achieved after it reaches the surface and is attained by the postemergent expansion of unsclerotized cuticle in the wings, legs, and body produced by a complex expansion behavior that includes walking, grooming, intaking air into the gut, and contracting muscles. Free rather than confined leg movement, as when burrowing, ends the extrication program

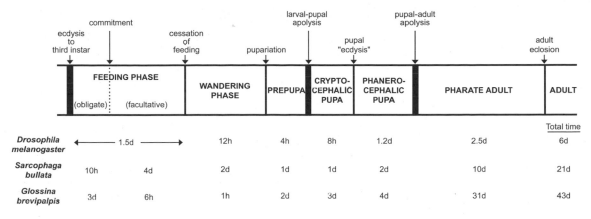

Species	FEEDING PHASE (obligate)	FEEDING PHASE (facultative)	WANDERING PHASE	PREPUPA	CRYPTO-CEPHALIC PUPA	PHANERO-CEPHALIC PUPA	PHARATE ADULT	Total time
Drosophila melanogaster	← 1.5d →		12h	4h	8h	1.2d	2.5d	6d
Sarcophaga bullata	10h	4d	2d	1d	1d	2d	10d	21d
Glossina brevipalpis	3d	6h	1h	2d	3d	4d	31d	43d

Fig. 10.9. Timetable of principal developmental and behavioral events from the beginning of the third larval instar to adult eclosion in three species of higher Diptera. Vertical arrows indicate landmarks for the beginning and end of each developmental stage. Times below each stage are the approximate durations of each interval at 25°C for the higher flies, *Drosophila melanogaster*, *Neobellieria bullata* (Sarcophagidae; *bullata* was transferred from the genus *Sarcophaga* to the genus *Neobellieria*), and *Glossina brevipalpis* (Glossinidae). (With permission from the *Annual Review of Entomology*, Volume 39 © 1994, by Annual Reviews www.AnnualReviews.org.)

and induces the expansion program (Denlinger and Zdárek, 1994). General sclerotization, induced by release of bursicon, occurs only after expansion is complete and only after the ptilinum is retracted for the last time and the lips of its invagination are sealed to form the frontal lunule.

10.2.2 MESODERMAL STRUCTURES
Larval skeletal and visceral muscles, hemocytes, the male and female primary reproductive exit system, dorsal vessel, and fat body differentiate from mesoderm during embryogenesis (8.1) and undergo more or less profound changes during metamorphosis of holometabolous insects.

10.2.2.1 MUSCULATURE
Five kinds of transformation have been observed to occur in insect skeletal muscles during metamorphosis (Whitten, 1968) (Fig. 10.10):
- Larval muscles pass unchanged in structure and function to the adult (A in Fig. 10.10).
- Larval muscles are respecified to form imaginal muscles, which may or may not have the same function (B).
- Larval muscles are destroyed and not replaced (C).
- Larval muscles are destroyed but replaced by differentiation of imaginal muscles having similar functions (D).
- New, imaginal muscles develop de novo from muscle progenitor cells or myoblasts (E).

In ametabolous, paurometabolous, and hemimetabolous insects, the first and second kinds of transformations predominate, while in neometabolous and holometabolous insects the last three do.

New, imaginal muscles can develop either through cleavage of larval fibers to act as myoblast fusion targets for formation of adult muscles or by de novo accumulation and differentiation of myoblasts.

In *Manduca sexta*, larval leg motor neurons survive degeneration of their larval target muscles and are respecified to innervate new muscles that differentiate from myoblasts during development of adult legs (Leudeman and Levine, 1996). If imaginal myoblasts, derived from developing thoracic legs of early pupae, are cultured in vitro with neurons from anywhere in the body, they first become spindle-shaped and then aggregate into multinucleate contractile myotubes. However, if only myoblasts are present, such myotubes fail to form. If labeled bromodeoxyuridine (BrdU) is added to co-cultures of myoblasts and neurons from anywhere in the nervous system, its uptake by myoblasts is enhanced, suggesting that neurons promote proliferation of myogenic cells. However, medium without neurons is not effective in myotube induction even if it is first conditioned by their temporary presence. Finally, if 20E is added to the medium, BrdU incorporation is also enhanced, even in the absence of neurons, indicating a role for this hormone in imaginal muscle development.

These observations are reinforced by recent work on the ventral diaphragm of this insect, which covers the ventral nerve cord (VNC) (Fig. 8.1C), consists of connective tissue and muscle fibers, and undergoes a greater or lesser change in configuration during metamorphosis. In pupae of *M. sexta*, the adult diaphragm develops through proliferation, aggregation, and differentiation of myoblasts associated with the transverse nerve of each VNC ganglion, with both processes controlled by release of 20E (Champlin et al., 1999). Tonic exposure to concentrations of 20E between 40 ng/ml and 1 µg/ml stimulate myoblast proliferation by activating an ecdysteroid-dependent control point in G2 of the myoblast cell cycle, and proliferation can be turned on or off simply by adjusting the concentration of ecdysteroid above or below these thresholds. However,

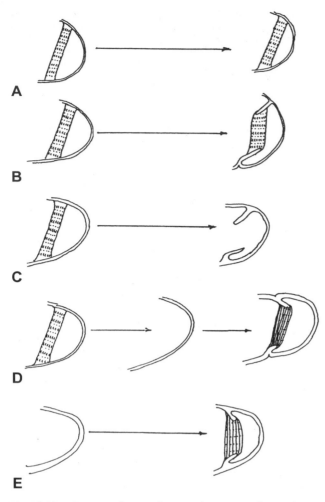

Fig. 10.10. Degrees of muscular transformation observed in various insects during metamorphosis. See text for details.

concentrations of 20E above 1 μg/ml irreversibly inhibit myoblast proliferation and, instead, induce the cells to differentiate into muscle fibers. Both the stop to proliferation and the stimulus to differentiation are inhibited by applying the juvenile hormone (JH) analog methoprene (Fig. 12.11F), even though methoprene has no effect on ecdysteroid-dependent proliferation. Finally, premature exposure to high (>1 μg/ml) concentrations of 20E in the absence of JH triggers precocious differentiation. These same ecdysteroid thresholds influence the proliferation and differentiation of cells in the optic lobes (10.2.4.1.2) and compound eyes (10.2.4.2.3) during metamorphosis of this species.

In higher flies, the imaginal, thoracic, skeletal muscles differentiate from loose, mesenchymal, adepithelial cells within the thoracic imaginal discs, and the abdominal muscles differentiate from those associated with the peripheral nerves in paired dorsal, lateral, and ventral clusters in each segment (Lawrence, 1992; Bate, 1993; Hartenstein, 1993; Roy and VijayRaghavan, 1999). Development of peripheral nerves and skeletal muscles

during metamorphosis seems to be reciprocal. For example, in *D. melanogaster*, the larval mesothoracic, intersegmental nerves and motor neurons are remodeled to innervate the dorsolongitudinal and two pairs of dorsoventral flight muscles of the adult, the dorsolongitudinal muscles using modified larval muscles as templates to develop while the dorsoventral muscles arise de novo from imaginal myoblasts in the imaginal discs (Fernandes and VijayRaghavan, 1993; Rivlin et al., 2000). If development of these muscles is delayed or permanently eliminated, adult-specific branching of motor neurons continues, but higher-order, synaptic branching fails to occur until the delayed muscle fibers reach a state of development suitable for their innervation (Fernandes and Keshishian, 1998). If these nerves are destroyed, the number of imaginal myoblasts available decreases and formation of dorsoventral flight muscles is blocked. However, the dorsolongitudinal flight muscles continue to develop unless deprived of their larval template fibers, in which case they too require innervation to differentiate.

An additional complication in higher flies is that special eclosion muscles, which function to facilitate escape of the adult from the puparium and its movement to the soil surface, differentiate during metamorphosis but then degenerate (Whitten, 1968; Bate, 1993; Chapman, 1998). In pharate adults of the sarcophagid *Neobellieria bullata*, for example, in addition to ptilinal muscles (10.2.1), there are four pairs of thoracic muscles and many abdominal muscles that function during eclosion but later degenerate. The thoracic muscles lose their contractility within 30 min of eclosion, their high glycogen content within 4 h, and their myosin ATPase and succinate dehydrogenase activity between 12 and 24 h, after which they degenerate (Bothe and Rathmayer, 1994). Results of neck ligation and muscle isolation experiments suggest that two signals induce this degeneration: the first, which is released about 2 h before eclosion and probably is EH, induces slow degeneration; the second, released 20–30 min after eclosion, causes loss of muscle contractility but is neither bursicon nor a direct neuronal command (bursicon extract did not affect contractility of these muscles).

10.2.2.2 HEMOCYTES

Larvae of higher flies have five types of hemocytes formed in their *lymph glands* (large, paired hematopoietic cells associated with the dorsal vessel [Fig. 8.7C]), of which one type, the *granulocytes* (plasmatocytes), undergoes major changes in structure and function during metamorphosis (Fig. 10.11: top, hemocyte type 5) (Whitten, 1968). Early in this process, the granulocytes change their adhesion and phagocytic properties and undergo nuclear multiplication. They then become giant multinucleate spherule cells or macrophages that phagocytose fragments of degenerating larval tissues and later certain imaginal tissues, and recycle their components to adult tissues developing from imaginal discs and histoblasts. These large birefringent cells are easily seen by

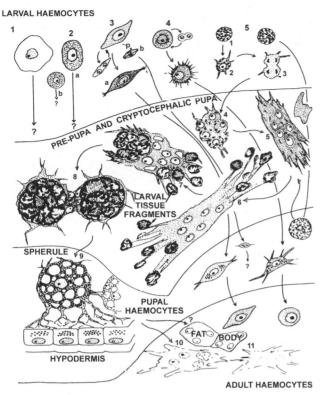

LARVAL HAEMOCYTES

PRE-PUPA AND CRYPTOCEPHALIC PUPA

LARVAL TISSUE FRAGMENTS

SPHERULE

PUPAL HAEMOCYTES

FAT BODY

HYPODERMIS

ADULT HAEMOCYTES

Fig. 10.11. Summary of the changes occurring in larval hemocytes during metamorphosis in pupae of *Neobellieria bullata* (Diptera: Sarcophagidae). 1. Hyaline cell. 2–4. Plasmatocytes. 5. Granular hemocytes. Granular hemocytes undergo nuclear division to form large multinucleate masses that phagocytose degenerating larval tissues and become pupal "spherules" or macrophages. These persist into the pharate adult, where they selectively ingest specific imaginal epidermal cells at specific times. Fusiform (3) and round (4) plasmatocytes are carried through metamorphosis and appear to function in the adult. (Reprinted from J. M. Whitten. Metamorphic changes in insects. In W. Etkin and L. I. Gilbert (eds.), *Metamorphosis: a problem in developmental biology*, p. 75, Fig. 7. Appleton-Century-Crofts, N.Y. © 1968.)

phase or interference contrast microscopy within living whole mounts of legs of sarcophagid or muscid pupae. While some larval plasmatocytes are carried through to the adult (Fig. 10.11: top, hemocyte types 3 and 4), the macrophages seem to degenerate (Whitten, 1968; Lanot et al., 2001).

10.2.2.3 DORSAL VESSEL

In *D. melanogaster* and other higher flies, the dorsal vessel is carried through to the adult but ceases to function in the midpupa because of dissolution and regeneration of contractile elements of the heart and alary muscles (Whitten, 1968; Sehnal, 1985a, 1985b). In addition, the dorsal vessel changes its course through the body to accommodate the changed body shape of the adult (Fig. 7.19) compared to the larva (Figs. 9.9D, 10.8). It then develops a conical chamber in the first and second abdominal segments, a layer of longitudinal muscles on the ventral surface of the heart in the abdomen, and adds a fourth pair of ostia, which become valvelike in the adult (Curtis et al., 1999). The dorsal vessel is reinnervated, and accessory pulsatile organs differentiate in association with the developing antennae, wings, and legs.

10.2.2.4 FAT BODY

In exopterygotes, larval fat body cells are usually carried over into the adult with little change except for an increase in the ploidy of their nuclei, but in endopterygotes, the fat body is reorganized during metamorphosis (Sehnal, 1985a, 1985b). In *D. melanogaster* and other higher flies, the larval fat body dissociates into individual cells and degenerates by day 3 of adult life, to be replaced by differentiation of imaginal fat cells apparently from mesenchyme within the imaginal discs (Schnal, 1985a, 1985b; Hartenstein, 1993; Richard et al., 1993). If corpora allata cells are removed from the ring gland of sarcophagid larvae, the larval fat body is retained, suggesting release of JH causes its degeneration. However, Richard and coworkers (1993) were unable to demonstrate fat body histolysis in vitro in the presence of JH, of larval ring glands, or of adult ovaries and concluded that fat body lysis is not mediated by JH but by some as-yet-unidentified factor(s) possibly originating in the ovary.

10.2.3 STRUCTURES OF MIXED GERM LAYER ORIGIN

Anlagen of the reproductive system differentiate in the embryo from cells of mixed germ-line, ectodermal, and mesodermal origin (8.4.1), and those of the gut differentiate from ectodermal and endodermal cells (8.2).

10.2.3.1 REPRODUCTIVE SYSTEMS

At hatch, the reproductive system consists of two undifferentiated mesodermal, gonadal rudiments, each connected to the back of sternum 9 (male) or 7 (female) by a delicate primary exit duct rudiment of similar origin (Fig. 8.53A: left and right). These may or may not end posteriorly in an expanded genital ampulla that, in males, is usually attached to a median ectodermal thickening, the *genital disc* (Fig. 8.53B).

10.2.3.1.1 MALES

A median, unpaired ejaculatory duct rudiment invaginates anteriorly from the genital disc at the back of sternum 9 between the ampullae of the vasa deferentia and carries them forward into the hemocoel (Fig. 10.12A–D). Depending on species, accessory gland primordia (*mesadenes*) later evaginate from walls of the ampullae and others (*ectadenes*), from those of the ejaculatory duct (Fig. 10.12E, G).

The male external genitalia or phallus (1.1.4) differentiates from a pair of ectodermal primary phallic lobes evaginating posteriorly from the genital disc immediately behind the genital ampullae and on either side of the ejaculatory duct rudiment at the back of sternum

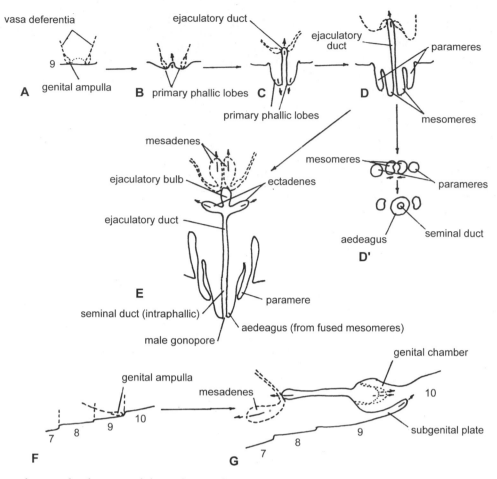

Fig. 10.12. Postembryonic development of the male reproductive exit system and external genitalia in an exopterygote insect in ventral aspect (A–E) and in sagittal (F and G) and transverse (D') sections (highly diagrammatic). Structures indicated with a dashed line are of mesodermal origin, and those with an unbroken line are of ectoderm. Only ectodermal structures are lined or covered by cuticle. Numerals indicate abdominal segments.

9 (Fig. 10.12B, C). These divide longitudinally into inner mesomeres and outer parameres. The mesomeres fuse together mesially to form the rudiment of the aedeagus, except for a median seminal duct that remains unoccupied between the mesomeres and extends posteriorly from the posterior opening of the ejaculatory duct through the phallus to the male gonopore at its tip (Fig. 10.12D, D', E). Posterior growth of sternum 9 under the genitalia then encloses them within a genital chamber (Fig. 10.12G). The reproductive system assumes its final form and function with the deposition and differential sclerotization of adult cuticle on or within its ectodermal portions. The cuticle of the phallus is usually deposited by the cells in its retracted, telescoped configuration (e.g., Heming, 1970). In addition, the innervation and differentiation of genital muscles and of secretory cells must be completed before the insect is able to inseminate a female; these processes are not usually completed until the end of the premating period, well after adult emergence (Matsuda, 1976).

10.2.3.1.2 FEMALES

A median, unpaired common oviduct rudiment invaginates anteriorly from ectoderm at the back of sternum 7 between the posterior ends of the lateral oviduct rudiments (its posterior opening is the primary gonopore); the rudiment of the spermatheca, from the back of 8; and that of the accessory gland(s), from the back of 9 (Fig. 10.13A, B) (2.1 describes the adult system). The gonopore is then shifted posteriorly to the back of sternum 8 as the secondary gonopore by formation and closure, from front to back, of a midventral longitudinal groove (Fig. 10.13B, B', C), and the openings of all three structures are enclosed within a bursa copulatrix or genital chamber by posterior growth of sternum 7 or 8 to form a subgenital plate (Fig. 10.13D). The genital chamber opens to the outside posteriorly through the vulva, often into the base of the ovipositor (Figs. 10.13D, 10.14C) (Matsuda, 1976).

In females having a valvular ovipositor (Fig. 2.4), primordia of the paired anterior ventral valves grow out

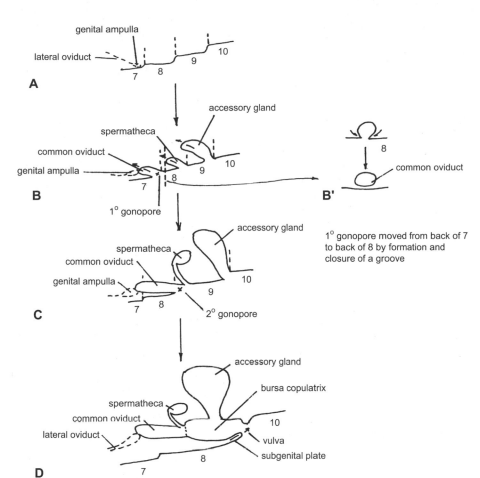

Fig. 10.13. Postembryonic development of the female reproductive exit system in an exopterygote insect in sagittal (A–D) and transverse (B') section (highly diagrammatic). Structures indicated with a dashed line are of mesodermal origin, and those with an unbroken line are of ectoderm. Only ectodermal structures are lined or covered by cuticle. Numerals indicate abdominal segments.

posteriorly from ectoderm at the back of sternum 8 on either side of the spermathecal rudiment, and primordia of the dorsal and lateral valves, from the back of 9 on either side of the accessory gland rudiment (Fig. 10.14B, C). They assume their final form and function with deposition and differential sclerotization of adult cuticle and with innervation and differentiation of their musculature (Matsuda, 1976).

In ectognathous apterygotes and in exopterygotes of both sexes, development of reproductive primordia usually begins just before or just after hatch, with differentiation continuing until the end of the preoviposition (female) or premating (male) periods, well after adult emergence. However, in neometabolous insects, development is delayed until later instars (Matsuda, 1976) and, in endopterygotes, it is delayed until metamorphosis, when there is also an increasing tendency, in higher dipterans, for the genital discs of adjacent genital segments to fuse longitudinally and for the ectodermal, secondary exit system to substitute for the mesodermal, primary one. For example, the single median genital disc in *D. melanogaster* larvae (Fig. 10.33) comprises the fused discs of segments 8–11 in females and of 9–11 in males, and, except for the gonads, gives rise to the entire reproductive system, the external genitalia, abdominal segment 8, and the posterior part of the hindgut (Epper, 1983; Hartenstein, 1993). In addition, primordia of both male and female components are present in the genital discs of both sexes (Epper, 1983). The ovary rudiments likewise divide into ovarioles in the pupa and pharate adult (King et al., 1968) rather than in the embryo or larva as in members of basal and intermediate lineages (8.4.1.3).

10.2.3.2 ALIMENTARY CANAL
In exopterygotes and in members of basal clades of endoptcrygotes, the entire gut increases in size throughout postembryogenesis through increases in cell number, size, and ploidy while midgut epithelial cells are partially or completely replaced at each molt from replacement cells scattered next to the basal lamina or by proliferation from regenerative crypts (Sehnal, 1985a, 1985b). Engelhard and coworkers (1991) used the proteinase dispase to separate the intact midgut epithelium of third, fourth, and fifth instar caterpillars of thc noctuid *Trichoplusia ni* from its underlying connective and muscle tissue layer and made living, whole mount preparations of them. They then characterized its cell types (columnar, goblet, differentiating, and stem cells) by double fluorescent labeling of F-actin and nuclei

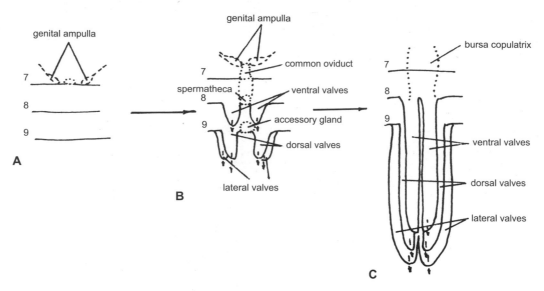

Fig. 10.14. Postembryonic development of the valvular ovipositor in a female exopterygote insect in ventral aspect (highly diagrammatic). Structures indicated with a dashed line are of mesodermal origin, and those with an unbroken line are of ectoderm. Only ectodermal structures are lined or covered by cuticle. Numerals indicate abdominal segments.

and compared their numbers by digital image analysis. Changes during larval development included changes in cell division rate and differentiation, the former greatly decreasing from fourth to fifth instar.

In higher flies, the larval gut is replaced by development of a new, imaginal gut at metamorphosis, as summarized in Fig. 10.15. The imaginal foregut originates both by posterior proliferation of cells from the labial discs at the anterior end of the larval foregut and by anterior proliferation from an imaginal ring of cells at its inner end. The imaginal hindgut originates in similar fashion from the genital disc and another imaginal ring at its inner end. As proliferation proceeds, the larval cells are replaced by imaginal ones that spread underneath them and slough them into the body cavity, where they degenerate and are engulfed by macrophages.

The larval midgut epithelium is replaced by imaginal cells proliferating from imaginal midgut precursor cells in hundreds of small *histoblasts* or islands scattered throughout the midgut and is sloughed into the lumen. It eventually forms a "yellow body" and at adult eclosion exits the body as the meconium along with nitrogenous waste voided from the Malpighian tubules (9.6.2.3).

The Malpighian tubules are carried through to the adult in most investigated exopterygotes, beetles, and flies and continue to function throughout metamorphosis (Sehnal, 1985a, 1985b). However, in some lepidopterans, apical cryptonephridia of the tubules associated with the hindgut degenerate and the remainder is reconstructed (Ryerse, 1979). In larvae of the honeybee *A. mellifera* and other eusocial hymenopterans fed by workers within wax or paper cells in the hive or nest, the anus and the openings between the midgut, Malpighian tubules, and hindgut are sealed by membrane to prevent

waste from contaminating the food with which the cells are provisioned (Snodgrass, 1956: 176–177, 197–198). After the larvae have completed their development and are sealed within their cells in preparation for metamorphosis, these membranes disintegrate, and for the first time, accumulated waste is voided from their bodies. The larval Malpighian tubules then degenerate and are replaced by a new set of imaginal tubules differentiating from primordia between their bases in the apex of the hindgut.

10.2.4 ECTODERMAL STRUCTURES

The ectoderm gives rise to the central, peripheral, and stomatogastric nervous sytems; the neuroendocrine sytem; the oenocytes; tracheal tubes; salivary glands and body wall epidermis; the foregut and hindgut; and primordia of the secondary exit system and external genitalia during embryogenesis (8.3).

10.2.4.1 CENTRAL NERVOUS SYSTEM

The central nervous system (CNS) of insects consists of the VNC and brain (Fig. 8.10), which, together with the endocrine system, control the body's activities. Thus, metamorphic changes in the CNS correlate with and influence changes in other organ systems (Whitten, 1968). Because the VNC undergoes metamorphosis differently and separately from the brain, it is treated separately.

10.2.4.1.1 VENTRAL NERVE CORD

Larval and imaginal VNCs usually differ, the VNC of the imaginal system being more concentrated than that of the larva because of rostral fusion of adjacent ganglia during metamorphosis to service the developing flight

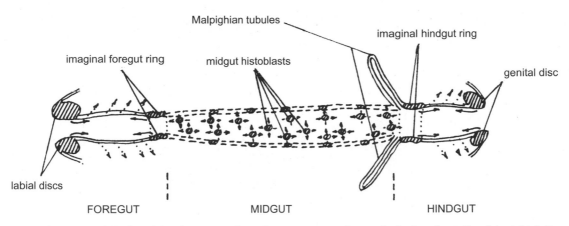

Fig. 10.15. Replacement of the larval by the imaginal gut during metamorphosis of a higher fly. Cells of the labial discs and imaginal foregut ring, of the imaginal hindgut ring and genital disc, and of the midgut histoblasts proliferate to replace the larval epithelium, respectively, of the foregut, midgut, and hindgut. Larval cells of the foregut and hindgut are sloughed into the hemocoel where they are phagocytosed by macrophages; those of the midgut are sloughed into its lumen where they persist along with nitrogenous waste from the Malpighian tubules to form the meconium voided on adult eclosion.

and genital muscles (Fig. 10.16A, B). This change in ganglionic position results from a proliferation of sheath (perineurial) cells about the VNC, breakdown of its neurilemma, and migration and redistribution of glial cells and ganglia (Whitten, 1968; Edwards, 1969; Pipa, 1973). In higher flies, in which the larval VNC is a single compact mass, the abdominal peripheral nerves shorten by shortening their perineurial sheath cells and pull the fused thoracicoabdominal ganglionic mass farther back into the thorax, the interganglionic connectives between the subesophageal and thoracic ganglia lengthening (Fig. 10.16C, D) (Whitten, 1968; Hartenstein, 1993; Truman et al., 1993). In other insects in which the imaginal VNC is more concentrated than the larval one, the connectives between adjacent ganglia shorten (Fig. 10.16A, B).

Holometabolous insects in advanced lineages differ from hemimetabolous ones in having two major bouts of neurogenesis, one in the embryo and the other in the late larval and pupal stages (Truman et al., 1993). During these periods, larval neurons either are respecified to perform new, adult functions or degenerate to be replaced by a new batch of strictly imaginal neurons that originate by proliferation from persisting embryonic neuroblasts (NBs) (Levine, 1986; Truman et al., 1993). For example, in *D. melanogaster* only 5%–10% of larval neurons in the VNC are carried over to the adult, most of these being respecified to function either as adult premotor interneurons or as motor neurons (Truman et al., 1993). Degenerating larval neurons are phagocytosed by glial cells in both the brain and the VNC just after pupariation (Cantera and Technau, 1997).

Prokop and Technau (1991) showed that embryonic and postembryonic NBs in thoracic ganglia of *D. melanogaster* have the same origin in the embryo. They transplanted single neuroectodermal cells from donor embryos triple labeled early in gastrulation into unla-

beled host embryos at the same stage of development (7.1.4 for methods). They then followed development of these transplants from early gastrulation until the late third instar and found the postembryonically derived neurons to always cluster with the embryonically derived ones. Labeling with BrdU at various times during embryogenesis suggested that the embryonic NB itself, rather than one of its progeny, resumes proliferation after hatch as the postembryonic NB. Purely embryonic clones, producing only larval neurons, were also observed. Of the 53 NBs estimated to occur in each embryonic neuromere of each segment, 47 thoracic and 6 abdominal NBs were shown to reactivate in the larva.

The question then arises as to what inhibits mitosis of embryonic NBs in older embryos and in first and second instar larvae of *D. melanogaster*. The answer seems to be the glial cells (Ebens et al., 1993). In *anachronism* (*ana*) mutants, quiescent NBs in the brain begin to divide precociously. The protein encoded by this gene is secreted by larval glial cells and seems to inhibit proliferation of these NBs until the third instar. It is not known how this inhibition is lifted, but it may be by release of 20E (Truman et al., 1993).

Not surprisingly, the number and distribution of glial cells in the VNC also change during metamorphosis (Awad and Truman, 1997). A newly hatched first instar of *D. melanogaster* has 40–50 midline glial cells in each ganglion of its VNC (Fig. 8.29). By localizing beta-galactosidase (β-GAL) expression in enhancer trap lines and by monitoring BrdU incorporation in pulse-chase experiments, Awad and Truman (1997) found these cells to proliferate to about 230 in the late third instar and to spread dorsoventrally throughout the midline. Just after pupariation, following release of the puparial peak of 20E (Fig. 12.20), these cells cease dividing, and for the first half of the pupal stage there is no change in their number or position. However, from 50% to 80% of

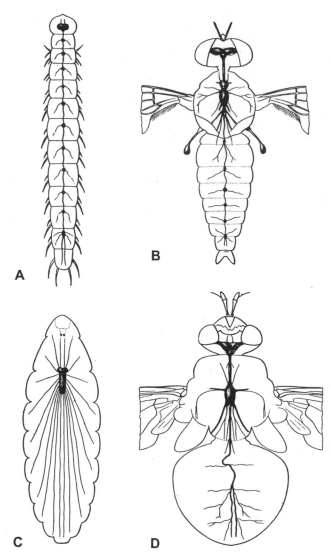

Fig. 10.16. The central nervous sytem of a larva (A) and adult (B) of *Bibio hortulanus* (Diptera: Bibionidae) and *Phryna vanessae* (C and D) (Diptera: Muscidae) in dorsal aspect. Rostral fusion of ventral nerve cord ganglia as in *B. hortulanus* is characteristic of less specialized insects having complete metamorphosis, whereas in higher flies (C and D), the larval system is more condensed than that of the adult. (Reprinted with permission from L. F. Henneguy. *Les insectes. Morphologie, reproduction, embryogénie*, p. 483, Figs. 473, 474; p. 485, Figs. 477, 478. Masson et Cie Editeurs, Paris. © 1904 Masson SA.)

metamorphosis, they all degenerate. Awad and Truman (1997) speculated that these pupal glial cells may stabilize the growth of outgrowing imaginal neurites crossing the midline of the VNC or running along its sides, as they do in the developing embryo (Fig. 8.22).

Origin of Sexual Differences in the Ventral Nerve Cord
In adult insects, sexual differences exist in the neuronal pattern of posterior ganglia. They are associated with differences in skeletomusculature and innervation of the male and female reproductive systems and external genitalia. In *D. melanogaster*, these differences arise through sex-specific division of a unique set of postembryonic NBs during larval and early pupal development (Truman et al., 1993). Animals mutant for several sex-determining genes (4.2) were analyzed to determine the genetic regulation of NB commitment to male or female division pattern and the time when these decisions are made. In *double sex* (*dsx*) mutants, these NBs failed to undergo postembryonic mitosis in either male or female larvae; normally they are determined to adopt male or female fates at the end of the first instar.

Posterior Migration of Neurons between Ventral Ganglia during Metamorphosis
In the subesophageal and thoracic ganglia of last (fifth) instar caterpillars of *Agrotis segetum* (Noctuidae) and *Manduca sexta* (Sphingidae), bilaterally paired clusters of immature imaginal neurons and associated glial cells migrate posteriorly through the interganglionic connectives to enter the next ganglion posteriorly (Cantera et al., 1995). Migration commences at the onset of metamorphosis when the clusters gradually separate from other neurons in the cortex, enter the anterior ends of the connectives, and reach the next ganglion 3 d later, at 26°C. During migration, each cluster is completely enveloped by a single giant glial cell spanning the length of the connective, and during its movement, each neuron extends a long neurite posteriorly into the next ganglion. On its arrival, each neuron rapidly intermingles with surrounding neurons, presumably synapses with them, and soon is no longer recognizable. What the functional significance and phylogenetic distribution of this migration might be are unknown.

10.2.4.1.2 BRAIN
In holometabolous insects, the brain greatly enlarges during larval development and at metamorphosis undergoes profound changes in its optic lobes relating to replacement of larval stemmata (Fig. 8.27) by imaginal compound eyes (Fig. 10.17) (Nordlander and Edwards, 1969a, 1969b; Hofbauer and Campos-Ortega, 1990; Meinertzhagen and Hanson, 1993; Salecker et al., 1998); in its *corpora pedunculata* (mushroom bodies) relating to development of complex, imaginal behavioral patterns (Nordlander and Edwards, 1968; Technau and Heisenberg, 1982; Truman et al., 1993; Kurusu et al., 2000); and in its antennal lobes relating to differences in the number and function of larval and imaginal antennal sensilla (Nordlander and Edwards, 1968; Rospars, 1988).

Optic Lobes
In both exopterygote (Anderson, 1978b; Mouze, 1978; Stark and Mote, 1981) and endopterygote insects, interneurons of the three imaginal neuropiles of each optic lobe—the lamina, medulla, and lobula/lobula plate—arise through proliferation of two separate aggregations of NBs within the larval brain, the outer (OOA)

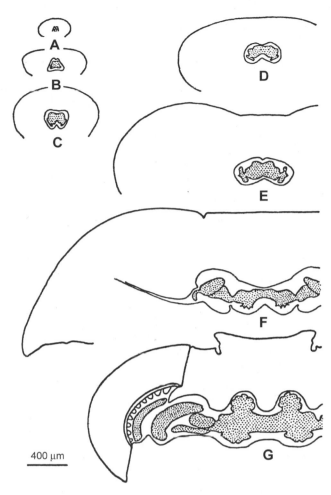

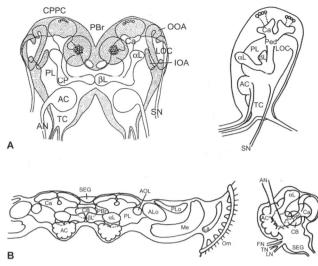

Fig. 10.18. Summary of the metamorphosis of the brain in *D. plexippus*. A. Third instar larval brain from the front (left) and side (right; optic lobes omitted). B. Adult brain from above (left) and side (right; optic lobe omitted). AC, antennal lobe; AN, antennal nerve; αL, alpha lobe of corpus pedunculatum; AOL, accessory optic lobe; ALo, anterior lobula; βL, beta lobe of corpus pedunculatum; Ca, calyx of corpus pedunculatum; CB, central body; CP, corpus pedunculatum; CPPC, corpus pedunculatum proliferation center; FN, frontal nerve; IOA, inner optic anlage; La, lamina ganglionaris; LN, labral nerve; LOC, larval optic lobe; Me, Medulla; Om, ommatidia of compound eye; OOA, outer optic anlage; PBr, protocerebral bridge; Ped, peduncle of corpus pedunclatum; PL, protocerebral lobe; PLo, posterior lobula; SEG, subesophageal ganglion; SN, stemmatal nerve; TC, tritocerebrum; TN, tegmental nerve. (A left reprinted with permission from R. H. Nordlander and J. S. Edwards. Postembryonic brain development in the monarch butterfly, *Danaus plexippus plexippus* L. III. Morphogenesis of centers other than the optic lobes. *W. Roux's Arch. Entwicklungs mech. Organ.* 164: 247–260. © 1970 Springer-Verlag.) (A right and B from Morphology of the larval and adult brains of the monarch butterfly, *Danaus plexippus plexippus*, L., R. H. Nordlander and J. S. Edwards, *J. Morphol.*, Copyright © 1968 Wiley-Liss, Inc. Reprinted by permission of Wiley-Liss, Inc., a subsidiary of John Wiley & Sons, Inc.)

Fig. 10.17. Frontal sections of the head and brain of the monarch butterfly, *Danaus plexippus* (Lepidoptera: Danaidae), at successive stages of development. Neuropile is represented by stippling, and cortex containing neuronal cell bodies, by unstippled areas. A. Newly hatched first instar. B. Early second instar. C. Early third instar. D. Early fourth instar. E. Early fifth instar. F. Early pupa. G. Adult. Note the spectacular expansion of the optic lobes in E–G associated with differentiation of the imaginal compound eyes. (Reprinted with permission from R. H. Nordlander and J. S. Edwards. Postembryonic brain development in the monarch butterfly, *Danaus plexippus plexippus* L. I. Cellular events during brain morphogenesis. *W. Roux's Arch. Entwicklungs mech. Organ.* 162: 197–217. © 1969 Springer-Verlag.)

and inner (IOA) optic anlagen (Fig. 10.18A), as summarized in Fig. 10.19 for the monarch butterfly, *Danaus plexippus* (Nordlander and Edwards, 1969b). Asymmetrical division of NBs in each outer anlage (area A in Fig. 10.19C) generates ganglion mother cells (GMCs) on either side (areas B and D) that divide once to generate interneurons for both the lamina cortex and lamina (area E) and for part of the medulla cortex and medulla (area C) (Fig. 10.19A–C). The inner anlage produces interneurons for the rest of the medulla and for the posterior lobula cortex and lobula/lobula plate (Fig. 10.19A, B) in the directions shown by the arrows in Fig. 10.19.

The total number of NBs, GMCs, and neurons appearing during optic lobe development in larvae and early pupae of *D. melanogaster* is summarized in Fig. 10.20 for both outer and inner anlagen (Hofbauer and Campos-Ortega, 1990). Note that the number of NBs increases during larval development, peaks in the late third instar, and rapidly decreases early in pupal development.

By culturing lobes of *M. sexta* in vitro and monitoring mitotic rate by incorporation of BrdU, Champlin and Truman (1998a) showed proliferation of optic anlagen to depend on continuous exposure to either ecdysone or 20E above a threshold of 90–100 ng/ml (they could turn it on and off repeatedly by shifting the hormone concentration above and below this level, the NBs arresting in G2 of the cell cycle at subthreshold concentrations). They also demonstrated lobe development to have two

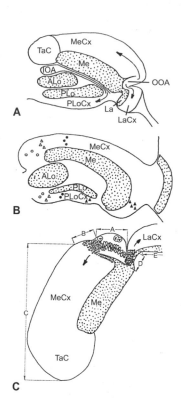

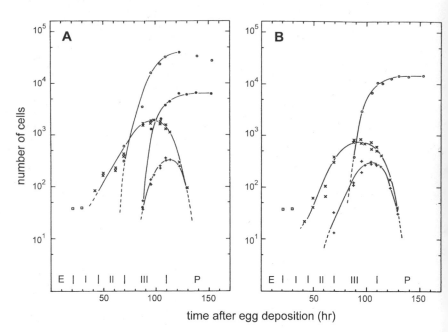

Fig. 10.20. Cell counts of neuroblasts and ganglion cells (including both ganglion mother cells and neurons) within the optic lobes during larval and early pupal development of *D. melanogaster*. A. Outer optic anlage and the cells it generates. ×, neuroblasts of entire anlage; +, neuroblasts giving rise to lamina cells; open circles, medulla cells; closed circles, lamina cells. B. Inner optic anlage and the cells it generates. ×, all neuroblasts of the anlage (including epithelium and strains); +, neuroblasts of the strains; open circles, ganglion cells originating from the lateral proliferation zone. In both diagrams, the two squares indicate neuroblasts of the early optic anlagen (inner and outer anlagen not yet distinguishable). E, embryogenesis; I, II, and III, larval instars; P, prepupa and pupa. (Reprinted with permission from A. Hofbauer and J. A. Campos-Ortega. Proliferation pattern and early differentiation of the optic lobes in *Drosophila melanogaster*. *W. Roux's Arch. Dev. Biol.* 198: 264–274. © 1990 Springer-Verlag.)

Fig. 10.19. Diagrammatic frontal sections of the developing optic lobe in brains of *D. plexippus*, showing how neuroblasts of the outer (OOA) and inner (IOA) optic anlagen give rise to interneurons of the lamina (La, LaCx), medulla (Me, MeCX), and anterior (ALo) and posterior (PLo, PLoCx) lobula (anterior is up and lateral, to the right). A. Early pupa. Arrows indicate the direction in which new interneurons are generated during metamorphosis. B. Late pupa. The diagram showing the final position of medulla and posterior lobula cortex interneurons produced at four different developmental stages is based on autoradiography using thymidine-methyl-H³ (H³-Td-R). Symbols indicate cells formed at the beginning of the third (open circles), fourth (open triangles), and fifth larval instars (filled circles) and at pupation (filled triangles). C. Summary of activity in the outer optic anlage as seen in frontal section. A. Neuroblast division in the anlage. Regions B and D contain interphase and dividing ganglion mother cells and newly formed interneurons, and C and E contain older differentiating interneurons. Cells of the lamina cortex emanate from the lateral edge of the anlage, and those of the medulla cortex emanate from its medial edge. (Reprinted with permission from R. H. Nordlander and J. S. Edwards. Postembryonic brain development in the monarch butterfly, *Danaus plexippus plexippus* L. II. The optic lobes. *W. Roux's Arch. Entwicklungs mech. Organ.* 163: 197–220. © 1969 Springer-Verlag.)

ecdysteroid-dependent phases: an earlier one in which moderate levels of ecdysteroids stimulate proliferation and a later one in which high concentrations trigger a wave of cell death, marking completion of the proliferative phase (compound eye development has similar ecdysteroid-dependent phases) (10.2.4.2.3).

That developmental events are similar in optic lobes of the grasshopper *Schistocerca gregaria* (Anderson, 1978b), the dragonfly *Aeshna cyanea* (Mouze, 1978), the cockroach *Periplaneta americana* (Stark and Mote, 1981), the butterfly *D. plexippus* (Nordlander and Edwards, 1969b), the moth *M. sexta* (Monsma and Booker, 1996a, 1996b), and the fly *D. melanogaster* (Meinertzhagen and Hanson, 1993) suggests that they are probably the same in all pterygote insects having well-developed compound eyes. The important induc-

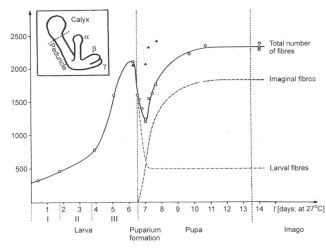

Fig. 10.21. Total number of Kenyon cell axons in the caudal peduncle of the corpus pedunculatum (mushroom body; inset) of *D. melanogaster* during postembryogenesis. Axons are shown as open circles and cell bodies (relative counts), as triangles. The axons of most larval cells (though not their nuclei) degenerate shortly after pupariation (dash-dot line) and are replaced by imaginal axons (dashed line) differentiating from progeny of imaginal neuroblasts. A bundle of about 500 larval axons persists centrally within the peduncle. (Reprinted with permission from *Nature* (G. Technau and M. Heisenberg. Neural reorganization during metamorphosis of the corpora pedunculata in *Drosophila melanogaster*. 295: 405–407) © 1982 Macmillan Magazines Limited.)

tive role of photoreceptor axons in optic lobe development is summarized in 10.2.4.2.4.

Corpora Pedunculata
Corpora pedunculata are complex paired structures that are often called the mushroom bodies (MBs). They are located on either side of the pars intercerebralis in the brain in arthropods and annelids (Figs. 8.24, 10.18) and function in sensory discrimination, integration of sensory input with behavior, and olfactory learning and memory (Strausfeld et al., 1995; Zars et al., 2000). In adult *D. melanogaster*, each consists of a dorsal group of about 2500 interneurons, the *Kenyon cells*, within each protocerebral lobe, nestled within a flattened cup of neuropile, the *calyx*, from which a stalk (peduncle) projects ventrally and medially before dividing into alpha (α and α'), beta (β and β'), and gamma (γ and γ') lobes (Figs. 8.24; 10.18; 10.21, inset) (Truman et al., 1993; Ito et al., 1997; Zars et al., 2000).

In *D. melanogaster*, the Kenyon cells and glia of each MB derive as separate clones from only four NBs (out of a total of 85 NBs in the larval brain) that begin to divide at the time of hatch and continue to bud off GMCs and neurons until the end of metamorphosis. At least three types of neurons are produced by each NB: the first type prior to the third instar and projecting into the gamma lobe, and the second and third types after puparium formation and projecting into the α' and β' lobes. Just before

pupariation, the axons (though not the somata) of most larval interneurons begin to degenerate and are replaced by differentiation of imaginal ones (Fig. 10.21) (Technau and Heisenberg, 1982; Truman et al., 1993; Raabe and Heisenberg, 1996; Lee et al., 1999).

By monitoring the expression pattern of 19 different GAL4 enhancer trap strains marking various subsets of MB cells, Ito and colleagues (1997) showed that each clone contributes collectively to formation of the MB. As well, partial ablation of one or more NBs by hydroxyurea showed each one to be capable of autonomously generating all structures within this synaptic center. The *eyeless* (*ey*), *twin-of-eyeless*, and *dachshund* (*dac*) genes are expressed in MB cells during their development. Mutations in *ey* (Kurusu et al., 2000) and *dac* (Martini et al., 2000) completely disrupt development of MB neuropile, and those in *dac* disrupt axonal projection.

Development of MBs in the larvae and pupae of worker honeybees (*Apis mellifera*) is similar except that these structures are relatively much larger and more complex and develop from a larger number of NBs (about 2000 just before pupation) (Farris et al., 1999). Each adult MB differentiates from four clusters of no more than 45 NBs present at the time of hatch. During MB ontogeny, subpopulations of Kenyon cells, differing in projection pattern, position, and immunohistochemistry, are born at different but overlapping times, the final complement being in place by midpupation. In adults, the cell bodies of the Kenyon cells in each calyx are concentrically arranged, the outer rows being the oldest, having been pushed to the periphery by later-born ones that remain closer to the center of proliferation, a pattern reflected in their projection pattern and immunohistochemistry.

Differentiation of MBs in *A. mellifera* is intimately related to that of antennal glomeruli within the antennal lobes and of synaptic centers in the optic lobes, as the Kenyon cells provide a template for the ramification of olfactory and visual projection neurons (Schröter and Malun, 2000).

Antennal Lobes
Paired antennal lobes within the deutocerebrum of the brain in both juvenile and adult insects contain a species-specific number of synaptic centers or *glomeruli*, the targets of olfactory axons projecting into the brain from the antennal sensilla (Fig. 10.18). In adults of *D. melanogaster*, each lobe contains 40 glomeruli of characteristic shape, size, and position (Laissue et al., 1999). These develop through gradual substitution and reorganization of elements in the larval lobes (Stocker et al., 1995). Pulse-chase labeling of dividing cells in larval and pupal brains with BrdU showed that some adult interneurons in each lobe derive from a single lateral NB that begins to divide early in instar 1, but most descend from additional NBs initiating division in later larval stages, with a peak of 10–12 NBs in the late third instar. Glial cell proliferation within each lobe reaches its peak 12 h APF, and these cells assist in organizing formation of the glomeruli (Oland and Tolbert, 1989).

The origin and differentiation of antennal interneurons and formation of imaginal antennal glomeruli during metamorphosis depend on entry and synaptic contact of olfactory afferents from newly differentiating imaginal antennal sensilla (Oland et al., 1998), and the lobes fail to differentiate if the antennal discs are removed (Levine, 1986). Growth cones of extending olfactory axons are attracted to the lobes by soluble or surface-bound cues and, just before entering them, associate with other axons having a common target and a common odor specificity. When differentiation is complete, the fly's brain contains a complete map of the peripheral olfactory system (Vosshall et al., 2000).

When a male antennal disc of *M. sexta* is transplanted into the head capsule of a female caterpillar whose own antennal discs have been removed, the female antennal lobe is innervated by male-specific antennal afferents. These induce formation of a *macroglomerular complex* (MGC), a male-specific structure whose interneurons respond to exposure of the adult male antenna to molecules of female-produced sex pheromones (Schneiderman and Hildebrand, 1985).

Using laser-scanning, confocal microscopy and various fluorescent staining techniques, Rössler and coworkers (1998) showed the MGCs of normal males of *M. sexta* to differentiate early in metamorphosis in only 3 d at 25°C following ingrowth of the first olfactory axons from developing adult antennae. Adult females have two female-specific glomeruli in each of their antennal lobes that develop at the same time and in the same relative position as the MGCs of males, but their function is unknown (Rössler et al., 1998).

Glia

As the larval brain metamorphoses into that of the adult (Figs. 10.17, 10.18, 10.22), changes might be expected to occur in the number and distribution of glial cells producing factors that promote the outgrowth of imaginal neurites and guide their growth cones. Hähnlein and Bicker (1997) examined this in metamorphosing neuropiles of the honeybee *A. mellifera* brain, using a tagged antibody raised against the glia-specific Repo protein of *D. melanogaster* and immunocytochemistry to locate glial somata. In early larval instars, a continuous layer of glial cells defines the boundaries of all growing neuropiles within the brain, but intrinsic glial nuclei are almost absent. During metamorphosis, glial cells migrate to identifiable locations within the corpora pedunculata, antennal lobes, and central body from the larval cortex, probably by following guidance cues on the surface of particular neuronal fibers. In fact, the growth of these glial cells may prepattern these neuropilar boundaries, as in the developing larval brain (8.3.1.2).

Neuroendocrine System

Changes in the relative position of the corpora cardiaca (CC) and corpora allata (CA) between larva and adult of *M. sexta* brain are illustrated in Fig. 10.22, and those in the location and projection patterns of some neurosecre-

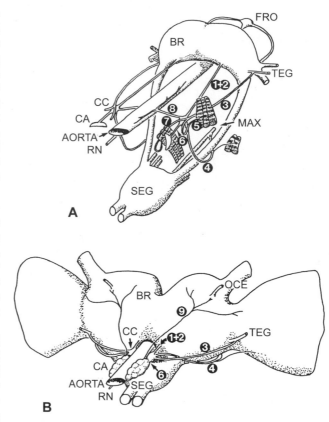

Fig. 10.22. Schematic diagrams of the cerebral neurosecretory system in the larva (A) and adult (B) of *Manduca sexta* (Lepidoptera: Sphingidae). Numbers refer to various nerves associated with the corpora allata (CA) and corpora cardiaca (CC). BR, brain; FRO, frontal ganglion; MAX, maxillary nerve; OCE, ocellar nerve (adult only); RN, recurrent nerve; SEG, subesophageal ganglion; TEG, tegumentary nerve; 1 + 2, nervi corporis cardiaci 1 + 2; 3, nervus corporis cardiaci 3 (NCC-3); 4, nervus corporis ventralis; 5, posterolateral branch of NCC-3; 6, nervus corporis allati; 7, posteromedial branch of NCC-3; 8, nervi corporis cardiaci-nervus recurrens + aorticus; 9, nervus corporis dorsalis. (Metamorphosis of the cerebral neuroendocrine system in the moth *Manduca sexta*, P. F. Copenhaver and J. W. Truman, *J. Comp. Neurol.*, Copyright © 1986 Wiley-Liss, Inc. Reprinted by permission of Wiley-Liss, Inc., a subsidiary of John Wiley & Sons, Inc.)

tory cells (NSCs) are shown in Fig. 10.23 (Copenhaver and Truman, 1986).

The larval brain contains several discrete groups of NSCs projecting to the CC-CA complex and to a variety of more peripheral structures (Fig. 10.23: top). During metamorphosis, the cerebral neuroendocrine system undergoes extensive reorganization, including reduction or loss of many larval neurons and the repositioning of other neurons and their dendritic fields (compare Fig. 10.23: top and bottom). Eventually, five separate types of NSCs can be recognized within the adult brain, each with a distinctive pattern of dendritic arborizations within the brain and of terminal neurohemal processes projecting to the CC, CA, aorta, or a combination of these regions (Fig. 10.23: bottom). However, despite

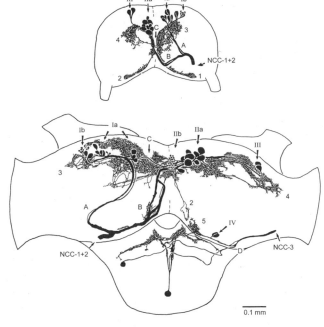

Fig. 10.23 Camera lucida drawings of whole mount preparations illustrating somata, projection patterns, and dendritic arborizations (1–5) of group I, II, and III neurosecretory cells in the brain of *M. sexta* as revealed by cobalt chloride backfilling. Top: Brain of fifth instar as seen from behind, showing neurosecretory cells associated with the nervi corporis cardiaci 1 + 2. Bottom: Brain and subesophageal ganglion of adult as seen from above and behind, illustrating neurosecretory cells of the left nervi corporis cardiaci 1 + 2 and the right nervus corporis cardiaci 3. Notice the extensive repositioning of cell clusters and their arborizations and the appearance of group IIb cells near the dorsal midline. (Metamorphosis of the cerebral neuroendocrine system in the moth *Manduca sexta*, P. F. Copenhaver and J. W. Truman, *J. Comp. Neurol.*, Copyright © 1986 Wiley-Liss, Inc. Reprinted by permission of Wiley-Liss, Inc., a subsidiary of John Wiley & Sons, Inc.)

these changes, most central elements of the larva are retained in the adult.

Normal adult exopterygote and endopterygote insects are unable to molt because their prothoracic glands degenerate prior to or during adult emergence (9.4.2, 9.4.3). In *D. melanogaster*, prothoracic gland cells of the ring gland gradually lose their ability to synthesize and secrete ecdysteroids, separate from the rest of the gland, and degenerate just after adult emergence, while the fused CA undergo ultrastructural changes associated with changes in JH synthesis and in their position relative to the brain (Dai and Gilbert, 1991).

In, *M. sexta*, the ability of prothoracic gland cells to synthesize ecdysteroids reaches a peak on day 4 after pupation, decreases drastically by day 6, and reaches basal levels by day 8, the processes of prothoracic gland degeneration being initiated on day 6 (Dai and Gilbert, 1997). Comparable changes occur in the CA, CC, and prothoracic glands of other insects (Hanton et al., 1993; Dorn, 1998).

10.2.4.2 PERIPHERAL NERVOUS SYSTEM

The peripheral nervous system (PNS) consists of the sense organs, their axonal projections to ganglia of the CNS, and the axons of motor neurons projecting to effector organs within the body, the latter two within the peripheral nerves (Fig. 8.16). Its embryogenesis was discussed in 8.3.2.

10.2.4.2.1 SENSE ORGANS OTHER THAN EYES

In many insects, both larvae and adults have a species-specific pattern of sense organs over the surface of their bodies that differs from each other, is genetically determined, and is of considerable use in identification (see keys and illustrations in Stehr, 1987b, 1991; Borror et al., 1989; and Daly et al., 1998). However, in some members of ametabolous and of basal and intermediate exopterygote lineages, such as the blood-feeding reduviid *Rhodnius prolixus*, the setal pattern becomes more complex at each molt and seems to result from each sensillum having a "zone of inhibition," inside of which other sensilla are unable to develop. As the insect grows, these sensilla gradually separate from one another, uncovering uninhibited areas of epidermis and making room for others to develop. Wigglesworth (1965) suggested that this might result from absorption of some essential nutrient or inductor substance by a developing sensillum but more probably results from sequential action of proneural and other genes, at least some of which are evolutionarily conserved (compare Fig. 8.30) (the same genes are involved in sense organ precursor cell [SOP] specification) (Gerhart and Kirschner, 1997: 273–285; P. Simpson et al., 1999; Wülbeck and Simpson, 2000).

The amount of sensory remodeling occurring during metamorphosis in holometabolous insects is extensive. For example, in *M. sexta*, the axons of all sense organs in each larval leg project to the ganglion in the VNC in four primary nerve branches (Consoulas, 2000). Using nerve tracing techniques, birth date labeling with BrdU, confocal microscopy, and electrophysiology, Consoulas showed most larval sensory neurons to degenerate during the larval-pupa transition and to be replaced by a new set of adult neurons that originate and differentiate during the pupal stage. At larva-pupal apolysis, the imaginal leg epidermis proliferates within the larval cuticle, encircles the larval sensory neurons, and walls them off from their target sensilla. However, those neurons innervating the larval leg's trichoid sensilla, and its femoral and tibial chordotonal organs, continue to exhibit electrical activity throughout metamorphosis, despite the absence of their sensilla, and persist into the adult, where they have similar functions. When the new, adult set of sensory organs differentiate, their axons contact and fasciculate with these persisting larval neurons, so that five of seven imaginal leg nerve branches develop on preexisting larval trajectories.

In *D. melanogaster* and other higher flies, all larval sensilla, except for paired Keilin's organs on the venter of each thoracic segment, degenerate, their cuticular components are shed at the larval-pupal apolysis, and a new set of some 5000 imaginal sensilla develops in three separate waves from SOPs in the imaginal discs and histoblasts (Hartenstein and Posakony, 1989; Fristrom and Fristrom, 1993; Jan and Jan, 1993; Shepherd and Smith, 1996; Gerhart and Kirschner, 1997: 273–285). The first wave arises late in the third instar and provides progenitors for large mechanoreceptive bristles on the adult head and thorax, for chordotonal organs in the legs and wings, for some campaniform sensilla on the wings, and for chemoreceptors on the proboscis, legs, and wing margin. These SOPs undergo asymmetrical differentiative mitoses identical to those occurring in embryos (Fig. 8.33). Their sensory axons either pioneer tracts to the CNS, to be followed by the growth cones of those of later-appearing sensilla, or fasciculate with persisting larval nerves to find their way.

After larval-pupal apolysis, eversion of the head and appendages, and spread and fusion of imaginal epidermis from the imaginal discs (Fig. 10.8B–F), these early sensilla become surrounded by proneural regions of imaginal epidermis in the head and thorax, from which a second set of SOPs for head, thorax, and appendages emerges. These SOPs undergo differential mitosis in the pupa 16–24h APF, their axons following tracts to the CNS pioneered by sensilla of the first wave. The last imaginal sensilla to differentiate are those of the abdomen, beginning about 32 h APF.

Many experiments have been devised to learn how axons of the peripheral sensilla find their appropriate synaptic targets within the CNS during metamorphosis. A particularly effective approach has been to generate sensilla in abnormal (ectopic) places and then to examine their projection patterns (Anderson, 1981; Stocker and Lawrence, 1981). This can be done either by surgically grafting pieces of epidermis from one place to another in a young insect before or after sensilla have developed, or by use of homeotic mutations (7.2.1.3). Results of these experiments usually show such sensilla to project to sites in the CNS appropriate to their new locations, although their nerve of entry into the CNS and their terminal arborizations within it may differ. For example, homeotic tarsal neurons in the antennal mutant *spineless-aristopedia* (*ss^a*) project to the antennal lobe of the brain and to the posterior portion of the subesophageal ganglion (Fig. 10.24C), as do those of normal *antennal* sense organs (Fig. 10.24A, B), but not to thoracic centers, as do those of wild-type legs (Fig. 10.24D) (Stocker and Lawrence, 1981).

10.2.4.2.2 COMPOUND EYES IN
DROSOPHILA MELANOGASTER
The compound eyes of adult insects and of some apterygote and most exopterygote juveniles consist of ommatidia of characteristic number, size, and structure (Paulus, 1979; Caveney, 1998). Although their postembryogenesis has been examined descriptively and experimentally in larvae and adults of many species, particularly in dragonflies (Sherk, 1978a, 1978b; Mouze, 1979), cockroaches (Nowel, 1980, 1981a, 1981b; Shelton et al.,

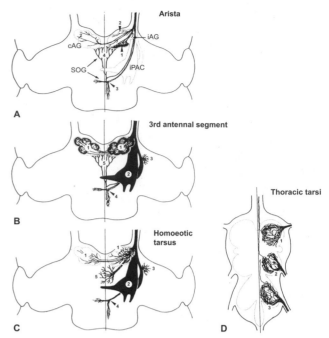

Fig. 10.24. Diagram of the adult brain of *D. melanogaster* in ventral aspect, summarizing sensory projection patterns from normal and genetically transformed antennae. A. Projection pattern from a wild-type arista: 1. Arborization of sensory axons in the ipsilateral (same side) antennal glomerulus (iAG). 2. Same, in the periphery of both iAG and in the contralateral (opposite side) antennal glomerulus (cAG). 3. Same, of small axons in the subesophageal ganglion (SOG). 4. Single large-diameter axon projecting bilaterally into the anterior SOG. iPAC, ipsilateral posterior antennal center. B. Projection pattern from the third antennal segment: 1. Glomerular arrangement of terminals in both iAG and cAG. 2. Axons projecting into the iPAC respect its boundaries. 3. Fibers projecting into the ventrolateral protocerebrum. 4 and 5. Small and large fibers in the SOG show the same branching patterns as aristal dye fills. C. Projection patterns from a homeotic tarsus (*spineless-aristapedia* [*ss^a*] mutant): 1. Axons terminate in iAC and cAC just after they enter. 2. Terminals in iPAC respect its boundaries identically to those from antennal segment III. 3. Arborization of axons in the ventrolateral protocerebrum. 4. Arborization of small fibers in the SOG. 5. Additional ectopic arborization extending from the ventral iAG into the anterior SOG. D. Wild-type sensory projections of sensilla on tarsi of the first (1), second (2), and third (3) legs are located mostly in the ipsilateral half of the corresponding thoracic ganglion. (Reproduced with permission from R. F. Stocker and P. A. Lawrence. Sensory projections from normal and homoeotically transformed antennae in *Drosophila. Dev. Biol.* 82: 224–237. © 1981 Academic Press, Inc.)

1983), locusts (Anderson, 1978a, 1978b), and moths (Maxwell and Hildebrand, 1981; Egelhaaf et al., 1988; Champlin and Truman, 1998b), that of *D. melanogaster* is by far the best understood (reviewed in Tomlinson, 1988; Lawrence, 1992; Dickson and Hafen, 1993; Wolff and Ready, 1993; Freeman, 1997; and Gerhart and Kirschner, 1997).

In adults of *D. melanogaster*, each compound eye consists of a two-dimensional array of some 750 (aver-

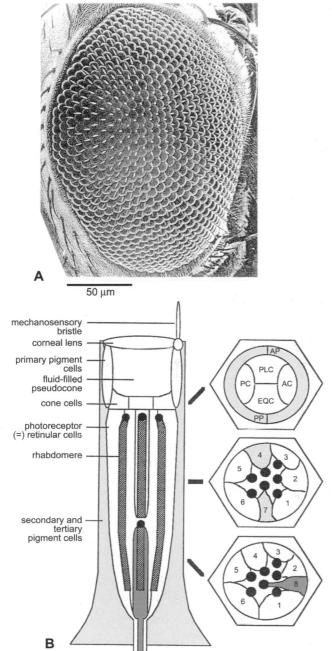

A

50 µm

mechanosensory bristle
corneal lens
primary pigment cells
fluid-filled pseudocone
cone cells
photoreceptor (=) retinular cells
rhabdomere
secondary and tertiary pigment cells

B

Fig. 10.25. A. Scanning electron micrograph of a *D. melanogaster* compound eye. Each facet represents a single ommatidium, with mechanosenory hairs projecting between them. Anterior is to the right and dorsal upward. B. Schematic diagram of a single ommatidium of *D. melanogaster* in sagittal section, showing its principal parts. The rhabdomeres of photoreceptor cells R1–R6 extend the length of the ommatidium. R7 provides a central rhabdomere apically and R8, the same basally. The pigment cells optically isolate each ommatidium from adjacent ones. (Reproduced with permission from A. Tomlinson. Cellular interactions in the developing *Drosophila* eye. *Development* 104: 183–193. © 1988 Company of Biologists, Ltd.)

age: female, 772; male, 732) ommatidia with mirror-image symmetry above and below a horizontal equator (Fig. 10.25A) (García-Bellido, 1972). Each ommatidium contains 25 cells of seven types: 8 photoreceptor (retinular) cells (R1–R8), 4 cone cells forming the pseudocone of the *dioptric* (light-focusing) apparatus, 2 m cells of unknown function, and 11 pigment cells (2 primary, 6 secondary, and 3 tertiary) (these last are shared with adjacent ommatidia) that optically isolate each ommatidium from its fellows. There is also an associated mechanoreceptor hair of four cells between adjacent ommatidia (Fig. 10.26A, B) (Tomlinson, 1988).

Each compound eye differentiates from an eye-antennal disc during the third larval and pupal stages (Fig. 10.26A). Until the third instar, the eye part of the disc consists of a single layer of undifferentiated, proliferating ectodermal cells covered by peripodial membrane. Over a period of about 2 d at 25°C, the assembly of ommatidial cell clusters proceeds in an orderly progressive wave from back to front across the disc as it is increasing eightfold in size to a total of some 20,000 cells. Closely associated with this process is a vertically oriented, morphogenetic furrow (MF) that moves from back to front across the disc, beginning at the optic nerve (Fig. 10.26B).

The MF is preceded by a parallel, asynchronous wave of mitosis (I) and is followed by a synchronous wave (II) (Fig. 10.27). Cells anterior to the furrow are undifferentiated, while those behind it assemble into ommatidial clusters in two steps—one in front of the second wave of mitosis (to form preclusters of what will become photoreceptor cells R8, R2, R5, R3, and R4), the other behind it (to add photoreceptor cells R1, R6, and R7 and the cone and pigment cells to each cluster)—at the rate of one new ommatidial column every 2 h. Assembly of each ommatidium is stereotyped such that each of its 25 cells has precise, invariable contacts with other cells in the cluster (Fig. 10.27: bottom). Development of mirror-image symmetry of ommatidia above and below the equator of each eye follows progression of the furrow (Fig. 10.30A).

Origin and Anterior Propagation of the Morphogenetic Furrow

Initial steps in pattern formation in the eye disc of *Drosophila* begin several cell rows ahead of the MF and are positively regulated by secreted signal proteins and by a proneural transcription factor encoded by the *atonal* (*ato*) gene (Fig. 10.28) (Brown et al., 1995). Three other regulatory proteins functioning to suppress neuronal development in other tissues—Extramacrochaete (Emc), Hairy (H), and Patched (Ptc)—are also expressed in front of the MF (Fig. 10.28). In *emc⁻ h⁻* clones, the MF and differentiated eye field advance up to eight ommatidial rows ahead of those of adjacent wild-type tissue, suggesting that MF progression and neuronal differentiation are negatively regulated by a combination of these anteriorly expressed proteins.

Continued anterior progression of the MF depends on

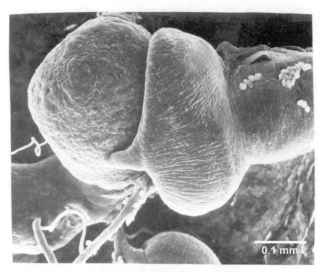

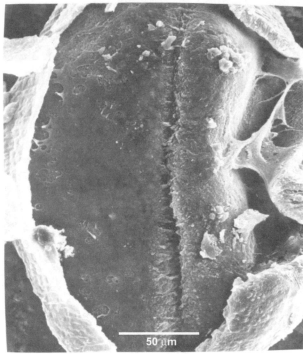

Fig. 10.26. Top: Scanning electron micrograph of the eye-antennal disc of a third instar larva of *D. melanogaster* in lateral aspect. Anterior is to the right and dorsal upward. The eye portion of the disc is connected by the optic nerve to one of the cerebral hemispheres. The ventral nerve cord curves away to the left. Tubelike structures are tracheae and nerves. To the right, several hemocytes are adhering to the surface of the peripodial membrane of the antennal portion of the disc. Bottom: Scanning electron micrograph of the eye disc, with the peripodial membrane partly removed. Anterior is to the right. The vertical groove is the morphogenetic furrow, which is moving from left to right. Cells behind the furrow aggregate into ommatidial clusters. (Reproduced with permission from D. F. Ready, T. E. Hanson, and S. Benzer. Development of the *Drosophila* retina, a neurocrystalline lattice. *Dev. Biol.* 53: 217–240. © 1976 Academic Press, Inc.)

production of the secreted Hedgehog (Hh) protein by ommatidial clusters emerging behind the MF (Fig. 10.28) (Strutt et al., 1995; Heberlein et al., 1995; Jarmen, 1996). Hh molecules disperse anteriorly and trigger expression of Decapentaplegic (Dpp) protein in cells entering the MF anteriorly; Dpp is also secreted and also diffuses anteriorly. Loss of Dpp production occurs in cells emerging from the MF posteriorly, except in the central cell of each cluster, which becomes photoreceptor R8.

Undifferentiated cells in front of the MF are competent to respond to Hh and Dpp but are inhibited from spontaneous activation by the Emc, H, Ptc, and protein kinase A (PKA) proteins they contain (Gerhart and Kirschner, 1997). The combination of Hh and Dpp overcomes this inhibition and induces cells just ahead of the MF (Fig. 10.28)

- to activate *hedgehog* (*hh*) expression to keep the Hh signal wave moving;
- to synchronize their cell cycles in G1;
- to undergo the shape change making the MF (apical constriction and cell shortening caused by expression of the gene *act up* [*acu*] preventing the *profilin* gene from inducing actin polymerization [Benlali et al., 2000]); and
- to initiate differentiation of ommatidia, the first step of which is to establish the spacing of prospective R8 cells far enough apart so that each is surrounded by some 25 uncommitted cells (spacing depends on activation of the *ato* gene).

Before the MF appears, Dpp is expressed around the rim of the eye disc, where it maintains and induces its own expression. However, after it does, expression of the segment-polarity gene *wingless* (*wg*) inhibits Dpp production in the disc except adjacent to the optic nerve (Fig. 10.30B). In *wg* mutants, an MF develops all around the rim and rushes inward, eventually generating a muddled clump of ommatidia (Cho and Choi, 1998). Dpp and Wingless (Wg) proteins are mutually inhibitory, and loss of Dpp function results in expansion of Wg expression and vice versa (Chanut and Heberlein, 1997). Loss of function of either the *hh* or *dpp* gene results in failure of the MF to form or advance and of ommatidia to differentiate. As the MF approaches the anterior margin of the disc, *dpp* expression is reduced and the fold stops moving and disappears (Chanut and Heberlein, 1997).

Recently Cho and colleagues (2000) and Gibson and Schubiger (2000) showed that additional Hh signaling by cells of the peripodial membrane (Fig. 10.26B) across the lumen of the disc to the disc cells contributes to forward progression of the MF and to cellular proliferation and pattern formation of disc cells. As the MF moves forward over the disc, peripodial cells immediately above it extend microtubule-based filopodia across the lumen and into the fold. If Hh secretion by cells of this membrane is lost or if the membrane is destroyed, defects appear in the surface pattern of the resulting compound eye.

Expression of these genes seems to be coordinated by secretion of the molting hormone 20E and by activation of the ecdysteroid signaling pathway, as genetic disrup-

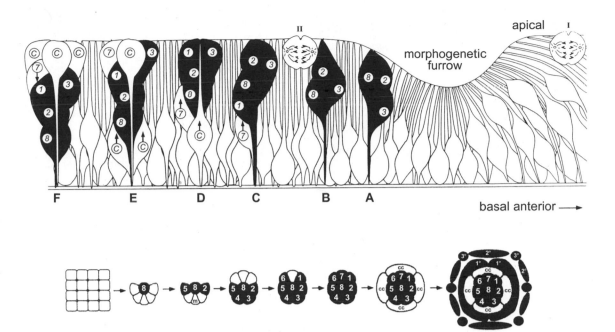

Fig. 10.27. Passage of the morphogenetic furrow across the eye disc from posterior (left) to anterior (right) is associated with the aggregation of disc cells into ommatidial clusters. These diagrams illustrate the process in *D. melanogaster*, but recent comparative work has shown it to be similar in the moth *Manduca sexta*. Top: Frontal section through the eye disc reconstructed from serial section electron micrographs and autoradiographic analyses of cell division. Arrows indicate direction of nuclear movement. Each of cell pairs R2/R5, R3/R4, R1/R6, and the cone cell (C) is represented by a single cell. All cells except for dividing ones extend across the epithelium; dividing cells round up at the apical surface. Notice that proliferation occurs both in front of (I) and behind (II) the morphogenetic furrow (from Wolff and Ready, 1993: 1314, Fig. 19). Bottom: Schematic diagram showing epithelial cells of the eye disc assembling into a single ommatidal unit (from Dickson and Hafen, 1993: 1330, Fig. 2B). (Reprinted with permission from M. Bate and A. Martinez Arias (eds.), *The development of* Drosophila melanogaster. Vol. 2. © 1993 Cold Spring Harbor Laboratory Press.)

tion of either ecdysone release (with an *ecdysoneless'* [*ecd'*] mutation) or response (with a *Broad Complex* [*BR-C*] mutation) disrupts MF progression (Brennan et al., 1998). In addition, transcription of both the ecdysteroid receptor (EcR) (12.4.1.5.1) and of the Z1 isoform of the BR-C protein (discussed in 12.4.1.5.2) is localized in cells within and close to the MF.

Orthologs of many of these genes function during compound eye development in larvae of the grasshopper *Schistocerca gregaria* and the red flour beetle *Tribolium castaneum*, but with different temporal and spatial patterns of expression (Friedrich and Benzer, 2000).

Specification of Cell Fate within the Ommatidium
Except for R8, the 25 cells of each ommatidial cluster are still undetermined as to fate at the time each cluster forms behind the MF (Fig. 10.27: bottom). Acquisition of photoreceptor cell identity within the cluster occurs sequentially in pairs of cells in the order R8 → R2 + R5 → R3 + R4, in preclusters in front of the second wave of mitosis, followed by R1 + R6 → R7 → cone cells → pigment cells in those behind it (Fig. 10.27: bottom).

A particular cell's identity within each cluster depends on its contacts with neighboring cells, particularly R8 (Moses, 1991), and on its state of determination when it receives signals from these cells (Freeman,

1997). For example, information cueing the precursor of receptor cell R7 for the R7 developmental pathway is presented by cells with which that cell is in contact: R1, R6, and R8 (Fig. 10.27: bottom, fifth and sixth from left); the same is true of other photoreceptor cells. The entire process of cell determination within a cluster takes about 25 h at 25°C.

In 1991, Moses proposed a sequential induction model for cell fate assignment within the ommatidium in which each cell type, as it is determined, expresses a specific cell-surface protein or ligand. Adjacent precursor cells possess receptors for these ligands and are thus induced to adopt the next appropriate cell fate. For example, control of photoreceptor cell R7 determination depends on the action of at least two regulatory genes: *sevenless* (*sev*) and *bride-of-sevenless* (*boss*) (Fig. 10.29). The Bride-of-sevenless (Boss) protein is expressed only by photoreceptor cell R8 and acts as a ligand for the transmembrane receptor protein Sevenless (Sev) in cell R7 encoded by the *sev* gene. Production of Boss protein induces transient production of Sev protein, which has a tyrosine kinase domain (such enzymes activate a signal transduction pathway within a cell, in this instance, by adding phosphate to Downstream of receptor kinase [Drk] protein). If this interaction does not occur, cell R7 differentiates as a nonneural, epidermal cell. Many steps

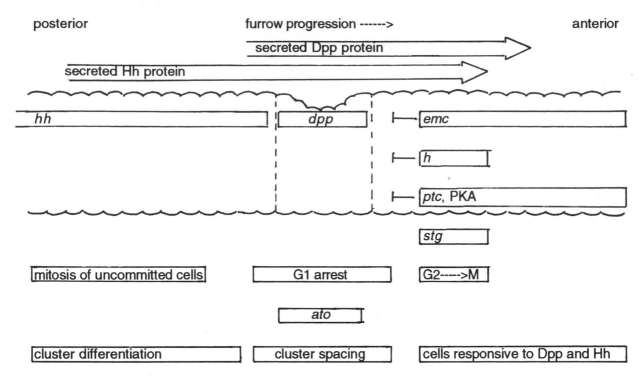

Fig. 10.28. Diagrammatic frontal section of the eye disc of *D. melanogaster*, showing gene expression occurring during passage of the morphogenetic furrow from back (left) to front (right). See text for details. (Adapted with permission from J. Gerhart and M. Kirschner. *Cells, embryos, and evolution. Toward a cellular and developmental understanding of phenotypic variation and evolutionary adaptability*, p. 227, Fig. 6-15. © 1997 Blackwell Science, Ltd.)

in this signaling pathway have recently been worked out, as indicated in Fig. 10.29.

Other photoreceptor cells within the ommatidium are determined in the same way by R8-activating genes in other photoreceptor pairs (*rough* in R2/R5; *sev* in R2/R5, R1/R6, and R7; and *seven-up* in R1/R6), with at least 27 genes being involved, including those of the *Notch/Delta* signaling system (8.3.1.4). Preliminary work on development of moth (*Ephestia kühniella*) ommatidia (Egelhaaf et al., 1988) indicates that its cell specification may be similar. Recruitment of photoreceptor cell R7 closes access to the R8 Boss signal so that additional, surrounding, undifferentiated cells are unable to assume a photoreceptor identity; instead, they are specified as cone and pigment cells (Fig. 10.27: bottom, right).

Normal differentiation of microvilli within the photoreceptive rhabdomere of each photoreceptor of an ommatidium (Fig. 10.25B) depends on presence of the photopigment rhodopsin (Chang and Ready, 2000). This acts by way of the rhodopsin guanosine triphosphatase Drac 1 to induce development of an actin cytoskeleton, the rhabdomere terminal web or stalk, which functions to prevent collapse of the microvilli into photoreceptor cytoplasm.

Because most other types of insect sensilla are known to develop by a sensillum-specific sequence of asymmetrical mitoses from a single SOP (Fig. 8.33 [Bate, 1978]), it was long thought that such was also true for

cells of each ommatidium (e.g., Weber, 1966: Fig. 28; Kühn, 1971: 360, 361). As just described, it is obvious that this is not so. If the second mitotic wave behind the MF is blocked by introducing the human cyclin-dependent kinase inhibitor $p^{21\,CIP1/WAF\,1}$ to the disc, each cell type normally added to the precluster after the second wave is still added (de Hooij and Hariharan, 1995). However, resulting ommatidia, though having all their photoreceptor, cone, and primary pigment cells, lack one or more of their other cells, indicating that the second wave is required to generate sufficient cells for full ommatidial differentiation.

Ommatidial Spacing

In compound eyes of *D. melanogaster* and other insects, interommatidial spaces separate adjacent ommatidia and result in their forming a hexagonal array (Fig. 10.25A). Cagan and associates (summarized by Roush [1997]) showed that as the MF moves forward, the cells immediately behind it, and destined to form new ommatidia, produce a protein signal that prevents epidermal cells immediately in front of them from forming another ommatidium.

The *ato* gene mentioned earlier (Fig. 10.28) encodes a signal transcription factor specifying recipient cells to become R8 photoreceptors (Fig. 10.27: bottom, second from left). This gene is also expressed just in front of the advancing MF. When the MF reaches these *ato-*

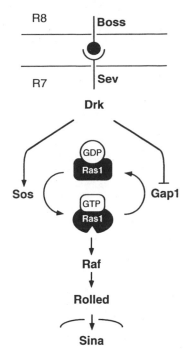

Fig. 10.29. Model of the Sevenless (Sev) signal transduction pathway in the eye disc of *D. melanogaster* specifying identity of photoreceptor R7 in an ommatidium. Arrows represent the hierarchy of the pathway as determined by genetic interactions and do not imply direct molecular interactions of gene products except for Bride-of-sevenless (Boss). (Reprinted with permission from B. Dickson and E. Hafen. Genetic dissection of eye development in *Drosophila*. In M. Bate and A. Martinez Arias (eds.), *The development of* Drosophila melanogaster. Vol. 2, p. 1348, Fig. 7. © 1993 Cold Spring Harbor Laboratory Press.)

expressing cells, *ato* is switched off immediately in those cells destined to become interommatidial space but a little later in prospective R8 cells (Fig. 10.28).

The signal to turn off *ato* expression is provided by the Ras signal transduction pathway. In prospective R8 cells, the Ras signal is weakened by presence of another protein encoded by the *rhomboid (rho)* gene. Ras and Rhomboid together induce expression of the Argos protein, which interferes with function of the Ras pathway and further slows down the shutdown of *ato* in R8 cells. However, Argos also prevents this same slowdown from occurring in the row of cells immediately in front of the R8 cells. Without Rhomboid, the Ras pathway cannot produce new Argos, so more distant cells don't receive Argos early, allowing them to produce a new row of ommatidia, which, in turn, activate new Argos to continue the cycle (Roush, 1997).

Specification of the Dorsoventral Axis and of
Ommatidial Symmetry above and below the Equator
The compound eye of adult *D. melanogaster* is divided by a horizontal equator into dorsal and ventral halves having mirror-image symmetry in their ommatidia, as

indicated by rhabdomere position in transverse sections (Fig. 10.30A). This boundary seems to be established before the MF arises, in much the same way as the wing margin is in wing discs (Fig. 11.10A) but with the dorsoventral (D-V) axis reversed, and to function in organizing subsequent morphogenesis of the eye disc (Fig. 10.30B) (Royet and Finkelstein, 1997; Cho and Choi, 1998; Domínguez and de Celis, 1998; Tomlinson and Struhl, 1999; Cavodeassi et al., 1999).

The signaling protein Fringe (Fng) is expressed only in the ventral half of the undifferentiated eye disc from the late first to mid second instars because it is inhibited in the dorsal half by proteins of the *Iroquois* gene complex (*mirror* [*mirr*], *araucon*, and *caupolican*). This creates a horizontal Fng boundary (Fig. 10.30B). Absence of an Fng boundary disrupts the equatorial expression of the Notch-signaling proteins Delta (Dl) and Serrate (Ser) and prevents Notch (N) expression at the D-V midline normally required for growth and ommatidial formation (Fig. 10.30B).

The correct specification of photoreceptors R3 and R4 in each ommatidial cluster above and below the equator (Figs. 10.25B: right; 10.30A: right) is essential to establishing their D-V polarity, with cell R3 receiving the polarizing signal by way of the receptor protein Frizzled (Fz) before or at higher levels than in the R4 cell and thereby generating the difference between them (Cooper and Bray, 1999). The stepwise sequence proposed is shown in Fig. 10.30C, with Fz signaling required to specify R3 by way of Dishevelled (Dsh) and with Delta (Dl) and Notch (N) signaling required to specify R4.

10.2.4.2.3 COMPOUND EYE POSTEMBRYOGENESIS
IN OTHER INSECTS
Ommatidial assembly is similar in investigated crustaceans (*Triops cancriformes*) (Melzer et al., 2000; Harzsch and Walossek, 2001) and exopterygotes, but begins in the embryo (e.g., the stick insect *Carausius morosus* [Such, 1978]) and at the anterior, not the posterior, margin of the eye. Compound eye development also involves continuous increase in ommatidial size and number throughout postembryogenesis rather than just at metamorphosis. In dragonfly larvae, for example, newly hatched first instars have 7 to about 270 ommatidia depending on genus, and adults have almost 8000 in each compound eye (Ando, 1962: 93–94; Sherk, 1978a). New ommatidia originate from a proliferation zone in the dorsal, anterior, and ventral margins of the eye and move posteriorly over its surface as new units are added, so that those of the first instar are at the back of the eye in adults (Sherk, 1978b; Mouze, 1979). In addition, ommatidia formed during larval development must function in both water (larvae) and air (adult) (Sherk, 1978a).

In each compound eye of the cockroach *Periplaneta americana* (Nowel, 1980, 1981a, 1981b; Shelton et al., 1983) and the locust *Schistocerca gregaria* (Anderson, 1978b), development is similar, the number of ommatidia increasing from about 100 in first instars to more

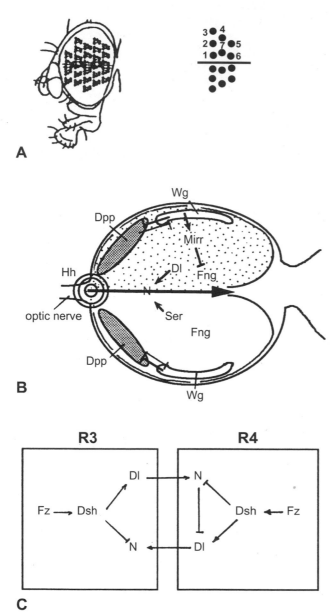

A

B

C

Fig. 10.30. Specification of dorsoventral (D-V) polarity in the compound eye of *D. melanogaster* during eye disc development. A. Diagram of the left compound eye showing the mirror-image orientation of dorsal and ventral ommatidia above and below the equator, as indicated by the position of the rhabdomeres of photoreceptors R1–R7 in transverse section (right); photoreceptor R8 is below the plane of section. B. Model for the role of Fringe (Fng) protein in organizing the D-V boundary of the eye disc prior to formation of the morphogenetic furrow. Mirror (Mirr) and other *Iroquois* proteins (Araucan and Caupolican) can inhibit expression of Fng and so provide a mechanism for its ventral specificity. Serrate (Ser) and Delta (Dl) proteins activate Notch (N) expression at the equator. Intersection of equatorial N signaling and of an unknown factor at the posterior margin of the disc at its juncture with the optic nerve (circles) is critical for eye morphogenesis, as the eye fails to develop in the absence of an equatorial boundary. Wingless (Wg) suppresses expression of Decapentaplegic (Dpp), and vice versa, to prevent formation of an ectopic morphogenetic furrow around the disc margin, and regulates Mirr expression and planar polarity. C. Proteins specifying identity of photoreceptors R3 and R4 above and below the equator. Dsh, Dishevelled; Fz, Frizzled. (A reprinted with permission from *Nature* (M. Domínguez and J. F. de Celis. A dorsal/ventral boundary established by Notch controls growth and polarity in the *Drosophila* eye. 396: 276–278) © 1998 Macmillan Magazines Limited.) (B adapted with permission from *Nature* (K.-O. Cho and K.-W. Choi. Fringe is essential for mirror symmetry and morphogenesis in the *Drosophila* eye. 396: 272–276) © 1998 Macmillan Magazines Limited.) (C adapted with permission from *Nature* (M. T. D. Cooper and S. J. Bray. Frizzled regulation of Notch signaling polarizes cell fate in the *Drosophila* eye. 397: 526–530) © 1999 Macmillan Magazines Limited.)

kühniella (Egelhaaf et al., 1988), *M. sexta* (Champlin and Truman, 1998b), and the red flour beetle *Tribolium castaneum* (Friedrich et al., 1996) indicates events similar to those occurring in *D. melanogaster*.

In fifth (last) instar larvae and pupae of *M. sexta*, secretion of ecdysteroids differentially governs two aspects of compound eye development (Champlin and Truman, 1998b) just as they do during ventral diaphragm (10.2.2.1) and optic lobe (10.2.4.1.2) development in this species (Champlin and Truman, 1998a). And just as in *D. melanogaster* (Brennan et al., 1998), tonic exposure to moderate levels of either ecdysone or 20E is required for forward progression of the MF across the eye disc and for proliferation, aggregation, and cell type specification in ommatidial clusters behind the MF. All can be reversibly started or stopped simply by adjusting ecdysteroid levels above or below a critical threshold concentration.

Tonic exposure to levels of ecdysteroid 17-fold higher than that required for MF progression induces the ommatidia to differentiate, and premature exposure to such concentrations, at any time during eye development, causes precocious differentiation. In these prepupae and pupae, the MF arrests irreversibly, and

than 3500 in adult cockroaches. In both insects, new ommatidia are recruited from an anterior budding zone that can regenerate from differentiated ommatidia if it is destroyed, specifically from dedifferentiated pigment cells of peripheral ommatidia. Also, in interspecific ommatidial chimeras produced by grafting budding zones between larval eyes of the cockroaches *Gromphadorhina portentosa* and *Leucophaea maderae*, in which equivalent cell types of each species are cytologically distinguishable, Nowel (1980) showed that the cells of each unit, rather than descending from a single stem cell as do other sense organs (Fig. 8.33), are formed by aggregation of adjacent epithelial cells in the budding zone, with fate depending on position, just as in *D. melanogaster* eyes. Preliminary experimental work on eye discs of the Mediterranean flour moth *Ephestia*

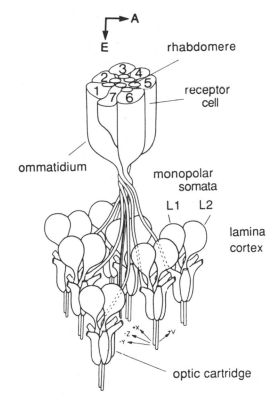

Fig. 10.31. The projection of receptor axons from a single ommatidium to six adjacent cartridges in the lamina of a fly's optic lobe. The axons of R1–R6 twist through 180 degrees as they project to the lamina before diverging to their respective lamina cartridges (only L1 and L2 of the lamina interneurons are shown). R1–R6 in all adjacent ommatidia also contribute axons to these cartridges through identical projections so that each cartridge receives not one but six receptor terminals. (Reprinted with permission from I. A. Meinertzhagen and T. E. Hanson. The development of the optic lobe. In M. Bate and A. Martinez Arias (eds.), *The development of* Drosophila melanogaster. Vol. 2, p. 1368, Fig. 2C. © 1993 Cold Spring Harbor Laboratory Press.)

ommatidial clusters behind the furrow differentiate to form a normal but miniature compound eye.

10.2.4.2.4 INFLUENCE OF PHOTORECEPTOR AXONS ON DIFFERENTIATION OF THE OPTIC LOBE
In insects having compound eyes, the three neuropiles of each optic lobe—the lamina, medulla, and lobula (Fig. 10.18B)—are of exceedingly complex, modular construction and consist, respectively, of an array of first-, second-, and third-order interneurons that in each center are *retinotopic* (i.e., they are arranged in columns corresponding to the ommatidial array innervating the lobe) (Meinertzhagen and Hanson, 1993). In the lamina these columns form optic cartridges, the axons of each of the six outer photoreceptor cells in each ommatidium, R1–R6, projecting to a different but adjacent group of five interneurons, L1–L5, in the lamina (Fig. 10.31). The

axons of photoreceptors R7 and R8, however, project to the medulla (Anderson, 1978a; Meinertzhagen and Hanson, 1993).

In third instar larvae of *D. melanogaster*, the wave of morphogenesis in the eye disc (Fig. 10.27) is translated into a wave of innervation in the optic lobe because innervation of the lamina by photoreceptor axons is required for normal differentiation of the lamina and, to a lesser extent, the medulla (Meinertzhagen and Hanson, 1993; Salecker et al., 1998; Huang and Kunes, 1998; Fortini and Bonini, 2000). As each ommatidial cluster differentiates behind the MF (Fig. 10.27), the axons of its photoreceptor cells project through the optic nerve to GMCs (lamina precursor cells) budded off from NBs on the lamina side of the outer optic anlage (Figs. 10.19; 10.32: arrow 2). As first shown by Power in 1943, signals from these axons coordinate the production of lamina interneurons from the GMCs and their assembly into columns.

As growth cones of the photoreceptor cell axons contact the GMCs, they release the signal protein Hh, which induces them to enter the S phase of the cell cycle (they were arrested in G1) and to each divide into two lamina interneurons. This step is controlled by the Hh-dependent transcriptional regulator Cubitus interruptus (Ci) (Huang and Kunes, 1998). Following GMC mitosis, the photoreceptor growth cones then release Spitz (Spi) protein, which induces these interneurons to aggregate into columns and to differentiate. At onset of metamorphosis, the growth cone of each of photoreceptors R1–R6 then leaves its original laminal column; extends to a specific neighboring column, a process that depends on the release of nitric oxide by lamina cells; and establishes a precise pattern of synaptic contacts with laminal interneurons L1–L5 in the adult optic cartridge (Fig. 10.31). This sequence of events is repeated upon arrival of photoreceptor axons from each new ommatidium. That the axons of photoreceptors R1–R6 project only to the lamina and not to the medulla, like those of R7 and R8, is due to expression of the gene *brakeless* (*bks*); in loss-of-function mutants, they too project to the medulla (Senti et al., 2000).

In addition to NBs and GMCs, the outer optic anlage produces glial cells that also depend on innervation by photoreceptor R1–R6 axons to migrate to and to differentiate within appropriate sites within the lamina (Salecker et al., 1998). In addition, retinal basal glial cells, originating in the optic stalk, migrate into the eye disc, where they enable young, growing photoreceptor axons behind the MF to grow into the optic stalk (Rangarajan et al., 1999).

Experiments on optic lobe metamorphosis in dragonfly (Mouze, 1978), cockroach (Nowel, 1981a, 1981b), and locust (Anderson, 1978b) larvae reveal both similarities and differences. If the lamina is destroyed, the retina grows and differentiates normally. If the retina is destroyed, the production of new lamina interneurons may continue autonomously, but they fail to differentiate and subsequently degenerate.

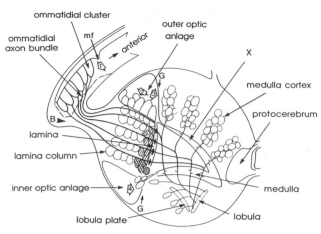

Fig. 10.32. Frontal section through a late third instar optic lobe of *D. melanogaster*, showing the relationship between cell generation and fiber pathway formation during optic lobe metamorphosis (the size of the lamina has been enlarged for clarity and the neuropiles reduced in size). Cells from the morphogenetic furrow (mf) assemble into new ommatidia (arrow, direction 1). Photoreceptor axon bundles from newly formed ommatidial clusters fasciculate on previously projected axons to reach the outer optic anlage. Cells are contributed to the lamina cortex from one side of the outer anlage (arrow, direction 2) and for the medulla cortex on the other side (arrow, direction 3). On the lamina-forming arm of the outer anlage, large neuroblasts generate ganglion mother cells and interneurons of the lamina columns. These sit atop a supraneuropile layer (crosshatched cells). Proximal to the lamina (i.e., toward the right), the diagram is very schematic. Proliferation from the inner optic anlage is poorly understood and is shown as occurring in a single direction (arrow, direction 4). The corresponding axon processes and their growth cones generate a plexus associated with their particular cortex, and these will form the neuropiles of the adult lamina, medulla, and lobula/lobula plate. Compare with Fig. 10.19. (Reprinted with permission from I. A. Meinertzhagen and T. E. Hanson. The development of the optic lobe. In M. Bate and A. Martinez Arias (eds.), *The development of Drosophila melanogaster*. Vol. 2, p. 1387, Fig. 10F. © 1993 Cold Spring Harbor Laboratory Press.)

10.2.4.2.5 DORSAL OCELLI

Adult and juvenile bristle tails (Archaeognatha), mayflies (Ephemeroptera), and stoneflies (Plecoptera) and most fully winged pterygote adults have two or three dorsal ocelli on the vertex of the head between the compound eyes. In adult *D. melanogaster*, each ocellus contains about 80 photoreceptors with very short axons synapsing with, and projecting to the brain by, four giant interneurons within a common ocellar nerve (García-Alonso et al., 1996). These three ocelli differentiate from a median and two lateral optic placodes originating from epidermis in the dorsal margin of each eye-antennal disc (Jürgens and Hartenstein, 1993; Fristrom and Fristrom, 1993). The median ocellus has a paired origin that becomes one on dorsal fusion of antennal parts of the discs following eversion. Just before head eversion at 12–13 h after pupariation (Fig. 10.8C–E), each lateral pla-

code and each half of each median placode, collectively, produce about 50 transient ocellar neurons whose axons project to the brain through an extracellular matrix containing Laminin A that covers and connects the disc epithelium and brain (García-Alonso et al., 1996). After head eversion, 12 ocellar interneurons originate in the brain and project their axons to the ocelli along these pioneers, which then die. In Laminin A mutants, the pioneers display path-finding defects leading to defects in the ocellar nerve.

10.2.4.3 EPIDERMIS

In addition to the secondary exit system and external genitalia of males and females (Figs. 10.12–10.14), the foregut and hindgut (Fig. 10.15), and ultimately, the CNS (10.2.4.1) and PNS (10.2.4.2), the ectoderm gives rise to the appendages (apterygotes, exopterygotes, and basal endopterygotes) and imaginal discs and histoblasts (intermediate and higher endopterygotes), and to the body wall, tracheal system, salivary glands, and various pheromone-dispensing structures.

10.2.4.3.1 IMAGINAL DISCS AND BODY WALL EPIDERMIS

During metamorphosis in *D. melanogaster*, the head and thorax of the adult differentiate from 10 pairs of invaginated imaginal discs, and the abdomen differentiates from 24 (male) or 28 (female) pairs of uninvaginated imaginal histoblasts and a single median invaginated genital disc, each primordium having a characteristic size, shape, location, cell number, and, in the discs, pattern of folds facilitating its individual recognition (Fig. 10.33) (Ursprung and Nöthiger, 1972; Madhavan and Schneiderman, 1977; Jürgens and Hartenstein, 1993; Hartenstein, 1993; Fristrom and Fristrom, 1993).

Cells of these primordia are determined during embryogenesis, the discs arising on each side of the embryo on the anteroposterior (A-P) compartment boundaries of their respective segments (Figs. 8.43: 5; 8.44). Depending on disc identity, they invaginate into the hemocoel at varying times after hatch and grow slowly by mitosis throughout larval development while maintaining their attachment to larval epidermis via their peripodial stalks (Fig. 10.34B). After larval-pupal apolysis and on commencement of pupal cuticle deposition, their peripodial stalks shorten and widen and the wing and leg discs elongate, evert, and fuse with each other (Figs. 10.8C–E, 10.34, 10.35). The wings, halteres, and thoracic nota and pleura develop from the prothoracic, wing, and haltere discs, and the thoracic sterna, ventral pleura, and legs develop from three pairs of leg discs (Fig. 10.33).

The cephalic discs evert and fuse with each other (Fig. 10.8F) (Jürgens and Hartenstein, 1993). The clypeolabral discs give rise to the clypeolabrum; the eye-antennal discs, to the antennae, proximal proboscis, maxillary palpi, head capsule, and compound eyes; and the labial discs, to the distal proboscis and labellar lobes. The reproductive exit ducts, external genitalia, analia, and

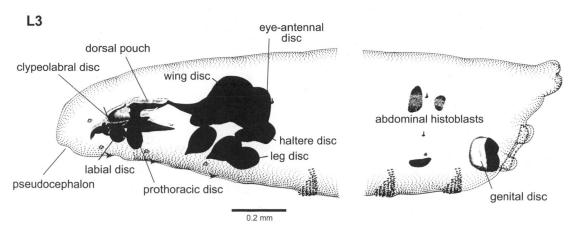

L3

clypeolabral disc
dorsal pouch
eye-antennal disc
wing disc
abdominal histoblasts
haltere disc
leg disc
labial disc
pseudocephalon
prothoracic disc
genital disc

0.2 mm

Fig. 10.33. Lateral aspect of a third instar larva of *D. melanogaster*, showing the principal imaginal discs and histoblasts (those of abdominal segments 1–6 are omitted). (Reprinted with permission from V. Hartenstein. *Atlas of* Drosophila *development*, p. 24. © 1993 Cold Spring Harbor Laboratory Press.)

part of the hindgut differentiate from the genital disc (Fristrom and Fristrom, 1993).

Morphogenesis of each thoracic disc can be divided into three processes, each relatively independent of each other (Figs. 10.34B–D; 10.35B, C): (1) elongation and shaping, (2) eversion, and (3) spreading and fusion of peripheral disc tissues to form a continuous imaginal epidermis (Fristrom and Fristrom, 1993). Elongation begins about 6 h before pupariation and ends about 6 h APF. In each leg disc, tarsal segments 2–5, the distal tibia and basitarsus, the proximal tibia and distal femur, and the proximal femur, trochanter, coxa, and body wall tend to develop slightly differently and somewhat independently of each other.

Following its eversion at larval-pupal apolysis, each wing disc passes through four steps, each occurring twice, once in the pharate pupa and again in the pharate adult (Fig. 10.8C–H) (Waddington, 1941; Fristrom and Fristrom, 1993):

- Apposition: Basal surfaces of dorsal and ventral wing epithelia come together following wing eversion.
- Adhesion: Basal junctions form between opposed wing epithelia except at lacunae, which become the veins.
- Expansion: Wing area increases as a result of the cell flattening.
- Separation: Dorsal and ventral wing epithelia separate by increases in hemostatic pressure and by production of a bulky extracellular matrix but remain connected by the basal junctions and by slender epidermal processes containing longitudinally disposed microtubules and microfilaments.

After eversion, marginal cells at the leading edge of the proximal, notal part of each wing disc (Fig. 10.35A) fuse middorsally to those of the other wing disc in a process called *thorax closure*. This fusion occurs by a mechanism seemingly redeployed from that inducing embryonic dorsal closure (8.1.4), as it involves expres-

sion of many of the same genes (Zeitlinger and Bohmann, 1999). After deposition of imaginal cuticle, and just before adult eclosion, the wing epidermal cells begin to degenerate.

Nardi and coworkers (1991) similarly examined eversion and differentiation of wing discs in the wing dimorphic lymantriid moth *Orgyia leucostigma*. In this species, males are fully winged but females are brachypterous (short winged). Even so, the wing pads of male and female pupae are of similar size. In female pupae, a period of programmed cell death occurs over 2 d following disc eversion that removes distal wing epidermis. If wing pads from female pupae are transferred into male pupae of the same age, their cells still degenerate, indicating that female wing discs were programmed earlier in development to self-destruct during metamorphosis.

It should be emphasized that invagination is not a universal feature of imaginal discs in endopterygotes, since megalopterans, raphidiopterans, neuropterans, mecopterans, and some basal coleopterans, lepidopterans, hymenopterans, and dipterans lack invaginated wing discs in the penultimate and even in the last larval instar, the imaginal legs and antennae differentiating from cells within the larval appendages (references in Svácha, 1992). Recent detailed analyses of wing (Quennedey and Quennedey, 1990) and imaginal leg (Huet and Lenoir-Rousseaux, 1976) development in last instar larvae and pupae of the tenebrionid beetle *Tenebrio molitor* provide good examples.

Unlike the discs, the abdominal histoblasts do not invaginate or proliferate during larval development and therefore, are still small patches of 6–15 diploid cells at the time of pupariation (Fig. 10.36A) (Bautz, 1971; Madhavan and Schneiderman, 1977). From 3 to 15 h APF, their cells proliferate rapidly, but the histoblasts maintain their size while depositing pupal cuticle

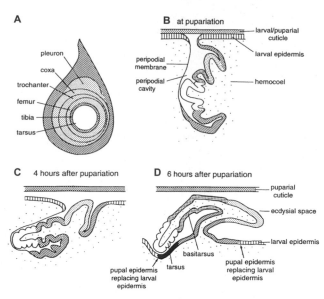

Fig. 10.34. Development of the adult leg in *D. melanogaster*. A. Fate map of the leg disc. B–D. Sagittal sections through the disc at various times after pupariation, showing elongation and eversion of the leg from the peripodial cavity. See text for details. (From R. F. Chapman. 1998. *The insects. Structure and function*, p. 384, Fig. 15.21. 4th ed. With the permission of Cambridge University Press; adapted from Fristrom and Fristrom, 1993: 871, 848.)

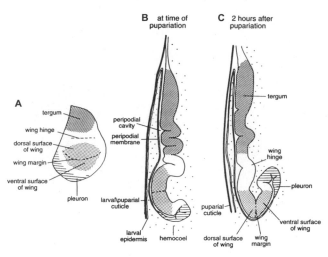

Fig. 10.35. Development of the adult wing in *D. melanogaster*. A. Fate map of a wing disc. B. Sagittal section through the disc at the time of pupariation. C. Section through the disc 2 h after pupariation. (From R. F. Chapman. 1998. *The insects. Structure and function*, p. 385, Fig. 15.21. 4th ed. With the permission of Cambridge University Press; adapted from Fristrom and Fristrom, 1993: 874.)

(Fristrom and Fristrom, 1993). At 18 h APF, they begin to spread between the larval cuticle and epidermis (Fig. 10.36B) and to cast the epidermal cells into the hemocoel, where they degenerate and are phagocytosed by macrophages (Bautz, 1975). Eventually they fuse, totally replacing the larval epidermis of each abdominal segment by 40 h APF (Fig. 10.36C). If such histoblasts are selectively destroyed by gamma irradiation, some larval cells survive metamorphosis and deposit imaginal cuticle at the same time as imaginal epidermis (Madhavan and Schneiderman, 1977, 1984).

10.2.4.3.2 DEVELOPMENT OF PIERCING AND SUCKING MOUTHPARTS IN MOSQUITOES

In odonates, whiteflies, and male scale insects (9.4.2) and in advanced endopterygotes (9.4.3), the mouthparts of larva and adult differ substantially in structure and function. For example, adult fleas (Siphonaptera) and the females of certain flies (psychodids, culicids, simuliids, ceratopogonids, tabanids, rhagionids, asilids, some muscids, and glossinids) have piercing and sucking mouthparts very different from those of their larvae (Matsuda, 1965; Chaudonneret, 1990), the one being replaced by the other during metamorphosis.

In mosquitoes, the larval mouthparts are equipped with feeding brushes and are specialized for setting up feeding currents and for filtering microorganisms out of the water column (Fig. 10.37A, B). In females, replacement, by imaginal piercing and sucking mouthparts

begins in the pharate pupa where the stylet-like labrum, hypopharynx, mandibles and maxillae, and the elongate labium of the adult begin to differentiate from epidermis within equivalent larval structures following larval-pupal apolysis (Fig. 10.37C, D) (Snodgrass, 1959; Christophers, 1960: chapter 15). They rapidly elongate, are ensheathed by pupal cuticle, are glued throughout their lengths to the underside of the body (Fig. 10.37E–G) (mosquito pupae are adecticous obtect and can swim effectively; 9.6.2.1), and are released from pupal cuticle upon adult emergence (Fig. 10.37H). These changes are accompanied by major changes in innervation and musculature that, to my knowledge, have yet to be investigated.

10.2.4.3.3 GIANT PULVILLAR NUCLEI AND THE DEPOSITION OF IMAGINAL CUTICLE

In calypterate flies, the distal segment or pretarsus of each adult leg consists of a pair of pretarsal claws or *ungues*, each subtended by a large "foot pad" or *pulvillus* covered ventrally by thousands of short, spatulate tenent hairs (Fig. 10.38D). Proximally and midventrally, an unguitractor plate and apodeme are attached to these structures into which is inserted a pretarsal depressor muscle. Contraction of this muscle flexes the pretarsus and allows the insect to walk effectively, with claws on rough surfaces and with pulvilli on smooth.

Formation of the pretarsus during metamorphosis is of developmental interest for two reasons: (1) emergence of claws and pulvilli from imaginal pretarsal epidermis is associated with programmed cell death, and (2) the entire dorsal cuticle of each pulvillus is deposited by

Fig. 10.36. Development, at 25°C, of imaginal histoblasts in the posterior half of segment 2 and in segment 3 into adult abdominal epidermis during metamorphosis of *Calliphora erythrocephala* (Diptera: Calliphoridae). The segments are shown in posterolateral aspect, with the larval and pupal cuticles omitted. A. At white puparium stage. B. At 48 h after puparium formation. C. At 72 h after puparium formation. l, border between tergal and sternal cells. (Reproduced from A.-M. Bautz. Chronologie de la mise en place de l'hypoderme imaginal de l'abdomen de *Calliphora erythrocephala* Meigen (Insecte, Diptère, Brachycère). *Arch. Zool. Exp. Gen.* 112: 157–178. © 1971.)

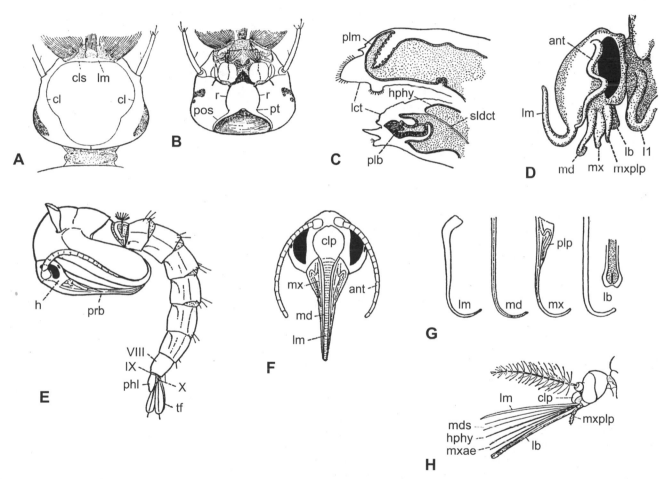

Fig. 10.37. Metamorphosis of the mouthparts in female mosquitoes (Diptera: Culicidae). A. *Aedes aegypti*. Dorsal aspect of head of fourth instar larva. B. Same as A, in ventral aspect. Fan rays on the mouthparts function to produce feeding currents. C. *Culex* sp., exuvial pharate pupa. Median sagittal section through anterior part of head, showing pupal labrum (plm) and labium (plb) inside larval cuticle (lct). D. Same as C but with larval cuticle removed to show lengthening adult mouthparts. E. *Aedes aegypti*. Lateral aspect of a male pupa, showing head (h) and proboscis (prb). F. *Aedes aegypti*. Frontal aspect of head of pharate adult female, showing proboscis. G. Same as F, but with individual mouthparts removed to show their sheaths of pupal cuticle. H. Lateral aspect of head of adult female. ant, antenna; cls, clypeolabral sulcus; cl, cleavage line (ecdysial suture) in larval head capsule; clp, clypeus; hphy, hypopharynx; lb, labium; lm, labrum; md, mandible; mx, maxilla; mxplp, maxillary palp; r, grooves in ventral head capsule; phl, phallus; plp, palp; pos, postoccipital suture; pt, posterior tentorial pit; sldct, salivary duct; tf, tail fin. (Reprinted from R. E. Snodgrass. 1959. The anatomical life of the mosquito. *Smithson. Misc. Collect.* 139 (8): 1–87.)

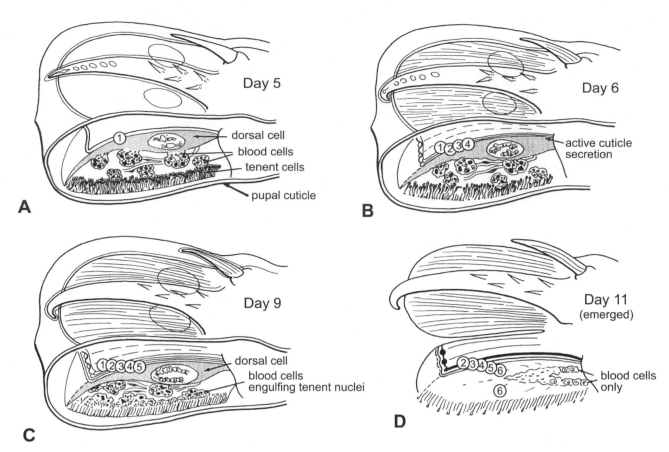

Fig. 10.38. Four metamorphic stages in the adult pretarsus of *Neobellieria bullata* (Diptera: Sarcophagidae). Note the two pulvilli, one in dorsal aspect and the other in sagittal section, and one of the two claws (ungues). The persisting pupal cuticle (A–C) is shed on adult emergence (D). A. Day 5. Pupal-adult apolysis has occurred, the ecdysial membrane (layer 1) persisting about the adult pretarsus. Note the six epidermal nuclei in the claw and the dorsal, giant and ventral, tenent cells with intervening hemocoel and blood cells. B. Day 6. Cuticle layers 2–4 have been deposited. C. Day 9. Five layers of cuticle have been deposited, the cells have withdrawn from the claw, and the tenent cell nuclei have degenerated and are being engulfed by hemocytes. D. Day 11, newly emerged adult. Layer 6 (endocuticle) has been deposited, and the giant cells have degenerated and have been phagocytosed by hemocytes, which are withdrawing into the distal tarsomere. (Reprinted with permission from J. M. Whitten. Coordinated development in the foot pad of the fly *Sarcophaga bullata* during metamorphosis: changing puffing patterns of the giant cell chromosomes. *Chromosoma* 26: 215–244. © 1969 Springer-Verlag.)

only two giant epidermal cells, each containing banded polytene chromosomes in their nuclei similar to those of larval salivary glands in *D. melanogaster* (Figs. 12.17, 12.18).

Metamorphosis of the sarcophagid *Neobellieria bullata* requires 12.5 d at 25°C (Bultmann and Clever, 1969). Following pupal-adult apolysis, each leg is surrounded by a sheath of pupal cuticle devoid of sense organs that is spatulate at its tip to encompass the developing adult pretarsus (Figs. 10.8H, 10.38A–C) (Whitten, 1969b, 1969c). Cell division continues in the pretarsal epidermis after apolysis, with six cells on each side emerging dorsally to form the distal part of each claw, one cell laterally and one medially on either side of the claw base to form the dorsal surface of each pulvillus, and thousands of small cells ventrally to form the tenent hairs (Fig. 10.38A). Gradual emergence of claws and pulvilli

from imaginal epidermis is caused by programmed cell death (8.6.1) between them exactly as occurs during digit formation in vertebrate embryos (Whitten, 1969a) and during formation of the distally dissected wings of the plume moth *Oidaematophorus hirosakianus* (Pterophoridae) (Yoshida et al., 1998). Growth of the four dorsal pulvillar cells in each pretarsus is at the expense of thousands of other dorsal epidermal cells, which degenerate shortly after pupal-adult apolysis (Whitten, 1969b). By day 4 the pretarsus has assumed its final shape, and by day 6 its cells have begun to deposit imaginal cuticle (Fig. 10.38B).

Unlike those of the larval salivary glands, which degenerate following pupariation, the giant foot pad cells function throughout the cuticular pharate adult stage (Fig. 10.38A–C) and degenerate during adult eclosion (Fig. 10.38D) (Whitten, 1969b), enabling investigators

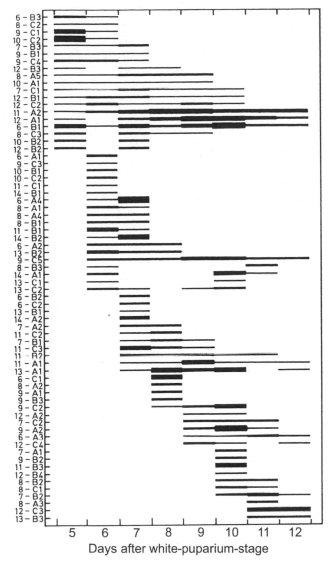

Fig. 10.39. Changes in puffing (gene expression) patterns at individual loci during pharate adult development in the giant pulvillar cells of *N. bullata*. All puffs in sections 6A–16B of chromosome B were examined at 24 h intervals. The thickness of the bars indicates average puff size in all four pulvillar cells of a single pretarsus. Symbols at left refer to puff positions in the chromosome map. (Reprinted with permission from H. Bultmann and U. Clever. Chromosomal control of foot pad development in *Sarcophaga bullata*. I. The puffing pattern. *Chromosoma* 28: 120–135. © 1969 Springer-Verlag.)

(Whitten, 1969c; Bultmann and Clever, 1969, 1970; Clever and Bultmann, 1972; Roberts et al., 1976; Carruthers and Roberts, 1979) to associate the puffing (gene-expressing) pattern in their polytene chromosomes with imaginal cuticle deposition (Figs. 10.39, 10.40). Bultmann and Clever (1969, 1970) used uptake of [14C]thymidine to monitor DNA synthesis, of [3H]uridine for RNA synthesis, of [14C]glucose for chitin synthesis, of [14C]leucine for sclerotin synthesis, and of

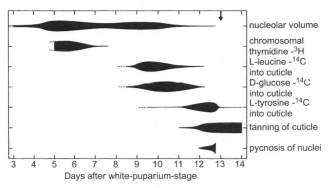

Fig. 10.40. Metabolic changes in the pulvilli of *N. bullata* during adult cuticle deposition. The arrow indicates adult eclosion. (Reprinted with permission from H. Bultmann and U. Clever. Chromosomal control of foot pad development in *Sarcophaga bullata*. II. Cuticle formation and tanning. *Dev. Biol.* 22: 601–621. © 1970 Academic Press, Inc.)

[14C]tyrosine for sclerotization to associate a specific sequence of 71 puffs on chromosome B with various stages of cuticle deposition (Figs. 10.39, 10.40). A similar analysis by Trepte (1976) on polytene chromosomes of hair- and socket-forming cells of the large scutellar bristles in pharate adults of another sarcophagid, *Sarcophaga barbata*, had similar results.

10.2.4.3.4 TRACHEAL SYSTEM

As in the embryo (Fig. 8.41), the imaginal tracheal system originates as 10 tracheal metameres, one each in the mesothorax and metathorax and in each of the first eight abdominal segments (Whitten, 1968; Manning and Krasnow, 1993). One-half of each metamere originates from a tracheal histoblast associated with the spiracle on each side of each segment, with tracheal nodes developing where branches of the two sides of the same segment or of two adjacent segments meet. At ecdysis, the remains of the cuticular lining of the tracheal system of the previous instar break at these nodes and are shed from the spiracle from which that portion of the metamere originated. Nardi (1990) used a fluorescently tagged monoclonal antibody to label a protein, probably Fasciclin II, characteristic of both commissural and longitudinal nodal cells in larvae and pupae of *M. sexta*.

Where developing organs of two successive instars differ, as at the larval-pupal or pupal-adult molts, further ramification of the system occurs to service these tissues by proliferation, spread, and hollowing out of tracheoblasts originating from epithelium at the tips of extant branches, to form new tracheoles; the cross-sectional area of the more proximal tubes increases in direct proportion to the new area serviced (Locke, 1958). In fact, it is now known that oxygen-starved cells in newly developing parts of the body secrete Branchless protein, which functions as a chemoattractant to guide new terminal branches to them (Jarecki et al., 1999).

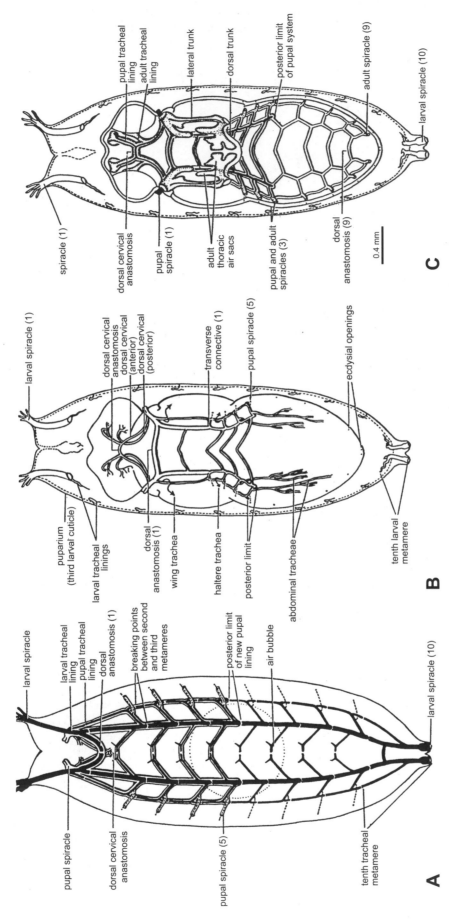

Fig. 10.41. Metamorphosis of the tracheal system in *D. melanogaster* in dorsal aspect (highly diagrammatic). A. Cuticular pharate pupa (prepupa of drosophilists) just after larval–pupal apolysis (compare with Fig. 10.8B). The posterior half of the larval system, from tracheal metamere 6 back, degenerates while the pupal system is deposited by tracheal epithelium surrounding the larval system anteriorly. Notice the tracheal nodes or fusion points in each metamere of the larval system, where the lining of the cuticle breaks prior to larval–pupal apolysis. B. Phanerocephalic pupa within the puparium (compare with Fig. 10.8G and H). The pupal system is functional while the cuticular lining of the larval system has been shed and is attached to the inner surface of the puparium at the nonfunctional spiracles marking the tracheal-placodes from which they invaginated in the embryo (see Fig. 8.41). Notice the tracheal branches growing posteriorly into the abdomen to form a new, adult abdominal tracheal system. C. Cuticular pharate adult within puparial and pupal cuticles, with the adult system being deposited around the pupal system (larval tracheal metamere 10 is not replaced). Imaginal airsacs are beginning to expand within the head and thorax to service the developing adult brain and flight muscles. (Reprinted from J. M. Whitten. Metamorphic changes in insects. In W. Etkin and L. I. Gilbert (eds.), *Metamorphosis: a problem in developmental biology*, pp. 48–105. Appleton-Century-Crofts, N.Y. © 1968.)

This method of tracheal deposition and replacement occurs in all pterygote insects at each molt regardless of how many pairs of spiracles the insect has, even in the juveniles of aquatic insects having gills instead of spiracles, and is good evidence for the terrestrial origin of aquatic insects (Pritchard et al., 1993).

In exopterygotes and in basal endopterygotes, the juvenile tracheal system functions in the adult, with additional branches and air sacs developing to service the imaginal brain and the flight and genital muscles (Whitten, 1968). However, in intermediate and higher endopterygotes, the pupa has unique respiratory problems because its tracheal system, formed at the time of the larval-pupal molt, must service larval tissues that will later degenerate, larval tissues persisting throughout metamorphosis and developing imaginal tissues not present when the system develops (Whitten, 1968). For example, in *D. melanogaster* and other higher flies, tracheal cells of the larval system from metamere 6 back do not secrete pupal cuticle and degenerate (Fig. 10.41A) (Whitten, 1968; Hartenstein, 1993; Manning and Krasnow, 1993). The pupal system differentiates during late larval and early pupal development from tracheal histoblasts situated in the abdomen at the position of the embryonic tracheal placodes from which the larval system originally developed (Fig. 8.41), and in the head and thorax, and from others along the dorsal cervical and the second dorsal transverse anastomoses (Fig. 10.41B) (Whitten, 1968; Hartenstein, 1993). In addition, it is characterized by grapelike bunches of coiled tracheoles suspended in the hemocoel, respiratory exchange, at first, apparently occurring from tracheole to hemolymph rather than from tracheole to cell (Whitten, 1968). With gradual development of imaginal organs, these tracheolar bunches grow out to service them.

The imaginal system (Fig. 10.41C) develops from the pupal system with further growth of tracheal tubes. A new abdominal system, much like that of the larva except for absence of metamere 10, develops from abdominal histoblasts while the principal tracheae of the head and thorax expand into a complex system of air sacs about the brain and flight muscles.

10.2.4.3.5 SALIVARY GLANDS

In pupae of the blowfly *Calliphora erythrocephala*, the cells of the two larval salivary glands are sloughed into the body cavity by proliferation and spread of imaginal cells underneath them from an imaginal ring surrounding the duct of each gland proximally. They then degenerate and are phagocytosed by hemocytes (Berridge et al., 1976). Imaginal cells proliferate throughout the 6-d larval period and until day 5 of the 10-d pupal period at 25°C but are unable to secrete fluid in response to 5-hydroxytryptamine stimulation until 2 h after adult emergence.

In lepidopterans, larval glands have been demonstrated to have three fates depending on species (Hakim, 1976). In bombycids the larval glands degenerate during metamorphosis; in saturniids they persist and secrete a dilute salt solution; and in noctuids and sphingids they synthesize products assisting in digestion.

10.2.4.3.6 PHEROMONE SYSTEMS

In many insect species, the female, male, or both have special glands that elaborate sex attractants functioning to attract conspecifics of the opposite sex, these characteristically differentiating during metamorphosis. Egelhaaf and colleagues (1992) followed in great detail the origin and differentiation of the male pheromone system of the arctiid moth *Creatonotos transiens* from paired epidermal anlagen at the anterior border of abdominal segment 8. The male pheromone in this insect is hydroxydandaidal and is attractive to adults of both sexes. Pyrrolizidine alkaloids ingested by the larva with its food are precursors to this pheromone but, in addition, act as a morphogen quantitatively controlling growth of the male abdominal scent organs (*coremata*) and thus their final size and hair number. Access to the diverse literature on this subject can be found in Egelhaaf et al., 1992.

11/ Specification of the Adult Body Pattern

Now that we have some knowledge of the structural changes occurring during metamorphosis, it is possible to consider observations and concepts relating to pattern formation in the imaginal epidermis and appendages and some recent progress in understanding the molecular genetics of wing pattern formation in butterflies. Some of the earliest evidence to shed light on the proximate causes of these processes came from studies on how juvenile insects regenerate lost appendages, so I discuss this first.

11.1 Wound Healing

The ability of a young insect to regenerate a damaged or missing appendage is a special kind of wound healing (Bullière and Bullière, 1985). Insects of practically any age repair wounds in their integument by means of the following steps (Fig. 11.1) (Wigglesworth, 1965):
- The wound closes temporarily through coagulation of hemocytes to form a scab.
- As damaged epidermal cells degenerate, they release polypeptide "wound factors," which attract surrounding epidermal cells to the wound margins (Fig. 11.1A: c).
- These congregate about the wound's edges, resulting in their source, peripheral areas becoming sparsely populated (Fig. 11.1A: d).
- The aggregated cells spread across the wound (Fig. 11.1A: a), close it, and deposit a new cuticle under the edges of the old, its surface pattern providing a record of this migration (Fig. 11.1B: g).
- Simultaneously, mitosis occurs in cells of the thinned-out peripheral areas and restores their original density (Fig. 11.1A: d).

11.2 Regeneration

Many of these steps also occur during regeneration of appendages or appendage parts removed at an early instar (Bullière, 1972; Truby, 1983; Bullière and Bullière, 1985). The ability of an insect to regenerate a missing appendage or appendage part is an example of postembryonic regulation and, except in apterygotes, occurs only in juveniles during molts, since this is the only time when the epidermis separates from cuticle and

when new cuticle is produced (Wigglesworth, 1965). The number of molts required depends on when during the intermolt the appendage was removed. If removed early in an instar, before the secretion of 20-OH ecdysone (20E) (12.5), then regeneration can be accomplished in one molt; if removed afterward, more than one is required (Bullière and Bullière, 1985). The regenerate will increase in size at each subsequent molt but usually will remain smaller than the uninjured appendages unless removed from a sufficiently early instar. In addition, removing an appendage will often stimulate or prolong a molt. In fact, juvenile odonates and stick insects can voluntarily sever a leg at the trochanterofemoral joint by contracting specific muscles there in a process called autotomy, much as do lizards with their tails (Wigglesworth, 1965).

The base of each appendage is surrounded by a regenerative (individuation) field whose epidermal cells are capable of dedifferentiating, proliferating into a blastema, and then regenerating an appendage after it is removed (Bullière, 1972; Truby, 1983). The same occurs if one or more distal segments of an appendage are removed; the epidermis at the cut end of the most distal, remaining segment forms a blastema and regenerates the missing ones.

The capacity to regenerate resides exclusively in the epidermis (Wigglesworth, 1965). If the appropriate ganglion of the ventral nerve cord is removed at the same time as the appendage it innervates, a young insect will regenerate over one or more molts an appendage that is normal in every way except that it lacks nerves and muscles and is nonfunctional. If only the appendage is removed, it will be reinnervated and remusculated as it regenerates, innervation being required for differentiation of skeletal muscle within it (see 10.2.2.1 for a summary of the role of the nervous system in normal muscle differentiation). Once an appendage has regenerated completely, it will be of almost normal size and function.

11.2.1 THE ROLE OF MORPHOGENS IN REGENERATION
Examination of the body of an insect or one of its appendages reveals that the long axes of its setae point either posteriorly (body) or distally (appendages). What might be the basis for this polarization? Results of countless experiments on a variety of insects have

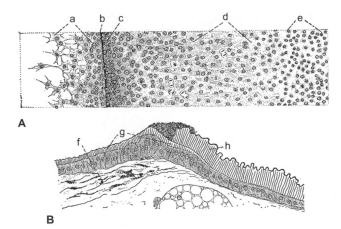

A

B

Fig. 11.1. Wound healing in an adult of *Rhodnius prolixus* (Hemiptera: Reduviidae). A. Surface view of the epidermis 4 d after a piece of cuticle was excised. a, epidermal cells spreading over the gap; b, margin of wound; c, cells piled up along margin of wound; d, zone depleted of cells by cellular migration to the wound; mitosis occurring; e, unchanged epidermal cells. B. Sagittal section through margin of a wound after 3 wk. f, new epidermis established under excised region; g, new cuticle extending outward under the old cuticle; h, old cuticle at margin of wound. (Reprinted from *The principles of insect physiology*, 6th ed., 1965, p. 99, V. B. Wigglesworth, Fig. 75, © 1965 Kluwer Academic Publishers, with kind permission from Kluwer Academic Publishers.)

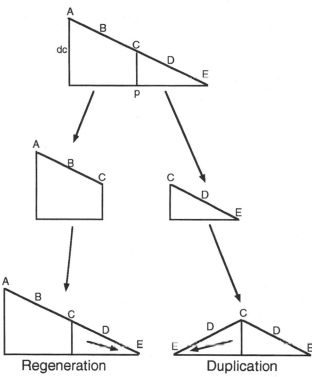

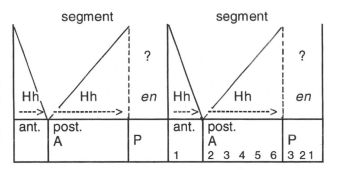

Regeneration   Duplication

Fig. 11.2. Regeneration and duplication explained in terms of a gradient of developmental capacity (dc) that varies with position within a segment (p). If the segment is cut, those positions higher in the gradient (A–C) can regenerate those lower down (D and E). Points low on the gradient can only duplicate themselves. (Adapted from Russell, 1980: 113, Fig. 7.)

shown regeneration of sclerites, tracheae, appendages, and segments to occur as if their cells were governed by action of an anteroposterior (A-P; body) or proximodistal (P-D; tracheae, appendages) morphogen gradient within the epidermis (Lawrence, 1992: chapter 6). The polarity in hair pattern seems to mark the direction of slope of a linear morphogen gradient that extends either posteriorly (body) or distally (tracheae, appendages) and is repeated in each segment. It seems to be generated at the anterior (proximal) margin of a segment and to be removed at its posterior (distal) margin, the former functioning as a source for the morphogen and the latter as a sink (Fig. 11.2). The factors in this gradient behave exactly as do individual sand grains, in a sand gradient and its slope can be altered predictably by damaging, removing, transplanting, rotating, and replacing pieces of abdominal, appendicular, or tracheal integument (Lawrence, 1992).

As shown in Fig. 11.2, regulation toward normality following experimental interference can occur by either regeneration or duplication, the cells exposed at the free epithelial surface resulting from removal of part of a segment regenerating structures corresponding only to lower levels in the gradient. Thus, the part removed can only duplicate itself.

Recently, Struhl and others (1997) demonstrated that this model of gradient distribution may be too simple, at least for establishing A-P polarity in abdominal segments of *D. melanogaster* embryos (Fig. 11.3). As discussed in 7.2.1.2, the segment-polarity gene *engrailed*

Fig. 11.3. The gradients of Hedgehog (Hh) secretion within two adjacent segments of a *D. melanogaster* embryo. See text for details. A, anterior compartment; P, posterior compartment. (Constructed from data in Struhl et al., 1997.)

(*en*) is expressed in the posterior (P) but not in the anterior (A) compartment of each segment. Hedgehog (Hh) protein is secreted by posterior cells, in response to *en* expression, and spreads not only anteriorly into the A compartment of its own segment but also posteriorly partway into the A compartment of the segment behind to form opposing gradients that then organize cell pattern and polarity within the A compartment.

Anterior and posterior cells within the A compartment respond in opposite ways to the presence of Hh by expressing different combinations of downstream genes and forming different cell types. They also form cuticular denticle belts that in anterior cells, point down the Hh gradient and in posterior cells, point up the Hh gradient, so that all structures in the A compartment point posteriorly (Fig. 11.3). Also, when Stuhl and colleagues (1997) expressed *hh* ectopically in the middle of each A compartment, its cells were transformed into an ectopic P compartment that secreted Hh and reorganized cellular pattern and polarity around it in exactly the same way.

Lawrence and coworkers (1999) have since evaluated this model experimentally in the developing adult abdomen, with results that support and expand it. As they indicate, many of these results were unexpected based on previous transplant work (reviewed in Lawrence, 1992: chapter 6) and suggest that our understanding of the role of gradients and compartments in patterning insect segments is still incomplete.

11.2.2 THE POLAR COORDINATE HYPOTHESIS

Results of regeneration experiments on appendages of a variety of animals from insect to mouse can be explained by use of an ingenious model formulated by French and associates (1976) (see also Bryant et al., 1977). The polar coordinate hypothesis is two-dimensional because most regeneration, at least in insects, involves changes in a simple sheet of epidermal cells. According to this model, each cell within an appendage segment has its position specified with respect to both its circumferential and radial (proximodistal) location within a set of polar coordinates. In each leg segment, the outer circle (A in Fig. 11.4) represents its base or proximal margin, while the center (E) represents its tip. In an imaginal disc, circle A represents the disc margin and E, the apex of the appendage that forms from it.

The two coordinates have the following characteristics (Fig. 11.4):

- The radial component is specified by a morphogen gradient and is identified alphabetically, A–E. The circular component is not in a gradient but instead forms a complete circle of meridians, 1–12.
- When a piece of integument is removed from a segment, the cut edges of both the segment and the piece heal together, thus bringing into contact cells with different radial and/or circular positional values.
- When cells with different positional values come together, they respond by proliferating. The resulting cells take up proximodistal (Fig. 11.5A) or circular (Fig. 11.5B) values, intermediate to those brought together, by intercalary regeneration using the shorter of the two routes according to the shortest intercalation rule. Thus, a fragment containing less than half the circumferential values will always intercalate a mirror-image duplicate of itself, while one containing more than half these values will always regenerate the missing fragment.
- For distal structures to be regenerated, a complete circle of proximal positional values is needed according to the complete circle rule (Fig. 11.5B: d).

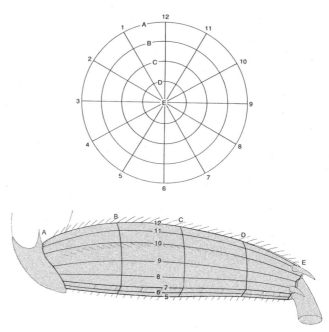

Fig. 11.4. The polar coordinate system represented as a disk (top). Twelve circumferential meridians are identified numerically and five radial values, alphabetically. As indicated in the cockroach (*Periplaneta americana*) femur (bottom), the model represents proximal-to-distal positional values in its periphery-to-center divisions and dorsoventral and anteroposterior values in its circumferential divisions. (Reprinted with permission from P. J. Bryant, S. V. Bryant, and V. French. Biological regeneration and pattern formation. *Sci. Am.* 237(1): 66–81. © 1977 Scientific American.)

When this model was published, it stimulated much additional experimentation to test its validity, and it was soon found that certain results could not be explained by the model as formulated above. Among these was the observation that some distal outgrowth could occur from proximal sites consisting of only partial circles (i.e., they violated the complete circle rule) (Bryant et al., 1981). Bryant and colleagues suggested this observation to be expected of a mechanism where new cells at the growing, distal tip are specified circumferentially by intercalation between cells at the most distal preexisting level, and where each new cell is forced to adopt a more distal positional value if its assigned value would otherwise be identical to that of a preexisting adjacent cell.

11.2.3 THE REACTION-DIFFUSION HYPOTHESIS

According Meinhardt's hypothesis (1982), positional information within an appendage or imaginal disc is provided in the form of morphogen gradients emanating from compartment boundaries (Fig. 11.6). He proposed that spontaneously organized patterning might be produced by coupling a short-range autocatalytic process to a long-range inhibitory process caused by diffusion of the molecules involved. Such a reaction-diffusion system could generate borders between neighboring cells having different states of determination.

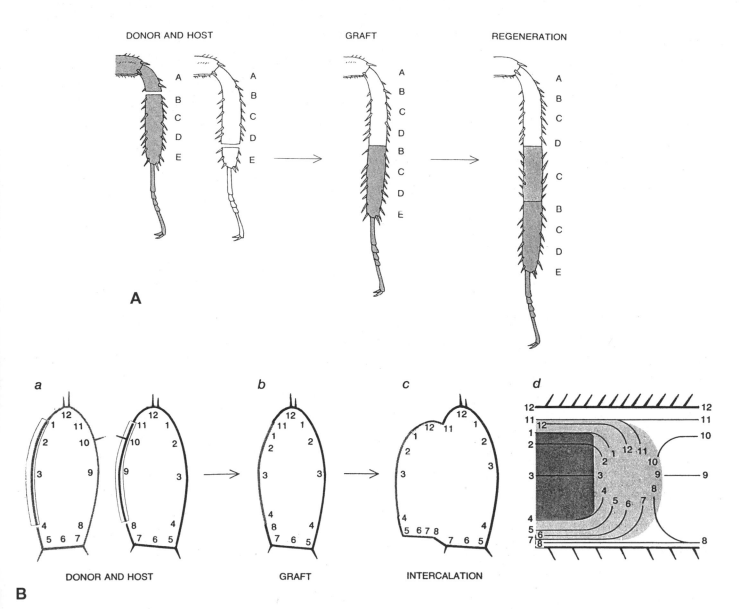

Fig. 11.5. A. Results of transplant experiment on nymphal legs of the cockroach *Periplaneta americana*, demonstrating the existence of proximodistal values. When a proximal-level portion of a tibia from a donor is grafted onto a distal-level portion of a tibia on a host, the confrontation induces intercalary regeneration and generates a reversed, intermediate organization level (C between D and B) where bristles point proximally rather than distally. B. Transplant experiment demonstrating circumferential values. A donor graft to a host (a) replaces the circumferential values between 8 and 11 with values 1 through 4 (b). Intercalary growth (c) on both sides of the graft results in formation of a complete circumference of values at the ends of the graft. A lateral view (d) shows the distal circumference diagrammatically. An intercalated circumference is equivalent to a circumference formed by amputation and produces whatever leg structure would normally exist distal to the circumference. (Reprinted with permission from P. J. Bryant, S. V. Bryant, and V. French. Biological regeneration and pattern formation. *Sci. Am.* 237(1): 66–81. © 1977 Scientific American.)

Meinhardt's model combines positional information (acquired by cellular interpretation of thresholds in a morphogen gradient; Figs. 7.7, 7.17B) and compartmentalization (based on differential expression of selector genes; Fig. 7.17A, B), with cooperation between gene expression domains at their borders specifying the source of specific morphogen gradients. Formation of a boundary between two compartments, say, A and P, would generate a morphogen that would diffuse in oppo-site directions from it, in this example, into compartments A and P. The distance of particular cells within each compartment from the border would determine their positional value and influence the downstream genes they would later express. A similar process could generate the dorsoventral (D-V) boundary and its positional coordinates.

To generate positional information in the proximodistal axis, Meinhardt proposed that a third interac-

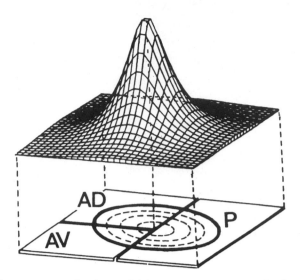

Fig. 11.6. Meinhardt's model for generating positional information in the proximodistal axis of an appendage primordium. A leg disc of *D. melanogaster* is divided into three compartments—anterodorsal (AD), anteroventral (AV), and posterior (P)—which cooperate to produce the morphogen. A high production rate is possible only where cells of all three compartments are juxtaposed at the intersection of the three compartment boundaries. The local concentration of the resulting, cone-shaped morphogen gradient provides the positional information specifying that a particular segment be formed, for example, the concentric pattern of the leg disc (see Fig. 10.34A). A certain minimum concentration of the morphogen is required for a cell to be specified as part of the leg disc. (Reprinted with permission from H. Meinhardt. *Models of biological pattern formation*, p. 80, Fig. 92. © 1982 Academic Press, Ltd.)

tion, between A-P and D-V morphogens, could create a local singularity at the point of intersection of these compartment boundaries. The distalmost structures of an appendage would be produced in the center of the disc at the highest concentration in the morphogen gradients, and the most proximal structures, at the lowest concentration at its margins (Fig. 11.6). This model predicts that distal transformation will occur in a disc if a new (ectopic) intersection of A-P and D-V boundaries is created experimentally, a prediction fully supported by results of experiments on gene expression in *D. melanogaster* wing and leg discs (11.3). Thus, according to this model, distal transformation requires only the intersection of a minimum of three compartments, not a complete circumference as in the polar coordinate hypothesis.

11.2.4 THE CARTESIAN COORDINATE HYPOTHESIS
Since imaginal discs are actually three- rather than two-dimensional, Russell (1983) suggested that positional information would more likely be specified by means of a Cartesian coordinate system with three orthogonal axes: D-V, A-P, and proximodistal. Positional information along each axis would be established by the con-

centration gradient of a different morphogen. Every cell on the disc surface would thus have a unique positional value (x, y, z) determined by its position in relation to the three independent morphogen gradients, enabling the disc surface to be envisioned as a sphere enclosing an origin with the positional value 0, 0, 0.

The advantage of this model over both the polar coordinate and reaction-diffusion models is that a single rule can account for both regeneration and duplication of disc fragments: values in each coordinate are assigned by interpolation whenever normally non-adjacent cells are juxtaposed. By converting such values to those of the polar coordinate model, Russell showed the application of this rule to automatically generate the shortest intercalation and complete circle values of that model. Thus, an apparently complex cellular behavior can be explained by a physically simple process of diffusive smoothing in chemical gradients.

Like the polar coordinate and reaction-diffusion hypotheses, this model is able to predict the outcome of all grafting and regeneration/duplication experiments on appendage primordia: a disc fragment containing the origin (0, 0, 0) will regenerate, while one that does not will duplicate.

This model can likewise be related to compartmentalization (7.1.6). Recall that progressive expression of selector genes was thought to divide the wing disc orthogonally into four compartments: anterodorsal, anteroventral, posterodorsal, and posteroventral (Fig. 7.16), each characterized by cells having a different state of determination. The boundaries between these compartments are geometrically related to a Cartesian coordinate system and might be established along threshold values for each of the three morphogens. Both this hypothesis and that of Meinhardt (1982) combine ideas on morphogen gradients, compartmentalization, and positional information to explain duplication, and regeneration, and both are fully supported by results of more recent molecular genetic studies.

11.3 Genetic and Molecular Basis of Leg and Wing Formation

Since 1980, much has been learned about the genetic and molecular bases for the origin and differentiation of imaginal discs in *D. melanogaster* and for the hypotheses just summarized. To date, all discs investigated have been shown to use the same three secreted signaling molecules—Hedgehog (Hh), Wingless (Wg) (both the products of segment-polarity genes: 7.2.1.2), and Decapentaplegic (Dpp; product of the dominant, zygotic D-V gene: 7.2.1.1.1)—to specify their characteristics, the interrelationships between them having been selected to accommodate the particular characteristics of each disc.

Specification of disc regions is now known to occur at two levels: (1) expression of regulatory pattern-control genes sets up the expression patterns of the signaling molecules and the rules governing their interrelationships, and (2) expression of these same or of additional regulatory genes confers upon the epidermis the ability to respond to these signals in a way specific for each disc.

11.3.1 ANTIGENIC DIFFERENCES AROUND THE CIRCUMFERENCE AND BETWEEN THE BASE AND APEX OF LEG SEGMENTS

Bullière and coworkers (1982) provided the first direct evidence of molecular differences in cells around the circumference and between the base and apex of the segments of an insect appendage. They immunized mice with membrane preparations of epidermal cells taken from the internal and external face of the femur and the apex and base of the tibia of the metathoracic legs of cockroach (*Blaberus craniifer*) nymphs. By indirect immunofluorescence, they found that the labeled antibody probes bound preferentially to the epidermal membranes from which their antigens originated, demonstrating that such membranes from different parts of each leg segment do, indeed, differ in their antigenic properties both around a segment and along its proximodistal axis, exactly as hypothesized in the polar coordinate model (11.2.2).

11.3.2 GENETIC SPECIFICATION OF CIRCUMFERENTIAL AND RADIAL (PROXIMODISTAL) AXES IN LEG AND WING DISCS

Results of experiments on the leg and wing discs of *D. melanogaster* larvae have begun to provide a molecular-genetic explanation for appendage development and for the polar coordinate, reaction-diffusion, and Cartesian coordinate models summarized earlier.

11.3.2.1 LEG DISCS

Many of the selector genes specifying segmental pattern in *D. melanogaster* embryos (7.2.1.2, 7.2.1.3) are now known to do the same during appendage specification prior to and during metamorphosis (Bryant, 1993; Cohen, 1993; Couso et al., 1993; Fristrom and Fristrom, 1993).

As discussed in 7.2.1 and 8.3.6, thoracic imaginal discs arise at the parasegmental boundaries and are subdivided into compartments owing to localized expression of transcription factors encoded by the dorsoventral, segmentation, and homeotic genes. Interactions between cells within each compartment then establish organizing centers for generating spatial pattern and for promoting cellular proliferation and outgrowth of developing appendages (Basler and Struhl, 1994; Fernando et al., 1994; Held et al., 1994; Gerhart and Kirschner, 1997).

In each leg disc, cells in the P compartment express the *en*, *invected* (*inv*), and *hh* genes (Fig. 11.7A: left) (Gerhart and Kirschner, 1997: 564). The secreted Hh protein is translated and spreads forward into the anterodorsal (A-D) and anteroventral (A-V) compartments of the disc. Cells receiving Hh via the receptor Patched in the A compartment adjacent to the A-P com-

partment boundary respond by secreting the Dpp protein in the A-D compartment and the Wg protein in the A-V compartment (Fig. 11.7A: center). Both proteins diffuse fore and aft into both A and P compartments and establish gradients in mirror-image symmetry, with their high point in the secreting cells (Neumann and Cohen, 1997b). The distal organizer for the leg arises in the center of the disc where the A-P and D-V boundaries intersect and where both Dpp and Wg proteins are secreted (Fig. 11.7A: right). In these cells, activation of the regulatory genes *Distal-less* (*Dll*; expressed first, distal to the presumptive coxa, but later only in the tibia and tarsus [Panganiban, 2000]) and *aristaless* (*al*; expressed at the presumptive apex) establishes the proximodistal axis of the leg. Note the close similarity of these events to that postulated by Meinhardt (1982) in his reaction-diffusion model (Fig. 11.6). Inhibitory and stimulatory interactions between the products of these and additional genes are illustrated in Fig. 11.7B.

Following its specification in the embryo, each leg disc is gradually divided into proximal, intermediate, and distal domains in the third instar by expression of the proteins Homothorax (Hth), Escargot (Esg), Dachshund (Dac), and Distal-less (Dll) (Fig. 11.8A) (Abu-Shaar and Mann, 1998; Wu and Cohen, 1999; Goto and Hayashi, 1999). Hth is required for nuclear localization of Extradenticle (Exd), a cofactor facilitating expression of several *HOM* genes (7.2.1.3). Wg and Dpp act together to define the extent of all three domains by inducing expression of Dac and Dll distally (which, at first, mutually inhibit each other) and by restricting expression of Hth and Esg to the proximal domain (Fig. 11.8A). In addition, Hth inhibits Dpp and Wg activity and so restricts expression of Dac and Dll and other downstream genes (e.g., *H15* and *optomotor blind* [*omb*]) to non-*hth*-expressing cells. Once these domains are established, proximal, intermediate, and distal cells sort themselves out such that cells forced to express Hth are unable to mix with distal cells and distal cells expressing Dll are unable to mix with proximal cells (cells in a narrow band between each pair of domains later express both genes; Fig. 11.8A). These antagonistic relationships help to convert the Dpp and Wg activity gradients into discrete domains of gene expression along the proximodistal axis of the legs (Fig. 11.8A).

After proximal and distal domains have been established in the leg disc, the Notch (N) signaling pathway (8.3.1.4) plays a fundamental role in leg segmentation (Fig. 11.8B) (Bishop et al., 1999; Rauskolb and Irvine, 1999; Casares and Mann, 2001). Coincident expression of the ligands Serrate (Ser) and Delta (Dl) in a ring at the apex of each leg segment activates *Enhancer-of-split* (*E(spl)*) expression in cells of the proximal end of the next, more distal segment by way of the N receptor to induce formation of a joint, while the proteins Fringe (Fng) and Dishevelled (Dsh) inhibit expression of Ser and Dl in proximal cells in each segment (Fig. 11.8B).

Specification of the tarsal and pretarsal segments is influenced by activation of the genes *BarH1* and *BarH2*

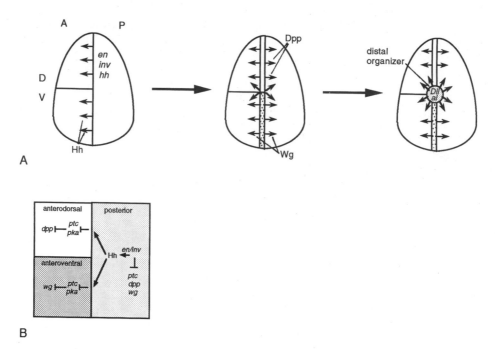

Fig. 11.7. Intercellular signaling and proximodistal organization of the leg disc in third instar larvae of *D. melanogaster*. A. The leg disc is divided into anterodorsal (A-D), anteroventral (A-V), and posterior (P) compartments (left). The genes *engrailed* (*en*), *invected* (*inv*), and *hedgehog* (*hh*) are expressed in cells of the P compartment, and Hh protein is translated and diffuses forward into the A-D and A-V compartments. Cells receiving the Hh signal immediately in front of the anterior-posterior boundary respond by expressing the *decapentaplegic* (*dpp*) or *wingless* (*wg*) genes and secreting Dpp (A-D) or Wg (A-V) protein, respectively, in the directions indicated (center). The distal organizer (right) forms at that site where Dpp and Wg proteins diffuse together at the center of the disc. *Distal-less* (*Dll*), *aristaless* (*al*), and other downstream genes are expressed within the organizer and establish the proximodistal axis of the leg, which then grows out from the leg disc. B. Intercellular circuitry with Hh signaling the derepression of *dpp* in nearby cells of the A-D compartment and of *wg* in those of the A-V compartment. (Adapted with permission from J. Gerhart, and M. Kirschner. *Cells, embryos, and evolution. Toward a cellular and developmental understanding of phenotypic variation and evolutionary adaptability*, p. 564, Fig. 10-28. © 1997 Blackwell Science, Ltd.)

beginning early in the third instar (Kojima et al., 2000). Juxtaposition of cells along the proximodistal axis differentially expressing these genes later in this instar generates the segmental boundaries between distal tarsal segments and influences the function of other genes specifying tarsomere identity.

Fig. 11.9 (Bryant, 1993) illustrates the expression pattern of additional genes functioning to specify segments and bristle pattern in *D. melanogaster* leg discs. Notice that *achaete* and *hairy* have a circumferential expression pattern in the disc; the others have radial patterns, exactly as postulated by French and associates (1976) in their polar coordinate hypothesis (Fig. 11.4).

Sequential expression of the genes just summarized was revealed by

- following, throughout larval and pupal development, the distribution of tagged cDNA probes to their transcripts by in situ hybridization and of labeled antibodies to their proteins;
- by use of beta-galactosidase (β-GAL) reporter gene systems (see Fig. 7.34); and
- by analyzing disc phenotypes in larvae in which one or more of the genes were mutant or were expressed in the wrong compartment.

Collectively, they and other genes constitute the principal coordinating system for development of all tubular appendages in *D. melanogaster*, including antennae, mouthparts, genitalia, and analia. They begin to be expressed in the extended germ band (polypod, phylotypic) stage in the embryo, and provide the basis for postphylotypic processes contributing to the evolution of appendage diversity within the Hexapoda (13.3.8). Orthologs of many of these genes are similarly expressed during leg development in embryos and larvae of the grasshopper *Schistocerca gregaria* (Jockusch et al., 2000) and in embryos of representative Crustacea and Chelicerata (Abzhanov and Kaufman, 2000b).

11.3.2.2 WING DISCS
Most regulatory genes and signaling proteins involved in specifying development of the leg discs do the same for wing discs in *D. melanogaster*, the differences resulting from the wing being an extended, flattened vane rather than a blind-ended tube (Fig. 11.10) (Williams et al., 1993, 1994; Shubin et al., 1997; Gerhart and Kirschner, 1997).

Proximodistal outgrowth of the wing is organized by cells along the D-V margin of the wing disc. The Wg and

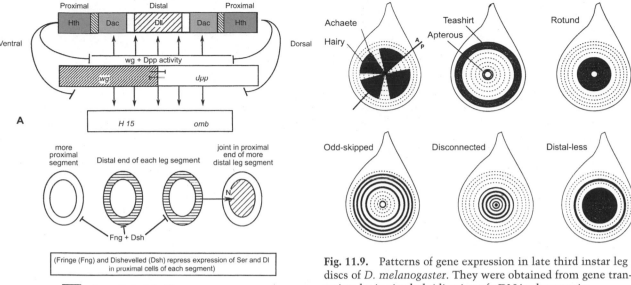

Fig. 11.8. Formation of multiple antagonistic domains along the proximodistal axis of a leg disc during larval development in *D. melanogaster*. A. Diagram showing how expression of *wingless* (*wg*) and *decapentaplegic* (*dpp*) in cells anterior to the anterior-posterior compartment boundary (compare with Fig. 10.34B) induce expression of Distal-less (Dll) and Dachshund (Dac) to specify distal domains but are inhibited by Homothorax (Hth) proximally. *omb, optomotor blind*. B. Diagram of transverse sections of developing leg segments. Coincident expression of the ligands Serrate and Delta in a distal ring in each leg segment activate *Enhancer-of-split* expression in epidermis of the proximal end of the next, more distal segment by way of the Notch (N) receptor to induce formation of a joint. (Fringe and Dishevelled inhibit expression of Ser and Dl in proximal cells of each segment.) (A adapted with permission from M. Abu-Shaar and R. S. Mann. Generation of multiple antagonistic domains along the proximodistal axis during *Drosophila* leg development. *Development* 125: 3821–3830. © 1998 Company of Biologists, Ltd.) (B adapted with permission from S. A. Bishop, T. Klein, A. Martinez Arias, and J. P. Couso. Composite signaling from *Serrate* and *Delta* establishes leg segments in *Drosophila* through Notch. *Development* 126: 2993–3003. © 1999 Company of Biologists, Ltd.)

Vestigial (Vg) proteins are expressed first in the wing disc and are crucial for distinguishing between notum/pleuron and wing blade and for compartmentalizing dorsal (D) and ventral (V) wing surfaces (Fig. 10.35A). The *apterous* (*ap*) gene is expressed only in the D compartment (Fig. 11.10A: left). In response to receipt of Apterous (Ap) protein, cells of this compartment secrete Ser protein, which induces local expression of the *wg* gene in a row of cells at the D-V boundary by way of the gene product interactions summarized in the box in Fig. 11.10A. When Wg protein is translated, it diffuses proximally into both D and V compartments in mirror-image gradients, with their high point in the

Fig. 11.9. Patterns of gene expression in late third instar leg discs of *D. melanogaster*. They were obtained from gene transcripts by in situ hybridization of cDNAs, by protein immunolocalization, or by staining for beta-galactosidase produced by a reporter gene. The patterns for *achaete* and *hairy* are from the pupal leg. A/P, anterior-posterior compartment boundary. (Reprinted with permission from P. J. Bryant. The polar coordinate model goes molecular. *Science* 259: 471–472. Copyright 1993 American Association for the Advancement of Science.)

Wg-secreting cells; this becomes the wing margin (Fig. 11.10A: center left) (Neumann and Cohen, 1997a, 1997b). Wg protein acts directly and at long range to define expression of *Dll* and *vestigial* (*vg*) in a concentration-dependent manner that is required for wing blade outgrowth (Neumann and Cohen, 1997a, 1997b).

Fig. 11.10B summarizes the interaction between the A and P compartments. As in the leg disc (Fig. 11.7A: left), posterior cells express the *en* and *inv* genes and hence the *hh* gene (Shubin et al., 1997). Hh protein is secreted into the A compartment, where it induces cells just in front of the A-P boundary to secrete Dpp protein. This diffuses fore and aft into both the A and P compartments in mirror-image gradients, with their high point in the Dpp-secreting cells, and influences expression of various downstream genes such as *Spalt* (*Sal*) and *optomotor blind* (*omb*) (Fig. 11.10B: center left) (Neumann and Cohen, 1997a, 1997b; Gritzan et al., 1999). Hh also specifies the positions of most sense organ precursor cells (SOPs) in the notum and some in the wing veins (Mueller et al., 1997). Close to the A-P boundary, SOPs are also specified by Hh, and late-signaling Hh, after inducing *dpp* expression, is responsible for specifying vein 3 and the scutellum of the notum and determining the distance between veins 3 and 4. The box in Fig. 11.10B shows the interaction between various genes in the two compartments based on the Hh-mediated inhibition of Patched/protein kinase A in the A compartment, which in the absence of Hh inhibits *dpp* gene expression.

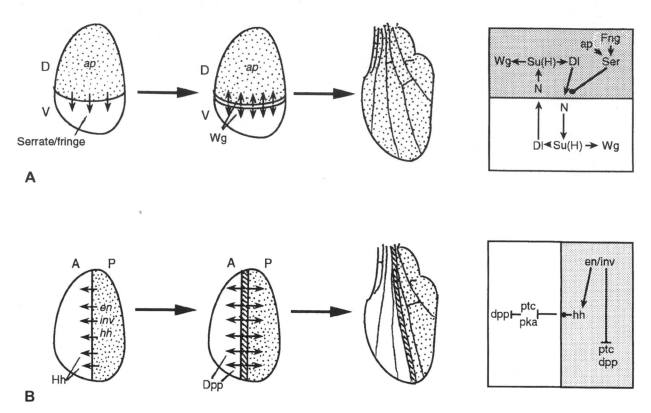

Fig. 11.10. Intercellular signaling and pattern specification in third instar wing discs of *D. melanogaster*. A. Interactions in the dorsoventral (D-V) axis. The gene *apterous* (*ap*) is expressed in the dorsal compartment (D). Dorsal cells secrete the Serrate (Ser) protein (left, short arrows), leading to local expression of *wingless* (*wg*) in cells of the D-V compartment boundary and secretion of Wg protein, which diffuses in both directions (middle, short arrows) to create two gradients in mirror-image symmetry, with their high point at the wing margin. The fully developed wing is shown on the right. The box on the right summarizes the intercellular circuitry of dorsal and ventral cells, based on the Notch (N)-Delta (Dl) network used in many other tissues. Ser activates the N receptor of a ventral cell, thereby stimulating *wg* expression via the Suppressor of hairy (Su [H]) transcription factor. B. Interactions in the anteroposterior (A-P) axis. Cells in the posterior (P) compartment express the *engrailed* (*en*) and *invected* (*in*) genes, and hence the *hedgehog* (*hh*) gene. The Hedgehog (Hh) protein is secreted to cells just in front of the A-P compartment boundary (A, left). These cells respond by expressing the *decapentaplegic* (*dpp*) gene and secreting Decapenta-plegic (Dpp) protein, which diffuses fore and aft to create two gradients in mirror-image symmetry, with their high point at the A-P boundary (center, arrows). The fully developed wing is shown on the right. The box on the right shows the intercellular circuitry based on Hh-mediated inhibition of the Patched (Ptc)/protein kinase A (PKA) inhibition of *dpp* expression. (Adapted with permission from J. Gerhart and M. Kirschner. *Cells, embryos, and evolution. Toward a cellular and developmental under-standing of phenotypic variation and evolutionary adaptability*, p. 556, Fig. 10-29. © 1997 Blackwell Science, Ltd.)

Expression of *wg* is also required for development of the wing hinge at the base of the wing, including the tegula and axillary sclerites, but does not activate *vg*, as it does in the wing blade (Casares and Mann, 2000; Azpiazu and Morata, 2000). Instead, Wg activates expression of the *hth* gene in the hinge, which is required for its development. In addition, Hth protein limits where wing blade development can initiate along the D-V boundary of the disc. The *teashirt* (*tsh*) gene is coexpressed with *hth* throughout wing disc development and collaborates with *hth* to repress *vg* expression and to block wing blade development. Note that proximodistal expression patterns of these genes are similar to those occurring in the leg discs (Figs. 11.7–11.9). Presence of *Iroquois* Complex (*Iro*-C) homeodomain proteins allows cells in the most proximal dorsal part of each wing disc

to develop as mesothoracic notum; if *Iro*-C is absent, these cells assume a more distal fate and develop as a wing hinge (Diaz del Corral et al., 1999).

Orthologs of many of these genes are involved in specifying limb outgrowth and differentiation in vertebrate limb buds, as has been summarized by others (Gaunt, 1997; Gerhart and Kirschner, 1997; Shubin et al., 1997; Johnson and Tabin, 1997; Wolpert, 1998; Carroll et al., 2001), suggesting such limbs to be homologous. However, after establishing an explicit, hierarchical set of criteria for assessing the role of convergence and descent in explaining these genetic regulatory similarities (their number, type, and phylogenetic distribution), Tabin and coauthors (1999) concluded that there has been no continuity of structure from which insect and vertebrate appendages could be derived and that therefore, they are not homologous. There is, however, good evidence for continuity in the genetic information required for constructing body wall outgrowths in representatives of several phyla, which must date back to a Precambrian common ancestor of Protostomes and Deuterostomes.

11.3.2.2.1 WHY DO WINGS NORMALLY GROW TO THE CORRECT SIZE?

A normal winged adult insect invariably has wings and legs that are proportional in size to its body, and the question arises as to how this proportionality is achieved. Edgar (in Day and Lawrence, 2000) manipulated genes in *Drosophila* larvae that affect the mitotic rate of cells in the wing discs, and found that if they divided too fast, the wing had too many cells and, if too slow, it had too few cells; the resulting wings, however, were of normal size regardless of cell number. Thus, if cell number is altered, the fly responds with a change in cell size. Similarly, if cell size is altered, the fly responds with a change in cell number. These observations suggested that this fly can measure the length and volume of its wing and, presumably, of its other appendages, and can determine when they should stop growing regardless of cell number or size. As summarized earlier and in Fig. 11.10, wing cell type is determined by distal-proximal concentration gradients of Wg and Dpp protein produced, respectively, at the D-V and just in front of the A-P compartment boundaries. In mutant flies having more than normal numbers of Wg- and Dpp-producing cells, extra wing tip patches result and induce neighboring cells to grow out as an extra wing from the wing margin, which again grows to normal adult size (Day and Lawrence, 2000).

Before the wing blade portion of a wing disc starts to grow out (Fig. 10.35C), it has steep gradients of Dpp and Wg protein, with the highest concentrations at the intersection of the D-V and A-P boundaries and the lowest concentrations at its edges (as in Fig. 11.6). However, as the wing blade grows out, both gradients are stretched, so that the concentrations of Dpp and Wg detected by each wing cell at each position on the length of the wing would drop. Thus, wing cells seem to measure the local slope or steepness of the Dpp and Wg gradients as an indication of their distance from the wing tip, to stop growing when the slope is sufficiently gradual, and to respond in the same way regardless of size or number of cells (Day and Lawrence, 2000).

11.3.2.2.2 SPECIFICATION OF WING VEINS

Each adult wing in *D. melanogaster* has five principal longitudinal veins, I–V, and two cross veins through which neurons and trachea pass and through which hemolymph circulates (Fig. 7.13: middle). According to the Comstock-Needham system of wing vein nomenclature, these veins are, respectively, R_1, R_{2+3}, R_{4+5}, M_1, and CuA_1 (C, Sc [0], $A_1 + CuA_2$ [VI], and A_2 are small and have not been analyzed). The position of each future vein in the wing disc is specified in the third instar and is first

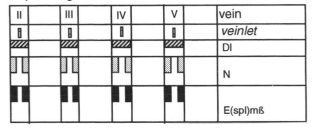

Fig. 11.11. Expression pattern of *veinlet*, Delta (Dl), Notch (N), and Enhancer-of-split (E(spl)mβ) in the wing disc and pupal wing of *D. melanogaster*. These genes and their products specify the pattern of longitudinal wing veins II–V (i.e., R_{2+3}, R_{4+5}, M_1, CuA_1) and interveins and also the boundary cells between these two domains. (Adapted with permission from J. F. de Celis, S. Bray, and A. García-Bellido. Notch signaling regulates *veinlet* expression and establishes boundaries between veins and interveins in the *Drosophila* wing. *Development* 124: 1919–1928. © 1997 Company of Biologists, Ltd.)

evident 3–4h after puparium formation as spaces between the dorsal and ventral wing epithelia, with veins II and V in the dorsal epithelium and III and IV in the ventral.

Longitudinal vein width and spacing is influenced by activity of the Delta/Notch signaling pathway (de Celis et al., 1997; Gerhart and Kirschner, 1997: 273–285; P. Simpson et al., 1999), while *rhomboid* (rho) mediates vein formation by way of the epidermal growth factor receptor Spitz and its ligand, Torpedo (Stark et al., 1999). Expression of the Notch-ligand Delta (Dl) is restricted to developing veins and, at first, coincides with places where Notch (N) transcription is lower (Fig. 11.11: wing disc). This asymmetrical distribution of ligand and receptor leads to activation of Notch on both sides of each vein within a territory of Delta-expressing cells (Fig. 11.11: pupal wing), which leads to establishment of boundary cells that separate the veins from adjacent interveins. In the interveins, *Enhancer-of-split* (E(spl)mβ) is activated and transcription of the vein-promoting gene *veinlet* is repressed (Fig. 11.11: wing disc), thus restricting vein differentiation. The expression of many other genes is also required (e.g., Blair, 1999; de Celis and Barrio, 2000).

That differences in wing shape and venation used in keys to identify adult flies in other families (e.g., Borror

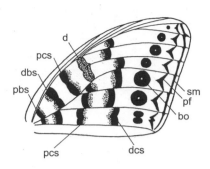
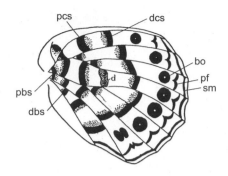

Fig. 11.12. Ground plan color pattern of the wing in nymphalid butterflies (Lepidoptera). The pattern elements from proximal to distal on the wing are as follows: pbs and dbs, proximal and distal bands of basal symmetry system; pcs and dcs, proximal and distal bands of central symmetry system; d, discal spot; bo, border ocelli; pf, parafocal elements; sm, submarginal bands. (Reprinted with permission from H. F. Nijhout. A comprehensive model for colour pattern formation in butterflies. *Proc. R. Soc. Lond. B Biol. Sci.* 239: 81–113. © 1990, The Royal Society.)

et al., 1989; McAlpine et al., 1981, 1987) often resemble the phenotypes of wing mutations induced in *D. melanogaster*, the phyletic phenocopy paradigm, suggests that changes in expression of orthologs of these genes could have contributed to the origin of such differences during dipteran diversification (Stark et al., 1999). However, Stark and coauthors (1999) emphasize (p. 122) that equating gene expression pattern with morphology is wrong since it fails to include consideration of phylogeny.

11.3.2.2.3 WING COLOR PATTERN IN LEPIDOPTERA

The exquisite color patterns on the wings of butterflies and moths consist of a species-specific mosaic of colored scales on both wing surfaces, each scale being the product of a single hair-forming cell (Nijhout, 1991; Ghiradella, 1998). The color of each scale depends on its pigment content, its three-dimensional cuticular structure, and the way it reflects or refracts light.

Fig. 11.12 is the ground plan wing color pattern for nymphalid butterflies as proposed by Nijhout (1990). Its principal pattern elements consist, from base to apex, of the proximal and distal bands of the basal symmetry system, proximal and distal bands of the central symmetry system, discal spot, border ocelli, parafocal elements, and submarginal bands.

The development of pattern within each wing "cell" (an area of membrane delimited by veins and not to be confused with the eukaryote cell as discussed elsewhere in this book) is largely independent of that in other "cells." The overall pattern consists of a serial repetition of homologous pattern elements from "cell" to "cell," each varying independently from the others in position, size, and shape.

Since 1978, Nijhout and colleagues have investigated wing pattern development by manipulating the wing discs of caterpillars and the wing pads of pupae of the buckeye butterfly *Precis coenia* and other nymphalids and have compared in great detail the wing patterns of 2208 species in 330 genera of this family (Nijhout, 1990). Fig. 11.13 illustrates the principal variations in pattern

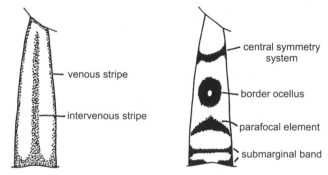

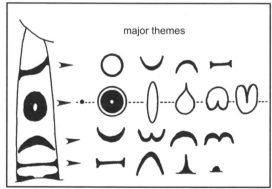

Fig. 11.13. Major pattern themes in the nymphalid ground plan wing color pattern. The top two panels show the extreme forms that a wing "cell" pattern can take: the intervenous stripe system (left) and the pattern elements system (right). Intervenous stripes always occur on the wing "cell" midline and vary only in length and width. The pattern element system exhibits the most diverse modifications. Each pattern element can vary independently from the others in position, size, shape, and color (though their relative positions stay the same). The bottom panel illustrates the principal shape themes encountered for each pattern element throughout the Nymphalidae. (Reprinted with permission from H. F. Nijhout. A comprehensive model for colour pattern formation in butterflies. *Proc. R. Soc. Lond. B Biol. Sci.* 239: 81–113. © 1990, The Royal Society.)

elements identified in this survey. In 1991, Nijhout published a beautiful book summarizing not only his findings but also those of previous investigators.

This wing pattern fulfills a number of design principals, with most pattern elements sharing the following characteristics:

- They are repeated from "cell" to "cell."
- They remain entirely within the boundaries of a single "cell" except for border ocelli.
- They are usually truncated or deflected near the veins.
- Most are bilaterally symmetrical, with their line of symmetry running down the midline of the "cell."
- All circular pattern elements have their origin on the "cell" midline.
- The form of most pattern elements is largely independent of the size and shape of the "cell" in which they are contained.
- The submarginal bands and parafocal elements run parallel to the wing margin.

Experimental Evidence

From his observations, Nijhout concluded that all wing patterns observed in these butterflies derive from the selective expression, suppression, displacement, or distortion of the pattern elements illustrated in Figs. 11.12 and 11.13. Results of his experiments suggest wing color pattern in lepidopterans to arise by a two-step process: (1) one that determines the distribution of discrete signaling sources or organizers for color pattern (begins in the penultimate larval instar), and (2) one by which these sources induce pattern differentiation in their surrounding epidermal cells (begins in last larval instar and ends early in the pharate adult). Point sources of an inductive signal serve as organizers for circular pattern elements such as the border ocelli and are always located in one, two, or three locations along the wing "cell" midline (Fig. 11.14). When a point source at the center of a presumptive ocellus is removed by cautery, the ocellus fails to develop, and when the source is transplanted to a different, more distal location in the wing membrane, it induces development of a new ocellus in host tissue surrounding its new location.

The wing margin serves as an organizer for pattern elements in its vicinity (Fig. 11.14). Parafocal elements and submarginal bands usually occur at a characteristic distance from the margin, their position being independent of that of other pattern elements in that "cell." In addition, these elements are reorganized parallel to the new margin after surgical removal of distal portions of a wing disc.

Wing veins serve as boundaries for the pattern of each "cell," and this accounts for the serial repetition of identical patterns in adjoining "cells" and for the fact that most wing patterns are sharply dislocated in adjacent "cells" (Fig. 11.12). Wing veins also act as organizers or line sources for the determination of pattern elements (Fig. 11.14), and they interact with other pattern elements (Fig. 11.15). The wing veins carry tracheae and sensory axons, and in the pupal and adult wing, they

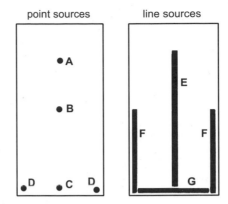

Fig. 11.14. Nijhout's toolbox of source and sink locations required for color pattern determination in butterflies. This diagram summarizes the locations where sources need to be placed to generate the major pattern themes seen in the border ocelli and parafocal elements of butterflies. By selecting sources indicated by one or, at most, two letters in this figure, and by additionally assuming that sources at the wing veins and distal margin (bottom edge of diagrams) could, alternatively, also be sinks, one can simulate diffusion gradients whose contours resemble one or more of the major pattern themes found in these elements in nature. (Reprinted with permission from H. F. Nijhout. A comprehensive model for colour pattern formation in butterflies. *Proc. R. Soc. Lond. B Biol. Sci.* 239: 81–113. © 1990, The Royal Society.)

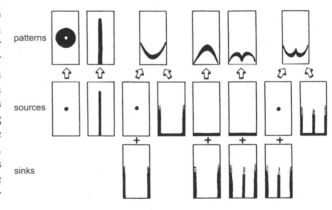

Fig. 11.15. Examples of simple source and sink distributions that generate concentration contours resembling some of the principal pattern themes in the parafocal elements and border ocelli shown in Fig. 11.13. (Reprinted with permission from H. F. Nijhout. A comprehensive model for colour pattern formation in butterflies. *Proc. R. Soc. Lond. B Biol. Sci.* 239: 81–113. © 1990, The Royal Society.)

carry hemolymph in a slow unidirectional flow. Hemolymph is the exclusive source of material input into the wing. The wing margin originates as a peripheral lacuna.

The Role of Diffusion

Results of Nijhout's experiments suggest species-specific color patterns to arise through cellular interactions within the wing epithelium by way of gap junctions between adjacent epidermal cells. For example, during border ocellus development, the pattern is determined by signaling activity of a small group of cells at its center, the focus. Determination of the outer margin of an eyespot in the butterfly *P. coenia* spreads out from this focus within the "cell" of a wing at about 0.27 mm ($\pm$30 cell diameters)/d at 29°C with a diffusion coefficient of $1.5 \times 10^{-8} \, \text{cm}^2\text{s}^{-1}$. This value is approximately that of a substance with a molecular mass of about 1000 Da diffusing through the gap junctions between adjacent wing epidermal cells. These values are consistent with a diffusional mechanism for cell-cell communication and with a reaction-diffusion model to account for wing pattern formation (see Meinhardt, 1982, for a detailed explanation).

A Model for Wing Pattern Formation

Based on his findings, Nijhout developed a model for wing pattern formation based on simple diffusion and threshold mechanisms. He proposed a pattern of source-sink distributions that could generate the diversity of patterns in the wing "cells" he observed in his survey (Fig. 11.13), tested the effectiveness of his resulting source-sink "toolbox" (Fig. 11.14) by computer simulation, and re-created many of the same color patterns observed in various species of Nymphalidae (Nijhout, 1990, 1991). These simulations resulted in a two-gradient model being proposed: a simple, additive relation between the two gradients suffices to generate most of the diversity of patterns observed in nymphalid wings (Fig. 11.15). He also discovered that the required positions of the sources and sinks of his toolbox (Fig. 11.14) emerge readily from Meinhardt's (1982) lateral inhibition model for reaction-diffusion. In its final form, Nijhout's model has two steps (Fig. 11.15): (1) a reaction-diffusion system that sets up a source-sink pattern followed by (2) a simple diffusion of a morphogen from these sources to form the patterns observed.

Genetics of Wing Color Pattern Formation

Sheppard and colleagues (1985) demonstrated about 40 genetic loci that affect wing color pattern in two species of *Heliconius* butterflies. These included genes that could

- switch the color of a given pattern element from red to yellow and from red to black;
- displace pattern elements proximally or distally along the wing "cell" midline;
- affect the size of single pattern elements; or
- influence the appearance or disappearance of discrete pattern elements without affecting the color or form of other elements of the color pattern.

Some of the genes and molecular processes underlying Nijhout's model were recently identified and found to be the same as those specifying wing pattern in *Drosophila* (Fig. 11.10) (Carroll et al., 1994; Nijhout, 1994a; Brakefield and French, 1995; French and Brakefield, 1995; Brakefield et al., 1996; Keys et al.,

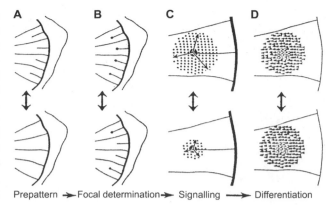

Prepattern → Focal determination → Signalling → Differentiation

Fig. 11.16. Eyespot formation and diversity are regulated progressively in four stages, as shown in two species (contrast top and bottom rows). A. In the larva, a prepattern of gene activity creates the potential focal pattern reflected by early *Distal-less* (*Dll*) expression, which occurs only in eyespot-bearing species. B. Foci are determined and *Dll* expression is stabilized only in specific wing subdivisions. The number of wing foci differs between species and between upper and lower wing surfaces. C. In the pupa, signaling from the focus (black circle) induces surrounding cells, with differently sized eyespots being controlled by the size of the focus. D. Induced cells later differentiate into scales of different colors depending on their distance from the focus and their position in the wing membrane. (Reprinted with permission from *Nature* (P. M. Brakefield, J. Gates, D. Keys, F. Kesbeke, P. J. Wijngaarden, A. Monteiro, V. French, and S. B. Carroll. Development, plasticity and evolution of butterfly eyespot patterns. 384: 236–242). © 1996 Macmillan Magazines Limited.)

1999; summarized in Nijhout, 1996; Carroll, 1997; Brakefield and French, 1999). Carroll and coworkers (1994) cloned *P. coenia* orthologs of the *Drosophila* genes *ap, inv, wg, scalloped* (*sd*), *dpp*, and *Dll*; produced labeled cDNA and antibody probes derived from them; and used them to examine gene expression in fifth instar wing discs of *P. coenia* caterpillars. They found the butterfly wing pattern to be organized by two spatial coordinate systems: one specifying positional information with respect to the whole wing field and practically the same as in *D. melanogaster* (Fig. 11.10), and the other, superimposed on the first, involving expression of several of the same genes and operating within each wing "cell" to elaborate the discrete pattern elements described by Nijhout (Figs. 11.12, 11.13). For example, the line sources generated by Nijhout's model (Fig. 11.14) are at least partly established by expression of Wg, Dpp, and Dll proteins beginning at the wing margin and diffusing inward almost exactly as predicted by Nijhout's model (Fig. 11.16).

The ontogeny of Dll protein distribution is particularly interesting, as this protein originates from the wing margin, diffuses inward along the midline of each "cell" (Fig. 11.16A), expands at the proximal tip of each midline (Fig. 11.16B), and eventually buds off as a proximal spot in 200–300 epidermal cells (Fig. 11.16C: black,

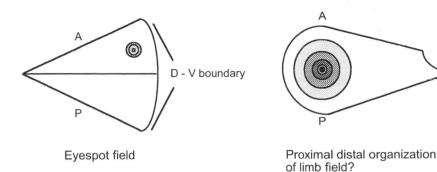

Eyespot field

Proximal distal organization
of limb field?

Fig. 11.17. Eyespot represented as concentric circles of pigment surrounding a focus (clear spot) (left). The expression of *Distal-less* (*Dll*) is within the eyespot. The circular organization of the leg disc of a third instar *D. melanogaster* larva (right) is represented as concentric rings of cells. Central cells, determined to form the distalmost structures of the leg, express *Dll*. Compare with Fig. 11.9. (Adapted with permission from S. B. Carroll, J. Gates, D. N. Keys, S. W. Paddock, G. E. F. Panganiban, J. E. Selegue, and J. W. Williams. Pattern formation and eyespot determination in butterfly wings. *Science* 265: 109–114. Copyright 1994 American Association for the Advancement of Science.)

central circle)—a developmental history identical to the model of ocellar focus origin generated in Nijhout's reaction-diffusion simulations (Figs. 11.14, 11.15). Only posterior eyespots are marked by Dll expression in both fore and hind wings (such spots are by far the most common in nymphalids) and appear to be generated by a process similar to, and perhaps evolved from, that governing proximodistal pattern formation in insect appendages—"a matter of teaching an old gene a new trick" (Fig. 11.17) (Carroll, 1997: 31).

In addition, Carroll and Brunetti (in Pennisi, 2000) identified genes that determine the size, pattern, and perhaps the color of eyespots after their specific location has been established. Again, they found some of these genes to be the same as those functioning earlier in development: *en*, *Spalt* [*Sal*], and *Dll*. In wing discs of *Bicyclus anynana*, they found all three genes to be expressed in the central white spot, *Sal* and *Dll* in the black ring around it but only *en* in the outer ring. However, in eyespots of *P. coenia*, all three genes are expressed in both the central spot and its surrounding ring, with expression of a fourth, still unidentified gene characterizing the outer ring. And the gene expression pattern was different again in the wing discs of two other nymphalid butterflies. Thus, the genes specifying eyespot characteristics seem to differ in closely related butterflies, a flexibility tolerated, according to Carroll, because the details of the body pattern have already been established. Such flexibility enables environmental factors such as temperature or predators to rapidly select for subtle changes in wing spot number and pattern both in members of different species and in members of different populations or generations of the same species.

Gene expression involved in eyespot development is now known to occur in four stages during postembryogenesis (Fig. 11.16) (Brakefield and French, 1995; French and Brakefield, 1995; Brakefield et al., 1996; Nijhout, 1996):

- stage A (prepattern in early fifth instar larva): Dll broadly expressed in wing.
- stage B (focal determination; late fifth instar larva): Dll expression restricted to eyespots (expression of the genes *cyclops* and *spotty* is required: *cyclops* mutants vary in the number and position of eyespots on the hind wing; *spotty* mutants develop extra eyespots in the fore wings).
- stage C (signaling; early pupa): Activation and spread of focal signal (*Bigeye* expression is required and may involve response of cells to receipt of focal signals).
- stage D (differentiation): Determination of eyespot color and differentiation of scales. These are not properties of the signal but rather are determined by the cells receiving the signal (foci transplanted to different distal locations on the wing epithelium induce eyespots of different color but initiate no response when transplanted proximally).

Events occurring in stages A and B control the number and position of eyespots, and those in C and D, their size, color, and polyphenism.

Finally, the role of selection in influencing the function of such regulatory genes in specifying wing pattern is vividly illustrated by species exhibiting different wing patterns at different times of year (summarized in Brakefield and French, 1999). The African butterfly *B. anynana* has two color patterns that differ according to time of year. Butterflies developing in the warm, wet season have large eyespots on the underside of their wings and those during the cool, dry season have tiny ones. When Brakefield and Reitsma (1995; cited in Carroll, 1997) released butterflies with large eyespots during the dry season, they were preferentially preyed upon by birds and lizards.

The proximate cause of small eyespots in dry-season forms is lower temperatures, which reduce the expression of *Dll*. When Brakefield and associates (1996) reared

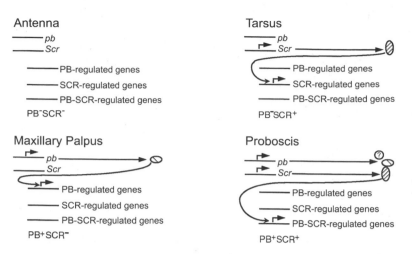

Fig. 11.18. Summary of the role of the homeotic genes *proboscipedia* (*pb*) and *Sex combs reduced* (*Scr*) in specifying adult antennal, tarsal, maxillary palpus, and proboscis in *D. melanogaster* (compare with Fig. 7.36: Hexapoda). (Adapted with permission from A. Percival-Smith, J. Weber, E. Gilfoyle, and P. Wilson. Genetic characterization of the role of the two HOX proteins: Proboscipedia and Sex Combs Reduced, in determination of adult antennal, tarsal, maxillary palp and proboscis identities in *Drosophila melanogaster*. *Development* 124: 5049–5062. © 1997 Company of Biologists, Ltd.)

two populations of *B. anynana* at a constant temperature and selected them for large or small eyespots, in less than 20 generations members of these lines lost the ability to develop alternate forms, indicating how rapidly selection can act on butterfly populations living under changing environmental conditions by affecting expression of regulatory genes.

11.4 Genetic Specification of Other External Imaginal Primordia during Development of *Drosophila melanogaster*

Knowledge of how genes influence the formation of adult external structures other than eyes, legs, and wings in *D. melanogaster* larvae is of recent vintage and, again, has been found to involve expression of many of the same genes: *en*, *inv*, *hh*, *wg*, *dpp*, *hth*, and *Dll*, along with the homeotic genes.

11.4.1 ANTENNAL VERSUS LEG DEVELOPMENT
During the evolution of terrestrial hexapods from a marine, crustacean-like ancestor, changes in the pattern of *HOM* gene expression or in the response of their downstream target genes are thought to have influenced the diversification of originally identical appendages (see Figs. 13.6 and 13.7). In adults of *D. melanogaster*, antennae and legs are serially homologous and differ from each other in expression of the *HOM* gene *Antennapedia* (*Antp*). Expression of *Antp* in posterior T1 to anterior T3 (Fig. 13.7: Hexapoda) promotes leg identity by repressing distal expression of two homeobox-containing, antenna-determining genes, *extradenticle* (*exd*) and *hth* (Casares and Mann, 1998; SiDong et al., 2000). If *exd*, *hth*, or *Dll* is mutant, the antenna develops as a leg whether *Antp* is expressed or not. Normally, *hth* and *exd* are expressed in most antennal cells, while in leg (and wing) discs, expression of both genes is restricted to proximal cells (Fig. 11.8A). *Antp* represses expression of *hth* distally, with the result that Extradenticle (Exd) protein is unable to enter epidermal nuclei to bind to the control sites of the downstream genes required for antennal development.

Expression of both *hth* and *Dll* activates expression of *Sal* in the antennae, and coexpression of these genes can induce the formation of ectopic antennae in derivatives of the head, wing, leg, or genital discs (SiDong et al., 2000). Thus, *hth* is an antennal selector gene, and *Antp* promotes leg development by repressing expression of Hth, entry into epidermal nuclei by Exd protein, and antennal development.

In the absence of *Antp* and *hth* activity in the mesothoracic legs and of *hth* activity in the antennae, both types of appendages develop into a ground-state appendage having only two parts: a normal, five-segmented tarsus and a proximal part consisting, in the leg, of the fused coxa, trochanter, femur, and tibia (Casares and Mann, 2001). This observation suggested to Casares and Mann (2001) that the ground state of appendages in *Drosophila*, in the absence of selector gene function, is two segmented, that this could be its ancestral state in all arthropods, and that legs first arose without *HOM* or *hth* gene input.

11.4.2 ANTENNAE, TARSI, MAXILLARY PALPUS, AND PROBOSCIS
Differential expression of the *HOM* genes *proboscipedia* (*pb*) and *Sex combs reduced* (*Scr*) in different head and thoracic segments (Fig. 13.7: Hexapoda) appears to specify identity of adult antennal, maxillary palpus, proboscis, and tarsal appendages during postembryogenesis of *D. melanogaster* according to the scenario illustrated in Fig. 11.18 (Percival-Smith et al., 1997).

11.4.3 ABDOMINAL TERGA, STERNA, AND PLEURA
Each of larval abdominal segments 1–7 in *D. melanogaster* contains three pairs of imaginal histoblasts that were established during embryogenesis (Fig. 8.43: 17): the anterodorsal, posterodorsal, and ventral histoblast nests (Fig. 11.19A). These proliferate and fuse during metamorphosis to form each adult abdominal segment (Fig. 10.36). The imaginal tergum, pleura, and sternum of each segment are specified during the pupal stage by expression of the same genes as those involved in leg and wing specification (Kopp et al., 1999). Expression of Wg

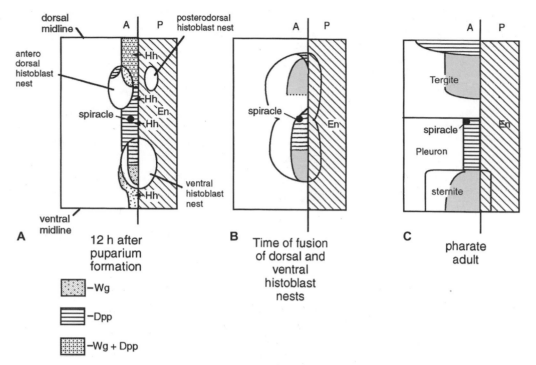

Fig. 11.19. Gene expression in an abdominal segment of *D. melanogaster*, in lateral aspect, at three different times during metamorphosis, showing proliferation and fusion of imaginal histoblast nests (compare with Fig. 10.36). A, anterior compartment; Dpp, Decapentaplegic protein; En, Engrailed protein; Hh, Hedgehog protein (secreted); P, posterior compartment; Wg, Wingless protein (secreted). (Adapted with permission from A. Kopp, R. K. Blackman, and I. Duncan. Wingless, Decapentaplegic and EGF receptor signaling pathways interact to specify dorso-ventral pattern in the adult abdomen of *Drosophila*. *Development* 126: 3495–3507. © 1999 Company of Biologists, Ltd.)

and Dpp is activated in the posterior margin of the A compartment by Hh signaling from the P compartment (Fig. 11.19A). *Drosophila* epidermal growth factor receptor (DER) acts synergistically with Wg to specify tergite and sternite identity, whereas their activity is inhibited by Dpp signaling, which promotes pleural identity (Fig. 11.19B, C). Dpp signaling at the dorsal midline controls D-V patterning within the tergite and promotes cuticular pigmentation medially (Fig. 11.19C).

11.4.4 THE GENITAL DISC

The genital disc of *D. melanogaster* larvae of both sexes contains primordia for both male and female and gives rise to the analia and to the male and female internal and external genitalia (10.2.3.1). In the embryo, the disc is specified by expression of the posterior pattern-control genes (7.2.1) such as *caudal* (Moreno and Morata, 1999) and, like other discs, is divided into A and P compartments (Fig. 11.20) (Chen and Baker, 1997; Casares et al., 1997; Sánchez et al., 1997). Similarly, expression of *wg* and *dpp* in the A compartment is induced by secretion of Hh protein, which in turn is activated by expression of *en* and *inv* in the P compartment. Mutual inhibition between Wg and Dpp proteins in the A compartment restricts their distributions, respectively, to medial and lateral portions of the genital disc except where they overlap.

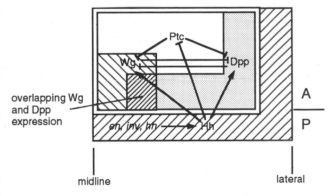

Fig. 11.20. Gene expression in the right half of the genital disc of a larva of *D. melanogaster* (it is identical in the left half but in mirror-image symmetry). Note that the same genes are involved in determination of the legs (Fig. 11.7), wings (Fig. 11.10), and wing eyespots (Fig. 11.17). Dpp, Decapentaplegic protein; *en*, *engrailed*; *hh*, *hedgehog*; Hh, Hedgehog protein; *inv*, *invected*; Ptc, Patched receptor protein; Wg, Wingless protein. (Adapted with permission from L. Sánchez, F. Casares, N. Gorfinkel, and I. Guerro. The genital disc of *Drosophila melanogaster*. II. Role of the genes *hedgehog*, *decapentaplegic* and *wingless*. *Dev. Genes Evol.* 207: 229–241. © 1997 Springer-Verlag.)

The sex determination gene *doublesex* (4.2.1) determines which of the genital primordia in each disc will differentiate (Sánchez et al., 2001). In female embryos and larvae, presence of Doublesex (F) induces differentiation of female genitalia and inhibits that of male genitalia by inhibiting induction of *dpp* expression by Hh in male primordia. In males, presence of Doublesex (M) induces development of male genitalia and inhibits that of female genitalia by blocking Hh-induced expression of *wg* in female primordia.

11.5 Effects of Symbiotic Microorganisms

Recently, it was demonstrated that certain host organisms develop normally only in the presence of associated symbiotic microorganisms. In certain insects, metabolic and regulatory products encoded by a symbiont's genome are required for normal postembryogenesis of the host's reproductive system (Moran and Telang, 1998).

12/ Hormones, Molting, and Metamorphosis

The structural events occurring during insect molting and metamorphosis (chapter 10) and regeneration (chapter 11) are mediated by blood-borne factors or hormones synthesized in and released from specific endocrine organs upon receipt of information from the sense organs by way of the central nervous system (CNS). In this chapter, I comment on the discovery, structure, and function of these hormones. A far more extensive treatment is provided in the superb book by Nijhout (1994b).

12.1 Classic Experiments Revealing How Hormones Induce Molting

The role of hormones in animal development was first demonstrated experimentally by Gudernatsch in 1912 and 1914 when he discovered that thyroid tissue will accelerate metamorphosis when it is fed to frog tadpoles (Kopéc, 1922). Observations on other vertebrates soon revealed the existence of many hormones, some of which proved to be widely distributed (Wigglesworth, 1985). As a result, some of them were evaluated in insect larvae in vain attempts to influence their metamorphosis. The lack of success and the earlier demonstration (by Oudemans [1899] and others) that insects lack sex hormones had many biologists believing that insects did not use hormones to develop and reproduce. What follows is a brief summary of some early attempts to understand how insects control molting and metamorphosis. Wigglesworth (1985) provided a detailed and critical historical account of this topic, and Bodenstein (1971) provided reprints of key papers with commentary and biographical information.

Working with caterpillars of the gypsy moth *Lymantria dispar* (Lepidoptera: Lymantriidae), S. Kopéc (1922) showed the following:

- If a fifth (last) instar caterpillar was tightly ligated into halves about the anterior abdomen with a hair loop 0–7 d after the molt into this instar, only the anterior half pupated. If ligated after day 7 (i.e., at the end of a critical period [CP] of 7 d), both halves did (the ventral nerve cord [VNC] was not severed by these ligations).
- If the brain of a caterpillar was removed before the end of the CP, pupation did not occur; if removed after day 7, it did.

- If the VNC of a caterpillar was severed before the end of the CP, pupation still occurred.

These results convinced Kopéc that a blood-borne "metamorphosis factor" must be released from the brain into the blood 7–10 d after the last larval molt to induce metamorphosis.

D. Bodenstein (1933) transplanted bristles, thoracic legs, and abdominal prolegs from donor caterpillars of the nymphalid butterflies *Vanessa urticae* and *V. io* into the body cavities of host larvae of the same or of different species and of the same age or older or younger. He found that all implants molted with the host, that implants could molt more frequently than they would have normally, and that they were ejected out of the body when host larvae molted to the pupal stage. Based on these results, Bodenstein concluded that blood-borne factors induce the molt and determine their number, that probably two molting factors control the kind of cuticle produced, and that these factors are not species-specific.

In a series of simple but elegant experiments, V. B. Wigglesworth (1934, 1936, 1940, 1985) investigated the control of molting in larvae of the blood-feeding reduviid bug *Rhodnius prolixus*. This bug has the following life history: $1 \to 2 \to 3 \to 4 \to 5 \to$ adult, with a blood meal required for each molt. When a bug was decapitated between 0 and 4 (in instars 1–4) or 0 and 7 (in instar 5) d after a blood meal, the insect lived for several months but would not molt. When decapitated after day 4 (instars 1–4) or day 7 (instar 5), the bug did molt, although it was unable to shed the remains of its previous cuticles. Therefore, a CP of 4 or 7 d must exist between the time of the blood meal and the time of release of a brain hormone (BH) into the blood from what Wigglesworth first thought was the corpus allatum (the two corpora allata [CA] are fused in this species; Fig. 12.1A: e) causing the molt. He also demonstrated, by severing the VNC before the end of the CP, that release of BH results from expansion of the abdomen due to uptake of the blood meal and resulting passage of information to the brain along the VNC.

When two fourth instar larvae were decapitated, one before and the other after the CP, and were then joined via a glass capillary tube, both molted.

By implanting the pars intercerebralis of "active" brains (i.e., those removed from donor larvae after the

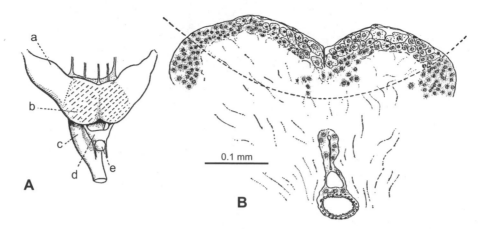

Fig. 12.1. A. Dorsal aspect of the brain of *Rhodnius prolixus* (Hemiptera: Reduviidae). The hatched area of a hormone-secreting protocerebrum induces molting when implanted into a juvenile host lacking its own brain. a, optic lobe; b, protocerebrum; c, subesophageal ganglion; d, fused corpora cardiaca; e, fused corpora allata. B. Transverse section through the posterior part of the protocerebrum, showing the somata of neurosecretory cells on either side of the midline, identified by their fuchsinophil inclusions. The portion above the dotted line was removed and implanted. (Reprinted from *The principles of insect physiology*, 6th ed., 1965, p. 59, V. B. Wigglesworth, Fig. 41. © 1965 Kluwer Academic Publishers, with kind permission from Kluwer Academic Publishers.)

CP) into the abdomens of larvae decapitated before the end of the CP, Wigglesworth showed that that part of the brain containing neurosecretory cells (NSCs) was the source of BH (Fig. 12.1). Wigglesworth's decision to do this experiment was influenced by the neuroanatomist B. Hanström, who had suggested to Wigglesworth, at a scientific meeting in Copenhagen in 1937, that NSCs rather than the CA might secrete BH. Hanström (1938) later demonstrated their presence in the brains of the bug sent to him by Wigglesworth. This led Wigglesworth to remove and transplant the pars intercerebralis (Wigglesworth, 1985).

V. Hachlow (1931) transected the bodies of moth pupae at various levels, sealed the cut ends of the resulting fragments, and found only those containing the thorax to molt to the adult, suggesting that something in the thorax is required for molting to occur. In a similar study on caterpillars of the sphingid moth *Deilephila* sp., but using transplantation techniques, Plagge (1938) found that active brains would induce pupation when transplanted into intact, fifth (last) instar caterpillars debrained before the CP but not when placed into isolated abdomens of such larvae. Thus, something in the head or thorax, in addition to the brain, is required to induce the molt into the pupa.

S. Fukuda (1940, 1942), while working on caterpillars and pupae of the commercial silkworm *Bombyx mori* (Lepidoptera: Bombycidae), noted paired strings of cells in either side of the prothorax that swelled and subsided immediately before a molt (compare Fig. 12.2) (these "granulated vessels" had been beautifully illustrated in 1762 by P. Lyonet in dissections of caterpillars of the goat moth *Cossus ligniperda* [Cossidae]). When Fukuda extirpated either the brain or the prothoracic glands (PGs) before the end of the CP, he prevented molt-

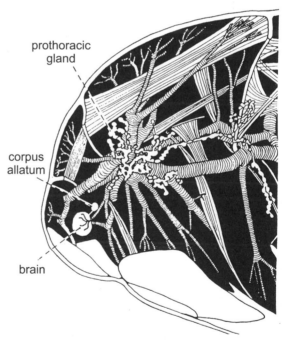

Fig. 12.2. Dissection of the head and thorax of an early pupa of *Hyalophora cecropia* (Lepidoptera: Saturniidae), showing the brain, corpus allatum, prothoracic gland, tracheal system, and some muscles on the right side. Fat body and other tissues have been omitted. (Reprinted with permission from A. Kühn. *Lectures on developmental physiology*, p. 339, Fig. 432. © 1971 Springer-Verlag.)

ing. Also, implanting active PGs caused a molt in a host larva whose own brain had been removed before the end of the CP. Therefore, two factors must be required to induce molting in this insect:

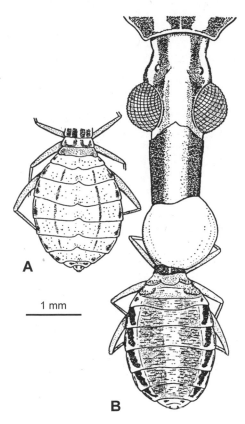

Fig. 12.3. A. Thorax and abdomen of a normal second instar larva of *R. prolixus*. B. Precocious "adult" produced when a first instar, decapitated larva is joined with paraffin wax to the tip of the head of a molting fifth instar larva. (Reprinted from *The principles of insect physiology*, 6th ed., 1965, p. 68, V. B. Wigglesworth., Fig. 49, © 1965 Kluwer Academic Publishers, with kind permission from Kluwer Academic Publishers.)

Brain NSCs → factor A → epidermis → molt
Prothoracic gland (PG) → factor B↗

On reading these results, H. Piepho (1942) then suggested that BH of the wax moth *Galleria mellonella* might induce the PG to secrete a molting hormone.

C. M. Williams (1947, 1948, 1952) used newly molted, diapausing pupae of the giant saturniid moth *Hyalophora cecropia* in his experiments. He cut off the abdomens of these pupae and sealed each shut with a plastic coverslip glued over the wound (such abdomens would not metamorphose into adult abdomens). When he implanted an active PG into such abdomens, they molted into adult abdomens, some even emitting sex attractants and depositing eggs. The brain alone could not initiate development in such diapausing, isolated abdomens, for no molt followed when an active (chilled) brain was implanted. Neither did a molt occur when an inactive PG from a diapausing pupa was implanted. However, an active brain from a chilled pupa plus an inactive PG induced molting.

Therefore, by the end of 1952, the sequence of factors known to induce molting in insects was as follows:

Active brain → factor A → PG →
factor B → epidermis → molt

This principal sequence of events is accepted to this day (Fig. 12.6) but with much added detail on the control and mode of action. Factor A is now known to be the prothoracicotropic hormone (PTTH) and factor B, the molting hormone (MH), 20-OH ecdysone (20E).

12.2 Classic Experiments Revealing How Hormones Induce Metamorphosis

Juvenile hormone (JH), synthesized in and released from cells of the CA, is now known to cause metamorphosis by inhibiting the epidermis from depositing adult cuticle in response to release of MH. Wigglesworth first demonstrated this between 1936 and 1940 in larvae and adults of *R. prolixus*. A young (instar 1 or 2) larva decapitated before the end of the CP and joined in parabiosis to a fifth instar decapitated after the CP underwent precocious metamorphosis (i.e., it developed some adult characteristics; Fig. 12.3). When a third instar had only its brain removed immediately after the CP, it molted to a fourth instar. But when the brain and the CA were removed at this time, it molted into a precocious adult. When an active CA, removed from a fourth instar larva, was implanted into the abdomen of a fifth instar, the latter molted into a giant sixth instar, rather than into an adult (Fig. 12.4). An adult joined in parabiosis to a fifth instar after the CP produced new adult cuticle under the old, but if active CA from younger larvae were introduced, then the new cuticle produced showed partial reversion to juvenile characteristics.

These results indicated that an inhibitory factor, secreted by cells of the CA, when present in the hemolymph, both maintains juvenile characteristics and inhibits the epidermis from depositing imaginal cuticle when a molt is induced by release of the molting factor. (The CA were first illustrated by Lyonet [1762] in larvae of *C. ligniperda*; were named by C. Janet in 1899, based on their presence in the ant *Myrmica rubra*; and were suggested to be endocrine organs by A. Nabert in 1913 [Wigglesworth, 1985].)

J. J. Bounhiol (1938), using caterpillars and pupae of *B. mori*, showed that when the CA are removed from second instar caterpillars, they molt into tiny pupae (0.025 gm) or larva/pupa intermediates and, eventually, into tiny, sterile adults.

H. Piepho (1942), working on caterpillars of the wax moth *G. mellonella* (Lepidoptera: Pyralidae), showed the following:

• When a piece of integument (epidermis + cuticle) is removed from a young donor caterpillar, say, a second instar, and transplanted into the hemocoel of an older host larva, say, a fourth instar, epidermal cells at the margins of the implant proliferate

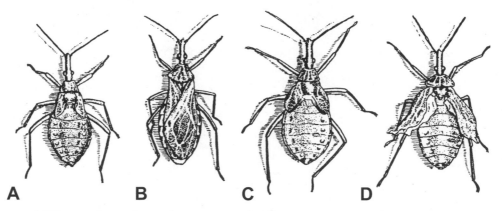

Fig. 12.4. A. Normal fifth instar larva of *R. prolixus*. B. Normal adult. C. Giant sixth instar larva produced by implanting an active (secreting) corpus allatum from a fourth instar into the abdomen of a fifth instar. D. A sixth instar larva/adult intermediate produced in the same way. (Reprinted from *The principles of insect physiology*, 6th ed., 1965, p. 70, V. B. Wigglesworth. Fig. 51, © 1965 Kluwer Academic Publishers, with kind permission from Kluwer Academic Publishers.)

to form a hollow vesicle within the body cavity of the host, with its piece of cuticle in its lumen (Fig. 12.5A, B).

- When the host molts to a fifth instar, the implant sheds its second instar cuticle into its interior and deposits fifth (not third) instar cuticle inside the vesicle (Fig. 12.5C); when it molts to the pupa, the implant sheds its fifth instar cuticle and deposits pupal cuticle (Fig. 12.5D); and when the host molts to an adult, the implant deposits adult cuticle complete with scales about the cast-off larval (second and fifth) and pupal cuticles (Fig. 12.5E).
- This same implant can then be removed from the adult host and cut into small pieces, each piece reintroduced into another young larva, and the implanted epidermis again induced to secrete larval, pupal, and adult cuticles, a process that can be continued indefinitely but not forever (9.7.8).

(*Note*: In this experiment, the transplants were replicated several times, and at each host molt, several hosts were killed and their implants removed, sectioned, and examined for the kind of cuticle produced. For example, transplants might be made into 50 hosts, with 10 hosts being killed and examined at each molt. Since all nervous connections between the piece of integument and the donor's CNS were severed when it was removed and placed in the body cavity of the host, only blood-borne agents could affect its epidermis. Possible secondary innervation from the host's CNS was shown not to have occurred when test implants were examined carefully in histological sections.)

Based on the results of these experiments, Piepho concluded that circulating blood-borne factors (hormones) both induce the molt and determine the kind of cuticle produced; the epidermis can deposit larval, pupal, and adult cuticle at any age; the epidermis can revert to producing larval and pupal cuticle after it has deposited adult cuticle; and the number of molts is not controlled by the epidermis. With subsequent work on

this species, Piepho (1951) developed the idea that high concentrations of CA factor cause the epidermis to deposit larval cuticle in response to MH; intermediate concentrations, pupal cuticle; and no CA factor, adult cuticle. That is, its action is concentration-dependent.

Based on all these results, Wigglesworth (1952) concluded the following:

- In juveniles of *R. prolixus*, the CA factor is secreted in the first four larval instars prior to molts. It both influences the epidermis to deposit larval cuticle and inhibits it from depositing adult cuticle when it responds to release of the molting factor.
- In the fifth instar, CA factor release is inhibited by nervous input from the brain, so that adult cuticle can be deposited in response to molting factor alone.
- In holometabolous insects, CA factor is released in relatively smaller amounts in the last larval instar, and this results in deposition of pupal rather than imaginal cuticle by the epidermis.
- Young epidermis tends to be more responsive to the presence of CA factor than does older epidermis (Fig. 12.6).

The results demonstrated conclusively that the CA were the source of a hormone, now known as JH, serving to maintain larval characteristics and to inhibit differentiation of adult characteristics (it cannot do so without MH). This action has since been demonstrated by use of similar methods in representatives of many insect orders but not in apterygote hexapods or in certain *neotenous* (sexually mature but of juvenile phenotype) insects such as stick insects (13.3.3).

In apterygote hexapods, adults resemble juveniles except in size and continue to molt after becoming reproductively mature (Fig. 9.3: Archaeognatha). In addition, their cuticle is generally similar, though not identical, to that of their juveniles. As in adults of other investigated insects, JH acts as a gonadotropin and is required for vitellogenesis in females (2.3.2) and influ-

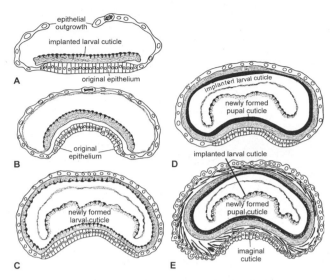

Fig. 12.5. Development of integumental transplants introduced into the hemocoel of the wax moth *Galleria mellonella* (Lepidoptera: Pyralidae) caterpillars. Transverse sections of transplants removed from the host are shown at various times following transfer. A and B. Formation of a hollow epidermal vesicle just after transfer. C. When the host caterpillar molts to the next larval instar, the implant sheds its old cuticle into its lumen and produces a new larval cuticle along with the host. D. When the host molts into the pupa, the transplant sheds its larval cuticle into its lumen and deposits pupal cuticle. E. When the host molts into the adult, the transplant sheds its pupal cuticle and deposits adult cuticle complete with scales. (Reprinted with permission from A. Kühn. *Lectures on developmental physiology*, p. 346, Fig. 443. © 1971 Springer-Verlag.)

ences spermatogenesis in males (1.3.7.2). It does not function in metamorphosis because it is not present during adult molts. Thus, an adult female alternates between depositing eggs and molting, each molt occurring in the absence of JH (Nijhout, 1994b: 149).

Based on these observations, the Czech endocrinologist V. J. A. Novák (1975) suggested that JH first evolved as a gonadotropin (but see 9.6.2.5.11) and was later "captured" to function in inhibiting adult development during the evolution of incomplete and complete metamorphosis (9.6.2.4). Wigglesworth (1936) had first suggested that gonadotropin and JH were different hormones, and Pfeiffer (1945) was the first to suggest that they might be the same (Wigglesworth, 1985).

Present dogma concerning the interaction of molecules of PTTH, 20E, and JH in controlling molting and metamorphosis in holometabolous insects is summarized in Fig. 12.6 for *Manduca sexta* (Riddiford, 1993b), though the real story is far more complex (Nijhout, 1994b). Neither PTTH, 20E, nor JH is species-specific, since interspecies transplants of appropriate hormone-secreting endocrine organs or injection of purified hormones have similar effects on juveniles of different pterygote species.

12.3 Neurosecretory Cells and Endocrine Organs

Specialized NSCs within ganglia of the CNS in insects are known to synthesize the developmental neuropeptide hormones PTTH, allatotropin, allatostatin, eclosion hormone (EH), diapause homone (DH), and bursicon (Scharrer and Scharrer, 1945; Raabe, 1982; Orchard and Loughton, 1985; Sedlak, 1985; Nässl, 1993; Nijhout, 1994b; Braünig, 1998; Schooneveld, 1998).

The somata of these cells occur in the cortex of the brain and all ganglia of the VNC in numbers and locations characteristic for the species, and most release their product into the hemolymph at sites some distance from their target cells (Fig. 12.7) (Raabe, 1982). In fresh dissections of the CNS in insect physiological saline solution viewed under incident light, they appear opalescent blue-white due to scattering of light by their characteristic inclusions, the neurosecretory granules (NSGs) (Fig. 12.8). These are 100–300 nm across, consist of molecules of neurohormone bound to a carrier protein and enclosed within a membrane, and are responsible for the selective staining of different NSCs by dyes specific for them such as paraldehyde fuchsin, azocarmine, and chrome hematoxylin (Panov, 1980).

The axon of a typical NSC projects to a characteristic release (perisympathetic, neurohemal) organ usually remote from its cell body, where it branches profusely to end in swollen terminals from which the neurohormone is released into the hemolymph (Figs. 12.8, 12.9). Both neurohormones and NSG proteins are synthesized in the cell body of the NSC on polysomes associated with the rough endoplasmic reticulum, are secreted into its lumen, are concentrated in cisternae of the Golgi apparatus, and are released as Golgi vesicles (NSGs) that pass down its axon to its terminals at speeds of up to 400 mm/d. They are released into the blood by exocytosis (Fig. 12.8), such release being associated with action potentials like those of typical neurons during transmitter release. This movement is by fast axonal transport along microtubular tracts (slow axonal transport at speeds of 1–5 mm/d is also known for smaller compounds). Fast retrograde transport along axonal microtubules toward the cell body from the terminals at rates of 200–300 mm/d carries aging cytoplasmic organelles such as mitochondria, cell membrane, and endocytotic vesicles back to the cell body.

Movement of NSGs down the axons of NSCs is by action of motor proteins such as the *kinesins*, which bind to microtubules within the cells in a stepwise fashion and haul proteins toward either their negative or positive end (Vale and Milligan, 2000). The motors have a "head on a stalk" organization and move along the microtubular tract by forceful interaction of the heads with the surface of the tubule.

Westbrook and colleagues (1991) developed computer-assisted methods for reconstructing the three-dimensional architecture and projection patterns of NSCs in brains of *M. sexta* after marking the cells with doubly labeled monoclonal antibodies raised against

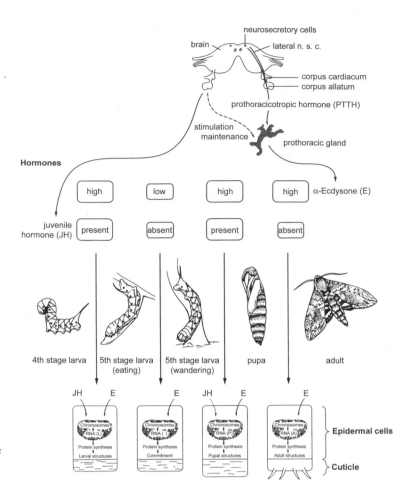

Fig. 12.6. The principal endocrine tissues in the tobacco hornworm *Manduca sexta*, and the classic roles of prothoracicotropic hormone, ecdysone (E), and juvenile hormone (JH) in molting and metamorphosis. n. s. c., neurosecretory cells. (Reprinted with permission from L. M. Riddiford. Hormone receptors and the regulation of insect metamorphosis. *Receptor* 3: 203–209. © The Humana Press, Inc.)

PTTH and EH (see 10.2.4.1.2). Cell bodies of two pairs of group III lateral neurosecretory cells (L-NSC III), which synthesize the "big" (28-kDa) molecular variant of PTTH stimulating each molt, are located dorsolaterally in each protocerebral hemisphere (Fig. 12.10A, C), while those of the group II medial neurosecretory cells (M-NSC IIa$_2$), which probably synthesize the "small" (6-kDa) variant of PTTH, are located anteriorly in the mediodorsal protocerebrum (Fig. 12.10B). Both types project to the CA, where they arborize into neurohemal terminals but with different patterns (Fig. 12.10A, B). Cell bodies of four group II ventromedial neurosecretory cells (VMNSCs) (Fig. 12.10C), synthesizing EH, project into the VNC via the circumesophageal connectives and along its length to the terminal abdominal ganglion to end in perisympathetic release organs associated with the hindgut.

The position, innervation, size, and cellular structure of corpora cardiaca (CC), CA, and PGs vary depending on species and stage of development (Raabe, 1982; Sedlak, 1985; Nijhout, 1994b; Cassier, 1998; Schooneveld, 1998).

12.4 The Developmental Hormones

The presently known developmental hormones in insects include ecdysteroids, PTTH and other ecdys-

iotropins, the JHs, EH, pre-ecdysis and ecdysis-triggering hormones, bursicon, allatotropin, allatostatins, and DH (Fig. 12.11).

12.4.1 20-OH ECDYSONE

20E (beta-ecdysone, ecdysterone) is the functional molting hormone of insects and other arthropods (Fig. 12.11B) (Rees, 1995).

12.4.1.1 CHEMICAL STRUCTURE AND BIOSYNTHESIS

Its immediate precursor, ecdysone (alpha-ecdysone) (Fig. 12.11A), was isolated, crystallized, and shown to be a steroid in pupae of the commercial silkworm *Bombyx mori* by Butenandt and Karlson in 1954 (they extracted 25 mg of ecdysone from 500 kg of pupae!), and its final structure was published by Huber and Hoppe in 1965. Insects can synthesize the molecule from cholesterol, which is required in their diets (Richards, 1981; Rees, 1985, 1995), but actual demonstration in vitro of ecdysteroid synthesis by PG cells from [^{14}C]-labeled cholesterol was not achieved until 1974 (by Chino et al. in *B. mori*). The reason for the 9-y delay is that cholesterol is converted to ecdysone only when offered to PG cells as cholesterol-containing lipoproteins in solution in the blood.

20E is the active molting hormone of insects and is

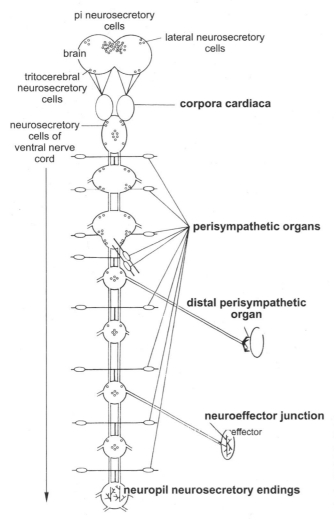

Fig. 12.7.   The distribution of neurosecretory cell bodies and the release sites of neurohormones in the central nervous system of an insect (the circumesophageal connectives are omitted). pi, pars intercerebralis. (Reprinted with permission from M. Raabe. *Insect hormones*, p. 54, Fig. 20. Plenum Press, N.Y. © 1982 Kluwer Academic/Plenum Publishers.)

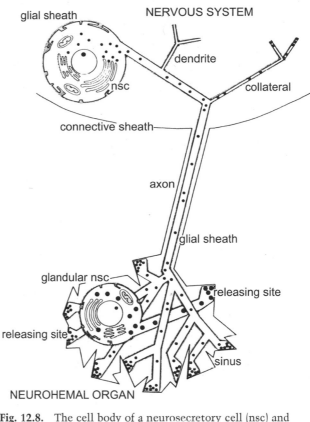

Fig. 12.8.   The cell body of a neurosecretory cell (nsc) and its projection to a neurohemal organ. The large and small dots are neurosecretory granules. Note the intrinsic glandular neurosecretory cell within the neurohemal organ. (Reprinted with permission from M. Raabe. *Insect hormones*, p. 3, Fig. 1. Plenum Press, N.Y. © 1982 Kluwer Academic/Plenum Publishers.)

100- to 200-fold more active in promoting molting than is ecdysone (King, 1972; Richards, 1981). In larvae of *Manduca sexta*, 3-dehydroxyecdysone secreted by PG cells is changed to ecdysone in the blood. Ecdysone is then metabolized to 20E in cells of the epidermis, fat body, midgut epithelium, and other target tissues (Bidmon et al., 1992). There are undoubtedly differences in its biosynthesis in different species (Nijhout, 1994b).

The biosynthetic pathway leading to ecdysone and its subsequent degradation has been worked out in detail, and some of the intermediates have also been shown to have hormonal action (Horn and Bergamasco, 1985; Rees, 1985; Smith, 1985; Koolman and Karlson, 1985; Sakurai and Gilbert, 1990). Ecdysteroids can exist in several different forms that differ in the number of

—OH groups and, when in solution in the hemolymph, are usually bound to an ecdysteroid-binding protein (Fig. 12.14) (Richards, 1981).

Injection of serum raised in rabbits against ecdysone (antiecdysone antibody) into wandering, last instar larvae of the wax moth *Galleria mellonella* will delay metamorphosis, presumably by binding to the active sites of ecdysteroid receptor proteins in ecdysone-receptive tissues of this insect (Fig. 12.14) (Birkenbeil and Eckert, 1991). The serum also binds to cells of the PG, where it labels sites of ecdysone release.

12.4.1.2 REGULATION OF SECRETION BY THE PROTHORACIC GLANDS

Regulation of ecdysteroid secretion by the PGs is very complex, at least in *M. sexta*, in which it is best understood (Fig. 12.12) (Smith, 1985; Bollenbacher, 1988; Nijhout, 1994b). In glands of *Rhodnius prolixus*, its synthesis and release before a molt has a circadian (24-h) rhythm both in vivo and in vitro (Vafopoulou and Steel, 1999).

Ecdysteroid secretion by PGs can also be influenced

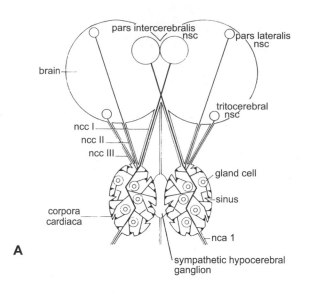

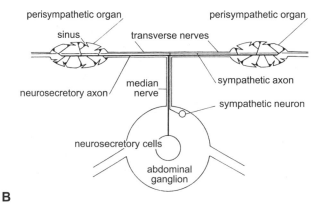

Fig. 12.9. A. A pair of insect corpora cardiaca, the storage sites and release organs for neuropeptide hormones synthesized within neurosecretory cells (nsc) in the brain. ncc I, II, and III, nervi corporis cardiaci I, II, and III; nca 1, nervi corporis allati 1. B. A pair of insect perisympathetic organs associated with a ganglion in the ventral nerve cord. These function as the release organs for neuropeptide hormones synthesized within neurosecretory cells of these ganglia. The interganglionic connectives are omitted. (Reprinted with permission from M. Raabe. *Insect hormones*, p. 38, Fig. 11. Plenum Press, N.Y. © 1982 Kluwer Academic/Plenum Publishers.)

by external factors. For example, larvae of the tobacco budworm *Heliothis virescens* (Noctuidae) fail to pupate if parasitized by the braconid wasp *Cardiochilis nigriceps*, because their PGs are inactivated in the last instar (Pennachio et al., 1998). The same will occur if an unparasitized caterpillar is injected with fluid from the calyx of the wasp's lateral oviduct or with venom, both of which contain a polydnavirus.

12.4.1.3 DEGENERATION OF THE PROTHORACIC GLANDS AT ADULT EMERGENCE IN EXOPTERYGOTES AND ENDOPTERYGOTES
Degeneration of the PGs before or after adult emergence is the proximate reason why adults of most exoptery-

gotes and endopterygotes do not molt (9.7.8) (in apterygotes, which continue to molt as adults, the glands persist and continue to function [Gabe, 1953]). Based on results of transplantation and parabiosis experiments, Wigglesworth (1955) concluded that degeneration of these glands in *R. prolixus* resulted from their passage through the fifth (last juvenile) instar in the absence of JH plus release of an additional, unidentified factor present at the time of adult emergence. In fact, Smith and Nijhout (1983) showed topical application of JH to inhibit the breakdown of PGs in several hemimetabolous and holometabolous insects, an observation recently confirmed in pharate adults of *M. sexta* by Dai and Gilbert (1998).

At 25°C and 60%–70% relative humidity, fragmentation of PG DNA normally begins on day 5 of pupal-adult development in this insect and peaks 1 d later. Injection of JH into young pupae not only prevents degeneration of PG cells for up to 11 d but also maintains their ability to synthesize ecdysteroids (control PGs had degenerated by this time), suggesting that JH both maintains PGs and stimulates their cells to secrete ecdysteroids (Fig. 12.12: bottom, E, F/G).

In the milkweed bug *Oncopeltus fasciatus* (Hemiptera: Lygaeidae) reared at 27°C, the PGs normally degenerate on day 7 of the fifth instar (Smith and Nijhout, 1983). If removed on day 2 and cultured in Grace's medium with 2.5 µg/ml of 20E, their cells begin to degenerate after 24-h exposure. If removed at daily intervals following the molt into the fifth instar and cultured without 20E, only those removed on day 3 or later degenerated, the same time at which endogenous secretion of ecdysteroids commences in this instar in vivo. Smith and Nijhout (1983) concluded that secretion of 20E in the absence of JH stimulates the degeneration of PGs and that no additional factor is required. Earlier they had shown sensitivity of PG cells to JH to decline sharply during day 3 of the fifth instar and to disappear entirely by day 4 (Smith and Nijhout, 1982). The same events likely occur in other exopterygote and endopterygote insects.

12.4.1.4 OTHER SOURCES OF ECDYSTEROIDS
The classic scheme (Fig. 12.6) has the PGs as the principal source of ecdysteroids. However, they are now known to be synthesized in the sheath cells of adult testes of several lepidopterans and flies (1.3.7.2.3), in the follicle cells of the ovaries (2.2.1.4), in larval epidermal cells, and in larval cells of other tissues in the abdomen (Nakanishi et al., 1972; Hsiao et al., 1975; Richards, 1981; Delbeque et al., 1990; Sakurai et al., 1991; Lafont, 1997; Sláma, 1998) and to be widely distributed in bacteria, fungi, ferns, higher plants, and many invertebrates and vertebrates including humans (Richards, 1981; Nijhout, 1994b; Sláma, 1998). In fact, Sláma cited evidence from a host of older papers to suggest that the widely accepted brain-PG-ecdysteroid sequence (Fig. 12.6) is misleading. For example, many workers (e.g., Nakanishi et al., 1972; Hsiao et al., 1975; Delbeque et al.,

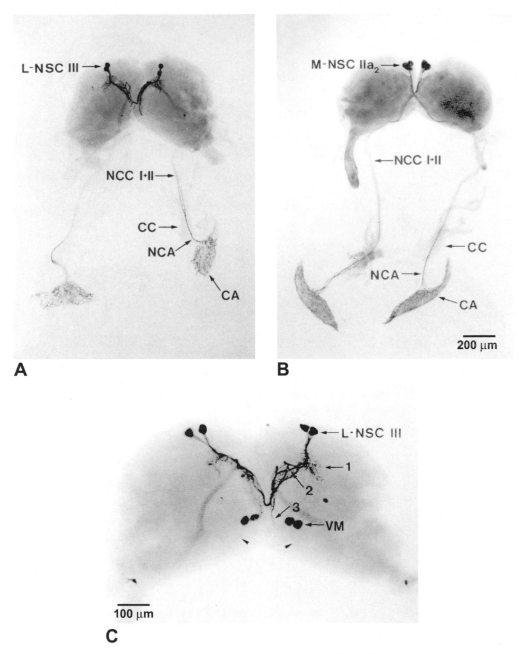

Fig. 12.10. Photomicrographs of whole mount preparations of the brains of 3-d-old, fifth instar larvae of *Manduca sexta* immunostained to show the neurosecretory cells and their projections (in A and B) to the corpora allata (CA) (in this species and in other lepidopterans, the release sites for brain neuropeptides are in the CA rather than the corpora cardiaca [CC]). C. Photomicrographs at a deeper plane through the brain, showing the position of the four ventromedian neurosecretory cells (VM). L-NSC III, group III lateral neurosecretory cells; M-NSC IIa₂, group II medial neurosecretory cells; NCA, nervi corporis allati; NCC I + II, nervi corporis cardiaci I + II. (Three-dimensional architecture of identified cerebral neurosecretory cells in an insect, A. L. Westbrook, M. E. Haire, W. R. Kier, and W. E. Bollenbacher, *J. Morphol.*, Copyright © 1991 Wiley-Liss, Inc. Reprinted by permission of Wiley-Liss, Inc., a subsidiary of John Wiley & Sons, Inc.)

1990; Sakurai et al., 1991; Lafont, 1997; Sláma, 1998) have noted that removal of the PGs has no effect on the molt cycle of certain insects. Also, the endogenous peaks of ecdysteroid release occurring before molts and during metamorphosis (Fig. 12.25) were said by Sláma (1998) to usually originate in nonfeeding periods as polar metabo-lites of sterols retrieved from old disintegrating tissues from previous instars, the contribution of the PGs being negligible (Fig. 12.13). In fact, he considered ecdysteroids not to meet the definition of a hormone but instead to represent peripheral feedback tissue factors resulting from the breakdown of larval tissues. Remobilization

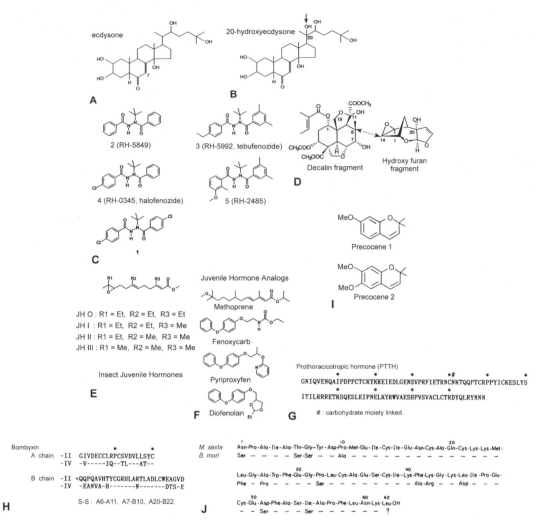

Fig. 12.11. The principal developmental hormones in insects and artificially synthesized insecticidal compounds interfering with molting and metamorphosis. A. Ecdysone. B. 20-OH ecdysone (20E), the active form of ecdysone. C. Five bisacylhydrazines: synthetic, nonsteroidal agonists of 20E (from Dhadialla et al., 1998: 549, Fig. 1). D. Azadirachtin from the Neem tree. E. The four principal juvenile hormones (JH). Et, ethyl group; Me, methyl group (from Dhadialla et al., 1998: 557, Fig. 3). F. Four juvenile hormone analogs (from Dhadialla et al., 1998: 557, Fig. 3). G. Amino acid sequence of *Bombyx mori* (Lepidoptera: Bombycidae) prothoracicotropic hormone. H. Amino acid sequences of the two components of *B. mori* bombyxin. I. Precocenes I and II from the bedding plant *Aegeratum houstonianum* (Asteraceae). J. Comparison of amino acid sequences of the eclosion hormones (EH) of *M. sexta* and *B. mori*. (A, B, C, E, and F reprinted with permission from the *Annual Review of Entomology*, Volume 43 © 1998, by Annual Reviews www.AnnualReviews.org. D reprinted with permission from Aldhous 1992. Copyright 1992 American Association for the Advancement of Science. G and H reproduced with permission from Nagasawa 1992. © 1992 Birkhäuser Verlag. I reprinted with permission from Bowers et al. 1976. Copyright 1976 American Association for the Advancement of Science. J reprinted with permission from Truman 1990. Japanese Scientific Society Press. © 1990 Springer-Verlag.)

from a central ecdysteroid pool then functions in the differentiation of adult tissues during metamorphosis: his transport and reutilization theory (Fig. 12.13).

Results of classic experiments (12.1, 12.2) resulting in development of our present perception (Fig. 12.6) of how brain, PGs, and 20E control molting and metamorphosis are convincing but based on study of too few species, just as is most current work in insect endocrinology (Nijhout, 1994b). In addition, many of the model species in use have been in culture for generations without infusion of wild germ plasm. Thus, our current understand-

ing of the brain-PG-ecdysteroid axis may not provide an accurate picture of how hormones control molting and metamorphosis throughout the hexapods, as illustrated by the exceptions Sláma (1998) mentioned and as Nijhout (1994b) detailed.

12.4.1.5 MODE OF ACTION

20E is a typical steroid hormone (Fig. 12.11B). Because it is slightly hydrophobic, it is usually bound reversibly to a carrier protein while in the hemolymph (Fig. 12.14) (Richards, 1981; O'Conner, 1985; Horodyski, 1996).

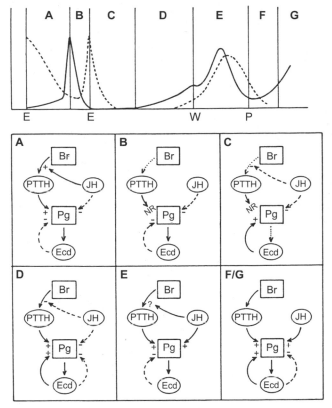

Fig. 12.12. The endocrine control of prothoracic gland secretion in *M. sexta*. Upper panel: Regulation of secretory activity of the prothoracic glands. Changes in juvenile hormone (JH) titer (dashed line) and secretory activity of the prothoracic glands (solid line) for the fourth instar (A and B), fifth instar (C–E), and early pupal stage (F and G). E, larva ecdysis; W, onset of wandering; P, pupation. Lower panel: Interendocrine control of prothoracic gland activity. Solid arrows with plus sign and dashed arrows with minus sign indicate activation and inhibition, respectively. Dotted arrows indicate no secretion of hormone. Br, brain; Ecd, ecdysteroids; NR indicates that the prothoracic glands are not responsive to prothoracicotropic hormone (PTTH). (Reprinted with permission from E. Ohnishi and H. Ishizaki (eds.), *Molting and metamorphosis*, p. 96, Fig. 6. Japanese Scientific Society Press. © 1990 Springer-Verlag.)

When released from this protein, each molecule diffuses through the plasma membrane of its target cell (it is lipophilic) and binds reversibly to an ecdysteroid-specific receptor protein (EcR) within the nucleus. Binding of hormone to the receptor causes an allosteric change in its three-dimensional configuration, enabling the complex to bind with high affinity to a specific DNA sequence, the *ecdysone-response element* (EcRE), where it is believed to act as a transcriptional enhancer (Harshman and James, 1998). The protein products of these activated genes, the primary response proteins, may in turn bind to the regulatory sites of other, downstream, regulatory genes, activate them, and so produce a delayed secondary response, thereby extending the ini-

tial effect of the hormone in a gene expression cascade (Fig. 12.15).

12.4.1.5.1 ECDYSTEROID RECEPTOR PROTEINS

Proteins in the cell membrane, cytosol, or nucleus of target cells that function to detect a hormone molecule and transduce its signal are called *receptors* (Yund and Osterbur, 1985). The *D. melanogaster* gene *EcR* has been cloned and is known to encode an ecdysone receptor (Bidmon and Silter, 1990; Koelle et al., 1991). EcR protein binds both to 20E and to specific DNA sequences in the chromosomes and is found in the nuclei of all target tissues examined at each 20E-responsive gene and at each developmental stage marked by a peak of 20E release. In addition, there are at least three isoforms of this receptor—EcR-A, EcR-B1, and EcR-B2—that influence which genes can be expressed, and each of these in turn can be regulated by the hormonal milieu in which it acts (Thummel, 1995; Lezzi et al., 1999).

Two of these receptors, EcR-A and EcR-B1, are expressed in the CNS of *D. melanogaster* and *M. sexta*, and Truman and colleagues (1994) used labeled, isoform-specific antibodies to track their prevalence throughout postembryogenesis. They found most larval neurons of both insects to express high levels of EcR-B1 at the onset of metamorphosis when they are losing their larval characteristics in response to release of ecdysteroid at the pupal commitment peak (Fig. 12.25) (during larval molts these neurons have no detectable EcR receptors and do not respond to ecdysteroids). During the pupal-adult transformation, most of these neurons switch to EcR-A expression and transform into their adult form, while a subset of larval neurons, innervating muscles functioning in adult emergence, hyperexpress EcR-A and die after adult emergence. Imaginal neuroblasts express EcR-B1 during the last larval instar, but the imaginal neurons they generate express only EcR-A. Thus, EcR-A seems to predominate in maturing cells, and EcR-B1, in those proliferating or regressing.

As of 1995, over 150 different nuclear receptor proteins in the steroid superfamily had been identified in animals, but only 10, including the three forms of EcR, have so far been discovered in insects (Segraves, 1991; Jindra, 1994; Mangeldorf and Evans, 1995; Hannon and Hill, 1997; Lezzi et al., 1999). They function in embryos, larvae, pupae, and adults; regulate development, cell differentiation, metamorphosis, and physiology; and are probably encoded by genes descending from a single ancestral gene that arose before the divergence of vertebrates and invertebrates.

A typical nuclear receptor protein includes the following elements (Lezzi et al., 1999):

$$N \underline{\quad \overset{\text{A/B} \quad \text{C} \quad \text{D} \quad \text{E} \quad \text{F}}{\hphantom{xxxxxxxxxxxxxxxxxxxxxxxxxxxxxxxxxx}} } C$$

where A/B is the N-terminal region; C, the DNA-binding domain; D, the variable hinge region; E, the conserved, ligand-binding domain; and F, the variable

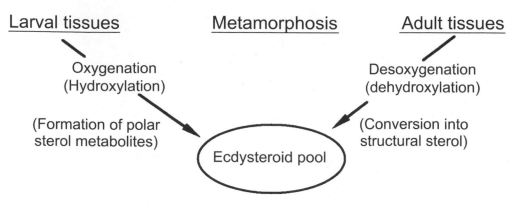

Fig. 12.13. Sterol use during insect metamorphosis as hypothesized by Sláma (1998). (Adapted with permission from K. Sláma. The prothoracic gland revisited. *Ann. Entomol. Soc. Am.* 91: 168–174. © 1998 Entomological Society of America.)

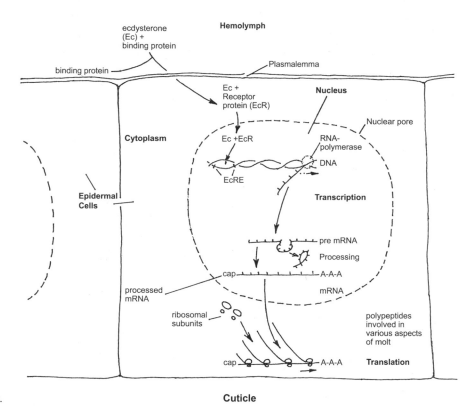

Fig. 12.14. Mechanism of ecdysteroid action in an epidermal cell. EcRE, ecdysteroid response element.

C-terminal. The DNA-binding domain has two highly conserved zinc fingers that bind the ligand-receptor complex to the ecdysteroid response sequence on the DNA of the target cell, while the ligand-binding domain functions in ligand recognition, ensuring both specificity and selectivity.

In *D. melanogaster*, the EcR protein must heterodimerize with the Ultraspiracle (USP) protein for DNA binding and transactivation of downstream genes to occur (Fig. 12.16) (Thomas et al., 1993; Yao et al., 1993). Thus, both USP and EcR-A are required in late third instar larvae for appropriate developmental and transcriptional responses to the 20E pulse triggering puparium formation (Fig. 12.20) (Hall and Thummel,

1998). Also, both proteins bind to the DNA of at least eight somatic cell types within the ovary rudiments of third instar larvae of this fly and correlate with ovarian differentiation during metamorphosis; if either *Ecr* or *ultraspiracle* (*usp*) genes are mutant, such morphogenesis is defective (Hodin and Riddiford, 1998).

Despite the vast phylogenetic distance separating vertebrates and insects, the molecular mechanisms by which steroids act to regulate gene expression are closely conserved, as summarized by Jindra (1994) and as shown by demonstrating that *D. melanogaster* and blowfly (*Lucilia cuprina*) EcRs can act as ecdysteroid-dependent transcription factors in cultured mammalian cells (Hannon and Hill, 1997). The vertebrate and tick

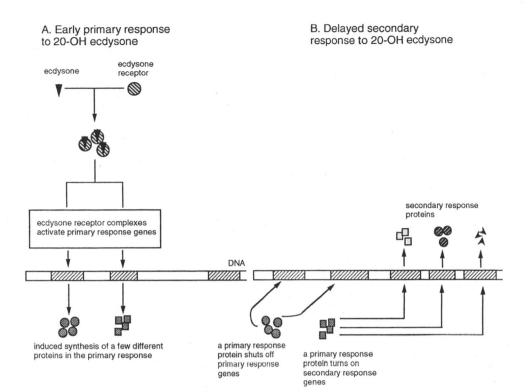

A. Early primary response to 20-OH ecdysone

B. Delayed secondary response to 20-OH ecdysone

ecdysone

ecdysone receptor

ecdysone receptor complexes activate primary response genes

DNA

induced synthesis of a few different proteins in the primary response

a primary response protein shuts off primary response genes

a primary response protein turns on secondary response genes

secondary response proteins

Fig. 12.15. Early primary response (A) and delayed secondary response (B) to 20-OH ecdysone release in cells of *D. melanogaster*. Some of the primary response proteins activate secondary response genes, whereas others turn off the primary response genes in a feedback loop. (© 1994 From *Molecular biology of the cell*, 4th ed., by B. Alberts et al. Adapted by permission of Routledge, Inc., part of The Taylor & Francis Group.)

ortholog of USP is the retinoid X receptor (RxR) (Palmer et al., 1999).

As of 1997, ten members of the steroid receptor gene superfamily had been identified in *D. melanogaster*. They encode proteins possessing both DNA- and ligand-binding domains, with some of these having orthologs in *Chironomus tentans*, *Aedes aegypti*, and *L. cuprina* (Diptera); in *Bombyx mori*, *Galleria mellonella*, *Choristoneura fumiferana*, and *M. sexta* (Lepidoptera) (Hannon and Hill, 1997); and in the crab *Uca pugilator* (Crustacea) (Durica et al., 1999) and the tick *Amblyomma americanum* (Chelicerata) (Palmer et al., 1999). However, only EcR has had its ligand identified specifically as 20E. The others function principally in specifying body pattern in embryos or photoreceptor identity in eye discs (10.2.4.2.2). The regulation and role of nuclear receptor proteins during molting and metamorphosis is reviewed in Riddiford et al., 1999, and Lezzi et al., 1999.

12.4.1.5.2 THE 20-OH ECDYSONE–INDUCED CASCADE OF GENE EXPRESSION AT METAMORPHOSIS
The first evidence to suggest that 20E induces molting by influencing differential gene expression was the discovery of chromosome puffs in polytene chromosomes of the salivary gland cells of various flies, including *D. melanogaster*.

Polytene Chromosomes, Chromosomal Puffing, and Differential Gene Expression
Polytene chromosomes in larval cells of dipterans and in salivary gland cells of the collembolan *Bilobella grassei* (Cassagnau, 1971) result when DNA and chromosome replication occur in the absence of cell division and when all copies of each chromosome remain side by side to create a single giant chromosome (Fig. 12.17A) (Alberts et al., 1994: 348–352).

Polytene chromosomes were first described and illustrated by Balbiani in 1881 in salivary gland cells of *Chironomus* (Diptera: Chironomidae) larvae and were recognized as being in interphase by Heitz and Bauer in 1933. In his drawings, Balbiani included doughnut-shaped "Balbiani rings" about the chromosomes in several positions. Beerman (1952) subsequently showed these rings to differ in position in relation to the cross bands in homologous chromosomes of salivary gland, Malpighian tubule, midgut, and rectal cells of *C. tentans* larvae (Fig. 12.18A). Based on this evidence, Beerman proposed that a particular sequence of puffs represents a corresponding pattern of gene activity and that gene expression differs in different cell types.

It is now known that chromosome puffs represent regions in the polytene chromosomes in which the chromatin strands are separated into loops (Fig. 12.17B). By use of [^{3}H]uridine autoradiography, Beerman's group

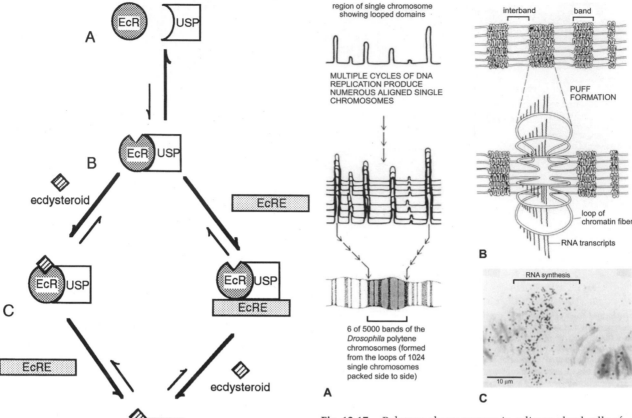

Fig. 12.16. Model illustrating the binding of the ecdysteroid–ecdysteroid receptor (EcR)–Ultraspiracle (USP) heterodimer complex to the ecdysteroid response element (EcRE) in the DNA of *D. melanogaster*. A. The subunits are incapable of high-affinity ligand or DNA binding but are in equilibrium with the unstable EcR/USP heterodimer. B. Although the EcR/USP complex is unstable, this interaction presumably induces a conformational change in EcR (as indicated by a notch) so that the complex becomes capable of high-affinity binding to ecdysteroid and to EcREs. C. Binding of ligand or EcREs or both to EcR/USP favors formation of stable EcR/USP complexes to create a functional complex (D) that tranduces the ecdysteroid signal. (Adapted with permission from *Nature* (T.-P. Yao, B. M. Forman, Z. Jlang, L. Cherbas, J.-D. Chen, M. McKeown, P. Cherbas, and R. M. Evans. Functional ecdysone receptor is the product of *EcR* and *ultraspiracle* genes. 366: 476–479) © 1993 Macmillan Magazines Limited.)

Fig. 12.17. Polytene chromosomes in salivary gland cells of *D. melanogaster*. A. Diagram showing how the bands in polytene chromosomes are thought to be generated by the side-to-side packing of homologous looped domains in the chromosome. B. A highly schematic view of puff formation in a polytene chromosome. C. Autoradiogram of a single puff. The portion shown is undergoing RNA synthesis and has therefore become labeled with [³H]uridine, as indicated by the deposition of silver grains. (© 1989 From *Molecular biology of the cell*, 3rd ed., by B. Alberts et al. Adapted by permission of Routledge, Inc., part of The Taylor & Francis Group.)

were found to change as development of the insect progressed (Fig. 12.18B), some of these puffs being hormone-dependent and others not.

Early, Intermediate, and Late Ecdysteroid-Inducible Genes

Clever and Karlson (1960) were the first to demonstrate that ecdysteroid injection induces chromosome puffing in larval salivary gland chromosomes of *C. tentans*. In 1974, Ashburner and colleagues proposed a model to explain how release of 20E prior to the larval-pupal molt might control puff formation and metamorphosis in salivary gland cells of *D. melanogaster* (Fig. 12.19). They proposed that the ecdysone/receptor complexes (ER) resulting from 20E release bind to the ecdysone response elements of early (EGs), intermolt (IGs), and late (LGs) ecdysone-responsive genes, activating expression of the EGs and inhibiting that of the IGs and LGs. Proteins of

showed these puffs to represent sites of mRNA synthesis or gene expression (Fig. 12.17C) that were both stage- and tissue-specific, the different puffs transcribing different transcripts. When the puffing patterns of homologous chromosomes in cells of different larval tissues were compared at the same time, their patterns were found to differ (Fig. 12.18A). In addition, the puffing pattern of chromosomes in nuclei of a specific tissue

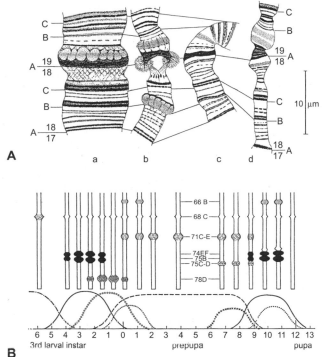

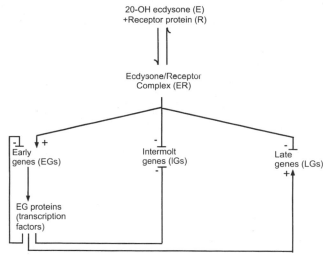

Fig. 12.19. Ashburner and colleagues' (1974) model of ecdysteroid action during the onset of insect metamorphosis.

Fig. 12.18. A. Polytene chromosome I of *Chironomus tentans* (Diptera: Chironomidae), showing differences in the puffing pattern in regions 18A to 19C in salivary gland (a), Malpighian tubule (b), hindgut (c), and midgut (d) cells and the varying levels of ploidy in each cell type (from Kühn, 1971: 483, Fig. 614a–d; after Beerman, 1952). B. The changing pattern of puffs in a region of chromosome III in salivary gland cells of *D. melanogaster* during the late third larval instar and early pupa. (A and B reprinted with permission from A. Kühn. *Lectures on developmental physiology*, p. 483, Fig. 614, and p. 485, Fig. 616 after Becker, 1959. © 1971 Springer-Verlag.)

the EGs were suggested to inhibit their own further expression via a feedback loop, to further inhibit expression of the IGs, and to stimulate expression of the LGs, the products of the latter genes causing metamorphosis to occur.

Six early ecdysone-inducible genes in *D. melanogaster* have since been shown to encode DNA-binding proteins that both repress their own synthesis and induce expression of some 103 downstream late puff genes over a 10-h period at 25°C, exactly as proposed in the model (Apple and Fristrom, 1990; Andres and Thummel, 1992; Jindra, 1994). The sequential appearance of late puff proteins is then thought to direct the tissue-specific changes associated with early stages of metamorphosis.

One of the early puff genes; *E74* (this designation represents its position on the left arm of chromosome 3; see Fig. 12.18B: 74EF), has two transcription units designated *E74A* and *E74B* that have separate promotors (Fletcher and Thummel, 1995). The mRNA of *E74B* appears 15–30 min after addition of 20E; that of *E74A*

appears 1 h later (this is because the *E74A* transcription unit is longer [60 kb] than that of *E74B* [20 kb]). In addition, the *E74A* and *E74B* promotors respond to different concentrations of 20E. Both E74A and E74B proteins are transcription factors that enter the nuclei of target cells, bind to the control regions of downstream late puff genes, and influence their expression.

Recessive, loss-of-function mutations in *E74A* and *E74B* do not affect the transcription of ecdysone-sensitive, primary response genes during late larval and early prepupal development but do affect many secondary response genes (Fletcher and Thummel, 1995). *E74B* mutants are defective in puparium formation and head eversion and die as prepupae or as cryptocephalic pupae (Fig. 10.8C–E), while *E74A* mutants pupariate normally and die either as prepupae or as pharate adults (Fletcher et al., 1995). Most secondary puffs are only modestly affected by mutation in *E74B*, but a subset of such puffs are submaximally induced in *E74A* mutants.

A second early ecdysone-inducible gene, *E75* (Fig. 12.18B: 75B), encodes at least three different proteins (E75A, E75C, E75G), each potentially multifunctional and a member of the steroid hormone receptor superfamily. The gene has been sequenced in larvae of *M. sexta* and *D. melanogaster* (Segraves and Woldin, 1993). Expression of *EcR*, *E74*, *E75*, and the *Broad Complex* (see below) is required for normal egg chamber maturation in *D. melanogaster* during stage 10 of oogenesis (Fig. 2.13A, B) (Buszczak et al., 1999).

The *Broad Complex*

The *Broad Complex* (*BR-C*) of *Drosophila* constitutes an X-chromosome–linked "early" gene residing within an "early" ecdysone-inducible puff at chromosome location 2B5 and having a major role in activating the structural reorganization that occurs during metamorphosis

(Karim et al., 1993; von Kalm et al., 1994). Larvae with *BR-C* deleted or mutant fail to initiate metamorphosis and die as wandering third instar larvae (Restifo and White, 1991, 1992). Study of *BR-C* mutants suggests that this locus encodes transcription factors necessary for proper regulation both of the intermolt period (the IGs) and of early and late gene expression during metamorphosis. The complex has been cloned and shown to encode a family of four DNA-binding proteins, each containing a pair of zinc fingers—Z1, Z2, Z3, and Z4—the functions of these tending to partially overlap (Bayer et al., 1997).

Phenotypic analysis of *BR-C* mutants shows the complex to have three complementation groups: *broad* (*br*), *reduced bristles on the palpus* (*rbp*), and *lethal 1(1) 2 Bc* (*l(1)2Bc*).

- Expression of *br⁺* is required for activation of the dihydroxyphenylalanine (DOPA) decarboxylase gene in the sclerotization pathway (Fig. 10.3) and for imaginal disc eversion and elongation (Figs. 10.8C–H; 10.34; 10.35). Elongation is partly caused by contraction of an actin-myosin ring in individual epidermal cells, with both actin and myosin genes being activated by the protein of *br⁺*, Z2.
- Expression of *rbp⁺* is required for development of specific adult bristles (its function provided by protein Z1 and partially by Z4).
- Expression of *l(1)2Bc⁺* is required for normal fusion of imaginal disc derivatives following their eversion at larval-pupal apolysis and for proper elongation of appendages (its function provided by protein Z3 and partially by Z2).

Internally, wild-type expression of *BR-C* is required for imaginal visual system differentiation; metamorphosis of the CNS; fusion of left and right brain hemispheres; degeneration of larval and development of adult salivary glands, proventriculus, and midgut; differentiation of the ovaries; and normal development of the adult flight muscles (Restifo and White, 1991, 1992; Jiang et al., 1997; Brennan et al., 1998; Buszczak et al., 1999). In addition, binding of *BR-C* proteins to regulatory site DNA of the cell death genes *reaper* (*rpr*) and *head involution defective* (*hid*) in *D. melanogaster* larvae (8.6.1) induces their expression just before destruction of larval salivary gland and midgut cells, while the anti–cell death gene *diap 2* is repressed (Jiang et al., 1997).

Zhou and coworkers (1998) isolated a cDNA ortholog of the *D. melanogaster BR-C* gene from *M. sexta* and showed it to encode homologs of the *Drosophila* Z2, Z3, and Z4 proteins, their DNA-binding domains having, respectively, 93%, 100%, and 85% sequence similarity to those of *Drosophila*. *BR-C* transcripts were not observed in abdominal epidermis during larval molts. However, three transcripts appeared at the end of the feeding stage in fifth instars at a time when epidermis is exposed to a small peak of 20E without JH and becomes committed to pupal development (the pupal commitment peak) (Fig. 12.25). They also induced transcription of *BR-C* in vitro in day 2 fifth instar larval epidermis

by applying ecdysteroids without JH and with dose-response and time courses similar to in vivo induction of pupal commitment. These observations indicate that JH can repress expression of *BR-C* in this moth.

In a screen for additional ecdysone-inducible puffs in *D. melanogaster*, D'Avino and Thummel (1998) discovered a mutation, *crooked legs* (*crol*), that causes defects in adult head eversion and leg morphogenesis and, eventually, pupal death. The wild-type gene is activated by release of 20E at the onset of metamorphosis; encodes at least three isoforms, each having a zinc finger DNA-binding domain; and seems to act upstream of *EcR*, *E74*, *E75*, and *BR-C* since these genes are submaximally expressed in *crol* mutants.

By analyzing *usp* mutant clones of cells in cultured wing discs of *Drosophila*, Schubiger and Truman (2000) demonstrated that in the absence of USP protein, early ecdysone-responsive genes such as *EcR*, *DHR3*, and *E75B* fail to upregulate in response to release of 20E, but that other genes that are normally expressed later such as *β-Ftz-F1* and the Z1 form of the *Broad Complex* (*BR-C-Z1*) are expressed precociously and induce precocious metamorphosis (monitored by following sensory neuron formation and axon outgrowth in the discs). They also showed that *BR-C-Z1* expression and early metamorphosis became steroid-*independent* in *usp* mutant clones.

Temporal Regulation of Ecdysteroid-Responsive Genes
Relationships between 20E release and gene expression prior to and at the onset of metamorphosis in *D. melanogaster* are illustrated in Fig. 12.20 (Thummel, 1996). Approximately 15 larval intermolt puffs, activated by the nuclear receptor DHR78 (Fisk and Thummel, 1998), regress in response to the 20E pulse triggering pupariation at 120 h (some of these encode glue proteins sticking the puparium to the substrate [Fig. 12. 21]), and at least 6 early puffs are directly induced. Transcription factors translated from these early genes then induce expression of about 103 late puff genes. In addition, a few mid prepupal puffs appear in response to the drop in ecdysone titer following puparium formation (Fig. 12.20). The prepupal (pharate pupal) pulse at 130 h reinduces the early-late puffing sequence. Fig. 12.21 summarizes the expression sequence of some of the best studied of these genes, and Fig. 12.22, the stimulatory and inhibitory effects of the proteins of some of the principal, primary response genes at each of these stages.

To identify and analyze the expression pattern of still more genes functioning in metamorphosis, White and associates (1999) assayed whole third instar larvae and prepupae of *D. melanogaster* from 18 h before puparium formation to 12 h after and constructed high-density DNA microarrays containing thousands of gene sequences representing 30%–40% of the genome of this fly. They grouped the genes according to similarity in expression pattern by use of pairwise correlation statistics and self-organizing maps and found them to fall into two categories: (1) those that are expressed at least 18 h

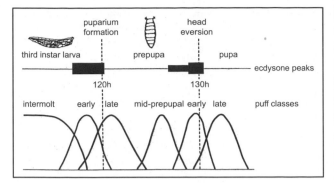

Fig. 12.20. Ecdysone regulation of polytene chromosome puffing during the onset of metamorphosis in *D. melanogaster*. Ecdysone pulses are shown at the top, with the magnitude of each pulse represented by the width of the bar. Developmental time proceeds from left to right, starting at the mid third instar when the polytene chromosomes can first be visualized. The sequence of intermolt, early, late, and mid prepupal puffs is represented at the bottom. (Reprinted from *Trends Genet.*, 12, Thummel, C. S. Flies on steroids—*Drosophila* metamorphosis and the mechanisms of steroid hormone action., pp. 306–310. Copyright 1996, with permission of Elsevier Science.)

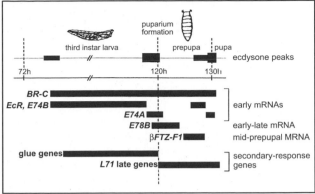

Fig. 12.21. Temporal patterns of ecdysone-regulated gene expression during the onset of metamorphosis in *D. melanogaster*. Ecdysone pulses are at the top, with the magnitude of each pulse represented by the width of the bar. Developmental time proceeds from left to right, with the major ecdysone-triggered transitions marked by vertical dotted lines. The one on the left represents the second to third instar larval molt; the one on the right, head eversion and the prepupal-pupal transition (see Fig. 10.8C–E). Horizontal bars show the timing of primary response, midpupal, and secondary response gene transcription. (Reprinted from *Trends Genet.*, 12, Thummel, C. S. Flies on steroids—*Drosophila* metamorphosis and the mechanisms of steroid hormone action., pp. 306–310. Copyright 1996, with permission of Elsevier Science.)

before puparium formation but then fall to low or undetectable levels during the late third instar larval ecdysone pulse (these are potentially repressed by ecdysone); and (2) those that are expressed at low or undetectable levels before the late larval pulse and are potentially induced by it.

They were able to assign the expression patterns of many of these genes, including those mentioned above, to all the developmental pathways known to function in early metamorphosis and to additional ones, hitherto unknown, and to identify many new genes of unknown function. Such genomic approaches to understanding metamorphosis have the potential to include all functioning genes and, together with molecular and genetic approaches, will greatly facilitate our understanding of metamorphosis and other aspects of development.

12.4.1.6 PRACTICAL ASPECTS

Knowledge of the chemistry and mode of action of 20E during insect ontogeny has resulted in the development of the synthetic, nonsteroidal bisacylhydrazine compounds (Fig. 12.11C) that kill pest insects by interacting with the EcR/USP receptor complex in a manner similar to 20E (Fig. 12.16) (Dhadialla et al., 1998). The insecticidal activity of the first of these, RH-5849, was discovered in 1983 by Rhom and Haas Company toxicologists, who showed it to halt feeding and to stimulate premature apolysis and cuticle synthesis in larvae of *M. sexta*.

RH-5992 (tebufenozide), a synthetic variant of RH-5849 announced in 1992 and marketed as Mimic, Confirm, and Romdan, is widely used against the larvae of pest lepidopterans. It acts to trigger a molt but not complete it (the new cuticle fails to sclerotize, trapping the insect within the remains of its old cuticle and prevent-

ing it from feeding). Unlike RH-5849, it is nontoxic to insects in other orders, including a wide range of parasitoids and predators. RH-0345 (halofenozide), marketed as Mach 2, is sold as a soil systemic and is effective against a wide spectrum of soil-dwelling larvae of turfgrass and ornamental pests. RH-2485, announced in 1996, is more active than tebufenozide against a wider spectrum of pest lepidopterans.

Azadirachtin, a tetranortriterpenoid (Fig. 12.11D), is extracted from seeds of the Neem tree *Azadirachta indica* of India and Burma (Mordue and Blackwell, 1993). Its two fragments, hydroxy furan and decalin, have different effects, the former deterring insect feeding and the latter disrupting growth and development (Aldhous, 1992). Decalin seems to block release of PTTH and allatotropins (see below), so that ecdysone and JH are not synthesized and released and molting fails to occur. Azadirachtin is known to be effective against over 200 pest species in many orders, inhibits chitin synthesis, disrupts sexual communication, induces sterility, decreases gut motility, and does not seem to affect insect parasitoids and predators. A plant systemic containing azadirachtin and sold as Margosan-O was released by W. R. Grace Company in 1992.

Finally, O'Reilly and Miller (1989) showed the *egt* gene of the alfalfa looper *Autographa californica* (Lepidoptera: Noctuidae) nuclear polyhedrosis virus (AcMNPV) to encode an ecdysteroid uridine-5'-diphosphate (UDP)-glucuronosyl transferase that

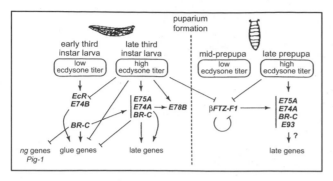

Fig. 12.22. Summary of the multiple, ecdysone-triggered regulatory hierarchies directing onset of metamorphosis in *D. melanogaster*. (Reprinted from *Trends Genet.*, 12, Thummel, C. S. Flies on steroids—*Drosophila* metamorphosis and the mechanisms of steroid hormone action., pp. 306–310. Copyright 1996, with permission of Elsevier Science.)

catalyzes transfer of glucose from UDP-glucose to ecdysteroids, thereby inactivating them. The enzyme seems to be secreted into the hemolymph by virus-infected cells and blocks molting in larvae of the fall armyworm *Spodoptera frugiperda* (Lepidoptera: Noctuidae).

12.4.2 PROTHORACICOTROPIC HORMONE

PTTH is the classic brain hormone of the older literature (12.1), whose release into the hemolymph starts the molting process by inducing cells of the PGs to synthesize and release ecdysone (Fig. 12.6) (Bollenbacher and Granger, 1985). In larvae of the moths *B. mori* and *M. sexta*, PTTH constitutes a family of structurally related neuropeptides belonging to the vertebrate growth factor superfamily that are classified according to size as either small (about 7 kDa) or large PTTH (about 28 kDa) (Bollenbacher et al., 1984; Vafopoulou and Steele, 1997). In *B. mori*, large PTTH is first translated from the transcript of the single *PTTH* gene as a 224–amino acid precursor containing a 109–amino acid subunit (Kawakami et al., 1990; Nagasawa, 1992). Two subunits are then linked by disulfide bonds and glycosylated before or after posttranslational cleavage to generate the homodimeric, mature PTTH (Fig. 12.11G). The prothoracicotropic potency of these PTTHs changes as the insects develop, suggesting their regulatory role to be stage-specific (Bollenbacher et al., 1984, 1993; Westbrook and Bollenbacher, 1990; Smith, 1993).

PTTH from newly emerged adults of *B. mori* induces a dose-dependent response and production of ecdysteroids by PGs of fifth instar larvae of the blood-feeding reduviid bug *Rhodnius prolixus*, suggesting that both function and conformation of PTTH molecules may be conserved between distantly related insects (Vafopoulou and Steele, 1997).

As mentioned in 12.3, large PTTH molecules are usually associated with NSGs in two pairs of dorsolaterally situated NSCs within the brain of *M. sexta*, the group III lateral NSCs (L-NSC III), and small PTTH molecules are

associated with NSGs in the group II medial NSCs (M-NSC IIa₂; Fig. 12.10A, B) (Gibbs and Riddiford, 1977; Agui et al., 1979; Westbrook et al., 1991). Both types are synthesized on polysomes of the rough endoplasmic reticulum within the cell bodies of these cells (Fig. 12.8). In *B. mori* only a single NSC on each side appears to be involved (Kawakami et al., 1990). The PTTH molecules are released into the blood with a 24-h periodicity (Fujishita and Ishizaki, 1981; Vafopoulou and Steele, 1999) from the neurohemal terminals of the axon branches of these cells by exocytosis, in the CC of most insects (Fig. 12.9A) but from the CA in *M. sexta*, *B. mori*, and other lepidopterans (Fig. 12.10A, B) (Agui et al., 1980; Nagasawa, 1992; Nijhout, 1994b).

Long-term perfusions and short-term batch incubations of brains and PGs of the cockroach *Periplaneta americana* with melatonin will induce release of PTTH, apparently the first evidence for a neurohemal release effect in an insect nervous system (Richter et al., 1999).

The reasons why the chemical identity of PTTH was so late in coming compared with that of ecdysone and JH are that brain medial NSCs are smaller, harder to get at, and more difficult to manipulate than PGs, CC, or CA; because there are other, similar NSCs in the same area of the brain, synthesizing other hormones; and because some NSCs synthesize more than one hormone (Chapman, 1998). Use of immunocytochemistry and radioimmuno assay has largely circumvented these problems. Monoclonal antibodies, raised in mice against PTTH, are now available for *M. sexta* (O'Brien et al., 1988) and *B. mori* (Mizoguchi et al., 1989).

12.4.2.1 MODE OF ACTION

PTTH molecules are lipophobic, like other neuropeptide hormones, and act on PG cells to synthesize ecdysteroids by way of a membrane-bound receptor protein and the second messenger, cyclic AMP (cAMP), as summarized in Fig. 12.23 for *M. sexta* (Vedickis et al., 1976; Smith and Cambert, 1985; Smith, 1993, 1995; Nijhout, 1994b; Chapman, 1998). They stimulate the synthesis of at least three proteins, one of which, beta-tubulin, may influence the dynamics of microtubule-dependent secretion or interorganelle movement of ecdysteroid precursors (Rybczynski and Gilbert, 1995). In vivo, beta-tubulin synthesis, resulting from increased transcription and translation in response to PTTH, was detected in some PG cells after only 5–10 min of exposure. The response of PGs to PTTH has a 24-h periodicity throughout larval and adult development of *R. prolixus* just like that of PTTH release, with high responsiveness at dusk and none during the day (Vafopoulou and Steele, 1999).

12.4.3 OTHER ECDYSIOTROPINS

Recently, a number of additional peptides, extracted from various insects, have been shown to induce secretion of 20E by PG cells. One, from larval hindguts of the European corn borer moth *Ostrinia nubialis* and the gypsy moth *Lymantria dispar*, stimulates production of

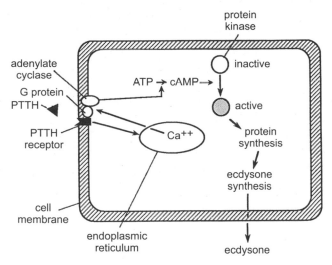

Fig. 12.23. Mode of action of prothoracicotropic hormone (PTTH) in inducing ecdysone synthesis and release in a prothoracic gland cell of *M. sexta* by way of the second messenger cyclic AMP. (From R. F. Chapman. 1998. *The insects. Structure and function*, p. 531, Fig. 21.13. 4th ed. Adapted with the permission of Cambridge University Press.)

ecdysteroids by PGs of *L. dispar* (Gelman et al., 1991) (this hormone may be responsible for the results observed by Sakurai ct al. [1991]; 12.4.1.4). It is a small peptide with a molecular size of 500–1500 Da and appears to act on PG cells the same way as PTTH, by way of the second messenger cAMP (compare Fig. 12.23). A third ecdysiotropin, extracted from gypsy moth brains, stimulates the synthesis of ecdysteroids by cells of its larval and pupal testes (1.3.7.2.3) (Wagner et al., 1997) and is a small peptide of 19 amino acids.

Bombyxin, extracted from brains of the commercial silkworm *B. mori*, consists of two dissimilar peptides, A and B (of 20 and 28 amino acids, respectively), held together by disulfide bonds. It is a member of the vertebrate insulin superfamily (Fig. 12.11H) (Nagasawa, 1992; Vafopoulou and Steel, 1997). Its functions are unknown in *B. mori* but it can stimulate PGs of the saturniid *Samia cynthia* and of the hemipteran *R. prolixus* to produce ecdysteroids, though not as effectively as PTTH does in the latter insect (Nagasawa, 1992; Vafopoulou and Steel, 1997). It is synthesized in four pairs of mid-dorsal NSCs in the brain projecting to the CA and is encoded on multiple copies of several bombyxin genes.

In blowflies, control of ecdysone synthesis by PGs seems to involve interactions between two ecdysiotropins of 5 and 16 kDa and an ecdysiostatin of 11 kDa (Hua et al., 1997).

12.4.4 ALLATOTROPIN

Results of many experiments on a variety of insects had earlier suggested that secretion of a factor from the brain induces synthesis and release of JH by cells in the CA (Nijhout, 1994b). Kataoka and coworkers (1989) extract-

ed such a hormone from 10,000 pharate adult heads of *M. sexta* and found it to be a 13–amino acid neuropeptide (Gly-Phe-Lys-Asn-Val-Glu-Met-Met-Thr-Ala-Arh-Gly-Phe-NH$_2$), to stimulate secretion of JH by CA cells in adults, and to accelerate the heart beat in pharate adults. Since then, expression of the *Mas-allatotropin* (*Mas-AT*) gene has been identified in two NSCs in the frontal ganglion that project their axons down the recurrent nerve to the gut, and in the terminal abdominal ganglion of larvae (Bhatt and Horodyski, 1999).

12.4.5 JUVENILE HORMONES

JHs, sesquiterpenes synthesized in cells of the CA, are best known for their role as primary gonadotropins in adult female insects (2.3.2.1) and for inhibiting metamorphosis in molts induced by 20E (Gilbert, 1976). However, they have numerous other effects, including a role in caste development in eusocial insects (12.8.2), in phase change in migratory locusts (12.8.1.3), in production of aphid morphs (12.8.1.2), and in the development of exaggerated structures such as the horns of scarab beetles (12.8.1) (Nijhout, 1994b; Jones, 1995; Emlen and Nijhout, 2000). They arc known to exist in at least five forms that differ in their side chains, stage of secretion, and phylogenetic distribution (Fig. 12.11E) (Richards, 1981; Nijhout, 1994b; Riddiford, 1994, 1996; Riddiford and Willis, 1996).

In *Schistocerca gregaria* nymphs (Orthoptera: Acrididae) and probably in other insects, the left and right CA can synthesize differing amounts of JH, with that producing the most JH being unpredictable (differences of up to 1000-fold have been recorded [Tobe, 1977]). However, the total amount of JH produced by both CA at any one time is the same and is characteristic for species and stage of development.

12.4.5.1 CHEMICAL STRUCTURE AND BIOSYNTHESIS
JH I was first chemically identified in 1967 by Röller and colleagues in larvae of the saturniid moth *Hyalophora cecropia* (Fig. 12.11E). JH 0 is known from *M. sexta* embryos; JH I and II, from various lepidopteran larvae; and JH III, from adults of 12 species of insects in six orders (Orthoptera, Blattodea, Isoptera, Coleoptera, Lepidoptera, and Diptera) (Fig. 12.11E) (Riddiford, 1996). These forms differ in physiological activity and in concentration in the hemolymph. It has been suggested (Novák, 1975) that JH III probably evolved first as a gonadotropin, the others being derived from it later through modification of their side groups after the origin of incomplete and complete metamorphosis. Farnesoic acid and methyl farnesoate, immediate precursors of JH III, but not JH III itself, are synthesized in and released from the mandibular organs of the crustacean *Procambarus clarkii* (Cusson et al., 1991).

12.4.5.2 MAINTENANCE OF JUVENILE HORMONE TITER IN THE HEMOLYMPH
Environmental factors of various kinds, depending on species, sex, and stage of development, stimulate spe-

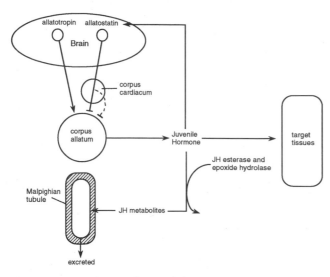

Fig. 12.24. Regulation of juvenile hormone titer in insects involves a balance between synthesis in the corpora allata and degradation in the hemolymph and secretion by the Malpighian tubules. (From R. F. Chapman. 1998. *The insects. Structure and function*, p. 580, Fig. 21.11. 4th ed. Adapted with the permission of Cambridge University Press.)

cific NSCs in the brain to synthesize and release allatotropin (Fig. 12.24) (de Kort and Granger, 1981; Feyereisen, 1985; Nijhout, 1994b; Chapman, 1998). Molecules of this hormone in turn stimulate cells of the CA, by way of second messengers, to synthesize and release free JH into the blood (Schooley and Barker, 1985; Rachinsky and Tobe, 1996). These molecules can be quickly degraded (half-life of 30–300 min) to JH acids (subsequently passed to and excreted by the Malpighian tubules) by action of JH-specific esterases and an epoxide hydrolase produced in the fat body (de Kort and Granger, 1981; Feyereisen, 1985; Hammock, 1985; Chapman, 1998). However, degradation of JH does not usually occur because it is generally bound to and protected from degradation by JH-binding proteins or lipoproteins (Goodman and Chang, 1985; Prestwich et al., 1996). This binding also prevents inappropriate uptake of JH and regulates its distribution within the body.

In larval instars of investigated species, JH is present in the blood throughout the instar, with its concentration tending to be high early in each instar and low toward its end, although there is much variation between species (Fig. 12.25: larva 4) (de Kort and Granger, 1981; Riddiford, 1990, 1993a, 1993b, 1994; Nijhout, 1994b; Chapman, 1998). In *M. sexta*, JH disappears on day 4 of the last (fifth) larval instar due to inhibition of JH synthesis and release by CA cells by the peptide hormone allatostatin (12.4.6) and by destruction of JH already in the blood by JH-specific esterases (Figs. 12.24; 12.25: larva 5) (pupal commitment caused by

ecdysteroid release occurs at this time) (Riddiford, 1990). A small release of JH occurs on day 9 of the fifth instar just after larval-pupal apolysis at the same time as a larger release of 20E (the pupal differentiation peak). JH is absent throughout the pupal stage (Fig. 12.25: pupa) but is synthesized again just after adult emergence (Hebda et al., 1994) and subsequently has a major role in gamete and pheromone production in members of both sexes (1.3.7.2.3, 2.3.2.1). Both qualitative and quantitative changes in JH concentration occur during postembryogenesis, and there are large differences between species in the absolute concentration of circulating JH (de Kort and Granger, 1981).

As mentioned earlier (12.4.1.3), one reason why hemimetabolous and holometabolous adults do not molt is that their PGs degenerate before (exopterygotes) or after (endopterygotes) adult emergence (another reason is that the adult exoskeleton lacks an ecdysial line). In larvae, the PGs are maintained by presence of JH. As first demonstrated by Bounhiol (1938) (12.2), when the CA are removed from an early larval instar, it molts to a tiny precocious pupa and then an adult, and its PGs degenerate. In addition, new, implanted PGs degenerate too, indicating that their degeneration is not under direct nervous control. Only "activated" PGs are competent to degenerate and will do so only if they function at least once in the absence of JH (the small pupal commitment peak of ecdysteroid on day 5 of the fifth instar in *M. sexta*; Fig. 12.25).

12.4.5.3 MODE OF ACTION

JH allows larval molting in response to release of 20E but prevents the switching of gene expression necessary for metamorphosis (Nijhout, 1994b; Jones, 1995; Riddiford, 1996; Restifo and Wikan, 1998). Its mode of action was long thought to be similar to that of ecdysteroids (Fig. 12.14), but the chief evidence for this is in the induction of vitellogenin synthesis in the fat body (2.3.2.1) (Nijhout, 1994b; Jones, 1995). Putative cytosolic and nuclear receptor proteins for JH have been identified in cells of many tissues (Riddiford and Hiruma, 1991; Riddiford and Truman, 1993; Riddiford et al., 1999), and Palli and coworkers (1991) showed that JH binds specifically to a 29-kDa protein (rjP29) in nuclei of larval *M. sexta* epidermal cells. However, this observation was shown to be erroneous by use of photoaffinity analogs and to be due to the presence of contaminating esterases (Riddiford, 1996).

Putative JH receptors are present in epidermal nuclei of *M. Sexta* during all immature instars but are not necessarily accompanied by JH (Fig. 12.25) (Riddiford, 1992; Nijhout, 1994b). In addition, results of experiments by Nijhout and his students on several insect species (reviewed in Nijhout, 1994b) suggest each instar of each species to have characteristic and relatively short JH-sensitive periods, presumably due to the presence of JH receptors, with JH only able to induce its effects on target tissues during these periods (Figs. 12.25, 12.26). In addition, the presence of JH receptors probably differs in

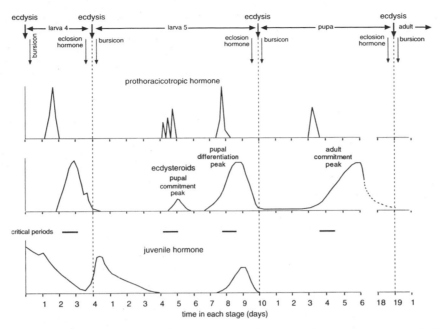

Fig. 12.25. Changes in hormone concentration regulating molting and metamorphosis in *M. sexta*. In the larva 4–larva 5 molt (and in all previous molts), juvenile hormone (JH) is present during a critical period of sensitivity to this hormone (JH-sensitive period) and inhibits the epidermis from depositing pupal or adult cuticle. After this molt, a brief pulse of ecdysteroid in the absence of JH (the pupal commitment peak) inhibits expression of the larval genetic program and activates genes involved in the transition from larva to pupa. JH is again present at a second critical period, just after larva-pupal apolysis, when a second, larger pulse of ecdysteroid is released (the pupal differentiation peak). This induces eversion and differentiation of imaginal discs after the larval-pupal ecdysis and other aspects of metamorphosis. A third, sustained pulse of ecdysteroid from day 3 to day 18 in the pupal stage, again without JH (the adult commitment peak), inhibits expression of the pupal transition genes and activates the imaginal ones. Eclosion hormone is released just before and bursicon just after each ecdysis in all molts. The secretion of pre-ecdysis– and ecdysis-triggering hormones and of crustacean cardioactive peptide, which are also involved (Fig. 12.28), is not shown. (From R. F. Chapman. 1998. *The insects. Structure and function*, p. 396, Fig. 15.30. 4th ed. Adapted with the permission of Cambridge University Press.)

different tissues at different times during development (Emlen and Nijhout, 2000).

JH is known to regulate the synthesis of pigment and cuticle proteins by epidermal cells in *M. sexta* larvae but has no effect on the ecdysteroid-induced expression of genes required for production of new cuticle at larval molts (Marsh, 1993). The presence of JH also inhibits the activation of genes required for transition from larval to pupal or adult development (Fig. 12.26). Also, if JH or JH agonists such as methoprene (Fig. 12.11F) are fed or topically applied to *D. melanogaster* larvae near the onset of metamorphosis, they disrupt the metamorphic reorganization of the CNS, salivary glands, and musculature in a dose-dependent manner and in a way similar to that caused by mutations in the *Broad Complex* (12.4.1.5), indicating a possible inhibitory role for JH in *BR-C* expression (Restifo and Wikan, 1998). In fact, JH I is now known to inhibit expression of *BR-C* during larval molts in *M. sexta* (Zhou et al., 1998).

By examining, by scanning electron microscopy, the JH analog–induced cuticles of larval-adult, larval-pupal, and pupal-adult intermediates of the bug *Pyrrhocoris apteris* and of the moths *Galleria mellonella* and *M.*

sexta, Sláma and Weyda (1997) discovered the response of individual epidermal cells to the presence of these compounds to be all or none, the cuticles of intermediates always being a mosaic of individual larval, pupal, or adult elements deposited by individual cells (Fig. 12.27). Response of each epidermal cell to the presence of JH must thus be autonomous and depend on the presence of JH receptors; the proportion of cells doing so depends on the time of application, within an instar, of a minimum concentration of the analog. Their observations refute Piepho's (1951) long-accepted concept of concentration-dependent JH action (12.2).

Jones (1995) suggested our lack of success in nailing down the mode of action of JH to be due to its diverse effects (her Table 1) and to the design of the experiments, which has been influenced too much by findings on vertebrate steroid function. The presence of multiple binding sites for different transcription factors on the regulatory sites of JH-sensitive genes suggests that their regulation may involve coordinate action of multiple regulatory molecules as in segment specification in *Drosophila* embryos (7.2.1.4).

JH analogs (Fig. 12.11F), first developed as third-

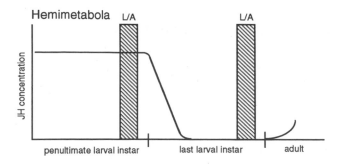

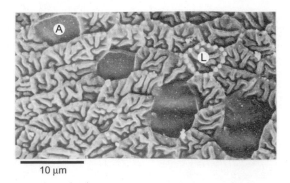

10 µm

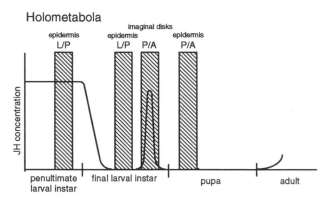

Fig. 12.27. Scanning electron micrograph of a small portion of the third abdominal tergite of a larval-adult intermediate (adultoid) of *Pyrrhocoris apterus* (Hemiptera: Pyrrhocoridae) produced by juvenoid application. Shown is a mosaic of larval (L) and adult (A) cuticular patches, each of which is the product of a single epidermal cell. This illustrates the all-or-none response of epidermal cells to the presence of ecdysteroids and juvenile hormone. (Reprinted with permission from K. Sláma and F. Weyda. The all-or-none rule in morphogenetic action of juvenile hormone on insect epidermal cells. *Proc. R. Soc. Lond. B. Biol. Sci.* 264: 1463–1470. © 1997, The Royal Society.)

Fig. 12.26. Typical juvenile hormone (JH) titer profiles during the end of larval life in hemimetabolous (based on *Schistocerca gregaria*) and holometabolous insects. Hypothetical JH-sensitive periods are indicated by the hatched bars. In exopterygotes, there is a JH-sensitive period near the end of each larval instar, during which the commitment to larval versus adult commitment is controlled (L/A). In endopterygotes, there is a similar JH-sensitive period for larval versus pupal commitment (L/P) near the end of each larval instar except the last. The two JH-sensitive periods in the final larval instar (L/P, P/A) are based on findings in *M. sexta* (see Fig. 12.25). There is an additional JH-sensitive period for pupal versus adult determination (P/A) early in the pupal stage at the onset of adult development. Note that in both types of insect, the change in gene expression from larva to adult (Hemimetabola), larva to pupa, or pupa to adult (Holometabola) can occur only in the absence of JH. (Nijhout, H. F. *Insect hormones.* Copyright © 1994 by Princeton University Press. Reprinted by permission of Princeton University Press.)

generation pesticides in the 1960s (Williams, 1967), mimic the action of JH at times when natural JH titer is low, and induce development of monstrous larval-pupal or pupal-adult intermediates and sterility in adult females (Dhadialla et al., 1998). In fact, Cusson and Reddy (2000) suggest that we can rejuvenate studies in insect pest management by

• isolating and cloning JH receptors;
• characterizing polydnavirus and entomopoxvirus gene products responsible for inhibiting host metamorphosis;
• isolating and cloning allatostatins (see below) and

designing synthetic allatostatic polypeptides inhibiting JH synthesis;
• characterizing JH biosynthesis enzymes specific to pest lepidopterans;
• characterizing the regulatory material introduced into host insects by parasitoid wasps of the genus *Chelonus* (Hymenoptera: Braconidae), which induce precocious metamorphosis;
• characterizing a tree-resistance mechanism involving anti-JH effects resulting in the failure of female pests to produce and deposit eggs;
• cloning JH esterase and epoxide hydrolase genes; and
• producing recombinant baculoviruses that overexpress these JH-degradative enzymes.

12.4.6 ALLATOSTATINS

For some years, experimental evidence had suggested the existence of factors that inhibit the synthesis and release of JH from CA cells (Fig. 12.24) (Hoffmann et al., 1999). In 1991, Kramer and colleagues purified a small peptide hormone, allatostatin, from head extracts of *M. sexta* that strongly inhibited JH synthesis in vitro by CA cells of fifth instar and adult females of this insect and of the moth *Heliothis virescens*. The peptide has 15 amino acids and is synthesized in lateral NSCs of the brain (Fig. 12.24) (Riddiford, 1992). Since it had no effect on CA cells in female cockroaches (*Periplaneta americana*), grasshoppers (*Melanoplus sanguinipes*), and mealworms (*Tenebrio molitor*), the molecules or their receptors must differ between distantly related species.

Since then, five additional allatostatins of 8–18 amino acids have been isolated from the cockroach *Diploptera*

Hormones, Molting, and Metamorphosis

punctata by immunocytochemistry. These may participate in the function of antennal pulsatile organs and hindgut in addition to inhibiting JH synthesis by CA cells (Stay and Woodhead, 1993; Pratt et al., 1997). Specific lateral NSCs of the brain deliver these peptides to the CA through NCC II, where they bind to two types of allatostatin receptors; a branch of NCC III transports them to muscles of the pulsatile organs (Fig. 12.24) (Pratt et al., 1997) (see Figs. 12.9A and 12.10A,B). Additional allatostatins have been identified from *P. americana*, *Blattella germanica* (Blattodea), *Gryllus bimaculata* (Orthoptera), *Aedes aegypti* (Diptera), *Apis mellifera* (Hymenoptera), and a crustacean, and they will no doubt be found to be universal in arthropods (Stay, 2000). Expression of an allatostatin gene has so far been observed only in particular brain NSCs and in midgut endocrine cells (Pratt et al., 1997; Hoffmann et al., 1999).

12.4.7 PRECOCENES AS ANTIALLATOTROPINS

In 1976, Bowers and associates discovered that bedding plants of the species *Ageratum houstonianum* (Asteraceae) produce chemicals, the precocenes, having anti-JH activity. If these are imbibed by feeding insects of a variety of species, they cause precocious metamorphosis of larvae and sterilize adult females (i.e., they produce effects identical to those induced by removal of the CA). The precocenes are in two forms: 1 and 2 (Fig. 12.11I) (Bowers, 1985). Precocene 2, with an additional methoxy group, is 10-fold more active than precocene 1 and, when applied topically to an insect, causes irreversible degeneration of CA cells (Unnithan et al., 1977; Pener et al., 1978). Apparently, the precocene is transformed into a highly reactive epoxide when imbibed by the insect.

12.4.8 ECLOSION HORMONE

EH was first demonstrated experimentally by Truman and Riddiford in 1970 in pupae of the saturniid moths *Hyalophora cecropia* and *Antheraea pernyi* by means of brain and CC transplants. For many years, it was considered to directly induce ecdysial behavior (Truman, 1985, 1990). In larvae of *M. sexta*, it is synthesized in two ventromedian pairs of neurosecretory cells (VMN-SCs) in the brain (Fig. 12.10C) that project, via the circumesophageal connectives through the VNC, to neurohemal areas on the hindgut (Truman and Copenhaver, 1989). However, in pharate adults, it is released from the CC in response to falling concentrations of 20E (Fig. 12.25), the change in release site resulting from restructuring in the VMNSCs during metamorphosis (Horodyski, 1996).

In *M. sexta*, EH is a neuropeptide of 62 amino acids (Fig. 12.11J) and is encoded by a single-copy *EH* gene containing three exons and three introns (Horodyski et al., 1989). During expression, exon I is not translated. Exon II encodes a 26–amino acid signal sequence and amino acids 1–4 of the EH peptide, and exon III encodes the remaining 58 amino acids. Transcripts of the gene are present in larval, diapausing pupal, and developing adult brains, and the hormone translated from them is released into the hemolymph as a prohormone, the signal sequence being snipped off to activate it.

12.4.8.1 MODE OF ACTION

The effectiveness of *M. sexta* EH in inducing molting behavior has been evaluated in bioassays involving 12 other species in seven orders (Blattodea, Orthoptera, Hemiptera, Coleoptera, Trichoptera, Lepidoptera, and Diptera) and was shown to be positive in all instances (Horodyski, 1996). Also, there is 70% sequence homology between the EH of *M. sexta* and *D. melanogaster* and 80% with that of *Bombyx mori* (Fig. 12.11J).

For many years, EH was considered to act directly on motor neurons in ganglia of the abdominal nerve cord by way of the second messenger cGMP to induce ecdysial behavior (Truman, 1985, 1990; Nijhout, 1994b). However, this action, at least in larvae and pupae of *M. sexta*, is now known to also involve secretion of a second pre-ecdysis–triggering hormone (PETH) and a third ecdysis-triggering hormone (ETH) by 18 large Inka cells attached to cuticle below each spiracle (Zitnan et al., 1996, 1999; Ewer et al., 1997; Predel and Eckert, 2000).

In the giant silkworm *Antheraea pernyi*, adult eclosion occurs on day 19 after pupation. Synthesis of EH begins on day 6, reaches a peak on day 12, and remains constant until eclosion. On day 19, EH molecules move to the CC in response to dropping concentrations of 20E (Fig. 12.25) and are released 2.5 h before eclosion at 100-fold the concentration required to induce ecdysial behavior. The hormone has a half-life of about 45 min in the blood. A similar peak occurs in every larval instar just before ecdysis (Fig. 12.25).

In addition to its role in inducing ecdysial behavior, EH also turns off pupal behavioral patterns and turns on adult ones; induces degeneration of pupal motor neurons and larval abdominal intersegmental muscles; plasticizes newly deposited imaginal cuticle prior to eclosion; and acts to permit release of the sclerotization hormone, bursicon, from parasympathetic organs associated with other NSCs in ganglia of the VNC (Figs. 12.7, 12.9B) (Reynolds, 1985; Truman, 1985).

12.4.9 *MANDUCA SEXTA* PRE-ECDYSIS– AND ECDYSIS–TRIGGERING HORMONES

While investigating immunohistological staining patterns before and after ecdysis in *M. sexta* caterpillars, Zitnan and coworkers (1996) observed 18 large (up to 250 μm in diameter) Inka cells, each associated with two other cells to form an epitracheal gland attached to the underside of the tracheal trunk, below each spiracle on each side of the prothorax and of abdominal segments 1–8. These cells were found to synthesize and release a 26–amino acid peptide, *M. sexta* ETH (or Mas-ETH), seemingly in response to receipt of molecules of EH. When they injected Mas-ETH into pharate larvae, pupae, or adults of *M. sexta*, they initiated pre-ecdysial behavior in 2–10 min. They also found isolated CNSs to respond to Mas-ETH but not to EH unless they were

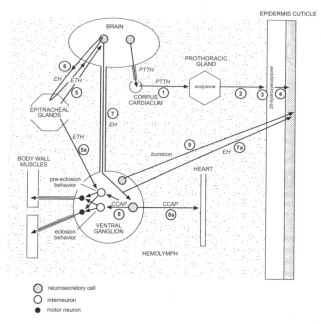

EPIDERMIS CUTICLE

BRAIN

PROTHORACIC
GLAND

PTTH

PTTH

CORPUS
CARDIACUM

ecdysone

20-hydroxyecdysone

EH ETH

EPITRACHEAL
GLANDS

ETH

EH

ETH

BODY WALL
MUSCLES

pre-eclosion
behavior

bursicon

EH

HEART

CCAP

CCAP

eclosion
behavior

VENTRAL
GANGLION

HEMOLYMPH

◨ neurosecretory cell

○ interneuron

● motor neuron

CONTROL OF APOLYSIS AND CUTICLE PRODUCTION
1 PTTH stimulates synthesis and release of ecdysone
2 ecdysone in hemolymph
3 ecdysone hydroxylated at tissues
4 20-hydroxyecdysone regulates genes producing cuticle

CONTROL OF ECDYSIS
5 ecdysis triggering hormone causes release of eclosion hormone
5a ecdysis triggering hormone switches on pre-eclosion hormone behavior
6 positive feedback loop between ETH and EH results in massive release of EH
7 central release of EH causes release of CCAP
7a EH acting via hemolymph plasticizes cuticle
8 CCAP switches on eclosion behavior and switches off pre-eclosion behavior

CONTROL OF EXPANSION AND SCLEROTIZATION
8a CCAP acting via hemolymph increases heartbeat
9 bursicon first plasticizes cuticle, then switches on cuticular sclerotization

Fig. 12.28. Summary of how hormones regulate the molt from pupa to adult based mostly on study of *M. sexta*. The abbreviations of hormones are italicized and are as follows: CCAP, crustacean cardioactive peptide; EH, eclosion hormone; ETH, ecdysis-triggering hormone (includes pre-ecdysis–triggering hormone, PETH); PTTH, prothoracicotropic hormone. (From R. F. Chapman. 1998. *The insects. Structure and function*, p. 397, Fig. 15.31. 4th ed. Reprinted with the permission of Cambridge University Press.)

accompanied by the tracheal system. This suggested the following sequence:

$$EH \rightarrow \text{Inka cells} \rightarrow \text{Mas-ETH} \rightarrow \text{ecdysial}$$
$$\text{behavior} \rightarrow \text{ecdysis}$$

However, additional study by Ewer (1997) and Zitnan (1999) and their respective associates showed instead that Inka cells and the source of EH, the VMNSCs (Fig. 12.10C), mutually excite each other—Mas-ETH inducing VMNSCs to release EH (Fig. 12.28: 6), and EH inducing Inka cells to release Mas-ETH (Fig. 12.28: 5)—the reciprocal excitation causing a massive EH/Mas-ETH surge in the blood. *M. sexta* PETH (or Mas-PETH), encoded by the same gene as Mas-ETH, induces pre-ecdysis I, and Mas-ETH induces pre-ecdysis II behavior by

way of ganglia in the VNC (Fig. 12.28: 5a) This phasic signal then tonically activates a segmentally repeated network of NSCs in CNS ganglia that synthesize and release crustacean cardioactive peptide (CCAP) (Fig. 12.28: 8) (Chapman, 1998; Zitnan et al., 1999). Release of CCAP switches off both pre-eclosion behaviors and induces eclosion behavior by way of interneurons and motor neurons in ganglia of the VNC (Fig. 12.28: 8). The motor neurons innervate the skeletal muscles (Fig. 12.28), foregut, and alary muscles of the dorsal vessel (Fig. 12.28: 8a), suggesting CCAP induces air swallowing and the circulatory changes accompanying ecdysis.

Prior to onset of ecdysis behavior, rising 20E levels at the end of each molt (Fig. 12.25) induce expression of the *EcR-B1* and *ETH* genes in the Inka cells and the synthesis of ecdysteroid receptors, Mas-PETH, and Mas-ETH (Zitnan et al., 1999). Subsequent ecdysteroid decline (Fig. 12.25) is required for peptide release, which is induced by release of EH and initiates the three behavioral patterns in specific order (Kingan and Adams, 2000):

Mas-PETH → pre-ecdysis I
Mas-ETH → pre-ecdysis II and, indirectly, ecdysis.

12.4.10 BURSICON

In 1935, Fraenkel demonstrated that cuticle sclerotization in newly emerged blowflies, *Calliphora erythrocephala*, could be delayed for over 24 h by forcing them to dig through sawdust using their ptilinum and various somatic movements (10.2.1). Once removed from the sawdust, they quickly expanded their wings and sclerotized their exoskeletons, presumably because inhibitory sense organs were no longer being stimulated by elements in the medium through which they burrowed (Denlinger and Zdárek, 1994).

That a factor additional to 20E was required for sclerotization was independently demonstrated by Cottrell and by Fraenkel and Hsaio in 1962. This factor was named bursicon by the second group in 1965. Release of bursicon is induced by the insect swallowing air. This in turn is part of the ecdysial behavior induced by release of EH, PETH, ETH, and CCAP (Figs. 12.25, 12.28).

12.4.10.1 MODE OF ACTION

When newly emerged flies are decapitated or tightly ligatured about their necks within a few minutes of emergence (i.e., before the end of the CP for bursicon release), they are unable to sclerotize their cuticle (though that of sensilla and joints is sclerotized before emergence [Fig. 10.4]). If this ligature is made after the end of the CP, they can. When blood from tanning flies is injected into flies decapitated or ligatured before the end of the CP, the block to sclerotization is lifted.

Bursicon is a protein of 30–60 kDa and probably exists in several forms (Reynolds, 1985). It is synthesized in four pairs of NSCs in each abdominal ganglion of *M. sexta* larvae and is released in massive quantities 3–30 min after eclosion in the sarcophagid fly *Neobellieria*

Hormones, Molting, and Metamorphosis

bullata at 100-fold the concentration required to initiate tanning; it is no longer detectable 11 h after eclosion (Fig. 12.25). However, judging from the results just summarized, bursicon must be synthesized by NSCs in the brain of adult flies.

In larvae of the mealworm *Tenebrio molitor* (Coleoptera: Tenebrionidae), the hormone occurs in two bands at 30 and 45 kDa (Kaltenhauser et al., 1995). Bursicons of other insects are similar and may be so for all arthropods (Kostron et al., 1995).

In addition to inducing sclerotization of cuticle after ecdysis, bursicon is known to plasticize adult cuticle at emergence (Fig. 12.28: 9) (EH does the same before ecdysis), trigger renewed adult cuticle deposition after ecdysis (it stops just before ecdysis), and causes the wing epidermis to degenerate (these cells maintain the shape of the adult wings during inflation and prevent their cuticle from ballooning) (Reynolds, 1985). Bursicon activates the enzyme tyrosinase in epidermal cells and thus, hydroxylation of tyrosine to DOPA in the sclerotization pathway by way of the second messenger cAMP (Fig. 10.3) (Reynolds, 1985; Hopkins and Kramer, 1992) (Fig. 10.1). Its ability to promote cuticle sclerotization is blocked by actinomycin D and puromycin, inhibitors, respectively, of the transcription and translation of RNA, so it must also have an effect on gene expression. In *M. sexta*, the epidermis does not acquire the ability to respond to bursicon until 9h before the hormone is released, presumably because of the absence of bursicon receptors. The hormone probably has a similar function in all insects (Kostron et al., 1995).

12.4.11 DIAPAUSE HORMONE

As discussed in 8.6.4.3, DH, a peptide of 24 amino acids, is synthesized in NSCs in the subesophageal ganglion of female *B. mori* pupae and induces embryos of the next generation to enter diapause (Denlinger, 1985; Nagasawa, 1992).

12.5 The Interaction of Hormones during Molting and Metamorphosis

The regulation of molting and metamorphosis by PTTH, 20E, JH, EH, and bursicon (but not PETH, ETH, and CCAP) is summarized in Fig. 12.25 for the tobacco hornworm, *Manduca sexta* (Richards, 1981; Riddiford, 1985, 1990, 1993a, 1993b; Bollenbacher, 1988; Riddiford and Truman, 1993; Nijhout, 1994b; Chapman, 1998). During each larval instar, the epidermis deposits a larval cuticle and specific larval pigments, owing to the presence of JH at critical (JH-sensitive) times during the molt cycle (Fig. 12.25: larva 4). JH also permits the growth of imaginal discs and maintains a strictly larval epidermis.

Metamorphosis is preceded by a fall in JH titer on day 4 of the fifth instar (Fig. 12.25: larva 5). This decrease in JH is followed by a pulse of 20E, the pupal commitment peak, on day 5, which in epidermal and other larval tissues switches the developmental program of the cells

from larval to pupal (no JH is present at this time), probably by inducing expression of the early and late genes indicated in Figs. 12.20–12.22. A second, larger peak of 20E immediately after larval-pupal apolysis on days 8 and 9 then triggers the onset of pupal differentiation. JH appears again transiently during this second pupal differentiation peak to prevent precocious adult differentiation of imaginal discs. These peaks in hormone release also influence the death or remodeling of larval motor neurons and muscles and the differentiation of imaginal ones (10.2.2.1, 10.2.4.1).

Following larval-pupal ecdysis, there is a third, sustained and much larger peak of 20E, also in the absence of JH, the adult commitment peak, which switches the epidermal cells from a pupal to an adult program (Fig. 12.25: pupa). This peak of 20E stimulates adult differentiation up until day 6 but thereafter inhibits it, this inhibition increasing with advancing age. The decrease in 20E titer commencing on day 10 causes certain additional larval interneurons to degenerate and stimulates release of EH, PETH, ETH, and CCAP (Fig. 12.28). Release of EH in turn induces release of bursicon. Note that in all molts, there is a peak of EH release just before each ecdysis, followed immediately by a peak of bursicon just after.

Hypothetical comparisons of JH titers and putative JH-sensitive periods in hemimetabolous and holometabolous insects (Fig. 12.26) indicate that the same probably occurs in hemimetabolous insects without the complication of a pupal stage (Nijhout, 1994b; Truman and Riddiford, 1999).

12.6 The Role of Critical Size in Molting and Metamorphosis

That critical size, in addition to hormones, is involved in controlling molting and metamorphosis was demonstrated ingeniously by Nijhout and his colleagues in larvae of *M. sexta* and the milkweed bug *Oncopeltus fasciatus* (references in Nijhout, 1994b).

12.6.1 MOLTING

Under optimal conditions, the number of molts is usually characteristic for the species (Table 9.4). However, when conditions are suboptimal, the number and timing of molts may vary. Normally, an insect molts only as it grows, suggesting that some process associated with growth triggers each molt (Nijhout, 1994b: 62). Since each molt is induced by secretion of PTTH (Fig. 12.25), control of molt timing and frequency must depend on processes controlling the synthesis and secretion of this hormone by NSCs in the brain.

In each larval instar of *Rhodnius prolixus* and other blood-feeding bugs, each molt is induced by uptake of a full blood meal (12.1). A smaller than normal meal or a sequence of small meals will not induce a molt, whereas a full meal of physiological saline solution will, suggesting that size of meal, not its nutritional value, is

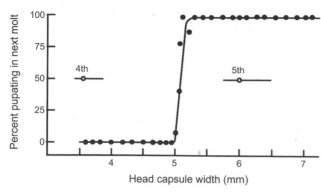

Fig. 12.29. The relationship between larval size (indexed by head capsule width) and the percentage of the test population pupating at the next molt in *M. sexta*. Larval size was manipulated by varying nutrition. Fourth instar larvae with head capsule widths larger than 5 mm molt to the fifth instar and pupate at the next molt, whereas those with smaller head capsules molt to another larval instar. Horizontal bars indicate the mean and range of head capsule widths in normal fourth (penultimate) and fifth (last) instar larvae. (Nijhout, H. F. *Insect hormones*. Copyright © 1994 by Princeton University Press. Reprinted by permission of Princeton University Press.)

what induces the molt (abdominal stretch receptors or nerves are stimulated mechanically by abdominal expansion due to uptake of the meal and pass impulses up the VNC to the brain, causing release of PTTH).

Larvae of *O. fasciatus* will molt only after they reach a critical weight in each instar, which is determined by maximum expansion of its cuticle following the previous molt. In fifth (last) instars, ecdysis to adult occurs 6–7 d after this weight is achieved but can be induced earlier in a larva below this weight by injecting saline solution into its abdomen, which creates the stretch necessary for a molt. The critical weight for the same instar can differ between individuals of a species, particularly when they are grown under suboptimal conditions, and depends on their body size at the beginning of the instar; a smaller fifth instar, for example, will have a lower critical weight than a larger one.

Larvae of holometabolous insects also have critical weights and sizes for each molt that likewise depend on larval size achieved following cuticular expansion at the beginning of the instar (references in Nijhout, 1994b). However, in these, the passage of time can override critical weight as a molting stimulus (recall that starved caterpillars of the clothes moth *Tineola bisselliella* can be forced to molt up to 40 times while decreasing in size [9.7.8]).

12.6.2 SIZE, GROWTH, AND DIFFERENTIATION OF WING IMAGINAL DISCS IN LEPIDOPTERANS

Larvae of the nymphalid butterfly *Precis coenia* normally have five larval instars (Miner et al., 2000). During the fifth instar, the wing discs grow continuously and at the same rate as the body. However, if such a larva is

starved, both it and its wing discs cease growing within 4 h, indicating that growth of the two is coupled. Topical application of the JH analog methoprene (Fig. 12.11F) to a normal fifth instar inhibits the growth of wing discs in a dose-dependent manner, even though the larva keeps growing normally.

Toward the end of the fifth instar, the wing discs begin to differentiate. In normally growing larvae, this differentiation does not begin until the discs have reached a critical size. If fifth instars are starved at any time after disc differentiation has begun, differentiation continues at a normal rate even though the discs and larvae are no longer growing. In addition, if fifth instars are starved before their wing discs begin to differentiate, they will not begin to do so spontaneously. Finally, if a fifth instar is treated topically with the analog JH after disc differentiation has begun, such differentiation is inhibited in a dose-dependent manner. Based on these observations, Miner and colleagues (2000) concluded that differentiation of wing discs only begins when they pass a critical size, initiation of wing disc differentiation and its continuation are controlled independently, and growth and continued differentiation of the discs can occur only in the absence of JH.

12.6.3 METAMORPHOSIS

Under optimal conditions, individuals of most species, after completing a species-specific number of juvenile molts, metamorphose into adults of a characteristic size (Nijhout, 1994b). However, if conditions are suboptimal, the number of larval molts can vary, as can adult size after metamorphosis. Because size increase in insects is exponential, with most growth occurring in the last larval instar (9.7.1), regulation of final adult size is largely determined in this instar. For example, when a fourth instar larva of *M. sexta* reaches a critical weight of about 5 gm with a head capsule width of 5.1 mm, it molts to a fifth instar (Figs. 12.29; 12.30: Roman numerals). Following the molt, its CA gradually cease releasing JH (Fig. 12.25) (the larva can continue to grow during this period). Disappearance of JH on day 4 is followed by two sequential releases of PTTH and ecdysteroids on days 5 and 7–9 (the pupal commitment and differentiation peaks), the second accompanied by a brief release of JH and occurring just after larval-pupal apolysis, and by larval-pupal ecdysis on day 10 (Fig. 12.25). Adult size is determined by the weight attained on day 5 of the fifth instar at the time the first secretion of PTTH begins in the fifth instar and is a function of the size of the cuticle deposited at the larva 4–larva 5 molt.

Normal larvae of *M. sexta* pass the threshold size for metamorphosis between instars 4 and 5 and pupate after the fifth instar (Fig. 12.30: Roman numerals), whereas those developing under suboptimal conditions grow more slowly and exhibit a broader range in size such that only a portion of fourth instars are above the threshold size for metamorphosis (Fig. 12.30: Arabic numerals) (Nijhout, 1994b). Those below the threshold size undergo one or more additional larval molts and pass through

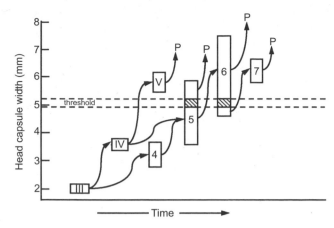

Fig. 12.30. Summary of the fate of larvae of *M. sexta* growing under various conditions. Boxes indicate the size range (indexed by head capsule width) in each instar. Roman numerals represent the values in larvae growing under optimal conditions and Arabic numerals, larvae growing under suboptimal conditions. P, pupate. See text for details. (Nijhout, H. F. *Insect hormones.* Copyright © 1994 by Princeton University Press. Reprinted by permission of Princeton University Press.)

a sixth and sometimes a seventh instar before pupating (it remains undetermined how the threshold size for metamorphosis is measured by the larvae). Based on knowledge of differences in metamorphosis in other insects, Nijhout (1994b) suggested that different species may have evolved variations on an ancestral mechanism by which they determine which larval instar is to be the last.

12.7 Endocrine Control of Molting and Metamorphosis in Other Arthropods

Ecdysteroids are now known to induce molting in representative chelicerates, crustaceans, myriapods, and nematodes in addition to hexapods (Chang et al., 1993; Loeb, 1993; Durica et al., 1999), and probably also in onychophorans, tardigrades, nematomorphs, priapulids, kinorhynchs, and loriciferans, lineages collectively referred to as the Ecdysozoa and thought to share a common ancestor based on recent comparison of homologous gene nucleotide sequences (Schmidt-Rhaesa and Bartolomaeus, 1998; Coast and Webster, 1998; Conway Morris, 2000). In both insects and crustaceans, synthesis and release of ecdysteroids are likewise regulated by brain-derived neuropeptides, and there is a drop in JH (or methyl farnesoyate or farnesoic acid in crustaceans) just before a molt. In crustaceans, a molt-inhibiting hormone continually inhibits ecdysteroid synthesis and secretion by the Y-organ in the eye stalk (Spazioni et al., 1999), while in the tick *Ornithoderus parkeri*, ecdysone is synthesized in the epidermis and hydroxylated to 20E in the fat body (Oliver and Dotson, 1993).

12.8 Hormones, Polymorphism, and Polyphenism

Hormones, particularly JH, influence the development of various forms in polymorphic insects by acting during species- and morph-specific JH-sensitive periods (Hardie and Lees, 1985; Nijhout, 1994b; Emlen, 2000; Emlen and Nijhout, 2000). A species is *polymorphic* if one or both sexes occur in two or more forms sufficiently distinct to be recognized without morphometric analysis, occur regularly, and make up a reasonable proportion of the population, or, as in eusocial insects, are essential to survival of the species (Richards, 1961). In some cases these morphs share the same genotype but develop by different, mutually exclusive routes induced by environmental signals acting at specific switch points during development. Such alternative polymorphism is referred to as polyphenism, the occurrence of two or more distinct phenotypes that can be induced in individuals of the same genotype by action of environmental factors (Hardie and Lees, 1985), and has probably evolved through action of selection on phenotypic plasticity (West-Eberhard, 1989). The sequential appearance of different forms during the life cycle of a species, as discussed in chapter 9, is *sequential polymorphism*.

12.8.1 POLYPHENISM IN NONSOCIAL INSECTS

In nonsocial insects, polyphenism can be associated with *size*, *sex*, or both (Emlen, 2000; Emlen and Nijhout, 2000), or with more or less elaborate life cycles involving the appearance of forms specialized for migration or dispersal (*dispersal polymorphism*) (Zera and Denno, 1997). For example, male dung beetles of the species *Onthophagus taurus* (Coleoptera: Scarabaeidae) either have or lack a pair of long, slender, dorsally recurved horns on the back of the head, the presence of which is correlated with body size: males with a pronotal width of 4.9 mm or greater have the horns, whereas smaller males have rudimentary horns or lack them entirely (Emlen and Nijhout, 1999). Emlen and Nijhout (1999) found these beetles to have a developmental switch late in the third (last) larval instar that activates horn growth only if the larva has acquired sufficient resources during feeding before metamorphosis. In larger males, horn growth begins with apolysis of dorsal epidermis from head capsule cuticle in the third instar, followed by rapid proliferation of horn cells during the pharate pupa to form what are essentially horn imaginal discs. During a short, JH-sensitive period late in this instar, topical application of the JH analog methoprene (Fig. 12.11F) to the head can switch the form of developing males: 80% of small, normally hornless males receiving methoprene during this period were induced to develop horns during the pharate pupa. They also observed a small peak of ecdysteroid at this time in both female and small male larvae that was absent in larger male larvae destined to form horns.

The development and evolution of exaggerated structures in polyphenic insects is complex and is critically discussed by Emlen (2000) and Emlen and Nijhout (2000) in relation to JH concentration, the distribution and concentration of putative JH receptor proteins in the cell membranes or nuclei of epidermal cells, scaling relationships (the covariation of trait magnitude with overall body size); and "reaction norms" (the range of possible forms that individuals with the same genotype could express were they reared across a range of environmental growth conditions). Emlen (2001) demonstrated also that position and size of horns in each of 161 species of onthophagine scarab correlate negatively with size of adjacent structures as if they competed for resources during their ontogeny—antennae competing with horns on the front of the head; compound eyes, with those at the back of the head; and wings, with those on the prothorax. The nature and magnitude of the trade-off depend on the ecology of each species and on the relative contributions of antennae, eyes, and wings to adult fitness. He concluded that such trade-offs should exist (1) when the enlarged structures develop coincident with the rest of the body and (2) when resources are limiting during at least part of this period, as in the pupal stage of scarabs.

12.8.1.1 CRICKETS

There is much experimental evidence from the study of crickets and aphids to indicate that elevated titers of JH, during a critical period of development, completely or partially block the normal development of wings, wing muscles, and associated structures in wing polymorphic species, resulting in apterous or brachypterous morphs (reviewed in Zera and Denno, 1997). Unfortunately, the only insect for which detailed hormonal information is available is the cricket *Gryllus rubens*, in which JH III (Fig. 12.11E) applied to long wing–destined penultimate or last instar juveniles results in brachypterous (short-winged) adults. In addition, prospective brachypters have higher JH titers than do prospective macropters, and the latter have higher JH esterase activity, all suggesting that JH degradation is reduced in last instar juveniles destined to become short-winged, that reduced degradation delays the drop in JH titer and that prolongation of an elevated JH titer inhibits normal development of wings and associated structures (there was no difference in JH synthesis between prospective long- and short-winged morphs).

12.8.1.2 APHIDS

Development of winged and wingless aphid morphs is influenced by population density, nutrition, photoperiod, and temperature, the apterae being regarded as "more larval" (neotenous) because they lack the sensorimotor apparatus (wings, specialized pterothorax, wing muscles, dorsal ocelli, and placoid antennal sensilla) of alates (5.5.2). This suggests that apterae may be influenced by an extended high titer of JH during postembryogenesis, and such seems to be true (Hardie and Lees, 1985; see 13.3.4). Apterous juveniles of *Brevicoryne brassicae*, for example, have larger CA than alate juveniles and are produced under long days and high temperatures, a response

that can be mimicked by applying JH I (Fig. 12.11E) to third (penultimate) larval instars. Also, topical application of precocene 2 (Fig. 12.11I) to adult *Acyrthosiphon pisum* causes their CA cells to degenerate and leads to production of alate progeny (12.4.7), an effect partially reversed by topically applying JH I (MacKauer et al., 1979, in Hardie and Lees, 1985). However, Zera and Denno (1997) provide examples of other aphid species in which topical application of JH either has no wing-inhibiting effects or seems to stimulate wing development.

Short days and lower temperatures in late summer and fall induce production of males and oviparae by gynoparae, whereas high temperatures and long days in summer inhibit development of these morphs. In females of *Megoura viciae*, sensitivity to photoperiod seems to reside in nine photoreceptors located in the anterior ventral neuropile of the protocerebrum, as indicated by their binding of labeled antibodies to various invertebrate opsins and phototransduction proteins (Gao et al., 1999).

Alternation of host plants by heteroecious aphids (Fig. 5.7C) is believed to result from change both in the physiological condition of the host and in the host preference of the aphids; the long-day, alate virginoparae prefer the herbaceous host, whereas the short-day gynoparae and oviparae prefer the woody one (Hardie and Lees, 1985). In some instances, these host preferences could be reversed by reversing the photoperiod under which potential gynoparae, oviparae, and virginoparae developed. Also, topical application of JH I to third and fourth instars juvenilized resulting adults (made them more like virginoparae) and increased their acceptance of bean hosts in direct relation to the amount they were juvenilized.

12.8.1.3 LOCUSTS

In 1921, Boris Uvarov published a classic paper on the migratory locust *Locusta migratoria* in which he showed both sexes of this species to exist in two principal phases, *solitaria* and *gregaria*. These phases differ extensively in structure, color, reproductive development and fecundity, physiology, behavior, and ecology (Hardie and Lees, 1985; Pener, 1991). When reared under crowded conditions in the lab, young hoppers develop into adults of the more mobile, *gregaria* phase, whereas isolated hoppers become adults approaching the *solitaria* phase (Fig. 12.31), and those in small loose groups develop into adults of the *transiens* phase between the two (Uvarov, 1921). The crowding response seems to begin in embryogenesis, to spread over all immature instars over several generations, and to have a strong maternal component. Implantation of active CA into hoppers developing under crowded conditions causes them to develop some *solitaria* characteristics, and removal of CA from fifth instar *solitaria* hoppers causes them to assume some *gregaria* characteristics. Also, JH titers tend to be higher in fourth and fifth instar *L. migratoria* reared in isolation than in groups. All these

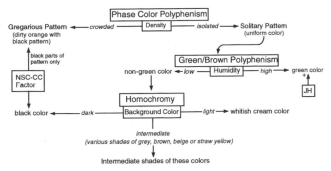

Fig. 12.31. Regulation of color pattern polyphenism in solitary and gregarious phases of the migratory locust *Locusta migratoria* (Orthoptera: Acrididae). Three aspects of color pattern are controlled independently: (1) pigment distribution, which can be bold and contrasting, or homogeneous and unpatterned; (2) if unpatterned, whether the overall color is green or not; and (3), if not green, whether the color is light, dark, or intermediate. Neurosecretory hormones (NSC-CC) and juvenile hormone (JH) affect this polyphenism at two different points. (Nijhout, H. F. *Insect hormones.* Copyright © 1994 by Princeton University Press. Reprinted by permission of Princeton University Press.)

observations suggest that extended JH titers during postembryogenesis influence development of the *solitaria* phase, which is considered to be more juvenile than the *gregaria* phase.

A detailed summary of exquisite behavioral experiments executed by Simpson and his students (S. J. Simpson et al., 1999) showed complex interactions between tactile, visual, and olfactory stimuli to induce gregarization, with tactile stimulation of mechanosensory hairs on the hind legs being the most important. Gregarization also increases with increasing population size relative to food abundance and increasing concentration of food resources.

Tawfik and coworkers (1999), working with another polyphenic locust, *Schistocerca gregaria*, found significant differences in the concentration of conjugated ecdysteroids (as 20E equivalents) (see 2.2.1.4) in the yolk of newly deposited eggs of *solitaria* (14 ng/egg) and *gregaria* (89 ng/egg) females and in newly hatched nymphs (*gregaria* nymphs contain fivefold more ecdysteroids than *solitaria* nymphs). These differences suggest physiological differences in oogenesis in solitary and crowded females.

In a series of carefully designed experiments on *S. gregaria*, McCaffery and colleagues (1998) showed that both crowding of solitary-reared females at the time of oviposition and high egg pod density promote the gregarization of first instars hatching from their eggs. When newly deposited, presumptively gregarious eggs are separated from their pods, the first instars that hatch are solitary, indicating that a gregarization factor associated with the egg pod promotes gregarization. This substance proved to emanate from the foam plugs of egg pods produced by crowded females but not from those of

solitary females. Saline extracts of the foam were found to contain an active factor that promoted gregarization of nymphs developing from newly deposited eggs of solitary females and from those of gregarious females earlier separated and washed to remove the factor. Since solitary eggs were only affected by the factor for up to 1 d after oviposition, it must have a short half-life or induce changes in gene expression in very early embryos that influence how they subsequently develop.

Preliminary characterization of the factor suggests it to be small (<3 kDa), water-soluble, and heat-sensitive; to consist of four components; and to be produced in the accessory glands of crowded females during oviposition. Synthesis of the factor in the glands is induced, in turn, by tactile stimulation of mechanosensory hairs on the hind legs of crowded females.

Regulation of the various color morphs in *L. migratoria* is summarized in Fig. 12.31 (Nijhout, 1994b). A "dark color–inducing neurohormone" of 11 amino acids synthesized in NSCs in the brain of *S. gregaria* is released from the CC and induces melanin formation in *gregaria* larvae (Tawfik et al., 1999).

12.8.2 POLYMORPHISM IN EUSOCIAL INSECTS
In eusocial insects, polymorphism is associated with dispersal and reproduction and with a division of labor within the social group. Such insects (Wilson, 1971) live in highly organized colonies; exhibit cooperation among individuals of the same species in care of their young; exhibit a division of labor in which large numbers of more or less sterile individuals assist a few fertile individuals; and have an overlap of generations such that offspring assist parents. Insect societies having all these characteristics have evolved independently at least 13 times (once in the termites, 10 times in the bees, and twice in the ants) (Wilson, 1971).

12.8.2.1 TERMITES
The incredible success of termites (order Isoptera) in the tropics and subtropics depends on maintaining a suitable balance in caste numbers in their colonies appropriate to colony size and food availability (Wilson, 1971; Hardie and Lees, 1985). Rapid replacement of the members of each caste in response to colony trauma results via input from a reservoir of juveniles. Each juvenile is able to develop into a member of any caste in response to receipt of morph-influencing pheromones produced by mature members of that caste and circulated through the colony by food exchange (trophallaxis). Colonies of species in the basal families have fewer castes but more flexible individual development, whereas those of the advanced family Termitidae have a more elaborate and rigid caste system in which development of the members of each caste is controlled by hormones acting at a few key points in the life cycle (Hardie and Lees, 1985; Nijhout, 1994b).

In the primitive, dry-wood termite *Kalotermes flavicollis* (Kalotermidae) investigated by Lüscher and students, mature colonies consist of adult males

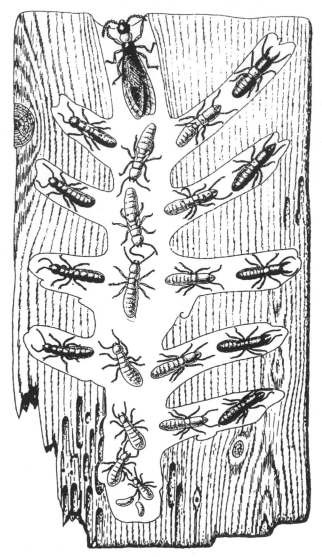

Fig. 12.32. Caste development in the primitive, dry-wood termite *Kalotermes flavicollis* (Isoptera: Kalotermidae). At the bottom are eggs and first instar nymphs. The nymphs molt five to seven times to reach the pseudergate stage (the termite in the middle of the drawing). This stage can molt repeatedly without further growth or differentiation, but at any molt it can also transform either into a supplementary reproductive (to the left of the drawing) or, through a presoldier instar, into a soldier (to the right). It can also change, by way of an intermediate stage with external wing pads, into a primary reproductive (at the top of the drawing). Supplementary reproductives and soldiers can also originate from this intermediate stage. (Reprinted with permission from M. Lüscher. The termite and the cell. *Sci. Am.* 188(5): 74–78 © 1953 Scientific American.)

(kings) and females (queens), replacement (secondary) reproductives, nymphs, soldiers, presoldiers, larvae, and pseudergates ("false workers") of both sexes, members of the latter functioning as workers but capable of undergoing an indefinite number of stationary molts (Fig. 12.32). Balance between the numbers of each caste is controlled by release of an inhibitory pheromone from the anus of fully differentiated caste members. This pheromone inhibits development of pseudergates into additional members of that caste, the sensitive period of the pseudergate to the soldier-inhibiting pheromone occurring late in the instar and that for induction of supplementary reproductives occurring early. These pheromones work by influencing the secretion of JH by the CA of recipient pseudergates, as was demonstrated by injecting active CA or JH analogs. The effect depends on both dosage and time of application during the intermolt but tends to induce recipients to molt into soldiers (the switch to supplementary reproductive requires the complete absense of JH) (Nijhout, 1994b).

JH-sensitive periods in the pseudergate controlling these caste determination switches seem to coincide with the pheromone-sensitive periods mentioned above, as shown in a model devised by Nijhout and Wheeler (1982). Fig. 12.33 shows hypothetical JH profiles during a pseudergate instar for animals molting to a soldier, a replacement reproductive, or an adult or remaining as a pseudergate (Nijhout, 1994b: 186). Three JH-sensitive periods are postulated to occur in the instar: the first controlling development of reproductive organs and genitalia in prospective reproductives; the second controlling development of wings and pigmentation in these adults; and the third controlling development of the head, pigments, and mandibles of soldiers. High JH titer during the first and second periods inhibits commitment to sexual and imaginal development. High JH titer during the third period induces development of soldier characteristics, whereas low JH at this time maintains the characteristics of the pseudergate. Thus (Fig. 12.33),

- to remain a pseudergate, JH titer must be high during the first two JH-sensitive periods and low during the third;
- to develop into a soldier, JH titer must be high throughout the instar;
- to become an adult, JH titer must be low throughout the instar; and
- to become a replacement reproductive, JH titer must be high during the second JH-sensitive period, so that reproductive but no other adult characters develop.

Termite species in the "higher" families tend to have both major (large) and minor (small) workers and soldiers but no pseudergate stage (Nijhout, 1994b: 187–188). At the end of larval life, each larva metamorphoses into a terminal worker, a soldier, or a reproductive, differentiation of the last two being influenced by release of inhibitory pheromones by soldiers and queens. As in basal termites, the switch to soldier development requires high JH titer but the JH-sensitive period is early in the instar (the hormonal cues stimulating worker and reproductive differentiation are still unidentified).

12.8.2.2 ANTS

Mature ant colonies typically contain winged haploid (in their germ line) males and two female castes, the queen and workers. The workers are monomorphic in some

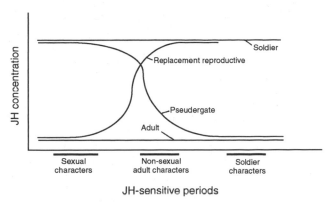

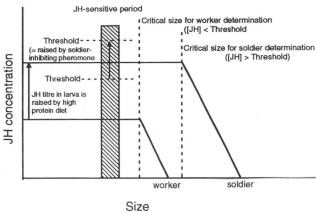

Fig. 12.33. Nijhout and Wheeler's (1982) model for the juvenile hormone (JH)–mediated control of caste determination in lower termites. Each curve represents a hypothetical JH titer profile during the pseudergate instar. Horizontal bars show the timing of JH-sensitive periods for sexual, nonsexual adult, and soldier characteristics. Pseudergates destined to become replacement reproductives or soldiers molt first to an intermediately differentiated presoldier or nymphal stage, respectively (compare with Fig. 12.32). See text for details. (Nijhout, H. F. *Insect hormones.* Copyright © 1994 by Princeton University Press. Reprinted by permission of Princeton University Press.)

Fig. 12.34. Wheeler and Nijhout's (1984) model for the control of worker/soldier polyphenism during the last larval instar of the ant *Pheidole bicarinata* (Hymenoptera: Formicidae). Under a low-protein diet, the juvenile hormone (JH) titer remains low and JH secretion stops when larvae reach a particular (low) critical size; such larvae metamorphose into workers. With a high-protein diet, the JH titer is higher, and if it rises above a threshold, the critical size at which JH secretion stops is revised upward. Such larvae grow for a longer time, attain a larger critical size, and stop secreting JH. Soldier genes are expressed, and the larva metamorphoses into a soldier. If too many soldiers are present in a colony, the JH threshold for soldier induction is increased by release of an inhibitory pheromone, so that even high titers of JH remain below threshold. The critical size remains low, and the larvae metamorphose into workers. (Nijhout, H. F. *Insect hormones.* Copyright © 1994 by Princeton University Press. Reprinted by permission of Princeton University Press.)

species and polymorphic (small, intermediate, and large or just small and large) in others, this polymorphism being either continuous or discontinuous (Hardie and Lees, 1985; Nijhout, 1994b). In species of *Myrmica*, *Pheidole*, and *Formica*, which are the best investigated, factors influencing the development of fertilized eggs into queens, soldiers, and workers can act either before or after hatch, with developmental plasticity being retained by developing larvae. The first fertilized eggs deposited in spring by a queen develop as queens or workers; later ones develop principally into workers. High-protein diets favor differentiation of queens; less rich, more liquid diets favor that of workers. The queen can also influence caste determination by modifying the behavior of nurse workers caring for larvae: in species with polymorphic workers, well-fed larvae develop into soldiers and less well-fed ones into minims.

Application of JH to fertilized eggs during an embryonic JH-sensitive period induces many of them to develop into queens. A similar application to a queen results in an increased proportion of her fertilized eggs metamorphosing into additional queens (Nijhout, 1994b).

Treatment of female worker larvae of *Pheidole bicarinata* by JH analogs causes them to develop into soldiers, the JH-sensitive period occurring during days 4–6 of the last larval instar (Fig. 12.34) (Nijhout, 1994b). If a larva receives no JH during this period, it reaches its normal critical size, pupates, and metamorphoses into a worker. However, if JH is present at this time, the critical size for metamorphosis is reprogrammed, and the larva grows to a larger critical size, pupates, and metamorphoses into a soldier having characters not present in smaller workers.

When too many soldiers are present in a colony, a smaller proportion of female larvae switches to soldier development until the original proportion of soldiers is restored (Nijhout, 1994b: 184). This results from production of a soldier-inhibiting pheromone by already present soldiers that prevents female larvae from switching to soldier development during their JH-sensitive periods. If larvae are fed a high-protein diet, their CA release enough JH during the JH-sensitive period to raise its titer above the threshold required for soldier determination. The soldier-inhibiting pheromone depresses the sensitivity threshold to JH during the JH-sensitive period. Thus, the proportion of soldiers in a colony depends on a balance between mean JH titer (a complex function of food availability and number of foraging workers) and the sensitivity threshold to JH, which is determined by the soldier-inhibiting pheromone produced by soldiers already present in the colony (Nijhout, 1994b: 185).

12.8.2.3 BEES

Bee species range from solitary to highly eusocial (Michener, 2000). In semisocial colonial halictines, queens and workers are similar and caste determination is imaginal and established by queen dominance, whereas in bumblebee and honeybee colonies, queens and workers differ structurally. Each fertilized egg has the potential to develop into either a queen or a worker depending on how the larvae are fed by nurse bees (i.e., caste determination is trophogenic).

In annual colonies of bumblebees, the first queens usually emerge at the end of the growing season but may appear earlier if the functioning queen dies or is removed. In some species, the queen may produce a contact pheromone that influences worker nursing behavior and thus larval nutrition, and in colonies of *Bombus terrestris*, larvae seem to be determined as workers in the first instar (Röseler and Röseler, 1979).

In longer-lived colonies of the stingless meliponine bee *Trigona*, transfer of larvae from large queen cells into smaller worker cells and vice versa provided no evidence of qualitative differences in larval feeding by workers (Velthuis, 1976).

In the honeybee *Apis mellifera*, caste determination is trophogenic (Hardie and Lees, 1985). Brood cells in the hive remain open until day 5 of postembryogenesis (fifth instar), and larvae are fed selectively by nurse bees depending on cell shape and size (Fig. 12.35). Larvae developing in large, queen cells (these hang down vertically from the lower margin of the comb, with their openings downward) are fed continuously on royal jelly (high in sugar, pantothenic acid, and biopterin) and emerge as adults on day 12 or 13 at a colony temperature of 34–35°C. Those developing in smaller, worker cells (these are within the comb, hexagonal, and slope upward 15 degrees to horizontal, with their openings outward) are switched to a diet of worker jelly during the fourth instar on day 3 and don't emerge until day 17 or 18. Prior to pupating, but after their cells are capped, fifth instars of both incipient castes orient themselves so that their heads are toward the opening of the cells, worker larvae depending on the texture of the cap, queen larvae on gravity, to do so.

Switching fertilized eggs and female larvae between queen and worker cells revealed that they remain competent to develop into either caste until the end of the fourth instar at 3–3.5 d after hatch but have become committed to develop into one caste or the other after the molt into instar 5 (Fig. 12.35). In smaller, younger colonies, the queen inhibits production of queen cells and additional queens by producing a contact pheromone in her mandibular glands, *queen substance* ([E]-9-oxodec-2-enoic acid), that is spread by trophallaxis throughout the colony.

Secretion of JH is critical to queen-worker divergence during larval development in honeybees. Topical application of JH I to 3.5-d-old fourth instar worker larvae induced development of worker-sized adults with queenlike characteristics. However, JH III, not JH I, is the functional JH in honeybees (Fig. 12.11E), and hemolymph titers are known to be about 10-fold higher in presumptive queens than in worker larvae at the time of caste determination on d 3.5, probably due to stimulation of CA by increased sugar intake. JH synthesis by CA cells increases early in the fifth instar and peaks

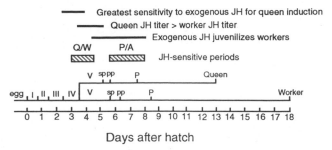

Fig. 12.35. Control of queen/worker polyphenism in the domestic honeybee *Apis mellifera* (Hymenoptera: Apidae). Queen and worker developmental pathways diverge after the fourth instar (IV). The pupal period (P) of a queen is shorter than that of a worker. The hatched bars indicate juvenile hormone (JH)–sensitive periods for the queen/worker (Q/W) and the pupal/adult (P/A) commitment switches. The black bars indicate the various effects of exogenous JH on these polyphenisms. (Nijhout, H. F. *Insect hormones*. Copyright © 1994 by Princeton University Press. Reprinted by permission of Princeton University Press.)

shortly before the larva spins its cocoon, reaching a rate 26 times that found in worker larvae. This period of synthesis corresponds to a JH-sensitive period postulated by Nijhout and Wheeler (1982) for queen/worker (Q/W) switching and precedes by 1 d a second JH-sensitive period required for pupal/adult (P/A) switching (Fig. 12.35). However, by this time, JH secretion by CA cells in both queen and worker larvae has declined to undetectable levels.

12.8.2.4 EUSOCIAL WASPS

In eusocial species of Vespidae, a maximum of two female castes occur, queen and worker, with caste differences ranging from the prominent dimorphism of *Vespula* spp. to those distinguished only by their place in the dominance hierarchy and their degree of ovarian development (e.g., *Polistes* spp.) (Hardie and Lees, 1985). In the former, larval transfer between worker and queen cells shows that fertilized eggs and female larvae develop as either workers or queens depending on how they are fed by nurse wasps. Application of JH can induce ovarial development and aggressive behavior in *Polistes annularis* females, and dominant females have larger CA and produce more JH than subordinate females.

In summary, though JHs obviously play a dominant role in the induction of polymorphism in eusocial insects, little is known about the conditions under which the CA are induced to secrete them or how they interact with other components of the endocrine system. Also, although both metamorphosis (sequential polymorphism) and polyphenism are JH-dependent, the timing of JH release in polymorphic insects is shifted so that it no longer coincides with metamorphosis (Figs. 12.33–12.35).

All polymorphic insects probably evolved from monomorphic ancestors in which JH regulated both ovarian maturation and metamorphosis. Thus, JH was probably captured independently in each polymorphic lineage to function in inducing polymorphism via selection on phenotypic plasticity (West-Eberhard, 1989). However, in honeybees, JH promotes queen determination only in the fourth instar, where its gonadotropic function persists, while in adult queens and workers, JH has practically lost this function (Hardie and Lees, 1985).

12.9 Future Developments

Incredible progress is being made in our understanding of the chemistry and mode of action of insect hormones (Nijhout, 1994b: 212–214). Isolation of the neuropeptide hormones in pure form has enabled them to be used to generate monoclonal antibodies in rabbits that are specific to them. With such probes available, the hormones can be analyzed by radioimmunoassay, a highly sensitive and specific technique for quantitatively examining the cellular dynamics of circulating hormones. Availability of probes for these hormones coupled to fluorescent markers enables cells containing hormone molecules to be localized by immunocytochemical techniques. Finally, intensive efforts are being made to reveal the signaling pathways by which insect hormones cause changes in gene expression in cells of their target tissues.

As emphasized by Nijhout (1994b: 212) and Emlen and Nijhout (2000), insect growth hormones function as switching signals that allow central control over developmental events and constitute a system through which their timing "can be flexibly regulated to meet the needs of an animal in a variable environment" (Nijhout, 1994b: 212). This system has the potential to uncouple "a relatively inflexible causal chain of developmental events into a centrally controlled set of developmental modules" (Nijhout, 1994b: 212) that allow these components to diverge in form and function to meet the requirements of constantly changing environments.

13/ Ontogeny and Hexapod Evolution

The details of development presented in this book provide a base for reflecting on how change in ontogeny might have influenced hexapod diversification. To consider such change productively requires us to recall how selection acts on populations to generate new species and higher taxa and to have some knowledge of the insect fossil record.

13.1 The Origin of Taxonomic Units or Taxa

Most systematists place the species they discover and describe within a hierarchical classification system that reflects our understanding of how we believe past evolutionary processes have generated the diversity we see today. The system is erected by carefully examining individual specimens and grouping together those sharing characteristics (including homologous nucleotide sequences in mitochondrial and nuclear genes) thought to be derived or specialized. The usual steps are as follows:

- Individual specimens are grouped together into taxa (singular: taxon), which in biological classification ideally should be "natural"—that is, their members should be related by descent from a nearest common ancestor. The fundamental category of study is the species.
- Each taxon is then assigned to a category within the hierarchy.
- Related taxa are grouped together to form taxa of higher rank, which are then assigned higher levels in the hierarchy. Thus, species, genus, family, and order are categories within the hierarchy, while *D. melanogaster*, *Drosophila*, Drosophilidae, and Diptera, respectively, are the names of natural groups or taxa.

13.1.1 THE ROLE OF VARIATION AND NATURAL SELECTION

In bisexual metazoans, genetic variability, housed within the gene pool of each species and originating through mutation, recombination, random assortment, introgression, and transposon "jumping" (Davies et al., 2000), is transformed by fertilization and individual ontogenies into the varying phenotypes of juvenile and adult offspring on which selection acts to generate evolutionary change. Natural selection is the unequal distribution of genotypes to the gene pool of the next generation through differential mortality and differences in reproductive success (Mayr and Ashlock, 1991).

13.1.2 SPECIATION

The following definitions (Mayr and Ashlock, 1991) are appropriate for hexapods, since most of them are known to be bisexual (chapter 5 considers exceptions):

- *species* are groups of interbreeding natural populations that are reproductively isolated from other such groups;
- *populations* are individuals of a given locality that potentially form a single interbreeding community; and
- *speciation* is the origin of genetic incompatibility between populations caused by acquisition of reproductive isolating mechanisms.

Based on the universality of the genetic code, most biologists would agree that extant bisexual species of metazoans probably descend through countless speciation events from a single marine ancestor species that lived deep in the Precambrian. Because formation of two or more species from one probably requires many years and generations, we have rarely observed it occurring in nature, even though we have produced reproductively isolated, laboratory populations of several *Drosophila* species by artificial selection (e.g., Higgie et al., 2000). The common feature of all postulated modes of speciation is the physical separation, in time or space, of some members of a species from the remainder for long enough that action of differential selection can generate different genotypes. Such genotypes are incompatible with that of the parental species, result in reproductive isolation between members of the two groups, and delimit now separate, new species. In areas of subsequent overlap in the distributions of such species, there is then selection to reinforce mating barriers between their members.

Evidence from comparative study of a variety of animals suggests that speciation could occur by allochronic (different time) or anagenetic, allopatric (different place) or geographic, sympatric (same place), stasipatric, and/or ecological speciation, as fully discussed by Grant (1963), Mayr (1963, 1982), Mayr and Ashlock (1991), Moore (1993), and many others. The allopatric mode probably is the most widespread.

The genes responsible for establishing reproductive isolation between members of two diverging populations of insects have just begun to be identified and could function before (prezygotic), during (zygotic), or after (postzygotic) fertilization or at two or more times. A recent insect example may be the homeobox gene *Odysseus* (*Ods*) in *Drosophila*, which likely controls transcription of downstream genes functioning in spermiogenesis (Ting et al., 1998). As described in 7.2.1.4, the homeodomain, in proteins of the homeobox genes, is a DNA-binding motif whose amino acid sequence is more or less conserved among members of diverse species. In most *Drosophila* species this domain of *Ods* is similarly conserved. However, in each member of the *D. melanogaster* sibling species complex (*D. mauritiana*, *D. simulans*, *D. sechellia*, and *D. melanogaster*), the amino acid sequences of the domain have rapidly diverged and may be responsible for the sterility of male hybrids between *D. mauritiana* and *D. simulans*. Thus, male offspring resulting from hybridization between males and females of different species within this complex are probably fertile only when the amino acid sequences of the Ods homeodomains and the binding sites in the regulatory sequences of their target gene(s) are compatible.

13.1.3 THE ORIGIN OF HIGHER TAXA: SALTATIONISM VERSUS GRADUALISM

A higher taxon is any monophyletic group of species or a single species, at any rank within the taxonomic hierarchy, that is separated from other taxa of the same rank by a gap in character distributions greater than any found within the taxon (Mayr and Ashlock, 1991). By convention, the size of the gap is proportional to the hierarchical position of the taxon (i.e., the higher the taxon in the hierarchy, the larger the gap), and it results from a combination of structural, ecological, physiological, and behavioral differences between members of two separate but related lineages. In addition, members of a given higher taxon are behaviorally and structurally equipped to occupy a distinctive portion of the environment called its *adaptive zone*. A higher taxon is more or less divergent depending on its rank, and most higher taxa include a nested set of lower-ranked monophyletic taxa, also more or less divergent from each other. For example, the class Insecta encompasses a number of orders; most orders, a number of families; most families, a number of genera; and most genera, a number of species.

Darwin (1859) considered new species to arise gradually through slow accumulation of advantageous traits and suggested that if continued indefinitely, such processes could lead eventually to the appearance of new higher taxa, although he acknowledged that such change might occur more rapidly during speciation than between cladistic events. For most of Darwin's contemporaries, this concept was difficult to accept, even by those convinced of the importance of natural selection in evolution. For example, though T. H. Huxley was a principal defender of Darwin's ideas, he was unable to accept the gradual emergence of higher levels of structural organization and instead proposed that new higher groups arise by major evolutionary jumps known as *saltations*.

This idea was supported by the observation that members of each higher taxon share a distinctive body plan consisting of an integrated assemblage of characters that function together as co-adaptations (this is why a knowledgeable taxonomist can assign an individual specimen by eye to its correct order, family, and genus). Saltationists argued that gradual evolution of such a distinctive, co-adapted suite of characters was impossible because no single component could be adaptive unless it functioned with the others. Neither could they have functioned effectively until all were fully developed. Their gradual evolution was thus not possible, since such characters could not have been selected individually until all were fully functional. Therefore, they must have arisen more or less simultaneously in a single major jump to a new functional stage that selection could then improve. If this were true, then intermediates between higher taxa could not have existed, and their absence in the fossil record is real and not a result of its deficiencies. According to this idea, most higher taxa must have arisen by sudden evolutionary jumps, as evidenced by the following observations on the fossil record as then known:

- In general, when members of a lineage first appear in the fossil record, they are already distinct from those of related lineages and already have the principal body plan characters considered adaptive for that group.
- Intermediate fossils are almost never found.
- There are large gaps in the fossil record, representatives of the major taxa generally appearing suddenly and often in great diversity.

For many years, saltationists were unable to invoke a mechanism for these sudden jumps other than special creation. But when Mendel's works were rediscovered in 1900 (Moore, 1993), some biologists attributed the sudden appearance of unusual variations in progeny to mutation. For example, the physiological geneticist Richard Goldschmidt (1940) suggested that "hopeful monsters" might occasionally arise during development via macromutation (Dietrich, 2000). Most such monsters would be ill adapted and would perish without passing on their traits, but occasionally one might be instantly adapted for a new way of life such that a new higher taxon of organisms might eventually descend from it (note the similarity of such thinking to that of some modern exponents of evolutionary developmental biology [Budd, 1999; Conway Morris, 2000]).

Several weaknesses are apparent in this explanation:
- The probability that fitness could be increased immediately by a single macromutation, while maintaining functional integrity, would seem to be zero ("Fisher's principle" [Fisher, 1930]). For example, four-winged *Drosophila* flies caused by mutation in the homeotic gene *Ultrabithorax* are incapable of flight because they lack functional wing muscles in the metathorax (Roy et al., 1997) and because their flight mechanism has evolved for two, not four, wings (Budd, 1999).

- The number of hopeful monsters required to generate the known diversity of metazoan higher taxa at all levels in the taxonomic hierarchy would be astronomical.
- There is no evidence for such monstrosities from nature, either living or in the fossil record.
- Conspecific males and females must be structurally, genetically, and behaviorally compatible to reproduce successfully.

Thus, for saltation to be effective in generating new body plans, at least two compatible monsters of opposite sex must originate simultaneously and relatively close to each other within the distribution area of a species—both similarly adapted for life in their environment in spite of their independent origin and drastically changed structure—and they must find each other and have compatible genitalia, behavioral repertoires, and genomes appropriate for producing viable offspring. Could such improbable coincidences occur as repeatedly as they must?

The saltation theory was based on lack of evidence, that is, on the absence of intermediates in the fossil record. In fact, such gaps are more apparent than real, and a number of transitional forms are well documented and continue to be discovered in some lineages. Examples include the evolution of amphibians from fish, of reptiles from amphibians, of mammals and birds from reptiles, and of horses, whales, and hominids from less derived ancestors (Raff, 1996).

Following widespread acceptance of the evolutionary synthesis (Moore, 1993), the idea of hopeful monsters was replaced by Darwin's gradualism, and it became generally accepted that both new species and higher taxa could arise by gradual accumulation of small adaptations accompanied by continuing speciation and extinction (Mayr, 1963, 1982). (Eldredge and Gould's [1972] "punctuated equilibrium" model differs from gradualism only in that major evolutionary change in structure is considered to occur principally during periods of rapid speciation, to be followed by long periods of evolutionary stasis.)

13.1.4 THE ORIGIN OF HIGHER TAXA: ADAPTIVE ZONES AND KEY INNOVATIONS

The great vertebrate paleontologist G. G. Simpson (1953: chapter 6) observed that major biomes are usually divided into smaller regions in which characteristic modes of adaptation are required of their inhabitants. These regions are interconnected by intergrading habitats, so that each major realm contains many smaller regions that differ from each other in topography and climate and in edaphic and biotic factors, each presenting specific adaptive challenges to the organisms inhabiting them. As mentioned already, the members of each higher taxon have a particular body plan permitting them to use one or more of these environmental patches. The range of patches occupied by members of a taxon plus the way they are adapted to use them is called the adaptive zone of that taxon.

The limits of a particular adaptive zone depend on the

taxon occupying it, and the zone does not exist until it is evolved into. For example, until insect flight evolved, the air was not used actively in locomotion by any animal. But "air" is not an adaptive zone. Flight is part of the adaptive zone of flying animals and implies the existence of skeletomuscular complexes within their bodies—the wings and their articulations, innervation, muscles, and behavioral repertoires—allowing them to actively use the air.

The adaptations enabling entry into and exploitation of a new adaptive zone are called *key innovations*. They usually comprise a complex, co-adapted suite of characters and are part of the adaptive zone. Such innovations also are referred to as *evolutionary novelties* because they constitute a unique set of derived characters placing members of a taxon within its adaptive zone.

Saltationists had concluded that it was impossible to pass from one adaptive zone to another without saltation, but to a gradualist, such major jumps are unnecessary. So, how could members of a lineage gradually acquire the new, co-adapted suite of evolutionary novelties enabling them to enter a new adaptive zone, when their ability to do so would seem to be constrained by the requirements that (1) the complex of key innovations would have to be built up slowly by a series of small changes, and (2) the species would have to be adapted to the environment at all times while making the transition?

Adaptive zones are united more or less continuously by a series of *transition zones* in which behavioral and structural characteristics evolve and act as exaptations (preadaptations) allowing movement into a new adaptive zone. An *exaptation* is a structure or behavior evolved to perform one function that can be selected to perform another without losing its original one. An example is the evolution of flying birds from terrestrial ancestral reptiles:

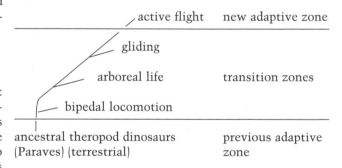

ancestral theropod dinosaurs (Paraves) (terrestrial) — previous adaptive zone

The exaptation in this example is the skeletomusculature of the forelimbs, its innervation, and the behavioral "wiring" in the central nervous system controlling its function. During the transition, the individuals of each species are completely adapted to the environment in which they are living in all life history stages, and each transition zone is simply an extension of the previous adaptive zone until a new adaptive zone is entered. It is also usual that speciation occurs during the transition, so that more species come out than went in.

Successful entry into a new adaptive zone, called the *critical* or *nodal point* (Kristensen, 1999), is often followed by a major radiation of that taxon as its members adapt to a wide range of new, unoccupied habitats. This *postadaptive zone* is where selection perfects the key innovations and, by increased rates of speciation, fills the habitats available with new species. The organisms in such descendant lineages are therefore alike because of the key innovations, constituting their body plans, inherited from their common ancestor, and constitute a new higher taxon.

Note that identifying probable cause for the success of a particular species-rich clade among the ground plan characteristics of its members is difficult and subjective, and the idea that key innovations originate at the base of a radiation in species diversity is controversial (Hunter, 1998; Masters and Rayner, 1998). For example, most known diversity within the Endopterygota (Fig. 0.2) is attributable to a small number of subordinate taxa within only 4 of its 11 orders: staphyliniform and cucujiform (mostly polyphagan) beetles, cyclorrhaphan flies, ditrysian lepidopterans, and apocritan wasps (Kristensen, 1999: Fig. 9).

13.2 Key Innovations and the Insect Fossil Record

If key innovations are critical to the origin and diversification of new higher taxa, then what evidence in the hexapod fossil record indicates which ones are most important and when these might have arisen?

13.2.1 KEY INNOVATIONS AND THE ADAPTIVE RADIATION OF THE HEXAPODS

Eight key innovations likely played a decisive role in the diversification of the supraordinal taxa of insects, and it is important to establish their identity and when they might have arisen. The inferred time of origin for each of these novelties is indicated on a phylogeny of the extant and extinct orders of hexapods in Fig. 13.1.

1. Internal fertilization (see 3.3.2): This is required for successful fertilization of eggs in species living in desiccating terrestrial environments.
2. Six legs: The presence of six legs promoted tripodal locomotion, stability, and adaptive radiation into diverse habitats (Dickinson et al., 2000).
3. Dicondylic mandibles: Members of the relatively species-poor, basal, apterygote clades Protura, Collembola, and Diplura have entognathous, protrusible mouthparts, while archaeognaths have ectognathous, monocondylic mandibles that have a single articulation point with the head capsule (Matsuda, 1965; Machida, 2000; but see Koch, 2001, in which detailed examinations of mandibular function in representatives of these taxa reveal a hitherto unappreciated diversity in mechanism). In most other insects, the mandible has two articulation points, a new, anterior articulation having evolved to impart a transverse rocking movement to the mandibles during adduction and abduction (Wheeler et al., 2001). This innovation has enabled descendant species to proliferate into terrestrial and freshwater habitats as general feeders, as pollen and leaf feeders, as leaf miners, as stem and wood borers, and as predators. Of course, one or the other of these articulation points or the entire mandible has been lost secondarily in species having specialized feeding mechanisms (e.g., lapping, sucking, mopping, piercing/sucking, grasping).
4. Ovipositors (see 3.9.2): These appendages facilitated entry into new habitats because of the numerous substrates into which eggs could be deposited by females. Specialization of ovipositors was followed by selection for specialized egg membranes to support embryonic survival in these new sites.
5. Wings (see 9.6.2.5.9 and 9.6.2.5.10): Insects were the first animals to fly actively. The origin of functional wings, at the latest in the Lower Carboniferous but probably in the Devonian (Fig. 13.1), undoubtedly contributed to pterygote diversification by increasing the ability of adults to disperse, find food and mates, and escape from flightless predators; for 140 million years, until the appearance of pterosaurs and the first birds in the Upper Jurassic, insects were the only actively flying animals (Sereno, 1999).
6. Wing flexing: All extant winged insects are either paleopterous or neopterous. Mayflies (Ephemeroptera) and damselflies and dragonflies (Odonata) are generally considered to be *paleopterous*: their principal flapping muscles are inserted more or less directly into the wing bases and their fore and hind wings alternate during flight and are extended vertically or laterally from the body when not in use. All other winged insects are *neopterous*: their principal flapping muscles are inserted indirectly into the walls of the thorax, their fore and hind wings usually function as a unit during flight, and they are flexed flat and lengthwise over the abdomen when not in use (Brodsky, 1994).

The names Paleoptera (old wing) and Neoptera (new wing) suggest the latter mechanism to be derived from the former and to be more specialized. However, there is little doubt that both methods of flight are highly specialized in extant members of both lineages, that of paleopterans for an aerial existence in which wing flexing is unnecessary and that of neopterans for a terrestrial existence in which wings are out of the way when not in use. Both seem very different from the mechanisms suggested for Paleozoic pterygotes based on examination of fossilized wing bases (Fig. 13.1) (9.6.2.5.9) and on theoretical analyses of wing loading, pterothorax mass total mass ratios, aspect ratios, shape descriptors, and their functional significance (Wootton and Kukalová-Peck, 2000).

The ability of insects to flex their wings lengthwise and flat over the back (in neopterans, owing to the presence of an off-line, third axillary sclerite and its associated muscles) was a major evolutionary novelty and appears to have arisen twice independently among insects in the Lower Carboniferous (Kukalová-Peck, 1978, 1991). However, the mechanism of wing folding

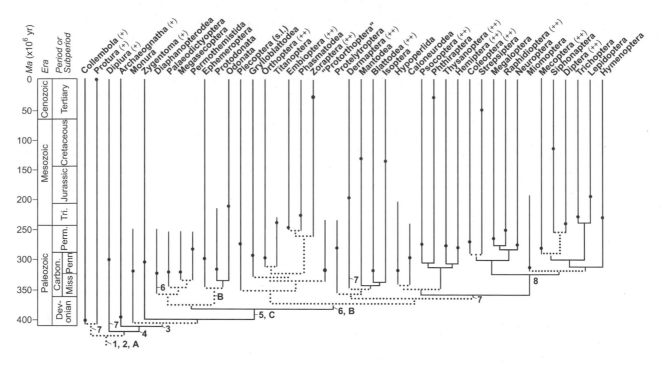

Key Innovations Contributing to Hexapod Diversification
1. Origin of internal fertilization.
2. Origin of six functional legs.
3. Origin of dicondylic mandibles.
4. Origin of a valvular ovipositor in females.
5. Origin of wings and of active flight.
6. Origin of wing flexing.
7. Nurse cell origin.
8. Evolution of complete metamorphosis.

Changes in Hormone Function and in Expression of Regulatory Genes
A. *HOM-C* genes: *lab, pb, Hox3, Dfd, Scr, Antp II Ubx, abdA, AbdB*
B. Cis-regulatory sites for receipt of *HOM-C* transcription factors inhibiting wing development on prothoracic and abdominal segments.
C. Juvenile hormone functioning in metamorphosis.

Fig. 13.1. Evolutionary relationships of presently recognized extant and fossil hexapod orders in relation to the geological time scale as inferred by Labandeira (1998a). Dots on vertical branches indicate the earliest documented fossils for each order (most branches extend below the dots to indicate extensive evolution prior to the deposition of the first fossil discovered). Solid horizontal lines indicate well-corroborated relationships, and dotted lines, more tentative ones. On this diagram I have indicated the probable time of origin of the eight key innovations most influencing hexapod diversification and information on expression of homeotic genes and of significant changes in hormone function. Order names indicated by a plus sign contain primarily litter-dwelling species, and those indicated by two plus signs contain significant numbers of such species, usually in basal lineages. The wide phylogenetic distribution of this lifestyle suggests it to be plesiotypic for hexapods. (Adapted with permission from the *Annual Review of Earth and Planetary Sciences*, Volume 26 © 1998 by Annual Reviews www.AnnualReviews.org.)

appears different in diaphanopteroids, and only the neopteran lineage diversified to the present (Fig. 13.1). The advantages to an adult of being able to flex its wings are obvious: it allows entry into microhabitats closed to paleopterous and tetrapod predators and allows it to hide from them in foliage, under logs and litter, and so on, without damaging the wings.

7. Nurse cells (see 2.2.7.1): Oogonia-derived nurse cells within the ovaries of female hexapods having meroistic ovarioles function to amplify the genome and to increase the rate of RNA synthesis and egg production during oogenesis. Their known phylogenetic distribution (Figs. 2.22, 13.1) correlates with ordinal diversity (Fig. 0.2) and may have contributed to its development.

8. Complete metamorphosis (see 9.6.2.4): The significance of this kind of postembryonic development is

reflected in the tremendous success of certain lineages within 4 of the 11 extant orders whose members have it (some 88.6% of known insect species are holometabolous) (Kristensen, 1999). The usual explanation for its adaptive advantage is that internal wing development allowed juveniles to specialize for mining or boring in solid substrates (wood, stems, under bark, in leaves, etc.) closed to their adults and to all stages of exopterygote and apterygote insects, and allowed the juveniles and adults to diverge in form, food, and habit and thereby avoid competing with each other.

The subsequent appearance of additional, more restricted evolutionary novelties has undoubtedly contributed to diversification within supraordinal, ordinal, and subordinal taxa. Examples are the piercing and sucking mouthparts of juvenile and adult hemipterans

(8.3.8.2), thysanopterans (8.3.8.1), and some lice (8.3.8.3) and of adult fleas and some flies (10.2.4.3.2); the sclerotized elytra in beetles; the puparium of higher flies (10.2.1); and the narrow abdominal petiole of apocritan wasps (Kristensen, 1999). Fossil evidence for the evolution of insect feeding mechanisms is fully discussed by Labandeira (1997). Unfortunately, the fossil record does not explicitly document the exact time of origin for any of these innovations, each being well established and highly specialized by time of first occurrence (Fig. 13.1).

A *fossil* is a preserved organism, a part of an organism, or the preserved work of an organism (ichnofossil or trace fossils). For hexapods, the record is far from complete because only a minuscule fraction of the insects that ever lived were fossilized, only under unusual circumstances was a complete individual fossilized, and only a tiny proportion of insects fossilized have been found and described. Because of biases in preservation, discovery, and collection, the remains of insects living in low-lying swamp or riparian habitats are disproportionately represented. This is because rapid deposition in fine silt, volcanic ash, or mud in swamps or in deep quiet lakes with low oxygen concentration generated the best fossils until the appearance in the Jurassic of conditions suitable for preservation in fossilized resins (amber) (Poinar, 1992).

13.2.2 EARTH HISTORY AND THE ORIGIN OF THE HEXAPOD ORDERS

The sun, Earth, and planets of our solar system are considered to have condensed from the solar nebula about 4.6 billion years ago (Cloud, 1988). The oldest fossils yet discovered are of cyanobacterial filaments from western Australia that are about 3.5 billion years old. Thus, for most of its existence, Earth has supported life.

Geologists have developed a chronology of Earth history by considering the composition of the rocks, the order of their formation, and the fossils they contain. The upper layers of an undisturbed sequence of sediments are assumed to be younger than the lower layers. Layers at different places on Earth are compared and matched by examining their fossils: when these are the same, the strata in which they are found are also assumed to be the same. Absolute age is assigned by dating using various radioactive isotopes, but estimates of the temporal extent of each period continue to change as new methods are introduced and because one must use different isotopes for formations of different ages (Renne et al., 1998). The times indicated below are from Hecht (1995) and Kenrick and Crane (1997).

The longest time divisions of the geological record are *eons*. The sequence of rocks in the most recent eon, the Phanerozoic, from 543 million years ago (Ma) to present, contains most known fossils and is understood in such detail that it has been divided into *eras*, the eras into *periods*, and the periods into *subperiods*, *epochs*, and *stages* or *regional sequences* (Fig. 13.1). Of earlier eons, only the Proterozoic (Precambrian; from 2.5 billion years

ago to 543 Ma) is known well enough to allow its tentative division into eras. The fossil record of eukaryotes is in the Phanerozoic, with the exception of Neoproterozoic (1 billion years ago to 543 Ma) impressions of some peculiar invertebrates at Ediacara in southern Australia and some 20 other places (from 570 to 535 Ma) (see summaries in Raff, 1996; Conway Morris, 1998a, 1998b, 1998c, 2000; and Martin et al., 2000), some putative 590-million-year-old (Myr) cnidarian embryos and larvae from the Doushantuo Formation of China (Chen et al., 2000), some perfectly preserved, 570-Myr, three-dimensional embryos of triploblastic but taxonomically unassignable animals from China (Xiao et al., 1998; Gould, 1998), numerous ichnofossils of similar age, and some billion-year-old algae.

Recent estimates of divergence times for animals, plants, and fungi and for basal animal phyla, based on consideration of differences in homologous nucleotide sequences for 75 independently acting nuclear genes, are from Wang and colleagues (1999):

- a three-way split (unresolved) for plants, animals, and fungi: 1576 ± 80 Ma;
- basal animal phyla (Eumetazoa: Porifera, Cnidaria, Ctenophora): 1200–1500 Ma;
- nematode/higher animal divergence: 1173 ± 79 Ma; and
- chordate/arthropod divergence: 993 ± 46 Ma.

These authors thus have the earliest extant animal phyla originating deep in the Proterozoic more than 400 million years before their first known fossil occurrence (problems with this analysis are discussed by Conway Morris [2000]).

13.2.2.1 PALEOZOIC ERA

Immediately following the Precambrian begins the Paleozoic (early life) era, the first era in geological history from which substantial fossil information has been gained. It began about 543 Ma, lasted until 251 Ma, and is divided into six periods.

13.2.2.1.1 CAMBRIAN PERIOD (543–506 MA)

The Cambrian is characterized by the apparently sudden appearance of the fossils of diverse marine animals representing most extant phyla, the so-called Cambrian explosion, with trilobites, chelicerates, crustaceans, and lobopods (onychophorans) represented by numerous species (Fortey et al., 1997; Conway Morris, 1998a, 2000; Valentine et al., 1999; Knoll and Carroll, 1999).

At this time, the positions of the principal landmasses have been inferred to be those shown in Fig. 13.2A (Scotese, 1997a), with what are now Africa, South America, Antarctica, Australia, and India forming a single landmass, Gondwana, extending from the equator to the South Pole and with the remaining continents separated by large expanses of sea, the Panthalassic and Iapetus Oceans (Scotese, 1997b). As time passed, these continents slowly drifted toward each other, and epicontinental seas gradually transgressed the land until they covered half its surface. Throughout Earth's history,

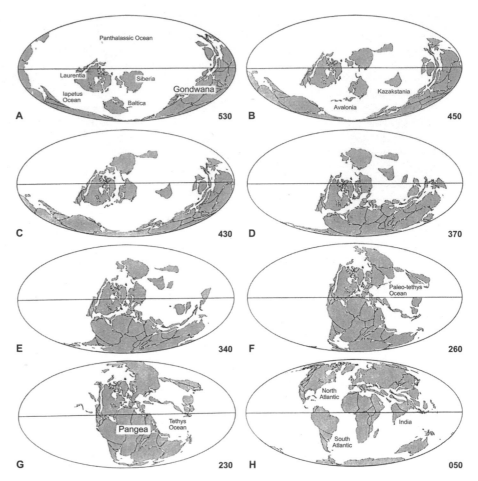

Fig. 13.2. Paleogeographic reconstructions of continental position during the Paleozoic, Mesozoic, and Lower Cenozoic. Numbers in the lower right indicate the age of each reconstruction (millions of years ago; Ma). A. Middle Cambrian. B. Upper Ordovician. C. Lower Silurian. D. Upper Devonian. E. Middle Carboniferous. F. Middle Permian. G. Middle Triassic. H. Middle Eocene. (Reprinted with permission from C. R. Scotese. *Phanerozoic plate tectonic reconstructions.* 7th ed. Paleomap Progress Rep. 36. Department of Geology University of Texas, Arlington. © 1997 The PALEOMAP Project.)

such changes in the relative positions of continents, together with quasi-periodic changes in Earth's orbital geometry (Zachos et al., 2001), have influenced the advance and retreat of glaciers, the transgression and regression of seas, ocean currents, global climate, and ultimately, the diversification of the marine and terrestrial biota (Cloud, 1988). The Cambrian is thought to have been warm and uniform except near the South Pole (Scotese, 1997b).

13.2.2.1.2 ORDOVICIAN PERIOD (506–439 MA)
Although the Ordovician period (Fig. 13.2B) began with extensive epicontinental seas, toward its end, mountain building took place and the seas began to withdraw. The climate was probably similar to that of today, with marine species of practically all animal phyla currently recognized being represented by the end of this period. The first terrestrial plants (Embryophytes) had probably evolved by the Middle Ordovician (Kenrick and Crane, 1997).

13.2.2.1.3 SILURIAN PERIOD (439–408.5 MA)
During the Silurian, epicontinental seas continued to withdraw and the climate, again, was probably similar to

that of today (Fig. 13.2C). Early terrestrial plant communities (liverworts, hornworts, mosses [Kenrick and Crane, 1997]) and terrestrial mitelike trigonotarbids, millipedes, and centipedes (Gray and Shear, 1992) are known from the Upper Silurian, indicating that terrestrial animal communities were well established by this time. Since fossils of Lower Devonian archaeognaths (bristletails) and collembolans (springtails) are known, internal fertilization (key innovation 1), the six-legged body plan (2), and well-developed valvular ovipositors in females (4) all probably originated sometime in the Silurian (Fig. 13.1). Also, phylogenetic relationships suggest that dicondylous mandibles (3) may likewise have orginated in the Upper Silurian (Fig. 13.1).

13.2.2.1.4 DEVONIAN PERIOD (408.5–362.5 MA)
Physically and climatically, the Devonian was probably similar to the Silurian, with the position of the continents as shown in Fig. 13.2D. Terrestrial plant life had diversified extensively, so that by the Middle Devonian, vast forests of tree ferns and horsetails had appeared. The first seed plants, primitive conifers, appear in the fossil record at the end of the Devonian. These became more diverse as the period advanced, providing food,

shelter, and more stable living conditions for terrestrial animals.

Of particular importance to early hexapods was the influence of forests on characteristics of the land's surface: generating litter, modifying and intensifying soil formation, and increasing the humidity of the forest floor, thereby increasing habitat heterogeneity and stimulating hexapod diversification (Villani et al., 1999).

The earliest known hexapods are from the Lower Devonian: the collembolan *Rhyniella praecursor* from the Rhynie chert in Scotland (Fig. 13.1). Archaeognath fragments are known from the Lower Devonian of Quebec and the Middle Devonian of Gilboa, New York, together with fossil mites, amblypygids, opilionids, scorpions, pseudoscorpions, centipedes, arthropleurids, trigonotarbids, and spiders (Figs. 13.1, 13.4) (Gray and Shear, 1992). Fossil Protura have yet to be discovered (Fig. 13.1), undoubtedly because of their small size, delicate structure, and soil-dwelling habits, but phylogeny suggests that they have surely been around since at least the Devonian if not earlier. And the Diplura, Monura (extinct), and Zygentoma, too, probably originated in the Devonian or earlier, even though the oldest fossil representatives of these taxa are from the Upper Carboniferous (Fig. 13.1).

13.2.2.1.5 CARBONIFEROUS PERIOD (362.5–286 MA)
North American geologists generally recognize two named periods—Mississippian and Pennsylvanian—whereas geologists from elsewhere recognize only the Carboniferous, divided into two subperiods—the Lower and Upper Carboniferous. In the Lower Carboniferous, the continents are believed to have had the positions shown in Fig. 13.2E, with the climate warm and moist. There is also evidence that the oxygen content of the atmosphere, presently 21%, was much higher during the Upper Carboniferous and Permian (35%), possibly due to photosynthesis by plants in the rich widespread swamp forests present at the time, and that this might have influenced the evolution of giant Carboniferous insects (Conway Morris, 1995; Berner et al., 2000). Terrestrial plant life was similar to that of the Upper Devonian, with some tree ferns, horsetails, and lycopods reaching heights of 35 m. Forests were well established and production of new organic matter exceeded its decay, contributing to formation of the widespread coal beds deposited at this time.

At least two additional hexapod innovations probably arose, at the latest, in the Lower Carboniferous—wings (5) and the ability to flex them lengthwise over the back (6)—even though we have no fossil insects from this period to prove it (Figs. 13.1, 13.4). In fact, Kukalová-Peck (1991) suggested that all major pterygote stem groups had probably arisen by the end of this period: those of the extinct Paleodictyoptera and allies and of ephemeroids, odonatoids, plecopteroids, orthopteroids, blattoids, hemipteroids, and even endopterygotes based on the origin of key innovation 7, complete metamorphosis (Fig. 13.1).

The first diverse insect fossils are preserved in Upper Carboniferous strata (Figs. 13.1, 13.4), with representatives of 17 orders from many sites including the Commetary Shales of Europe; coal strata in Germany and Poland, in the Czech and Slovak Republics, and in Pennsylvania; and from mudstone concretions from Mazon Creek, Illinois (Shabica and Hay, 1997). At the time, all these areas were located in a warm humid tropical belt on or adjacent to the equator (Fig. 13.2E).

Knowledge of the fossils of this subperiod is summarized by Kukalová-Peck (1991), is described in detail by Carpenter (1992a, 1992b), and is illustrated in Fig. 13.4, while an imaginative description of the organisms' presumed lifestyles is provided by Shear and Kukalová-Peck (1990). Upper Carboniferous insects appear to have been extraordinarily diverse, and many had ecological roles differing from those of today. Though present (Labandeira, 1998a), herbivory seems not to have been as prevalent as today, with most primary productivity being funneled through detritivores. This indicates that the leaf litter habitat is probably ancestral in hexapods, a prediction supported by the widespread distribution of this lifestyle in members of basal clades within many extant hexapod lineages (Fig. 13.1).

13.2.2.1.6 PERMIAN PERIOD (286–251 MA)
By the end of the Devonian (Fig. 13.2D), the more northern continents had begun to approach Gondwana to form a single landmass (Scotese, 1997a). Consolidation of continents into the supercontinent Pangaea ("all land") continued throughout the Lower (Fig. 13.2E) and Upper Carboniferous, to reach completion in the Permian (Fig. 13.2F). Cooling occurred throughout the Upper Carboniferous, and by its end, the southern continents (South America, South Africa, Antarctica, Australia, and India) were experiencing a major ice age (Scotese, 1997b). The northern continents (Laurentia and Eurasia) lay near the equator and continued to experience warm, moist conditions.

The Lower Permian was marked by continued cooling and glaciation in the Southern Hemisphere, by gradual uplift of land resulting from the consolidation of continents, and by withdrawal of large inland seas from continental interiors due to formation of southern glaciers. This drained many of the coal-forming swamps in both Northern and Southern Hemispheres and resulted in the development of extensive deserts. Seed plants (gymnosperms), which first appeared in the Upper Devonian, had diversified throughout the Carboniferous to dominate forests of the Permian.

If the stem groups of extant orders are included as separate lineages, 30 orders of insects are known from Permian deposits, of which representatives of 10 have yet to be found, though probably will be, in Upper Carboniferous strata (Fig. 13.1). Representatives of all but two of these are living today.

The end of the Permian witnessed the greatest period of organismal extinction in the known geological record—80%–95% of Permian species, according to Erwin (1994)—although its magnitude and pattern varied tremendously among different higher taxa. Many invertebrate lineages, including the trilobites, became extinct, as did many of the early terrestrial plant taxa, while others were markedly reduced in numbers of species (Erwin, 1994).

Among the insects, many Upper Carboniferous orders declined in diversity (Figs. 13.1, 13.4). Of these, all but the Ephemeroptera, Plecoptera, Orthoptera, Blattodea, and Miomoptera disappeared before the end of the Permian, while the Protodonata persisted into the Triassic in greatly reduced diversity before becoming extinct.

Most orders that were eventually to become extinct had done so by the end of the Permian or not long thereafter (Figs. 13.1, 13.4). For this reason, the insect fauna after the Permian was beginning to assume a modern appearance, and the only order to originate after the Permian and to become extinct was the Titanoptera, which did both in the Triassic. The subsequent fossil record amounts to a continuing, progressive addition of previously unrecorded orders until a modern fauna is reached.

Since it contained both Upper Carboniferous and new orders of insects, the Permian fauna was more diverse than that of the Upper Carboniferous (Fig. 13.1). "In terms such as diversity of form and the association of generalized and specialized species, the fauna of the Permian was probably the most diverse in the history of the insects" (Carpenter, 1977: 69).

Why did so many groups become extinct at the end of the Permian? The decline in both fauna and flora was too irregular to be due to a single cause (Erwin, 1994), but recent uranium/lead zircon data on the ratio of carbon 13 to carbon 12 in rocks from Meishan in South China (Bowring et al., 1998) suggest that the event probably lasted less than 165,000 years (251 ± 0.3 Ma) and was likely caused by a catastrophic influx of carbon 12. A contributing factor was surely the gradual uplift of land resulting from formation of Pangaea (Fig. 13.2F). This caused the drainage of inland seas and coal swamps and the onset of more arid and variable conditions. A large number of habitats must have been reduced in extent or wiped out completely, events placing considerable strain on the basically tropical- and subtropical-adapted swamp faunas of the Upper Carboniferous.

Associated with this change in climate were extensive changes in community structure. For instance, by the end of the Permian, gymnosperms dominated the forests, placing a new set of environmental constraints on hexapods adapted to life in the forest before (Labandeira, 1998a). Not only did this result in a wholly different group of plants to feed on, requiring different mouthparts (Labandeira, 1997) and behavior, but also the entire aspect of the forest had changed, including canopy characteristics, leaf and tree shape, amounts and types of cover, and litter composition. New insect forms evolving in response to these changed environmental conditions must have competed with representatives of the more archaic orders persisting from the Upper Carboniferous, perhaps contributing to their demise.

13.2.2.2 MESOZOIC ERA

We enter the Mesozoic era (251–65 Ma) with 23 known hexapod orders (3 apterygote, 2 paleopterous, 12 neopterous and exopterygote, and 6 endopterygote) (Figs. 13.1, 13.4). With the great extinctions of the Permian, the Paleozoic came to a close about 251 Ma. At this time, the climate was cool and dry and the land was still united into the single supercontinent of Pangaea, which was high and dry (Fig. 13.2F). Gymnosperms and seed ferns dominated the land plants and the reptiles had already begun to diversify. The Mesozoic is divided into three periods: Triassic, Jurassic, and Cretaceous (Fig. 13.1).

13.2.2.2.1 TRIASSIC PERIOD (251–208 MA)

Highlands resulting from mountain building in the Permian prevented reinvasion of seas for much of the Triassic, and the climate warmed throughout the period to levels much higher than today. The supercontinent Pangaea began to split up in the Middle Triassic to yield a northern supercontinent, Laurasia, and a southern supercontinent, Gondwanaland, separated by the Tethys Sea (Fig. 13.2G) (Scotese, 1997a). Appearance of these two landmasses early in the Mesozoic defined centers of diversification for many family- and genus-level taxa of extant insects, and the relative isolation of the faunas of these two landmasses is a major aspect of the biogeographical analysis of the insect groups arising in the Lower Mesozoic.

The Triassic flora was much like that of the Permian, its major elements being gymnosperms (pines, cedars, and firlike forms) and tree ferns. However, little is known of the insects of the Triassic, since their fossils are rare (Carpenter, 1992b) (a Triassic Lagerstätte recently described from the Virginia–North Carolina border seems to be an exception [Fraser et al., 1996] and contains superbly preserved fossils of many insects including a possible thysanopteran). This lack of information seems to have resulted from the conditions of deposition prevailing in this period: most sediments were coarse grained, and the relative dryness of the period seems to have presented fewer situations suitable for fossilization than did conditions of other periods.

Orders appearing for the first time in the Triassic fossil record are Odonata, Phasmatodea, Titanoptera (giant cricket-like insects that probably originated in the Upper Carboniferous), Thysanoptera (?), Trichoptera, Diptera, and Hymenoptera (Figs. 13.1, 13.4). Diptera are represented by fossils of the structurally more primitive lower flies (Nematocera), and Hymenoptera, by the equally primitive sawflies (Symphyta) (Carpenter, 1992b).

In addition to fossils of flies and sawflies, there is copious fossil evidence of their feeding activity in the form of leaf damage, leaf mines, bore holes in stems and wood, and galls, coprolites (fossil frass), and gut contents (Scott et al., 1992; Labandeira, 1997, 1998a). Some petrified tree trunks from this period show evidence of the feeding galleries of scolytid and buprestid beetles, a good indication that wood-boring beetles were diverse by the Triassic.

13.2.2.2.2 JURASSIC PERIOD (208–145 MA)

The physical conditions of the Jurassic were probably much like those of the Triassic. The division of Pangaea that had begun in the Triassic (Fig. 13.2G) had continued, so that by the end of the Jurassic, the continents had the positions shown in Fig. 13.3 (150 Ma). Africa and South America were beginning to separate from Antarctica and Australia but were still connected to each other.

The Jurassic climate was much warmer than today and sea level was some 200 m higher (there is no evidence for glaciation in polar regions at this time) (Scotese, 1997b). Some fossil trees show annual rings in transverse section, indicating that climate was becoming seasonal. In addition, there was apparently a more than 95% species-level turnover of the Triassic-Jurassic megaflora, perhaps caused by a fourfold increase in atmospheric carbon dioxide and an associated 3–4°C "greenhouse" warming (McElwain et al., 1999).

Even though the Triassic-Jurassic boundary marks a major faunal mass extinction, apparently there was little change in the Triassic insect fauna. The first fossil Dermaptera and Lepidoptera are from the Jurassic (Figs. 13.1, 13.4), and although Kukalová-Peck (1991), Carpenter (1992a), Labandeira and Sepkoski (1993), and Labandeira (1998a) showed Thysanoptera as arising in the Permian (*Liassothrips longipennis*), the earliest probable fossil thrips is from the Triassic (Fraser et al., 1996). Fossil representatives of some modern families appear first in the Jurassic (Carpenter, 1992b).

13.2.2.2.3 CRETACEOUS PERIOD (145–65 MA)

The continents had continued to drift apart, so that by the end of this period (Fig. 13.3: 80 Ma), South America was separated from Africa and North America from Europe, except for a narrow connection through what is now Greenland, by the rapidly widening Atlantic Ocean. India had split from Antarctica and was beginning to move north for an eventual collision with Asia. The arrival of exotic terranes (e.g., Wrangellia and Stikinia) along the western coast of North America correlated with commencement of the Rocky Mountain orogeny (Scotese, 1997b).

The Cretaceous was warm and moist worldwide, owing to an absence of glaciers at the poles, to rapid seafloor spreading, and to the presence of extensive continental seaways due to sea levels 100–200 m higher than at present. Warm water from equatorial regions could thus be transported northward, warming the polar regions (Scotese, 1997b). The flora was uniform throughout the world during the Lower Cretaceous, with subtropical forests extending to 70°N and S and coal-

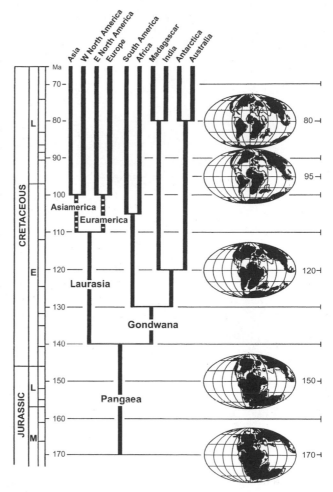

Fig. 13.3. Temporally calibrated area cladogram showing the breakup of Pangaea into 10 major land areas by the end of the Cretaceous. The large area of sea to the right in the bottom three maps is the Tethys Sea separating Laurasia from Gondwana. In the maps, black is dry land and unshaded outline are shallow seas. (Reprinted with permission from P. C. Sereno. The evolution of dinosaurs. *Science* 284: 2137–2147. Copyright 1999 American Association for the Advancement of Science.)

forming swamps existing on most continents and at most latitudes.

The most important floristic event of the Cretaceous was the explosive radiation of the flowering plants, or angiosperms, beginning about 130 Ma and correlating, probably in a coevolutionary mode, with that of insects specializing on flowering plants, especially Orthoptera, Hemiptera, Coleoptera, Diptera, Lepidoptera, and Hymenoptera (Fig. 13.4) (Labandeira and Sepkoski, 1993; Crane et al., 1995; Labandeira, 1998a, 1998b).

Little was known of Cretaceous insects until the discovery of extensive amber deposits in Canada (Clear Lake, Manitoba; Medecine Hat, Alberta), Siberia, and Lebanon (Poinar, 1992). Although such fossils are biased toward small insects and those likely to land on tree

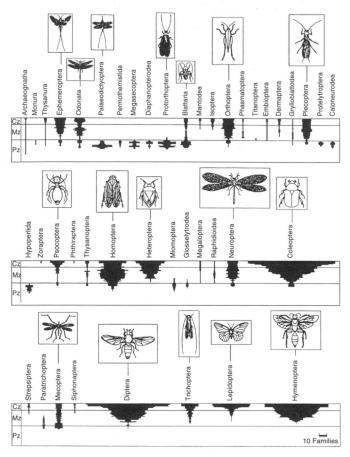

Fig. 13.4. Spindle diagrams displaying the diversity of known fossil families within extant and extinct insect orders (entognath hexapods [Protura, Collembola, and Diplura] are omitted) in stratigraphic stages of the Phanerozoic. Boxed illustrations (not to scale) depict typical adult representatives of some orders. Flowering plants make their fossil appearance about two-thirds of the way up the band for the Mesozoic (just above the "M" in Mz). Pz, Paleozoic (Silurian through Permian); Mz, Mesozoic; Cz, Cenozoic. (Reprinted with permission from Labandeira and Sepkoski 1993. Copyright 1993 American Association for the Advancement of Science.)

trunks oozing resin, Cretaceous amber is particularly useful in understanding the evolution of family-level taxa, since many basal members of extant families are often found in it (Poinar, 1992). Therefore, most Cretaceous insects can be placed in modern families and even in modern genera.

Fossil representatives of Mantodea, Isoptera, and Siphonaptera and of numerous basal families of Lepidoptera are found for the first time in Cretaceous amber (Figs. 13.1, 13.4) (Poinar, 1992). Of special interest is a fossil vespoid wasp nest and a primitive ant, indicating that eusociality had developed in both Isoptera and Hymenoptera by this period.

As mentioned, the Cretaceous is of special interest because of the proliferation of insect pollinators.

Hughes and Smart (1967) and many botanists had postulated earlier that the most primitive angiosperms were similar to extant magnolias and had large, showy, fragrant blossoms, which today are well adapted to insect pollination, particularly by beetles. This required that *entomophily* (insect pollination) be primitive in angiosperms and *anemophily* (wind pollination) be derived by reduction. Most modern insect pollinators are attracted to flowers by their pollen, color, shape, size, and scent and must have appropriate sensilla, olfactory receptor proteins, and behavioral repertoires to respond.

Good fossil evidence suggests that insect pollination of angiosperms did not become widespread until the Lower Cretaceous and then probably by transfer from ancestors that had fed on the pollen of cycads, Ephedrales, and conifers (Crepet et al., 1991; Labandeira, 1998a, 1998b). At first, there must have been little specialization of flowers to attract insects (the oldest-known putative angiosperm, from the Upper Jurassic of Liaoning, China, *Archaefructus liaoningensis*, had carpels enclosing ovules but apparently no other flower parts [Sun et al., 1998]). The close pollination relationships between insects and angiosperms existing today apparently did not arise until the Upper Cretaceous and Cenozoic (Labandeira, 1998a, 1998b).

The end of the Cretaceous was another time of major extinction, resulting in complete extinction for many taxa such as the dinosaurs (except for birds [Sereno, 1999]) but apparently affecting neither angiosperms nor insects, both of which were rapidly diversifying. These massive extinctions brought an end to the Cretaceous and to the Mesozoic about 65 Ma. A possible cause of the extinctions was an asteroid or comet, some 10 km in diameter, that impacted Earth on the north coast of Yucatan in the Gulf of Mexico, creating the Chicxulub crater (Alvarez, 1997; Morgan et al., 1997).

At the close of the Cretaceous, the climate was becoming seasonal owing to continental collision and mountain building: India with Asia, to form the Himalayas and Tibetan plateau; Spain with France, the Pyrenees; Italy with France and Switzerland, the Alps; Greece and Turkey with the Balkans, the Hellinide and Dinaride Mountains; Arabia with Iran, the Zagros Mountains; and Australia with Indonesia. There was an increase in the world's ocean basins (Scotese, 1997b), and angiosperms and insects were diversifying, often in a coevolutionary mode. By the end of this period, the insect fauna was essentially modern, and most extant families were present and represented by many extant genera.

13.2.2.3 CENOZOIC ERA

The Cenozoic (65 Ma–present) is divided by most geologists into two periods: Tertiary (65–1.9 Ma), which is further divided by some into the Paleogene and Neogene periods, and Quarternary (1.9–0.01 Ma) (Zachos et al., 2001). Because the fossil and stratigraphic record is less disturbed in these more recent deposits, these periods can be studied with more resolution and are divided into epochs and regional sequences:

Tertiary Period	*Quaternary Period*
Paleogene	Pleistocene (1.9–0.01 Ma)
Palcocene (65–54.5 Ma)	Holocene (0.01 Ma–
Eocene (54.5–33.5 Ma)	present)
Oligocene (33.5–23.8 Ma)	
Neogene	
Miocene (23.8–5.5 Ma)	
Pliocene (5.5–1.9 Ma)	

By the Lower Tertiary, Australia and New Zealand had split from Antarctica, and by the Middle Tertiary, the Indian plate had contacted the Asian plate to begin uplift of the Himalayas and Tibetan plateau (Fig. 13.2H), as is continuing today. At the close of the Tertiary, the continents had assumed positions essentially the same as those existing today. The relative times and sites of continental separation during the Lower Tertiary are important for understanding the distributions of modern insect groups. Important elements of the modern insect fauna evolved in the Tertiary, with most extant genera and some extant species of insects originating during the Paleocene, Eocene, and Oligocene.

13.2.2.3.1 TERTIARY PERIOD

Paleocene-Eocene Epochs
The climate during the Paleocene and Eocene is thought to have been mild (in North America, palms occurred as far north as North Dakota and forests of *Metasequoia* sp. on Axel Heiberg Island at 80°N). The nature and complexity of the insect fauna of the late Eocene and early Oligocene are indicated by fossils preserved in Baltic amber.

Oligocene Epoch
A general cooling trend in the Oligocene continued to the end of the Tertiary. Of particular interest is an important fossil-bearing deposit in Florissant, Colorado. Its compression fossils are preserved in fine volcanic ash, which appears to have been deposited in a lake basin. Wilson (1978a, 1978b) provided a fine analysis of these faunas in northwestern North America.

Miocene-Pliocene Epochs
The climate dried, grasslands spread, and many grassland-inhabiting groups of insects diversified during the Miocene and Pliocene. The climate continued to cool and dry, accompanied by the onset of glaciation at the North and South Poles (Scotese, 1997b).

13.2.2.3.2 QUARTERNARY PERIOD

Pleistocene Epoch
Although the end of the Pleistocene marked the extinction of many vertebrate taxa, that of some megafauna undoubtedly caused by humans (Roberts et al., 2001;

Alroy, 2001), few insects appear to have disappeared and relatively little speciation seems to have occurred in most lineages. The most important effect of this epoch seems to have been on insect distribution; it was at this time that the modern distributions of insect species were established. The epoch ended about 10,000 years ago with the melting of the last great ice sheets.

With this background in Earth history and hexapod fossils, we can now consider the possible interactions between ontogeny and phylogeny in hexapod evolution and when these interactions might have occurred.

13.3 Ontogeny and Hexapod Evolution

I began this book by stating that for a descendant to differ from its parents, it must alter the way it develops. Thus, development must play a major role in evolutionary change since it produces the variants on which selection acts to generate new lineages. Here, I present a brief summary of how our present ideas on insect development and evolution have arisen.

13.3.1 THE "BIOGENETIC LAW" OF MÜLLER-HAECKEL
The idea that during its ontogeny an organism repeats, in a shortened way, the past evolutionary history of its adult ancestors dates to Aristotle's "Great Chain of Being": a linear ordering of living things from lowest to highest according to their structural complexity and level of perfection, with humans at the top (Raff, 1996: 293). In the late 18th and early 19th centuries, a group of German naturalists established the concept of *Naturphilosophie*, according to which all nature is in flux from lower organisms to higher ones, and from initial chaos to man, with the animal kingdom envisaged as a single "superorganism" stopping earlier or later in its development to generate the forms we now see as representative "lower" or "higher" species (Gould, 1977).

Naturphilosophen prominent in laying the groundwork for recapitulation included J. F. Meckel and L. Oken (De Beer, 1962; Gould, 1977). In 1811, Meckel published a 60-page essay in which he considered the similarities between the embryonic stages of higher animals and the adults of lower animals (it included 57 pages of perceived examples of recapitulation organ by organ), whereas Oken (1847) proposed that "the animal kingdom is only a dismemberment of its highest member Man" and that "all development begins with a primal zero and progresses to complexity by the successive addition of organs in a determined sequence" (Gould, 1977: 40).

From this kind of thinking, Fritz Müller and Ernst Haeckel gradually developed their "biogenetic law," the idea that developmental stages during an individual's ontogeny represent the adult forms of ancestral animals added sequentially to the end of its ancestral ontogeny. Two processes were thought to act concurrently in this progression: (1) Phylogenetic change within a lineage of animals was postulated to proceed by each addition contributing new structures to the adult members of the descendant lineage (*terminal addition*). (2) As terminal addition proceeded, development was accelerated so that features added earlier in the history of the lineage were pushed back into earlier developmental stages of descendant individuals (*condensation*). For example, some polypod-stage insect embryos often bear evanescent abdominal appendages (Fig. 6.22B) that are thought to represent the functional legs of the adult of an ancestral, multilegged arthropod pressed back into an earlier stage of development (compare Fig. 13.14).

Both terminal addition and condensation must have occurred if the mechanism of the "law" was to be explained. Otherwise, the ontogenies of descendants would have become longer and longer with addition of each succeeding stage, and there was no evidence that such was so. Thus, according to Haeckel (1866: p. 300), "ontogeny is the short and rapid recapitulation of phylogeny.... During its own rapid development ... an individual repeats the most important changes in form evolved by its ancestors during their long and slow paleontological development." However, Haeckel soon discovered that this seemingly logical relationship between ontogeny and phylogeny was obscured by the occurrence of specialized characters in juveniles of many taxa. Characters added by terminal addition and inherited in proper sequence were said to be *palingenetic* and to reflect the course of evolution, while those added to juvenile stages as adaptations to local conditions or inherited out of sequence were said to be *cenogenetic* and to confound one's ability to reconstruct the history of lineages (Gould, 1977: 82). In 1905, Haeckel reformulated his law to include juvenile adaptations: "The rapid and brief ontogeny is a condensed synopsis of the long and slow history of the stem: this synopsis is the more faithful and complete in proportion as palingenesis has been preserved by heredity and cenogenesis has not been introduced by adaptation" (1905: 14).

Because Haeckel was the preeminent biologist of his time in Germany, he exerted immense influence on 19th- and early-20th-century biology (Oppenheimer, 1955), and as might be expected, his ideas permeate the writings of late-19th- and early-20th-century students of insect structure and development (read, for example, Korschelt and Heider's [1899] review of invertebrate ontogeny).

According to Oppenheimer (1955), De Beer (1962), Gould (1977), and many other historians of biology, Haeckel's emphasis on the biogenetic law delayed the experimental analysis of ontogeny because phylogeny was considered to be its cause. However, his influence began to wane with the rise of experimental embryology begun by Roux (1888; discussed in Gould, 1977), with the rediscovery of Mendel's laws of heredity in 1900 (Gould, 1977), and with the gradual realization that many of Haeckel's supposed facts and interpretations either were wrong, had many exceptions (De Beer, 1962),

or were manipulated to bolster his scheme (Gould, 2000).

Experimental embryologists ignored Haeckel's historical explanations and sought proximate causes for ontogeny in the intrinsic and extrinsic factors controlling animal development. No evidence was found for the existence of terminal addition or condensation, and most biologists now accept that ontogeny influences phylogeny more than phylogeny influences ontogeny: "Ontogeny does not recapitulate phylogeny, it creates it" (Garstang, 1922: 81). Instead, what recapitulation really means is that it is easier to find ancestral character states early in development and derived ones later, an observation used today to assist in polarizing character states in phylogenetic analysis (the *ontogenetic method*; Mayr and Ashlock, 1991). Even though biologists no longer accept the concept of recapitulation, evidence from comparative, descriptive study of animal ontogeny is "providing an increasingly precise description of the results of phylogenetic evolution" (De Beer, 1962: 172).

13.3.2 THE INFLUENCE OF VON BAER AND DARWIN

A devastating challenge to recapitulation was actually available long before Haeckel published his ideas. In *Entwicklungsgeschichte der Thiere: Beobachtung und Reflexion* (1828), K. E. von Baer compared the embryology of the domestic chicken with embryos of other vertebrates. In fact, discovery of the similarity between these embryos, after completing their gastrulation and neurulation, was actually the first recognition of the phylotypic stage (Gerhart and Kirschner, 1997: 300). Further, von Baer enunciated four "laws of development," collectively the antithesis of recapitulation and still influencing us today in how we interpret the relationship between development and evolution (e.g., Raff, 1996; Arthur, 1997):

- In development within the egg, the general characters appear before the special characters.
- From the more general characters, the less general and finally the special characters are developed.
- During its development, an animal departs more and more from the form of animals in other taxa.
- Younger stages in the development of an animal are not like the adult stages of other animals lower in the great chain of being, but are like their younger stages (De Beer, 1962: 3).

Darwin (1859: 449) used developmental evidence to assist in identifying homologous structures and to derive diverse organisms from a common stock: "Community in embryonic structure reveals community of descent." (A *homologous* feature is one derived from the same or equivalent feature of the nearest common ancestor [Mayr and Ashlock, 1991: 418].) Müller (1864), based on his detailed comparative knowledge of crustacean larvae, suggested that individual development provides clues to ancestral history, whereas Haeckel, a staunch Darwinist, had concluded that phylogeny caused ontogeny.

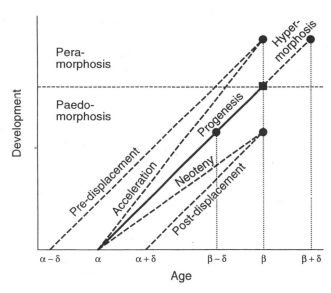

Fig. 13.5. Possible heterochronic changes occurring during evolution as modified from Alberch and colleagues (1979). The degree of development (ordinate) is a measurement of an organ or another measurement of shape or size. The solid line denotes the ontogenetic trajectory of the ancestor and the square, its adult condition. Dashed lines and filled circles represent descendant ontogenies and adults. Acceleration is an increase and neoteny, a decrease in the rate of development, as visualized in the steeper or flatter slopes of the corresponding descendant trajectories. Early onset of growth (at time $\alpha-\delta$ instead of α) is predisplacement, whereas $\alpha + \delta$ is postdisplacement. Progenetic descendants follow the same ontogenetic trajectory as their ancestors but terminate development early (at time $\beta-\delta$). Hypermorphic descendants extend the ancestral trajectory (termination at time $\beta + \delta$). Paedomorphic forms are below and peramorphic forms, above the horizontal dashed line. (Reprinted with permission from C. Klingenberg and J. R. Spence. Heterochrony and allometry: lessons from the water-strider genus *Limnoporus*. *Evolution* 47: 1834–1853. © 1993 Society for the Study of Evolution.)

13.3.3 THE ROLE OF HETEROCHRONY IN ONTOGENY AND PHYLOGENY

For decades after the evolutionary synthesis of the 1930s and 1940s (Mayr, 1982; Moore, 1993), evolutionary and developmental biologists ignored each other (Raff, 1996). However, Garstang (1922), De Beer (1962), Gould (1977), Alberch and colleagues (1979), McKinney and McNamara (1991), and McNamara (1997), among others, considered the observed relationship between developmental stages and phylogeny to be real and to result from action of heterochrony during the ontogeny of members of descendant lineages. *Heterochrony* is phyletic change in the onset or timing of development such that appearance or rate of development of a feature in a descendant ontogeny is either retarded or accelerated relative to appearance or rate of development of that feature in an ancestral ontogeny (De Beer, 1962; Gould, 1977; Alberch et al., 1979; Klingenberg and Spence, 1993; McNamara, 1997; Klingenberg, 1998).

As shown in Fig. 13.5, retardation can lead either to paedomorphosis, by slowing somatic development such that a descendant individual becomes sexually mature while still of juvenile phenotype (neoteny), or to recapitulation (hypermorphosis, peramorphosis), by slowing sexual maturation such that adult structure is more evolved at the time of sexual maturation than in the ancestor. Acceleration can also lead to paedomorphosis, by increasing the rate of sexual maturation such that a descendant individual is still of juvenile phenotype when sexually mature (progenesis), or to hypermorphosis, by increasing the rate of somatic development so that the descendant organism, again, is more evolved at the time of sexual maturation. Early onset of growth of an organ primordium in a descendant ontogeny as compared with that of the ancestor is predisplacement, while late onset is postdisplacement.

According to this model, change in size or shape of a descendant of any age could result from quantitative change in the time of onset or rate of either somatic or germ-line development. Also, a new character could appear at any stage in ontogeny and, by action of heterochrony, be accelerated or retarded so as to appear earlier or later in a descendant ontogeny. Thus, one can plot size and shape as independent variables operating during the history of a lineage (i.e., the two can become dissociated in ontogeny) (Gould, 1977; Alberch et al., 1979), and one can compare an extant or an extinct (as a fossil) descendant with its supposed ancestor (as a fossil) if they are at the same developmental stage (though assigning a fossil both to a particular taxon and to a particular stage in a life cycle can require assumptions unacceptable to an experimental biologist).

13.3.3.1 HETEROCHRONY AND LIFE HISTORY

Gould (1977) related the processes of acceleration and retardation in evolution to the life histories of extant forms. For example, he envisaged paedomorphosis (Fig. 13.5) to be an escape from overspecialization of the adult, and hypermorphosis to be a means of adding new structures to the adult (Gould, 1977). He also suggested progenesis to be a possible life history mechanism for r-selected species: r-selection acts on populations below the carrying capacity of their habitats to raise their maximum rate of increase (r) and favors the evolution of species with high reproductive potential, good colonizing ability, and short life cycles, as in aphids (5.5.2) and paedogenetic gall midges (5.5.3).

In the same way, neoteny (Fig. 13.5) was suggested to be a life history mechanism for K-selected species: K-selection acts to maintain population levels at the carrying capacity of their habitats (K is the number of individuals that the environment can support) and usually favors production of a few, slowly developing young, well adjusted to a stable environment.

13.3.3.2 HETEROCHRONY AND MACROEVOLUTION

According to Gould (1977), both progenesis and neoteny could also contribute to the evolution of new higher taxa. Such could occasionally arise by progenesis because selection on imaginal structure would relax in a developmental context, mixing juvenile and imaginal characters: precocious sexual maturity usually leaves some somatic characters juvenile while accelerating the development of others more directly associated with reproduction such as external genitalia and legs. In a similar vein, neoteny could provide a flexible alternative to the hypermorphic overspecialization of the adult that usually accompanies delay in adult somatic maturation (De Beer, 1962; Gould, 1977).

Heterochronic changes are regulatory and require alteration in the time of appearance or rate of development of a developmental program already in place (Gould, 1977). For example, expression of the pair-rule gene *hairy* (Fig. 7.26) is heterochronic in early embryos of three closely related species of *Drosophila* (Kim et al., 2000). Thus, change in expression of one or more genes regulating the timing of development of a particular organ could be important in generating new differences in structure and function in members of a descendant lineage. Such differences could act as evolutionary novelties and facilitate entry into a new adaptive zone. In fact, regulatory genes such as *lin-14* that control the timing of development of particular cell lineages have been identified in the nematode *Caenorhabditis elegans*. Mutations in such heterochrony genes cause certain sets of cells in mutant worms to develop normally but relatively earlier or later in ontogeny than in wild-type worms (see Alberts et al., 1994: 1074; and Raff, 1996: 276–277).

Raff (1996: 427) acknowledged the importance of heterochrony in contributing to evolutionary dissociation of developmental events between ancestor and descendant but suggested that since "development flows along a time axis, almost any alteration will manifest itself as a change in timing," such as mutation in a variety of early expressed, pattern-control genes that do not themselves influence timing (7.2.1). He thus considered most examples of heterochrony to result from rather than to contribute to major changes in ontogeny.

13.3.4 HORMONES, ABNORMAL METAMORPHOSIS, AND MACROEVOLUTION

Possibly influencing heterochrony, and especially pertinent to insects with their complex, hormonally controlled postembryonic development (chapter 12), was Matsuda's (1987) proposal that changes in descendants could result from disturbance in the balance of the various growth hormones influencing ontogeny, particularly juvenile hormone (JH). As we have already seen, JH is a female gonadotropin (2.3.2.1), can influence the rate of spermatogenesis in males (1.3.7.2.3), can inhibit the differentiation of adult characteristics during postembryonic development of exopterygotes and endopterygotes (12.2, 12.4.5, 12.5), and plays a critical role in the development of polymorphisms and polyphenisms (12.8). When the relatively high JH concentrations of earlier instars persist into later development, neoteny can

result, as in aphid virginoparae (12.8.1.2). Such high concentrations of JH could inhibit 20-OH ecdysone (20E) from inducing expression of early ecdysteroid-inducible genes such as the *Broad Complex* (12.4.1.5.2) (Restifo and Wikan, 1998), leading either to inhibition of metamorphosis or to changes in it. When reduction or loss of structures occurs during postembryogenesis, as in the failure of certain imaginal characters to develop in neotenous or progenetic species, compensatory and possibly adaptive development of other structures may occur (*material compensation*) to generate animals with a new structural organization—"hopeful monsters" in Goldschmidt's (1940) view—but simultaneously, in large numbers, and of both sexes.

Hormonal balance in insects can be perturbed simultaneously in large numbers of juvenile individuals by action of a variety of environmental factors; the resulting "abnormal" metamorphoses could be advantageous and, according to Matsuda (1987), have the potential to influence macroevolutionary change. In insects, two common results of material compensation are excessive development of ovaries and legs (particularly hind legs) and reduction or loss of wings and associated structures in adults, all resulting from higher than normal JH concentrations during late postembryogenesis. Such new structures in large numbers of individuals could function as evolutionary novelties and allow entry into a new adaptive zone. At the same time, the loss of previous imaginal structures, such as those associated with flight, could end a structural homology in adults of such a lineage.

The evolution of Phthiraptera (lice) from a psocopteroid ancestor might provide an example of such a process. Wing polymorphism can be induced in adults of certain psocids by changes in temperature or population density during postembryogenesis (Matsuda, 1987). These environmental factors presumably act by prolonging into later instars the high juvenile concentration of JH. This has yet to be demonstrated in psocids but has been in aphids and in some other wing polymorphic species such as termites and stick insects (12.2, 12.8.2.1). Perhaps several conspecific psocids settled in the nest or on the body of a bird or mammal where the stimulus for maintenance of elevated JH concentrations in their older offspring was always present (compare 2.4, no. 13), leading to constant aptery in them. In fact, adults of several psocid species in the genera *Lepinotis* (Trogiidae) and *Liposcelis* (Liposcelidae) are always apterous and do commonly occur in the nests of and, phoretically, on the bodies of certain mammals and birds (Lyal, 1985). Lice and liposcelids share 12 derived character states, mostly losses, suggesting the family Liposcelidae to be the sister group of the Phthiraptera and rendering the remaining Psocoptera paraphyletic. The common ancestor of Liposcelidae and Phthiraptera is suggested to have been *nidicolous* (nest dwelling) and facultatively ectoparasitic. Since winglessness was advantageous in this new, ectoparasitic adaptive zone, the insects persisted and diversified, their descendants being selected to specialize further for an ectoparasitic existence, as detailed by Lyal (1985: 160–161). All these specializations may have ultimately originated through inhibition of normal adult differentiation (i.e., neoteny) by prolonging high JH concentrations into later instars and could have initiated the lineage we now recognize as the Phthiraptera. A similar scenario could explain the evolution of the highly specialized adults of fleas (Siphonaptera) from a mecopteroid ancestor (Kristensen, 1999).

Unfortunately, there is a Lamarckian "inheritance-of-acquired-characters" slant to Matsuda's ideas. He also ignored the value of having a well-established phylogeny on which to map the direction of evolutionary change in ontogeny and the need to assess age, developmental stage, and size independently to determine how a metamorphic pattern might have been altered (McCune, 1988).

13.3.5 THE EFFECTS OF PHYLETIC CHANGE IN SIZE

Selection for decrease in phyletic size, especially extreme decrease in size or *miniaturization*, is also important in generating novel structural characteristics during evolution (Hanken and Wake, 1993). For example, a critical limit to smaller size in female insects could be their eggs, since there is a lower limit to egg size, probably about 0.18 by 0.06 mm, as in certain thrips, beyond which they contain insufficient yolk to support the embryogenesis of a fully formed, self-sufficient juvenile. Of course, egg size can decrease below this limit, but only by one of two ways: by dispensing with yolk and compensating for it by supplying nutrients directly to the embryo from the hemolymph of the mother via a trophamnion, as in viviparous aphids (5.5.2) and paedogenetic midge larvae (5.5.3), or by placing the egg in the body cavity of another animal or into the yolk of its egg, as in endoparasitoid wasps (6.7.2.1). As selection for small body size continues, the number of eggs that can be accommodated and matured within the female's abdomen is reduced until the minimum of one egg at a time is reached, as in certain ptiliid beetles (adults of some species are less than 0.5 mm long [Dybas and Dybas, 1981]). In other words, the relative size of an egg is inversely proportional to the body size of its mother (Cobben, 1968: Figs. 273, 274; Hinton, 1981; Andersen, 1982: Fig. 573); the same is true of the central nervous system and sense organs (Hanken and Wake, 1993). To cope with the relatively huge eggs of such tiny females, there must be compensatory changes in the structure of the abdomen and reproductive system so that an egg can still be developed, ovulated, fertilized, and deposited.

Such may have occurred during the evolution of the semiaquatic bugs (Hemiptera: infraorder Gerromorpha), in which the reproductive exit system of females is characterized by an intricate, sperm-manipulating gynatrial complex (Fig. 3.17) (Andersen, 1982; Heming-van Battum and Heming, 1986). This complex probably arose originally to accommodate the fertilization of the pro-

portionately larger eggs of a lineage of bugs ancestral to Gerromorpha that was being selected for small body size. In members of descendant lineages, it has been further modified in characteristic ways, possibly due to selection acting to increase the role of the female in controlling fertilization of eggs (3.7). Proportionately large eggs are metabolically expensive to produce, fertilize, and deposit, so that it profits a female to accept sperm from only the biologically fittest males. Because of the gynatrial complex, females of *Hebrus pusillus* (Hebridae) seem able to control intromission (only the most persistent males succeed), insemination (the female seems to draw sperm out of the exit system of the male during copulation), and fertilization (the site of sperm storage differs from the sites of insemination and fertilization, and females control movement of sperm from site to site). These developments in females in turn have resulted in epigamic selection on males for characteristics enabling the females' sperm-manipulating mechanisms to be circumvented. For example, after the long sperm of the male of *H. pusillus* (they are longer than the body of the female) have entered the tubular spermatheca of the female, they reverse themselves within it so that their heads face its opening, possibly giving them an advantage over later-arriving sperm in fertilizing eggs (Heming-van Battum and Heming, 1989). The diversity in structure of the exit systems in male and female semiaquatic bugs (Andersen, 1982) and their adaptive significance (Heming-van Battum and Heming, 1986) can be understood throughout the infraorder in the context summarized above.

13.3.6 THE INFLUENCE OF CONSERVED CELLULAR PROCESSES

Until the recent use of homologous nucleotide sequences in mitochondrial and nuclear genes to infer phylogeny (Hillis and Moritz, 1990) and the investigation of pattern-control gene distribution and function in embryos (7.2.1) of animals other than *Drosophila*, the discoveries of cell biology had little impact on metazoan or hexapod evolutionary biology. This was because until recently, few evolutionary biologists had any interest or expertise in molecular biology and because most cell biologists are reductionist: "to understand cell function, knowledge of evolution is much less important than knowledge of chemistry" (Gerhart and Kirschner, 1997: x). In their remarkable book *Cells, Embryos, and Evolution*, the molecular biologists Gerhart and Kirschner attempted to replace Dobzhansky's (1973) dictum that "nothing in biology makes sense except in the light of evolution" with "nothing in biology makes sense except in the light of cell biology." In a marvelous exercise in well-substantiated, bottom-up speculation, they proposed the profuse radiation of metazoans since the Proterozoic to reside within the "robust flexibility" of animal development, that is, in the ability of a central set of conserved cellular processes to direct a much wider range of developmental processes that in turn have facilitated the evolution of diverse phenotypes via selection on populations. In other words, selection has generated and maintained a capacity to evolve in animal ontogeny by using, again and again, the same cellular processes to regulate a host of different functions.

The low specificity of regulatory protein interactions in animal development (e.g., cytoskeletal, receptor, transcription factor, and signaling proteins) has generated a fundamental flexibility that has facilitated the origin, selection, and evolution of new regulatory roles. Exploratory behavior, such as that exhibited by growing microtubules (1.3.4, 2.2.2.6, 3.6.2, 6.9.1, 7.2.2), and growth cone filopodia of developing neuronal (8.3.1), muscle (8.1.2), and tracheolar cells (8.3.4, 10.2.2.1), has allowed complex structures such as the nervous, skeletomuscular, and tracheal systems to develop with a minimum of explicit instruction. Thus, mechanisms operating in animal cells and embryos are inherently flexible since they respond to a wide variety of signals, regulate a wide variety of cellular processes, and can thus generate a wide variety of phenotypes. At the same time, they are sufficiently robust by their ubiquity to accommodate any external or internal perturbations that might arise. Thus, new regulatory interactions can evolve relatively easily without upsetting existing ones already operating within an embryo. These interactions can influence the development of phenotype and, through selection, the diversification of descendants. In fact, the principal adaptive value of animal behavior, according to Gerhart and Kirschner, is to ensure entry into new habitats providing new opportunities for the selection of new mechanisms.

As discussed throughout this book and as summarized later (13.3.8–13.3.10), all these cellular processes are known to function during development of *D. melanogaster*, with recent comparative work on embryos and larvae of other hexapods suggesting that subtle differences in some of them correlate with differences in larval and adult body plans and have had a role in the development of hexapod diversity.

Nevertheless, consideration of these mechanisms alone cannot address the origin of diversity, as the selective forces shaping animal bodies do not act directly on cells, but on whole organisms through survival of the fittest. Thus, the ideas of Gerhart and Kirschner (1997) are incomplete, "resolutely internalist" (Wray, 1998: 284), and problematic in terms of both functional continuity and population genetics (Wilkins, 1998; Budd, 1999).

13.3.7 THE ROLE OF DEVELOPMENTAL CONSTRAINT
Much evidence suggests that the individual variation on which selection acts is not unlimited but is constrained by history and by action of a variety of extrinsic and intrinsic factors (Raff, 1996; Gerhart and Kirschner, 1997; Moore and Willmer, 1997; Arthur, 1997; Emlen, 2000). In his chapter 9, Raff (1996) considered these constraints to be imposed respectively by physics, genomic organization, and resistance to reorganization:

• Physics: Results of countless biomechanical studies on living animals show that the physical laws of gravity, air and water flow, strength of materials, surface-volume relationships, and efficiency of energy conversion influence the possible shapes assumed by organisms and how they are constructed and perform (Vogel, 1988; Dickinson et al., 2000; Thomas et al., 2000). Such rules also affect how tissues and cells behave during ontogeny, as in the formation of sheets (like the insect epidermis) or tubes (like a trachea, gonadal exit duct, or foregut) of cells or the cellular diffusion of morphogens (7.1.3, 7.2.1, 11.2.1, 11.3.2.2.3).

• Genome organization: Genetic constraint on variation can result from differences in genome size (affects cell size and division rate and recombination during meiosis) and on the presence of neutral or "selfish" DNA: the presence of transposons within the genome can affect mutation rate, inversion, and translocation and the expression of neighboring regulatory genes.

• Resistance to reorganization: Finally, it has long been recognized that integrated developmental interactions are more or less resistant to modification or reorganization by selection (*canalization*) (Waddington, 1942; Raff, 1996; Arthur, 1997; Gellon and McGinnis, 1998). The phyletic history of a lineage has entailed the establishment of a particular, specialized ontogeny for the basic body plan of members of that taxon (first observed in the polypod or extended germ band stage in insects and ultimately enabling its members to be part of a particular adaptive zone) such that starting variation is no longer available to allow the origin of diverse new body plans. Any subsequent major evolutionary change must either follow existing developmental mechanisms, constraining the evolution of future possibilities to those of that developmental mode, or jettison them, as in certain direct-developing echinoderms (Raff, 1996).

A good example of this resistance in insects is the complex of pattern-control genes (maternal effect, segmentation, and *HOM*) and signaling proteins (Dpp, Hh, Wg, etc.) specifying development of body plans in *Drosophila* embryos and larvae and influencing expression of the downstream regulatory and structural genes generating the phenotype (7.2.1, 11.3). Though orthologs of many of these genes and their proteins are evolutionarily conserved and function in the development of other animals (Figs. 7.32, 7.36, 7.38), they have duplicated, diverged, dissociated, and been redeployed for additional roles in members of these descendant lineages, at the same time losing others. Thus, whereas the *HOM* genes of *Drosophila* specify segmental identity in *Drosophila* and in other arthropod embryos, the orthologous *Hox* genes of vertebrates specify that of brain rhombomeres, mesodermal somites, and vertebrae (Fig. 7.38) (Raff, 1996; Shubin et al., 1997; Gerhard and Kirschner, 1997).

Arthur (1997) rearticulated von Baer's laws (13.3.2) to include the role of pattern-control genes and canalization in appearance of new body plans. His "inverted developmental cone model" envisaged the apex of the cone as the single-cell zygote and its flared top as the organs and specialized cell types of the adult. He speculated that mutations affecting early stages in its ontogeny would be reflected in widespread phenotypic change in its adult descendants, while those affecting later stages would have lesser effects. Arthur then used this model to explain the failure of new, phylum-level body plans to originate since the Cambrian. He proposed that action of selection over geological time tends to generate homeostatic mechanisms in increasingly complex, descendant ontogenies that protect developmental processes from genetic and environmental perturbation and that tend to stabilize body plans. The only genes having the potential to produce large-scale phenotypic change are those acting early in ontogeny. These are difficult to identify, as few regulatory genes are only early acting; most are expressed at different times throughout ontogeny (e.g., most homeoprotein-encoding genes; 7.2.1, 11.3).

Davidson and coworkers (1995) and Peterson and Davidson (2000) proposed that the Neoproterozoic progenitors of extant animal phyla were small and simple with relatively few cell types, and Arthur (1997) suggested that they exhibited little canalization during development. Mutations affecting early development of such simple organisms could produce viable imaginal phenotypes well outside the normal range of variation, phenotypes that would be impossible today because of the highly integrated nature of development in most post-Cambrian metazoans and because of the competition that surviving mutant individuals would face from better-adapted, wild-type compatriots. Such radical mutations might occasionally have been viable during the late Proterozoic only because internal homeostatic mechanisms had a lesser role than now and because marine habitats at that time were relatively empty ecologically. These made the origin of new body plans possible in the late Precambrian. During and following the Cambrian, however, ecological space quickly filled up, and subsequent canalization inflicted internal constraints on phenotypic change that prevented the emergence of additional new body plans.

13.3.8 MODULARITY

Systematists commonly observe that different parts of the body, in both juveniles and adults, have different rates of evolutionary change within members of a lineage, owing to the action of "mosaic evolution" (Mayr and Ashlock, 1991: 252). Its occurrence suggests both that metazoan ontogeny might be modular (i.e., compartmentalized) and that selection might act relatively independently on different body parts to generate module-specific changes in the body plans of descendants (West-Eberhard, 1989; Raff, 1996; Gerhart and Kirschner, 1997).

The basic body plan of a developing animal is first recognizable in the *phylotypic stage* (the polypod or extended germ band stage in insect embryos), a highly conserved period in embryogenesis in which the germ layers and organ primordia typical of its phylum and class have been established but not those defining membership in lower-level taxa (Sander, 1976b; Cohen and Massey, 1983; Raff, 1996; Gerhart and Kirschner, 1997; Arthur, 1997; Scholtz, 1998). Although it is not functional (organogenesis has just begun), the phylotypic stage has in place a series of spatially organized *phylotypic processes* (germ layers, compartments, and reliably distributed signal proteins) that "can activate, orient, place and scale" the more local, postphylotypic development of modules to generate the anatomical differences characterizing members of lower-level taxa (Gerhart and Kirschner, 1997: 511, 573).

According to Sander (1976b), the phylotypic stage is the most stable of any that proceed or follow it, and is when embryos of disparate species within a higher taxon most closely resemble each other. Raff (1996: 204, 206) suggested this stage to have been conserved in evolution because of the "complex interactions among what will eventually be discrete, non-contiguous modules," that is, those occurring between the transcription factors of *Drosophila* pattern-control genes specifying the segments and compartments and their identity (7.2.1). Gerhart and Kirschner (1997) argued that conservation of this stage was likely due to its "flexibility, versatility, and robustness," since it represents the basic body plan characteristic of a higher taxon on which subsequent development can add the characteristics of members in lower taxa. Finally, Slack (1997) suggested it has stayed the same because selection has had little effect on this stage of embryogenesis. All are plausible but untested scenarios.

Modules established in the phylotypic stage continue to develop in later embryogenesis and after hatch. However, because they no longer interact extensively with each other (compartment boundaries prevent this from occurring), they can undergo relatively independent modification

- through broad or subtle change in expression of the *HOM* genes underlying their specification and identity, resulting, possibly, from subtle but rapid change in their cis-regulatory modules (Figs. 13.6, 13.7) (Carroll, 2000);
- through change in regulatory interactions between *HOM* genes (7.2.1.3.4);
- by recruitment of genes to assume additional functions within previously established compartments (e.g., recruitment of the *hedgehog* [*hh*] regulatory circuit to function in eyespot determination in butterfly wings [Keys et al., 1999]);
- by changes in the regulation or function of downstream target genes (Fig. 13.6: 4) (Gellon and McGinnis, 1998; Guss et al., 2001); or
- by changes in the distribution, timing, and sensitivity of JH-sensitive periods in the cells of particular compartments during postembryonic development, a primary basis for the development of polyphenisms (12.8) (Emlen, 2000; Emlen and Nijhout, 2000).

Variation in such change during development of individuals of a stem species could influence the phenotypes

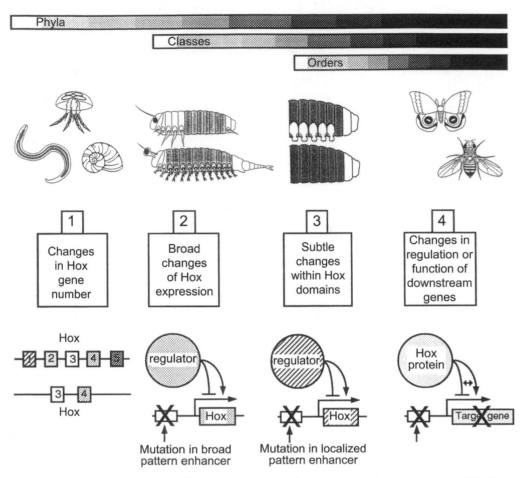

Fig. 13.6. Evolutionary changes in *Hox* gene pathways arranged according to presumed impact on animal body structure. The numbered boxes list some of the changes that *Hox* pathways can undergo during evolution. Diagrams below the boxes indicate likely molecular mechanisms mediating such changes and those above, the modifications in body structure that result. Graded bars at the top suggest a rough correlation between taxonomic level and variation in *Hox* pathways. (Shaping animal body plans in development and evolution by modulation of *Hox* expression patterns, G. Gellon and W. McGinnis, *Bioessays*, Copyright © 1998 Wiley-Liss, Inc. Adapted by permission of Wiley-Liss, Inc., a subsidiary of John Wiley & Sons, Inc.)

of the organ or organ parts developing from that module and provide a platform for selection. Similar changes might partially explain the origin of the multiplicity of tagmata and appendages in arthropods (Fig. 13.7); of cephalic and thoracic horns in polyphenic, male scarabs (12.8); of mouthparts (Fig. 13.8), legs, and wing venational (Stark et al., 1999) and pigmentation patterns (11.3.2.2); and of external genitalia characterizing adults of each insect order and family. The independent development and selection of modules has been suggested to be an important source of structural disparity within members of higher taxa such as the Insecta (Gerhart and Kirschner, 1997: chapter 10) but is only beginning to be investigated comparatively with the vigor recently directed at the *HOM* genes (Stark et al., 1999; Emlen, 2000, 2001; Emlen and Nijhout, 2000; Guss et al., 2001).

Prior to development of its phylotypic stage, an early, preblastoderm stage embryo of *D. melanogaster* is a single system that has its axes gradually specified through interactions, within the periplasm, of transcription fac-

tor gradients. These result from sequential expression of maternal and zygotic dorsoventral (Fig. 7.20), terminal, and anteroposterior (Fig. 7.24) genes, beginning while the egg is still developing within its follicle (Fig. 2.20). Because the interactions occur throughout the periplasm, early stages of embryogenesis, like postphylotypic ones, are unstable and can be acted on by selection. They also can come to differ extensively from one species to another as, for example, do cleavage patterns (6.2) and blastoderm (6.3), germ band (6.7), germ layer (6.8), and segment (6.10) formation in embryos of different insect species and as do early developmental modes in closely related, directly versus indirectly developing cnidarian (Martin, 1997) and sea urchin embryos (Raff, 1996: 223–237). The latter have "set-aside" cells (imaginal rudiments) that function like the imaginal discs in *Drosophila* larvae, to generate the adult during metamorphosis (Davidson et al., 1995; Wray, 1997; Peterson and Davidson, 2000).

Raff (1996: 208) summarized these differences

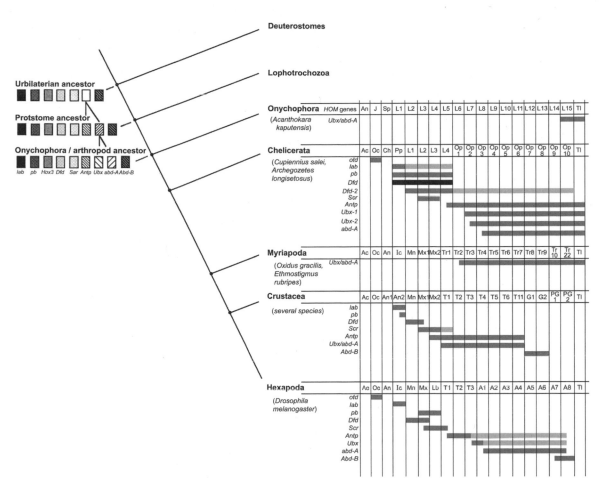

Fig. 13.7. *HOM* gene expression and evolution of body plan in the onychophoran-arthropod clade. The last common ancestor of protostomes and deuterostomes possessed at least seven divergent *HOM* genes (top row of boxes). The common ancestor of the protostomes (middle row of boxes) possessed an additional, central class gene (hatched to upper right) that gave rise to the *Ultrabithorax* (*Ubx*) and *abdominal-A* (*abd-A*) genes in the onychophoran-arthropod ancestor. Diversification of this clade has evolved around a conserved set of nine *HOM* genes (bottom row of boxes: *lab*, labial; *pb*, proboscipedia; *Hox 3*; *Dfd*, Deformed; *Scr*, Sex combs reduced; *Antp*, Antennapedia; *Ubx*; *abd-A*; and *Abd-B*, Abdominal-B). Extant onychophorans and arthropods differ in tagmata and in segment number, identity, and structure even within each class. Within each species, expression of different *HOM* genes influences the structure of body parts. The relative boundaries of *HOM* gene expression have diverged among arthropods and often correlate with transitions in appendage structure and function from one tagma to another. The heteronomy and diversification of arthropod body plans correlate with diversification of *HOM* gene regulation and illustrate, in general, that regulatory changes in conserved pattern-control genes underlie body plan evolution. Body segments: A1, etc., abdominal segment 1, etc.; Ac, acron; An, antennal; Ch, cheliceral; G1, G2, genital; Ic, intercalary; J, jaws; L1, etc, leg (or lobopod) segment 1, etc.; Mn, mandibular; Mx1, first maxillary; Mx2, second maxillary; Oc, ocular; Op1, etc., opisthosomal segment 1, etc.; Pg1, etc., postgenital segment 1, etc.; Pp, pedipalpal; Sp, slime papilla; T1, etc., thoracic segment 1, etc.; Tl, telson; Tr1, etc., trunk segment 1, etc. (Left half adapted with permission from A. H. Knoll and S. B. Carroll. Early animal evolution: emerging views from comparative biology and geology. *Science* 284: 2129–2137. Copyright 1999 American Association for the Advancement of Science. Right half from Fig. 7.36.)

between early, mid, and late stage embryos in a "developmental hourglass model" in which development rather than sand flows from top to bottom (Fig. 13.9). According to this model, early (with no modules) and late stage embryos (with numerous, independently developing modules) can accommodate change more readily than can the highly interlinked, phylotypic stage.

As both Raff (1996) and Gerhard and Kirschner (1997) emphasized, the regulatory genes specifying development of body pattern also act in a modular fashion, their sequential expression being reflected in the sequential appearance of an increasing number of ever smaller compartments in the embryo and larva as it develops (7.2.1, 11.3). And recent comparative work on species in other phyla has shown their development to be modular also and to likewise be under control of orthologous pattern-control genes (Gerhart and Kirschner, 1997: chapter 7; Gellon and McGinnis, 1998), referred to collectively by

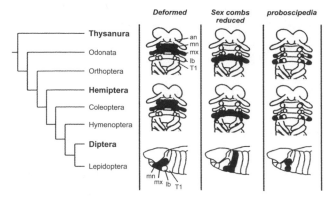

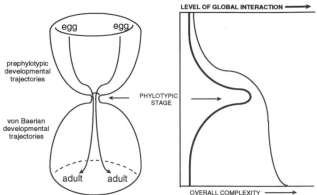

Fig. 13.8. Expression pattern of the homeotic genes *Deformed* (*Dfd*), *Sex combs reduced* (*Scr*), and *proboscipedia* (*pb*) in gnathal segments of polypod (extended germ band) stage embryos of *Thermobia domestica* (Zygentoma: Lepismatidae), *Oncopeltus fasciatus* (Hemiptera: Lygaeidae) (both in ventral aspect), and *D. melanogaster* (Diptera: Drosophilidae; in lateral aspect). Larvae of *T. domestica* have biting and chewing mouthparts; those of *O. fasciatus*, piercing and sucking mandibular and maxillary stylets (see Fig. 8.49); and in those of *D. melanogaster*, the head is later invaginated into the front of the body and the mouthparts are reduced to mouth hooks (see Fig. 8.45). Note the subtle difference in expression in each gene in embryos of each species. The inferred phylogeny of the most diverse insect orders is shown on the left (the authors use the old order name Thysanura, which includes both Zygentoma and Archaeognatha). an, antenna; lb, labium; mn, mandible; mx, maxilla; T1, prothorax. (Reproduced with permission from the authors from A. Popadíc, A. Abzhanov, D. Rusch, and T. C. Kaufman. Understanding the genetic basis of morphological evolution: the role of homeotic genes in the diversification of the arthropod bauplan. *Int. J. Dev. Biol.* 42: 453–461. © 1998 University of the Basque Country Press.)

Fig. 13.9. Raff's "developmental hourglass" model. The arrows represent variant developmental trajectories of two species passing through the hourglass, converging on the phylotypic stage and diverging again. The volume of the hourglass represents probability space, and its width, at any level, the probability that an evolutionary change can be successfully integrated into a developmental pathway at that level. Early (with no modules) and late stage embryos (with numerous, independently developing modules) can accommodate change more readily than can the highly interlinked phylotypic stage. The graph on the right plots the overall increase in complexity of an individual embryo during its ontogeny (i.e., as it passes through the hourglass) and the level of interaction among modules. Interaction increases as the embryo approaches the phylotypic stage, and then declines as modules become increasingly autonomous in later development. (Adapted with permission from R. A. Raff. *The shape of life. Genes, development, and the evolution of animal form*, p. 208, Fig. 6.7. © 1996 The University of Chicago Press.)

Slack and associates (1993) as the zootype (see also Raff, 1996: 179–190 and chapters 10 and 12).

A third level of modularity is introduced if one considers the possible roles in evolution of change in the regulatory modules of the control sequences influencing the expression of pattern-control genes such as *even-skipped* (*eve*) (Fig. 7.39; 7.2.1.4) (Carroll, 2000; Carroll et al., 2001; Davidson, 2001):

- Individual regulatory modules in a gene control sequence can act and evolve independently of others.
- Regulation of the expression of such genes is combinatorial and depends on the identity, number, and affinity of the transcription factors binding to their regulatory sequences (e.g., Guss et al., 2001).
- The DNA-binding specificities of these factors are sufficiently imprecise that the affinity and number of binding sites for each factor can evolve relatively rapidly. For example, the binding sites for the four transcription factors controlling expression of *eve* stripe 2 in *Drosophila* embryos (Bicoid, Hunchback, Giant, and Krüppel) have diverged between closely related species.

Carroll (2000) also wondered whether the same genetic mechanisms underlying intraspecific variation and

interspecific differences are sufficient to account for the origin of the large differences in body plan characteristic of higher taxa.

Evolutionary biologists such as Budd (1999) considered all such bottom-up scenarios to be saltational (13.1.3) and problematic and suggested instead that action of selection for morphological change in individuals drives evolutionary change in the expression of body patterning genes rather than vice versa (Fig. 13.10). According to his "homeotic takeover" model (Fig. 13.11), mutations in the pattern-control genes may have been selected because they increased efficiency in building a particular body plan but only after selection for morphological innovation and reorganization had already occurred. Fig. 13.11 illustrates the selective modification of a pair of posterior, feeding appendages of a hypothetical arthropod for a locomotory function while allowing them to remain functionally integrated with appendages of the head. The subsequent shift in *Hox* gene expression then occurs, increasing the efficiency of this development without disrupting the function of other locomotory and feeding appendages. Significantly, a similar developmental shift in expression of the *HOM* gene *Sex combs reduced* (*Scr*) was recently shown to correlate with a leg to maxilliped

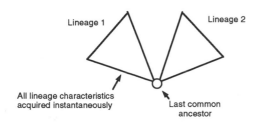

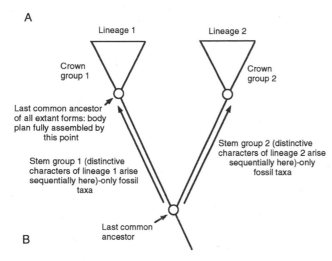

Fig. 13.10. Differing perceptions on how the common body plans characterizing members of higher taxa evolve. A. In the bottom-up, "saltational" ("hopeful monster") scenario preferred by most developmental geneticists, all body plan characteristics of the members of a phylum or other higher taxon appear suddenly in descendants of the common ancestor of the group, owing to a sudden change in *Hox* gene expression pattern in each one. B. In the top-down, "gradualist" model preferred by most systematists, paleontologists, morphologists, and population geneticists, the crown groups (lineages 1 and 2) are as in A. However, each is considered to descend from a common but now extinct stem species. Major body plan characteristics accumulate gradually and progressively in the stem species of each group after their reproductive isolation. (Does evolution in body patterning genes drive morphological change—or vice versa?, G. E. Budd, *Bioessays*, Copyright © 1999 Wiley-Liss, Inc. Adapted by permission of Wiley-Liss, Inc., a subsidiary of John Wiley & Sons, Inc.)

transformation during embryogenesis of the isopod crustacean *Porcellio scaber* (compare Fig. 13.7 [Abzhanov and Kaufman, 1999a]).

This scenario could provide one mechanism for the role of phenotypic plasticity in generating new species and higher taxa with modified body plans (West-Eberhard, 1989). *Phenotypic plasticity* is the ability of a single genotype to produce more than one alternative form of morphology, physiological state, or behavior in response to environmental conditions (West-Eberhard, 1989: 249) (examples include the polyphenisms exhibited by crickets, aphids, locusts, and scarab beetles discussed in 12.8). West-Eberhard (1989) suggested that

phenotypic evolution proceeds first by phenotypic plasticity followed by mutation and genetic stabilization of particularly advantageous phenotypes, so that changes in phenotype again lead to changes in genotype during animal diversification.

13.3.9 EXPRESSION OF HOMEOTIC AND *DISTAL-LESS* GENES AND THE EVOLUTION OF ARTHROPOD LIMBS

Although extant arthropods (chelicerates, myriapods, crustaceans, and hexapods) share many characteristics (exoskeleton, jointed appendages, a reduced coelom, hemocoel, etc.), members of each class differ substantially from those of other classes in number of body segments and tagmata and in number, position, structure, and function of their limbs (Fig. 13.14). These similarities and differences have generated vigorous debate for more than a century about whether these animals descend from a single Precambrian ancestor that was already an arthropod, or polyphyletically from a number of different, nonarthropod ancestors, members of each lineage independently achieving the arthropod grade of organization (Gorsberg, 1990; Stys and Zrzavy, 1994; Raff, 1996; Fryer, 1996, 1998; Moore and Willmer, 1997; Shubin et al., 1997; Gerhart and Kirschner, 1997; various authorities in Fortey and Thomas, 1998; Conway Morris, 1998a; and Deuve, 2001).

Molecular evidence is increasingly supporting the existence of a common genetic mechanism underlying segmental identity and limb development, not only in arthropods but also in other appendage-bearing metazoans (see 7.2.1.2.1, 7.2.1.3.4, 7.2.1.5). The many differences among chelicerate, myriapod, crustacean, and insect embryos and larvae indicate that differences in tagmata, segmental identity, and appendage plan correlate with differences in expression of particular *HOM* genes (Fig. 13.7), and also of the downstream genes influencing limb outgrowth (e.g., *Distal-less* [*Dll*]) and differentiation (Fig. 13.6).

For example, absence of *Ubx* and *abd-A* expression in anterior thoracic segments of embryos of 13 species in nine orders of Crustacea correlates with modification of their appendages into feeding maxillipeds associated with the head (compare Fig. 13.7) (Averof and Patel, 1997). And in embryos of the terrestrial isopod *P. scaber*, modification in expression of an ortholog of the *HOM* gene *Sex combs reduced* (*PsScr*) is also involved (Abzhanov and Kaufman, 1999a). In early embryos of *P. scaber*, transcripts of *PsScr* appear in both the second maxillary and first thoracic segments but are translated into protein only in maxillary segment 2. However, in late embryos, PsScr protein also appears in the first thoracic segment and correlates with a change of its appendages into maxillipeds (compare Fig. 13.7).

Limb branching in crustacean and in some chelicerate (e.g., *Limulus polyphemus*) embryos is a second-order phenomenon, affected by expression of the *Dll* gene, that initiates distal outgrowth both of unbranched limbs in fruit fly (Figs. 7.40, 7.41, 11.7A, 11.8A, 11.9), arachnid, and centipede embryos and of each branch of

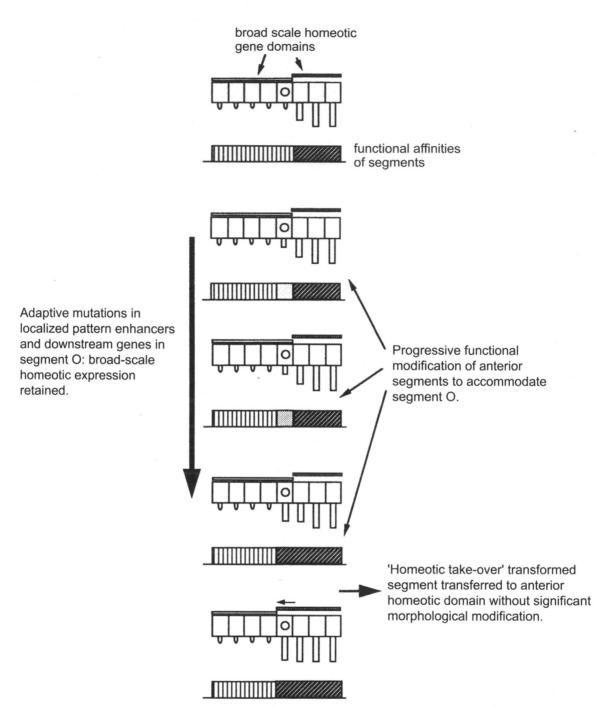

broad scale homeotic
gene domains

functional affinities
of segments

Adaptive mutations in
localized pattern enhancers
and downstream genes in
segment O: broad-scale
homeotic expression
retained.

Progressive functional
modification of anterior
segments to accommodate
segment O.

'Homeotic take-over' transformed
segment transferred to anterior
homeotic domain without significant
morphological modification.

Fig. 13.11. Budd's "homeotic takeover" model: structural "anticipation" of homeotic transformation in the evolution of arthropod trunk segments. Schematic diagrams represent trunk segments and their appendages related to the expression domains of two homeotic genes. Graphs depict level of functional integration: diagonally hatched → vertically hatched, functional affinity of adjacent segments. Modulation of local homeotic gene regulation and downstream gene expression enables gradual adaptive modification of segment O and its appendages while also allowing them, at all times, to be functionally integrated with adjacent appendages. Shifts in large-scale homeotic expression domains (caused by mutation in high-level regulation) can then occur and be selected based on efficiency grounds without structural or functional disruption. (Does evolution in body patterning genes drive morphological change—or vice versa?, G. E. Budd, *Bioessays*, Copyright © 1999 Wiley-Liss, Inc. Adapted by permission of Wiley-Liss, Inc., a subsidiary of John Wiley & Sons, Inc.)

biramous limbs in crustacean embryos (Panganiban et al., 1995). *Dll* is not expressed distally in leg 4 primordia of the mite *Archegozetes longisetosus*, probably because prelarvae and larvae of mites have only three pairs of functional legs (Thomas and Telford, 1999). All differences in limb branching correspond to differences in the way expression of *Dll* is regulated by the *HOM* genes during embryogenesis and postembryogenesis. For example, each branch of each limb of embryos and larvae of the crustaceans *Artemia franciscana* and *Mysidopsis bahia* expresses *Dll* at its tip. The exact pattern of expression even varies between body segments: branches of the cephalic appendages of *M. bahia* grow simultaneously from a single group of *Dll*-expressing cells, while those of thoracic limbs develop sequentially from separate cell groups. This suggests that the limb branches themselves cannot distinguish members of any arthropod taxon based on *Dll* expression. In fact, orthologs of *Dll* are expressed in the distal portion of developing antennae, jaws, slime papillae, and lobopods of onychophoran embryos (*Peripatopsis capensis, Acanthokara kaputensis*); in distal portions of the parapodia of *Chaetopterus variopedatus* (Annelida); in the ampulla of ascidians; and in the CNS and epidermis of the lancelet *Branchiostoma* (Cephalochodata). The five or six *Dll* orthologs in vertebrate embryos are expressed, within the mesoderm. These findings collectively suggest that *Dll* expression was probably present in the common ancestor of all metazoans (DiSilvestro, 1997; Panganiban et al., 1997; Grenier et al., 1997).

Differences in *HOM* gene expression in representative embryos of each arthropod class are shown in Fig. 13.7, which also illustrates the increasing molecular evidence supporting the validity of the previously recognized subphylum Mandibulata (Myriapoda + [Crustacea + Hexapoda]) (Knoll and Carroll, 1999).

13.3.10 FOSSILS, PATTERN-CONTROL GENES, KEY INNOVATIONS, AND HEXAPOD EVOLUTION

As summarized in 13.2, hexapods have a known fossil record extending from the Lower Devonian (Fig. 13.1); myriapods, from the Upper Silurian; and trilobites, crustaceans, and chelicerates, from the Lower Cambrian (Fryer, 1996, 1998; Fortey et al., 1997; Dunlop and Selden, 1998; Shear, 1998; Conway Morris, 1998a; Walossck and Müller, 1998; Valentine et al., 1999). Comparisons between homologous nucleotide sequences of nuclear genes, together with morphological, developmental, phylogenetic, and fossil evidence, suggest to many recent workers (e.g., Labandeira and Beall, 1990; Kukalová-Peck, 1991; Carpenter 1992a, 1992b; Labandeira and Sepkoski, 1993; Pritchard et al., 1993; Carroll, 1995; Raff, 1996; Grenier et al., 1997; Fortey et al., 1997; Scholtz, 1998; Popadíc et al., 1998a; Knoll and Carroll, 1999) that arthropods are monophyletic and descend from a common, segmented, arthropodized ancestor that lived in the Late Precambrian and had jointed biramous limbs and a homonomous (undifferentiated) trunk (Fig. 13.14).

Fryer (1996, 1998) and Moore and Willmer (1997), however, building on the exquisite comparative, functional morphological studies of Sidnie Manton (summarized in 1977), the embryological fate map reconstructions of Anderson (1973), and recent discoveries in paleontology (Briggs and Fortey, 1989), pattern-control genes (7.2.1), and eye development (8.3.2.4.4), believed such a conclusion to greatly underestimate the role of convergence in arthropod evolution. Fryer proposed instead that all arthropod classes originated independently in the Upper Precambrian from a monophyletic assemblage of soft-bodied, segmented, hemocoel- and lobopod-possessing wormlike ancestors, with arthropodization having occurred independently within each lineage (Fig. 13.12). He considered the gap of 144 million years between the earliest known trilobite, chelicerate, and crustacean fossils, all marine, and the earliest known terrestrial, hexapod fossils to be real and to result from descendants of the soft-bodied, pre-Devonian ancestors of each extant hexapod class (Collembola, Protura, Diplura, Archaeognatha, and Insecta) having arthropodized independently (Fig. 13.13).

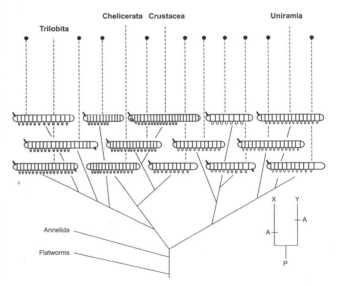

Fig. 13.12. Polyphyletic origin of the arthropods from a number of different but related segmented worms having a hemocoel and lobopods (affinities of the latter are arbitrary). Although the genome of each major lineage in the Cambrian differed, members of the various lineages would have shared major elements because of their common ancestry. Filled circles indicate extinct lineages. The inset shows the possible development of two arthropod lineages, X and Y, from a prearthropod ancestor, P, with independent arthropodization occurring in each (A). Members of the two lineages might be expected to have retained many similarities in the nucleotide sequences of their genomes such that use of molecular methods would be unable to detect the point at which either lineage became arthropodized. (Reprinted with permission from G. Fryer, Reflection on arthropod evolution. *Biol. J. Linn. Soc.* 58: 1–55. © 1996 Academic Press, Ltd.)

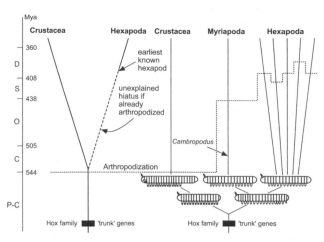

Fig. 13.13. Comparison between monophyletic (left) and polyphyletic (right) concepts of the subphylum Mandibulata (with or without Myriapoda), showing shared occurrence of the *Hox* genes, which must have arisen before arthropodization. On the right, all five extant hexapod classes (Protura, Collembola, Diplura, Archaeognatha, and Insecta, according to Manton [1977]) are shown arising independently from a single prearthropod ancestor. Note the time gap of more than 100 million years between the earliest known fossils of crustaceans and those of myriapods and hexapods (late arthropodization, as indicated by the dotted line, is congruent with the polyphyletic hypothesis). C, Cambrian; D, Devonian; O, Ordovician; P-C, Precambrian; S, Silurian. (Reprinted with permission from G. Fryer. Reflection on arthropod evolution. *Biol. J. Linn. Soc.* 58: 1–55. © 1996 Academic Press, Ltd.)

These two hypotheses have yet to be reconciled (Fortey and Thomas, 1998; Conway Morris, 1998a, 1998b, 1998c, 2000; Jenner and Schram, 1999; Klass and Kristensen, 2001), but the monophyletic hypothesis seems to have more support. Both hypotheses are similar in postulating an early origin for progenitors of the pattern-control genes that specify metazoan body plans, the descendant genes operating today in members of each lineage having derived by successive duplication, divergence, dissociation, and redeployment from a more simple, ancestral regulatory platform originating in the Neoproterozoic (Figs. 7.38, 13.7) (Hall, 1992; Carroll, 1995; Davidson et al., 1995; Kerr, 1995; Wray, 1995; Fryer, 1996; Gilbert et al., 1996; Valentine et al., 1996; Raff, 1996; Erwin et al., 1997; Shubin et al., 1997; Grenier et al., 1997; Gerhart and Kirschner, 1997; Gellon and McGinnis, 1998; Popadíc et al., 1998a; Conway Morris, 1998c; Knoll and Carroll, 1999; Peterson and Davidson, 2000).

As outlined in 13.2, key innovations in the diversification of supraordinal taxa of hexapods were, arguably, the origin of internal fertilization, hexapody, dicondylic mandibles, and ovipositors in the Silurian; of wings and wing flexing and of ovariolar nurse cells in the Upper Devonian or Lower Carboniferous; and of complete metamorphosis in the Lower or Upper Carboniferous (Fig. 13.1). Since extant ectognath insects have 20 body

segments (6 head, 3 thoracic, 11 abdominal), the basic insect body plan had probably originated, at the latest, by the end of the Upper Silurian (Fig. 13.1) (Kukalová-Peck, 1991). Order-level differences in body plan have probably originated since via speciation, extinction, and mosaic evolution, influenced by mutation and selection, which affect postphylotypic processes within the cells of each compartment in postpolypod stage embryos and larvae and influence, in turn, the expression of downstream structural genes (Fig. 13.6) (Graba et al., 1997; Gellon and McGinnis, 1998). For example, mouthpart primordia of polypod stage embryos of known hexapods are similar (Fig. 13.8) and only begin to differentiate into the highly specialized, invaginated, cephalopharyngeal complex of higher flies (Fig. 8.45) or into the piercing and sucking mouthparts of thrips (Fig. 8.47B), bugs (Fig. 8.49), and lice (8.3.8.3) during later embryogenesis. This is possibly due to action of differential selection and to subsequent change in expression of the genes regulating ontogeny within cells of the gnathal segments giving rise to these appendages (8.3.8.2.1). For example, the *HOM* gene *Scr* is expressed in the labial segment in *Thermobia domestica*, *Oncopeltus fasciatus*, *Tribolium castaneum*, and *D. melanogaster* embryos where it contributes to medial fusion of the second maxillae to form the labium (Fig. 13.8); this characteristic distinguishes hexapods and some myriapods from other arthropods and differentiates this segment from both the mandibular and first maxillary segments (Rogers and Kaufman, 1997; Popadíc et al., 1998). The gene *proboscipedia* (*pb*) is expressed in both the maxillary and labial appendages in *T. domestica* and *T. castaneum* embryos, which are similar to one another, but only in the labial segment in *O. fasciatus* embryos, where these appendages are quite different (Fig. 13.8).

In members of subordinal taxa, change in expression of the genes controlling determination and differentiation of cells within each terminal compartment, and possibly relating to changes in JH secretion and tissue sensitivity, will probably be found to occur still later in development, as discussed earlier and as implied in von Baer's 170-y-old laws of development (13.3.2), an aspect just beginning to be investigated at the genetic and molecular levels (Warren et al., 1994; Powell, 1996; Shubin et al., 1997; Gerhart and Kirschner, 1997; Graba et al., 1997; Gellon and McGinnis, 1998; Emlen, 2000).

The probable time of origin of the eight evolutionary novelties that, I believe, have most influenced diversification of supraordinal taxa of hexapods is indicated in Fig. 13.1. These are discussed more fully below.

13.3.10.1 THE ORIGIN OF INTERNAL FERTILIZATION
The selective advantages of internal over external fertilization for terrestrial animals are obvious and were discussed in 3.3.2. To my knowledge, nothing is known concerning the genetic and developmental basis for this switch, but it is intriguing that paired clusters of epidermal cells on either side of the ventral midline in segments T1–A9 in *Manduca sexta* embryos temporarily

express the same antibody as cells of the primary exit ducts and those giving rise to the secondary exit system and external genitalia in this insect (Fig. 8.53A: center). Could these clusters represent an ancestral, but no longer functional, developmental feature in insect embryos inherited from a polygoneate marine ancestor (crustacean?) that perhaps reproduced by external fertilization?

13.3.10.2 THE ORIGIN OF SIX LEGS

According to Shubin (1997) and Panganiban (1997) and their respective coworkers, the paired, jointed appendages of arthropods may descend from the simple, unjointed appendages of a marine ancestor that resembled the lobopodia of extant onychophorans and tardigrades, the transition to jointed appendages having occurred before the Cambrian (Fig. 13.14). Next came external segmentation and sclerotization and jointed, biramous appendages, each of the latter perhaps resulting from basal fusion of an outer, gill-like lateral lobe with a medioventral lobopodium, as appears to be present in the Cambrian lobopod *Opabinia regalis*. Further diversification of early arthropods was accompanied by changes in the number, pattern, and function of limbs (Shubin et al., 1997; Panganiban et al., 1997; Gerhart and Kirschner, 1997; Walossek and Müller, 1998; Scholtz, 1998); by increase in the number of limb types (up to 10 in hexapods); and by the subsequent appearance of unbranched (uniramous) appendages in the ancestors of those chelicerates, myriapods, crustaceans, and hexapods assuming a terrestrial existence.

In *Drosophila* embryos, each pair of thoracic appendages is specified by expression of a different combination of homeotic genes (Fig. 13.7):

- *Sex combs reduced* (*Scr*) in the anterior prothorax (T1);
- *Antennapedia* (*Antp*) in posterior T1, in T2, and in anterior T3; and
- *Ultrabithorax* (*Ubx*) in posterior T3.

However, development of antennae on the head does not require expression of homeotic genes but instead is specified by that of the gaplike genes *orthodenticle* (*otd*), *empty spiracles* (*ems*), and *buttonhead* (*btd*) (Fig. 7.41). In addition, total loss of homeotic gene function in embryos of the beetle *Tribolium castaneum* results in development of a larva bearing a pair of antennae on each head, thoracic, and abdominal segment, whereas combined inhibition of *Deformed* (*Dfd*), *pb*, and *Scr* function in *Oncopeltus fasciatus* embryos causes the mandibles to develop as distal antennal segments and the maxillae and labium to develop as antennae (7.2.1.3). Thus, the potential to develop paired appendages seems to be present in all segments of the insect embryo, the differential expression of homeotic (and in the pregnathal head, gaplike) genes influencing only whether appendages form and what they develop into (Fig. 13.7) (Shubin et al., 1997; Knoll and Carroll, 1999; Casares and Mann, 2001).

HOM gene expression in embryos and larvae of other arthropods leads to the appearance of different but

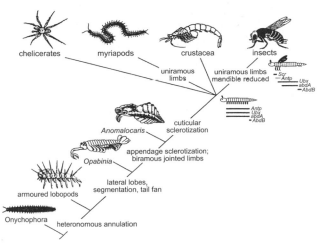

Fig. 13.14. The lobopod-arthropod transition and diversification of arthropod limb patterns according to Shubin and colleagues (1997). Various Cambrian fossil lobopods that may represent increasing degrees of arthropodization are depicted. *Opabinia* is shown in partial cutaway view to reveal the lateral lobes and ventral lobopodia. The relationship among arthropod classes is shown as an unresolved polytomy (see Fig. 13.7 for an alternate hypothesis). The ancestral euarthropod was probably fully sclerotized with jointed biramous appendages and a homonomous trunk. Single branched (uniramous) limbs evolved in terrestrial arthropods. Tagmatization and diversification of individual limbs in extant insects involved regulatory changes in *HOM* gene expression domains along the anteroposterior axis compared with tagmatization in the ancestor in which trunk appendages and *HOM* gene expression were thought to be mostly identical (*HOM* gene expression diagrams in upper right; see also Figs. 13.6 and 13.7). (Reprinted with permission from *Nature* (N. Shubin, C. Tabin, and S. Carroll. Fossils, genes and the evolution of animal limbs. 388: 639–648) © 1997 Macmillan Magazines Limited.)

serially homologous limbs, suggesting the increase in appendage diversity in onychophorans, chelicerates, myriapods, crustaceans, and hexapods to have involved diversification in *HOM* gene regulation and function in arthropod trunk segments or in the regulatory modules of the downstream genes they influence (Figs. 13.6, 13.7) (Carroll, 1995; Shubin et al., 1997; Gerhart and Kirschner, 1997; Gellon and McGinnis, 1998; Scholtz, 1998). For example, in embryos and larvae of the branchiopod crustacean *Artemia franciscana*, expression of orthologs of the *HOM* genes *Antp*, *Ubx*, and *abd-A* is somewhat coincident and the structure of all thoracic appendages the same, while in those of *Drosophila*, expression of the same genes differentiates among T1, T2, T3, and A1 (Fig. 13.7).

The abdomens of adult pterygote insects are legless and result from repression of *Dll* and of limb formation by products of the *Ubx*, *abd-A*, and *Abdominal-B* (*Abd-B*) genes (Fig. 13.7; 7.2.1.3), an event perhaps related to the origin of hexapody (Fig. 13.1). That abdominal appendages temporarily appear in the polypod embryos of many basal hexapods and in some endopterygotes and that they persist and differentiate into functional styli and eversible vesicles in immature and adult diplurans, archaeognaths, and silverfish and as prolegs in larval lepidopterans, mecopterans, and sawflies (see 6.10, 9.6.1) would suggest that such repression is not complete. In fact, in embryos of the butterfly *Precis coenia*, *Ubx* and *abd-A* are expressed in the abdominal segments until 20% of embryogenesis but are repressed in paired ventral regions of abdominal segments 1, 3–5, and 9 to allow development of embryonic abdominal appendages into pleuropodia and larval prolegs (7.2.1.3.4). Finally, orthologs of *Ubx* and *abd-A* are also expressed in all but the first opisthosomal segments of chelicerate embryos, in all but the first trunk segments of myriapod embryos, and in the thoracic segments of some crustacean embryos, but they do not repress *Dll* expression and appendage formation (Fig. 13.7; see 7.2.1.3.4).

Diversity in arthropod limb type correlates with differences in expression of the proximodistal regulatory gene *Dll*. Orthologs of this gene control not only distal outgrowth of limbs and limb branches during embryogenesis and postembryogenesis in insects, other arthropods, and vertebrates, but also the ampullae and siphons of tunicates, the tube feet of echinoderms, the parapodia of annelids, and the lobopodia of onychophorans (Shubin et al., 1997; Panganiban et al., 1997; Grenier et al., 1997). Significantly, *Dll* is not expressed in the embryonic mandibles of hexapods and only at the beginning of their development in myriapod and direct-developing crustacean embryos (7.2.1.5; Grenier et al., 1997; Scholtz, 1998), refuting Manton's (1977) concept of a whole limb mandible in mandibulate arthropods. This conclusion is supported by Machida's recent (2000) homologization of the mandibular condyle, stipes, incisor, and molar of the archaeognath *Pedetontus unimaculatus*, respectively, with the cardo, stipes, galea, and lacinia of the maxilla, the mandibular palpus having been lost.

In arthropod embryos, phylopod, polyramous, biramous, uniramous, and chelate limbs all originate on the ventrolateral parasegmental boundary within each limb-bearing segment (8.3.6) but differ in their dorsoventral branch points (Shubin et al., 1997). This suggests that shifts in the signals or in the target sites of these genes along the dorsoventral axis of the body wall (7.2.1.1.1) and appendage could have influenced the development of polyramy, biramy, and uniramy and that additions or reductions in branch number could easily evolve.

13.3.10.3 THE ORIGIN OF DICONDYLIC MANDIBLES

As summarized in 13.2.1, the origin of dicondylic mandibles has probably contributed in a major way to insect diversification (Fig. 13.1). Unfortunately, nothing is known about the expression of homeotic or downstream genes in embryonic heads of any apterygote hexapod except *Thermobia domestica* (Zygentoma), which has dicondylic mandibles. In structural detail, mandibular ontogeny in monocondylic hexapods is identical to that in dicondylic ones until development of the single

articulation with the head capsule late in embryogenesis (Machida, 2000).

13.3.10.4 THE ORIGIN OF THE OVIPOSITOR IN FEMALES

Fossil evidence of archaeognaths in the Lower Devonian (13.2.2.1) suggests the ground plan valvular ovipositor of insects to have arisen, at the latest, in the Upper Silurian (Figs. 3.20, 13.1). Comparative evidence from development (Fig. 10.14) and structure (Matsuda, 1976) implies that the valves derive from paired embryonic abdominal appendages of abdominal segments 8 and 9 only in orthopterans, those of other females originating independently of these appendages from sternal cells situated medial to them either before or after the embryonic appendages have withdrawn. Nevertheless, recent molecular analysis of genital disc specification in male and female *Drosophila* larvae (Casares et al., 1997; Sánchez et al., 1997; Estrada and Sánchez-Herrero, 2001) shows their components to be specified by many of the same genes and signaling molecules as those specifying components of the appendages in other segments (Fig. 11.20) and for them to develop into legs or antennae in the absence of *Abd-B* expression.

Kukalová-Peck (1991: Fig. 6.3), based on her extensive knowledge of Paleozoic fossils, considered the valvular ovipositor to have arisen from the ancestral, 11-segmented legs of the eighth and ninth abdominal segments: the gonangula from subcoxae of the legs of abdominal segment 9, the first and second valves (gonapophyses), respectively, from trochanteral endites of appendages of segments 8 and 9, and the basal plates or gonocoxites (valvifers) and gonoplacs (third valvulae) from the fused coxae and trochanters of these appendages in segment 9. I find this assignment to be unconvincing because the evidence for it is weak and speculative.

The genetic and molecular basis of ovipositor specification and development is unknown because females of *D. melanogaster*, the best-known insect genetically and developmentally, have secondarily lost the valvular ovipositor and replaced it with extensible, terminal abdominal segments constituting the oviscapt and developing from the genital disc. Nevertheless, I suspect that molecular examination of ovipositor development in cricket or grasshopper larvae will soon occur and that it will involve the function of orthologs of many of the same genes as in the genital discs of *Drosophila* (11.4.4).

13.3.10.5 THE ORIGIN OF WINGS

The origin of wings some time in the Devonian (Fig. 13.1) (Kukalová-Peck, 1991; Carpenter, 1992a, 1992b) has influenced insect diversification to such an extent that today the pterygotes comprise some 99.5% of known hexapod and about 75% of known animal species (Daly et al., 1998). Until recent discoveries on their genetic and molecular basis in *Drosophila* (8.3.6, 11.3.2.2), speculation about the origin of wings was based on comparative morphology and paleontology and focused mostly on whether the wings were novelties unique to insects (e.g., the "paranotal theory") or modified versions of ancestral structures present in other arthropods (e.g., the "limb-exite theory" of Kukalová-Peck [1991]). According to Kukalová-Peck, each functional insect wing arose from an exite borne by the most proximal, epicoxal segment of a polyramous, 11-segmented leg of an aquatic pterygote ancestor through the following transformation series:

- First, the limb exites were modified into flaplike structures on all thoracic and abdominal segments to facilitate respiration and propulsion in water, as in the abdominal gills of fossil (Fig. 9.16) and extant mayfly nymphs.
- Next, some of these insects, both subadult and adult, might have used such protowings of the thoracic segments in surface skimming, as do certain extant adult stoneflies (Marden and Kramer, 1994; Thomas et al., 2000) and mayflies (Ruffieux et al., 1998), thereby acquiring the mechanical strength, flexibility (sclerotized veins, corrugation, flexion lines, etc.), skeletomusculature, and innervation preadapting them to function in active flight.
- These adaptations were then perfected by selection, resulting in the ability to fly actively.

With subsequent evolution and diversification of pterygote lineages, taxa were generated whose members bore three pairs (some extinct paleodictyopterans), two pairs (most species), or only one pair of thoracic wings (male scale insects, strepsipterans, dipterans) (Figs. 13.1, 13.18).

13.3.10.5.1 WING EVOLUTION IN INSECTS: THE MOLECULAR EVIDENCE

As shown in Fig. 13.7, the identity of the three thoracic segments in embryos of *D. melanogaster* is specified by expression of the homeotic genes *Scr*, *Antp*, and *Ubx*, the wing-bearing mesothorax being specified by *Antp* and the haltere-bearing metathorax by *Antp* and *Ubx*. Expression of orthologs of these genes is practically identical in embryos of the apterygote insect *Thermobia domestica* (Zygentoma), even though its adults are primitively wingless, suggesting that these genes do not specify wing development (Peterson et al., 1999).

Support for this suggestion was provided by Carroll and coworkers (1995), who examined the influence of mutations in these genes on wing and haltere disc development in *D. melanogaster* embryos and larvae. They found disc formation in this fly not to be promoted by the products of any of these genes but instead to be repressed in the wingless segments by expression of different *HOM* genes. This suggests that wings first arose without *HOM* gene involvement in a pterygote ancestor having a *HOM* gene expression pattern similar to that of *D. melanogaster* (Fig. 13.7) and *T. domestica*. At different stages of pterygote evolution, wing formation then came under the negative control of individual *HOM* genes (perhaps as summarized in Fig. 13.6) to produce

the diversity of wing numbers known in fossil and extant adults (Carroll et al., 1995; Shubin et al., 1997).

Carroll and coworkers (1995) considered formation of wing primordia to be a ground plan condition in pterygote segments (Fig. 13.1) and ectopic wing disc formation in homeotic mutants of *Drosophila* to result from de-repressing the wing development program in cells of those segments in which they occurred. They found

- *Scr* mutants to have an additional pair of wing discs in the prothorax (T1);
- *Antp* mutants to develop normally;
- *Ubx* mutants to have the haltere discs of the mesothorax enlarged into wing discs plus additional ectopic discs;
- *abd-A⁻* mutants to have a few *snail* (*sna*)-expressing cells in segments A1–A7; and
- *Ubx⁻/abd-A⁻* mutants to develop wing discs in segments T2–A7.

Thus, *Drosophila* embryos do not require *Antp*, *Scr*, *Ubx*, and *abd-A* function for the formation of adult wings from embryonic primordia, and it is significant that fossil evidence shows wings or protowings to be present on all thoracic and abdominal segments in the juveniles of certain Paleozoic pterygotes (Fig. 9.16) (Kukalová-Peck, 1991).

So how did the *HOM* genes come to have a role in repressing development of wings in prothoracic and abdominal segments in most pterygote insects? Recent comparative evidence indicates that *Hox* genes probably originated in the Neoproterozoic and have a role in specifying body plan in all metazoans (Fig. 7.38). They thus originated long before the origin of wings in hexapods and were present and expressed in ancestral Paleozoic hexapods both lacking wings and having thoracic and abdominal winglets (Fig. 13.1). Subsequent to wing origin, certain downstream genes involved in wing formation acquired cis (nearby)-regulatory elements responding to inhibitory binding of transcription factors of the *HOM* genes. Ensuing modifications to the original pterygote wing or protowing pattern could thus have been achieved (Fig. 13.6) by the evolution of Scr-, Antp-, Ubx-, and abd-A-responsive, cis-regulatory elements in downstream genes involved in wing development (see Gellon and McGinnis, 1998); and/or by subtle changes in the expression domains of these homeotic repressors (such differences in *Ubx* and *abd-A* expression are known in 13 species of Crustacea, where they correlate with one to three thoracic segments being transformed into additional, functioning cephalic segments [13.3.8]). The evolution of Scr-responsive elements thus led to elimination of prothoracic wings, and that of Ubx- and abd-A-responsive elements led to elimination of abdominal wings and, in flies, to reduction and modification of the metathoracic wings into halteres (Fig. 13.7).

The evolution of changes in *HOM* gene regulation (Fig. 13.7) and in regulatory interactions between homeotic transcription factors and the genes involved in appendage formation (Fig. 13.6) may have contributed also to the diversification of arthropod body plans and to

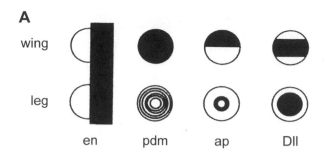

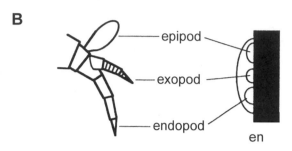

Fig. 13.15. Gene expression in insect and crustacean appendage primordia. A. *Drosophila* wing and leg discs, showing the expression domains of *engrailed* (*en*), *nubbin/pdm* (*pdm*), *apterous* (*ap*), and *Distal-less* (*Dll*). Dorsal is up, anterior is to the left, and proximodistal is perpendicular to the page, with distal regions in the center and proximal regions at the periphery of each disc. B. A three-branched crustacean limb (left) showing its anteroposterior organization relative to expression of the *engrailed* ortholog (right). (Reprinted with permission from *Nature* (M. Averof and S. M. Cohen. Evolutionary origin of insect wings from ancestral gill. 385: 627–630) © 1997 Macmillan Magazines Limited.)

the origin of different numbers and positions of wing pairs in members of certain insect higher taxa such as male scale insects, strepsipterans, and dipterans (Carroll et al., 1995; Gellon and McGinnis, 1998).

13.3.10.5.2 MOLECULAR EVIDENCE THAT THE INSECT WING IS HOMOLOGOUS TO THE DISTAL EPIPODITE OF CRUSTACEAN BIRAMOUS APPENDAGES

Averof and Cohen (1997) isolated orthologs of two genes having wing-specific functions in *Drosophila* embryos and larvae, *pdm* and *apterous* (*ap*), from embryos and larvae of the brine shrimp *Artemia franciscana* and the crayfish *Pasifasticus lenisculus*. Their expression patterns during embryonic and larval development provide molecular support for Kukalová-Peck's (1991) hypothesis that the insect wing evolved from a gill-like exite (specifically, the distal epipodite) already present in the last common marine ancestor of mandibulate arthropods (Fig. 13.16) (but see Fryer, 1998).

The segment-polarity gene *engrailed* (*en*) is expressed in the posterior compartment of both leg and wing primordia in insect embryos and larvae, and its ortholog is expressed in this compartment in primordia of the biramous appendages of crustacean embryos (Fig. 13.15).

Two *pdm* genes are present in *Drosophila*, *pdm1* and *pdm2*. Gene *pdm 1* is expressed throughout the prospective wing and has been implicated in wing blade specification (it is also expressed weakly as a set of rings in the leg discs) (Fig. 13.15A). The protein of the *pdm* gene of *Artemia* embryos is Af-PDM. Averof and Cohen raised antibodies against this protein, tagged them, and used them as probes to discover where this protein is expressed in *Artemia* embryos and larvae. At first it is expressed all over each appendage (Fig. 13.16c) but soon is restricted to a dorsal lobe of the appendage that will later differentiate into the distal epipodite (Fig. 13.16d, e). Fig. 13.16a shows the fully differentiated appendage, which has osmoregulatory and respiratory functions.

In *Drosophila* larvae, the *ap* gene is expressed specifically in the dorsal epidermis of developing wings (Figs. 11.10A, 13.15A) and as a ring in the fourth tarsal segment of leg discs (Figs. 11.9, 13.15A). Using the same methods as above, Averof and Cohen showed expression of the protein for *ap* in *Artemia* embryos (Af-AP) to be practically identical to that of Af-PDM (Fig. 13.16f–h) except that it is also expressed in limb muscles (Fig. 13.16h). However, it is not restricted to the dorsal surface of the distal epipodite as in wing primordia, since this appendage is not dorsoventrally flattened.

Expression of crayfish *pdm* in crayfish embryos is similar to that in *Artemia* embryos (Fig. 13.16i–k) but resembles that in the leg discs of *Drosophila* embryos (Fig. 13.15A) in that it is also expressed in rings in crayfish endopodites (Fig. 13.16j, k), the branch to which the insect leg is probably homologous. These patterns of gene expression are widely distributed in crustaceans, as *Artemia* is a branchiopod and the crayfish, a malacostracan, two widely divergent clades of crustaceans.

Based on this molecular evidence, Averof and Cohen (1997) (Fig. 13.17) postulated that distinct, structural progenitors of legs and epipodites/wings were present in the presumably aquatic, last common ancestor of mandibulate arthropods; that the wing is a direct, evolutionary derivative of the distal epipodite; and that these were supposedly repressed in members of the myriapod and apterygote clades as a terrestrial adaptation. However, this scenario depends on acceptance of the phylogenetic relationships of mandibulate arthropods as being those postulated in Fig. 13.17, when these are still in dispute (compare Figs. 13.7, 13.12–13.14 and the arguments presented by the various authorities in Deuve, 2001). Repression of distal epipodites/wings in myriapods and apterygotes could be construed as a loss synapomorphy uniting these taxa as monophyletic. Or, if the apterygote orders are considered to be monophyletic ([[[Protura + Collembola] + Diplura] + [Archaeognatha + Zygentoma]] + Pterygota), then this loss could have occurred twice: once in the apterygote lineage and once in the myriapods. And if the hexapods are polyphyletic (Fig. 13.13), then the epipodites/wings would have had to have been repressed seven times independently: once in the myriapods and once each in proturan, collembolan, dipluran, archaeognath, monuran (extinct), and zygentoman lin-

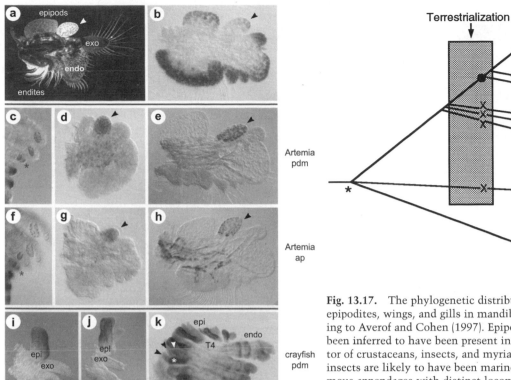

Fig. 13.16. Photomicrographs of PDM and Apterous (AP) protein expression in developing crustacean appendages. a. A thoracic appendage of *Artemia franciscana* (Af) (Branchiopoda) showing the position of the proximal and distal epipodites, the exopod (exo), endopod (endo), and endites. Dorsal is up and proximal is to the left. b. Expression of Distal-less (Dll) protein in all outgrowing branches of a developing larval thoracic appendage in *A. franciscana*. c–e. Expression of Af *pdm*. f–h. Expression of Af *ap*. c and f. Dorsal aspect of a series of thoracic appendage primordia in the body of a developing *A. franciscana* larva. Development of posterior primordia (at the bottom) trails that of anterior primordia. The stage at which expression becomes restricted to the distal epipodite is marked by an asterisk. d, e, g, and h. Individual limb primordia oriented as in a. d and g. Young limb buds just after appearance of distinct lobes (the same age as those in c and f marked with an asterisk). e and h. Same, but more mature. Arrowheads mark distal epipodites. i–k. Expression of *pdm* in the thoracic limbs of the crayfish *Pacifastacus leniusculus* (Malacostraca) at ±70% of embryonic development. i. Limb of first thoracic segment (T1). j. Limb of T3. k. Limbs of T4–T8 viewed in dorsal aspect. Epipodites of T2 and T3 are visible anteriorly. Epipodites, exopods, and endopods are labeled where present (expods absent from T4 to T8). Note strong staining in a single, distal and posterior epipodite per limb (except T8) marked by an asterisk in T6 (other epipodites are labeled with arrowheads). (Reprinted with permission from *Nature* (M. Averof and S. M. Cohen. Evolutionary origin of insect wings from ancestral gill. 385: 627–630) © 1997 Macmillan Magazines Limited.)

Fig. 13.17. The phylogenetic distribution of respiratory epipodites, wings, and gills in mandibulate arthropods according to Averof and Cohen (1997). Epipodites (asterisk) have been inferred to have been present in the last common ancestor of crustaceans, insects, and myriapods. The ancestors of insects are likely to have been marine and to have had biramous appendages with distinct locomotory legs and respiratory gills. With the transition to terrestrial life, the epipodite gills gradually lost their respiratory and osmoregulatory function and were lost independently in myriapods and in early apterygote hexapods (x). These structures were retained in modified form (filled circle) in the lineage leading to pterygote insects, perhaps initially as gills in aquatic larvae (and as in the abdominal appendages of fossil and extant mayfly larvae; see Fig. 9.16) and later as wings. The evolutionary relationships among crustaceans, myriapods, and insects are controversial and are presented here as an unresolved trichotomy. (Reprinted with permission from *Nature* (M. Averof and S. M. Cohen. Evolutionary origin of insect wings from ancestral gill. 385: 627–630) © 1997 Macmillan Magazines Limited.)

eages. Acceptance of this hypothesis would suggest that the ancestors of each of these lineages were aquatic and had distal epipodites, and that each lineage independently assumed a terrestrial lifestyle. The most recent molecular evidence for arthropods suggests the relationship Onychophora + [Chelicerata + [Myriapoda + [Crustacea + Hexapoda]]] (Fig. 13.7) (Abzhanov et al., 1999; Schultz and Regier, 2000; Cook et al., 2001), in which case the epipodites/wings would have been repressed only in the apterygotes. This example shows how important a sound phylogeny is to correctly interpreting the evolutionary history of a particular adaptation.

The differences between fore and hind wings in adults of many orders are probably influenced by subtle differences in the expression patterns of *Antp* and *Ubx*, by changes in the cis-regulatory sequences of downstream genes influenced by their transcription factors, or by divergence in wing patterning genes (Fig. 13.6).

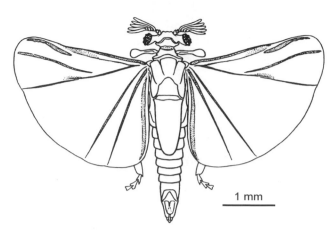

Fig. 13.18. Adult male strepsipteran, *Coriophagus rieki* (Halictophagidae), showing mesothoracic halteres and metathoracic wings, just the opposite of *D. melanogaster* and other dipterans and perhaps due to a switch in the segmental expression of *Ubx* and *abd-A*. (Reprinted with permission from J. Kathirithamby. Strepsiptera. In CSIRO (ed.), *The insects of Australia. A textbook for students and research workers.* 2nd ed. Vol. 2, p. 684, Fig. 36.1. Melbourne University Press and Cornell University Press © 1991. Commonwealth Scientific and Industrial Research Organisation, Division of Entomology.)

In male twisted wings (Strepsiptera), the halteres are borne by the mesothorax and the functional wings, by the metathorax (Fig. 13.18), exactly the opposite of their location in flies (Fig. 7.19) and in male coccoid hemipterans (Fig. 9.6E); the latter have functional mesothoracic wings and metathoracic hamulohalteres (Carver et al., 1991). Whiting and Wheeler (1994) and Whiting (1998) proposed this reversal to have arisen through anteroposterior reversal in expression of *Antp* and *Ubx*. This postulate together with cladistic analysis of homologous nucleotide sequences in 18S and 28S rDNA genes (Whiting and Wheeler, 1994) and in 18S rRNA (Chalwatzis et al., 1996) and of both these and morphological characters (Whiting et al., 1997; Whiting, 1998; Wheeler et al., 2001) suggested that Strepsiptera constitutes the sister group of the Diptera, to form the monophyletic taxon Halteria, rather than of the Coleoptera, which is the more widely accepted hypothesis (Fig. 13.1). Kristensen (1999) was not convinced.

13.3.10.6 THE ORIGIN OF WING FLEXING
Ability to flex the wings flat and lengthwise over the abdomen has apparently arisen at least twice during insect evolution: once in the extinct, Upper Carboniferous Diaphanopterodea and once in the stem species of the Neoptera, in which it involved the origin of the third axillary sclerite and associated musculature and innervation (Fig. 13.1). The idea that wing flexing is ancestral is refuted by the wings being extended in compression fossils of most Carboniferous wing-bearing insects (Kukalová-Peck, 1991; Carpenter, 1992a, 1992b).

Unfortunately little is known about the origin of the third axillary sclerite or about the genetic and molecular basis for its specification in *D. melanogaster* larvae and pupae. In this insect, the wing hinge, including the tegula and axillary sclerites, differentiates from cells in the wing disc between what will eventually differentiate into notum, pleuron, and wing blade (Fig. 10.35A, C) (Bryant, 1975), and is induced to do so by expression of the Wingless (Wg), Homothorax (Hth), and Teashirt (Tsh) proteins in this part of the disc (11.3.2.2). However, we have yet to understand how the third axillary is specified.

13.3.10.7 THE ORIGIN OF NURSE CELLS
Functional, oogonia-derived nurse cells have arisen independently at least four times within the ovarioles of female hexapods: once each in the stem groups of Collembola + Protura, Diplura, Dermaptera, and Paraneoptera + Holometabola (Figs. 2.22, 13.1). The germaria of polytrophic ovarioles contain one or more stem oogonia (Fig. 2.11) and are characterized by sequential proliferation but incomplete cytokinesis of each of a sequence of stem cell–generated cystoblasts into clones of interconnected cystocytes (Fig. 2.12); by differentiation of most cystocytes into nurse cells (Fig. 2.9C, D); by amplification of each nurse cell genome; and by their collective takeover of germ-line gene expression from the oocyte's chromosomes (Table 2.2). Telotrophic ovarioles, probably derived from ancestral, polytrophic ones (Fig. 2.21), have a similar developmental history but are characterized by two or more cystocytes per clone developing into oocytes (Figs. 2.9E, F; 2.14) and by retention of nurse cells within the germarium to form a tropharium (Figs. 2.3B, 2.14, 2.16A, 2.17B).

Our interpretation of nurse cell origin depends on whether cluster formation of cystocytes is ancestral, as it is for spermatogonia in males, or derived (2.2.7.1). If it is ancestral, then all genes responsible for cystocyte cluster formation (i.e., those functioning in maintaining the ring canals and fusomes during each cell cycle) must be blocked in female insects with panoistic ovarioles and independently reactivated four times in members of the taxa indicated above, along with genes influencing nurse cell and oocyte origin, divergence, and function (2.5). If it is derived, then these gene complexes must have evolved independently at least four times; of course, there are subtle differences in development in the females of different lineages having meroistic ovarioles (2.2.2). Parsimony would favor the former scenario.

13.3.10.8 THE ORIGIN OF COMPLETE METAMORPHOSIS
As discussed in 9.6.2.4, structural divergence between young and adult that is greater than a certain, relatively small, amount necessitates the presence of a last, transitional pupal instar in which a larva changes into an adult (Fig. 9.2). Presence of the pupal stage in turn has facilitated divergent juvenile adaptation to a variety of habitats closed to the young of apterygote and exopterygote

insects and the radiation of holometabolous insects to the point that they now comprise some 88% of known insect species (Fig. 0.2).

At least three innovations may have contributed to the origin of complete metamorphosis in insects: (1) a shift in the secretion of JH to an earlier stage in embryogenesis, to temporarily inhibit the differentiation of first instar nymphal characteristics and to promote differentiation of pronymphal (larval) characteristics (9.6.2.5.11); (2) redeployment of the originally gonadotrophic (1.3.7.2.3, 2.3.2.1) JH for a metamorphic function (9.6.2.5.11, 12.2, 12.4.5) in most exopterygote and endopterygote insects (Fig. 13.1); and (3) the origin of a complex, 20E-induced cascade of early, intermediate, and late regulatory genes, which, with the periodic absence of JH, mediated the switch from larval to pupal to adult developmental programs (12.4.1.5.2). Although best understood in *D. melanogaster* and *Manduca sexta*, comparative evidence indicates that similar genes have similar roles in other endopterygotes. Only additional comparative work on representative species in other endopterygote and neometabolous lineages will allow us to discover how widespread these genes are and whether they constitute a key innovation in the origin of holometabolous insects.

Finally, selection for change in the timing of JH release and in the tissue distribution of JH-sensitive periods (JH receptors?) has provided the basis for the evolution of intraspecific polymorphisms and polyphenisms (12.8).

References

Abmayr, S. M., M. S. Erickson, and B. A. Bowe. 1995. Embryonic development of the larval body wall musculature in *Drosophila melanogaster*. *Trends Genet.* 11: 153–159.

Abrams, J. M., K. White, L. I. Fessler, and H. Steller. 1993. Programmed cell death during *Drosophila* embryogenesis. *Development* 117: 29–43.

Abu-Hakima, R., and K. G. Davey. 1975. Two actions of juvenile hormone on the follicle cells of *Rhodnius prolixus* Stål. *Can. J. Zool.* 53: 1187–1188.

———. 1977. The action of juvenile hormone on follicle cells of *Rhodnius prolixus in vitro*: the effect of colchicine and cytochalasin B. *Gen. Comp. Endocrinol.* 32: 360–370.

Abu-Shaar, M., and R. S. Mann. 1998. Generation of multiple antagonistic domains along the proximodistal axis during *Drosophila* leg development. *Development* 125: 3821–3830.

Abzhanov, A., and T. C. Kaufman. 1999a. Novel regulation of the homeotic gene *Scr* associated with a crustacean leg-to-maxilliped appendage transformation. *Development* 126: 1121–1128.

———. 1999b. Homeotic genes and the arthropod head: expression patterns of the *labial*, *proboscipedia* and *Deformed* genes in crustaceans and insects. *Proc. Natl. Acad. Sci. U.S.A.* 96: 10224–10229.

———. 2000a. Crustacean (malacostracan) Hox genes and the evolution of the arthropod trunk. *Development* 127: 2239–2249.

———. 2000b. Homologs of *Drosophila* appendage genes in the patterning of arthropod limbs. *Dev. Biol.* 227: 673–689.

Abzhanov, A., A. Popadíc, and T. C. Kaufman. 1999. Chelicerate Hox genes and the homology of arthropod segments. *Evol. Dev.* 1: 77–89.

Adams, M. D., et al. (125 authors). 2000. The genome sequence of *Drosophila melanogaster*. *Science* 287: 2185–2195.

Adams, T. S. 1970. Ovarian regulation of the corpus allatum in the house fly, *Musca domestica*. *J. Insect Physiol.* 16: 349–370.

Agui, N., W. E. Bollenbacher, N. A. Granger, and L. I. Gilbert. 1980. Corpus allatum is release site for insect prothoracicotropic hormone. *Nature* 285: 669–670.

Agui, N., N. Granger, L. I. Gilbert, and W. E. Bollenbacher. 1979. Cellular localization of the insect prothoracicotropic hormone: *in vitro* assay of a single neurosecretory cell. *Proc. Natl. Acad. Sci. U.S.A.* 76: 669–670.

Akam, M. 1987. The molecular basis for metameric pattern in the *Drosophila* embryo. *Development* 101: 1–22.

———. 1991. Wondrous transformation. *Nature* 399: 282.

Alberch, P., S. J. Gould, G. F. Oster, and D. B. Wake. 1979. Size and shape in ontogeny and phylogeny. *Paleobiology* 55: 296–317.

Alberts, B., D. Bray, J. Lewis, M. Raff, K. Roberts, and J. D. Watson. 1989. *Molecular biology of the cell*. 2d ed. Garland Publishers.

———. 1994. *Molecular biology of the cell*. 3rd ed. Garland Publishers.

Aldhous, P. 1992. Neem chemical: the pieces fall in place. *Science* 258: 893.

Alexander, R. D. 1964. The evolution of mating behaviour in arthropods. In K. C. Highnam (ed.), *Insect reproduction*, pp. 78–94. Symp. R. Entomol. Soc. Lond. 2. Royal Entomological Society of London.

Allan, B. B., and W. E. Balch. 1999. Protein sorting by directed maturation of Golgi compartments. *Science* 285: 63–66.

Alroy, J. 2001. A multispecies overkill simulation of the end-Pleistocene megafaunal mass extinction. *Science* 292: 1893–1896.

Alvarez, W. 1997. *T. rex and the crater of doom*. Princeton University Press.

Andéol, Y. 1994. Early transcription in different animal species: implication for transition from maternal to zygotic control in development. *W. Roux's Arch. Dev. Biol.* 204: 3–10.

Andersen, N. M. 1982. *The semiaquatic bugs (Hemiptera, Gerromorpha). Phylogeny, adaptations, biogeography and classification*. Entomonograph Vol. 3. Scandinavian Science Press.

Andersen, S. A. 1990. Sclerotization of insect cuticle. In E. Ohnishi and H. Ishizaki (eds.), *Molting and metamorphosis*, pp. 133–155. Japanese Scientific Society Press.

Anderson, A. 1992. The evolution of sexes. *Science* 257: 324–326.

Anderson, D. T. 1962. The embryology of *Dacus tryoni* (Frogg.) [Diptera, Trypetidac], the Queensland fruitfly. *J. Embryol. Exp. Morphol.* 10: 248–292.

———. 1972a. The development of hemimetabolous insects. In S. J. Counce and C. H. Waddington (eds.), *Developmental systems: insects*. Vol. l, pp. 95–165. Academic Press.

———. 1972b. The development of holometabolous insects. In S. J. Counce and C. H. Waddington (eds.), *Developmental systems: insects*. Vol. l, pp. 165–242. Academic Press.

———. 1973. *Embryology and phylogeny in annelids and arthropods*. Pergamon Press.

Anderson, H. 1978a. Postembryonic development of the visual system of the locust, *Schistocerca gregaria*. I. Patterns of growth and developmental interactions in the retina and optic lobe. *J. Embryol. Exp. Morphol.* 45: 55–83.

———. 1978b. Postembryonic development of the visual system of the locust, *Schistocerca gregaria*. II. An experimental investigation of the formation of the retina-lamina projection. *J. Embryol. Exp. Morphol.* 46: 147–170.

———. 1981. Projections from sensory neurons developing at ectopic sites in insects. *J. Embryol. Exp. Morphol.* 65 (Suppl.): 209–224.

Anderson, J. F., and W. R. Horsfall. 1963. Thermal stress and anomalous development of mosquitoes (Diptera: Culicidae). I. Effect of constant temperature on dimorphism of adults of *Aedes stimulans. J. Exp. Zool.* 154: 67–107.

——. 1964. Dimorphic development of transplanted juvenile gonads of mosquitoes. *Science* 147: 1183–1184.

Anderson, K. 1995. One signal, two body axes. *Science* 269: 489–490.

Anderson, M. 1994. *Sexual selection.* Princeton University Press.

Ando, H. 1962. *The comparative embryology of Odonata with special reference to the relict dragonfly* Epiophlebia superstes *Selys.* Japan Society for the Promotion of Science.

Ando, H., and Cz. Jura (eds.). 1987. *Recent advances in insect embryology in Japan and Poland.* Arthropodan Embryological Society of Japan. ISEBU.

Ando, H., and K. Miya (eds.). 1985. *Recent advances in insect embryology in Japan.* Arthropodan Embryological Society of Japan. ISEBU.

André, H. M. 1989. The concept of stase. In H. M. André and J.-Cl. Lions (eds.), *Ontogeny and the concept of stase,* pp. 3–14. Agar.

Andres, A. J., and C. S. Thummel. 1992. Hormones, puffs and flies: the molecular control of metamorphosis by ecdysone. *Trends Genet.* 8: 132–138.

Apple, R. T., and J. W. Fristrom. 1990. Regulation of pupal gene expression during metamorphosis in *Drosophila melanogaster.* In E. Ohnishi and H. Ishizaki (eds.), *Molting and metamorphosis,* pp. 223–237. Japanese Scientific Society Press.

Arabaki, N., T. Myoshi, and H. Noda. 2001. *Wolbachia*-mediated parthenogenesis in the predatory thrips *Franklinothrips vespiformis* (Thysanoptera, Insecta). *Proc. R. Soc. Lond. B Biol. Sci.* 268: 1011–1016.

Arendt, D., and K. Nübler-Jung. 1994. Inversion of dorsoventral axis? *Nature* 371: 26.

——. 1996. Common ground plans in early brain development in mice and flies. *Bioessays* 18: 255–259.

——. 1999. Comparison of early nerve cord development in insects and vertebrates. *Development* 126: 2309–2325.

Arnquist, G. 1998. Comparative evidence for the evolution of genitalia by sexual selection. *Nature* 393: 784–786.

Artavanis-Tsakonas, S., K. Matsuno, and M. E. Fortini. 1995. Notch signaling. *Science* 268: 225–232.

Artavanis-Tsakonas, S., M. D. Rand, and R. J. Lake. 1999. Notch signaling: cell fate control and signal integration in development. *Science* 284: 770–776.

Arthur, W. 1997. *The origin of animal body plans: a study in evolutionary developmental biology.* Cambridge University Press.

Ashburner, M., C. Chihara, P. Meltzer, and G. Richards. 1974. On the temporal control of puffing activity in polytene chromosomes. *Cold Spring Harb. Symp. Quant. Biol.* 38: 655–662.

Askew, R. R. 1988. *The dragonflies of Europe.* Harley Books, Martens.

Astaurov, B. L. 1972. Experimental model of the origin of bisexual polyploid species in animals (the hypothesis of indirect origin of polyploid animals via parthenogenesis and hybridization). *Biol. Zentralbl.* 91: 137–150.

Audy, J. R., F. J. Rodovsky, and P. H. Vercammen-Grandjean. 1972. Neosomy: radical intrastadial metamorphosis associated with arthropod symbiosis. *J. Med. Entomol.* 9: 487–494.

Austin, A. D., and T. O. Browning. 1981. A mechanism for movement of eggs along insect ovipositors. *Int. J. Insect Morphol. Embryol.* 10: 93–108.

Averof, M. 1998. Origin of the spider's head. *Nature* 395: 436–437.

Averof, M., and S. M. Cohen. 1997. Evolutionary origin of insect wings from ancestral gill. *Nature* 385: 627–630.

Averof, M., and N. H. Patel. 1997. Crustacean appendage evolution associated with changes in Hox gene expression. *Nature* 388: 682–686.

Awad, T. A., and J. W. Truman. 1997. Postembryonic development of the midline glia in the CNS of *Drosophila*: proliferation, programmed cell death and endocrine regulation. *Dev. Biol.* 187: 283–297.

Azevedo, R. B. R., V. French, and L. Partridge. 1997. Life history consequences of egg size in *Drosophila melanogaster. Am. Nat.* 150: 250–282.

Azpiazu, N., and G. Morata. 2000. Function and regulation of *homothorax* in the wing imaginal discs of *Drosophila. Development* 127: 2685–2693.

Azpiazu, N., P. A. Lawrence, J.-P. Vincent, and M. Frasch. 1996. Segmentation and specification of the *Drosophila* mesoderm. *Genes Dev.* 10: 3183–3194.

Baccetti, B. 1979. Ultrastructure of sperm and its bearing on arthropod phylogeny. In A. P. Gupta (ed.), *Arthropod phylogeny,* pp. 609–644. Van Nostrand-Reinhold.

——. 1998. Spermatozoa. In F. W. Harrison and M. Locke (eds.), *Microscopic anatomy of invertebrates.* Vol. 11C. *Insecta,* pp. 843–894. Wiley-Liss.

Baccetti, B., and B. A. Afzelius. 1976. *The biology of the sperm cell.* Karger.

Baehrecke, E. H., and M. R. Strand. 1990. Embryonic morphology and growth of the polyembryonic parasitoid *Copidosoma floridanum* (Ashmead) (Hymenoptera: Encyrtidae). *Int. J. Insect Morphol. Embryol.* 19: 165–175.

Bailey, W. J., and J. Ridsdill-Smith (eds.). 1991. *Reproductive behaviour of insects.* Chapman and Hall.

Baker, R., and G. Schubiger. 1995. Ectoderm induces muscle-specific gene expression in *Drosophila* embryos. *Development* 121: 1387–1398.

Balbiani, E. G. 1881. Sur la structure du noyau des cellulaires salivaires chez les larves de *Chironomus. Zool. Anz.* 4: 637–641, 662–666.

Ball, E. E., and C. S. Goodman. 1985a. Muscle development in the grasshopper embryo. II. Syncytial origin of the extensor tibiae muscle pioneers. *Dev. Biol.* 111: 399–416.

——. 1985b. Muscle development in the grasshopper embryo. III. Sequential origin of the flexor tibiae muscle pioneers. *Dev. Biol.* 111: 417–424.

Ball, E. E., R. K. Ho, and C. S. Goodman. 1985. Muscle development in the grasshopper embryo. I. Muscles, nerves, and apodemes in the metathoracic leg. *Dev. Biol.* 111: 383–398.

Ball, E. E., H. G. de Couet, P. L. Horn, and J. M. Quinn. 1987. Hemocytes secrete basement membrane components in embryonic locusts. *Development* 99: 255–259.

Ballarino, J., M. Ma, T. Ding, and C. Lamison. 1991. Development of male-incubated ovaries in the gypsy moth, *Lymantria dispar. Arch. Insect Biochem. Physiol.* 16: 221–234.

Bango, P., and K. White. 2000. Regulation and execution of apoptosis during *Drosophila* development. *Dev. Dyn.* 218: 68–79.

Bantock, C. R. 1970. Experiments on chromosome elimination in the gall midge, *Mayetiola destructor. J. Embryol. Exp. Morphol.* 24: 257–286.

Barinaga, M. 1995. Focusing on the eyeless gene. *Science* 267: 1766–1767.

Basler, K., and G. Struhl. 1994. Compartment boundaries and the control of *Drosophila* limb pattern by hedgehog protein. *Nature* 368: 208–214.

Bastiani, M. J., H. G. de Couet, J. M. A. Quinn, R. O. Karstrom, K. Kotrla, C. S. Goodman, and E. E. Ball. 1992. Position-specific expression of the annulin protein during grasshopper embryogenesis. *Dev. Biol.* 154: 129–142.

Bate, C. M. 1976a. Embryogenesis of an insect nervous system. I. A map of the thoracic and abdominal neuroblasts in

Locusta migratoria. J. Embryol. Exp. Morphol. 35: 107–123.

——. 1976b. Pioneer neurons in an insect embryo. *Nature* 260: 54–56.

——. 1978. Development of sensory systems in Arthropods. In M. Jacobson (ed.), *Handbook of sensory physiology.* Vol. 9, pp. 1–53. Springer-Verlag.

——. 1993. The mesoderm and its derivatives. In M. Bate and A. Martinez Arias (eds.), *The development of Drosophila melanogaster.* Vol. 2, pp. 1013–1090. Cold Spring Harbor Laboratory Press.

Bate, C. M., and E. B. Grunewald. 1981. Embryogenesis of an insect nervous system. II: a second class of neuron precursor cells and the origin of intersegmental connectives. *J. Embryol. Exp. Morphol.* 61: 317–330.

Bate, C. M., and A. Martinez Arias. 1991. The embryonic origin of imaginal discs in *Drosophila. Development* 112: 755–761.

Bate, M., and A. Martinez Arias (eds.). 1993. *The development of Drosophila melanogaster.* Vols. 1 and 2. Cold Spring Harbor Laboratory Press.

Bateson, W. 1894. *Materials for the study of variation treated with especial regard to discontinuity in the origin of species.* Macmillan.

Bautz, A.-M. 1971. Chronologie de la mise en place de l'hypoderme imaginal de l'abdomen de *Calliphora erythrocephala* Meigen (Insecte, Diptère, Brachycère). *Arch. Zool. Exp. Gen.* 112: 157–178.

——. 1975. Growth and degeneration of the larval abdominal epidermis in *Calliphora erythrocephala* (Meigen) (Diptera: Calliphoridae) during larval life and metamorphosis. *Int. J. Insect Morphol. Embryol.* 4: 495–515.

Bayer, C. A., L. von Kalm, and J. W. Fristrom. 1997. Relationships between protein isoforms and genetic functions demonstrate functional redundancy at the *Broad Complex* during *Drosophila* metamorphosis. *Dev. Biol.* 186: 267–282.

Baylies, M. K., and M. Bate. 1996. *twist:* a myogenic switch gene in *Drosophila. Science* 272: 1481–1484.

Baylies, M. K., M. Bate, and M. R. Gomez. 1998. Myogenesis: a view from *Drosophila. Cell* 93: 921–927.

Beardsley, T. 1991. Smart genes. *Sci. Am.* 265(2): 86–95.

Becker, S., G. Pasca, D. Strumpf, L. Min, and T. Volk. 1997. Reciprocal signaling between *Drosophila* epidermal muscle attachment cells and their corresponding muscles. *Development* 124: 2615–2622.

Beerman, W. 1952. Chromomerenkon-stanz und spezifische Modifikationen der Chromosomen-Struktur in der Entwicklung und Organdifferenzierung von *Chironomus tentans. Chromosoma* 5: 139–198.

Bell, W. J. 1972. Yolk formation by transplanted cockroach oocytes. *J. Exp. Zool.* 181: 41–48.

Belles, X., P. Cassier, X. Cerda, N. Pascual, M. Andre, V. Rosso, and M. D. Pulachs. 1993. Induction of choriogenesis by 20-hydroxyecdysone in the German cockroach. *Tissue Cell* 25: 195–204.

Ben-Ari, E. T. 1999. Paternity battles. *Bioscience* 49: 951–956.

Benlali, A., I. Draskovic, D. J. Hazelett, and J. E. Treisman. 2000. *act up* controls actin polymerization to alter cell shape and restrict Hedgehog signaling in the *Drosophila* eye. *Cell* 101: 271–281.

Bennet-Clark, H. C. 1963. The relation between epicuticular folding and the subsequent size of an insect. *J. Insect Physiol.* 9: 43–46.

Bennettova, B., and G. Fraenkel. 1981. What determines the number of ovarioles in a fly ovary? *J. Insect Physiol.* 27: 403–410.

Bentley, D., and M. Caudy. 1983. Pioneer axons lose directed growth after selective killing of guide post cells. *Nature* 304: 62–65.

Bentley, D., and H. Keshishian. 1982a. Pioneer neurons and pathways in insect appendages. *Trends Neurosci.* 5: 364–367.

——. 1982b. Pathfinding by peripheral pioneer neurons in grasshoppers. *Science* 218: 1082–1088.

Bentley, D., H. Keshishian, M. Shankland, and A. Toroian-Raymond. 1979. Quantitative staging of embryonic development of the grasshopper, *Schistocerca nitens. J. Embryol. Exp. Morphol.* 54: 47–74.

Bergerard, J. 1972. Environmental and physiological control of sex-determination and differentiation. *Annu. Rev. Entomol.* 17: 57–74.

Berlese, A. 1913. Intorno alle metamorfosi degli insetti. *Redia* 9: 121–136.

Berlot, J., and C. S. Goodman. 1984. Guidance of peripheral pioneer neurons in the grasshopper: adhesive hierarchy of epithelial and neuronal surfaces. *Science* 223: 493–496.

Bernays, E. A. 1971. The vermiform larva of *Schistocerca gregaria* (Forskål): form and activity. *Z. Morphol. Tiere* 70: 183–200.

——. 1972a. The intermediate molt (first ecdysis) of *Schistocerca gregaria* (Forskål). *Z. Morphol. Tiere* 71: 160–179.

——. 1972b. The muscles of newly hatched *Schistocerca gregaria* larvae and their possible functions in hatching, digging and ecdysial movements (Insecta: Acrididae). *J. Zool.* 166: 141–158.

——. 1972c. Hatching in *Schistocerca gregaria* (Forskål). *Acrida* 1: 41–60.

——. 1986a. Diet-induced head allometry among foliage-chewing insects and its importance for graminivores. *Science* 231: 495–497.

——. 1986b. Evolutionary contrasts in insects: nutritional advantages of holometabolous development. *Physiol. Entomol.* 11: 377–382.

Bernays, E. A., and J. Hamai. 1987. Head size and shape in relation to grass-feeding in Acridoidea (Orthoptera). *Int. J. Insect Morphol. Embryol.* 16: 323–336.

Berner, R. A., S. T. Petsch, J. A. Lake, D. J. Beerling, B. N. Popp, R. S. Lane, E. A. Laws, M. B. Westley, N. Cassar, F. I. Woodward, and W. P. Quick. 2000. Isotope fractionation and atmospheric oxygen: implications for phanerozoic O_2 evolution. *Science* 287: 1630–1633.

Berridge, M. J., B. L. Gupta, J. L. Oschman, and B. J. Wall. 1976. Salivary gland development in the blowfly, *Calliphora erythrocephala. J. Morphol.* 149: 459–482.

Berrigan, D., and S. J. Locke. 1991. Body size and male reproductive performance in the flesh fly, *Neobelleiria bullata. J. Insect Physiol.* 37: 575–581.

Berríos-Ortiz, A., and R. B. Selander. 1979. *Skeletal musculature in larval phases of the beetle* Epicauta segmenta *(Coleoptera, Meloidae).* Ser. Entomol. W. Junk.

Berry, S. J. 1982. Maternal direction of oogenesis and early embryogenesis in insects. *Annu. Rev. Entomol.* 27: 205–227.

——. 1985. RNA synthesis and storage during insect oogenesis. In L. Browder (ed.), *Developmental biology. A comprehensive synthesis.* Vol. 1, *Oogenesis,* pp. 351–384. Plenum Press.

Bhat, K. M., and P. Schedl. 1997. Establishment of stem cell identity in the *Drosophila* germline. *Dev. Dyn.* 210: 371–382.

Bhatt, T. R., and F. M. Horodyski. 1999. Expression of the *Manduca sexta* allatotropin gene in cells of the central and enteric nervous system. *J. Comp. Neurol.* 403: 407–420.

Bidmon, H. J., and T. J. Silter. 1990. The ecdysteroid receptor. *Invert. Reprod. Dev.* 18: 13–28.

Bidmon, H. J., W. E. Stumpf, and N. A. Granger. 1992. Ecdysteroid receptors in the neuro-endocrine axis of a moth. *Experientia* 48: 42–47.

Bilder, D., and M. P. Scott. 1998. Hedgehog and Wingless induce metameric pattern in the *Drosophila* visceral mesoderm. *Dev. Biol.* 201: 43–56.

Bilinski, S. M. 1991. Are accessory nuclei involved in the establishment of developmental gradients in hymenopteran oocytes? Ultrastructural studies. *W. Roux's Arch. Dev. Biol.* 199: 423–426.

———. 1993. Structure of ovaries and oogenesis in entognathans (Apterygota). *Int. J. Insect Morphol. Embryol.* 22: 255–269.

Bircher, U., and E. Hauschteck-Jungen. 1997. The length of the sperm nucleus in *Drosophila obscura* group species is depending on the total length of the sperm. *Invert. Reprod. Dev.* 32: 225–229.

Birkenbeil, H., and M. Eckert. 1991. Temporal molt inhibition by anti-ecdysone serum injected into lepidopteran larvae. *Arch. Insect Biochem. Physiol.* 18: 195–201.

Birkhead, T. R. 2000. *Promiscuity: an evolutionary history of sperm competition and sexual conflict.* Faber and Faber.

Birkhead, T. R., and A. P. Møller (eds.). 1998. *Sperm competition and sexual selection.* Academic Press.

Bishop, S. A., T. Klein, A. Martinez Arias, and J. P. Couso. 1999. Composite signaling from *Serrate* and *Delta* establishes leg segments in *Drosophila* through Notch. *Development* 126: 2993–3003.

Blackith, R. E., R. G. Davies, and E. A. Moy. 1963. A biometric analysis of development in *Dysdercus fasciatus* Sign. (Hemiptera: Pyrrhocoridae). *Growth* 27: 317–334.

Blackman, R. L. 1985a. Spermatogenesis in the aphid *Amphorophora tuberculata* (Homoptera, Aphididae). *Chromosoma* 92: 357–362.

———. 1985b. Aphid cytology and genetics. In *Evolution and biosytematics of aphids*, pp. 171–237. Proc. Int. Aphidological Symp., Jablonna, 1981. Polska Akademia Nauk.

Blackman, R. L., and D. F. Hales. 1986. Behaviour of the X-chromosomes during growth and maturation of parthenogenetic eggs of *Amphorophora tuberculata* (Homoptera, Aphididae), in relation to sex determination. *Chromosoma* 94: 59–64.

Blair, S. S. 1999. *Drosophila* imaginal disc development: patterning the adult fly. In V. E. A. Russo, D. J. Cove, L. G. Edgar, R. Jaenisch, and F. Salamini (eds.), *Development. Genetics, epige-netics and environmental regulation,* pp. 347–370. Springer-Verlag.

Blair, S. S., and A. Ralston. 1997. Smoothened-mediated Hedgehog signaling is required for the maintenance of the anterior-posterior lineage restriction in the developing wing of *Drosophila. Development* 124: 4053–4063.

Blum, M. S., and N. A. Blum (eds.). 1979. *Sexual selection and reproductive competition.* Academic Press.

Bock, E. 1941. Wechselbeziehungen zwischen den Keimblättern bei der Organbildung von *Chrysopa perla* L. *W. Roux's Arch. Entwicklungsmech. Organ.* 141: 159–247.

Bodenstein, D. 1933. Beintransplantationen an Lepidopteren raupen. I. Transplantationen zur analyse der Raupen- und Puppenhaütung. *W. Roux's Arch. Entwicklungsmech. Organ.* 128: 564–583.

———. (ed.) 1971. *Milestones in developmental physiology of insects.* Papers in development and heredity. Appleton-Century-Crofts.

Bodmer, R. 1993. The gene *tinman* is required for specification of the heart and visceral muscles in *Drosophila. Development* 118: 719–729.

Bodnar, J. W. 1997. Programming the *Drosophila* embryo. *J. Theor. Biol.* 188: 391–445.

Boleli, I. C., Z. L. Paulino-Simões, and M. M. G. Bitondi. 1999. Cell death in ovarioles causes permanent sterility in *Frieseomelitta varia* worker bees. *J. Morphol.* 242: 271–282.

Bollenbacher, W. E. 1988. The interendocrine regulation of larval-pupal development in the tobacco hornworm, *Manduca sexta*: a model. *J. Insect Physiol.* 34: 941–947.

Bollenbacher, W. E., and N. A. Granger. 1985. Endocrinology of the prothoracicotropic hormone. In G. A. Kerkut and L. I. Gilbert (eds.), *Comprehensive insect physiology, biochemistry and pharmacology.* Vol. 7, pp. 109–151. Pergamon Press.

Bollenbacher, W. E., R. S. Gray, D. P. Muehleisen, S. A. Regan, and A. L. Westbrook. 1993. The biology of the prothoracicotropic hormone peptidergic neurons in an insect. *Am. Zool.* 33: 316–323.

Bollenbacher, W. E., E. J. Katahira, M. O'Brien, L. I. Gilbert, M. K. Thomas, N. Agui, and A. H. Baumhover. 1984. Insect prothoracicotropic hormone: evidence for two molecular forms. *Science* 224: 1243–1245.

Bolvar, J., J. R. Huynh, H. Lopez-Schier, G. Gonzalez, D. St. Johnston, and G. Gonzalez-Reyes. 2001. Centrosome migration into the *Drosophila* oocyte is independent of *BicD* and *egl*, and of the organization of the microtubule cytoskeleton. *Development* 128: 1889–1897.

Bonhag, P. F. 1959. Histological and histochemical studies on the ovary of the American cockroach, *Periplaneta americana* (L.). *Univ. Calif. Public. Entomol.* 16: 81–106; pl. 10–17.

Bonhag, P. F., and W. J. Arnold. 1961. Histology, histochemistry and tracheation of the ovariole sheaths in the American cockroach, *Periplaneta americana* (L.). *J. Morphol.* 108: 107–129.

Bonhag, P. F., and J. R. Wick. 1953. The functional anatomy of the male and female reproductive systems of the milkweed bug, *Oncopeltus fasciatus* (Dallas) (Heteroptera: Lygaeidae). *J. Morphol.* 93: 177–284.

Borden, J. H., and C. E. Slater. 1968. Induction of flight muscle degeneration by synthetic juvenile hormone in *Ips confusus. Z. Vergl. Physiol.* 61: 366–368.

Bordereau, C. 1982. Ultrastructure and formation of the physogastric termite queen cuticle. *Tissue Cell* 14: 371–396.

Borkowski, O. M. D., N. H. Brown, and M. Bate. 1995. Anterior-posterior subdivision and the diversification of the mesoderm in *Drosophila. Development* 121: 4183–4193.

Borror, D. J., C. A. Triplehorn, and N. F. Johnson. 1989. *An introduction to the study of insects.* 6th ed. Saunders.

Bossing, T., G. Udolph, C. Q. Doe, and G. M. Technau. 1996. The embryonic central nervous system lineages of *Drosophila.* I. Neuroblast lineages derived from the ventral half of the neuroectoderm. *Dev. Biol.* 179: 41–64.

Bothe, G. W. M., and W. Rathmayer. 1994. Programmed degeneration of the thoracic eclosion muscle in the flesh fly, *Sarcophaga bullata. J. Insect Physiol.* 40: 983–995.

Bounhiol, J. J. 1938. Recherches expérimentales sur le déterminisme de la métamorphose chez les Lépidoptères. *Bull. Biol. Suppl.* 24: 1–199.

Bour, B. A., M. Chakravarti, J. M. West, and S. M. Abmayr. 2000. *Drosophila* SNS, a member of the immunoglobulin superfamily that is essential for myoblast fusion. *Genes Dev.* 14: 1498–1511.

Bourtzis, K., and S. O'Neill. 1998. *Wolbachia* infections and arthropod reproduction. *Bioscience* 48: 287–293.

Bowers, W. S. 1985. Antihormones. In G. A. Kerkut and L. I. Gilbert (eds.),

Comprehensive insect physiology, biochemistry and pharmacology. Vol. 8, pp. 551–564. Pergamon Press.

Bowers, W. S., T. Ohta, J. S. Cleere, and P. A. Marsella. 1976. Discovery of insect anti-juvenile hormones in plants. *Science* 193: 542–547.

Bownes, M. 1992. Molecular aspects of sex determination in insects. In J. M. Crampton and P. Eggleston (eds.), *Insect molecular science*, pp. 76–100. 16th Symp. R. Entomol. Soc. Lond. Academic Press.

Bowring, S. A., D. H. Erwin, Y. G. Yin, M. W. Martin, K. Davidek, and W. Wang. 1998. U/Pb zircon geochronology and tempo of the end-Permian mass extinction. *Science* 280: 1039–1045.

Boyan, G. S., and J. L. D. Williams. 1997. Embryonic development of the pars intercerebralis/central complex of the grasshopper. *Dev. Genes Evol.* 207: 317–329.

Boyan, G. S., J. L. D. Williams, and H. Reichert. 1995a. Morphogenetic reorganization of the brain during embryogenesis in the grasshopper. *J. Comp. Neurol.* 361: 429–440.

Boyan, G., S. Therianos, J. L. D. Williams, and H. Reichert. 1995b. Axogenesis in the embryonic brain of the grasshopper *Schistocerca gregaria*: an identified cell analysis of early brain development. *Development* 121: 75–86.

Boyce, N. 1998. Monster sperm. *New Sci.* 158: 40–43.

Boyle, M., and S. DiNardo. 1995. Specification, migration and assembly of the somatic cells of the *Drosophila* gonad. *Development* 121: 1815–1825.

Boyle, M., N. Bonini, and S. DiNardo. 1997. Expression and function of *clift* in the development of somatic gonadal precursors within the *Drosophila* mesoderm. *Development* 124: 971–982.

Brakefield, P. M., and V. French. 1995. Eyespot development in butterfly wings: the epidermal response to damage. *Dev. Biol.* 168: 98–111.

———. 1999. Butterfly wings: the evolution of development of colour patterns. *Bioessays* 21: 391–401.

Brakefield, P. M., J. Gates, D. Keys, F. Kesbeke, P. J. Wijngaarden, A. Monteiro, V. French, and S. B. Carroll. 1996. Development, plasticity and evolution of butterfly eyespot patterns. *Nature* 384: 236–242.

Braünig, P. 1998. Networks of neurosecretory (neurohemal) endings. In F. W. Harrison and M. Locke (eds.), *Microscopic anatomy of invertebrates*. Vol. 11B. *Insecta*, pp. 539–549. Wiley-Liss.

Brendza, R. P., L. R. Serbus, J. B. Duffy, and W. M. Saxton. 2000. A function for kinesis I in the posterior transport of *oskar* mRNA and Staufen protein. *Science* 289: 2120–2122.

Brennan, C. A., M. Ashburner, and K. Moses. 1998. Ecdysone pathway is required for furrow progression in the developing *Drosophila* eye. *Development* 125: 2653–2664.

Bressec, C., and E. Hauschteck-Jungen. 1996. *Drosophila subobscura* females preferentially select long sperm for storage and use. *J. Insect Physiol.* 42: 323–328.

Brewer, L. R., M. Corzett, and R. Balhorn. 1999. Protamine-induced condensation and decondensation of the same DNA molecule. *Science* 286: 120–123.

Brewster, R., and R. Bodmer. 1996. Cell lineage analysis of the *Drosophila* peripheral nervous system. *Dev. Genet.* 18: 50–63.

Brian, M. V. 1980. Social control over sex and caste in bees, wasps, and ants. *Biol. Rev.* 55: 379–415.

Bridges, C. B. 1915. A linkage variation in *Drosophila*. *J. Exp. Zool.* 19: 1–21.

Briggs, D. E. G., and R. A. Fortey. 1989. The early radiation and relationships of the major arthropod groups. *Science* 246: 241–243.

Brinck, P. 1962. Die Entwicklung der Spermaübertragung der Odonaten. *Proc. XIth Int. Congr. Entomol. Vienna* 11: 715–718.

Britton, E. B. 1970. Coleoptera (beetles). In I. M. Mackerras (ed.), *The insects of Australia. A textbook for students and research workers*, pp. 495–620. Melbourne University Press.

Broadie, K. S., and M. Bate. 1993. Innervation directs receptor synthesis and localization in *Drosophila* embryo synaptogenesis. *Nature* 361: 350–353.

Brodsky, A. K. 1994. *The evolution of insect flight.* Oxford University Press.

Broihier, H. T., L. A. Moore, M. Van Doren, S. Newman, and R. Lehmann. 1998. *zfh-1* is required for germ cell migration and gonadal mesoderm development. *Development* 125: 655–666.

Brokaw, C. J. 1989. Direct measurements of sliding between outer doublet microtubules in swimming sperm flagella. *Science* 243: 1593–1596.

Brooks, M. A., and T. J. Kurtti. 1972. Male rudimentary ovaries: a case of cellular symbiosis in *Blattella germanica* (L.) (Dictyoptera: Blattellidae). *Int. J. Insect Morphol. Embryol.* 1: 169–179.

Browder, L. W. 1980. *Developmental biology.* Saunders College.

Brown, J. J., and M. Friedländer. 1995. Influence of parasitism on spermatogenesis in the codling moth, *Cydia pomonella* L. *J. Insect Physiol.* 41: 957–963.

Brown, N. L., C. A. Sattler, S. W. Paddock, and S. B. Carroll. 1995. *Hairy* and *emc* negatively regulate morphogenetic furrow formation in the *Drosophila* eye. *Cell* 30: 879–887.

Brown, S., M. De Camillis, K. Gonzalez-Charnice, M. Denell, R. Beeman, W. Nic, and R. Denell. 2000. Implications of the *Tribolium Deformed* mutant phenotype for the evolution of Hox gene function. *Proc. Natl. Acad. Sci. U.S.A.* 97: 4510–4514.

Brown, S., S. Holtzman, T. Kaufman, and R. Denell. 1999. Characterization of the *Tribolium Deformed* ortholog and its ability to directly regulate *Deformed* target genes in the rescue of a *Drosophila Deformed* null mutant. *Dev. Genes Evol.* 209: 389–398.

Brust, R. A. 1968. Temperature-induced intersexes in *Aedes* mosquitoes: comparative study of species from Manitoba. *Can. Entomol.* 100: 879–891.

Bryant, E. H. 1969. A system favoring the evolution of holometabolous development. *Ann. Entomol. Soc. Am.* 62: 1087–1091.

Bryant, P. J. 1975. Pattern formation in the imaginal wing disc of *Drosophila melanogaster*: fate map, regeneration and duplication. *J. Exp. Zool.* 193: 49–77.

———. 1993. The polar coordinate model goes molecular. *Science* 259: 471–472.

Bryant, P. J., S. V. Bryant, and V. French. 1977. Biological regeneration and pattern formation. *Sci. Am.* 237(1): 66–81.

Bryant, P. J., V. French, and S. V. Bryant. 1981. Distal regeneration and symmetry. *Science* 212: 993–1002.

Budd, G. E. 1999. Does evolution in body patterning genes drive morphological change—or vice versa? *Bioessays* 21: 326–332.

Bull, A. L. 1966. *Bicaudal*, a genetic factor which affects the polarity of the embryo in *Drosophila melanogaster*. *J. Exp. Zool.* 161: 221–241.

Bullière, D. 1972. Étude de la régénération d'appendice chez un insecte: stades de la formation des régénérats et rapports avec le cycle de mue. *Ann. Embryol. Morphol.* 5: 61–74.

Bullière, D., and F. Bullière. 1985. Regeneration. In G. A. Kerkut and L. I. Gilbert (eds.), *Comprehensive insect physiology, biochemistry and pharmacology.* Vol. 2, pp. 372–424. Pergamon Press.

Bullière, D., F. Bullière, K. Mounaji, M. de Reggi, and B. Gharib. 1982. Positional cell surface antigens in an insect appendage. *W. Roux's Arch. Dev. Biol.* 191: 222–227.

Bullock, T. H., and G. A. Horridge (eds.). 1965. *Structure and function of the nervous systems in invertebrates.* Vols. 1 and 2. W. H. Freeman.

Bulmer, M. G. 1982. Cyclical parthenogenesis and the cost of sex. *J. Theor. Biol.* 94: 197–207.

Bultmann, H., and U. Clever. 1969. Chromosomal control of foot pad development in *Sarcophaga bullata.* I. The puffing pattern. *Chromosoma* 28: 120–135.

——. 1970. Chromosomal control of foot pad development in *Sarcophaga bullata.* II. Cuticle formation and tanning. *Dev. Biol.* 22: 601–621.

Büning, J. 1993. Germ cell cluster formation in insect ovaries. *Int. J. Insect Morphol. Embryol.* 22: 237–253.

——. (ed.) 1994. *The insect ovary.* Chapman and Hall.

——. 1996. Review on ovary development and phylogeny of insects. In *Proceedings of the 20th International Congress of Entomology, Firenze,* p. 110.

——. 1998a. Reductions and new innovations dominate oogenesis of Strepsiptera (Insecta). *Int. J. Insect Morphol. Embryol.* 27: 3–8.

——. 1998b. The ovariole: structure, type, and phylogeny. In F. W. Harrison and M. Locke (eds.), *Microscopic anatomy of invertebrates.* Vol. 11C. *Insecta,* pp. 897–932. Wiley-Liss.

Burrows, M. 1996. *The neurobiology of an insect brain.* Oxford University Press.

Bursell, E. 1974. Environmental aspects—temperature. In M. Rockstein (ed.), *The physiology of Insecta.* Vol. 2, pp. 1–41. Academic Press.

Buszczak, M., M. R. Freeman, J. R. Carlson, M. Bender, L. Cooley, and W. A. Segraves. 1999. Ecdysone response genes govern egg chamber development during mid-oogenesis in *Drosophila. Development* 126: 4581–4589.

Butenandt, A., and P. Karlson, 1954. Über die Isolierung eines Metamorphosen-Hormons der Insekten in kristallisierten Form. *Z. Naturforsch.* 96: 389–391.

Callahan, C. A., J. L. Bonkovsky, A. L. Scully, and J. B. Thomas. 1996. *derailed* is required for muscle attachment site selection in *Drosophila. Development* 122: 2761–2767.

Campos-Ortega, J. A. 1993. Early neurogenesis in *Drosophila melanogaster.* In M. Bate and A. Martinez Arias (eds.), *The development of* Drosophila melanogaster. Vol. 2, pp. 1091–1129. Cold Spring Harbor Laboratory Press.

——. 1994. Cellular interactions in the developing nervous system of *Drosophila. Cell* 77: 969–977.

Campos-Ortega, J. A., and V. Hartenstein. 1985. *The embryonic development of* Drosophila melanogaster. 1st ed. Springer-Verlag.

——. 1997. *The embryonic development of* Drosophila melanogaster. 2nd ed. Springer-Verlag.

Cantera, R., and G. M. Technau. 1997. Glial cells phagocytose neuronal debris during the metamorphosis of the central nervous system in *Drosophila melanogaster. Dev. Genes Evol.* 206: 277–280.

Cantera, R., K. S. J. Thompson, E. Halberg, D. R. Nässl, and J. P. Bacon. 1995. Migration of neurons between ganglia in the metamorphosing insect ganglion. *W. Roux's Arch. Dev. Biol.* 205: 10–20.

Carayon, J. 1966. Traumatic insemination and the paragenital system. In R. L. Usinger (ed.), *Monograph of Cimicidae (Hemiptera-Heteroptera).* Vol. 7, pp. 81–166. Thomas Say Foundation. Entomological Society of America.

——. 1977. Insemination extra-génitale traumatique. In P. P. Grassé (ed.) *Traité de zoologie,* Vol. VIII 5A, pp. 351–390. Masson et Cie.

Carle, F. L. 1982. Evolution of odonate copulatory process. *Odonatologica* 11: 271–286.

Carpenter, F. M. 1977. Geological history and the evolution of insects. In *Proceedings of the XV International Congress of Entomology, Washington, D. C., 1976,* pp. 63–70.

——. 1992a. *Superclass Hexapoda.* Part R, *Arthropoda 4.* Vol. 3. *Treatise on invertebrate paleontology.* Geological Society of America and the University of Kansas.

——. 1992b. *Superclass Hexapoda.* Part R, *Arthropoda 4.* Vol. 4. *Treatise on invertebrate paleontology.* Geological Society of America and the University of Kansas.

Carroll, R. 1988. *Vertebrate paleontology and evolution.* W. H. Freeman.

Carroll, S. B. 1995. Homeotic genes and the evolution of arthropods and chordates. *Nature* 376: 479–485.

——. 1997. Genetics on the wing or How the butterfly got its spots. *Nat. Hist.* 107 (February): 28–32.

——. 2000. Endless forms: the evolution of gene regulation and morphological diversity. *Cell* 101: 577–580.

Carroll, S. B., J. Gates, D. N. Keys, S. W. Paddock, G. E. F. Panganiban, J. E. Selegue, and J. W. Williams. 1994. Pattern formation and eyespot determination in butterfly wings. *Science* 265: 109–114.

Carroll, S. B., J. K. Grenier, and S. D. Weatherbee. 2001. *From DNA to diversity. Molecular genetics and the evolution of animal design.* Blackwell Science.

Carroll, S. B., S. D. Weatherbee, and J. A. Langeland. 1995. Homeotic genes and the regulation and evolution of insect wing number. *Nature* 375: 58–61.

Carruthers, G., and B. Roberts. 1979. Studies on the differentiation and growth of the pretarsus in *Tricholioproctia impatiens* (Walker) (Diptera: Sarcophagidae). *J. Aust. Entomol. Soc.* 18: 305–314.

Carter, D., and M. Locke. 1993. Why caterpillars do not grow short and fat. *Int. J. Insect Morphol. Embryol.* 22: 81–102.

Carvalho, J. C., C. E. Vanario-Alonso, and S. E. Abdelhay. 1999. Specialized features of *Rhynchosciara americana* embryogenesis. *Int. J. Insect Morphol. Embryol.* 28: 309–319.

Carver, M., G. F. Gross, and T. E. Woodward. 1991. Hemiptera (bugs, leafhoppers, cicadas, aphids, scale insects, etc.). In CSIRO (ed.), *The insects of Australia. A textbook for students and research workers.* 2nd ed. Vol. 1, pp. 429–509. Melbourne University Press.

Casares, F., and R. S. Mann. 1998. Control of antennal versus leg development in *Drosophila. Nature* 392: 723–726.

——. 2000. A dual role for *homothorax* in inhibiting wing blade development and specifying proximal identities in *Drosophila. Development* 127: 1499–1508.

——. 2001. The ground state of the ventral appendage in *Drosophila. Science* 293: 1477–1480.

Casares, F., L. Sánchez, I. Guerro, and E. Sánchez-Herrero. 1997. The genital disc of *Drosophila melanogaster.* I. Segmental and compartment organization. *Dev. Genes Evol.* 207: 216–228.

Cassagnau, P. 1971. Les chromosomes salivaires polytènes chez *Bilobella grassei* (Denis) (Collemboles: Neanuridae). *Chromosoma* 35: 57–83.

Cassier, P. 1998. The corpora allata. In F. W. Harrison and M. Locke (eds.), *Microscopic anatomy of invertebrates.* Vol. 11C, *Insecta,* pp. 1041–1058. Wiley-Liss.

Castelli-Gair, J. 1998. The *lines* gene of *Drosophila* is required for specific functions of the abdominal-B Hox protein. *Development* 125: 1269–1274.

Cavallin, M. 1971. La polyembryonie substitutive et la problème de l'origine de la lignée germinale chez le phasme *Carausius morosus* Br. *C. R. Acad. Sci. Paris* 272: 462–465.

Cave, M. D. 1982. Morphological manifestations of ribosomal DNA amplification during insect oogenesis. In R. C. King, and II. Akai (eds.), *Insect ultrastructure*. Vol. l, pp. 86–117. Plenum Press.

Caveney, S. 1969. Muscle attachment related to cuticle architecture in Apterygota. *J. Cell Sci.* 4: 541–559.

——. 1998. Compound eyes. In F. W. Harrison and M. Locke (eds.), *Microscopic anatomy of invertebrates*. Vol. 11B, *Insecta*, pp. 423–445. Wiley-Liss.

Cavodeassi, F., R. Diez del Corral, S. Campuzano, and M. Dominguez. 1999. Compartments and organizing boundaries in the *Drosophila* eye: the role of the homeodomain Iroquois proteins. *Development* 126: 4933–4942.

Chalwatzis, N., J. Hauf, Y. van der Peer, R. Kinzelbach, and F. K. Zimmermann. 1996. 18S rRNA genes of insects: primary structure of the genes and molecular phylogeny of the Holometabola. *Ann. Entomol. Soc. Am.* 89: 788–803.

Champlin, D. T., and J. W. Truman. 1998a. Ecdysteroid control of cell proliferation during optic lobe neurogenesis in the moth *Manduca sexta*. *Development* 125: 269–277.

——. 1998b. Ecdysteroids govern two phases of eye development during metamorphosis of the moth, *Manduca sexta*. *Development* 125: 2009–2018.

Champlin, D. T., S. E. Reiss, and J. W. Truman. 1999. Hormonal control of ventral diaphragm myogenesis during metamorphosis of the moth, *Manduca sexta*. *Dev. Genes Evol.* 209: 265–274.

Chang, E. S., M. J. Bruce, and S. L. Tamone. 1993. Regulation of crustacean molting: a multihormonal system. *Am. Zool.* 33: 324–329.

Chang, H.-Y., and D. F. Ready. 2000. Rescue of photoreceptor degeneration in rhodopsin-null *Drosophila* mutants by activated Rac1. *Science* 290: 1978–1980.

Chang, W. S., K. Serikawa, K. Allen, and D. Bentley. 1992. Disruption of pioneer growth cone guidance *in vivo* by removal of glycosyl-phosphtidylinositol-anchored cell surface proteins. *Development* 114: 507–519.

Chanut, F., and N. Heberlein. 1997. Role of *decapentaplegic* in initiation and progression of the morphogenetic furrow in the developing *Drosophila* retina. *Development* 124: 559–567.

Chapman, M. J., M. F. Dolan, and L. Margulis. 2000. Centrioles and kinetosomes: form, function and evolution. *Q. Rev. Biol.* 75: 409–429.

Chapman, R. F. 1982. *The insects. Structure and function*. 3rd ed. Harvard University Press.

——. 1998. *The insects. Structure and function*. 4th ed. Cambridge University Press.

Chapman, T., L. F. Liddle, J. M. Kalb, M. F. Wolfner, and L. Partridge. 1995. Cost of mating in *Drosophila melanogaster* females is mediated by male accessory gland products. *Nature* 373: 241–244.

Charlesworth, B. 1991. The evolution of sex chromosomes. *Science* 251: 1030–1033.

Chasan, R., and K. V. Anderson. 1993. Maternal control of dorsal-ventral polarity and pattern in the embryo. In C. M. Bate and A. Martinez Arias (eds.), *The development of* Drosophila melanogaster. Vol. 1, pp. 387–424. Cold Spring Harbor Laboratory Press.

Chaudonneret, J. 1990. *Les pièces buccales des insectes. Thème et variations*. Édition Hors Série du Bull. Sci. Bourgogne. Dijon.

Chen, C.-L., D. J. Lampe, H. M. Robertson, and J. B. Nardi. 1997. Neuroglian is expressed in cells destined to form the prothoracic glands of *Manduca* embryos as they segregate from surrounding cells and rearrange during morphogenesis. *Dev. Biol.* 181: 1–13.

Chen, E. H., and B. S. Baker. 1997. Compartmental organization in the *Drosophila* genital discs. *Development* 124: 205–218.

Chen, J.-Y., P. Oliveri, C.-W. Li, G.-L. Zhou, F. Gao, J. W. Hagedorn, K. J. Peterson, and E. H. Davidson. 2000. Precambrian animal diversity: putative phosphatized embryos from the Doushantuo Formation of China. *Proc. Natl. Acad. Sci. U.S.A.* 97: 4457–4462.

Chen, T. T., and L. J. Hillen. 1983. Review article: expression of the vitellogenin genes in insects. *Gamete Res.* 7: 179–196.

Chiang, R. G., and K. G. Davey. 1988. A novel receptor capable of monitoring applied pressure in the abdomen of an insect. *Science* 241: 1665–1667.

Chino, H., S. Sakurai, T. Ohtaki, N. Ikekawa, H. Miyazaki, M. Ishibashi, and H. Abuki. 1974. Biosynthesis of α-ecdysone by prothoracic glands *in vitro*. *Science* 183: 529–530.

Cho, K.-O., and K.-W. Choi. 1998. Fringe is essential for mirror symmetry and morphogenesis in the *Drosophila* eye. *Nature* 396: 272–276.

Cho, K.-O., J. Chen, S. Izaddoost, and K.-W. Choi. 2000. Novel signaling from the peripodial membrane is essential for eye disc patterning in *Drosophila*. *Cell* 103: 331–342.

Christophers, S. R. 1960. *Aëdes aegypti (L.), the yellow fever moquito. Its life history, bionomics and structure*. Cambridge University Press.

Clark, A. G., D. J. Begun, and T. Prout. 1999. Female x male interactions in *Drosophila* sperm competition. *Science* 283: 217–220.

Clarke, K. U. 1957. On the increase in linear size during growth in *Locusta migratoria* L. *Proc. R. Entomol. Soc. Lond. Ser. A* 32: 35–39.

Clements, A. N. 1992. *The biology of mosquitoes*. 2nd ed. Vol. 1. *Development, nutrition and reproduction*. Chapman and Hall.

Clever, U., and H. Bultmann. 1972. Chromosomal control of foot pad development in *Sarcophaga bullata*. III. Requirement of RNA and protein synthesis for cuticle formation and tanning. *Cell Diff.* 1: 37–42.

Clever, U., and P. Karlson. 1960. Induktion von Puff-Veränderungen in den Speicheldrüsen-chromosomen von *Chironomus tentans* durch Ecdyson. *Exp. Cell. Res.* 20: 623–626.

Cloud, P. 1988. *Oasis in space. Earth history from the beginning*. W. W. Norton.

Coast, G. M., and S. G. Webster (eds.). 1998. *Recent advances in arthropod endocrinology*. Cambridge University Press.

Cobben, R. H. 1968. *Evolutionary trends in Heteroptera. Part I. Eggs, architecture of the shell, gross embryology and eclosion*. Pudoc.

——. 1978. *Evolutionary trends in Heteroptera. Part II. Mouthpart—structures and feeding strategies*. Meded. 78–5. Landbouwhogeschool.

Cohen, B., A. Simcox, and S. M. Cohen. 1993. Allocation of the thoracic imaginal primordia in the *Drosophila* embryo. *Development* 117: 597–608.

Cohen, J., and B. D. Massey. 1983. Larvae and the origins of major phyla. *Biol. J. Linn. Soc.* 19: 321–328.

Cohen, S., and G. Jürgens. 1991. *Drosophila* headlines. *Trends Genet.* 7: 267–272.

Cohen, S. M. 1993. Imaginal disc development. In M. Bate and A. Martinez Arias (eds.), *The development of* Drosophila melanogaster. Vol. 2, pp. 747–841. Cold Spring Harbor Laboratory Press.

Cohen, S. M., G. Brönner, F. Küttner, G. Jürgens, and H. Jäckle. 1989. *Distalless* encodes a homeodomain protein required for limb development in *Drosophila*. *Nature* 338: 432–434.

Colas, J.-F., J.-M. Launay, J.-L. Vonesch, P. Heckel, and L. Maroteaux. 1999. Serotonin synchronizes convergent extension of ectoderm with morphogenetic gastrulation movements in *Drosophila*. *Mech. Dev.* 87: 77–91.

Condic, M. L., and D. Bentley. 1989. Removal of basal lamina *in vivo* reveals growth cone-basal lamina adhesive interactions and axonal tension in grasshopper embryos. *J. Neurosci.* 9: 2678–2686.

Consoulas, C. 2000. Remodeling of the leg sensory system during metamorphosis of the hawkmoth, *Manduca sexta*. *J. Comp. Neurol.* 419: 154–174.

Conway Morris, S. 1995. Ecology in deep time. *Trends Ecol. Evol.* 10: 290–294.

——. 1998a. *The crucible of creation. The Burgess Shale and the rise of animals.* Cambridge University Press.

——. 1998b. Metazoan phylogenies: falling into place or falling to pieces? A palaentological perspective. *Curr. Opin. Genet. Dev.* 8: 662–667.

——. 1998c. Early metazoan evolution: reconciling palaentology and molecular biology. *Am. Zool.* 38: 867–877.

——. 2000. Evolution: bringing molecules into the fold. *Cell* 100: 1–11.

Cook, C. E., M. L. Smith, M. J. Telford, A. Bastianello, and M. Akam. 2001. Hox genes and the phylogeny of the arthropods. *Curr. Biol.* 11: 759–763.

Cook, P. A., and N. Wedell. 1999. Nonfertile sperm delay female re-mating. *Nature* 397: 486.

Cooley, L., and W. E. Theurkauf. 1994. Cytoskeletal functions during *Drosophila* oogenesis. *Science* 266: 590–596.

Cooper, M. T. D., and S. J. Bray. 1999. Frizzled regulation of Notch signaling polarizes cell fate in the *Drosophila* eye. *Nature* 397: 526–530.

Copeland, J. W. R., A. Nasiadka, B. H. Dietrich, and H. M. Krause. 1996. Patterning of the *Drosophila* embryo by a homeodomain-deleted Ftz polypeptide. *Nature* 379: 162–165.

Copenhaver, P. F. 1993. Origins, migration and differentiation of glial cells in the insect enteric nervous system from a discrete set of glial precursors. *Development* 117: 59–74.

Copenhaver, P. F., and P. H. Taghert. 1989a. Development of the enteric nervous system in the moth. I. Diversity of cell types and the embryonic expression of FMRFamide-related neuropeptides. *Dev. Biol.* 131: 70–84.

——. 1989b. Development of the enteric nervous system in the moth. II.

Stereotyped cell migration precedes the differentiation of embryonic neurons. *Dev. Biol.* 131: 85–101.

——. 1990. Neurogenesis in the insect enteric nervous system: generation of pre-migratory neurons from an epithelial placode. *Development* 109: 17–28.

——. 1991. Origins of the insect enteric nervous system: differentiation of the enteric ganglia from a neurogenic epithelium. *Development* 115: 1115–1132.

Copenhaver, P. F., and J. W. Truman. 1986. Metamorphosis of the cerebral neuroendocrine system in the moth *Manduca sexta*. *J. Comp. Neurol.* 249: 186–204.

Copenhaver, P. F., A. M. Horgan, and S. Combes. 1996. An identified set of visceral muscle bands is essential for the guidance of migratory neurons in the enteric nervous system of *Manduca sexta*. *Dev. Biol.* 179: 412–426.

Corbet, P. S. 1999. *Dragonflies: behavior and ecology of Odonata*. Cornell University Press.

Costa, M., D. Sweeton, and E. Wieschaus. 1993. Gastrulation in *Drosophila*: cellular mechanisms of morphogenetic movements. In M. Bate and A. Martinez Arias (eds.), *The development of* Drosophila melanogaster. Vol. 1, pp. 425–465. Cold Spring Harbor Laboratory Press.

Cottrell, C. B. 1962. The imaginal ecdysis of blowflies. The control of cuticular hardening and darkening. *J. Exp. Biol.* 39: 35–43.

Counce, S. J. 1973. The causal analysis of insect embryogenesis. In S. J. Counce and C. H. Waddington (eds.), *Developmental systems: insects*. Vol. 2, pp. 1–156. Academic Press.

Couso, J. P., M. Bate, and A. Martinez Arias. 1993. A *wingless*-dependent polar coordinate system in *Drosophila* imaginal discs. *Science* 259: 484–489.

Craig, S. F., L. B. Slobodkin, G. A. Wray, and C. H. Biermann. 1997. The "paradox" of polyembryony: a review of the cases and a hypothesis for its evolution. *Evol. Ecol.* 11: 127–143.

Crane, P. R., E. M. Friis, and K. R. Pedersen. 1995. The origin and early diversification of the angiosperms. *Nature* 374: 27–33.

Crepet, W. L., E. M. Friis, and K. C. Nixon. 1991. Fossil evidence for the evolution of biotic pollination. *Philos. Trans. R. Soc. Lond. B Biol. Sci.* 333: 187–195.

Crespi, B. J., D. A. Carmean, and T. W. Chapman. 1997. Ecology and evolution of galling thrips and their allies. *Annu. Rev. Entomol.* 42: 51–71.

Crozier, R. H. 1975. *Hymenoptera*. Ani-

mal cytogenetics. Vol. 3. *Insecta*. 7. Gebrüder Borntraeger.

CSIRO (ed.). 1991. *The insects of Australia. A textbook for students and research workers.* 2d ed. Vol. 1, pp. 125–140. Melbourne University Press.

Cuellar, O. 1977. Animal parthenogenesis. *Science* 197: 837–843.

Cumberledge, S. J., J. Szabad, and S. Sakonju. 1992. Gonad formation and development requires the *abd-A* domain of the bithorax complex in *Drosophila melanogaster*. *Development* 115: 395–402.

Curtis, N. J., J. M. Ringo, and H. B. Dowse. 1999. Morphology of the pupal heart, and associated tissues in the fruit fly, *Drosophila melanogaster*. *J. Morphol.* 240: 225–235.

Curtis, S. K., and D. B. Benner. 1991. Movement of spermatozoa of *Megaselia scalaris* (Diptera: Brachycera: Cyclorrhapha: Phoridae) in artificial and natural fluids. *J. Morphol.* 210: 85–99.

Curtis, S. K., D. B. Benner, and G. Musil. 1989. Ultrastructure of the spermatozoon of *Megaselia scalaris* Loew (Diptera: Brachycera, Cyclorrhapha: Phoridea: Phoridae). *J. Morphol.* 200: 47–61.

Cusson, M., and J. N. McNeil. 1989. Involvement of juvenile hormone in the regulation of pheromone release activities in a moth. *Science* 243: 210–212.

Cusson, M., and S. R. Reddy. 2000. Can juvenile hormone research help rejuvenate integrated pest management? *Can. Entomol.* 132: 263–280.

Cusson, M., K. J. Yagi, Q. Ding, H. Duve, A. Thorpe, J. M. McNeil, and S. S. Tobe. 1991. Biosynthesis and release of juvenile hormone and its precursors in insects and crustaceans. The search for a unifying arthropod endocrinology. *Insect Biochem.* 21: 1–6.

Dai, J. G., and L. I. Gilbert. 1991. Metamorphosis of the corpus allatum and degeneration of the prothoracic glands during the larval-pupal-adult transformation of *Drosophila melanogaster*. A cytophysiological analysis of the ring gland. *Dev. Biol.* 144: 309–326.

——. 1997. Programmed cell death of the prothoracic glands of *Manduca sexta* during pupal-adult metamorphosis. *Insect Biochem. Mol. Biol.* 27: 69–78.

——. 1998. Juvenile hormone prevents the onset of programmed cell death in the prothoracic glands of *Manduca sexta*. *Gen. Comp. Endocrinol.* 109: 155–165.

Dallai, R., and B. A. Afzelius. 1990. Microtubular diversity in insect spermatozoa: results obtained with a new fixative. *J. Struct. Biol.* 103: 164–179.

——. 1993. Substructure of the axoneme of pterygote insect spermatozoa: phylogenetic considerations. *Int. J. Insect Embryol. Morphol.* 22: 449–458.

Dallai, R., B. A. Afzelius, S. Lanzavecchia, and P. L. Bellon. 1991. Bizarre flagellum of thrips spermatozoa (Thysanoptera, Insecta). *J. Morphol.* 209: 343–347.

Dallai, R., P. P. Fanciulli, and F. Frati. 1999. Chromosome elimination and sex determination in springtails (Insecta, Collembola). *J. Exp. Zool.* 285: 215–225.

Daly, H. V., J. T. Doyen, and A. H. Purcell III. 1998. *Introduction to insect biology and diversity.* 2d ed. Oxford University Press.

Dambach, M., and H. Igelmund. 1983. Das Ei-Ablageverhalten von Grillen (Saltatoria: Grylloidea). *Entomol. Gen.* 8: 267–281.

Damen, W. G. M., and D. Tautz. 1998. A Hox class 3 orthologue from the spider *Cupiennius salei* is expressed in a Hox gene-like fashion. *Dev. Genes Evol.* 208: 586–590.

Damen, W. G. M., M. Hausdorf, E.-A. Seyfarth, and D. Tautz. 1998. A conserved mode of head segmentation in arthropods revealed by the expression of Hox genes in a spider. *Proc. Natl. Acad. Sci. U.S.A.* 95: 10665–10670.

Damen, W. G. M., M. Weller, and D. Tautz. 2000. Expression patterns of *hairy, even-skipped* and *runt* in the spider *Cupiennius salei* imply that these genes were segmentation genes in the basal arthropod. *Proc. Natl. Acad. Sci. U.S.A.* 97: 4515–4519.

Daniel, A., K. Dumstrei, J. A. Lengyel, and V. Hartenstein. 1999. The control of cell fate in the embryonic visual system by *atonal, tailless,* and EGFR signaling. *Development* 126: 2945–2954.

Danks, H. V. 1987. *Insect dormancy: an ecological perspective.* Biological Survey of Canada (Terrestrial Arthropods). Monograph series No. 1. National Museum of Natural Sciences, Ottawa.

Darlington, P. J., Jr. 1980. *Evolution for naturalists. The simple principles and complex reality.* Wiley-Interscience.

Darwin, C. 1859. *The origin of species by means of natural selection or the preservation of favored races in the struggle for life.* J. Murray.

——. 1874. *The descent of man, and selection in relation to sex.* 2d ed. A. L. Burt.

Davey, K. G. 1959. Spermatophore production in *Rhodnius prolixus. Q. J. Microsc. Sci.* 100: 289–322.

——. 1965. *Reproduction in the insects.* Oliver and Boyd.

——. 1974. Reproduction in the females of some hematophagous insects. *J. Med. Entomol.* 11: 40–45.

——. 1993. Hormonal integration of egg production in *Rhodnius prolixus. Am. Zool.* 33: 397–402.

Davidson, E. H. 2001. *Genomic regulatory systems. Development and evolution.* Academic Press.

Davidson, E. H., K. J. Peterson, and R. A. Cameron. 1995. Origin of bilaterian body plans: evolution of developmental regulatory mechanisms. *Science* 270: 1319–1325.

Davies, D. R., I. Y. Goryshin, W. R. Reznikoff, and I. Raymont. 2000. Three-dimensional structure of the Tn5 synaptic complex transposition intermediate. *Science* 289: 77–85.

D'Avino, P. P., and C. S. Thummel. 1998. *crooked legs* encodes a family of zinc finger proteins required for morphogenesis and ecdysone-regulated gene expression during *Drosophila* morphogenesis. *Development* 125: 1733–1745.

Davis, N. T. 1966. Reproductive physiology. In R. L. Usinger (ed.), *Monograph of Cimicidae (Hemiptera-Heteroptera).* Vol. 7, pp. 167–178. Thomas Say Foundation. Entomological Society of America.

Day, S. J., and P. A. Lawrence. 2000. Measuring dimensions: the regulation of size and shape. *Development* 127: 2977–2987.

De Beer, G. 1962. *Embryos and ancestors.* 3rd ed. Oxford University Press.

de Celis, J. F., and R. Barrio. 2000. Function of the *spalt/spalt-related* gene complex in positioning the veins in the *Drosophila* wing. *Mech. Dev.* 91: 31–41.

de Celis, J. F., S. Bray, and A. García-Bellido. 1997. Notch signaling regulates *veinlet* expression and establishes boundaries between veins and interveins in the *Drosophila* wing. *Development* 124: 1919–1928.

de Cuevas, M., and A. C. Spradling. 1998. Morphogenesis of the *Drosophila* fusome and its implications for oocyte specification. *Development* 125: 2781–2789.

Dedeins, F., F. Vavre, F. Fleury, B. Loppin, M. E. Hochberg, and M. Boulétreau. 2001. Removing symbiotic *Wolbachia* bacteria specifically inhibits oogenesis in a parasitic wasp. *Proc. Natl. Acad. Sci. U.S.A.* 98: 6247–6252.

Degrugillier, M. E. 1985. *In vitro* release of housefly, *Musca domestica* L. (Diptera: Muscidae), acrosomal material after treatments with secretion of female accessory gland and micropyle cap substance. *Int. J. Insect Morphol. Embryol.* 14: 381–391.

Degrugillier, M. E., and R. A. Leopold. 1973. Internal genitalia of the female house fly, *Musca domestica* L. (Diptera: Muscidae): analysis of copulation and oviposition. *Int. J. Insect Morphol. Embryol.* 2: 313–325.

——. 1976. Ultrastructure of sperm penetration of house fly eggs. *J. Ultrastruct. Res.* 56: 312–325.

de Hooij, J. C., and I. K. Hariharan. 1995. Uncoupling cell fate determination from patterned cell division in the *Drosophila* eye. *Science* 270: 983–985.

Dej, K. J., and A. C. Spradling. 1999. The endocycle controls nurse cell polytene chromosome structure during *Drosophila* oogenesis. *Development* 126: 293–303.

de Kort, C. A. D., and N. A. Granger. 1981. Regulation of the juvenile hormone titer. *Annu. Rev. Entomol.* 26: 1–28.

de Lange, T. 1998. Telomeres and senescence: ending the debate. *Science* 279: 334–335.

Delbeque, J.-P., K. Weidner, and K. H. Hoffmann. 1990. Alternative sites for ecdysteroid production in insects. *Invert. Reprod. Dev.* 18: 29–42.

De Loof, A. 1983. The meroistic insect ovary as a miniature electrophoresis chamber. *Comp. Biochem. Physiol.* 74A: 3–9.

Deng, W., and H. Lin. 1997. Spectrosomes and fusomes anchor mitotic spindles during asymmetric germ cell divisions and facilitate the formation of a polarized microtubule array for oocyte specification in *Drosophila. Dev. Biol.* 189: 79–94.

Denlinger, D. L. 1985. Hormonal control of diapause. In G. A. Kerkut and L. I. Gilbert (eds.), *Comprehensive insect physiology, biochemistry and pharmacology.* Vol. 8, pp. 353–412. Pergamon Press.

Denlinger, D. L., and J. Zdárek. 1994. Metamorphosis behavior in flies. *Annu. Rev. Entomol.* 39: 243–266.

——. 1997. A hormone from the uterus of the tsetse fly, *Glossina morsitans,* stimulates parturition and abortion. *J. Insect Physiol.* 43: 135–142.

De Robertis, E. M. 1997. The ancestry of segmentation. *Nature* 387: 25–26.

De Robertis, E. M., and Y. Sasai. 1996. A common plan for dorsoventral patterning in Bilataria. *Nature* 380: 37–40.

de Rosa, R., J. K. Grenier, T. Andreeva,

C. E. Cook, A. Adouette, M. Akam, S. B. Carroll, and G. Balavoine. 1999. Hox genes in brachiopods and priapulids and protostome evolution. *Nature* 399: 772–776.

DeSalle, R., and D. Grimaldi. 1993. Phylogenetic pattern and developmental process in *Drosophila*. *Syst. Biol.* 42: 458–475.

Deshpande, G., G. Calhoun, J. L. Yanowitz, and P. D. Schedel. 1999. Novel functions of *nanos* in down regulating mitosis and transcription during development of the *Drosophila* germline. *Cell* 99: 271–281.

Deuve, T. (ed.). 2001. Origin of the Hexapoda. *Ann. Soc. Entomol. Fr. N. S.* 37(1/2): 1–306.

Dewhurst, C. F. 1999. Changes in the morphology of the larval mandibles of African Armyworm, *Spodoptera exempta* (Walker) (Lepidoptera: Noctuidae). *Afr. Entomol.* 7: 261–264.

de Wilde, J., and J. Beetsma. 1982. The physiology of caste development in social insects. *Adv. Insect Physiol.* 16: 167–246.

de Wilde, J., and A. De Loof. 1973. Reproduction-endocrine control. In M. Rockstein (ed.), *The physiology of Insecta.* 2d ed. Vol. 1, pp. 97–115. Academic Press.

Dhadialla, T. S., G. R. Carlson, and D. P. Le. 1998. New insecticides with ecdysteroidal and juvenile hormone activity. *Annu. Rev. Entomol.* 43: 545–569.

Diaz del Corral, R., P. Aroca, J. L. Gómez-Skarmeta, F. Cavodeassi, and J. Modolell. 1999. The Iroquois homeodomain proteins are required to specify body wall identity in *Drosophila*. *Genes Dev.* 13: 1754–1761.

Dickinson, M. H., C. T. Farley, R. J. Full, M. A. R. Koehl, R. Kram, and S. Lehman. 2000. How animals move: an integrative view. *Science* 288: 100–106.

Dickson, B., and E. Hafen. 1993. Genetic dissection of eye development in *Drosophila*. In M. Bate and A. Martinez Arias (eds.), *The development of Drosophila melanogaster.* Vol. 2, pp. 1327–1362. Cold Spring Harbor Laboratory Press.

Dietrich, M. R. 2000. From hopeful monsters to homeotic effects: Richard Goldschmidt's integration of development, evolution and genetics. *Am. Zool.* 40: 738–747.

DiSilvestro, R. L. 1997. Out on a limb. *Bioscience* 47: 729–731.

Dittman, F., R. Ehni, and W. Engels. 1981. Bioelectric aspects of the hemipteran telotrophic ovariole (*Dysdercus inter-*

medius). *W. Roux's Arch. Entwicklungsmech. Organ.* 190: 221–225.

Dittman, F., R. Hörner, and E. Engels. 1984. Endopolyploidization of tropharium nuclei during larval development and the first gonocycle in *Dysdercus intermedius* (Heteroptera). *Int. J. Invert. Reprod. Dev.* 7: 279–290.

Dittman, F., D. G. Weiss, and A. Münz. 1987. Movement of mitochondria in the ovarian trophic cord of *Dysdercus intermedius* (Heteroptera) resembles nerve axonal transport. *W. Roux's Arch. Dev. Biol.* 196: 407–413.

Dobeus, L. L., J. S. Peterson, J. Treisman, and L. A. Raftery. 2000. *Drosophila bunched* integrates opposing DPP and EGF signals to set the operculum boundary. *Development* 127: 745–754.

Dobson, S., and M. Tanouye. 1997. The paternal sex ratio chromosome induces chromosome loss independently of *Wolbachia* in the wasp *Nasonia vetripennis*. *Dev. Genes Evol.* 206: 207–217.

Dobzhansky, T. 1973. "Nothing in biology makes sense except in the light of evolution." *Am. Biol. Teacher* 35: 125–129.

Doe, C. L. 1996. Spindle orientation and asymmetric localization in *Drosophila*: both *inscuteable*? *Cell* 86: 695–697.

Doe, C. L., and C. S. Goodman. 1985a. Early events in insect neurogenesis. I. Development and segmental differences in the pattern of neuronal precursor cells. *Dev. Biol.* 111: 193–205.

———. 1985b. Early events in insect neurogenesis. II. The role of cell interactions and cell lineage in the determination of neuronal precursor cells. *Dev. Biol.* 111: 206–219.

Domínguez, M., and J. F. de Celis. 1998. A dorsal/ventral boundary established by Notch controls growth and polarity in the *Drosophila* eye. *Nature* 396: 276–278.

Dorn, A. 1983. Hormones during embryogenesis of the Milkweed bug, *Oncopeltus fasciatus* (Heteroptera: Lygaeidae). *Entomol. Gen.* 8: 193–214.

———. 1998. Comparative structural aspects of development in neuroendocrine systems. In F. W. Harrison and M. Locke (eds.), *Microscopic anatomy of invertebrates.* Vol. 11C. *Insecta*, pp. 1059–1092. Wiley-Liss.

Dorn, A., L. I. Gilbert, and W. E. Bollenbacher. 1987. Prothoracicotropic hormone activity in the embryonic brain of the tobacco hornworm, *Manduca sexta*. *J. Comp. Physiol. B* 157: 279–283.

Dover, G. A. 1986. How to drive an egg. *Trends Genet.* 2: 300–302.

Downes, J. A., and W. W. Wirth. 1981. Ceratopogonidae. In J. F. McAlpine, B. V. Peterson, G. E. Shewell, H. J. Teskey, J. R. Vockeroth, and D. M. Wood (eds.), *Manual of nearctic Diptera.* Vol. 1, pp. 393–421. Monograph No. 27. Research Branch, Agriculture Canada.

Downes, W. L. 1987. The impact of vertebrate predators on early arthropod evolution. *Proc. Entomol. Soc. Wash.* 89: 389–406.

Driever, W. 1993. Maternal control of anterior development in the *Drosophila* embryo. In C. M. Bate and A. Martinez Arias (eds.), *The development of Drosophila melanogaster.* Vol. 1, pp. 301–324. Cold Spring Harbor Laboratory Press.

Drosopoulos, S. 1976. Triploid pseudogamous biotype of the leafhopper, *Muellerianella fairmairei*. *Nature* 263: 499.

———. 1978. Laboratory synthesis of a pseudogamous triploid "species" of the genus *Muellerianella* (Homoptera, Delphacidae). *Evolution* 32: 916–920.

Drummond-Barbosa, D., and A. Spradling. 2001. Stem cells and their progeny respond to nutritional changes during *Drosophila* oogenesis. *Dev. Biol.* 231: 265–278.

Duboule, D. 1994. *Guidebook to the homeobox genes.* Oxford University Press.

Duchek, P., and P. Rørth. 2001. Guidance of cell migration by EGF receptor signaling during *Drosophila* oogenesis. *Science* 291: 131–133.

Duman-Scheel, M., and N. H. Patel. 1999. Analysis of molecular marker expression reveals neuronal homology. *Development* 126: 2327–2334.

Dumser, J. B. 1980. The regulation of spermatogenesis in insects. *Annu. Rev. Entomol.* 25: 341–369.

Dumser, J. B., and K. G. Davey. 1974. Endocrinological and other factors influencing testis development in *Rhodnius prolixus*. *Can. J. Zool.* 52: 1011–1022.

———. 1975a. The *Rhodnius* testis: hormones, differentiation of the germ cells and duration of the molting cycle. *Can. J. Zool.* 53: 1673–1681.

———. 1975b. The *Rhodnius* testis: hormonal effects on germ cell divisions. *Can. J. Zool.* 53: 1682–1689.

Dumstrei, K., C. Nassif, G. Abboud, A. Aryai, and V. Hartenstein. 1998. EGFR signaling is required for the differentiation and maintenance of neural progenitors along the dorsal midline of the *Drosophila* embryonic head. *Development* 125: 3417–3426.

Dunlop, J. A., and P. A. Selden. 1998. The

early history and phylogeny of the chelicerates. In R. A. Fortey and R. H. Thomas (eds.), *Arthropod relationships*. Systematics Association Special Vol. 55, pp. 221–235. Chapman and Hall.

Durica, D. S., A. C.-K. Chung, and P. M. Hopkins. 1999. Characterization of *EcR* and *RxR* gene homologs and receptor expression during the molt cycle in the crab, *Uca pugilator*. *Am. Zool.* 39: 758–773.

Dybas, L. K., and H. S. Dybas. 1981. Coadaptation and taxonomic differentiation of sperm and spermathecae in featherwing beetles. *Evolution* 35: 168–174.

Eady, J. 1994. Sperm transfer and storage in relation to sperm competition in *Callosobruchus maculatus*. *Behav. Ecol. Sociobiol.* 35: 123–129.

Ebens, A. J., H. Garren, B. N. R. Cheyette, and S. L. Zipursky. 1993. The *Drosophila anachronism* locus: a glycoprotein secreted by glia inhibits neuroblast proliferation. *Cell* 74: 15–27.

Eberhard, W. G. 1985. *Sexual selection and animal genitalia*. Harvard University Press.

——. 1991. Copulatory courtship and cryptic female choice in insects. *Biol. Rev.* 66: 1–32.

——. 1996. *Female control: sexual selection by cryptic female choice*. Princeton University Press.

Edgar, B. A., and C. F. Lehner. 1996. Developmental control of cell cycle regulators: a fly's perspective. *Science* 274: 1646–1651.

Edmunds, G. F., Jr., and W. P. McCafferty. 1988. The mayfly subimago. *Annu. Rev. Entomol.* 33: 509–529.

Edwards, J. S. 1969. Postembryonic development and regeneration of the insect nervous system. *Adv. Insect Physiol.* 6: 97–137.

Edwards, J. S., and L. P. Tolbert. 1998. Insect neuroglia. In F. W. Harrison and M. Locke (eds.), *Microscopic anatomy of invertebrates*. Vol. 11B. *Insecta*, pp. 449–466. Wiley-Liss.

Egelhaaf, A., H. Altenfeld, and H.-U. Hoffmann. 1988. Evidence for the priming role of the central retinula cell in ommatidium differentiation of *Ephestia kuehniella*. *W. Roux's Arch. Dev. Biol.* 197: 184–189.

Egelhaaf, A., S. Rick-Wagner, and D. Schneider. 1992. Development of the male scent organ of *Creatonotos transiens* (Lepidoptera, Arctiidae) during metamorphosis. *Zoomorphology* 111: 125–139.

Eldredge, N., and S. J. Gould. 1972. Punctuated equilibria: an alternative to phyletic gradualism. In T. J. M. Schopf (ed.), *Models in paleobiology*, pp. 82–115. Freeman, Cooper.

Elledge, S. J. 1996. Cell cycle checkpoints: preventing an identity crisis. *Science* 274: 1664–1671.

Ellgaard, L., M. Molinari, and A. Helenius. 1999. Setting the standards: quality control in the secretory pathway. *Science* 286: 1882–1888.

Emlen, D. J. 2000. Integrating development with evolution: a case study with beetle horns. *Bioscience* 50: 403–418.

——. 2001. Costs and the diversification of exaggerated animal structures. *Science* 291: 1534–1536.

Emlen, D. J., and H. F. Nijhout. 1999. Hormonal control of male horn length dimorphism in the dung beetle, *Onthophagus taurus* (Coleoptera: Scarabaeidae). *J. Insect Physiol.* 45: 45–53.

——. 2000. The development and evolution of exaggerated morphologies in insects. *Annu. Rev. Entomol.* 45: 661–708.

Emlen, S. T., and L. W. Oring. 1977. Ecology, sexual selection, and the evolution of mating systems. *Science* 197: 215–223.

Engelhard, E. K., B. A. Keddie, and L. E. Volkman. 1991. Isolation of third, fourth, and fifth instar larval midgut epithelia of the moth, *Trichoplusia ni*. *Tissue Cell* 23: 917–928.

Engelmann, F. 1970. *The physiology of insect reproduction*. Pergamon Press.

——. 1979. Insect vitellogenin: identification, biosynthesis, and role in vitellogenesis. *Adv. Insect Physiol.* 14: 49–108.

Enserink, M. 1997. Thanks to a parasite, asexual reproduction catches on. *Science* 275: 1743.

Ephrussi, A., and R. Lehmann. 1992. Induction of germ cell formation by *oskar*. *Nature* 358: 387–392.

Ephrussi, A., L. K. Dickinson, and R. Lehmann. 1991. *oskar* organizes the germ plasm and directs localization of the posterior determinant *nanos*. *Cell* 66: 37–50.

Epper, F. 1983. Three-dimensional fate map of the female genital disc of *Drosophila melanogaster*. *W. Roux's Arch. Dev. Biol.* 192: 270–274.

Ermating, G., and A. Isaake. 2000. Ectotherms, temperature, and trade-offs: size and number of eggs in a carabid beetle. *Am. Nat.* 155: 804–813.

Ernst, K.-D. 1972. Die Ontogenie der basiconischen Riechsensillen auf der Antenne von *Necrophorus* (Coleoptera). *Z. Zellforsch.* 129: 217–236.

Erwin, D. H. 1994. The permo-triassic extinction. *Nature* 367: 231–236.

Erwin, D. H., J. Valentine, and D. Jablonski. 1997. The origin of animal body plans. *Am. Sci.* 85: 126–137.

Estrada, B., and E. Sánchez-Herrero. 2001. The Hox gene *Abdominal-B* antagonizes appendage development in the genital disc of *Drosophila*. *Development* 128: 331–339.

Ewer, J., S. C. Gammie, and J. W. Truman. 1997. Control of insect ecdysis by a positive feedback endocrine system: roles of eclosion hormone and ecdysis triggering hormone. *J. Exp. Biol.* 200: 869–881.

Fabrizio, J. J., G. Hime, S. K. Lemmon, and C. Bazinet. 1998. Genetic dissection of sperm individualization in *Drosophila melanogaster*. *Development* 125: 1833–1843.

Farrell, E. R., and H. Keshishian. 1999. Laser ablation of persistent *twist* cells in *Drosophila*: muscle precursor fate is not segmentally restricted. *Development* 126: 273–280.

Farris, S. M., G. E. Robinson, R. L. Davies, and S. E. Fahrbach. 1999. Larval and pupal development of the mushroom bodies in the honeybee, *Apis mellifera*. *J. Comp. Neurol.* 414: 97–113.

Fausto, A.-M., M. Carcupino, M. Mazzini, and F. Giorgi. 1994. An ultrastructural investigation on vitellophage invasion of the yolk mass during and after germ band formation in embryos of the stick insect, *Carausius morosus* Br. *Dev. Growth Diff.* 36: 197–207.

Fausto, A.-M., M. Mazzini, A. Cecchettini, and F. Giorgi. 1997. The yolk sac in late embryonic development of the stick insect *Carausius morosus* (Br.). *Tissue Cell* 29: 257–266.

Ferenz, H. J. 1993. Yolk protein accumulation in *Locusta migratoria* (R. & F.) (Orthoptera: Acrididae) oocytes. *Int. J. Insect Morphol. Embryol.* 22: 295–314.

Fernandes, J. J., and H. Keshishian. 1998. Nerve-muscle interactions during flight muscle development in *Drosophila*. *Development* 125: 1769–1779.

Fernandes, J., and K. VijayRaghavan. 1993. The development of indirect flight muscle innervation in *Drosophila melanogaster*. *Development* 118: 215–227.

Fernando, J.-D.-B., B. Cohen, and S. M. Cohen. 1994. Cell interaction between compartments establishes the proximo-distal axis of *Drosophila* legs. *Nature* 372: 175–179.

Feyereisen, R. 1985. Regulation of juvenile hormone titre: synthesis. In G. A.

Kerkut and L. I. Gilbert (eds.), *Comprehensive insect physiology, biochemistry and pharmacology.* Vol. 7, pp. 391–429. Pergamon Press.

Field, S. A., and A. D. Austin. 1994. Anatomy and mechanics of the telescopic ovipositor system of *Scelio* Latreille (Hymenoptera: Scelionidae) and related genera. *Int. J. Insect Morphol. Embryol.* 23: 135–158.

Finkelstein, R., and E. Boncinelli. 1994. From fly head to mammalian forebrain: the story of *otd* and *otx*. *Trends Genet.* 10: 310–315.

Finkelstein, R., and N. Perrimon. 1991. The molecular genetics of head development in *Drosophila melanogaster*. *Development* 112: 899–912.

Fisher, R. A. 1930. *The genetical theory of natural selection.* Clarendon Press.

——. 1931. The evolution of dominance. *Biol. Rev.* 6: 345–368.

Fisk, G. J., and C. S. Thummel. 1998. The DHR78 nuclear receptor is required for ecdysteroid signaling during the onset of *Drosophila* metamorphosis. *Cell* 93: 543–555.

Fitch, K. R., and B. T. Wakimoto. 1998. The paternal effect gene *ms(3)sneaky* is required for sperm activation and the initiation of embryogenesis in *Drosophila melanogaster*. *Dev. Biol.* 197: 270–282.

Fleming, R. J., Y. Gu, and N. A. Hukriede. 1997. *Serrate*-mediated activation of *Notch* is specifically blocked by the product of the gene *fringe* in the dorsal compartment of the *Drosophila* wing imaginal disc. *Development* 124: 2973–2981.

Fletcher, J. C., and C. S. Thummel. 1995. The *Drosophila E74* gene is required for the proper stage- and tissue-specific transcription of ecdysone-regulated genes at the onset of metamorphosis. *Development* 121: 1411–1421.

Fletcher, J. C., K. C. Burtis, D. S. Hogness, and C. S. Thummel. 1995. The *Drosophila E74* gene is required for metamorphosis and plays a role in the polytene chromosome puffing response to ecdysone. *Development* 121: 1455–1465.

Foe, V. E. 1989. Mitotic domains reveal early commitments of cells in *Drosophila* embryos. *Development* 107: 1–22.

Foe, V. E., and G. M. Odell. 1989. Mitotic domains partition fly embryos, reflecting early cell biological consequences of determination in progress. *Am. Zool.* 29: 617–652.

Foe, V. E., C. M. Field, and G. M. Odell. 2000. Microtubules and mitotic cycle phase modulate spatiotemporal distributions of F-actin and myosin II in *Drosophila* syncytial blastoderm embryos. *Development* 127: 1767–1787.

Foe, V. E., G. M. Odell, and B. A. Edgar. 1993. Mitosis and morphogenesis in the *Drosophila* embryo: point and counterpoint. In C. M. Bate and A. Martinez Arias (eds.), *The development of* Drosophila melanogaster. Vol. 1, pp. 149–300. Cold Spring Harbor Laboratory Press.

Forbes, A., and R. Lehmann. 1998. Nanos and Pumilio have critical roles in the development and function of *Drosophila* germline cells. *Development* 125: 679–690.

Forjanic, J. P., C.-K. Chen, H. Jäckle, and M. G. Gaitán. 1997. Genetic analysis of stomatogastric nervous system development in *Drosophila* using enhancer trap lines. *Dev. Biol.* 186: 137–154.

Fortey, R. 2000. *Trilobite! Eyewitness to evolution.* Harper Collins.

Fortey, R. A., and R. H. Thomas (eds.). 1998. *Arthropod relationships.* Systematics Association Special Vol. 55. Chapman and Hall.

Fortey, R. A., D. E. G. Briggs, and M. A. Wille. 1997. The Cambrian evolutionary "explosion" recalibrated. *Bioessays* 19: 429–434.

Fortini, M. E., and N. M. Bonini. 2000. Modeling human neurodegenerative diseases in *Drosophila*: on a wing and a prayer. *Trends Genet.* 16: 161–167.

Fowler, G. L. 1973. Some aspects of the reproductive biology of *Drosophila*: sperm transfer, sperm storage, sperm utilization. *Adv. Genet.* 17: 293–360.

Fox, C. W., and M. E. Czesak. 2000. Evolutionary ecology of progeny size in arthropods. *Annu. Rev. Entomol.* 45: 341–369.

Fraenkel, G. 1935. Observations and experiments on the blowfly (*Calliphora erythrocephala*) during the first day after emergence. *Proc. Zool. Soc. Lond.* 1935: 893–904.

Fraenkel, G., and G. Bhaskaran. 1973. Pupariation and pupation in cyclorrhaphous flies (Diptera). Terminology and interpretation. *Ann. Entomol. Soc. Am.* 66: 418–422.

Fraenkel, G., and C. Hsiao. 1962. Hormonal and nervous control of tanning in the fly. *Science* 138: 27–29.

Frank, L. H., and C. Rusklow. 1996. A group of genes required for maintenance of amnioserosa tissue in *Drosophila*. *Development* 122: 1343–1352.

Fraser, N. C., D. A. Grimaldi, P. E. Olsen, and B. Axsmith. 1996. A Triassic Lagerstätte from eastern North America. *Nature* 380: 615–619.

Freeman, M. 1997. Cell determination strategies in the *Drosophila* eye. *Development* 124: 261–270.

French, V., and P. M. Brakefield. 1995. Eyespot development in butterfly wings: the focal signal. *Dev. Biol.* 168: 112–123.

French, V., B. J. Bryant, and S. V. Bryant. 1976. Pattern regulation in epimorphic fields. *Science* 193: 969–981.

Friedländer, M. 1989. 20-Hydroxyecdysone induces glycogen accumulation within the testicular sheath during in vitro spermatogenesis renewal in diapausing codling moths (*Cydia pomonella*). *J. Insect Physiol.* 35: 29–39.

——. 1997. Mini review. Control of the eupyrene-apyrene sperm dimorphism in Lepidoptera. *J. Insect Physiol.* 43: 1085–1092.

Friedländer, M., and G. Benz. 1981. The eupyrene-apyrene dichotomous spermatogenesis of Lepidoptera. Organ culture study on the timing of apyrene commitment in the codling moth. *Int. J. Invert Reprod.* 3: 113–120.

——. 1982. Control of spermatogenesis resumption in post-diapausing larvae of the codling moth. *J. Insect Physiol.* 28: 349–355.

Friedländer, M., and J. J. Brown. 1995. Tebufenozide (Mimic®), a non-ecdysteroidal ecdysone agonist, induces spermatogenesis reinitiation in isolated abdomens of diapausing codling moth larvae (*Cydia pomonella*). *J. Insect Physiol.* 41: 403–411.

Friedländer, M., and H. Gitay. 1972. The fate of the normal-anucleated spermatozoa in inseminated females of the silkworm *Bombyx mori*. *J. Morphol.* 138: 121–130.

Friedländer, M., and S. Miesel. 1977. Spermatid anucleation during the normal and atypical spermiogenesis of the warehouse moth, *Ephestia cautella*. *J. Submicrosc. Cytol.* 9: 173–185.

Friedländer, M., and S. E. Reynolds. 1988. Meiotic metaphases are induced by 20-Hydroxyecdysone during spermatogenesis of the tobacco hornworm, *Manduca sexta*. *J. Insect Physiol.* 34: 1013–1019.

Friedländer, M., and J. Wahrman. 1972. The number of centrioles in insect sperm: a study of two kinds of differentiating silkworm spermatids. *J. Morphol.* 134: 383–397.

Friedländer, M., R. E. Jeger, and C. H. Scholtz. 1999. Intra-follicular musculature in *Ormogus freyi* (HAAF)

(Coleoptera: Trogidae). *Int. J. Insect Morphol. Embryol.* 28: 5–11.

Friedländer, M., A. Jeshtadi, and S. E. Reynolds. 2001. The structural mechanism of trypsin-induced intrinsic sperm motility in *Manduca sexta. J. Insect Physiol.* 47: 245–255.

Friedrich, M., and S. Benzer. 2000. Divergent *decapentaplegic* expression patterns in compound eye development and the evolution of insect metamorphosis. *J. Exp. Zool.* 288: 39–55.

Friedrich, M., I. Rambold, and R. R. Melzer. 1996. The early stages of ommatidial development in the flour beetle *Tribolium castaneum* (Coleoptera, Tenebrionidae). *Dev. Genes. Evol.* 206: 136–146.

Fristrom, D., and J. W. Fristrom. 1993. The metamorphic development of the adult epidermis. In M. Bate and A. Martinez Arias (eds.), *The development of* Drosophila melanogaster. Vol. 2, pp. 843–897. Cold Spring Harbor Laboratory Press.

Fryer, G. 1996. Reflection on arthropod evolution. *Biol. J. Linn. Soc.* 58: 1–55.

——. 1998. A defence of arthropod polyphyly. In R. A. Fortey and R. H. Thomas (eds.), *Arthropod relationships.* Systematics Association Special Vol. 55, pp. 23–33. Chapman and Hall.

Fujishita, M., and H. Ishizaki. 1981. Circadian clock and prothoracicotropic hormone secretion in relation to the larval-larval ecdysis rhythm of the saturniid *Samia cythia ricini. J. Insect Physiol.* 27: 121–128.

Fukuda, S. 1940. Induction of pupation in silkworm by transplanting the prothoracic gland. *Proc. Imp. Acad. Tokyo* 16: 414–416.

——. 1942. The hormonal mechanism of larval molting and metamorphosis in the silkworm. *J. Fac. Sci. Imp. Univ. Tokyo Sec. IV.* 6: 477–532; plates 11–15.

Fuller, M. T. 1993. Spermatogenesis. In M. Bate and A. Martinez Arias (eds.), *The development of* Drosophila melanogaster. Vol. 1, pp. 71–147. Cold Spring Harbor Laboratory Press.

Fullilove, S. L., and A. G. Jacobson. 1971. Nuclear elongation and cytokinesis in *Drosophila montana. Dev. Biol.* 26: 560–577.

Furlong, E. E. M., E. C. Andersen, B. Null, K. P. White, and M. P. Scott. 2001. Patterns of gene expression during *Drosophila* mesoderm development. *Science* 293: 1629–1633.

Furman, D. P., and E. P. Catts. 1970. *Manual of medical entomology.* 3rd ed. National Press Books.

Furtado, A. F. 1979. The hormonal control of mitosis and meiosis during oogenesis in a blood-feeding bug *Panstrongylus megistus. J. Insect Physiol.* 25: 561–570.

Gabe, M. 1953. Données histologiques sur les glandes endocrines céphaliques des quelques Thysanoures. *Bull. Soc. Zool. Fr.* 78: 177–193.

Gack, C., and K. Peschke. 1994. Spermathecal morphology, sperm transfer and a novel mechanism of sperm displacement in the rove beetle, *Aleochara curtula* (Coleoptera, Staphylinidae). *Zoomorphology* 114: 227–237.

Gadenne, C., M.-C. Dufour, F. Rossignol, J. M. Bécard, and F. Couillaud. 1997. Occurrence of non-stationary larval moults during diapause in the cornstalk borer, *Sesamia nonagrioides* (Lepidoptera: Noctuidae). *J. Insect Physiol.* 43: 425–431.

Gage, A. R., and C. J. Bernard. 1996. Male crickets increase sperm number in relation to competition and female size. *Behav. Ecol. Sociobiol.* 38: 349–353.

Gage, M. J. D. 1994. Associations between body size, mating pattern, testes size and sperm length. *Proc. R. Soc. Lond. B Biol. Sci.* 258: 247–254.

Gage, M. J. D., and P. A. Cook. 1994. Sperm size or numbers: effects of nutritional stress upon eupyrene and apyrene sperm production strategies in the moth *Plodia interpunctella* (Lepidoptera: Pyralidae). *Funct. Ecol.* 8: 594–599.

Gaino, E., and M. Mazzini. 1991. Aflagellate sperm in three species of Leptophlebiidae (Ephemeroptera). *Int. J. Insect Morphol. Embryol.* 20: 119–125.

Galliot, B., and D. Miller. 2000. Origin of anterior patterning. How old is our head? *Trends Genet.* 16: 1–5.

Gallitano-Mendel, A., and R. Finkelstein. 1997. Novel segment polarity gene interactions during embryonic head development in *Drosophila. Dev. Biol.* 192: 599–613.

Ganfornina, M. D., D. Sánchez, and M. J. Bastiani. 1996a. The role of the cell surface in neuronal pathfinding. *Bioscience* 46: 344–354.

——. 1996b. Embryonic development of the enteric nervous system of the grasshopper, *Schistocerca americana. J. Comp. Neurol.* 372: 581–596.

Gao, N., M. von Schantz, R. G. Foster, and J. Hardie. 1999. The putative brain photoperiodic photoreceptors in the vetch aphid, *Megoura viciae. J. Insect Physiol.* 45: 1011–1019.

García-Alonso, L., R. D. Fetter, and C. S. Goodman. 1996. Genetic analysis of *Laminin A* in *Drosophila*: extracellular matrix containing laminin A is required for ocellar axon pathfinding. *Development* 122: 2611–2621.

García-Bellido, A. 1972. Pattern formation in imaginal discs. In II. Urspung and R. Nöthiger (eds.), *The biology of imaginal discs*, pp. 59–91. Springer-Verlag.

García-Bellido, A., P. A. Lawrence, and G. Morata. 1979. Compartments in animal development. *Sci. Am.* 241(1): 102–110.

García-Bellido, A., P. Ripoll, and G. Morata. 1973. Developmental compartmentalization of the wing disc of *Drosophila. Nature* 245: 251–253.

Garstang, W. 1922. The theory of recapitulation: a critical restatement of the biogenetic law. *J. Linn. Soc. Zool.* 35: 81–101.

Gaunt, S. J. 1997. Chick limbs, fly wings and homology at the fringe. *Nature* 386: 324–325.

Gayon, J. 2000. History of the concept of allometry. *Am. Zool.* 40: 748–758.

Gehring, W. 1998. *Master control genes in development and evolution. The homeobox story.* Yale University Press.

Gehring, W., Y-Q. Qian, M. Billeter, K. Furukuba-Tokunaga, A. F. Schier, D. Rosendez-Perez, M. Affolter, G. Otting, and K. Wüthrich. 1994. Homeodomain-DNA recognition. *Cell* 78: 211–223.

Gehring, W. J. 1985. The molecular basis of development. *Sci. Am.* 253(4): 153–162.

Gehring, W. J., and Y. Hiromi. 1986. Homeotic genes and the homeobox. *Annu. Rev. Genet.* 20: 147–173.

Gellon, G., and W. McGinnis. 1998. Shaping animal body plans in development and evolution by modulation of *Hox* expression patterns. *Bioessays* 20: 116–125.

Gelman, D. B., B. S. Thyagaraja, T. J. Kelly, E. P. Masier, R. A. Bell, and A. B. Borkovec. 1991. The insect gut: a new source of ecdysiotropic peptides. *Experientia* 47: 247–253.

Geoffroy St.-Hilaire, É. 1822. Considérations générales sur la vertebre. *Mem. Mus. Hist. Nat.* 9: 89–119.

Gerber, G. H. 1970. Evolution of the methods of spermatophore formation in pterygote insects. *Can. Entomol.* 102: 358–362.

Gerhart, J. 2000. Inversion of the chordate body axis: are there alternatives? *Proc. Nat. Acad. Sci. U.S.A.* 97: 4445–4448.

Gerhart, J., and M. Kirschner. 1997. *Cells, embryos, and evolution. Toward a* cellular and developmental understanding of phenotypic variation and evolutionary adaptability. Blackwell Science.

Gessner, B., and F. Ruttner. 1977. Transfer der Spermatozoen in die Spermatheka der Bienenkönigin. *Apidologie* 8: 1–18.

Geyer, J. W. C. 1951. The reproductive organs of certain termites, with notes on the hermaphrodites of *Neotermes*. *Entomol. Mem. Dept. Agric. S. Afr.* 2: 233–325.

Ghiradella, H. 1998. Hairs, bristles and scales. In F. W. Harrison and M. Locke (eds.), *Microscopic anatomy of invertebrates*. Vol. 11A. *Insecta*, pp. 257–287. Wiley-Liss.

Ghysen, A., and C. Dambly-Chaudière. 2000. A genetic program for neural connectivity. *Trends Genet.* 16: 221–226.

Ghysen, A., C. Dambly-Chaudière, E. Aceves, L. Jan, and Y. Jan. 1986. Sensory neurons and peripheral pathways in *Drosophila* embryos. *W. Roux's Arch. Dev. Biol.* 195: 281–289.

Gibbs, D., and L. M. Riddiford. 1977. Prothoracicotropic hormone in *Manduca sexta*: localization by a larval assay. *J. Exp. Biol.* 66: 255–266.

Gibson, M. C., and G. Schubiger. 2000. Peripodial cells regulate proliferation and patterning of *Drosophila* imaginal discs. *Cell* 103: 343–350.

Giebultowicz, J. M., J. E. Joy, J. G. Riemann, and A. K. Raina. 1994. Changes in protein patterns in sperm and vas deferens during the daily rhythm of sperm release in the gypsy moth. *Arch. Insect Biochem. Physiol.* 27: 65–75.

Giebultowicz, J. M., J. G. Riemann, A. K. Raina, and R. L. Ridgway. 1989. Circadian system controlling release of sperm in the insect testes. *Science* 245: 1098–1100.

Giebultowicz, J. M., F. Weyda, E. F. Erbe, and W. P. Wergin 1997. Circadian rhythm of sperm release in the gypsy moth *Lymantria dispar*: ultrastructural study of transepithelial penetration of sperm bundles. *J. Insect Physiol.* 43: 1133–1147.

Gilbert, C. 1994. Form and function of stemmata in larvae of holometabolous insects. *Annu. Rev. Entomol.* 39: 323–349.

Gilbert, L. I. (ed.). 1976. *The juvenile hormones*. Plenum Press.

Gilbert, S. F., J. M. Optiz, and R. A. Raff. 1996. Resynthesizing evolutionary and developmental biology. *Dev. Biol.* 173: 357–372.

Gillooly, D. J., and H. Stenmark. 2001. A lipid oils the endocytosis machine. *Science* 291: 993–994.

Giniger, E., L. Y. Jan, and Y. N. Jan. 1993. Specifying the path of the intersegmental nerve of the *Drosophila* embryo: a role for Delta and Notch. *Development* 117: 431–440.

Giorgi, F., L. Yin, A. Cecchettini, and J. H. Nordin. 1997. The vitellin processing protease of *Blattella germanica* is derived from a pro-protease of maternal origin. *Tissue Cell* 29: 293–303.

Gnatzy, W., and J. Tautz. 1977. Sensitivity of an insect mechanoreceptor during molting. *Physiol. Entomol.* 2: 279–288.

Godt, D., and F. A. Lanski. 1995. Mechanisms of cell rearrangement and recruitment in *Drosophila* ovary morphogenesis and the requirement of *bric-à-brac*. *Development* 121: 173–187.

Goldschmidt, R. 1940. *The material basis of evolution*. Yale University Press.

Goldstein, B., and G. Freeman. 1997. Axis specification in animal development. *Bioessays* 19: 105–116.

Gómez, M. R., and M. Bate. 1997. Segregation of myogenic lineages in *Drosophila* requires Numb. *Development* 124: 4857–4866.

Gönczy, P., and S. DiNardo. 1996. The germ line regulates somatic cyst cell proliferation and fate during *Drosophila* spermatogenesis. *Development* 122: 2437–2447.

Gönczy, P., E. Matunis, and S. DiNardo. 1997. *bag-of-marbles* and *benign gonial cell neoplasm* act in the germline to restrict proliferation during *Drosophila* spermatogenesis. *Development* 124: 4361–4371.

Gönczy, P., S. Viswanathan, and S. DiNardo. 1992. Probing spermatogenesis in *Drosophila* with P-element enhancer detectors. *Development* 114: 89–98.

González-Reyes, A., and D. St. Johnston. 1998. The *Drosophila* A-P axis is polarized by the cadherin-mediated position of the oocyte. *Development* 125: 3635–3644.

González-Reyes, A., H. Elliott, and D. St Johnston. 1997. Oocyte determination and the origin of polarity in *Drosophila*: the role of the *spindle* genes. *Development* 124: 4927–4937.

Goodman, C. S. 1984. Landmarks and labels that help developing neurones find their way. *Bioscience* 34: 300–307.

Goodman, C. S., and M. J. Bastiani. 1984. How embryonic nerve cells find one another. *Sci. Am.* 251(6): 58–66.

Goodman, C. S., and C. Q. Doe. 1993. Embryonic development of the *Drosophila* central nervous system. In M. Bate and A. Martinez Arias (eds.), *The development of* Drosophila melanogaster. Vol. 1, pp. 1131–1206 Cold Spring Harbor Laboratory Press.

Goodman, C. S., and N. C. Spitzer. 1979. Embryonic development of identified neurones: differentiation from neuroblast to neurone. *Nature* 280: 208–214.

Goodman, C. S., M. J. Bastiani, C. Q. Doe, S. Du Lac, S. Helfand, J. Y. Kurada, and J. B. Thomas. 1984. Cell recognition during neuronal development. *Science* 225: 1271–1279.

Goodman, W. S., and E. S. Chang. 1985. Juvenile hormone cellular and hemolymph binding proteins. In G. A. Kerkut and L. I. Gilbert (eds.), *Comprehensive insect physiology, biochemistry and pharmacology*. Vol. 7, pp. 491–510. Pergamon Press.

Gorsberg, R. K. 1990. Out on a limb: arthropod origins. *Science* 250: 632–633.

Goss, J. 1903. Untersuchungen über die Histologie des Insektenovarium. *Zool. Jarhb. abt. Anat. Ontog.* 18: 71–186.

Goto, S., and S. Hayashi. 1997. Specification of the embryonic limb primordium by graded activity of Decapentaplegic. *Development* 124: 125–132.

———. 1999. Proximal to distal communication in the *Drosophila* leg provides a basis for an intercalary mechanism of limb patterning. *Development* 126: 3407–3413.

Gould, S. J. 1977. *Ontogeny and phylogeny*. Harvard University Press.

———. 1994. Common pathways of illumination. *Nat. Hist.* 103(12): 10–20.

———. 1997. As the worm turns. *Nat. Hist.* 107(1): 24–27, 68–75.

———. 1998. On embryos and ancestors. *Nat. Hist.* 107(6): 20–22, 58, 60–62, 64–65.

———. 2000. Abscheulich! (Atrocious!). *Nat. Hist.* 109(12): 42–49.

Govind, S., and R. Steward. 1991. Dorsoventral pattern formation in *Drosophila*. *Trends Genet.* 7: 119–125.

Graba, Y., D. Aragnol, and J. Pradel. 1997. *Drosophila* Hox complex, downstream targets and the function of homeotic genes. *Bioessays* 19: 379–388.

Graham, J. H., D. C. Freeman, and J. M. Emlen. 1993. Developmental stability: a sensitive indicator of populations under stress. In W. G. Landis, J. S. Hughes, and M. A. Lewis (eds.), *Environmental toxicology and risk assessment*, pp. 136–158. American Society for Testing and Materials.

Grandjean, F. 1938. Sur l'ontogénese des acariens. *C. R. Acad. Sci. Paris D* 206: 146–150.

Grant, V. 1963. *The origin of adaptations.* Columbia University Press.

Gray, J., and W. Shear. 1992. Early life on land. *Am. Sci.* 80: 442–456.

Grbíc, M., L. M. Nagy, S. B. Carroll, and M. Strand. 1996. Polyembryonic development: insect pattern formation in a cellularized environment. *Development* 122: 795–804.

Grbíc, M., L. M. Nagy, and M. R. Strand. 1998. Development of polyembryonic insects: a major departure from typical insect embryogenesis. *Dev. Genes Evol.* 208: 69–81.

Grbíc, M., D. Rivers, and M. R. Strand. 1997. Caste formation in the polyembryonic wasp, *Copidosoma floridanum* (Hymenoptera: Encyrtidae): *in vivo* and *in vitro* analysis. *J. Insect Physiol.* 43: 553–565.

Green, P., A. Y. Hartenstein, and V. Hartenstein. 1993. The embryonic development of the *Drosophila* visual system. *Cell Tissue Res.* 273: 583–598.

Greene, E. 1989. A diet-induced developmental polymorphism in a caterpillar. *Science* 243: 643–646.

Gregory, G. E. 1965. The formation and fate of the spermatophore in the African migratory locust, *Locusta migratoria migratorioides* Reiche and Fairmaire. *Trans. R. Entomol. Soc. Lond.* 117: 33–66.

Grenier, J. K., and S. B. Carroll. 2000. Functional evolution of the Ultrabithorax protein. *Proc. Natl. Acad. Sci. U.S.A.* 97: 704–709.

Grenier, J. K., T. L. Garber, R. Warren, P. M. Whitington, and S. Carroll. 1997. Evolution of the entire arthropod *Hox* gene complex predated the origin and radiation of the onychophoran/arthropod clade. *Curr. Biol.* 7: 547–553.

Gritzan, U., V. Hatini, and S. DiNardo. 1999. Mutual antagonism between signals secreted by adjacent Wingless and Engrailed cells leads to specification of complementary regions of the *Drosophila* parasegment. *Development* 126: 4107–4115.

Grodner, M. L., and W. L. Steffans. 1978. Evidence of a chemotactic substance in the spermathecal gland of the female boll weevil (Coleoptera: Curculionidae). *Trans. Am. Microsc. Soc.* 97: 116–120.

Grossniklauf, U., K. M. Cardigan, and W. J. Gehring. 1994. Three maternal coordinate systems co-operate in the patterning of the *Drosophila* head. *Development* 120: 3155–3171.

Grueber, W. B., and J. W. Truman. 1999. Development and organization of a nitric-oxide sensitive peripheral nerve plexus in larvae of the moth, *Manduca sexta. J. Comp. Neurol.* 404: 127–141.

Gullan, P. J., and P. S. Cranston. 1994. *The insects. An outline of entomology.* 1st ed. Chapman and Hall.

Günther, K. G. 1961. Funktionell-anatomische Untersuchung des männlichen Kopulations-apparates der Flöhe unter besonderer Berücksichtigung seiner postembryonalen Entwicklung (Siphonaptera). *Dtsch. Entomol. Z. N. F.* 8: 258–349.

Gura, T. 1997. Multiphoton imaging. Biologists get up close and personal with live cells. *Science* 276: 1988–1991.

Gurdon, J. B., S. Dyson, and D. T. St. Johnston. 1998. Cells' preception of position in a concentration gradient. *Cell* 95: 159–162.

Guss, K. A., C. E. Nelson, A. Hudson, M. E. Kraus, and S. B. Caroll. 2001. Control of a genetic regulatory network by a selector gene. *Science* 292: 1164–1167.

Gutzeit, H. O. 1985. The role of ecdysteroids in reproduction. In G. A. Kerkut and L. I. Gilbert (eds.), *Comprehensive insect physiology, biochemistry and pharmacology.* Vol. 1, pp. 205–262. Pergamon Press.

Gwynne, D. T., and L. W. Simmons. 1990. Experimental reversal of courtship roles in an insect. *Nature* 346: 172–173.

Haas, M. S., S. J. Brown, and R. W. Beeman. 2001a. Pondering the procephalon: the segmental origin of the labrum. *Dev. Genes Evol.* 211: 89–95.

——. 2001b. Homeotic evidence for the appendicular origin of the labrum in *Tribolium castaneum. Dev. Genes Evol.* 211: 96–102.

Haasen, B., and H. Vaessin. 1996. Regulatory interactions during early neurogenesis in *Drosophila. Dev. Genet.* 18: 12–27.

Hachlow, V. 1931. Zur Entwicklungsmechanik der Schmetterlinge. *W. Roux's Arch. Entwicklungsmech. Organ.* 125: 26–49.

Hackstein, J. H. P., H. Beck, R. Hochstenbach, H. Kremer, and H. Zaccharias. 1990. Spermatogenesis in *Drosophila hydei*: a genetic survey. *W. Roux's Arch. Dev. Biol.* 199: 251–288.

Hadfield, S. J., and J. M. Axton. 1999. Germ cells colonized by endosymbiotic bacteria. *Nature* 402: 482.

Haeckel, E. 1866. *Generelle Morphologie der Organismen: Allgemeine Grundzüge der organischen Formen-Wissenschaft, mechanisch begründet durch die von Charles Darwin reformirte Descendenz-Theorie.* 2 vols. G. Reimer.

——. 1905. *The evolution of man.* 2 vols. Transl. by J. McCabe from the 5th edition of *Anthropogenie.* Watts.

Haga, K. 1985. Oogenesis and embryogenesis of the idolothripine thrips, *Bactrothrips brevitubus* (Thysanoptera, Phlaeothripidae). In H. Ando and K. Miya (eds.), *Recent advances in insect embryology in Japan*, pp. 45–107. ISEBU.

Hagan, H. R. 1951. *Embryology of the viviparous insects.* Ronald Press.

Hagedorn, H. H. 1985. The role of ecdysteroids in reproduction. In G. A. Kerkut and L. I. Gilbert (eds.), *Comprehensive insect physiology, biochemistry and pharmacology.* Vol. 8, pp. 205–262. Pergamon Press.

Haget, A. 1969. Séparation expérimentale du soma et du derme de la gonade, par la transplantation des initiales germinales en position ecotopique, chez les embryons du Coléoptère *Leptinotarsa. C. R. Acad. Sci. Paris D* 269: 2226–2229.

——. 1977. L'Embryologie des Insectes. In P. P. Grassé (ed.), *Traité de zoologie.* Vol. VIII, fasc. 5B, pp. 1–262, 279–387. Masson et Cie.

Hähnlein, I., and G. Bicker. 1997. Glial patterning during postembryonic development of central neuropiles in the brain of the honeybee. *Dev. Genes Evol.* 207: 29–41.

Hakim, R. S. 1976. Structural changes occurring during transformation of labial gland cells to their adult form, in *Manduca sexta. J. Morphol.* 149: 547–566.

Halder, G., P. Callaerts, and W. J. Gehring. 1995. Induction of ectopic eyes by targeted expression of the *eyeless* gene in *Drosophila. Science* 267: 1788–1792.

Hales, D. F., and T. E. Mittler. 1983. Precocene causes male determination in the aphid *Myzus persicae. J. Insect Physiol.* 29: 819–823.

Hales, D. F., J. Tomiuk, K. Wärhmann, and P. Sunnucks. 1997. Evolutionary and genetic aspects of aphid biology: a review. *Eur. J. Entomol.* 94: 1–55.

Hall, B. K. 1992. *Evolutionary developmental biology.* Chapman and Hall.

——. 1998. Germ layers and the germ-layer theory revisited. *Evol. Biol.* 30: 121–186.

Hall, B. L., and C. S. Thummel. 1998. The R × R homologue of Ultraspiracle is an essential component of the *Drosophila* ecdysone receptor. *Development* 125: 4709–4717.

Hamilton, W. D., and G. M. H. Waites. 1990. *Cellular and molecular events in spermiogenesis.* Cambridge University Press.

Hammock, B. D. 1985. Regulation of juvenile hormone titre: degradation. In G. A. Kerkut and L. I. Gilbert (eds.), *Comprehensive insect physiology, biochemistry and pharmacology.* Vol. 7, pp. 431–472. Pergamon Press.

Handel, K., C. G. Grünfelder, S. Roth, and K. Sander. 2000. *Tribolium* embryogenesis: an SEM study of cell shapes and movements from blastoderm to serosal closure. *Dev. Genes Evol.* 210: 167–179.

Hanken, J., and D. B. Wake. 1993. Miniaturization of body size: organismal consequences and evolutionary significance. *Annu. Rev. Ecol. Syst.* 24: 501–519.

Hannon, G. M., and R. J. Hill. 1997. Cloning and characterization of LcEcR: a functional ecdysone receptor from the sheep blow fly *Lucilia cuprina. Insect Biochem. Mol. Biol.* 27: 479–488.

Hansen-Delkeskamp, E. 1969. Synthesis von RNS und Protein während der Oögenese und frühen Embryogenese von *Acheta domestica* L. *W. Roux's Arch. Entwicklungsmech. Organ.* 162: 114–120.

Hanström, B. 1938. Zwei Problems betreffs der hormalen Localization in Insektenkopf. *Acta. Univ. Lund N. F. Aut.* 2, 39: 1–17.

Hanton, W. K., R. D. Watson, and W. G. Bollenbacher. 1993. Ultrastructure of prothoracic glands during larval-pupal development of the tobacco hornworm, *Manduca sexta:* a reappraisal. *J. Morphol.* 216: 95–112.

Happ, G. M. 1992. Maturation of the male reproductive system and its endocrine regulation. *Annu. Rev. Entomol.* 37: 303–320.

Hardie, J. 1987. Juvenile hormone stimulation of oocyte development and embryogenesis in the parthenogenetic ovarioles of an aphid, *Aphis fabae. Int. J. Invert. Reprod. Dev.* 11: 189–202.

Hardie, J., and A. D. Lees. 1985. Endocrine control of polymorphism and polyphenism. In G. A. Kerkut and L. I. Gilbert (eds.), *Comprehensive insect physiology, biochemistry and pharmacology.* Vol. 8, pp. 441–490. Pergamon Press.

Hardy, R. W., K. T. Tokuyasu, D. L. Lindsley, and M. Gravito. 1979. The germinal proliferation center in the testis of *Drosophila melanogaster. J. Ultrastruct. Res.* 69: 180–190.

Harland, R. M. 2001. A twist on embryonic signalling. *Nature* 410: 423–424.

Harrat, A., S. Ihsan, and J. Schoeller-Raccoud. 1999. Development of the subesophageal body during embryogenesis without diapause in *Locusta migratoria* (Linnaeus) (Orthoptera: Acrididae). *Int. J. Insect Morphol. Embryol.* 28: 27–39.

Harris, W. A., and C. E. Holt. 1999. Slit, the midline repellant. *Nature* 398: 462–463.

Harshman, L. G., and A. A. James. 1998. Differential gene expression in insects: transcriptional control. *Annu. Rev. Entomol.* 43: 671–700.

Hartenstein, V. 1993. *Atlas of* Drosophila *development.* Published with M. Bate and A. Martinez Arias (eds.), *The development of* Drosophila melanogaster. Vol. 2. Cold Spring Harbor Laboratory Press.

Hartenstein, V., and Y. N. Jan. 1992. Studying *Drosophila* embryogenesis with P-lac Z enhancer trap lines. *W. Roux's Arch. Dev. Biol.* 201: 194–220.

Hartenstein, V., and J. W. Posakony. 1989. The development of adult sensilla on the wing and notum of *Drosophila melanogaster. Development* 107: 389–405.

Hartenstein, V., A. Lee, and A. W. Toga. 1995. A graphic digital database of *Drosophila* embryogenesis. *Trends Genet.* 11: 51–58.

Hartenstein, V., C. Nassif, and A. Lekeven. 1998. Embryonic development of the *Drosophila* brain. II. Pattern of glial cells. *J. Comp. Neurol.* 402: 32–47.

Hartenstein, V., G. M. Technau, and J. A. Campos-Ortega. 1985. Fate-mapping in wildtype *Drosophila melanogaster.* III. A fate map of the blastoderm. *W. Roux's Arch. Entwicklungsmech. Organ.* 194: 213–216.

Hartenstein, V., U. Tepass, and E. Gruszynski-de Feo. 1996. Proneural and neurogenic genes control specification and morphogenesis of stomatogastric nerve cell precursors in *Drosophila. Dev. Biol.* 173: 213–229.

Hartenstein, V., A. Younossi-Hartenstein, and A. Lekeven. 1994. Delamination and division in the *Drosophila* neuroectoderm: spatiotemporal pattern, cytoskeletal dynamics, and common control by neurogenic and segment polarity genes. *Dev. Biol.* 165: 480–499.

Hartl, D. L. 1971. Some aspects of natural selection in arrhenotokous populations. *Am. Zool.* 11: 309–315.

Harzsch, S., and D. Walossek. 2001. Neurogenesis in the developing visual system of the branchiopod crustacean *Triops longicaudatus* (LeConte, 1846): corresponding patterns of compound eye formation in Crustacea and Insecta? *Dev. Genes Evol.* 211: 37–43.

Hatakeyama, M., J. M. Lee, M. Sawa, and K. Oishi. 2000. Artificial reproduction in a hymenopteran insect, *Athalia rosae,* using eggs matured with heterospecific yolk proteins and fertilized with cryopreserved sperm. *Arch. Insect Biochem. Physiol.* 43: 137–144.

Haubruge, E., L. Arnaud, J. Mignon, and M. J. G. Gage. 1999. Fertilization by proxy: rival sperm removal and translocation in a beetle. *Proc. R. Soc. Lond. B Biol. Sci.* 266: 1183–1187.

Hayashi, F. 1999. Rapid evacuation of spermatophore contents and male post-mating behaviour in alderflies (Megaloptera: Sialidae). *Entomol. Sci.* 2: 49–56.

Hayashi, S., S. Herose, T. Metcalfe, and A. D. Shirras. 1993. Control of imaginal cell development by the *escargot* gene of *Drosophila. Development* 118: 105–115.

He, Y., T. Tanaka, and T. Miyata. 1995. Eupyrene and apyrene sperm and their numerical fluctuations inside the female reproductive tract of the armyworm, *Pseudaletia separata. J. Insect Physiol.* 41: 689–694.

Heald, R., R. Tournebize, T. Blank, R. Sandaltzopoulos, P. Becker, A. Hyman, and E. Karsenti. 1996. Self-organization of microtubules into bipolar spindles around artificial chromosomes. *Nature* 382: 420–425.

Hebda, C., J.-H. Yu, G. Bhaskaran, and K. H. Dahm. 1994. Reactivation of corpora allata in pharate adult *Manduca sexta. J. Insect Physiol.* 40: 849–858.

Heberlein, U., C. M. Singh, A. Y. Luk, and T. J. Donohoe. 1995. Growth and differentiation in the *Drosophila* eye coordinated by hedgehog. *Nature* 373: 709–711.

Hebert, P. D. N. 1987. Genotypic characteristics of cyclic parthenogens and the obligately asexual derivatives. In S. J. Stearns (ed.), *The evolution of sex and its consequences,* pp. 175–195. Birkhäuser Verlag.

Hecht, J. 1995. The geological timescale. Supplement in *New Scientist* 146 (May 20). Inside Science No. 81.

Hegner, R. W. 1908. The effects of removing the germ-cell determinants from the egg of some chrysomelid beetles. *Biol. Bull.* 16: 19–26.

Heitz, E., and H. Bauer. 1933. Beweise für die Chromosomen struktur der Kernschliefen in den Knäuelkernen von *Bibio hortulanus* L. (Cytologische Untersuchungen an Dipteren, I.). *Z. Zellforsch.* 17: 67–82.

Held, L. I. 1979. Pattern as a function of cell number and cell size on the second-leg basitarsus of *Drosophila*. *W. Roux's Arch. Dev. Biol.* 187: 105–127.

Held, L. I., Jr., M. A. Heup, J. M. Sappington, and S. D. Peters. 1994. Interactions of *decapentaplegic, wingless,* and *Distal-less* in the *Drosophila* leg. *W. Roux's Arch. Dev. Biol.* 203: 310–319.

Hellriegel, B., and G. Bernasconi. 2000. Female-mediated differential sperm storage in a fly with complex spermathecae, *Scatophaga stercoraria*. *Anim. Behav.* 59: 311–317.

Heming, B. S. 1970. Postembryonic development of the male reproductive system in *Frankliniella fusca* (Thripidae) and *Haplothrips verbasci* (Phlaeothripidae) (Thysanoptera). *Misc. Public. Entomol. Soc. Am.* 7(2): 235–272.

——. 1978. Structure and function of the mouthparts in larvae of *Haplothrips verbasci* (Osborn) (Thysanoptera, Tubulifera, Phlaeothripidae). *J. Morphol.* 156: 1–38.

——. 1979. Origin and fate of germ cells in male and female embryos of *Haplothrips verbasci* (Osborn) (Insecta, Thysanoptera, Phlaeothripidae). *J. Morphol.* 160: 323–343.

——. 1980. Development of the mouthparts in embryos of *Haplothrips verbasci* (Osborn) (Insecta, Thysanoptera, Phlaeothripidae). *J. Morphol.* 164: 235–263.

——. 1982. Structure and development of the larval visual system in embryos of *Lytta viridana* Leconte (Coleoptera, Meloidae). *J. Morphol.* 172: 23–43.

——. 1991. Order Thysanoptera. In F. W. Stehr (ed.), *Immature insects.* Vol. 2, pp. 1–21. Kendall/Hunt.

——. 1993a. Origin and fate of pleuropodia in embryos of *Neoheegeria verbasci* (Osborn) (Thysanoptera: Phlaeothripidae). In J. Bhatti (ed.), *Advances in Thysanopterology* (A volume in honour of the 80th birthday of Dr. A. Bournier). *Zoology (Delhi)* 4: 205–223.

——. 1993b. Structure, function, ontogeny and evolution of feeding in thrips (Thysanoptera). In C. W. Schaefer and R. A. B. Leschen (eds.), *Functional morphology of insect feeding,* pp. 3–41. Thomas Say Publications in Entomology. Entomological Society of America.

——. 1995. History of the germ line in male and female thrips (Thysanoptera). In B. L. Parker, M. Skinner, and T. Lewis (eds.), *Thrips biology and management,* pp. 505–535. Proc. 1993 Int. Conf. Thysanoptera. Towards Understanding Thrips Management, 28–30 Sept. 1993, Burlington, VT. Plenum Publishing.

——. 1996. Structure and development of larval antennae in embryos of *Lytta viridana* LeConte (Coleoptera: Meloidae). *Can. J. Zool.* 74: 1008–1034.

Heming, B. S., and E. Huebner. 1994. Development of the germ cells and reproductive primordia in male and female embryos of *Rhodnius prolixus* Stål (Hemiptera: Reduviidae). *Can. J. Zool.* 72: 1100–1119.

Heming-van Battum, K. E., and B. S. Heming. 1986. Structure, function and evolution of the reproductive system in females of *Hebrus pusillus* and *H. ruficeps* (Hemiptera, Gerromorpha, Hebridae). *J. Morphol.* 190: 121–167.

——. 1989. Structure, function and evolutionary significance of the reproductive system in males of *Hebrus ruficeps* and *H. pusillus* (Heteroptera, Gerromorpha, Hebridae). *J. Morphol.* 202: 281–323.

Henking, H. 1891. Üntersuchungen über die ersten Entwicklungsvorgänge in den Eiern der Insekten. *Z. Zool.* 51: 685–736.

Henneguy, L. F. 1904. *Les insectes. Morphologie, reproduction, embryogénie.* Masson et Cie.

Herth, W., and K. Sander. 1973. Mode and timing of body pattern formation (regionalization) in the early embryonic development of cyclorrhaphic dipterans (*Protophormia, Drosophila*). *W. Roux's Arch. Entwicklungsmech. Organ.* 172: 1–27.

Heslop-Harrison, G. 1958. On the origin and function of the pupal stadia in holometabolous insects. *Proc. Univ. Durham Phil. Soc. Ser. A* 13(8): 59–79.

Hewitt, G. M. 1979. *Orthoptera: grasshoppers and crickets. Animal cytogenetics.* Vol. 3. *Insecta. 1.* Gebrüder Borntraeger.

Higgie, M., S. Chenoweth, and M. W. Blows. 2000. Natural selection and the reinforcement of mate recognition. *Science* 290: 519–521.

Hillis, D. M., and C. Moritz (eds.). 1990. *Molecular systematics.* Sinauer Associates.

Hilversen, D. von, and O. von Hilversen. 1991. Pre-mating sperm removal in the bush cricket, *Metaplastes ornatus* Ramme 1931 (Orthoptera, Tettigonoidea, Phaneropteridae). *Behav. Ecol. Sociobiol.* 28: 391–396.

Hinds, M. J., and J. R. Linley. 1974. Changes in male potency with age after emergence in the fly, *Culicoides melleus*. *J. Insect Physiol.* 20: 1037–1040.

Hinton, H. E. 1948. On the origin and function of the pupal stage. *Trans. R. Entomol. Soc. Lond.* 99: 395–409.

——. 1955. On the structure, function and distribution of the prolegs of the Panorpoidea, with a criticism of the Berlese-Imms theory. *Trans. R. Entomol. Soc. Lond.* 106: 455–545.

——. 1963. The origin and function of the pupal stage. *Proc. R. Entomol. Soc. Lond.* 38: 77–85.

——. 1964. Sperm transfer in insects and the evolution of haemocoelic insemination. In K. C. Highnam (ed.), *Insect reproduction,* pp. 95–107. Symp. No. 2. Royal Entomology Society of London.

——. 1974. Accessory functions of seminal fluid. *J. Med. Entomol.* 11: 19–25.

——. 1981. *Biology of insect eggs.* Vols. 1–3. Pergamon Press.

Hinton, H. E., and I. M. MacKerras. 1970. Reproduction and metamorphosis. In I. M. Mackerras (ed.). *The insects of Australia. A textbook for students and research workers,* pp. 83–106. 1st ed. Melbourne University Press.

Hirata, J., H. Nakagoshi, Y. I. Nabeshima, and F. Matsuzaki. 1995. Asymmetric segregation of the homeodomain protein Prospero during *Drosophila* development. *Nature* 377: 627–630.

Hirth, F., S. Leuzinger, S. Therianos, T. Loop, W. J. Gehring, K. Furukubo-Tokunaga, and H. Reichert. 1996. Embryonic development of the *Drosophila* brain. In *Proceedings of the 20th International Congress of Entomology, Firenze,* p. 199.

Ho, K., O. M. Dunin-Berkowski, and M. Akam. 1997. Cellularization in locust embryos occurs before blastoderm formation. *Development* 124: 2761–2768.

Ho, R. K., E. E. Ball, and C. S. Goodman. 1983. Muscle pioneers: large mesodermal cells that erect a scaffold for developing muscles and motoneurons in grasshopper embryos. *Nature* 301: 66–69.

Hoage, T. R., and R. G. Kessel. 1968. An electron microscope study of the process of differentiation during spermatogenesis in the drone honey bee (*Apis mellifera* L.) with special reference to centriole replication and elimination. *J. Ultrastruct. Res.* 24: 6–32.

Hoch, M., K. Broadie, H. Jäckle, and H. Skaer. 1994. Sequential fates in a single cell are established by the neurogenic cascade in the Malpighian tubules of *Drosophila*. *Development* 120: 3439–3450.

Hodin, J., and L. M. Riddiford. 1998. The *ecdysone receptor* and *ultraspiracle* regulate the timing and progression of ovarian morphogenesis during *Drosophila* metamorphosis. *Dev. Genes Evol.* 208: 304–317.

——. 2000a. Different mechanisms underlie phenotypic plasticity and interspecific variation for reproductive characters in drosophilids (Insecta: Diptera). *Evolution* 54: 1638–1653.

——. 2000b. Parallel alterations in the timing of ovarian ecdysone receptor and Ultraspiracle expression characterizes the independent evolution of larval reproduction in two species of gall midges. *Dev. Genes Evol.* 210: 358–372.

Hofbauer, A., and J. A. Campos-Ortega. 1990. Proliferation pattern and early differentiation of the optic lobes in *Drosophila melanogaster*. *W. Roux's Arch. Dev. Biol.* 198: 264–274.

Hoffman, J. A., and M. Lagueux. 1985. Hormonal aspects of embryonic development. In G. A. Kerkut and L. I. Gilbert (eds.), *Comprehensive insect physiology, biochemistry and pharmacology*. Vol. 1, pp. 435–460. Pergamon Press.

Hoffman, J. A., F. C. Kafotis, C. A. Janeway Jr., and R. A. B. Ezekowitz. 1999. Phylogenetic perspectives in innate immunity. *Science* 284: 1313–1318.

Hoffmann, K. H., M. Meyering-Vos, and M. W. Lorenz. 1999. Allostatins and allatotropins: is the regulation of corpora allata activity their primary function? *Eur. J. Entomol.* 96: 255–266.

Holbrook, G. L., and C. Schal. 1998. Social influence on nymphal development in the cockroach, *Diploptera punctata*. *Physiol. Entomol.* 23: 121–130.

Holland, P. W. H., and J. Garcia-Fernández. 1996. Hox genes and chordate evolution. *Dev. Biol.* 173: 382–395.

Hölldobler, B., and E. O. Wilson. 1990. *The ants*. Belknap Press, Harvard University Press.

Holley, S. A., and E. L. Ferguson. 1997. Fish are like flies are like frogs: conservation of dorsal-ventral patterning mechanisms. *Bioessays* 19: 281–284.

Holley, S. A., P. D. Jackson, Y. Sasai, B. Lu, E. M. De Robertis, F. M. Hoffmann, and E. L. Ferguson. 1995. A conserved system for dorsal-ventral patterning in insects and vertebrates involving *sog* and *chordin*. *Nature* 376: 249–253.

Holmes, A. L., and J. S. Heilig. 2000. Fasciclin II and Beaten path modulate intercellular adhesion in *Drosophila* larval visual organ development. *Development* 127: 261–272.

Holt, G. G., and D. T. North. 1970. Effects of gamma radiation on the mechanisms of sperm transfer in *Trichoplusia ni*. *J. Insect Physiol.* 16: 2211–2222.

Honegger, H. W., B. Seibel, U. Kaltenhauser, and B. Bräunwig. 1992. Expression of Bursicon-like activity during embryogenesis of the locust *Schistocerca gregaria*. *J. Insect Physiol.* 38: 981–986.

Hopkins, T. R., and K. J. Kramer. 1992. Insect cuticle sclerotization. *Annu. Rev. Entomol.* 37: 273–302.

Horn, D. H. S., and R. Bergamasco. 1985. Chemistry of ecdysteroids. In G. A. Kerkut and L. I. Gilbert (eds.), *Comprehensive insect physiology, biochemistry and pharmacology*. Vol. 7, pp. 185–248. Pergamon Press.

Horodyski, F. M. 1996. Neuroendocrine control of insect ecdysis by eclosion hormone. *J. Insect Physiol.* 42: 917–924.

Horodyski, F. M., L. M. Riddiford, and J. W. Truman. 1989. Isolation and expression of the eclosion hormone gene from the tobacco hornworm, *Manduca sexta*. *Proc. Natl. Acad. Sci. U.S.A.* 86: 8123–8127.

Horsfall, W. R., and J. F. Anderson. 1961. Suppression of male characteristics of mosquitoes by thermal means. *Science* 133: 1830.

Horsfall, W. R., and M. C. Ronquillo. 1970. Genesis of the reproductive system of mosquitoes. II. Male of *Aedes stimulans* (Walker). *J. Morphol.* 131: 329–357.

Horsfall, W. R., M. C. Ronquillo, and W. J. Patterson. 1972. Genesis of the reproductive system of mosquitoes. IV: Thermal modification of *Aedes stimulans*. *Isr. J. Entomol.* 7: 73–84.

Howe, R. W. 1967. Temperature effects on embryonic development in insects. *Annu. Rev. Entomol.* 12: 15–42.

Hsiao, T. H., C. Hsiao, and J. De Wilde. 1975. Moulting hormone production in the isolated larval abdomen of the Colorado potato beetle. *Nature* 255: 727–728.

Hu, N., and J. Castelli-Gair. 1999. Study of the posterior spiracles of *Drosophila* as a model to understand the genetic and cellular mechanisms controlling morphogenesis. *Dev. Biol.* 214: 197–210.

Hua, Y.-F., R.-J. Jiang, and J. Koolman. 1997. Multiple control of ecdysone biosynthesis in blow fly larvae: interaction of ecdysterotropins and ecdysterostatins. *Arch. Insect Biochem Physiol.* 35: 125–134.

Huang, Z., and S. Kunes. 1998. Signals transmitted along retinal axons in *Drosophila*: hedgehog signal reception and the cell circuitry of lamina cartridge assembly. *Development* 125: 3753–3764.

Huber, R., and W. Hoppe. 1965. Die Kristall- und Molekülstrukturanalyse die Insektenverpuppungs hormon Ecdyson mit der automatisierten Factmolekülmethode. *Chem. Ber.* 98: 2403–2424.

Huebner, E. 1984. The ultrastructure and development of the telotrophic ovariole. In R. C. King and H. Akai (eds.), *Insect ultrastructure*. Vol. 2, pp. 3–48. Plenum Press.

Huebner, E., and W. Diehl-Jones. 1993. Nurse-cell oocyte interaction in the telotrophic ovariole. *Int. J. Insect Morphol. Embryol.* 22: 369–387.

——. 1998. Developmental biology of insect ovaries: germ cells and nurse cell-oocyte polarity. In F. W. Harrison and M. Locke (eds.), *Microscopic anatomy of invertebrates*. Vol. 11C. Insecta, pp. 957–993. Wiley-Liss.

Huet, C., and J. J. Lenoir-Rousseaux. 1976. Étude de la mise en place de la patte imaginale de *Tenebrio molitor*. 1. Analyse expérimentale des processus de restauration au cours de la morphogenèse. *J. Embryol. Exp. Morphol.* 35: 303–321.

Hughes, C. L., and T. C. Kaufman. 2000. RNAi analysis of *Deformed, proboscipedia*, and *Sex combs reduced* in the milkweed bug *Oncopeltus fasciatus*: novel roles for Hox genes in the hemipteran head. *Development* 127: 3683–3694.

Hughes, N. F., and J. Smart. 1967. Plant-insect relationships in the Paleozoic and later time. In S. B. Harland, C. H. Holland, M. R. House, N. F. Hughes, and A. B. Reynolds (eds.), *The fossil record*. pp. 107–117. Geological Society of London.

Hughes, T. D. 1980. The imaginal ecdysis of the desert locust, *Schistocerca gregaria*. IV. The role of the gut. *Physiol. Entomol.* 5: 153–164.

Hughes-Schrader, S. 1927. Origin and differentiation of the male and female germ cells in the hermaphrodite of *Icerya purchasi* (Coccidae). *Z. Zellforsch.* 6: 509–540.

Hunter, J. P. 1998. Key innovations and the ecology of macroevolution. *Trends Ecol. Evol.* 13: 281–282.

Hurst, L. D. 1992. Intragenomic conflict as an evolutionary force. *Proc. R. Soc. Lond. B Biol. Sci.* 248: 135–140.

Huxley, J. S. 1932. *Problems of relative growth*. Methuen.

Hyams, J. 1996. Look ma, no chromosomes. *Nature* 382: 397–398.

Hyams, J. S., and H. Stebbings. 1977. The distribution and function of microtubules in nutritive tubes. *Tissue Cell.* 9: 537–545.

Ikeda, Y., and R. Machida. 1998. Embryogenesis of the dipluran *Lepidocampa weberi* Oudemans (Hexapoda, Diplura, Campodcidac): external morphology. *J. Morphol.* 237: 101–115.

Ikeshima-Kataoka, H., J. B. Skeath, Y.-I. Nabeshima, C. Q. Doe, and F. Matsuzaki. 1997. Miranda directs Prospero to a daughter cell during *Drosophila* asymmetric divisions. *Nature* 390: 625–629.

Illmenscc, K. 1972. Developmental potencies of nuclei from cleavage, preblastoderm, and syncytial blastoderm transplanted into unfertilized eggs of *Drosophila melanogaster*. *W. Roux's Arch. Entwicklungsmech. Organ.* 170: 267–298.

Illmensee, K., and A. P. Mahowald. 1974. Transplantation of posterior polar plasm in *Drosophila*. Induction of germ cells in the anterior pole of the egg. *Proc. Natl. Acad. Sci. U.S.A.* 71: 1016–1020.

Ingham, P. W. 1988. The molecular genetics of embryonic pattern formation in *Drosophila*. *Nature* 335: 25–34.

Irvine, K. D., and E. Wieschaus. 1994. Cell intercalation during *Drosophila* germband extension and its regulation by pair-rule segmentation genes. *Development* 120: 827–841.

Ito, H. 1918. On the glandular nature of the corpora allata of the Lepidoptera. *Bull. Imp. Tokyo Ser. Coll.* 1: 64–108.

Ito, K., J. Urban, and G. M. Technau. 1995. Distribution, classification, and development of *Drosophila* glial cells in the late embryonic and early larval ventral nerve cord. *W. Roux's Arch. Dev. Biol.* 204: 284–307.

Ito, K., W. Awano, K. Suzuki, Y. Hiromi, and D. Yamamoto. 1997. The *Drosophila* mushroom body is a quadruple structure of clonal units each of which contains a virtually identical set of neurones and glial cells. *Development* 124: 761–777.

Ivanova-Kasas, O. M. 1972. Polyembryony in insects. In S. J. Counce and C. H. Waddington (eds.), *Developmental systems: insects.* Vol. 1, pp. 243–271. Academic Press.

Izunni, S., K. Yano, Y. Yamamoto, and S. Y. Takahashi. 1994. Yolk protein from insect eggs: structure, biosynthesis and programmed degradation during embryogenesis. *J. Insect Physiol.* 40: 735–746.

Jablonka, E., and M. J. Lamb. 1990. The evolution of heteromorphic sex chromosomes. *Biol. Rev.* 65: 249–276.

Jacobson, M. D., M. Weil, and M. C. Raff. 1997. Programmed cell death in animal development. *Cell* 88: 347–354.

Jagla, K., M. Frasch, T. Jagla, G. Dretzen, F. Bellard, and M. Bellard. 1997. *lady-bird*, a new component of the cardiogenic pathway in *Drosophila* required for diversification of heart precursors. *Development* 124: 34–79.

Jaglarz, M. K., and K. R. Howard. 1994. Primordial germ cell migration in *Drosophila melanogaster* is controlled by somatic tissue. *Development* 120: 83–89.

Jamieson, B. G. M. 1987. *The ultrastructure and phylogeny of insect sperm.* Cambridge University Press.

Jamieson, B. G. M., R. Dallai, and B. A. Afzelius. 1999. *Insects. Their sperm and phylogeny.* Science Publications.

Jan, N. J., and L. Y. Jan. 1993. The peripheral nervous system. In M. Bate and A. Martinez Arias (eds.), *The development of* Drosophila melanogaster. Vol. 2, pp. 1207–1244. Cold Spring Harbor Laboratory Press.

——. 2000. Polarity in cell division: what frames thy fearful asymmetry? *Cell* 100: 599–602.

Janet, C. 1899. Sur les nerfs cephaliques, les corpora allata et le tentorium de la fourmi *Myrmica rubra*. *Mém. Soc. Zool. Fr.* 12: 295–335.

Jang, J. K., L. Messina, M. B. Erdmann, T. Abel, and R. S. Hawley. 1995. Induction of metaphase arrest in *Drosophila* oocytes by chiasma-based kinetochore tension. *Science* 268: 1917–1919.

Jarecki, J., E. Johnson, and M. A. Krasnow. 1999. Oxygen regulation of airway branching in *Drosophila* is mediated by Branchless FGF. *Cell* 99: 211–220.

Jarmen, A. P. 1996. Epithelial polarity in the *Drosophila* compound eye: eyes left or right? *Trends Genet.* 12: 121–123.

Jenner, R. A., and F. R. Schram. 1999. The grand game of metazoan phylogeny: rules and strategies. *Biol. Rev.* 74: 121–142.

Jeschikov, J. J. 1941. Die Dottermenge im Ei und die Typen der Postembryonalen Entwicklung bei den Insekten. *Zool. Anz.* 134: 71–87.

Jiang, C., E. H. Baehrecke, and C. S. Thummel. 1997. Steroid regulated programmed cell death during *Drosophila* metamorphosis. *Development* 124: 4673–4683.

Jiménez, G., A. Guickst, A. Ephrussi, and J. Casanova. 2000. Relief of gene expression by Torso RTK signaling: role of *capicua* in *Drosophila* terminal and dorsoventral patterning. *Genes Dev.* 14: 224–231.

Jindra, M. 1994. Gene regulation by steroid hormones: vertebrates and insects. *Eur. J. Entomol.* 91: 163–187.

Jockusch, E. L., C. Nulsen, S. J. Newfeld, and L. M. Nagy. 2000. Leg development in flies versus grasshoppers: differences in *dpp* expression do not lead to differences in the expression of downstream components of the leg patterning pathway. *Development* 127: 1617–1626.

Johannsen, O. A., and F. H. Butt. 1941. *Embryology of insects and myriapods.* McGraw-Hill.

John, B. 1990. *Meiosis.* Cambridge University Press.

Johnson, E., and J. S. Berry. 1977. Cellular differentiation of a highly specialized insect secretory organ. *Differentiation* 8: 39–52.

Johnson, R. L., and C. J. Tabin. 1997. Molecular models for vertebrate limb development. *Cell* 90: 979–990.

Joly, D., C. Bressac, and D. Lachaise. 1995. Disentangling giant sperm. *Nature* 277: 202.

Joly, P. 1977. Le développement postembryonnaire des insectes. In P. P. Grasse (ed.), *Traité de Zoologie.* Vol.8-5A, pp. 409–637. Masson et Cie.

Jones, B. M. 1956a. Endocrine activity during insect embryogenesis. Function of the ventral head glands in locust embryos (*Locustana pardalina* and *Locusta migratoria*). *J. Exp. Biol.* 33: 174–185.

——. 1956b. Endocrine activity during insect embryogenesis. Control of events in development following the embryonic moult (*Locusta migratoria*, *Locustana pardalina*, Orthoptera). *J. Exp. Biol.* 33: 684–696.

Jones, B. W., R. D. Fetter, G. Tear, and C. S. Goodman. 1995. *glial cells missing*: a genetic switch that controls glial versus neuronal fate. *Cell* 82: 1013–1023.

Jones, G. 1995. Molecular mechanisms of action of juvenile hormone. *Annu. Rev. Entomol.* 40: 147–169.

Jones, J. C., and R. E. Wheeler. 1965. An analytical study of coitus in *Aedes aegypti* (Linnaeus). *J. Morphol.* 117: 401–424.

Jones, M. L., and B. S. Heming. 1979. Effects of temperature and relative humidity on embryogenesis in eggs of *Mamestra configurata* (Walker) (Lepidoptera: Noctuidae). *Quaest. Entomol.* 15: 257–294.

Jones, N. A., Y. M. Kuo, Y. H. Sun, and S. K. Beckendorf. 1998. The *Drosophila* Pax gene *eye gone* is required for embryonic salivary duct development. *Development* 125: 4163–4174.

Jones, R. T. 1978. The blood/germ cell barrier in male *Schistocerca gregaria*:

the time of its establishment and factors affecting its formation. *J. Cell Sci.* 31: 145–163.

Juberthie-Jupeau, L., J. Durand, and M. Cazals. 1983. Spermatogenèse comparée chez les coleoptères Bathysciinae souterrains. *Cytobios* 37: 187–208.

Junker, H. 1923. Cytologische Üntersuchungen an den Geschlechtsorgane der halbzwitteringen Steinfliege, *Perla marginata* (Panzer). *Arch. Zellforsch.* 17: 185–259.

Jura, C. 1972. Development of apterygote insects. In S. J. Counce and C. H. Waddington (eds.), *Developmental systems: insects*. Vol. 1, pp. 49–94. Academic Press.

Jura, C., and A. Krgyslofowicz. 1992. Initiation of embryonic development in *Tetrodontophora bielanensis* (Waga) (Collembola) eggs: meiosis, polyspermy, union of gametes and first cleavage. *Int. J. Insect Morphol. Embryol.* 21: 87–94.

Jürgens, G., and V. Hartenstein. 1993. The terminal regions of the body pattern. In C. M. Bate and A. Martinez Arias (eds.), *The development of* Drosophila melanogaster. Vol. 1, pp. 687–746. Cold Spring Harbor Laboratory Press.

Jürgens, G., R. Lehmann, M. Schardin, and C. Nüsslein-Volhard. 1996. Segmental organisation of the head in the embryo of *Drosophila melanogaster*. A blastoderm fate map of the cuticle structures of the larval head. *W. Roux's Arch. Dev. Biol.* 195: 359–377.

Kaltenhauser, U., J. Kellermann, K. Andersson, F. Lottspeich, and H. W. Honegger. 1995. Purification and partial characterization of bursicon, a cuticle sclerotizing neuropeptide in insects, from *Tenebrio molitor. Insect Biochem. Molec. Biol.* 25: 525–533.

Kalthoff, K. 1983. Cytoplasmic determinants in dipteran eggs. In W. R. Jeffrey and R. A. Raff (eds.), *Time, space and pattern in embryonic development*, pp. 313–348. Alan R. Liss.

Kalthoff, K., K.-G. Rau, and J. C. Edmond. 1982. Modifying effects of ultraviolet irradiation on the development of abnormal body patterns in centrifuged insect embryos (*Smittia* sp., Chironomidae, Diptera). *Dev. Biol.* 91: 413–422.

Kambysellis, M. P. 1968. Interspecific transplantation as a tool for indicating phylogenetic relationships. *Proc. Natl. Acad. Sci. U.S.A.* 59: 1166–1172.

———. 1970. Compatibility in insect tissue transplantations. I. Ovarian transplantations between *Drosophila* species endemic to Hawaii. *J. Exp. Zool.* 175: 169–180.

Kambysellis, M. P., and C. M. Williams. 1971a. *In vitro* development of insect tissues. I. A macromolecular factor prerequisite for silkworm spermatogenesis. *Biol. Bull.* 141: 527–540.

———. 1971b. *In vitro* development of insect tissues. II. The role of ecdysone in the spermatogenesis of silkworms. *Biol. Bull.* 141: 541–552.

Kandler-Singer, I., and K. Kalthoff, 1976. RNase sensitivity of an anterior morphogenetic determinant in an insect egg (*Smittia* spec. Chironomidae, Diptera). *Proc. Natl. Acad. Sci. U.S.A.* 73: 3739–3743.

Karim, F. D., G. M. Guild, and C. S. Thummel. 1993. The *Drosophila Broad Complex* plays a key role in controlling ecdysone-regulated gene expression at the onset of metamorphosis. *Development* 118: 977–988.

Karpen, G. H., M.-H. Le, and H. Le. 1996. Centric heterochromatin and the efficiency of achiasmatic disjunction in *Drosophila* female meiosis. *Science* 273: 118–122.

Karr, T. L. 1991. Intracellular sperm-egg interactions in *Drosophila*: a three-dimensional structural analysis of a paternal product in the developing egg. *Mech. Dev.* 34: 101–112.

———. 1996. Paternal investment and intracellular sperm-egg interactions during and following fertilization in *Drosophila. Curr. Top. Dev. Biol.* 34: 89–115.

———. 2001. Centrosome inheritance: a central "in-egg-ma." *Curr. Biol.* 11: 221–224.

Karr, T. L., and S. Pitnick. 1996. The ins and outs of fertilization. *Nature* 379: 405–406.

Kataoka, H., A. Toschi, J. P. Li, R. L. Carney, D. A. Schooley, and S. J. Kramer. 1989. Identification of an allatotropin from adult *Manduca sexta. Science* 243: 1481–1483.

Kathirithamby, J. 1991. Strepsiptera. In CSIRO (ed.), *The insects of Australia. A textbook for students and research workers.* 2d ed. Vol. 2, pp. 684–695. Melbourne University Press and Cornell University Press.

Katz, M. J. 1980. Allometry formula: a cellular model. *Growth* 44: 89–96.

Kauffman, S. A. 1981. Pattern formation in the *Drosophila* embryo. *Philos. Trans. R. Soc. Lond. B Biol. Sci.* 295: 567–594.

Kauffman, S. A., R. Shymko, and K. Trabert. 1978. Control of sequential compartment formation in *Drosophila. Science* 199: 259–270.

Kaufman, T. C., M. A. Seeger, and G. Olsen, 1990. Molecular and genetic organization of the Antennapedia gene complex of *Drosophila melanogaster. Adv. Genet.* 27: 309–362.

Kaulenas, M. S. 1992. *Insect accessory reproductive structures. Function, structure, development.* Springer-Verlag.

Kawakami, A., H. Kataoka, T. Oka, A. Mizoguchi, M. Kimura-Kawakami, T. Adachi, M. Iwami, H. Nagawawa, A. Suzuki, and H. Ishizaki. 1990. Molecular cloning of the *Bombyx mori* prothoracicotropic hormone. *Science* 247: 1333–1335.

Kawamura, N., N. Yamashiki, H. Satitoh, and K. Sahara. 2000. Peristaltic squeezing of sperm bundles in the late stage of spermatogenesis in the silkworm, *Bombyx mori. J. Morphol.* 246: 53–58.

Keil, T. A. 1997. Comparative morphogenesis of sensilla: a review. *Int. J. Insect Morphol. Embryol.* 26: 151–160.

Keil, T. A., and C. Steiner. 1990. Morphogenesis of the antenna of the male silkmoth *Antheraea polyphemus*, II. Differential mitoses of "dark" precursor cells create the anlagen of sensilla. *Tissue Cell.* 22: 705–720.

———. 1991. Morphogenesis of the antenna of the male silkmoth *Antheraea polyphemus*, III. Development of olfactory sensilla and the properties of hair-forming cells. *Tissue Cell* 23: 821–851.

Kenchington, W. 1969. The hatching thread of praying mantids: an unusual chitinous structure. *J. Morphol.* 129: 307–316.

Kenrick, P., and P. R. Crane. 1997. The origin and early evolution of plants on land. *Nature* 389: 33–39.

Kenyon, C. 1994. If birds can fly why can't we? Homeotic genes and evolution. *Cell* 78: 175–180.

Kerr, R. A. 1995. Embryos give clues to early animal evolution. *Science* 270: 1330–1331.

Keshishian, H. 1980. The origin and morphogenesis of pioneer neurons in the grasshopper metathoracic leg. *Dev. Biol.* 80: 388–397.

Kessel, R. G. 1961. Cytological studies of the sub-esophageal body cells and pericardial cells in the embryo of the grasshopper *Melanoplus differentialis differentialis* (Thomas). *J. Morphol.* 109: 289–322.

———. 1966. The association between microtubules and nuclei during spermiogenesis in the dragonfly. *J. Ultrastruct. Res.* 16: 293–304.

Keys, D. N., D. L. Lewis, J. E. Selegue, B. J. Pearson, L. V. Goodrich, R. J.

Johnson, J. Gates, M. P. Scott, and S. B. Carroll. 1999. Recruitment of a *hedgehog* regulatory circuit in butterfly eyespot evolution. *Science* 283: 532–534.

Kiger, A. A., H. White-Cooper, and M. T. Fuller. 2000. Somatic support cells restrict germline stem cell self-renewal and promote differentiation. *Nature* 407: 750–754.

Kim, J., J. Q. Kerr, and G.-S. Min. 2000. Molecular heterochrony in the early development of *Drosophila*. *Proc. Natl. Acad. Sci. U.S.A.* 97: 212–216.

Kimmel, C. B. 1996. Was Urbilateria segmented? *Trends Genet.* 12: 329–333.

King, D. S. 1972. Ecdysone metabolism in insects. *Am. Zool.* 12: 343–345.

King, F. J., and H. Lin. 1999. Somatic signaling mediated by fs(1)yb is essential for germline stem cell maintenance during *Drosophila* oogenesis. *Development* 126: 1833–1844.

King, R. C. 1970. *Ovarian development in* Drosophila melanogaster. Academic Press.

King, R. C., S. K. Aggarwal, and U. Aggarwal. 1968. The development of the female *Drosophila* reproductive system. *J. Morphol.* 124: 143–166.

King, R. C., J. D. Cassidy, and A. Rousset. 1982. The formation of clones of interconnected cells during gametogenesis in insects. In R. C. King and H. Akai (eds.), *Insect ultrastructure*. Vol. 1, pp. 3–31. Plenum Press.

King, R. W., R. J. Deshaies, J.-M. Peters, and M. W. Kirschner. 1996. How proteolysis drives the cell cycle. *Science* 274: 1652–1659.

Kingan, T. G., and M. E. Adams. 2000. Ecdysteroids regulate secretory competence in INKA cells. *J. Exp. Biol.* 203: 3011–3018.

Klag, J., and S. Bilinski. 1993. Öosome formation in two ichneumonid wasps. *Tissue Cell* 25: 121–128.

Klämbt, C. 1993. The *Drosophila* gene *pointed* encodes two ETS-like proteins which are involved in the development of the midline glial cells. *Development* 117: 163–176.

Klämbt, C., T. Hummel, T. Menne, E. Sadlowski, H. Scholz, and A. Stollwerk. 1996. Development and function of embryonic central nervous system glial cells in *Drosophila*. *Dev. Genet.* 18: 40–49.

Klass, K.-D., and N. P. Kristensen. 2001. The ground plan and affinities of hexapods: recent progress and open problems. *Ann. Soc. Entomol. Fr. (N. S.)* 37(1/2): 265–298.

Kleine-Schonnefeld, H., and W. Engels. 1981. Symmetrical pattern of follicle arrangement in the ovary of *Musca domestica* (Insecta, Diptera). *Zoomorphology* 98: 185–190.

Klingenberg, C. P. 1996a. Multivariate allometry. In L. F. Marcus et al. (eds.), *Advances in morphometrics*, pp. 23–49. Plenum Press.

——. 1996b. Individual variation of ontogenies: a longitudinal study of growth and timing. *Evolution* 50: 2412–2428.

——. 1998. Heterochrony and allometry: the analysis of evolutionary change in ontogeny. *Biol. Rev.* 73: 79–123.

Klingenberg, C. P., and H. F. Nijhout. 1998. Competition among growing organs and developmental control of morphological asymmetry. *Proc. R. Soc. Lond. B Biol. Sci.* 265: 1135–1139.

——. 1999. Genetics of fluctuating asymmetry: a developmental model of developmental instability. *Evolution* 53: 358–375.

Klingenberg, C. P., and J. R. Spence. 1993. Heterochrony and allometry: lessons from the waterstrider genus *Limnoporus*. *Evolution* 47: 1834–1853.

——. 1997. On the role of body size for life history evolution. *Ecol. Entomol.* 22: 55–68.

Klingenberg, C. P., and M. Zimmermann. 1992a. Dyar's rule and multivariate allometric growth in nine species of waterstriders (Heteroptera: Gerridae). *J. Zool. (Lond.)* 227: 453–464.

——. 1992b. Static, ontogenetic, and evolutionary allometry: a multivariate comparison in nine species of water striders. *Am. Nat.* 140: 601–620.

Klingenberg, C. P., G. S. McIntyre, and S. D. Zaklan. 1998. Left-right asymmetry of fly wings and the evolution of body axes. *Proc. R. Soc. Lond. B Biol. Sci.* 265: 1255–1259.

Klingler, M., and D. Tautz. 1999. Formation of embryonic axes and blastoderm pattern in *Drosophila*. In V. E. A. Russo, D. J. Cove, L. G. Edgar, R. Jaenisch, and F. Salamini (eds.), *Development. Genetics, epigenetics and environmental regulation*, pp. 311–330. Springer-Verlag.

Klose, M., and D. Bentley. 1989. Transient pioneer neurons are essential for formation of an embryonic peripheral nerve. *Science* 245: 982–984.

Knoll, A. H., and S. B. Carroll. 1999. Early animal evolution: emerging views from comparative biology and geology. *Science* 284: 2129–2137.

Kobayashi, M., and H. Ishikawa. 1993. Breakdown of indirect flight muscles of alate aphids (*Acyrthosiphon pisum*) in relation to their flight, feeding and reproductive behavior. *J. Insect Physiol.* 39: 549–554.

Kobayashi, Y., and H. Ando. 1988. Phylogenetic relationships among the lepidopteran and trichopteran suborders (Insecta) from the embryological standpoint. *Z. Zool. Syst. Evol. Forsch.* 26: 186–210.

Koch, M. 2001. Mandibular mechanisms and the evolution of hexapods. *Ann. Soc. Entomol. Fr. (N. S.)* 37(1/2): 129–174.

Koch, P., L. P. Pijnacker, and J. Kreke. 1972. DNA reduplication during meiotic prophase in the oocytes of *Carausius morosus* Br. (Insecta, Cheleutoptera). *Chromosoma* 36: 313–321.

Koelle, M. R., W. S. Talbot, W. A. Segraves, M. T. Bender, P. Cherbas, and D. S. Hogness. 1991. The *Drosophila EcR* gene encodes an ecdysone receptor, a new member of the steroid receptor superfamily. *Cell* 67: 59–77.

Kojima, T., M. Sato, and K. Saigo. 2000. Formation and specification of distal leg segments in *Drosophila* by dual *Bar* homeobox genes. *Development* 127: 769–778.

Koolman, J., and P. Karlson. 1985. Regulation of ecdysteroid titre: degradation. In G. A. Kerkut and L. I. Gilbert (eds.), *Comprehensive insect physiology, biochemistry and pharmacology*. Vol. 7, pp. 343–361. Pergamon Press.

Koolman, J., K. Scheller, and D. Bodenstein. 1979. Ecdysteroids in the adult male blowfly, *Calliphora vicina*. *Experientia* 35: 134–135.

Kopéc, S. 1922. Studies on the necessity of the brain for the inception of insect metamorphosis. *Biol. Bull.* 42: 323–342.

Kopp, A., R. K. Blackman, and I. Duncan. 1999. Wingless, Decapentaplegic and EGF receptor signaling pathways interact to specify dorso-ventral pattern in the adult abdomen of *Drosophila*. *Development* 126: 3495–3507.

Korschelt, E., and K. Heider. 1899. *Textbook of the embryology of invertebrates*. Vol. III. Swan, Sonnenschein.

Kostron, B. K., K. Marquardt, and H. W. Honegger. 1995. Bursicon: the cuticle sclerotizing hormone—comparison of its molecular mass in different insects. *J. Insect Physiol.* 41: 1045–1053.

Kramer, S. J., A. Toschi, C. A. Miller, H. Kataoka, G. B. Quistad, J. P. Li, R. L. Carney, and D. A. Schooley. 1991. Identification of an allatostatin from the tobacco hornworm, *Manduca sexta*. *Proc. Natl. Acad. Sci. U.S.A.* 88: 9458–9462.

Krause, G. 1939. Die Eitypen der Insekten. *Biol. Zentralbl.* 59: 495–536.

———. 1953. Die Aktionsfolge zur Gestaltung des Keimstreifs von *Tachycines* (Saltatoria), ins besondere das morphogenetische Konstruktionsbild bei *Duplicatas parallela*. *W. Roux's Arch. Entwicklungsmech. Organ.* 146: 275–370.

Kristensen, N. P. 1984. Studies on the morphology and systematics of primitive Lepidoptera (Insecta). *Steenstrupia* 10: 141–191.

———. 1991. Phylogeny of extant hexapods. In CSIRO (ed.), *The insects of Australia. A textbook for students and research workers*. 2d ed. Vol. 1, pp. 125–140. Melbourne University Press and Cornell University Press.

———. 1999. Phylogeny of endopterygote insects, the most successful lineage of living organisms. *Eur. J. Entomol.* 96: 237–253.

Krumlauf, R., and A. Gould. 1992. Homeobox cooperativity. *Trends Genet.* 8: 297–300.

Kubrakiewicz, J. 1997. Germ cell cluster organization in polytrophic ovaries of Neuroptera. *Tissue Cell* 29: 221–228.

Kubrakiewicz, J., and S. M. Bilinski. 1995. Extrachromosomal amplification of rDNA in oocytes of *Hemerobius* spp. (Insecta, Neuroptera). *Chromosoma* 103: 606–612.

Kühn, A. 1971. *Lectures on developmental physiology*. Transl. by Roger Milkman. Springer-Verlag.

Kukalová-Peck, J. 1978. Origin and evolution of insect wings and their relation to metamorphosis, as documented by the fossil record. *J. Morphol.* 156: 53–126.

———. 1987. New carboniferous Diplura, Monura and Thysanura, the hexapod ground plan and the role of the thoracic side lobes in the origin of wings (Insecta). *Can. J. Zool.* 65: 2327–2345.

———. 1991. Fossil history and the evolution of hexapod structures. In CSIRO (ed.), *The insects of Australia. A textbook for students and research workers*. 2d ed. Vol. 1, pp. 141–179. Melbourne University Press.

Künhe, H. 1972. Entwicklungsablauf und -stadien von *Micromalthus debilis* Le Conte (Col., Micromalthidae) aus einer Laboratoriumspopulation. *Z. Angew. Entomol.* 72: 157–168.

Künhe, H., and G. Becker. 1976. Zur Biologie und Ökologie von *Micromalthus debilis* Le Conte (Col., Micromalthidae). *Mat. Org.* 3: 447–461.

Kunkel, J. G. 1975. Cockroach molting. I. Temporal organization of events during molting cycle of *Blattella germanica* (L.). *Biol. Bull.* 148: 259–273.

Kunz, W., and K. H. Glätzer, 1980. Chromosome structure and function during gametogenesis, with special consideration of insect oogenesis and spermatogenesis. In R. L. Blackman, G. M. Hewitt, and M. Ashburner (eds.), *Insect cytogenetics*, pp. 51–64. Symp. R. Entomol. Soc. Lond. 10. Blackwell Science.

Kuo, Y. M., N. Jones, B. Zhou, S. Panzer, V. Larson, and S. K. Beckendorf. 1996. Salivary duct determination in *Drosophila*: roles of the EGF receptor signalling pathway and the transcription factors Fork head and Trachealess. *Development* 122: 1909–1917.

Kuroda, M. I., and A. M. Villeneuve. 1996. Promiscuous chromosomal proteins: complexes about sex. *Science* 274: 1633–1634.

Kurusu, M., T. Nagao, U. Walldorf, S. Flister, W. J. Gehring, and K. Furukubo-Tokunaga. 2000. Genetic control of development of the mushroom bodies, the associative learning centers in the *Drosophila* brain, by *eyeless*, *twin-of-eyeless* and *dachshund* genes. *Proc. Natl. Acad. Sci. U.S.A.* 97: 2140–2144.

Kusch, L., and R. Reuter. 1999. Functions for *Drosophila brachyenteron* and *forkhead* in mesoderm specification and cell signaling. *Development* 126: 3991–4003.

Labandeira, C. C. 1997. Insect mouthparts: ascertaining the paleobiology of insect feeding strategies. *Annu. Rev. Ecol. Syst.* 28: 153–193.

———. 1998a. Early history of arthropod and vascular plant associations. *Annu. Rev. Earth Planet. Sci.* 26: 329–377.

———. 1998b. The role of insects in Later Jurassic to Middle Cretaceous ecosystems. In S. G. Lucas, J. I. Kirkland, and J. W. Eastop (eds.), *Lower and middle cretaceous terrestrial ecosystems*, pp. 105–124. New Mexico Mus. of Nat. Hist. Sci. Bull. No. 14.

Labandeira, C. C., and B. S. Beall. 1990. Arthropod terrestriality. In S. J. Culver (ed.), *Arthropod paleobiology. Short courses in paleontology*, pp. 214–256. No. 3. Paleontological Society.

Labandeira, C. C., and J. J. Sepkoski Jr. 1993. Insect diversity in the fossil record. *Science* 261: 310–315.

Lacalli, T. 1995. Dorsoventral axis inversion. *Nature* 373: 110–112.

———. 1996. Dorsoventral axis inversion: a phylogenetic perspective. *Bioessays* 18: 251–254.

Lafont, R. 1997. Ecdysteroids and related molecules in animals and plants. *Arch. Insect Biochem. Physiol.* 35: 3–20.

Lagueux, M., C. Hetru, F. Goltzene, C. Kappler, and J. A. Hoffmann. 1979. Ecdysone titre and metabolism in relation to cuticulogenesis in embryos of *Locusta migratoria*. *J. Insect Physiol.* 25: 709–723.

Lagueux, M., M. Hirn, and J. A. Hoffmann. 1977. Ecdysone during ovarian development in *Locusta migratoria*. *J. Insect Physiol.* 23: 109–119.

Lai-Fook, J. 1982. Structural comparison between eupyrene and apyrene spermiogenesis in *Calpodes ethlius* (Hesperiidae, Lepidoptera). *Can. J. Zool.* 60: 1216–1230.

Laissue, P. P., C. Reiter, P. R. Hiesinger, S. Halter, K. F. Fischbach, and R. F. Stocker. 1999. Three-dimensional reconstruction of the antennal lobe in *Drosophila melanogaster*. *J. Comp. Neurol.* 405: 543–552.

Lamka, M. L., and H. D. Lipshitz. 1999. Role of the amnioserosa in germ band retraction of the *Drosophila melanogaster* embryo. *Dev. Biol.* 214: 102–112.

Lane, J. A., and H. Stebbing. 1994. Independent regulation of microtubule spacing and microtubule stability following redundancy of nutritive tubes in telotrophic ovaries in Hemiptera. *Int. J. Insect Morphol. Embryol.* 23: 297–309.

Lanot, R., J. Thiebold, M. Lagueux, G. Goltzene, and J. A. Hoffmann. 1987. Involvement of ecdysone in the control of meiotic reinitiation in oocytes of *Locusta migratoria* (Insecta, Orthoptera). *Dev. Biol.* 121: 174–181.

Lanot, R., D. Zachary, F. Holder, and M. Meister. 2001. Postembryonic haematopoiesis in *Drosophila*. *Dev. Biol.* 230: 243–257.

Lawrence, J. F. 1991. Micromalthidae (Archostemata). In F. Stehr (ed.), *Immature insects*. Vol. 2, pp. 300–302. Kendall/Hunt.

Lawrence, P. A. 1990. Compartments in vertebrates? *Nature* 344: 382–383.

———. 1992. *The making of a fly: the genetics of animal design*. Blackwell Science.

Lawrence, P. A., and G. Morata. 1994. Homeobox genes: their function in *Drosophila* segmentation and pattern formation. *Cell* 78: 181–189.

Lawrence, P. A., R. Bodmer, and J.-P. Vincent. 1995. Segmental patterning of heart precursors in *Drosophila*. *Development* 121: 4303–4308.

Lawrence, P. A., J. Casal, and G. Struhl. 1999. *hedgehog* and *engrailed*: pattern formation and polarity in the *Drosophila* abdomen. *Development* 126: 2431–2439.

Leary, R. F., and F. W. Allendorf. 1989. Fluctuating asymmetry as an indicator of stress: implications for conservation biology. *Trends Ecol. Evol.* 4: 214–217.

Lebestky, T., T. Chang, V. Hartenstein, and U. Banerjee. 2000. Specification of *Drosophila* hematopoietic lineage by conserved transcription factors. *Science* 288: 146–149.

Lee, T., A. Lee, and L. Luo. 1999. Development of the *Drosophila* mushroom bodies: sequential generation of three distinct types of neurons form a neuroblast. *Development* 126: 4065–4076.

Lefcort, F., and D. Bentley. 1987. Pathfinding by pioneer neurons in isolated opened, and mesoderm-free limb buds of embryonic grasshoppers. *Dev. Biol.* 119: 466–480.

Legay, J. M. 1977. Allometry and systematics: insect egg form. *J. Nat. Hist.* 11: 493–499.

Leinaas, H. P. 1983. Synchronized moulting controlled by communication in group-living Collembola. *Science* 219: 193–195.

Leopold, R. A. 1976. The role of male accessory glands in insect reproduction. *Annu. Rev. Entomol.* 21: 199–221.

Leopold, R. A., and M. E. Degrugillier. 1973. Sperm penetration of housefly eggs: evidence for involvement of a female accessory secretion. *Science* 181: 555–557.

Leopold, R. A., S. Meola, and M. E. Degrugillier. 1978. The egg fertilization site within the house fly, *Musca domestica* (L.) (Diptera: Muscidae). *Int. J. Insect Morphol. Embryol.* 7: 111–120.

Leptin, M., and B. Grunewald. 1990. Cell shape changes during gastrulation in *Drosophila*. *Development* 110: 73–84.

Leudeman, R., and R. B. Levine. 1996. Neurons and ecdysteroids promote the proliferation of myogenic cells cultured from the developing adult legs of *Manduca sexta*. *Dev. Biol.* 173: 51–68.

Leung, B., and M. R. Forbes. 1996. Fluctuating asymmetry in relation to stress and fitness: comparing efficacy of analysis involving multiple traits. *Ecoscience* 3: 400–413.

Leuzinger, S., F. Hirth, D. Gerlich, D. Acampara, A. Simeane, W. J. Gehring, R. Finkelstein, K. Furkubo-Tokunaga, and H. Reichert. 1998. Equivalence of the fly *orthodenticle* gene and the human *OTX* genes in embryonic brain development of *Drosophila*. *Development* 125: 1703–1710.

Levine, R. B. 1986. Reorganization of the insect nervous system during metamorphosis. *Trends Neurosci.* 9: 315–319.

Lewis, D. L., M. De Camillis, and R. L. Bennett. 2000. Distinct roles of the homeotic genes *Ubx* and *abd-A* in beetle embryonic abdominal appendage development. *Proc. Natl. Acad. Sci. U.S.A.* 97: 4504–4509.

Lewis, E. B. 1978. A gene complex controlling segmentation in *Drosophila*. *Nature* 276: 1–6.

Lezzi, M., T. Bergman, J.-F. Mouillet, and V. C. Hearich. 1999. The ecdysone receptor puzzle. *Arch. Insect Biochem. Physiol.* 41: 99–106.

Liang, Z., and M. D. Biggen. 1998. Eve and Ftz regulate a wide variety of genes in blastoderm embryos: the selector homeoproteins directly or indirectly regulate most genes in *Drosophila*. *Development* 125: 4471–4482.

Liebrich, W., P. J. Hanna, and O. Hess. 1982. Evidence of asynchronous mitotic cell divisions in secondary spermatogonia of *Drosophila*. *Int. J. Invert. Reprod.* 5: 305–310.

Lifschytz, E., and D. Hareven. 1977. Gene expression and the control of spermatid morphogenesis in *Drosophila melanogaster*. *Dev. Biol.* 58: 276–294.

Ligoxygakis, P., M. Strigini, and M. Awerd. 2001. Specification of left-right asymmetry in the embryonic gut of *Drosophila*. *Development* 180: 1171–1174.

Lilly, B., B. Zhao, G. Ranganayakulu, B. M. Paterson, R. A. Schulz, and E. N. Olson. 1995. Requirement of MADS domain transcription factor D-MEF-2 for muscle formation in *Drosophila*. *Science* 267: 688–693.

Lin, H., and S. Schagot. 1997. Neuroblasts: a model for asymmetric division of stem cells. *Trends Genet.* 13: 333–339.

Lin, H., L. Yue, and A. C. Spradling. 1994. The *Drosophila* fusome, a germ-line specific organelle, contains membrane skeletal proteins and functions in cyst formation. *Development* 120: 947–956.

Lin, Y.-J., L. Seroude, and S. Benzer. 1998. Extended life-span and stress resistance in the *Drosophila* mutant *methuselah*. *Science* 282: 943–946.

Lindsley, D. L., and K. T. Tokuyasu. 1980. Spermatogenesis. In M. Ashburner and T. R. F. Wright (eds.), *The genetics and biology of Drosophila*. Vol. 2D, pp. 225–294. Academic Press.

Linley, J. R. 1981a. Ejaculation and spermatophore formation in *Culicoides melleus* (Coq.) (Diptera: Ceratopogonidae). *Can. J. Zool.* 59: 332–346.

———. 1981b. Emptying of the spermatophore and spermathecal filling in *Culicoides melleus* (Coq.) (Diptera: Ceratopogonidae). *Can. J. Zool.* 59: 427–430.

Linley, J. R., and K. R. Simmons. 1983. Quantitative aspects of sperm transfer in *Simulium decorum* (Diptera: Simuliidae). *J. Insect Physiol.* 29: 581–584.

Locke, M. 1958. The coordination of growth in the tracheal system of insects. *Q. J. Microsc. Sci.* 99: 373–391.

———. 1985. A structural analysis of postembryonic development. In G. A. Kerkut and L. I. Gilbert (eds.), *Comprehensive insect physiology, biochemistry and pharmacology*. Vol. 2, pp. 87–149. Pergamon Press.

———. 1990. Epidermal cells. In E. Ohnishi and H. Ishizaki (eds.), *Molting and metamorphosis*, pp. 173–206. Japanese Scientific Society Press.

———. 1998. Epidermis. In F. W. Harrison and M. Locke (eds.), *Microscopic anatomy of invertebrates*. Vol. 11A. Insecta, pp. 75–138. Wiley-Liss.

———. 2001. The Wigglesworth Lecture: insects for studying fundamental problems in biology. *J. Insect Physiol.* 47: 495–507.

Locke, M., and P. Huie. 1981. Epidermal feet in insect metamorphosis. *Nature* 293: 733–735.

Lockwood, J. A., and R. N. Story. 1985. A bifunctional pheromone in the first instars of the southern green stink bug, *Nezara viridula* (L.) (Hemiptera: Pentatomidae): its characterization and interaction with other stimuli. *Ann. Entomol. Soc. Am.* 78: 474–479.

———. 1986a. Adaptive functions of nymphal aggregation in the southern green stink bug, *Nezara viridula* (L.). *Environ. Entomol.* 15: 739–750.

———. 1986b. Embryonic orientation in pentatomids: its mechanism and function in southern green stink bug (Hemiptera: Pentatomidae). *Ann. Entomol. Soc. Am.* 79: 963–970.

Loeb, M., C. W. Woods, E. P. Brandt, and A. B. Borkovec. 1982. Larval testes of the tobacco budworm: a new source of insect ecdysteroids. *Science* 218: 896–898.

Loeb, M. J. 1993. Hormonal control of growth and reproduction in arthropods: introduction to the symposium. *Am. Zool.* 33: 303–307.

Loeb, M. J., J. Kochansky, R. M. Wagner, and C. W. Woods. 1998. Structure-function analysis of testis ecdysiotropin. *Arch. Insect Biochem. Physiol.* 38: 11–18.

Lohs-Schardin, M. 1982. Dicephalic—a *Drosophila* mutant affecting polarity in follicle organization and embryonic patterning. *W. Roux's Arch. Dev. Biol.* 191: 28–36.

Lokki, J., and A. Saura. 1980. Polyploidy in insect evolution. In W. H. Lewis (ed.), *Polyploidy: biological relevance*, pp. 277–312. Plenum Press.

Longo, F. J. 1997. *Fertilization*. 2d ed. Chapman and Hall.

Lopez, A. J., and D. S. Hogness. 1991. Immunochemical dissection of the Ultrabithorax Homeo-protein family in *Drosophila melanogaster*. *Proc. Natl. Acad. Sci. U.S.A.* 88: 9924–9928.

Loppin, B., M. Docquier, F. Bonneton, and P. Couble. 2000. The maternal effect mutation *sésame* affects the formation of the male pronucleus in *Drosophila melanogaster*. *Dev. Biol.* 222: 392–404.

Lu, B., F. Roeglers, L. L. Jan, and Y. N. Yang. 2001. Adherens junctions inhibit asymmetric division in the *Drosophila* epithelium. *Nature* 409: 522–525.

Ludwig, P., J. L. D. Williams, E. Lodde, H. Reichert, and G. S. Boyan. 1999. Neurogenesis in the median domain of the embryonic brain of the grasshopper *Schistocerca gregaria*. *J. Comp. Neurol.* 414: 379–390.

Lüscher, M. 1953. The termite and the cell. *Sci. Am.* 188(5): 74–78.

Lutz, D. A., and E. Huebner. 1980. Development and cellular differentiation of an insect telotrophic ovary (*Rhodnius prolixus*). *Tissue Cell* 12: 773–794.

———. 1981. Development of nurse cell-oocyte interactions in the insect telotrophic ovary (*Rhodnius prolixus*). *Tissue Cell* 13: 321–335.

Lyal, C. H. C. 1985. Phylogeny and classification of the Psocodea, with particular reference to the lice (Psocodea: Pththiraptera). *Syst. Entomol.* 10: 145–165.

Lynch, M., and J. S. Conery. 2000. The evolutionary fate and consequences of duplicate genes. *Science* 290: 1151–1155.

Lyonet, P. 1762. *Traité anatomique de la chenille qui range le bois de saule*. La Haye.

Ma, X., D. Yuan, K. Diepold, T. Scarborough, and J. Ma 1996. The *Drosophila* morphogenetic protein Bicoid binds DNA cooperatively. *Development* 122: 1195–1206.

Mable, B. K., and S. P. Otto. 1998. The evolution of life cycles with haploid and diploid phases. *Bioessays* 20: 453–462.

Mach, J. M., and R. Lehmann. 1997. An Egalitarian-Bicaudal D complex is essential for oocyte specification and axis determination in *Drosophila*. *Genes Dev.* 11: 423–435.

Machida, R. 1981. External features of embryonic development of a jumping bristletail, *Pedetontus unimaculatus* Machida (Thysanura, Machilidae). *J. Morphol.* 168: 339–355.

———. 2000. Serial homology of the mandible and maxilla in the jumping bristletail, *Pedetontus unimaculatus* Machida, based on external embryology. *J. Morphol.* 245: 19–28.

Machida, R., and H. Ando. 1998. Evolutionary changes in developmental potentials of the embryo proper and embryonic membranes along with the derivative structures in Atelocerata with special reference to Hexapoda (Arthropoda). *Proc. Arthropod. Embryol. Soc. Jap.* 33: 1–13.

Mackerras, I. M. (ed.). 1970. *The insects of Australia. A textbook for students and research workers*. 1st ed. Melbourne University Press.

Madhavan, M. M. 1973. The dual origin of the nurse chamber in the ovarioles of the gall midge, *Heteropyza pygmaea*. *W. Roux's Arch. Entwicklungsmech. Organ.* 173: 164–168.

Madhavan, M. M., and H. A. Schneiderman. 1977. Histological analysis of the dynamics of growth of imaginal discs and histoblast nests during the larval development of *Drosophila melanogaster*. *W. Roux's Arch. Entwicklungsmech. Organ.* 183: 269–305.

———. 1984. Do larval epidermal cells possess the blueprint for adult pattern in *Drosophila*? *J. Embryol. Exp. Morphol.* 82: 1–8.

Mahowald, A. P. 1972. Oogenesis. In S. J. Counce and C. H. Waddington (eds.), *Developmental systems: insects*. Vol. 1, pp. 1–47. Academic Press.

———. 1992. Germ plasm revisited and illuminated. *Science* 255: 1216–1217.

Mäkel, M. 1942. Metamorphose und Morphologie des *Pseudococcus* Männchens mit besonderer Berücksichtung des Skelettmuskelsystems. *Zool. Jb. Anat. Ontog. Tiere* 67: 461–512.

Malzacher, P. 1968. Die Embryogenese des Gehirns paurometaboler Insekten. Untersuchungen an *Carausius morosus* und *Periplaneta americana*. *Z. Morphol. Tiere* 62: 103–161.

Mangeldorf, D. J., and R. M. Evans. 1995. The nuclear receptor superfamily: the second decade. *Cell* 83: 835–839.

Mann, T. 1984. *Spermatophores: development, structure, biochemical attributes and role in the transfer of spermatozoa*. Springer-Verlag.

Mannervik, M., Y. Nibu, H. Zhang, and M. Levine. 1999. Transcriptional coregulators in development. *Science* 284: 606–609.

Manning, G., and M. A. Krasnow. 1993. Development of the *Drosophila* tracheal system. In M. Bate and A. Martinez Arias (eds.), *The development of* Drosophila melanogaster. Vol. 1, pp. 609–685. Cold Spring Harbor Laboratory Press.

Manton, S. M. 1977. *The arthropoda: habits, functional morphology and evolution*. Clarendon Press.

Marchal, P. 1898. Le cycle évolutif de l'*Encyrtus fuscicolli*. *Soc. Entomol. Fr. Bull.* 1898, 109–111.

Marden, J. H., and M. G. Kramer. 1994. Surface-skimming stoneflies: a possible intermediate stage in insect flight evolution. *Science* 266: 427–430.

Marec, F. 1996. Synaptonemal complexes in insects. *Int. J. Insect Morphol. Embryol.* 25: 205–233.

Marec, F., J. Leutelt, W. Traut, and K. W. Wolf. 1993. Visualization of polyfusomes in gonads of a moth, *Ephestia kuehniella* Z. (Lepidoptera, Pyralidae) by a microspreading technique and electron microscopy. *Int. J. Insect Morphol. Embryol.* 22: 487–496.

Margaritis, L. H., and M. Mazzini. 1998. Structure of the egg. In F. W. Harrison and M. Locke (eds.), *Microscopic anatomy of invertebrates*. Vol. 11C. Insecta, pp. 995–1037. Wiley-Liss.

Markow, T. A., and P. F. Ankney. 1988. Insemination reaction in *Drosophila*: found in species whose males contribute material to oocytes before fertilization. *Evolution* 42: 1097–1101.

Marsh, J. 1993. Transforming frogs and flies. *Nature* 361: 116–117.

Marsh, M., and H. T. McMahon. 1999. The structural era of endocytosis. *Science* 285: 215–220.

Martin, M. W., D. V. Grazhdankin, S. A. Bowring, D. A. D. Evans, M. A. Fedonkin, and J. L. Kirschvink. 2000. Age of neoproterozoic bilatarian body and trace fossils, White Sea, Russia: implications for metazoan evolution. *Science* 288: 841–845.

Martin, V. J. 1997. Cnidarians, the jellyfish and hydras. In S. F. Gilbert and A. M. Raunio (eds.), *Embryology. Constructing the organism*, pp. 57–86. Sinaur Associates.

Martin-Bermudo, M. D. 2000. Integrins modulate Egfr signaling pathway to regulate tendon cell differentiation in the *Drosophila* embryo. *Development* 127: 2607–2615.

Martínez, M. I., and M. R. Cruz. 1999. Comparative morphological analysis of testes follicles in dung beetles (Coleoptera: Scarabaeidae: Scarabaeinae: Aphodiinae, Geotrupinae). *Proc. Entomol. Soc. Wash.* 101: 804–815.

Martinez Arias, A. 1993. Development and patterning of the larval epidermis of *Drosophila*. In M. Bate and A. Martinez Arias (eds.), *The development of Drosophila melanogaster.* Vol. 1, pp. 517–608. Cold Spring Harbor Laboratory Press.

Martini, S. R., G. Roman, S. Meuser, G. Mardon, and R. L. Davis. 2000. The retinal determination gene *dachshund*, is required for mushroom body cell differentiation. *Development* 127: 2663–2672.

Martoja, R. 1977. Les organes genitaux femelles. In P. P. Grassé (ed.), *Traité de zoologie.* Vol. VIII 5A, pp. 1–123. Masson et Cie.

Marx, J. 1994. Chromosome ends catch fire. *Science* 265: 1656–1658.

——. 1995. Helping neurones find their way. *Science* 268: 971–973.

Masters, J. C., and R. J. Rayner. 1998. Key innovations? *Trends Ecol. Evol.* 13: 281.

Matova, N., and L. Cooley. 2001. Comparative aspects of animal oogenesis. *Dev. Biol.* 231: 291–320.

Matsuda, R. 1965. *Morphology and evolution of the insect head.* Memoirs of the American Entomological Institute No. 4.

——. 1976. *Morphology and evolution of the insect abdomen.* Pergamon Press.

——. 1987. *Animal evolution in changing environments with special reference to abnormal metamorphosis.* Wiley Interscience.

Matunes, E., J. Tran, P. Gönczy, K. Caldwell, and S. DiNardo. 1997. *punt* and *schnurri* regulate a somatically derived signal that restricts proliferation of committed progenitors in the germline. *Development* 124: 4383–4391.

Matuszewski, B. 1982. *Diptera I, Cecidomyiidae. Animal cytogenetics.* Vol. 3. *Insecta.* 3. Gebrüder Borntraeger.

Maxwell, G. D., and J. G. Hildebrand. 1981. Anatomical and neurochemical consequences of deafferentation in the development of the visual system of the moth *Manduca sexta*. *J. Comp. Neurol.* 195: 667–680.

Mayr, E. 1963. *Animal species and evolution.* Belknap Press, Harvard University Press.

——. 1982. *The growth of biological thought.* Belknap Press, Harvard University Press.

Mayr, E., and P. D. Ashlock. 1991. *Principles of systematic zoology.* 2d ed. McGraw-Hill.

McAlpine, J. F., B. V. Peterson, G. E. Shewell, H. J. Teskey, J. R. Vockeroth, and D. M. Wood (eds.). 1981, 1987. *Manual of nearctic Diptera.* Vols. 1–3. Monograph No. 27. Research Branch, Agriculture Canada.

McCaffery, A. R., and V. A. McCaffery. 1983. Corpus allatum activities during overlapping cycles of öocyte growth in adult female *Melanoplus sanguinipes*. *J. Insect Physiol.* 29: 259–266.

McCaffery, A. R., S. J. Simpson, M. S. Islam, and P. Roessingh. 1998. A gregarizing factor present in the egg pod foam of the desert locust *Schistocerca gregaria*. *J. Exp. Biol.* 201: 347–363.

McCall, K., and H. Steller. 1998. Requirement for dcp-1 caspase during *Drosophila* oogenesis. *Science* 279: 230–234.

McClung, C. E. 1901. Notes on the accessory chromosome. *Anat. Anz.* 20: 220–226.

McCune, A. R. 1988. Ontogeny and phylogeny (review of Matsuda, R. 1987. *Animal evolution in changing environments with special reference to abnormal metamorphosis*). *Science* 239: 300–301.

McDonald, J. A., and C. L. Doe. 1997. Establishing neuroblast-specific gene expression in the *Drosophila* CNS: *huckebein* is activated by Wingless and Hedgehog and repressed by Engrailed and Gooseberry. *Development* 124: 1079–1087.

McElwain, J. C., D. J. Beerling, and F. I. Woodward. 1999. Fossil plants and global warming at the Triassic-Jurassic boundary. *Science* 285: 1386–1390.

McFarlane, C., and J. E. McFarlane. 1988. Sperm penetration and in vitro fertilization of the egg of the house cricket *Acheta domesticus*. *Int. J. Invert. Reprod. Dev.* 13: 171–182.

McGinnis, W., and M. Kuziora. 1994. The molecular architects of body design. *Sci. Am.* 270(2): 58–66.

McGinnis, W., M. Levine, E. Hafen, A. Kuroiwa, and W. J. Gehring. 1984. A conserved DNA sequence in homeotic genes of the *Drosophila* Antennapedia and bithorax complexes. *Nature* 308: 428–433.

McKearin, D. 1997. The *Drosophila* fusome, organelle biogenesis and germ cell differentiation: if you build it. *Bioessays* 19: 147–152.

McKee, B. D., and M. A. Handel. 1993. Sex chromosomes, recombination, and chromatin configuration. *Chromosoma* 102: 71–80.

McKim, K. S., and R. S. Hawley. 1995. Chromosomal control of meiotic cell division. *Science* 270: 1595–1600.

McKinney, M. L., and K. L. McNamara. 1991. *Heterochrony. The evolution of ontogeny.* Plenum Press.

McNamara, K. J. 1997. *Shapes of time: the evolution of growth and development.* Johns Hopkins University Press.

Meckel, J. F. 1811. *Entwurf einer Darstellung der zwischen dem Embryozustande der höheren Tiere und dem permanenten der niederen stattfindenen Parallel: Beyträge zur vergleichenden Anatomie,* Vol. 2. Carl Heinrich Reclam.

Meer, J. M. van der. 1988. The role of metabolism and calcium in the control of mitosis and ooplasmic movements in insect eggs: a working hypothesis. *Biol. Rev.* 63: 109–157.

Megraw, T. L., L.-R. Kao, and T. C. Kaufman. 2001. Zygotic development without functional mitotic centrosomes. *Curr. Biol.* 11: 116–129.

Megraw, T. L., K. Li, L.-R. Kao, and T. C. Kaufman. 1999. The centrosomin protein is required for centrosome assembly and function during cleavage in *Drosophila*. *Development* 126: 2829–2839.

Meier, P., and G. Evan. 1998. Dying like flies. *Cell* 95: 295–298.

Meier, P., A. Finch, and G. Evan. 2000. Apoptosis in development. *Nature* 407: 796–801.

Meier, R., M. Kotrba, and P. Ferrar. 1999. Ovoviviparity and viviparity in the Diptera. *Biol. Rev.* 74: 199–258.

Meier, T., F. Chabaud, and H. Reichert. 1991. Homologous patterns in the embryonic development of the peripheral nervous system in the grasshopper, *Schistocerca gregaria* and the fly *Drosophila melanogaster*. *Development* 112: 241–253.

Meinertzhagen, I. A., and T. E. Hanson. 1993. The development of the optic lobe. In M. Bate and A. Martinez Arias (eds.), *The development of Drosophila melanogaster.* Vol. 2, pp. 1363–1506. Cold Spring Harbor Laboratory Press.

Meinertzhagen, I. A., J. G. Emsley, and X. J. Sun. 1998. Developmental anatomy of the *Drosophila* brain: neuroanatomy as gene expression. *J. Comp. Neurol.* 402: 1–9.

Meinhardt, H. 1982. *Models of biological pattern formation.* Academic Press.

Melzer, R. R., C. Michalke, and U. Smola. 2000. Walking on insect paths? Early ommatidial development in the compound eye of the ancestral crustacean, *Triops cancriformes. Naturwissenschaften* 87: 308–311.

Mendel, J. (G.) 1866. Versuche über Pflanzen-hybriden. *Verh. Natur. Vereins Brünn* 4(1865): 3–57.

Mesnier, M. 1981. Study of the ovulation process in the stick insect, *Clitumnus extradentatus*. *J. Insect Physiol.* 27: 425–433.

Metzger, R. J., and M. A. Krasnow. 1999. Genetic control of branching morphogenesis. *Science* 284: 1635–1639.

Michener, C. D. 2000. *The bees of the world*. Johns Hopkins University Press.

Micholitsch, T., P. Krügel, and G. Pass. 2000. Insemination and fertilization in the seed bug *Lygaeus simulans*. *Eur. J. Entomol.* 97: 13–18.

Mickoleit, G. 1973. Über den Ovipositor der Neuropteroidea und Coleoptera und seine phylogenetische Bedeutung (Insecta, Holometabola). *Z. Morphol. Ökol. Tiere* 74: 37–64.

Mill, P. J. 1998. Tracheae and tracheoles. In F. W. Harrison and M. Locke (eds.), *Microscopic anatomy of invertebrates*. Vol. 11A. *Insecta*, pp. 303–336. Wiley-Liss.

Miller, D. J., and A. Miles. 1993. Homeobox genes and the zootype. *Nature* 365: 215–216.

Miller, O. L., Jr., and B. R. Beatty. 1969. Visualization of nucleolar genes. *Science* 164: 955–957.

Miller, P. L. 1990. Mechanisms of sperm removal and sperm transfer in *Orthetrum coerulescens* (Fabricius) (Odonata: Libellulidae). *Physiol. Entomol.* 15: 199–209.

Miner, A. L., A. J. Rosenberg, and H. F. Nijhout. 2000. Control of growth and differentiation of the wing imaginal disk of *Precis coenia* (Lepidoptera: Nymphalidae). *J. Insect Physiol.* 46: 251–258.

Mizoguchi, A., T. Oka, H. Kataoka., H. Nagasawa, A. Suzuki, and H. Ishizaki. 1989. Immunohistochemical localization of prothoracicotropic hormone-producing neurosecretory cells in the brain of *Bombyx mori*. *Dev. Growth Diff.* 32: 591–598.

Mohler, J., J. W. Mahaffey, E. Deutsch, and K. Vani. 1995. Control of *Drosophila* head segment identity by the bZIP Homeotic gene *cnc*. *Development* 121: 237–247.

Mojica, J. M., S. File-Emperador, and D. L. Bruck. 2000. Sperm bundle and spermatozoon ultrastructure in two species of the cardini group of *Drosophila*. *Invert. Reprod. Dev.* 37: 147–155.

Møller, A. P., and J. P. Swaddle. 1997. *Asymmetry, developmental stability and evolution*. Oxford University Press.

Monsma, S. A., and R. Booker. 1996a. Genesis of the adult retina and outer optic lobes of the moth, *Manduca sexta*. I. Patterns of proliferation and cell death. *J. Comp. Neurol.* 367: 10–20.

———. 1996b. Genesis of the adult retina and outer optic lobes of the moth, *Manduca sexta*. II. Effects of deafferentation and developmental hormone manipulation. *J. Comp. Neurol.* 367: 21–35.

Montagne, J., M. J. Stewart, H. Stocker, E. Hafen, S. C. Kozma, and G. Thomas. 1999. *Drosophila* S6 kinase: a regulator of cell size. *Science* 285: 2126–2129.

Montell, D. J. 1999. The genetics of cell migration in *Drosophila melanogaster* and *Caenorhabditis elegans* development. *Development* 126: 3035–3046.

Moore, J., and P. Willmer. 1997. Convergent evolution in invertebrates. *Biol. Rev.* 72: 1–60.

Moore, J. A. 1993. *Science as a way of knowing. The foundations of modern biology*. Harvard University Press.

Moore, L. A., H. T. Broihier, M. Van Doren, L. B. Lunsford, and R. Lehmann. 1998. Identification of genes controlling germ cell migration and embryonic gonad formation in *Drosophila*. *Development* 125: 667–678.

Moran, D. T., J. C. Rowley, S. N. Zill, and F. G. Verala. 1976. The mechanism of sensory transduction in a mechanoreceptor. *J. Cell Biol.* 71: 832–847.

Moran, N. A. 1992. The evolution of aphid life cycles. *Annu. Rev. Entomol.* 37: 71–92.

Moran, N. A., and A. Telang. 1998. Bacteriocyte-associated symbionts of insects. *Bioscience* 48: 295–304.

Mordue, A. J., and A. Blackwell. 1993. Review: Azadirachtin: an update. *J. Insect Physiol.* 39: 903–924.

Morell, V. 1994. Rise and fall of the Y chromosome. *Science* 263: 171–172.

Moreno, E., and G. Morata. 1999. *caudal* is the Hox gene that specifies the most posterior *Drosophila* segment. *Nature* 400: 873–877.

Morgan, J., M. Warner, J. Britton, R. Buffler, A. Camargo, G. Christeson, P. Denton, A. Hildebrand, R. Hobbs, H. Macintyre, G. Mackenzie, P. Maguire, L. Marin, Y. Nakamura, M. Pilkington, V. Sharpton, D. Snyder, G. Suarez, and A. Trejo. 1997. Size and morphology of the Chicxulub impact crater. *Nature* 390: 472–476.

Morgan, M. M., and A. P. Mahowald. 1996. Multiple signalling pathways establish both the individuation and the polarity of the oocyte follicle in *Drosophila*. *Arch. Insect Biochem. Physiol.* 33: 211–230.

Morgan, T. H., A. H. Sturtevant, H. J. Muller, and C. B. Bridges. 1915. *The mechanism of Mendelian heredity*. W. W. Norton.

Mori, H. 1979. Embryonic haemocytes: origin and development. In A. P. Gupta (ed.), *Insect haemocytes*, pp. 3–27. Cambridge University Press.

———. 1983. Origin, development, morphology, functions and phylogeny of the embryonic midgut epithelium in insects. *Entomol. Gen.* 8: 135–154.

———. 1985. Abortive anatrepsis and imperfect germ band formation in the capillary-coated eggs of the water strider, *Gerris paludum insularis* (Hemiptera: Gerridae). *Ann. Entomol. Soc. Am.* 78: 509–513.

Morisato, D., and K. V. Anderson. 1995. Signalling pathways that establish the dorsal-ventral pattern of the *Drosophila* embryo. *Annu. Rev. Genet.* 29: 371–399.

Moritz, G. 1997. Structure, growth and development. In T. Lewis (ed.), *Thrips as crop pests*, pp. 15–63. CAB International.

Morrow, E. H., and M. J. G. Gage. 2000. The evolution of sperm length in moths. *Proc. R. Soc. Lond. B Biol. Sci.* 267: 307–313.

Moses, K. 1991. The role of transcription factors in the developing *Drosophila* eye. *Trends Genet.* 7: 250–255.

Mouze, M. 1978. Rôle des fibres postrétiniennes dans la croissance du lobe optique de la larve d'*Aeshna cyanea* Müll. (Insect Odonate). *W. Roux's Arch. Entwicklungsmech. Organ.* 184: 325–350.

———. 1979. Étude cytologique de la genèse ommatidienne chez la larve d'un odonate anisoptère. *Rev. Can. Biol.* 38: 227–248.

Mueller, J. L., M. Calleja, J. Capdevila, and I. Guerro. 1997. Hedgehog activity, independent of Decapentaplegic, participates in wing disc patterning. *Development* 124: 1227–1237.

Mukai, M., M. Kashieawa, and S. Kobayashi. 1999. Induction of *indora* expression in pole cells by the mesoderm is required for female germ-line development in *Drosophila melanogaster*. *Development* 126: 1023–1029.

Mullen, J. R., and S. DiNardo. 1995. Establishing parasegments in *Drosophila* embryos: roles of *odd-skipped* and *naked* genes. *Dev. Biol.* 169: 295–308.

Müller, F. 1864. Für Darwin. In A. Müller

(ed.), *Fritz Müller. Werke, Briefs und Leben*, pp. 200–263. G. Fischer.

Müller, H.-A. J. 2000. Genetic control of epithelial cell polarity: lessons from *Drosophila. Dev. Dyn.* 218: 52–67.

Müller, H. J. 1940. Die Symbiose der Fulgoroiden (Homoptera-Cicadina). *Zoologica* 98: 1–220; pl. I–XXXVII.

——. 1951. Über das Schlüpfen der Zikaden (Homoptera auchenorrhyncha) aus dem Ei. *Zoologica* 103: 1–41; pl. I–XIV.

Muskavitch, M. A. T. 1994. Delta-Notch signaling and *Drosophila* cell fate choice. *Dev. Biol.* 166: 415–430.

Myat, M. M., and D. J. Andrew. 2000. Organ shape in the *Drosophila* salivary gland is controlled by regulated sequential internalization of the primordia. *Development* 127: 679–691.

Nabert, A. 1913. Die corpora allata der Insekten. *Z. Wiss. Zool.* 104: 181–358.

Nagasawa, H. 1992. Neuropeptides of the silkworm, *Bombyx mori. Experientia*, 48: 425–430.

Nagoo, T., K. Endo, H. Kawauchi, U. Walldorf, and K. Furukubo-Tokunaga. 2000. Patterning defects in the primary axonal scaffolds caused by the mutations of the *extradenticle* and *homothorax* genes in the embryonic *Drosophila* brain. *Dev. Genes Evol.* 210: 289–299.

Naisse, J. 1966a. Contrôle endocrinien de la différenciation sexuelle chez l'insecte *Lampyris noctiluca* (Coléoptère Malacoderme Lampyride). I. Rôle androgène des testicules. *Arch. Biol. (Liège)* 77: 139–201.

——. 1966b. Contrôle endocrinien de la différenciation sexuelle chez l'insecte *Lampyris noctiluca* (Coléoptère Malacoderme Lampyride). II. Phénomènes neurosécrétoires et endocrines au cours du développement postembryonaire chez le mâle et la femelle. *Gen. Comp. Endocrinol.* 7: 85–104.

——. 1966c. Contrôle endocrinien de la différenciation sexuelle chez *Lampyris noctiluca* (Coléoptère Lampyride). III. Influence des hormones de la pars intercerebralis. *Gen. Comp. Endocrinol.* 7: 105–110.

Nakanishi, K., H. Moriyama, T. Okauchi, S. Fujioka, and M. Koreeda. 1972. Biosynthesis of α- and β-ecdysones from cholesterol outside the prothoracic gland in *Bombyx mori. Science* 176: 51–52.

Namba, R., and J. S. Minden. 1999. Fate mapping of *Drosophila* embryonic mitotic domain 20 reveals that the larval visual system is derived from a subdomain of a few cells. *Dev. Biol.* 212: 465–476.

Nambu, J. R., J. O. Lewis, and S. T. Crewe. 1993. The development and functions of the *Drosophila* CNS midline cells. *Comp. Biochem. Physiol.* 104A: 399–409.

Nardi, J. B. 1990. Expression of a surface epitope on cells that link branches in the tracheal network of *Manduca sexta. Development* 110: 681–688.

——. 1993. Modulated expression of a surface epitope on migrating germ cells of *Manduca sexta* embryos. *Development* 118: 967–975.

Nardi, J. B., and E. G. Cattani. 1995. Expression of a cell surface protein during morphogenesis of the reproductive system in *Manduca sexta* embryos. Both moths and mammals have an indifferent stage of genital differentiation. *W. Roux's Arch. Dev. Biol.* 205: 21–30.

Nardi, J. B., G. L. Godfrey, and R. A. Bergstrom. 1991. Programmed cell death in the wing of *Orgyia leucostigme* (Lepidoptera: Lymantriidae). *J. Morphol.* 209: 121–131.

Nasiadka, A., and H. M. Karuse. 1999. Kinetic analysis of segmentation gene interactions in *Drosophila* embryos. *Development* 126: 1515–1526.

Nasiadka, A., A. Grill, and H. M. Karuse. 2000. Mechanisms regulating target gene selection by the homeodomain-containing protein Fushi tarazu. *Development* 127: 2965–2976.

Nasmyth, K. 1996. Viewpoint: putting the cell cycle in order. *Science* 274: 1643–1645.

Nassif, C., A. Noveen, and V. Hartenstein. 1998a. Embryonic development of the *Drosophila* brain. I. Pattern of pioneer tracks. *J. Comp. Neurol.* 402: 10–31.

Nassif, C., A. Daniel, J. A. Lengyel, and V. Hartenstein. 1998b. The role of morphogenetic cell death during *Drosophila* embryonic head development. *Dev. Biol.* 197: 170–186.

Nässl, D. R. 1993. Neuropeptides in the insect brain. *Cell Tissue Res.* 273: 1–29.

Ndiaye, M., X. Mattei, and O. T. Thiau. 1997. Maturation of mosquito spermatozoa during their transit throughout the male and female reproductive systems. *Tissue Cell* 29: 675–678.

Nelson, J. A. 1915. *The embryology of the honey bee.* Princeton University Press.

Neumann, C. J., and S. M. Cohen. 1997a. Long-range action of Wingless organizes the dorsal-ventral axis of the *Drosophila* wing. *Development* 124: 871–880.

——. 1997b. Morphogens and pattern formation. *Bioessays* 19: 721–729.

Newcomer, W. S. 1948. Embryological development of the mouthparts and related structures of the milkweed bug, *Oncopeltus fasciatus* (Dallas). *J. Morphol.* 82: 365–411.

Newmark, P. A., S. E. Mohr, L. Gong, and R. E. Bursell. 1997. *mago nashi* mediates the posterior follicle cell-to-oocyte signal to organize axis formation in *Drosophila. Development* 124: 3197–3207.

Nicklas, R. B. 1997. How cells get the right chromosomes. *Science* 275: 632–637.

Nieuwkoop, P. D., and L. A. Sutasurya. 1981. *Primordial germ cells in the invertebrates.* Cambridge University Press.

Nijhout, H. F. 1990. A comprehensive model for colour pattern formation in butterflies. *Proc. R. Soc. Lond. B Biol. Sci.* 239: 81–113.

——. 1991. *The development and evolution of butterfly wing patterns.* Smithsonian Institution Press.

——. 1994a. Genes on the wing. *Science* 265: 44–45.

——. 1994b. *Insect hormones.* Princeton University Press.

——. 1996. Focus on butterfly eyespot development. *Nature* 384: 209–210.

Nijhout, H. F., and D. J. Emlen. 1998. Competition among body parts in the development and evolution of insect morphology. *Proc. Natl. Acad. Sci. U.S.A.* 95: 3685–3689.

Nijhout, H. F., and D. E. Wheeler. 1982. Juvenile hormone and the physiological basis of insect polymorphism. *Q. Rev. Biol.* 57: 109–133.

Nilson, L. A., and T. Schupbach. 1998. Localized requirements for *windbeutel* and *pipe* reveal a dorsoventral prepattern within the follicular epithelium of the *Drosophila* ovary. *Cell* 93: 253–262.

Norbeck, B. A., Y. Feng, and J. L. Denburg. 1992. Molecular gradients along the proximo-distal axis of embryonic insect legs: possible guidance cues for pioneer axon growth. *Development* 116: 467–479.

Nordlander, R. H., and J. S. Edwards. 1968. Morphology of the larval and adult brains of the monarch butterfly, *Danaus plexippus plexippus*, L. *J. Morphol.* 126: 67–94.

——. 1969a. Postembryonic brain development in the Monarch butterfly, *Danaus plexippus plexippus* L. I. Cellular events during brain morphogenesis. *W. Roux's Arch. Entwick-lungsmech. Organ.* 162: 197–217.

——. 1969b. Postembryonic brain development in the Monarch butterfly, *Danaus plexippus plexippus* L. II. The optic lobes. *W. Roux's Arch. Entwicklungsmech. Organ.* 163: 197–220.

——. 1970. Postembryonic brain development in the Monarch butterfly, *Danaus plexippus plexippus* L. III. Morphogenesis of centers other than the optic lobes. *W. Roux's Arch. Entwicklungsmech. Organ.* 164: 247–260.

Noselli, S. 1998. JNK signaling and morphogenesis in *Drosophila*. *Trends Genet.* 14: 33–38.

Novák, V. J. A. 1975. *Insect hormones.* 2d English ed. Chapman and Hall.

Novák, V. J. A., and S. K. Zambre. 1974. To the problem of structure and function of the pleuropodia in *Schistocerca gregaria* Forskål embryos. *Zool. Jb. Physiol.* 78: 344–355.

Nowel, M. S. 1980. Ommatidium assembly and formation of the retina-lamina projection in interspecific chimeras of cockroach. *J. Embryol. Exp. Morphol.* 60: 345–358.

——. 1981a. Formation of the retina-lamina projection of the cockroach: no evidence for neuronal specificity. *J. Embryol. Exp. Morphol.* 62: 241–258.

——. 1981b. Postembryonic growth of the compound eye of the cockroach. *J. Embryol. Exp. Morphol.* 62: 259–275.

Nübler-Jung, K., and D. Arendt. 1994. Is ventral in insects dorsal in vertebrates? A history of embryological arguments favouring axis inversion in chordate ancestors. *W. Roux's Arch. Dev. Biol.* 203: 357–366.

Nüesch, H. 1987. Metamorphose bei Insekten: direkte und indirekte Entwicklung by Apterygoten and Exopterygoten. *Zool. Jahrb. (Anat.)* 115: 453–487.

Nur, U. 1971. Parthenogenesis in Coccids (Homoptera). *Am. Zool.* 11: 301–308.

——. 1980. Evolution of unusual chromosome systems in scale insects. In R. L. Blackman, G. M. Hewitt, and M. Ashburner (eds.), *Insect cytogenetics,* pp. 97–117. Symp. Roy. Ent. Soc. Lond. 10. Blackwell Science.

Nüsslein-Volhard, C. 1991. 1. Determination of the embryonic axes of *Drosophila*. In K. Roberts, E. Coen, C. Dean, J. Jones, K. Chater, R. Flavell, A. Wilkins, and N. Holder (eds.), *Molecular and cellular basis of pattern formation. Development* Suppl. 1: 1–10.

——. 1996. Gradients that organise embryo development. *Sci. Am.* 275(2): 54–61.

Nüsslein-Volhard, C., and E. Wieschaus. 1980. Mutations affecting segment number and polarity in *Drosophila. Nature* 287: 795–801.

Nüsslein-Volhard, C., H. G. Frohnhofer, and R. Lehmann. 1987. Determination of anteroposterior polarity in *Drosophila. Science* 238: 1675–1681.

Nylin, S., and K. Gotthard. 1998. Plasticity in life-history traits. *Annu. Rev. Entomol.* 43: 63–83.

O'Brien, M. A., E. J. Katahira, T. R. Flanagan, L. W. Arnold, G. Haughton, and W. E. Bollenbacher. 1988. A monoclonal antibody to the insect prothoracicotropic hormone. *J. Neurosci.* 8: 3247–3257.

O'Conner, J. D. 1985. Ecdysteroid action at the molecular level. In G. A. Kerkut and L. I. Gilbert (eds.), *Comprehensive insect physiology, biochemistry and pharmacology.* Vol. 8, pp. 85–98. Pergamon Press.

O'Farrell, P. H., B. A. Edgar, K. Lakich, and C. F. Lehner. 1989. Directing cell division during development. *Science* 246: 635–640.

Ohnishi, E., and H. Ishizaki (eds.). 1990. *Molting and metamorphosis.* Japanese Scientific Society Press.

Oken, L. 1847. *Elements of physiophilosophy.* Transl. by A. Tulk. Ray Society.

Oland, L. A., and L. P. Tolbert. 1989. Patterns of glial proliferation during formation of olfactory glomeruli in an insect. *Glia* 2: 10–24.

Oland, L. A., W. M. Pott, M. R. Heggins, and L. P. Tolbert. 1998. Targeted ingrowth and glial relationships of olfactory receptor axons in the primary olfactory pathway of an insect. *J. Comp. Neurol.* 398: 119–138.

Oldroyd, B. P., L. Halling, and T. E. Rinderer. 1999. Development and behaviour of anarchistic honeybees. *Proc. R. Soc. Lond. B Biol. Sci.* 266: 1875–1878.

Oliver, J. H., and E. M. Dotson. 1993. Hormonal control of molting and reproduction in ticks. *Am. Zool.* 33: 384–396.

Oliwenstein, L. 1993. By a thousand cuts. *Discover* 14(2): 24–25.

Oppenheimer, J. M. 1955. Problems, concepts and their history. In B. H. Willier, P. A. Weiss, and V. Hamburger (eds.), *Analysis of development,* pp. 1–24. W. B. Saunders.

Orchard, I., and B. G. Loughton. 1985. Neurosecretion. In G. A. Kerkut and L. I. Gilbert (eds.), *Comprehensive insect physiology, biochemistry and pharmacology.* Vol. 7, pp. 61–107. Pergamon Press.

O'Reilly, D. R., and L. K. Miller. 1989. A baculovirus blocks insect molting by producing ecdysteroid UDP-glucosyl transferase. *Science* 245: 1110–1112.

Orr, A. G., and R. L. Rutowski. 1991. The function of the sphragis in *Cressida cressida* (Fab) (Lepidoptera, Papilionidae): a visual deterrent to copulation attempts. *J. Nat. Hist.* 25: 703–710.

Orr-Weaver, T. L. 1994. Developmental modification of the *Drosophila* cell cycle. *Trends Genet.* 10: 321–327.

Osanai, M. 1996. Initiatorin, a serine-endopeptidase as inducer of sperm maturation of *Bombyx mori* (L.) (Lepidoptera: Bombycidae). In *Proceedings of the 20th International Congress of Entomology, Firenze,* p. 168.

Oudemans, J. T. 1899. Falter aus castrierten Raupen. *Zool. Jahrb. Syst.* 12: 132–133.

Overall, R., and L. F. Jaffe. 1985. Patterns of ionic current through *Drosophila* follicles and eggs. *Dev. Biol.* 108: 102–119.

Page, A. W., and T. L. Orr-Weaver. 1997. Activation of the meiotic divisions in *Drosophila* oocytes. *Dev. Biol.* 183: 195–207.

Painter, T. S. 1934. A new method for the study of chromosome aberrations and the plotting of chromosome maps in *Drosophila melanogaster. Genetics* 19: 176–188.

Palli, S. R., L. M. Riddiford, and K. Hiruma. 1991. Juvenile hormone and "retenoic acid" receptors in *Manduca* epidermis. *Insect Biochem.* 21: 7–15.

Palmer, A. R. 1996a. Waltzing with asymmetry. Is fluctuating asymmetry a powerful new tool for biologists or just an alluring new dance step? *Bioscience* 46: 518–532.

——. 1996b. From symmetry to asymmetry: phylogenetic patterns of asymmetry variation in animals and their evolutionary significance. *Proc. Natl. Acad. Sci. U.S.A.* 93: 14279–14286.

Palmer, M. J., M. A. Harmon, and V. Laudet. 1999. Characterization of EcR and RxR homologues in the ixodid tick, *Amblyomma americanum* (L.). *Am. Zool.* 39: 747–757.

Panganiban, G. 2000. *Distal-less* function during *Drosophila* appendage and sense organ development. *Dev. Dyn.* 218: 554–562.

Panganiban, G., A. Sebring, L. Nagy, and S. Carroll. 1995. The development of crustacean limbs and the evolution of arthropods. *Science* 270: 1363–1366.

Panganiban, G., S. M. Irvine, C. Lowe, H. Role, L. S. Corley, B. Sherbon, J. K. Grenier, J. F. Fallon, J. Kimble, M. Walker, G. A. Wray, B. J. Swalla, M. Q. Martindale, and S. B. Carroll. 1997. The origins and evolution of animal appendages. *Proc. Natl. Acad. Sci. U.S.A.* 94: 5162–5166.

Pankratz, M. J., and H. J. Jäckle. 1993. Blastoderm segmentation. In C. M.

Bate and A. Martinez Arias (eds.). *The development of* Drosophila melanogaster. Vol. 1, pp. 467–516. Cold Spring Harbor Laboratory Press.

Panov, A. A. 1980. Demonstration of neurosecretory cells in insect central nervous system. In N. J. Strausfeld and T. A. Miller (eds.), *Neuroanatomical techniques. Insect nervous system*, pp. 25–50. Springer-Verlag.

Panzer, S., D. Weigel, and S. K. Bechendorf. 1992. Organogenesis in *Drosophila melanogaster*: embryonic salivary gland determination is controlled by homeotic and dorsoventral patterning genes. *Development* 114: 49–57.

Papaj, D. R. 2000. Ovarian dynamics and host use. *Annu. Rev. Entomol.* 45: 423–448.

Park, Y. I., S. Shu, S. B. Ramaswamy, and A. Srinivasan. 1998. Mating in *Heliothis virescens*: transfer of juvenile hormone during copulation by male to female and stimulation of biosynthesis of endogenous juvenile hormone. *Arch. Insect Biochem. Physiol.* 38: 100–107.

Parker, G. A. 1970. Sperm competition and its evolutionary consequences in insects. *Biol. Rev.* 45: 525–558.

Parkhurst, S. M., and P. M. Meneely. 1994. Sex determination and dosage compensation: lessons from flies and worms. *Science* 264: 924–932.

Parsons, M. C. 1974. The morphology and possible origin of the hemipteran loral lobes. *Can. J. Zool.* 52: 189–202.

Passner, J. M., H. D. Ryoo, L. Shen, R. S. Mann, and A. K. Aggarwal. 1999. Structure of a DNA-bound Ultrabithorax-Extradenticle homeodomain complex. *Nature* 397: 714–719.

Patel, N. 1994. Developmental evolution: insights from studies of insect segmentation. *Science* 266: 581–590.

Patel, N. H., E. E. Ball, and C. S. Goodman. 1992. Changing role of *even-skipped* during the evolution of insect pattern formation. *Nature* 357: 339–342.

Patel, N. H., B. G. Condron, and K. Zinn. 1994. Pair-rule expression patterns of *even-skipped* are found in both short- and long-germ beetles. *Nature* 367: 429–434.

Patel, N. H., T. B. Kornberg, and C. S. Goodman. 1989. Expression of *engrailed* during segmentation in grasshopper and crayfish. *Development* 107: 201–212.

Paulus, H. F. 1979. Eye structure and the monophyly of the Arthropoda. In A. P. Gupta (ed.), *Arthropod phylogeny*, pp. 299–383. Van Nostrand Reinhold.

Pearson, K. G., G. S. Boyan, M. Bastiani, and C. S. Goodman. 1985. Heterogeneous properties of segmentally homologous interneurons in the ventral nerve cord of locusts. *J. Comp. Neurol.* 233: 133–145.

Pearson, M. J. 1974. The abdominal epidermis of *Calliphora erythrocephala* (Diptera). I. Polyteny and growth in the larval cells. *J. Cell Sci.* 16: 113–131.

Pener, M. P. 1991. Locust phase polymorphism and its endocrine relations. *Adv. Insect Physiol.* 23: 1–79.

Pener, M. P., L. Orshan, and J. de Wilde. 1978. Precocene II causes atrophy of corpora allata in *Locusta migratoria*. *Nature* 272: 350–353.

Pennachio, F., P. Falabella, and S. B. Vinson. 1998. Regulation of *Heliothis virescens* prothoracic glands by *Cardiochilis nigriceps* polydnavirus. *Arch. Insect Biochem. Physiol.* 38: 1–10.

Pennisi, E. 2000. An integrative science finds a home. *Science* 287: 570–572.

Percival-Smith, A., J. Weber, E. Gilfoyle, and P. Wilson. 1997. Genetic characterization of the role of the two HOX proteins: Proboscipedia and Sex Combs Reduced, in determination of adult antennal, tarsal, maxillary palp and proboscis identities in *Drosophila melanogaster*. *Development* 124: 5049–5062.

Peri, F., and S. Roth. 2000. Combined activities of Gurken and Decapentaplegic specify dorsal chorion structures of the *Drosophila* egg. *Development* 127: 841–850.

Perondini, A. L. P. 1998. Elimination of X chromosomes and the problem of sex determination in *Sciara ocellaris*. In R. N. Chatterjee and L. Sánchez (eds.), *Genome analysis in eukaryotes: developmental and evolutionary aspects*, pp. 149–166. Narosa Publishing.

Perotti, M.-E., F. Cattaneo, M. Pasini, and J. H. P. Hackstein. 1996. Sperm and egg surface components potentially involved in gamete interactions in *Drosophila melanogaster*. In *Proceedings of the 20th International Congress of Entomology, Firenze*, p. 115.

Peschke, K. 1986. Development, sex specificity, and site of production of aphrodisiac pheromones in *Aleochara curtula*. *J. Insect Physiol.* 32: 687–693.

Pesson, P. 1944. Contribution à l'étude morphologique et fonctionelle de la tête, de l'appareil buccal, et du tube digestif des femelles de coccides. In *Monographies publiées par les Stations et Laboratoires de Recherches Agronomiques*, pp. 1–226. Paris.

Pétavy, G. 1975. Involution des annexes embryonnaires dans l'oeuf de *Locusta migratoria migratorioides* R. et F. (Orthoptera: Acrididae): Morphologie et histologie. *Int. J. Insect Morphol. Embryol.* 4: 1–22.

——. 1976. Involution des annexes embryonnaires dans l'oeuf de *Locusta migratoria migratorioides* R. et F. (Orthoptera: Acrididae): Analyse cytologique. *Int. J. Insect Morphol. Embryol.* 5: 167–186.

Peterson, K. J., and E. H. Davidson. 2000. Regulatory evolution and the origin of bilatarians. *Proc. Natl. Acad. Sci. U.S.A.* 97: 4430–4433.

Peterson, M. D., A. Popadíc, and T. C. Kaufman. 1998. The expression of two *engrailed*-related genes in an apterygote insect and a phylogenetic analysis of insect *engrailed*-related genes. *Dev. Genes Evol.* 208: 547–557.

Peterson, M. D., B. T. Rogers, A. Popadíc, and T. C. Kaufman. 1999. The embryonic expression pattern of *labial*, posterior homeotic complex genes and the *teashirt* homologue in an apterygote insect. *Dev. Genes Evol.* 209: 77–90.

Petit, M. G., and M. P. Scott. 1992. Gene control systems affecting insect development. In J. M. Crampton and P. Eggleston (eds.), *Insect molecular science*, pp. 101–124. Academic Press.

Pfeiffer, I. W. 1945. The influence of the corpora allata over the development of nymphal characters in the grasshopper, *Melanoplus differentialis*. *Trans. Conn. Acad. Art Sci.* 36: 489–515.

Pflugfelder, O. 1936. Vergleichende-anatomische, experimentelle und embryologische Untersuchungen über das Nervensystem und die Sinnesorgane der Rhynchoten. *Zoologica* 93: 1–102; pl. I–XXV.

Phillips, D. M. 1970. Insect sperm: their structure and morphogenesis. *J. Cell Biol.* 44: 243–277.

——. 1974. *Spermiogenesis*. Academic Press.

Piepho, H. 1942. Untersuchungen zur Entwicklungsphysiologie der Insektenmetamorphose. Über die Puppenhäutung der Wacsmotte *Galleria mellonella* L. W. *Roux's Arch. Entwicklungsmech. Organ.* 141: 500–583.

——. 1951. Über die Lenkung der Insektenmetamorphose durch Hormone. *Verh. Dtsch. Zool. Ges.* (1951): 62–67.

Pijnacker, L. P. 1969. Automictic parthenogenesis in the stick insect *Bacillus rossius* Rossi (Cheleutoptera, Phasmidae). *Genetica* 40: 393–399.

Pijnacker, L. P., and M. A. Ferwerda. 1976. Experiments in blocking and unblocking of first meiotic metaphase in eggs of the parthenogenetic stick insect *Carausius morosus* Br. (Phasmida, Insecta). *J. Embryol. Exp. Morphol.* 36: 383–394.

Pijnacker, L. P., and J. Godeke. 1984. Development of ovarian follicle cells in the stick insect, *Carausius morosus* Br. (Phasmatodea), in relation to their function. *Int. J. Insect Morphol. Embryol.* 13: 21–28.

Pinet, J. M. 1969. Étude de la morphogenèse du stylet maxillaire de *Rhodnius prolixus* par des transplantations d'organes styligènes incomplets. *Arch. Zool. Exp. Gen.* 109: 613–641.

Pipa, R. L. 1973. Proliferation, movement, and regression of neurones during the postembryonic development of insects. In D. Young (ed.), *Developmental neurobiology of arthropods*, pp. 105–129. Cambridge University Press.

Pitnick, S. 1996. Investment in testes and the cost of making long sperm in *Drosophila*. *Am. Nat.* 148: 57–80.

Pitnick, S., and T. L. Karr. 1998. Paternal products and by-products in *Drosophila* development. *Proc. R. Soc. Lond. B Biol. Sci.* 265: 821–826.

Pitnick, S., and T. A. Markow. 1994a. Male gametic strategies: sperm size, testes size, and the allocation of ejaculate among successive mates by the sperm limited fly *Drosophila pachea* and its relatives. *Am. Nat.* 143: 785–819.

——. 1994b. Large-male advantages associated with costs of sperm production in *Drosophila hydei*, a species with giant sperm. *Proc. Natl. Acad. Sci. U.S.A.* 91: 9277–9281.

Pitnick, S., T. A. Markow, and G. S. Spicer. 1995a. Delayed male maturity is a cost of producing large sperm in *Drosophila*. *Proc. Natl. Acad. Sci. U.S.A.* 92: 10614–10618.

——. 1999. Evolution of multiple kinds of female sperm-storage organs in *Drosophila*. *Evolution* 53: 1804–1822.

Pitnick, S., G. S. Spicer, and T. A. Markow. 1995b. How long is a giant sperm? *Nature* 375: 109.

——. 1997. Phylogenetic examination of female incorporation of ejaculate in *Drosophila*. *Evolution* 51: 833–845.

Plagge, E. 1938. Weitere Untersuchungen über das Verpuppungshormon bei Schmetterlingen. *Biol. Zentralbl.* 58: 1–12.

Plautz, J. D., M. Kaneko, J. C. Hall, and S. A. Kay. 1997. Independent photoreceptive circadian clocks throughout *Drosophila*. *Science* 278: 1632–1635.

Pluta, A. F., A. M. Mackay, A. M. Ainsztein, I. G. Goldberg, and W. C. Earnshaw. 1995. The centromere: hub of chromosomal activities. *Science* 270: 1591–1594.

Pohlhammer, K. 1978. Die besamung der Eier bei der australischen Grille *Teleogryllus commodus* Walker (Insecta, Orthoptera). *Zool. Jb. Anat.* 99: 157–173.

Poinar, G. O., Jr. 1992. *Life in amber.* Stanford University Press.

Poirié, M., E. Niederer, and M. Steinmann-Zwicky. 1995. A sex-specific number of germ cells in embryonic gonads of *Drosophila*. *Development* 121: 1867–1873.

Pollock, D. A., and B. B. Normark. 2002. The life cycle of *Micromalthus debilis* Le Conte, 1878 (Coleoptera: Archostemata: Micromalthidae): historical review and evolutionary perspective. *J. Zool. Syst. Evol.* 40: 105–112.

Popadíc, A., A. Abzhanov, D. Rusch, and T. C. Kaufman. 1998a. Understanding the genetic basis of morphological evolution: the role of homeotic genes in the diversification of the arthropod bauplan. *Int. J. Dev. Biol.* 42: 453–461.

Popadíc, A., G. Panganiban, D. Rusch, W. A. Shear, and T. C. Kaufman. 1998b. Molecular evidence for the gnathobasic derivation of arthropod mandibles and for the appendicular origin of the labrum and other structures. *Dev. Genes Evol.* 208: 142–150.

Posakony, J. W. 1994. Nature versus nurture: asymmetric cell divisions in *Drosophila* bristle development. *Cell* 76: 415–418.

Poulson, D. 1945. Chromosomal control of embryogenesis in *Drosophila*. *Am. Nat.* 79: 340–363.

——. 1950. Histogenesis, organogenesis, and differentiation in the embryo of *Drosophila melanogaster* Meigen. In M. Demerec (ed.), *Biology of Drosophila*, pp. 168–274. Wiley.

Powell, J. R. 1996. Insect evolutionary biology: lessons from *Drosophila*. In *Proceedings of the 20th International Congress of Entomology, Firenze*, pp. 29–31.

Power, M. E. 1943. The effect of reduction in numbers of ommatidia upon the brain of *Drosophila melanogaster* *J. Exp. Zool.* 94: 33–71.

Poyarkoff, E. 1914. Essai d'une théorie de la nymphe des insectes holométaboles. *Arch. Zool. Exp. Gen.* 54: 221–265.

Pratt, G. E., G. C. Unnithan, K. F. Fak, N. R. Siegel, and R. Feyereisin. 1997. Structure-activity studies reveal two allatostatin receptor types in corpora allata of *Diploptera punctata*. *J. Insect Physiol.* 43: 627–634.

Pratt, S. A. 1968. An electron microscope study of Nebenkern formation and differentiation in spermatids of *Murgantia histrionica* (Hemiptera, Pentatomidae). *J. Morphol.* 126: 31–65.

Predel, R., and M. Eckert. 2000. Neurosecretion: peptidergic systems in insects. *Naturwissenschaften* 87: 343–350.

Presgraves, D. C., R. H. Baker, and G. S. Wilk. 1999. Coevolution of sperm and female reproductive tract morphology in stalk-eyed flies. *Proc. R. Soc. Lond. B Biol. Sci.* 266: 1041–1047.

Prestwich, G. D., H. Wojtasek, A. J. Lentz, and J. M. Rabinovich. 1996. Biochemistry of proteins that bind and metabolize juvenile hormones. *Arch. Insect Biochem. Physiol.* 32: 407–419.

Price, C. S. C. 1997. Conspecific sperm precedence in *Drosophila*. *Nature* 388: 663–666.

Price, C. S. C., K. A. Dyer, and J. A. Coyne. 1999. Sperm competition between *Drosophila* males involves both displacement and incapacitation. *Nature* 400: 449–452.

Pringle, J. A. 1938. A contribution to the knowledge of *Micromalthus debilis* LeC. (Coleoptera). *Trans. R. Entomol. Soc. Lond.* 87: 287–290.

Pritchard, G., M. H. McKee, E. M. Pike, G. J. Scrimgeour, and J. Zloty. 1993. Did the first insects live in water or in air? *Biol. J. Linn. Soc.* 49: 31–44.

Pritsch, M., and J. Büning. 1989. Germ cell cluster in the panoistic ovary of Thysanoptera (Insecta). *Zoomorphology* 108: 309–313.

Proctor, H. C. 1998. Indirect sperm transfer in arthropods: behavioral and evolutionary trends. *Annu. Rev. Entomol.* 43: 153–174.

Proctor, H. C., R. L. Baker, and D. T. Gwynne. 1995. Mating behaviour and spermatophore morphology: a comparative test of the female-choice hypothesis. *Can. J. Zool.* 73: 2010–2020.

Prokop, A., and G. Technau. 1991. The origin of postembryonic neuroblasts in the ventral nerve cord of *Drosophila melanogaster*. *Development* 111: 79–88.

Prout, T., and A. G. Clark. 2000. Seminal fluid causes temporarily reduced egg hatch in previously mated females. *Proc. R. Soc. Lond. B Biol. Sci.* 267: 201–203.

Quennedey, A. 1998. Insect epidermal gland cells: ultrastructure and morphogenesis. In F. W. Harrison and M. Locke (eds.), *Microscopic anatomy of invertebrates.* Vol. 11A. *Insecta,* pp. 177–207. Wiley-Liss.

Quennedey, A., and B. Quennedey. 1990. Morphogenesis of the wing anlagen in the mealworm beetle *Tenebrio molitor* during the last larval instar. *Tissue Cell,* 22: 721–740.

Quicke, D. L. J., M. G. Fitton, J. R. Tunstead, S. N. Ingram, and P. V. Gittens. 1994. Ovipositor structure and relationships within the Hymenoptera, with special reference to Ichneumonidae. *J. Nat. Hist.* 28: 635–682.

Quiring, R., U. Walldorf, U. Kloter, and W. J. Quiring. 1994. Homology of the *eyeless* gene in *Drosophila* to the *small eye* gene in mice and *aniridia* in humans. *Science* 265: 785–789.

Raabe, M. 1982. *Insect hormones.* Plenum Press.

——. 1986. Insect reproduction: regulation of successive steps. *Adv. Insect Physiol.* 19: 29–154.

Raabe, T., and M. Heisenberg. 1996. Structural and functional analysis of the *Drosophila* mushroom bodies. In *Proceedings of the 20th International Congress of Entomology, Firenze,* p. 204.

Rachinsky, A., and S. S. Tobe. 1996. Role of second messengers in the regulation of juvenile hormone production in insects with particular emphasis on calcium and phosphoinositide signaling. *Arch. Insect Biochem. Physiol.* 33: 259–282.

Raff, J. W., and D. M. Glover. 1989. Centrosomes and not nuclei initiate pole cell formation in *Drosophila* embryos. *Cell* 57: 611–619.

Raff, R. A. 1996. *The shape of life. Genes, development, and the evolution of animal form.* University of Chicago Press.

Raff, R. A., and T. C. Kaufman. 1983. *Embryos, genes and evolution.* Macmillan.

Raikhel, A. S., and T. S. Dhadialla. 1992. Accumulation of yolk proteins in insect oocytes. *Annu. Rev. Entomol.* 37: 217–251.

Raikhel, A. S., and E. S. Snigirevskaya. 1998. Vitellogenesis. In F. W. Harrison and M. Locke (eds.), *Microscopic anatomy of invertebrates.* Vol. 11C. *Insecta,* pp. 933–955. Wiley-Liss.

Raikhel, A. S., K. Miura, and W. A. Segraves. 1999. Nuclear receptors in mosquito vitellogenesis. *Am. Zool.* 39: 722–735.

Rangarajan, R., L. Gong, and U. Gaul. 1999. Migration and function of glia in the developing *Drosophila* eye. *Development* 126: 3285–3292.

Rauskolb, C., and K. D. Irvine. 1999. Notch-mediated segmentation and growth control of the *Drosophila* leg. *Dev. Biol.* 210: 339–350.

Ready, D. F., T. E. Hanson, and S. Benzer. 1976. Development of the *Drosophila* retina, a neurocrystalline lattice. *Dev. Biol.* 53: 217–240.

Reed, B. H., and T. L. Orr-Weaver. 1997. The *Drosophila* gene *morula* inhibits mitotic functions in the endo cell cycle and the mitotic cell cycle. *Development* 124: 3543–3553.

Rees, H. H. 1985. Biosynthesis of ecdysone. In G. A. Kerkut and L. I. Gilbert (eds.), *Comprehensive insect physiology, biochemistry and pharmacology.* Vol. 7, pp. 249–293. Pergamon Press.

——. 1995. Ecdysteroid biosynthesis and inactivation in relation to function. *Eur. J. Entomol.* 92: 9–39.

Regier, J. C., and F. C. Kafatos. 1985. Molecular aspects of chorion formation. In G. A. Kerkut and L. I Gilbert (eds.), *Comprehensive insect physiology, biochemistry and pharmacology.* Vol. 1, pp. 113–151. Pergamon Press.

Reichert, H., and G. Boyan. 1997. Building a brain: developmental insights in insects. *Trends Neurosci.* 20: 258–264.

Reichmann, V., K.-P. Rehorn, R. Reuter, and M. Leptin. 1998. The genetic control of the distinction between fat body and gonadal mesoderm in *Drosophila. Development* 125: 713–723.

Reichmann, V., U. Irion, R. Wilson, R. Grosskortenhaus, and M. Frasch. 1997. Control of cell fates and segmentation in the *Drosophila* mesoderm. *Development* 124: 2915–2922.

Reinitz, J., and D. H. Sharp. 1995. Mechanism of *eve* stripe formation. *Mech. Dev.* 49: 133–150.

Reith, F. 1925. Die Entwicklung des *Musca*—Eies nach Ausschaltung verschiedener Eibereiche. *Z. Wiss. Zool.* 126: 182–238.

Rempel, J. G. 1940. Intersexuality in Chironomidae induced by nematode parasitism. *J. Exp. Zool.* 84: 261–289.

——. 1975. The evolution of the insect head: the endless dispute. *Quaest. Entomol.* 11: 7–25.

Rempel, J. G., and N. S. Church. 1965. The embryology of *Lytta viridana* LeConte (Coleoptera: Meloidae). I. Maturation, fertilization and cleavage. *Can. J. Zool.* 43: 915–924.

——. 1969. The embryology of *Lytta viridana* LeConte (Coleoptera: Meloidae). V. The blastoderm, germ layers, and body segments. *Can. J. Zool.* 47: 1157–1171.

——. 1971. The embryology of *Lytta viridana* LeConte (Coleoptera: Meloidae). VII. Eighty-eight to 132 h, the appendages, the cephalic apodemes, and head segmentation. *Can. J. Zool.* 49: 1571–1581.

Renne, P. R., D. B. Karner, and K. R. Ludwig. 1998. Absolute ages aren't exactly. *Science* 282: 1840–1841.

Restifo, L. L., and K. White. 1991. Mutations in a steroid hormone-regulated gene disrupt the metamorphosis of the central nervous system in *Drosophila. Dev. Biol.* 148: 174–194.

——. 1992. Mutations in a steroid hormone-regulated gene disrupt the metamorphosis of internal tissues in *Drosophila*: salivary glands, muscle, and gut. *W. Roux's Arch. Dev. Biol.* 201: 221–234.

Restifo, L. L., and T. G. Wikan. 1998. A juvenile hormone agonist reveals distinct developmental pathways mediated by ecdysone-inducible *Broad Complex* transcription factors. *Dev. Genet.* 22: 141–159.

Retnakaran, A., and J. Percy. 1985. Fertilization and special modes of reproduction. In G. A. Kerkut and L. I. Gilbert (eds.), *Comprehensive insect physiology, biochemistry and pharmacology.* Vol. 1, pp. 231–293. Pergamon Press.

Reuter, R. 1994. The gene *serpent* has homeotic properties and specifies endoderm versus ectoderm within the *Drosophila* gut. *Development* 120: 1123–1135.

Reuter, R., and M. Leptin. 1994. Interacting functions of *snail, twist* and *huckebein* during the early development of germ layers in *Drosophila. Development* 120: 1137–1150.

Reynolds, S. E. 1985. Hormonal control of cuticle mechanical properties. In G. A. Kerkut and L. I. Gilbert (eds.), *Comprehensive insect physiology, biochemistry and pharmacology.* Vol. 8, pp. 335–351. Pergamon Press.

Reynolds, S. E., and R. I. Samuels. 1996. Physiology and biochemistry of insect molting fluid. *Adv. Insect Physiol.* 26: 157–232.

Rhyu, M. S., L. Y. Jan, and Y. N. Jan. 1994. Asymmetric distribution of Numb protein during division of the sensory organ precursor cell confers distinct fates to daughter cells. *Cell* 76: 477–491.

Rice, W. R. 1994. Degeneration of a non-recombining chromosome. *Science* 263: 230–232.

Richard, D. S., A. E. Armin, and L. I. Gilbert. 1993. A reappraisal of the hormonal regulation of larval fat body histolysis in female *Drosophila melanogaster*. *Experientia* 49: 150–156.

Richard-Mercier, N. 1982. Le contrôle de la différenciation sexuelle des insectes: étude chez une coleoptère chrysomelide, le doryphore. *Ann. Biol.* 21: 359–394.

Richards, A. G., and A. Miller. 1937. Insect development analyzed by experimental methods: a review. Part I. Embryonic stages. *J. N.Y. Entomol. Soc.* 45: 1–60.

Richards, G. 1981. Insect hormones in development. *Biol. Rev.* 56: 501–549.

Richards, O. W. 1961. An introduction to the study of insect polymorphism. *Symp. R. Entomol. Soc. Lond.* 1: 1–10.

Richter, K., E. Peschke, and D. Peschke. 1999. Effect of melatonin on the release of prothoracicotropic hormone from the brain of *Periplaneta americana* (Blattodea: Blattidae). *Eur. J. Entomol.* 96: 341–345.

Riddiford, L. M. 1985. Hormone action at the cellular level. In G. A. Kerkut and L. I. Gilbert (eds.), *Comprehensive insect physiology, biochemistry and pharmacology*. Vol. 8, pp. 37–84. Pergamon Press.

——. 1990. Hormonal control of sequential gene expression in lepidopteran epidermis. In E. Ohnishi and E. Ishizaki (eds.), *Molting and metamorphosis*, pp. 207–222. Japanese Scientific Society Press.

——. 1992. Molecular approaches to insect endocrinology. In J. M. Crampton and P. Eggleston (eds.), *Insect molecular science*, pp. 226–240. 16th Symp. R. Entomol. Soc. Lond. Academic Press.

——. 1993a. Hormones and *Drosophila* development. In M. Bate and A. Martinez Arias (eds.), *The development of Drosophila melanogaster*. Vol. 2, pp. 899–939. Cold Spring Harbor Laboratory Press.

——. 1993b. Hormone receptors and the regulation of insect metamorphosis. *Receptor* 3: 203–209.

——. 1994. Cellular and molecular actions of juvenile hormone. I. General considerations and premetamorphic actions. *Adv. Insect Physiol.* 24: 213–274.

——. 1996. Juvenile hormone: the status of its "status quo" action. *Arch. Insect Biochem. Physiol.* 32: 271–286.

Riddiford, L. M., and K. Hiruma. 1991. Juvenile hormone and retenoic acid receptors in *Manduca* epidermis. *Insect Biochem.* 21: 7–16.

Riddiford, L. M., and J. W. Truman. 1993. Hormone receptors and the regulation of insect metamorphosis. *Am Zool.* 33: 340–347.

Riddiford, L. M. and J. H. Willis (eds.). 1996. VI Conference on juvenile hormones. *Arch. Insect Biochem. Physiol.* 32: 265–674.

Riddiford, L. M., K. Hiruma, L. Lon, and B. Zhar. 1999. Regulation and role of nuclear receptors during larval molting and metamorphosis of Lepidoptera. *Am. Zool.* 39: 736–746.

Ridley, M. 1988. Mating frequency and fecundity in insects. *Biol. Rev.* 63: 509–549.

——. 1989. The incidence of sperm displacement in insects: four conjectures, one corroboration. *Biol. J. Linn. Soc.* 38: 349–367.

Riek, E. F. 1970. Hymenoptera (wasps, bees, ants). In I. M. Mackerras (ed.), *The insects of Australia. A textbook for students and research workers*, pp. 867–959. 1st ed. Melbourne University Press.

Riemann, J. G., and J. M. Giebultowicz. 1991. Secretion in the upper vas deferens of the gypsy moth correlated with the circadian rhythm of sperm release from the testes. *J. Insect Physiol.* 37: 33–62.

Riemann, J. G., and B. J. Thorson. 1971. Sperm maturation in the male and female genital tracts of *Anagasta kühniella* (Lepidoptera: Pyralidae). *Int. J. Insect Morphol. Embryol.* 1: 11–19.

Riemann, J. G., B. J. Thorson, and R. L. Ruud. 1974. Daily cycle of release of sperm from the testes of the Mediterranean flour moth. *J. Insect Physiol.* 20: 195–207.

Riese, J., G. Tremml, and M. Bienz. 1997. *D-fos*, a target gene of Decapentaplegic signaling with a critical role during *Drosophila* endoderm induction. *Development* 124: 3353–3361.

Ripparbelli, M. G., R. Stouthamer, and R. Dallai. 1998. Microtubule organization during the early development of the parthenogenetic development of the hymenopteran *Muscidifurax uniraptor*. *Dev. Biol.* 195: 89–99.

Rivera-Pomar, R., and H. Jäckle. 1996. From gradients to stripes in *Drosophila* embryogenesis: filling in the gaps. *Trends Genet.* 12: 478–483.

Rivera-Pomar, R., D. Niessing, U. Schmidt-Ott, W. J. Gehring, and H. Jäckle. 1996. RNA binding and translational suppression by Bicoid. *Nature* 379: 746–749.

Rivlin, P. K., A. M. Schneiderman, and R. Booker. 2000. Imaginal pioneers prefigure the formation of adult thoracic muscles in *Drosophila*

melanogaster. *Dev. Biol.* 222: 450–459.

Roberts, B., J. M. Whitten, and L. I. Gilbert. 1976. Patterns of incorporation of tritiated thymidine by the dorsal polytene foot-pad nuclei of *Sarcophaga bullata* (Sarcophagidae: Diptera). *Chromosoma* 54: 127–140.

Roberts, R. G., T. E. Flannery, L. K. Ayliffe, H. Yoshida, J. M. Olley, G. J. Prideaux, G. M. Laslett, A. Baynes, M. A. Smith, R. Jones, and B. L. Smith. 2001. New ages for the last Australian megafauna: continent-wide extinction about 46,000 years ago. *Science* 292: 1888–1891.

Roff, D. A. 1992. *The evolution of life histories: theory and analysis*. Chapman and Hall.

Rogers, B. T., and T. C. Kaufman. 1996. Structure of the insect head as revealed by the EN protein pattern in developing embryos. *Development* 122: 3419–3432.

——. 1997. Structure of the insect head in ontogeny and phylogeny: a view from *Drosophila*. *Int. Rev. Cytol.* 174: 1–84.

Rogers, B. T., M. D. Peterson, and T. C. Kaufman. 1997. Evolution of the insect body plan as revealed by the *Sex combs reduced* expression pattern. *Development* 124: 149–157.

Rogina, B., R. A. Reenan, S. P. Nilsen, and S. L. Helfanc. 2000. Extended life-span conferred by cotransporter gene mutations in *Drosophila*. *Science* 290: 2137–2140.

Röller, H., K. H. Dahm, C. C. Sweely, and B. M. Trost. 1967. The structure of the juvenile hormone. *Ang. Chem. Int. Ed. in English* 6: 179–180.

Ronquillo, M. C., and W. R. Horsfall. 1969. Genesis of the reproductive system of mosquitoes. I. Female of *Aedes stimulans* (Walker). *J. Morphol.* 129: 249–280.

Roosen-Runge, E. C. 1977. *The process of spermatogenesis*. Cambridge University Press.

Roote, C. E., and S. Zusman. 1995. Functions for PS integrins in tissue adhesion, migration and shape changes during early embryonic development in *Drosophila*. *Dev. Biol.* 169: 322–336.

Röseler, P.-F., and I. Röseler. 1979. Studies on the regulation of the juvenile hormone titre in bumble bee workers, *Bombus terrestris*. *J. Insect Physiol.* 24: 707–713.

Rospars, J. P. 1988. Structure and development of the insect antennodeutocerebral system. *Int. J. Insect Morphol. Embryol.* 17: 243–294.

Rössler, W., L. P. Tolbert, and J. G.

Hildebrand. 1998. Early formation of sexually dimorphic glomeruli in the developing olfactory lobe of the moth *Manduca sexta*. *J. Comp. Neurol.* 396: 415–428.

Roth, L. M. 1973. Inhibition of oocyte development during pregnancy in the cockroach *Eublaberus posticus*. *J. Insect Physiol.* 19: 455–469.

Roth, S., P. Jordan, and R. Karess. 1999. Binuclear *Drosophila* oocytes: consequences and implications for dorsal-ventral patterning in oogenesis and embryogenesis. *Development* 126: 927–934.

Rothschild, M. R. 1965. Fleas. *Sci. Am.* 213(6): 44–53.

Rothschild, M. R., B. Ford, and M. Hughes. 1970. Maturation of the male rabbit flea (*Spilopsyllus cuniculi*) and the oriental rat flea (*Xenopsylla cheopis*): some effects of mammalian hormones on development and impregnation. *Trans. Zool. Soc. Lond.* 32: 105–188.

Roush, W. 1997. Arraying the fly eye. *Science* 277: 640.

Roy, S., and K. VijayRaghavan. 1999. Muscle pattern diversification in *Drosophila*: the story of imaginal myogenesis. *Bioessays* 21: 486–498.

Roy, S., L. S. Shashidara, and K. VijayRaghavan. 1997. Muscles in the *Drosophila* second thoracic segment are patterned independently of autonomous homeotic gene function. *Curr. Biol.* 7: 222–227.

Royet, J., and R. Finkelstein. 1997. Establishing primordia in the *Drosophila* eye-antennal disc: the roles of *decapentaplegic*, *wingless* and *hedgehog*. *Development* 124: 4793–4800.

Rubenstein, E. C. 1979. The role of an epithelial occlusion zone in the termination of vitellogenesis in *Hyalophora cecropia* ovarian follicles. *Dev. Biol.* 71: 115–127.

Rudolph, K. M., G.-J. Liau, A. Daniel, P. Green, A. J. Coursey, V. Hartenstein, and J. A. Lengyel. 1997. Complex regulatory region mediating *tailless* expression in early embryonic patterning and brain development. *Development* 124: 4299–4308.

Ruffieux, L., J.-M. Elouard, and M. Sartori. 1998. Flightlessness in mayflies and its relevance to hypotheses on the origin of insect flight. *Proc. R. Soc. Lond. B Biol. Sci.* 265: 2135–2140.

Rugendorff, A., A. Younossi-Hartenstein, and V. Hartenstein. 1994. Embryonic origin and differentiation of the *Drosophila* heart. *W. Roux's Arch. Dev. Biol.* 203: 266–280.

Ruiz-Gómez, M., N. Coutts, and A. Price. 2000. *Drosophila* dumbfounded: a myoblast attractant essential for fusion. *Cell* 102: 189–198.

Ruiz-Gómez, M., S. Romani, C. Hartmann, H. Jäckle, and M. Bate. 1997. Specific muscle identities are regulated by *Krüppel* during *Drosophila* embryogenesis. *Development* 124: 3407–3417.

Rusch, J., and M. Levine. 1997. Regulation of a *dpp* target gene in the *Drosophila* embryo. *Development* 124: 305–311.

Russell, M. A. 1980. Imaginal discs. In R. Ransom (ed.), *A handbook of Drosophila development*, pp. 95–121. Elsevier Biomedical Press.

——. 1983. Positional information in imaginal discs. In P. Antonelli (ed.), *Mathematical essays on growth and emergence of form*, pp. 169–183. University of Alberta Press.

Rybczynski, R., and L. I. Gilbert. 1995. Prothoracicotropic hormone elicits a rapid, developmentally specific synthesis of β-tubulin in an insect endocrine gland. *Dev. Biol.* 169: 15–28.

Ryerse, J. S. 1979. Developmental changes in Malpighian tubule cell structure. *Tissue Cell.* 11: 533–551.

Sahota, T. S., and S. H. Farris. 1980. Inhibition of flight muscle degeneration by precocene II in the spruce bark beetle, *Dendroctonus rufipennis* (Kirby) (Coleoptera: Scolytidae). *Can. J. Zool.* 58: 378–381.

Sakurai, S., and L. I. Gilbert. 1990. Biosynthesis and secretion of ecdysteroids by the prothoracic glands. In E. Ohnishi and H. Ishizaki (eds.), *Molting and metamorphosis*, pp. 83–106. Japanese Scientific Society Press.

Sakurai, S., J. T. Warren, and L. I. Gilbert. 1991. Ecdysteroid synthesis and molting by the tobacco hornworm, *Manduca sexta*, in the absence of prothoracic glands. *Arch. Insect Biochem. Physiol.* 18: 13–36.

Salecker, I., and J. Boeckh. 1995. Embryonic development of the antennal lobes of a hemimetabolous insect, the cockroach *Periplaneta americana*: light and electron microscopic observations. *J. Comp. Neurol.* 352: 33–54.

Salecker, I., T. R. Clandinin, and S. L. Zipursky. 1998. Hedgehog and Spitz: making a match between photoreceptor axons and their targets. *Cell* 95: 587–590.

Salvini-Plawen, L. V., and E. Mayr. 1977. On the evolution of photoreceptors and eyes. In M. K. Hecht, W. C. Steere, and B. Wallace (eds.), *Evolutionary biology*. Vol. 10, pp. 207–263. Plenum Press.

Salzberg, A., and H. J. Bellen. 1996. Invertebrate vs. vertebrate neurogenesis: variations on the same theme? *Dev. Genet.* 18: 1–10.

Samakovilis, C., N. Hacohen, G. Manning, D. C. Sutherland, K. Guillemin, and M. A. Krasnow. 1996a. Development of the *Drosophila* tracheal system occurs by a series of morphologically distinct but genetically coupled branching events. *Development* 122: 1395–1407.

Samakovilis, C., G. Manning, P. Steneberg, N. Hacohen, R. Cantera, and M. A. Krasnow. 1996b. Genetic control of epithelial tube formation during *Drosophila* tracheal development. *Development* 122: 3531–3536.

Sánchez, I., F. Casares, N. Gorfinkiel, and I. Guerro. 1997. The genital disc of *Drosophila melanogaster*. II. Role of the genes *hedgehog*, *decapentaplegic* and *wingless*. *Dev. Genes Evol.* 207: 229–241.

Sánchez, I., N. Gorfinkiel, and I. Guerro. 2001. Sex determination genes control the development of the *Drosophila* genital disc, modulating the response to Hedgehog, Wingless and Decapentaplegic signals. *Development* 128: 1033–1043.

Sander, K. 1959. Analyse des ooplasmatischen Reaktionssystems von *Euscelis plebejus* Fall. (Cicadina) durch Isolieren und Kombinieren von Keimteilen. I. Die Differenzierungsleistungen vorderer und hinterer Eiteile. *W. Roux's Arch. Entwicklungsmech. Organ.* 151: 430–497.

——. 1960. Analyse des ooplasmatischen Reaktionssystems von *Euscelis plebejus* Fall. (Cicadina) durch Isolieren und Kombinieren von Keimteilen. II. Die Differenzierungsleistungen nach Verlagern von Hinterpolmaterial. *W. Roux's Arch. Entwicklungsmech. Organ.* 151: 660–707.

——. 1968. Mechanismen der Keimeseinrollung (Anatrepsis) im Insekten-Ei. *Verh. Deutsch. Zool. Ges.* 31: 81–89.

——. 1971. Pattern formation in longitudinal halves of leaf hopper eggs (Homoptera) and some remarks on the definition of "embryonic regulation." *W. Roux's Arch. Entwicklungsmech. Organ.* 167: 336–352.

——. 1976a. Morphogenetic movements in insect embryogenesis. In P. A. Lawrence (ed.), *Insect development*, pp. 35–52. R. Entomol. Soc. Lond. Symp. 8. Blackwell Science.

——. 1976b. Specification of the basic

body pattern in insect embryogenesis. *Adv. Insect Physiol.* 12: 125–238.

——. 1984. Embryonic pattern formation in insects: basic concepts and their experimental foundations. In G. M. Malacinski (ed.), *Pattern formation: a primer in developmental biology,* pp. 245–268. Macmillan.

——. 1986. The role of genes in ontogenesis—evolving concepts from 1886–1983 as perceived by an insect embryologist. In T. J. Horder, J. A. Witkowski, and C. C. Wylie (eds.), *A history of embryology,* pp. 363–395. Cambridge University Press.

——. 1997. Pattern formation in insect embryogenesis: the evolution of concepts and mechanisms. *Int. J. Insect Morphol. Embryol.* 25: 349–367.

Sander, K., H. Gutzeit, and H. Jäckle. 1985. Insect embryogenesis: morphology, physiology, genetical and molecular aspects. In G. A. Kerkut and L. I. Gilbert (eds.), *Comprehensive insect physiology, biochemistry and pharmacology.* Vol. 1, pp. 319–385. Pergamon Press.

Sappington, T. W., and A. S. Raikhel. 1998. Molecular characteristics of insect vitellogenins and vitellogenin receptors. *Insect Biochem. Mol. Biol.* 28: 277–300.

Satir, P. 1968. Studies on cilia. III. Further studies on the cilium tip and a "sliding filament" model of ciliary motility. *J. Cell Biol.* 39: 77–94.

Schaller, F. 1979. Significance of sperm transfer and formation of spermatophores in arthropod phylogeny. In A. P. Gupta (ed.), *Arthropod phylogeny,* pp. 587–608. Van Nostrand-Reinhold.

Scharrer, E., and B. Scharrer. 1945. Neurosecretion. *Physiol. Rev.* 25: 171–181.

Schatten, G. 1994. The centrosome and its mode of inheritance: the reduction of the centrosome during gametogenesis and its restoration during fertilization. *Dev. Biol.* 165: 299–335.

Scheiner, S. M. 1992. Grand synthesis in the making. *Science* 258: 1820–1822.

Schlupp, I., C. Marler, and M. J. Ryan. 1994. Benefit to male sailfin mollies of mating with heterospecific females. *Science* 263: 373–374.

Schmid, A., A. Chiba, and C. L. Doe. 1999. Clonal analysis of *Drosophila* embryonic neuroblasts: neural cell types, axon projections and muscle targets. *Development* 126: 4653–4689.

Schmidt, E. L., and C. M. Williams. 1953. Physiology of insect diapause. V. Assay of the growth and differentiation hormone of Lepidoptera by the method of tissue culture. *Biol. Bull* 105: 174–187.

Schmidt, H., C. Rickert, T. Bossing, O. Vef, J. Urban, and G. M. Technau. 1997. The embryonic central nervous system lineages of *Drosophila.* II. Neuroblast lineages derived from the dorsal part of the neuroectoderm. *Dev. Biol.* 189: 186–204.

Schmidt, O. 1980. Insect egg cortex isolated by microsurgery: specific protein pattern and uridine incorporation. *W. Roux's Arch. Entwicklungsmech. Organ.* 188: 23–26.

Schmidt-Ott, U., and G. M. Technau. 1992. Expression of *en* and *wg* in the embryonic head and brain of *Drosophila* indicated a refolded band of seven segment remnants. *Development* 116: 111–125.

——. 1994. Fate-mapping the procephalic region of the embryonic *Drosophila* head. *W. Roux's Arch. Dev. Biol.* 203: 367–373.

Schmidt-Ott, U., M. González-Gaitán, H. Jäckle, and G. M. Technau. 1994b. Number, identity and sequences of the *Drosophila* head segments as revealed by neural elements and their deletion patterns in mutants. *Proc. Natl. Acad. Sci. U.S.A.* 91: 8363–8367.

Schmidt-Ott, U., M. González-Gaitán, and G. M. Technau. 1995. Analysis of neural elements in head-mutant *Drosophila* embryos suggests segmental origin of the optic lobes. *W. Roux's Arch. Dev. Biol.* 205: 31–44.

Schmidt-Ott, U., K. Sander, and G. M. Technau. 1994a. Expression of *engrailed* in embryos of a beetle and five dipteran species with special reference to terminal regions. *W. Roux's Arch. Dev. Biol.* 203: 298–303.

Schmidt-Rhaesa, A., T. Bartolomaeus, C. Lemburg, U. Ehlers, and J. R. Garey. 1998. The position of the Arthropoda in the phylogenetic system. *J. Morphol.* 238: 263–285.

Schmucker, D., H. Jäckle, and U. Gaul. 1997. Genetic analysis of the larval optic nerve projection in *Drosophila. Development* 124: 937–948.

Schneiderman, A. M., and J. G. Hildebrand. 1985. Sexually dimorphic development of the insect olfactory pathway. *Trends Neurosci.* 6: 844–846.

Schnetter, M. 1934. Physiologische Untersuchungen über das Differenzierungszentrum in der Embryonalentwicklung der Honigbiene. *W. Roux's Arch. Entwicklungsmech. Organ.* 131: 285–323.

Schober, M., M. Schaefer, and J. A. Knoblich. 1999. Bazooka recruits Inscuteable to orient asymmetric cell divisions in *Drosophila* neuroblasts. *Nature* 402: 548–551.

Schöck, F., J. Reischl, E. Wimmer, H. Taubert, B. A. Purnell, and H. Jäckle. 2000. Phenotypic suppression of *empty spiracles* is prevented by *buttonhead. Nature* 405: 351–354.

Schoeller, J. 1964. Recherches descriptives et expérimentales sur la céphalogenèse de *Calliphora erythrocephala* (Meigen) au cours des développement embryonnaire et postembryonnaire. *Arch. Zool. Exp. Gener.* 103: 1–216.

Scholtz, G. 1998. Cleavage, germ band formation and head segmentation: the ground pattern of Euarthropoda. In R. A. Fortey and R. H. Thomas (eds.), *Arthropod relationships,* pp. 317–332. Systematics Association Special Vol. 55. Chapman and Hall.

Scholtz, G., B. Mittmann, and M. Gerberding. 1998. The pattern of *Distal-less* expression in the mouthparts of crustaceans, myriapods and insects: new evidence for a gnathobasic mandible and the common origin of Mandibulata. *Int. J. Dev. Biol.* 42: 801–810.

Schooley, D. A., and F. C. Barker. 1985. Juvenile hormone biosynthesis. In G. A. Kerkut and L. I. Gilbert (eds), *Comprehensive insect physiology, biochemistry and pharmacology.* Vol. 7, pp. 363–389. Pergamon Press.

Schooneveld, H. 1998. Neurosecretion. In F. W. Harrison and M. Locke (eds.), *Microscopic anatomy of invertebrates.* Vol. 11B. Insecta, pp. 467–486. Wiley-Liss.

Schrader, F. 1960a. Cytological and evolutionary implications of aberrant chromosome behavior in the harlequin lobe of some Pentatomidae (Heteroptera). *Chromosoma* 11: 103–128.

——. 1960b. Evolutionary aspects of aberrant meiosis in some Pentatominae (Heteroptera). *Evolution* 14: 498–508.

Schreiner, B. 1977. Vitellogenesis in the milkweed bug, *Oncopeltus fasciatus* Dallas (Hemiptera). A light and electron microscope investigation. *J. Morphol.* 151: 35–80.

Schröder, R., and K. Sander. 1993. A comparison of transplantable *bicoid* activity and partial *bicoid* homeobox sequences in several *Drosophila* and blow fly species (Calliphoridae). *W. Roux's Arch. Dev. Biol.* 203: 34–43.

Schröter, U., and D. Malun. 2000. Formation of antennal lobe and mushroom body neuropils during metamorphosis in the honeybee, *Apis mellifera. J. Comp. Neurol.* 422: 229–245.

Schubiger, M., and J. W. Truman. 2000.

The RxR ortholog USP suppresses early metamorphic processes in *Drosophila* in the absence of ecdysteroids. *Development* 127: 1151–1159.

Schuh, R. T., and P. Stys. 1991. Phylogenetic analysis of cimicomorphan family relationships (Heteroptera). *J. N.Y. Entomol. Soc.* 99: 298–350.

Schultz, J., and J. C. Regier. 2000. Phylogenetic analysis of arthropods using two nuclear protein-encoding genes supports a crustacean + hexapod clade. *Proc. R. Soc. Lond. B Biol. Sci.* 267: 1011–1019.

Schulz, C., R. Schröder, B. Hausdorf, C. Wolff, and D. Tautz. 1998. A *caudal* homologue in the short germ band beetle, *Tribolium* shows similarities to both the *Drosophila* and the vertebrate *caudal* expression patterns. *Dev. Genes Evol.* 208: 283–289.

Schüpbach, P. M., and R. Camenzind. 1983. Germ cell lineage and follicle formation in paedogenetic development of *Mycophila spyereri* Barnes (Diptera: Cecidomyiidae). *Int. J. Insect Morphol. Embryol.* 12: 211–223.

Schütt, C., and R. Nöthiger. 2000. Structure, function and evolution of the sex-determining system in dipteran insects. *Development* 127: 667–677.

Schwalm, F. E. 1965. Zell und Mitosenmuster der normalen und nach röntgenbestralung regulierunden Keimanlage von *Gryllus domesticus*. *Z. Morph. Okol. Tiere* 55: 915–1023.

———. 1988. *Insect morphogenesis*. Monogr. Dev. Biol. 20. Karger.

Scotese, C. R. 1997a. *Phanerozoic plate tectonic reconstructions*. Paleomap Progress Rep. 36. Department of Geology. University of Texas, Arlington.

———. 1997b. *Paleogeographic atlas*. Paleomap Progress Rep. 90-0497. Paleomap Project. University of Texas, Arlington.

Scott, A. C. 1941. Reversal of sex production in *Micromalthus*. *Biol. Bull.* 81: 420–431.

Scott, A. C., J. Stephanson, and W. C. Chaloner. 1992. Interaction and coevolution of plants and arthropods during the Palaeozoic and Mesozoic. *Philos. Trans. R. Soc. Lond. B. Biol. Sci.* 335: 129–165.

Scott, M. P., and A. J. Weiner. 1984. Structural relationships among genes that control development: sequence homology between the *Antennapedia*, *Ultrabithorax* and *fushi tarazu* loci of *Drosophila*. *Proc. Natl. Acad. Sci. U.S.A.* 81: 4115–4119.

Scudder, G. G. E. 1971. Comparative morphology of insect genitalia. *Annu. Rev. Entomol.* 16: 379–406.

Searle, J. B., and T. E. Mittler. 1981. Embryogenesis and the production of males by apterous viviparae of the green peach aphid *Myzus persicae* in relation to photoperiod. *J. Insect Physiol.* 27: 145–153.

Sedlak, B. J. 1985. Structure of endocrine glands. In G. A. Kerkut and L. I. Gilbert (eds.), *Comprehensive insect physiology, biochemistry and pharmacology*. Vol. 7, pp. 25–60. Pergamon Press.

Seecoomar, M., S. Aggarwal, K. Vani, G. Yang, and J. Mohler. 2000. *knot* is required for the hypopharyngeal lobe and its derivatives in the *Drosophila* embryo. *Mech. Dev.* 91: 209–215.

Segal, D., and T. Sprey. 1984. The dorsal/ventral compartment boundary in *Drosophila*: coincidence with the prospective operculum seam. *W. Roux's Arch. Dev. Biol.* 193: 133–138.

Segraves, W. A. 1991. Something old, something new: the steroid receptor superfamily in *Drosophila*. *Cell* 67: 225–228.

Segraves, W. A., and C. Woldin, 1993. The E75 gene of *Manduca sexta* and comparison with its *Drosophila* homolog. *Insect Biochem. Mol. Biol.* 23: 91–97.

Sehar, T. C., and M. Leptin. 2000. Tribbles, a cell-cycle brake that coordinates proliferation and morphogenesis during *Drosophila* gastrulation. *Curr. Biol.* 10: 623–629.

Sehnal, F. 1985a. Morphology of insect development. *Annu. Rev. Entomol.* 30: 89–109.

———. 1985b. Growth and life cycles. In G. A. Kerkut and L. I. Gilbert (eds.), *Comprehensive insect physiology, biochemistry and pharmacology*. Vol. 2, pp. 1–86. Pergamon Press.

Sehnal, F., P. Svácha, and J. Zrzavy. 1996. Evolution of insect metamorphosis. In L. I. Gilbert, J. R. Tata, and B. G. Atkinson (eds.), *Metamorphosis. Postembryonic reprogramming of gene expression in amphibian and insect cells*, pp. 3–58. Academic Press.

Seidel, F. 1924. Die Geschlechtsorgane in der embryonalen Entwicklung von *Pyrrhocoris apterus* L. *Z. Morphol. Okol. Tiere* 1: 429–506.

———. 1929. Untersuchungen über das Bildungsprinzip der Keimanlage bei der Libelle *Platycnemis pennipes*. I–IV. *W. Roux's Arch. Entwicklungsmech. Organ.* 119: 322–440.

Seiler, J. 1969. Intersexuality in *Solenobia triquetrella* F. R. and *Lymantria dispar* L. (Lepid.). Questions of determination. *Mon. Zool. Ital. (N. S.)* 3: 185–212.

Seimiya, M., and W. J. Gehring. 2000. The *Drosophila* homeobox gene *optix* is capable of inducing ectopic eyes by an *eyeless*-independent mechanism. *Development* 127: 1879–1886.

Selander, R. B. 1991. Meloidae (Tenebrionoidea). Blister beetles, oil beetles. In F. W. Stehr (ed.), *Immature insects*. Vol. 2, pp. 530–534. Kendall/Hunt.

Selivon, D., J. S. Morgante, and A. L. P. Perondini. 1997. Egg size, yolk mass extrusion and hatching behavior in two cryptic species of *Anastrepha fraterculus* (Wiedemann) (Diptera, Tephritidae). *Brazilian J. Genet.* 20: 587–594.

Selivon, D., J. S. Morgante, A. F. Ribeiro, and A. L. P. Perondini. 1996. Extrusion of masses of yolk during embryonary development of the fruit fly *Anastrepha fraterculus*. *Invert. Reprod. Dev.* 29: 1–7.

Sen, J., J. S. Goltz, L. Stevens, and D. Stein. 1998. Spatially restricted expression of *pipe* in the *Drosophila* egg chamber defines embryonic dorsal-ventral polarity. *Cell* 95: 471–481.

Senti, K.-A., K. Keleman, F. Eisenhaber, and B. J. Dickson. 2000. *brakeless* is required for lamina targeting of R1-R6 axons in the *Drosophila* visual system. *Development* 127: 2291–2301.

Sereno, P. C. 1999. The evolution of dinosaurs. *Science* 284: 2137–2147.

Seugnet, L., P. Simpson, and M. Haenlin. 1997. Transcriptional regulation of *Notch* and *Delta*: requirement for neuroblast segregation in *Drosophila*. *Development* 124: 2015–2025.

Sevala, V. M., and B. G. Loughton. 1992. Insulin-like peptides during oogenesis and embryogenesis in *Locusta migratoria*. *Invert. Reprod. Dev.* 21: 187–191.

Sgro, C. M., and L. Partridge. 2000. Evolutionary responses of the life history of wild-caught *Drosophila melanogaster* to two standard methods of laboratory culture. *Am. Nat.* 156: 341–353.

Shabica, C. W., and A. A. Hay (eds.). 1997. *Richardson's guide to the fossil fauna of Mazon Creek*. Northeastern Illinois. University.

Shankland, M., and D. Bentley. 1983. Sensory receptor differentiation and axonal pathfinding in the cercus of the grasshopper embryo. *Dev. Biol.* 97: 468–482.

Sharma, A. C., and M. Brand. 1998. Evolution and homology of the nervous system: cross phylum rescues of *otd/Otx* genes. *Trends Genet.* 14: 211–214.

Sharov, A. G. 1966. *Basic arthropodan stock*. Pergamon Press.

Sharp, J. G., G. C. Rogers, and J. M. Scholey. 2000. Microtubule motors in mitosis. *Nature* 407: 41–47.

Sharplin, J. 1965. Replacement of the tentorium of *Periplaneta americana* Linnaeus during ecdysis. *Can. Entomol.* 97: 947–951.

Shear, W. A. 1998. The fossil record and evolution of the Myriapoda. In R. A. Fortey and R. H. Thomas (eds.), *Arthropod relationships*, pp. 211–219. Systematics Association Special Vol. 55. Chapman and Hall.

Shear, W. A., and J. Kukalova-Peck. 1990. The ecology of Paleozoic terrestrial arthropods: the fossil evidence. *Can. J. Zool.* 68: 1807–1834.

Shelton, P. M. J., H.-D. Pfannenstiel, and E. Wachmann. 1983. Regeneration of the eye margin in *Periplaneta americana* (Insecta, Blattodea). *J. Embryol. Exp. Morphol.* 76: 9–25.

Shen, W., and G. Mordon. 1997. Ectopic eye development in *Drosophila* induced by directed *dachshund* expression. *Development* 124: 45–52.

Shepherd, D., and C. M. Bate. 1990. Spatial and temporal pattern of neurogenesis in the embryo of the locust (*Schistocerca gregaria*). *Development* 108: 83–96.

Shepherd, D., and S. A. Smith. 1996. Central projections of persistent larval sensory neurons prefigure adult sensory pathways in the CNS of *Drosophila*. *Development* 122: 2375–2384.

Shepherd, J. G. 1974. Activation of saturniid moth sperm by a secretion of the male reproductive tract. *J. Insect Physiol.* 20: 2107–2122.

Sheppard, P. M., J. R. G. Turner, K. S. Brown, W. W. Benson, and M. C. Singer. 1985. Genetics and the evolution of Muellerian mimicry in *Heliconius* butterflies. *Philos. Trans. R. Soc. Lond. B Biol Sci.* 308: 433–613.

Sherk, T. E. 1978a. Development of the compound eyes of dragonflies (Odonata). II. Development of the larval compound eyes. *J. Exp. Zool.* 203: 47–59.

——. 1978b. Development of the compound eyes of dragonflies (Odonata). IV. Development of the adult compound eyes. *J. Exp. Zool.* 203: 183–200.

Shi, L., T. C. Hsu, S. Pathak, and T. A. Granovsky. 1982. Behavior of centrioles in spermatogenesis of the Madagascar hissing cockroach. *J. Heredity* 73: 408–412.

Shimizu, I., S. Aoki, and T. Ishikawa. 1997. Neuroendocrine control of diapause hormone secretion in the silkworm, *Bombyx mori*. *J. Insect Physiol.* 43: 1101–1109.

Shingyogi, C., H. Higuchi, M. Yoshimura, E. Katayama, and T. Yanagida. 1998. Dynein arms are oscillating force generators. *Nature* 393: 711–714.

Shubin, N., C. Tabin, and S. Carroll. 1997. Fossils, genes and the evolution of animal limbs. *Nature* 388: 639–648.

SiDong, P. D., J. Chu, and G. Panganiban. 2000. Coexpression of the homeobox genes *Distal-less* and *homothorax* determines *Drosophila* antennal identity. *Development* 127: 209–216.

Silberglied, R. E., J. G. Shepherd, and J. L. Dickinson. 1984. Eunuchs: the role of apyrene sperm in Lepidoptera? *Am. Nat.* 123: 255–265.

Simpson, G. G. 1953. *The major features of evolution*. Columbia University Press.

Simpson, P., R. Woehl, and K. Usui. 1999. The development and evolution of bristle patterns in Diptera. *Development* 126: 1349–1364.

Simpson, S. J., A. R. McCaffery, and B. F. Hägele. 1999. A behavioural analysis of phase change in the desert locust. *Biol. Rev.* 74: 461–480.

Singer, M. A., M. Hortsch, C. S. Goodman, and D. Bentley. 1992. Annulin, a protein expressed at limb segment boundaries in the grasshopper embryo, is homologous to protein cross-linking transglutaminases. *Dev. Biol.* 154: 143–159.

Siva-Jothy, M. T. 1997. Odonate ejaculate structure and mating systems. *Odonatologica* 26: 415–437.

Siwinski, J. 1980. Sexual selection in insect sperm. *Fla. Entomol.* 63: 99–111.

Skaer, H. 1989. Cell division in Malpighian tubule development in *Drosophila melanogaster* is regulated by a single tip cell. *Nature* 342: 566–569.

——. 1993. The alimentary canal. In M. Bate and A. Martinez Arias (eds.), *The development of* Drosophila melanogaster. Vol. 2, pp. 941–1012. Cold Spring Harbor Laboratory Press.

——. 1998. Who pulls the string to pattern cell division in *Drosophila*? *Trends Genet.* 14: 337–339.

Skeath, J. B. 1999. At the nexus between pattern formation and cell type specification: the generation of individual neuroblast fates in the *Drosophila* embryonic central nervous system. *Bioessays* 21: 922–931.

Slack, J. M. W. 1983. *From egg to embryo. Determinative events in early development*. Developmental and Cell Biology 13. Cambridge University Press.

——. 1997. Book review. (Gerhart and Kirschner. 1997). *Nature* 387: 866–867.

Slack, J. M. W., P. W. H. Holland, and C. F. Graham. 1993. The zootype and the phylotypic stage. *Nature* 361: 490–492.

Sláma, K. 1998. The prothoracic gland revisited. *Ann. Entomol. Soc. Am.* 91: 168–174.

Sláma, K., and F. Weyda. 1997. The all-or-none rule in morphogenetic action of juvenile hormone on insect epidermal cells. *Proc. R. Soc. Lond. B Biol. Sci.* 264: 1463–1470.

Slifer, E. H. 1937. The origin and fate of the membranes surrounding the grasshopper egg; together with some experiments on the source of the hatching enzyme. *Q. J. Microsc. Sci.* 79: 493–506.

Small, S., A. Blair, and M. Levine. 1992. Regulation of *even-skipped* stripe 2 in the *Drosophila* embryo. *EMBO J.* 11: 4047–4057.

Smith, A. V., and T. L. Orr-Weaver. 1991. The regulation of the cell cycle during *Drosophila* embryogenesis: the transition to polyteny. *Development* 112: 997–1008.

Smith, E. L. 1969. Evolutionary morphology of external insect genitalia. *Ann. Entomol. Soc. Am.* 62: 1051–1079.

Smith, J. M. 1978. *The evolution of sex*. Cambridge University Press.

Smith, S. G. 1971. Parthenogenesis and polyploidy in beetles. *Am. Zool.* 11: 341–349.

Smith, S. G., and N. Virkki. 1978. *Coleoptera. Animal cytogenetics.* Vol. 3. *Insecta.* 5. Gebrüder Borntraeger.

Smith, S. L. 1985. Regulation of ecdysteroid titer: synthesis. In G. A. Kerkut and L. I. Gilbert (eds.), *Comprehensive insect physiology, biochemistry and pharmacology.* Vol. 7, pp. 295–341. Pergamon Press.

Smith, W. A. 1993. Second messengers and the action of prothoracicotropic hormone in *Manduca sexta*. *Am. Zool.* 33: 330–339.

——. 1995. Regulation and consequences of cellular changes in the prothoracic glands of *Manduca sexta* during the last larval instar: a review. *Arch. Insect Biochem. Physiol.* 30: 271–293.

Smith, W. A., and W. L. Cambert. 1985. Role of cyclic nucleotides in hormone action. In G. A. Kerkut and L. I. Gilbert (eds.), *Comprehensive insect physiology, biochemistry and pharmacology.* Vol. 8, pp. 263–299. Pergamon Press.

Smith, W. A., and H. F. Nijhout. 1982. Synchrony of juvenile hormone-sensitive periods for internal and external development in last-instar larvae of the milkweed bug, *Oncopeltus fasciatus*. *J. Insect Physiol.* 28: 797–803.

——. 1983. *In vitro* stimulation of cell death in the moulting glands of *Oncopeltus fasciatus* by 20-hydroxyecdysone. *J. Insect Physiol.* 29: 169–176.

Smithers, C. N. 1995. *Psilopsocus mimulus* Smithers (Psocoptera: Psilopsocidae), the first, known wood-boring psocopteran. *J. Aust. Entomol. Soc.* 34: 117–120.

Snodgrass, R. E. 1924. Anatomy and metamorphosis of the apple maggot, *Rhagoletis pomonella* Walsh. *J. Agric. Res.* 28(1): 1–36; pl. 1–6.

——. 1933. Morphology of the insect abdomen. Part II. The genital ducts and the ovipositor. *Smithson. Misc. Collect.* 89(8): 1–148.

——. 1935. *Principles of insect morphology*. McGraw-Hill.

——. 1954. Insect metamorphosis. *Smithson. Misc. Collect.* 122(9): 1–124.

——. 1956. *Anatomy of the honey bee*. Comstock, Cornell University Press.

——. 1957. A revised interpretation of the external reproductive organs of male insects. *Smithson. Misc. Collect.* 136(6): 1–60.

——. 1959. The anatomical life of the mosquito. *Smithson. Misc. Collect.* 139(8): 1–87.

Soller, M., M. Bownes, and E. Kubli. 1999. Control of oocyte maturation in sexually mature *Drosophila* females. *Dev. Biol.* 208: 337–351.

Sommer, R. J., and D. Tautz. 1993. Involvement of an orthologue of the *Drosophila* pair-rule gene *hairy* in segment formation of the short-germ band embryo of *Tribolium* (Coleoptera). *Nature* 361: 448–450.

Sonnenschein, M., and C. L. Hauser. 1990. Presence of only eupyrene spermatozoa in adult males of the genus *Micropteryx* Hübner and its phylogenetic significance (Lepidoptera: Zeugloptera: Micropterygidae). *Int. J. Insect Morphol. Embryol.* 19: 269–276.

Spane, E. P., and C. L. Doe. 1995. The *prospero* transcription factor is asymmetrically localized to the cell cortex during neuroblast mitosis in *Drosophila*. *Development* 121: 3187–3195.

Spane, E. P., C. Kopczynski, C. S. Goodman, and C. L. Doe. 1995. Asymmetric localization of Numb autonomously determines sibling neuron identity in the *Drosophila* CNS. *Development* 121: 3489–3494.

Spazioni, E., M. P. Mattson, W. L. Wang, and H. E. McDougald. 1999. Signaling pathways for ecdysteroid hormone synthesis in crustacean Y-organs. *Am. Zool.* 39: 496–512.

Spradling, A. C. 1993. Developmental genetics of oogenesis in *Drosophila*. In C. M. Bate and A. Martinez Arias (eds.), *The development of* Drosophila melanogaster. Vol. 1, pp. 1–70. Cold Spring Harbor Laboratory Press.

Sprenger, F., and C. Nüsslein-Volhard. 1993. The terminal system of axis determination in the *Drosophila* embryo. In C. M. Bate and A. Martinez Arias (eds.), *The development of* Drosophila melanogaster. Vol. 1, pp. 365–386. Cold Spring Harbor Laboratory Press.

Sreng, L., and A. Quennedey. 1976. Role of a temporary ciliary structure in the morphogenesis of insect glands. An electron microscope study of the tergal glands of male *Blattella germanica* L. (Dictyoptera, Blattellidae). *J. Ultrastruct. Res.* 56: 78–95.

Stanojevic, D., S. Small, and M. Levine. 1991. Regulation of a segmentation stripe by overlapping activators and repressors in the *Drosophila* embryo. *Science* 254: 1385–1387.

Stark, J., J. Bonacum, J. Remsen, and R. DeSalle. 1999. The evolution and development of dipteran wing veins: a systematic approach. *Annu. Rev. Entomol.* 44: 97–129.

Stark, R. J., and M. I. Mote. 1981. Postembryonic development of the visual system of *Periplaneta americana*. I. Pattern of growth and differentiation. *J. Embryol. Exp. Morphol.* 66: 235–255.

Stauber, M., H. Jäckle, and U. Schmidt-Ott. 1999. The anterior determinant *bicoid* of *Drosophila* is a derived Hox class 3 gene. *Proc. Natl. Acad. Sci. U.S.A.* 96: 3786–3789.

Stay, B. 2000. A review of the role of neurosecretion in the control of juvenile hormone synthesis: a tribute to Berta Scharrer. *Insect Biochem. Mol. Biol.* 30: 653–662.

Stay, B., and A. P. Woodhead. 1993. Neuropeptide regulators of insect corpora allata. *Am. Zool.* 33: 357–364.

Stay, B., T. Friedel, S. S. Tobe, and E. C. Mundell. 1980. Feedback control of juvenile hormone synthesis in cockroaches: possible role for ecdysone. *Science* 207: 898–900.

Stearns, S. C. 1992. *The evolution of life histories*. Oxford University Press.

Stebbings, H., and C. Hunt. 1983. Microtubule polarity in the nutritive tubes of insect ovarioles. *Cell Tissue Res.* 233: 133–141.

Stehr, F. W. 1987a. Order Lepidoptera. In F. W. Stehr (ed.), *Immature insects*. Vol. 1, pp. 288–596. Kendall/Hunt.

——. (ed.). 1987b. *Immature insects*. Vol. l. Kendall/Hunt.

——. (ed.) 1991. *Immature insects*. Vol. 2. Kendall/Hunt.

Steinmann-Zwicky, M. 1994. Sex determination of the *Drosophila* germ line: *tra* and *dsx* control somatic inductive signals. *Development* 120: 707–716.

Steller, H. 1995. Mechanisms and genes of cellular suicide. *Science* 267: 1445–1449.

Stern, C. 1936. Somatic crossing over and segregation in *Drosophila melanogaster*. *Genetics* 21: 625–730.

Stern, D. L., and D. J. Emlen. 1999. The developmental basis for allometry in insects. *Development* 126: 1091–1101.

Stevens, N. M. 1905. Studies in spermatogenesis with special reference to the "accessory chromosomes". *Public. Carnegie Inst.* 36: 1–32; 7 pl.

Stillman, B. 1996. Cell cycle control of DNA replication. *Science* 274: 1659–1664.

Stjernholm, F., and B. Karlsson. 2000. Nuptial gifts and the use of body resources for reproduction in the green-veined white butterfly, *Pieris napi*. *Proc. R. Soc. Lond. B Biol. Sci.* 267: 807–811.

St. Johnston, D. 1993. Pole plasm and the posterior group genes. In C. M. Bate and A. Martinez Arias (eds.), *The development of* Drosophila melanogaster. Vol. 1, pp. 325–363. Cold Spring Harbor Laboratory Press.

Stocker, R. F., and P. A. Lawrence. 1981. Sensory projections from normal and homoeotically transformed antennae in *Drosophila*. *Dev. Biol.* 82: 224–237.

Stocker, R. F., M. Tissot, and N. Gendre. 1995. Morphogenesis and cellular proliferation pattern in the developing antennal lobe of *Drosophila melanogaster*. *W. Roux's Arch. Dev. Biol.* 205: 62–72.

Stouthamer, R., J. A. J. Breeuwer, R. F. Luck, and J. H. Werren. 1992. Molecular identification of microorganisms associated with parthenogenesis. *Nature* 361: 66–68.

Strand, M. R., and M. Grbíc. 1997. The development and evolution of polyembryonic insects. *Curr. Topics Dev. Biol.* 35: 121–159.

Strausfeld, N. J. 1976. *Atlas of an insect brain*. Springer-Verlag.

Strausfeld, N. J., and I. A. Meinertzhagen. 1998. The insect neuron: types, morphologies, fine structure, and relationship to the architectonics of the insect nervous system. In F. W. Harrison and

M. Locke (eds.), *Microscopic anatomy of invertebrates*. Vol. 11B. *Insecta*, pp. 487–538. Wiley-Liss.

Strausfeld, N. J., E. K. Buschbeck, and R. S. Gomez. 1995. The arthropod mushroom body: its functional roles, evolutionary enigmas and mistaken identities. In O. Breidbach and W. Kutsch (eds.), *The nervous system of invertebrates: an evolutionary and comparative approach*, pp. 349–380. Birkhäuser.

Strøme, S. 1992. The germ of the issue. *Nature* 358: 368–369.

Struhl, G., D. A. Barbash, and P. A. Lawrence. 1997. Hedgehog organizes the pattern and polarity of epidermal cells in the *Drosophila* abdomen. *Development* 124: 1243–1254.

Strutt, D. I., V. Wiersdorff, and M. Mlodzik. 1995. Regulation of furrow progression in the *Drosophila* eye by cAMP-dependent protein kinase A. *Nature* 373: 705–709.

Sturm, H. 1992. Mating behaviour and sexual dimorphism in *Promesomachilis hispanica* Silvestri, 1923 (Machilidae: Archaeognatha, Insecta). *Zool. Anz.* 228: 60–73.

Sturtevant, A. H. 1929. The claret mutant type of *Drosophila simulans*: a study of chromosome elimination and cell lineage. *Z. Zool.* 135: 323–356.

Stutt, A. D., and T. Siva-Jothy. 2001. Traumatic insemination and sexual conflict in the bed bug *Cimex lectularius*. *Proc. Natl. Acad. Sci. U.S.A.* 98: 5683–5687.

Stys, P., and S. Bilinski. 1990. Ovariole types and the phylogeny of the hexapods. *Biol. Rev.* 65: 401–429.

Stys, P., and J. Zrzavy. 1994. Phylogeny and classification of extant Arthropoda: review of hypotheses and nomenclature. *Eur. J. Entomol.* 91: 257–275.

Stys, P., J. Zrzavy, and F. Weyda. 1993. Phylogeny of the Hexapoda and ovarian metamerism. *Biol. Rev.* 68: 365–379.

Such, J. 1978. Embryologie ultrastructurale de l'ommatidie chez le phasme *Carausius morosus* Br. (Phasmida: Lonchodidae): morphogenèse et cytodifferenciation. *Int. J. Insect Morphol. Embryol.* 2: 165–183.

Sugawara, T. 1993. Oviposition behaviour of the cricket *Teleogryllus commodus*: mechanosensory cells in the genital chamber and their role in the switch-over of steps. *J. Insect Physiol.* 39: 335–346.

Sun, G., D. L. Dilcher, S. Zheng, and Z. Zhou. 1998. In search of the first flower: a Jurassic angiosperm, *Archaefructus*, from Northeast China. *Science* 282: 1692–1695.

Suomalainen, E., A. Saura, and J. Lokke (eds.). 1987. *Cytology and evolution in parthenogenesis*. CRC Press.

Svácha, P. 1992. What are and what are not imaginal discs: re-evaluation of some basic concepts (Insecta, Holometabola). *Dev. Biol.* 154: 101–117.

Swan, A., and B. Suter. 1996. Role of *Bicaudal-D* in patterning the *Drosophila* egg chamber in mid-oogenesis. *Development* 122: 3577–3586.

Swanson, C. P., T. Merz, and W. J. Young. 1967. *Cytogenetics*. Prentice-Hall.

Szabo, S. P., and D. H. O'Day. 1983. The fusion of sexual nuclei. *Biol. Rev.* 58: 323–342.

Szöllosi, A. 1975. Electron microscope study of spermiogenesis in *Locusta migratoria* (Insect Orthoptera). *J. Ultrastruct. Res.* 50: 322–346.

——. 1982. Relationships between germ and somatic cells in the testes of locusts and moths. In R. C. King and H. Akai (eds.), *Insect ultrastructure*. Vol. 1, pp. 32–60. Plenum Press.

Szöllosi, A., and C. Marcaillou. 1977. Electron microscope study of the blood-testis barrier in an insect: *Locusta migratoria*. *J. Ultrastruct. Res.* 59: 158–172.

——. 1979. The apical cell of the locust testis: an ultrastructural study. *J. Ultrastruct. Res.* 69: 331–342.

Szöllosi, A., J. Riemann, and C. Marcaillou. 1980. Localization of the blood-testis barrier in the testis of the moth, *Anagasta kühniella*. *J. Ultrastruct. Res.* 72: 189–199.

Tabin, C. J., S. B. Carroll, and G. Panganiban. 1999. Out on a limb: parallels in vertebrate and invertebrate limb patterning on the origin of appendages. *Am. Zool.* 39: 650–663.

Takashima, S., and R. Murakami. 2001. Regulation of pattern formation in the *Drosophila* hindgut by *wg*, *hh*, *dpp*, and *en*. *Mech. Dev.* 101: 79–90.

Takeda, N., and T. Ohishi. 1976. Effect of a development-stimulating pheromone on the embryo of *Chironomus dorsalis*. *J. Insect Physiol.* 22: 1327–1330.

Takemura, Y., T. Kanda, and Y. Horie. 2000. Artificial insemination using cryopreserved sperm in the silkworm *Bombyx mori*. *J. Insect Physiol.* 47: 491–497.

Tamarelle, M. 1981. La formation et l'évolution de l'organe dorsal chez les embryons de cinq collemboles (Insecta: Apterygota). *Int. J. Insect Morphol. Embryol.* 10: 203–224.

Tanaka, S. 1993. Allocation of resources to egg production and flight muscle development in a wing dimorphic cricket, *Modicogryllus confirmatus*. *J. Insect Physiol.* 39: 493–498.

Tateno, M., Y. Nashida, and T. Adachi-Yamada. 2000. Regulation of JNK by Src during *Drosophila* development. *Science* 287: 324–327.

Tawfik, A. I., A. Vedrová, and F. Sehnal. 1999. Ecdysteroids during ovarian development and embryogenesis in solitary and gregarious *Schistocerca gregaria*. *Arch. Insect Biochem. Physiol.* 41: 134–143.

Tear, G. 1999. Neuronal guidance: a genetic prospective. *Trends Genet.* 15: 113–118.

Technau, G., and M. Heisenberg. 1982. Neural reorganization during metamorphosis of the corpora pedunculata in *Drosophila melanogaster*. *Nature* 295: 405–407.

Technau, G. M. 1987. A single cell approach to problems of cell lineage and commitment during embryogenesis. *Development* 100: 1–12.

Technau, G. M., and J. A. Campos-Ortega. 1986. Lineage analysis of transplanted individual cells in embryos of *Drosophila melanogaster*. III. Commitment and proliferative capabilities of pole cells and midgut progenitors. *W. Roux's Arch. Dev. Biol.* 195: 489–493.

Telfer, W. H. 1975. Development and physiology of the oocyte-nurse cell syncytium. *Adv. Insect Physiol.* 11: 223–319.

Telfer, W. H., and L. M. Anderson. 1968. Functional transformations accompanying the initiation of a terminal growth phase in the cecropia moth oocyte. *Dev. Biol.* 17: 512–535.

Telfer, W. H., E. Huebner, and D. S. Smith. 1982. The cell biology of vitellogenic follicles in *Hyalophora* and *Rhodnius*. In R. C. King and H. Akai (eds.), *Insect ultrastructure*. Vol. 1, pp. 118–149. Plenum Press.

Telford, M. J., and R. H. Thomas. 1998a. Expression of the homeobox genes shows chelicerate arthropods to retain their deutocerebral segment. *Proc. Natl. Acad. Sci. U.S.A.* 95: 10671–10675.

——. 1998b. Of mites and *zen*: expression studies in a chelicerate arthropod confirm that *zen* is a divergent Hox gene. *Dev. Genes Evol.* 208: 591–594.

Temin, G., M. Zander, and J. P. Roussel. 1986. Physico-chemical (GC-MS) measurements of juvenile hormone III titres during embryogenesis of *Locusta migratoria*. *Int. J. Invert. Reprod.* 9: 105–112.

Tepass, U. 1997. Epithelial differentiation in *Drosophila*. *Bioessays* 19: 673–682.

Tepass, U., and V. Hartenstein. 1994. Epithelium formation in the *Drosophila* midgut depends on the interaction of endoderm and mesoderm. *Development* 120: 579–590.

——. 1995. Neurogenic and proneural genes control cell fate specification in the *Drosophila* endoderm. *Development* 121: 393–405.

Tepass, U., L. I. Fessler, A. Aziz, and V. Hartenstein. 1994. Embryonic origin of hemocytes and their relationship to cell death in *Drosophila*. *Development* 120: 1829–1837.

Teskey, H. J. 1981. Key to families-larvae. In J. F. McAlpine, B. V. Peterson, G. E. Shewell, H. J. Teskey, J. R. Vockeroth, and D. M. Wood (eds.), *Manual of nearctic Diptera*. Vol. 1, pp. 125–147. Monograph No. 27. Research Branch, Agriculture Canada.

Tettamanti, M., J. D. Armstrong, K. Endo, M. Y. Yang, K. Furukubo-Tokunaga, K. Kaiser, and H. Reichert. 1997. Early development of the *Drosophila* mushroom bodies, brain centers for associative learning and memory. *Dev. Genes Evol.* 207: 242–252.

Therianos, S., S. Leuzinger, F. Hirth, C. S. Goodman, and H. Reichert. 1995. Embryonic development of the *Drosophila* brain: formation of commissural and descending pathways. *Development* 121: 3849–3860.

Theunissen, J. A. B. M. 1976. Aspects of gametogenesis and radiation pathology in the onion maggot, *Hylemya antiqua* (Meigen). I. Gametogenesis. *Meded. Landbhoog. Wageningen.* 76–3.

Theurkauf, W. E. 1994. Premature microtubule-dependent cytoplasmic streaming in *cappuccino* and *spire* mutant oocytes. *Science* 265: 2093–2096.

Theurkauf, W. E., B. M. Alberts, Y. N. Jan, and T. A. Jongens. 1993. A central role for microtubules in the differentiation of *Drosophila* oocytes. *Development* 118: 1169–1180.

Thomas, C., P. DeVries, J. Hardin, and J. White. 1996. Four-dimensional imaging: computer visualization of 3D movements in living specimens. *Science* 273: 603–607.

Thomas, H. E., H. G. Stunnenberg, and A. F. Stewart. 1993. Heterodimerization of the *Drosophila* ecdysone receptor with retenoid X receptor and ultraspiracle. *Nature* 362: 471–475.

Thomas, J. B., M. J. Bastiani, M. Bate, and C. S. Goodman. 1984. From grasshopper to *Drosophila*: a common plan for neuronal development. *Nature* 310: 203–207.

Thomas, M. B., K. A. Walsh, M. R. Wolf, B. A. Mcpheron, and J. H. Marden. 2000. Molecular phylogenetic analysis of evolutionary trends in stonefly wing structure and locomotor behavior. *Proc. Natl. Acad. Sci. U.S.A.* 97: 13178–13183.

Thomas, R. D. K., R. M. Shearman, and G. W. Stewart. 2000. Evolutionary exploitation of design options by the first animals with hard skeletons. *Science* 288: 1239–1242.

Thomas, R. H., and M. J. Telford. 1999. Appendage development in embryos of the oribatid mite *Archegozetes longisetosus* (Acari, Oribatei, Trhypochthoniidae). *Acta Zool.* 80: 193–200.

Thompson, D'Arcy W. 1917. *On growth and form.* Cambridge University Press.

Thompson, K. J., S. P. Sivanesan, H. R. Campbell, and K. J. Sanders. 1999. Efferent neurons and specialization of abdominal segments in grasshoppers. *J. Comp. Neurol.* 415: 65–79.

Thornhill, R. 1976. Sexual selection and paternal investment in insects. *Am. Nat.* 110: 153–163.

——. 1993. The allure of symmetry. *Nat. Hist.* 102(9): 30–36.

Thornhill, R., and J. Alcock. 1983. *The evolution of insect mating systems.* Harvard University Press.

Thummel, C. S. 1995. From embryogenesis to metamorphosis: the regulation and function of *Drosophila* nuclear receptor superfamily members. *Cell* 85: 871–877.

——. 1996. Flies on steroids-*Drosophila* metamorphosis and the mechanisms of steroid hormone action. *Trends Genet.* 12: 306–310.

Tihen, J. A. 1946. An estimate of the number of cell generations preceding sperm formation in *Drosophila melanogaster. Am. Nat.* 80: 389–392.

Ting, C.-T., S.-C. Tsaur, M.-L. Wu, and C.-I. Wu. 1998. A rapidly evolving homeobox at the site of a hybrid sterility gene. *Science* 282: 1501–1504.

Titschack, E. 1926. Untersuchungen über das Wachstum der Kleidermotte *Tineola biselliella* Hum. Gleichzeitig ein Beiträge zur Klärung der Insektenhuatung. *Z. Zool.* 128: 509–569.

——. 1930. Untersuchungen über das Wachstum, der Nahrungsverbrauch und die Eierzeugung. III. *Cimex lectularius* L. *Z. Morphol. Okol. Tiere* 17: 471–551.

Tobe, S. S. 1977. Asymmetry in hormone biosynthesis by insect endocrine glands. *Can. J. Zool.* 55: 1509–1514.

Tojo, K., and R. Machida. 1998. Early embryonic development of the mayfly *Ephemera japonica* McLachlan (Insecta, Ephemeroptera, Ephemeridae). *J. Morphol.* 238: 327–335.

Tomlinson, A. 1988. Cellular interactions in the developing *Drosophila* eye. *Development* 104: 183–193.

Tomlinson, A., and G. Struhl. 1999. Decoding vectorial information from a gradient: sequential roles of the receptors Frizzled and Notch in establishing planar polarity in the *Drosophila* eye. *Development* 126: 5725–5738.

Tram, U., and W. Sullivan. 2000. Reciprocal inheritance of centrosomes in the parthenogenetic hymenopteran *Nasonia vitripennis. Curr. Biol.* 10: 1413–1419.

Tran, J., T. J. Brenner, and S. DiNardo. 2000. Somatic control over the germline stem cell lineage during *Drosophila* spermatogenesis. *Nature* 407: 754–757.

Tremblay, E., and L. E. Caltagirone. 1973. Fate of polar bodies in insects. *Annu. Rev. Entomol.* 18: 421–441.

Tremblay, E., and D. Calvert, 1971. Embryosystematics in the aphidiines (Hymenoptera: Braconidae). *Bull. Lab. Entomol. Agric. "Filippo Silvestri" Portici* 29: 223–249.

Trendelenberg, M. F., W. W. Franke, and U. Scheer. 1977. Frequencies of circular units of nucleolar DNA in oocytes of two insects, *Acheta domesticus* and *Dytiscus marginalis*, and changes in nucleolar morphology during oogenesis. *Differentiation* 7: 133–158.

Trepte, H.-H. 1976. Das Puffmuster der Borstenapparat-Chromosomen von *Sarcophaga barbata. Chromosoma* 55: 137–164.

——. 1979. Rate of follicle growth, change in follicle volume and stages of macromolecular synthesis during ovarian development in *Musca domestica. J. Insect Physiol.* 25: 199–203.

Truby, P. R. 1983. Blastema formation and cell division during cockroach limb regeneration. *J. Embryol. Exp. Morphol.* 75: 151–164.

Truckenbrodt, W. 1979. The embryonic covers during blastokinesis and dorsal closure of the normal and of the actinomycin D-treated egg of *Odontotermes badius* (Hod.) (Insecta, Isoptera). *Zool. Jb. Anat.* 101: 7–18.

Truman, J. W. 1985. Hormonal control of ecdysis. In G. A. Kerkut and L. I. Gilbert (eds.), *Comprehensive insect physiology, biochemistry and pharmacology.* Vol. 8, pp. 413–440. Pergamon Press.

——. 1990. Neuroendocrine control of ecdysis. In E. Ohnishi and H. Ishizaki (eds.), *Molting and metamorphosis,* pp. 67–82. Japanese Scientific Society Press.

Truman, J. W., and E. E. Ball. 1998. Patterns of embryonic neurogenesis in a primitive wingless insect, the silverfish, *Ctenolepisma longicaudata:* comparison with those seen in flying insects. *Dev. Genes Evol.* 208: 357–368.

Truman, J. W., and P. E. Copenhaver. 1989. The larval eclosion hormone neurons in *Manduca sexta:* identification of the brain-proctodeal neurosecretory system. *J. Exp. Biol.* 147: 457–470.

Truman, J. W., and L. M. Riddiford. 1970. Neuroendocrine control of ecdysis in silkmoths. *Science* 167: 1624–1626.

——. 1999. The origins of insect metamorphosis. *Nature* 401: 447–452.

Truman, J. W., B. J. Taylor, and T. A. Ward. 1993. Formation of the adult nervous system. In M. Bate and A. Martinez Arias (eds.) *The development of* Drosophila melanogaster. Vol. 2, pp. 1245–1275. Cold Spring Harbor Laboratory Press.

Truman, J. W., W. S. Talbot, S. E. Fahrbach, and D. S. Hogness. 1994. Ecdysone receptor expression in the CNS correlates with stage-specific responses to ecdysteroids during *Drosophila* and *Manduca* development. *Development* 120: 219–234.

Tschinkel, W. R., and S. D. Porter. 1988. Efficiency of sperm use in queens of the fire ant, *Solenopsis invicta* (Hymenoptera: Formicidae). *Ann. Entomol. Soc. Am.* 81: 777–781.

Tschudi-Rein, K., and G. Benz. 1990a. Mechanisms of sperm transfer in female *Pieris brassicae* (Lepidoptera: Pieridae). *Ann. Entomol. Soc. Am.* 83: 1158–1164.

——. 1990b. The bipotentiality of lepidopteran spermatocytes questioned. *Bull. Soc. Entomol. Suisse* 63: 81–85.

Tuxen, S. L. (ed.). 1970. *Taxonomist's glossary of genitalia in insects.* 2d ed. Munksgaard.

Tuzet, O. 1977. Les spermatophores des insectes. In P. P. Grassé (ed.), *Traité de zoologie.* Vol. VIII 5A, pp. 277–330. Masson et Cie.

Udolph, G., K. Lüer, T. Bossing, and G. M. Technau. 1995. Commitment of CNS progenitors along the dorsoventral axis of *Drosophila* neuroectoderm. *Science* 269: 1278–1281.

Udolph, G., A. Prokop, T. Bossing, and G. M. Technau. 1993. A common precursor for glia and neurons in the embryonic CNS of *Drosophila* gives rise to segment-specific lineage variants. *Development* 118: 765–775.

Uemiya, H., and H. Ando. 1987. Embryogenesis of a springtail, *Tomocerus ishibashii* (Collembola, Tomoceridae): external morphology. *J. Morphol.* 191: 37–48.

Ueshima, N. 1979. *Hemiptera II: Heteroptera. Animal cytogenetics.* Vol. 3. Insecta. 6. Gebrüder Borntraeger.

Unnithan, G. C., K. K. Nair, and W. S. Bowers. 1977. Precocene-induced degeneration of the corpus allatum of adult females of the bug, *Oncopeltus fasciatus. J. Insect Physiol.* 23: 1081–1094.

Ursprung, H., and R. Nöthiger (eds.). 1972. *The biology of imaginal discs.* Springer-Verlag.

Uvarov, B. P. 1921. A revision of the genus *Locusta* L. (= *Pachytylus* Fieb.), with a new theory as to the periodicity and migrations of locusts. *Bull. Entomol. Res.* 12: 135–163.

Vafopoulou, K., and C. G. H. Steele. 1997. Ecdysteroidogenic action of *Bombyx* prothoracicotropic hormone and bombyxin on the prothoracic glands of *Rhodnius prolixus in vitro. J. Insect Physiol.* 43: 651–656.

——. 1999. Daily rhythm of responsiveness to prothoracicotropic hormone in prothoracic glands of *Rhodnius prolixus. Arch. Insect Biochem. Physiol.* 41: 117–123.

Vale, R. D., and R. A. Milligan. 2000. The way things move: looking under the hood of molecular motor proteins. *Science* 288: 88–95.

Valentine, J. W., D. H. Erwin, and S. D. Jablonski. 1996. Developmental evolution of metazoan body plans: the fossil evidence. *Dev. Biol.* 173: 373–381.

Valentine, J. W., D. Jablonski, and D. H. Erwin. 1999. Fossils, molecules and embryos: new perspectives on the Cambrian explosion. *Development* 126: 851–859.

van Doren, M., H. T. Broihier, L. A. Moore, and R. Lehmann. 1998. HMG-CoA reductase guides migrating primordial germ cells. *Nature* 396: 466–469.

Vedickis, W. V., W. E. Bollenbacher, and L. I. Gilbert. 1976. Insect prothoracic glands: a role for cyclic AMP in the stimulation of α-ecdysone secretion. *Mol. Cell. Endocrinol.* 5: 397–407.

Velthuis, H. H. W. 1976. Environmental, genetic and endocrine influences in stingless bee caste determination. In M. Lüscher (ed.), *Phase and caste determination in insects. Endocrine aspects,* pp. 35–53. Pergamon Press.

Vervoort, M., D. Zink, N. Pujol, K. Victoir, N. Duent, A. Ghysen, and C. Dambly-Chaudiere. 1995. Genetic determinants of sense organ identity in *Drosophila:* regulating interactions between *cut* and *poxn. Development* 121: 3111–3120.

Villani, M. G., L. L. Allee, A. Diaz, and P. S. Robbins. 1999. Adaptive strategies of edaphic arthropods. *Annu. Rev. Entomol.* 44: 233–256.

Vincent, A., J. T. Blankenship, and E. Wieschaus. 1997. Integration of the head and trunk segmentation systems controls cephalic furrow formation in *Drosophila. Development* 124: 3747–3754.

Vincent, J. F. V. 1975. How does the female locust dig her oviposition hole? *J. Entomol. (A)* 50: 175–181.

Vincent, S., J. L. Vonesch, and A. Giangrande. 1996. *glide* directs glial fate commitment and cell fate switch between neurones and glia. *Development* 122: 131–139.

Virkki, N. 1969. Sperm bundles and phylogenesis. *Z. Zellforsch.* 101: 13–27.

Visscher, S. N. 1971. Studies on the embryogenesis of *Aulocara elliotti* (Orthoptera; Acrididae). III. Influence of maternal environment and aging on development of the progeny. *Ann. Entomol. Soc. Am.* 64: 1057–1074.

Vogel, G. 1998. Doubled genes may explain fish diversity. *Science* 281: 1119–1120.

——. 2000. The come-hither, don't-touch-me proteins. *Science* 290: 2243–2244.

Vogel, S. 1988. *Life's devices.* Princeton University Press.

Volk, T. 1992. A new member of the spectrin superfamily may participate in formation of embryonic muscle attachments in *Drosophila. Development* 116: 721–730.

Volk, T., and K. VijayRaghavan. 1994. A central role for epidermal segment border cells in the induction of muscle patterning in the *Drosophila* embryo. *Development* 120: 59–70.

Vollmar, H. 1972. Die Einrollbewegung (Anatrepsis) des Keimstreifs im Ei von *Acheta domesticus* (Orthopteroidea, Gryllidae). *W. Roux's Arch. Dev. Biol.* 170: 135–151.

von Baer, K. E. 1828. *Entwicklungsgeschichte der Thiere: Beobachtung and Reflexion.* Bornträger.

von Dassow, G., E. Meir, E. M. Munro, and G. M. Odell. 2000. The segment polarity network is a robust developmental module. *Nature* 406: 188–192.

von Kalm, L., K. Crossgrove, D. von Seggern, G. M. Guild, and S. K. Beckendorf. 1994. The *Broad complex* directly controls a tissue-specific response to the steroid hormone ecdysone at the onset of *Drosophila* metamorphosis. *EMBO J.* 13: 3505–3516.

Vorhees, F. R., and W. R. Horsfall. 1971. Genesis of the reproductive system of mosquitoes. III. Supernumerary male genitalia. *J. Morphol.* 133: 399–408.

Vosshall, L. B., A. M. Wong, and R. Axel. 2000. An olfactory sensory map in the fly brain. *Cell* 102: 147–159.

Waddington, C. H. 1941. The genetic control of wing development in *Drosophila*. *J. Genet.* 41: 75–139.

——. 1942. Canalization of development and the inheritance of acquired characters. *Nature* 150: 563–565.

Wade, M. J., and N. W. Chang. 1995. Increased male fertility in *Tribolium confusum* beetles after infection with the intracellular parasite *Wolbachia*. *Nature* 373: 72–74.

Wagner, D. L., J. L. Loose, T. D. Fitzgerald, J. A. De Benedictis, and D. R. Davis. 2000. A hidden past: the hypermetamorphic development of *Marmara arbutiella* (Lepidoptera: Gracillariidae). *Ann. Entomol. Soc. Am.* 93: 59–64.

Wagner, R. M., M. J. Loeb, J. P. Kochansky, D. B. Gehnan, W. R. Lusby, and R. A. Bull. 1997. Identification and characterization of an ecdysiotropic peptide from brain extracts of the gypsy moth, *Lymantria dispar*. *Arch. Insect Biochem. Physiol.* 34: 175–189.

Wahli, W. 1988. Evolution and expression of vitellogenin genes. *Trends Genet.* 4: 227–232.

Walker, W. F. 1980. Sperm utilization strategies in nonsocial insects. *Am. Nat.* 115: 780–799.

Wallace, B. G. 1996. Signaling mechanisms mediating synapse formation. *Bioessays* 18: 777–780.

Walossek, D., and K. J. Müller. 1998. Cambrian "Orsten"-type arthropods and the phylogeny of Crustacea. In R. A. Fortey and R. H. Thomas (eds.), *Arthropod relationships*, pp. 139–153. Systematics Association Special Vol. 55. Chapman and Hall.

Wan, S., A.-M. Cato, and H. Skaer. 2000. Multiple signaling pathways establish cell fate and cell number in *Drosophila* malpighian tubules. *Dev. Biol.* 217: 153–165.

Wang, K. Y.-C., S. Kumar, and S. Blair Hedges. 1999. Divergence time estimates for the early history of animal phyla and for the origin of plants, animals and fungi. *Proc. R. Soc. Lond. B Biol. Sci.* 266: 163–171.

Warren, R. W., L. Nagy, J. Selegue, J. Gates, and S. Carroll. 1994. Evolution of homeotic gene regulation in flies and butterflies. *Nature* 372: 458–461.

Warrior, R. 1994. Primordial germ cell migration and the assembly of the *Drosophila* embryonic gonad. *Dev. Biol.* 166: 180–194.

Wasserman, J. D., and M. Freeman. 1998. An autoregulatory cascade of EGF receptor signaling patterns the *Drosophila* egg. *Cell* 95: 355–364.

Watler, D. 1982. Influence of social situation on food consumption and growth in nymphs of the house cricket, *Acheta domesticus*. *Physiol. Entomol.* 7: 343–350.

Watson, J. A. L. 1964. Moulting and reproduction in the adult firebrat, *Thermobia domestica* (Packard) (Thysanura, Lepismatidae). II. The reproductive cycle. *J. Insect Physiol.* 10: 399–408.

Weaver, R. J., G. E. Pratt, and J. R. Finney. 1975. Cyclic activity of the corpus allatum related to gonotrophic cycles in adult female *Periplaneta americana*. *Experientia*, 31: 597–598.

Weber, H. 1934. Die postembryonale Entwicklung der Aleurodinen (Hemiptera-Homoptera). Ein Beitrag zur Kenntnis der Metamorphosen der Insekten. *Z. Morphol. Okol. Tiere* 29: 268–305.

——. 1966. *Grundriss der Insektenkunde*. 2d ed. G. Fischer Verlag.

Weed-Pfeiffer, I. G. 1936. Removal of the corpora allata on egg production in the grasshopper, *Melanoplus differentialis*. *Proc. Soc. Exp. Biol. Med.* 34: 883–885.

Weeks, A. R., F. Marek, and J. A. J. Breeuwer. 2001. A mite species that consists entirely of haploid females. *Science* 292: 2479–2482.

Weismann. A. 1892. *Das Keimplasma: Eine Theorie der Vererbung*. Gustav Fischer.

Weisser, W. W., C. Braendle, and N. Minoretti, 1999. Predator-induced morphological shift in the pea aphid. *Proc. R. Soc. Lond. B Biol. Sci.* 266: 1175–1181.

Went, D. F. 1979. Paedogenesis in the dipteran insect *Heteropeza pygmaea*: an interpretation. *Int. J. Invert. Reprod.* 1: 21–30.

——. 1982. Egg activation and parthenogenetic reproduction in insects. *Biol. Rev.* 57: 319–344.

Went, D. F., and R. Camenzind. 1977. Hemolymph-dependent sex determination in a paedogenetic gall midge. *Naturwissenschaften* 64: 276.

Went, D. F., and G. Krause. 1973. Normal development of mechanically activated, unlaid eggs in an endoparasitic hymenopteran. *Nature* 244: 454–455.

Westbrook, A. L., and W. E. Bollenbacher. 1990. The prothoracicotropic hormone neuroendocrine axis in *Manduca sexta*: development and function. In E. Ohnishi and H. Ishizaki (eds.), *Molting and metamorphosis*, pp. 3–16. Japanese Scientific Society Press.

Westbrook, A. L., M. E. Haire, W. R. Kier, and W. E. Bollenbacher. 1991. Three-dimensional architecture of identified cerebral neurosecretory cells in an insect. *J. Morphol.* 208: 161–174.

West-Eberhard, M. J. 1989. Phenotypic plasticity and the origin of diversity. *Annu. Rev. Ecol. Syst.* 20: 249–278.

Wheeler, D. E., and H. F. Nijhout. 1984. Soldier determination in *Pheidole bicarinata*: inhibition by adult soldier. *J. Insect Physiol.* 30: 127–135.

Wheeler, W. C., M. Whiting, Q. D. Wheeler, and J. M. Carpenter. 2001. The phylogeny of the extant hexapod orders. *Cladistics* 17: 113–169.

White, K. P., S. A. Rifkin, P. Hurban, and D. S. Hogness. 1999. Microarray analysis of *Drosophila* development during metamorphosis. *Science* 286: 2179–2184.

White, M. J. D. 1973. *Animal cytology and evolution*. 3rd ed. Cambridge University Press.

White-Cooper, H., M. A. Schäfer, L. S. Alphey, and M. T. Fuller. 1998. Transcriptional and post-transcriptional control mechanisms co-ordinate the onset of spermatid differentiation with meiosis I in *Drosophila*. *Development* 125: 125–134.

Whiting, M. F. 1998. Phylogenetic position of the Strepsiptera: a review of molecular and morphological evidence. *Int. J. Insect Morphol. Embryol.* 27: 53–60.

Whiting, M. F., and W. W. Wheeler. 1994. Insect homeotic transformation. *Nature* 368: 696.

Whiting, M. F., J. C. Carpenter, Q. D. Wheeler, and W. C. Wheeler. 1997. The Strepsiptera problem: phylogeny of the holometabolous insect orders inferred from 18S and 28s rDNA sequences and morphology. *Syst. Biol.* 46: 1–68.

Whitington, P. M., and J. P. Bacon. 1998. The organization and development of the arthropod ventral nerve cord: insights into arthropod relationships. In R. A. Fortey and R. H. Thomas (eds.), *Arthropod relationships*, pp.

349–367. Systematic Association Special Vol. 55. Chapman and Hall.

Whitington, P. M., K.-K. Harris, and D. Leach. 1996. Early axonogenesis in the embryo of a primitive insect, the silverfish, *Ctenolepisma longicaudata*. *W. Roux's Arch. Dev. Biol.* 205: 272–281.

Whitington, P. M., T. Meier, and S. King 1991. Segmentation, neurogenesis, and formation of early axonal pathways in the centipede, *Ethmostigmus rubripes* (Brandt). *W. Roux's Arch. Dev. Biol.* 199: 349–363.

Whitten, J. M. 1968. Metamorphic changes in insects. In W. Etkin and L. I. Gilbert (eds.), *Metamorphosis: a problem in developmental biology*, pp. 48–105. Appleton-Century-Crofts.

——. 1969a. Cell death during early morphogenesis: parallels between insect limb and vertebrate limb development. *Science* 163: 1456–1457.

——. 1969b. Coordinated development in the fly foot: sequential cuticle secretion. *J. Morphol.* 127: 73–104.

——. 1969c. Coordinated development in the foot pad of the fly *Sarcophaga bullata* during metamorphosis: changing puffing patterns of the giant cell chromosomes. *Chromosoma* 26: 215–244.

Wigglesworth, V. B. 1934. The physiology of ecdysis in *Rhodnius prolixus* (Hemiptera). II. Factors controlling molting and "metamorphosis" *Q. J. Microsc. Sci.* 77: 191–222.

——. 1936. The function of the corpus allatum in the growth and reproduction of *Rhodnius prolixus* (Hemiptera). *Q. J. Microsc. Sci.* 79: 91–121.

——. 1940. The determination of characters at metamorphosis in *Rhodnius prolixus* (Hemiptera). *J. Exp. Biol.* 17: 201–222.

——. 1952. Hormone balance and the control of metamorphosis in *Rhodnius prolixus* (Hemiptera). *J. Exp. Biol.* 29: 620–631.

——. 1955. The breakdown of the thoracic gland in the adult insect *Rhodnius prolixus*. *J. Exp. Biol.* 32: 485–491.

——. 1956. Formation and involution of striated muscle fibres during the growth and moulting cycles of *Rhodnius prolixus* (Hemiptera). *Q. J. Microsc. Sci.* 97: 465–480.

——. 1964. *The life of insects*. Weidenfeld and Nicolson.

——. 1965. *The principles of insect physiology*. 6th ed. Methuen.

——. 1972. *The principles of insect physiology*. 7th ed. Chapman and Hall.

——. 1985. Historical perspectives. In G. A. Kerkut and L. I. Gilbert (eds.), *Comprehensive insect physiology, biochemistry and pharmacology*. Vol. 7, pp. 1–24. Pergamon Press.

Wilk, R., B. H. Reid, U. Tepass, and A. D. Lipshitz. 2000. The *hindsight* gene is required for epithelial maintenance and differentiation of the tracheal system in *Drosophila*. *Dev. Biol.* 219: 183–196.

Wilkins, A. S. 1998. Evolutionary developmental biology: where is it going? *Bioessays* 20: 783–784.

Williams, C. M. 1947. Physiology of insect diapause. II. Interaction between the pupal brain and prothoracic glands in the metamorphosis of the giant silkworm, *Platysamia cecropia*. *Biol. Bull.* 93: 89–98.

——. 1948. Physiology of insect diapause. III. The prothoracic glands in the cecropia silkworm, with special reference to their significance in embryonic and postembryonic development. *Biol. Bull.* 94: 60–65.

——. 1952. Physiology of insect diapause. IV. The brain and prothoracic glands as an endocrine system in the cecropia silkworm. *Biol. Bull.* 116: 323–338.

——. 1967. Third-generation pesticides. *Sci. Am.* 217(1): 13–17.

Williams, G. C. 1975. *Sex and evolution*. Princeton University Press.

Williams, J. A., S. W. Paddock, and S. B. Carroll. 1993. Pattern formation in a secondary field: a hierarchy of regulatory genes subdivides the developing *Drosophila* wing disc into discrete subregions. *Development* 117: 571–584.

Williams, J. A., S. W. Paddock, K. Vorwerk, and S. B. Carroll. 1994. Organization of wing formation and induction of a wing-patterning gene at the dorsal/ventral compartment boundary. *Nature* 368: 299–305.

Willis, J. H. 1986. The paradigm of stage-specific gene sets in insect metamorphosis: time for revision. *Arch. Insect Biochem. Physiol.* Suppl. 1: 47–57.

Wilson, E. B. 1905. The chromosomes in relation to the determination of sex in insects. *Science* 22: 500–502.

Wilson, E. O. 1971. *The insect societies*. Belknap Press, Harvard University Press.

Wilson, M. V. H. 1978a. Paleogene insect faunas of western North America. *Quaest. Entomol.* 14: 13–34.

——. 1978b. Evolutionary significance of North American Paleogene insect faunas. *Quaest. Entomol.* 14: 35–42.

Winner, E. A., A. Carleton, P. Harjes, T. Turner, and C. Desplan. 2000. *bicoid*-independent formation of thoracic segments in *Drosophila*. *Science* 287: 2476–2479.

Winston, M. L., and K. N. Sleesor. 1992. The essence of royalty: honey bee queen pheromone. *Am. Sci.* 80: 374–385.

Wolf, K. W. 1997. Centrosome structure is very similar in eupyrene and apyrene spermatocytes of *Ephestia kuehniella* (Pyralidae, Lepidoptera, Insecta). *Invert. Reprod. Dev.* 31: 39–46.

Wolf, R. 1972. The cytaster, a colchicine-sensitive migration organelle of cleavage nuclei in an insect egg. *Dev. Biol.* 62: 464–472.

Wolfe, S. L. 1981. *Biology of the cell*. 2d ed. Wadsworth.

Wolff, C., R. Schröder, C. Schulz, D. Tautz, and M. Klingler. 1998. Regulation of the *Tribolium* homologues of *caudal* and *hunchback* in *Drosophila*: evidence for maternal gradient systems in a short germ embryo. *Development* 125: 3645–3654.

Wolff, T., and D. F. Ready. 1993. Pattern formation in the *Drosophila* retina. In M. Bate and A. Martinez Arias (eds.), *The development of* Drosophila melanogaster. Vol. 2, pp. 1277–1325. Cold Spring Harbor Laboratory Press.

Wolpert, L. 1968. The french flag problem: a contribution to the discussion on pattern development and regulation. In C. H. Waddington (ed.), *Towards a theoretical biology*, pp. 125–133. Edinburgh University Press.

——. 1978. Pattern formation in biological development. *Sci. Am.* 239(4): 154–164.

——. (ed.). 1998. *Principles of development*. Current Biology Ltd., Oxford University Press.

Wong, J. T. W., W. T. C. Wu, and T. P. O'Conner. 1997. Transmembrane Semaphorin I promotes axon outgrowth in vivo. *Development* 124: 3597–3607.

Wood, G. C. 1981. Asilidae. In J. F. McAlpine, B. V. Peterson, G. E. Shewell, H. J. Teskey, J. R. Vockeroth, and D. M. Wood (eds.), *Manual of nearctic Diptera*. Vol. 1, pp. 549–573. Monograph No. 27. Research Branch, Agriculture Canada.

Woodward, T. E., J. W. Evans, and V. F. Eastop. 1970. Hemiptera (bugs, leafhoppers, etc.). In I. M. Mackerras (ed.), *The insects of Australia. A textbook for students and research workers*, pp. 387–457. 1st ed. Melbourne University Press.

Wootton, R. J., and J. Kukalová-Peck. 2000. Flight adaptations in Paleozoic Palaeoptera (Insecta). *Biol. Rev.* 75: 129–167.

Wray, G. A. 1995. Punctuated evolution in embryos. *Science* 267: 1115–1116.

——. 1997. Echinoderms. In S. F. Gilbert and A. M. Raunio (eds.), *Embryology. Constructing the organism*, pp. 309–329. Sinaur Associates.

——. 1998. A tough cell. (Review of Gerhart and Kirschner, 1997). *Evolution* 52: 291–294.

Wrensch, D. L. 1993. Evolutionary flexibility through haploid males or how chance favors the prepared genome. In D. L. Wrensch and M. A. Ebbert (eds.), *Evolution and diversity of sex ratio in insects and mites*, pp. 118–149. Chapman and Hall.

Wu, J., and S. M. Cohen. 1999. Proximodistal axis formation in the *Drosophila* leg: subdivision into proximal and distal domains by Homothorax and Distal-less. *Development* 126: 109–117.

Wu, L. H., and J. A. Lengyel. 1998. Role of *caudal* in hindgut specification and gastrulation suggests homology between *Drosophila* amnioproctodeal invagination and vertebrate blastopore. *Development* 125: 2433–2442.

Wülbeck, C., and P. Simpson. 2000. Expression of *achaete-scute* homologues in discrete proneural clusters on the developing notum of the medfly, *Ceratitis capitata*, suggest a common origin for the stereotyped bristle patterns of higher Diptera. *Development* 127: 1411–1420.

Wyatt, G. R. 1997. Juvenile hormone in insect reproduction—a paradox? *Eur. J. Entomol.* 94: 323–333.

Wyatt, G. R., and K. G. Davey. 1996. Cellular and molecular actions of juvenile hormone II. Roles of juvenile hormone in adult insects. *Adv. Insect Physiol.* 26: 1–156.

Wylie, C. 1999. Germ cells. *Cell* 96: 165–174.

Xiao, S., Y. Zhang, and A. H. Knoll. 1998. Three-dimensional preservation of algae and animal embryos in a Neoproterozoic phosphorite. *Nature* 391: 553–558.

Xie, T., and A. C. Spradling. 1998. *decapentaplegic* is essential for the maintenance and division of germline stem cells in the *Drosophila* ovary. *Cell* 94: 251–260.

——. 2000. A niche maintaining germ line stem cells in the *Drosophila* ovary. *Science* 290: 328–330.

Yajima, H. 1960. Studies on embryonic determination of the harlequin-fly, *Chironomus dorsalis*. I. Effects of centrifugation and of its combination with constriction and puncturing. *J. Embryol. Exp. Morphol.* 8: 198–215.

——. 1964. Studies on embryonic determination of the harlequin-fly, *Chironomus dorsalis*. II. Effects of partial irradiation of the egg by ulltra-violet light. *J. Embryol. Exp. Morphol.* 12: 89–100.

——. 1970. Study of the development of the internal organs of the double malformations of *Chironomus dorsalis* by fixed sectioned materials. *J. Embryol. Exp. Morphol.* 24: 287–303.

——. 1983. Induction of longitudinal double malformations by centrifugation or by partial UV irradiation of eggs of the chironomid species *Chironomus samoensis* (Diptera: Chironomidae). *Entomol. Gen.* 8: 171–191.

Yamashita, O., and L. S. Indrasith. 1988. Metabolic fates of yolk proteins during embryogenesis in arthropods. *Dev. Growth Diff.* 30: 337–346.

Yamauchi, H., and N. Yoshitake. 1982. Origin and differentiation of the oocyte-nurse cell complex in the germarium of the earwig, *Anisolabis maritima* Borelli (Dermaptera: Labiduridae). *Int. J. Insect Morphol. Embryol.* 11: 293–305.

Yao, T.-P., B. M. Forman, Z. Jiang, L. Cherbas, J.-D. Chen, M. McKeown, P. Cherbas, and R. M. Evans. 1993. Functional ecdysone receptor is the product of *EcR* and *ultraspiracle* genes. *Nature* 366: 476–479.

Yip, M. L. R., M. L. Lamka, and H. D. Lipshitz. 1997. Control of germ band retraction in *Drosophila* by the zinc-finger protein HINDSIGHT. *Development* 124: 2129–2141.

Yoshida, A., Y. Arita, Y. Sakamaki, K. Watanabe, and R. Kodama. 1998. Transformation from the pupal to adult wing in *Oidaematophorus hirosakianus* (Lepidoptera: Pterophoridae). *Ann. Entomol. Soc. Am.* 91: 852–857.

Young, J. H. 1953. Embryology of the mouthparts of Anoplura. *Microentomology* 18: 85–133.

Younossi-Hartenstein, A., P. Green, G.-J. Liaw, K. Rudolf, J. Lengyel, and V. Hartenstein. 1997. Control of early neurogenesis of the *Drosophila* brain by the head gap genes *tll, otd, ems* and *btd*. *Dev. Biol.* 182: 270–283.

Younossi-Hartenstein, A., C. Nassif, and V. Hartenstein. 1996. Early neurogenesis of the *Drosophila* brain. *J. Comp. Neurol.* 370: 313–329.

Younossi-Hartenstein, A., U. Tepass, and V. Hartenstein. 1993. Embryonic origin of the imaginal discs of the head of *Drosophila melanogaster*. *W. Roux's Arch. Dev. Biol.* 203: 60–73.

Yu, F., X. Morin, Y. Cai, X. Yang, and W. Chu. 2000. Analysis of *partner of inscuteable*, a novel player of *Drosophila* asymmetric divisions, reveals two distinct steps in Inscute-able apical localization. *Cell* 100: 399–409.

Yu, R., and S. W. Omholt. 1999. Early developmental processes in the honeybee (*Apis mellifera*) oocyte. *J. Insect Physiol.* 45: 763–767.

Yund, M. A., and D. L. Osterbur. 1985. Ecdysteroid receptors and binding proteins. In G. A. Kerkut and L. I. Gilbert (eds.), *Comprehensive insect physiology, biochemistry and pharmacology*. Vol. 7, pp. 473–490. Pergamon Press.

Zacharias, D., J. L. D. Williams, T. Meier, and H. Reichert. 1993. Neurogenesis in the insect brain: cellular identification and molecular characterization of brain neuroblasts in the grasshopper embryo. *Development* 118: 941–955.

Zacharuk, R. Y. 1985. Antennae and sensilla. In G. A. Kerkut and L. I. Gilbert (eds.), *Comprehensive insect physiology, biochemistry and pharmacology*. Vol. 6, pp. 1–69. Pergamon Press.

Zachos, J., M. Pagani, L. Sloan, E. Thomas, and K. Billups. 2001. Trends, rhythms, and aberrations in global climate 65 Ma to present. *Science* 292: 686–693.

Zars, T., M. Fischer, R. Schulz, and M. Heisenberg. 2000. Localization of a short-term memory in *Drosophila*. *Science* 288: 672–675.

Zeger, S. L., and S. D. Harlow. 1987. Mathematical models from laws of growth to tools for biologic analysis: fifty years of *Growth*. *Growth* 51: 1–21.

Zeh, D. W., J. A. Zeh, and R. L. Smith. 1989. Ovipositors, amnions and eggshell architecture in the diversification of terrestrial arthropods. *Q. Rev. Biol.* 64: 147–168.

Zeitlinger, J., and D. Bohmann. 1999. Thorax closure in *Drosophila*: involvement of Fos and the JNK pathway. *Development* 126: 3947–3956.

Zera, A. J., and R. F. Denno. 1997. Physiology and ecology of dispersal polymorphism in insects. *Annu. Rev. Entomol.* 42: 207–231.

Zhang, D., and R. B. Nicklas. 1996. 'Anaphase' and cytokinesis in the absence of chromosomes. *Nature* 382: 466–468.

Zhang, N., J. Zhang, K. J. Purcell, Y. Cheng, and K. Howard. 1997. The *Drosophila* protein Wunen repels migrating germ cells. *Nature* 385: 64–67.

Zheng, Z., A. Khoo, D. Fambrough Jr., L. Garza, and R. Booker. 1999. Homeotic gene expression in the wild-type and a homeotic mutant of the moth *Manduca sexta*. *Dev. Genes Evol.* 209: 460–472.

Zhou, B., K. Hiruma, T. Shinoda, and

L. M. Riddiford. 1998. Juvenile hormone prevents ecdysteroid-induced expression of the *Broad Complex* RNAs in the epidermis of the tobacco hornworm, *Manduca sexta. Dev. Biol.* 203: 233–244.

Zitnan, D., T. G. Kingan, J. L. Hermesman, and M. A. Adams. 1996. Identification of ecdysis-triggering hormone from an epitracheal endocrine system. *Science* 271: 88–91.

Zitnan, D., L. S. Ross, I. Zitanova, J. L. Hermesmen, S. S. Gill, and M. E. Adams. 1999. Steroid induction of a peptide hormone gene leads to orchestration of a defined behavioral sequence. *Neuron* 23: 523–535.

Zrzavy, J., and P. Stys. 1995. Evolution of metamerism in Arthropoda: developmental and morphological perspectives. *Q. Rev. Biol.* 70: 279–295.

Index

*Page numbers in **bold** indicate figures. The symbol † indicates an extinct taxon.*

A tubule, 9–10, **10**, **11**
abdominal-A (abd-A). See genes, homeotic
Abdominal-B (Abd-B). See genes, homeotic
Acanthokara kaputensis (onychophoran): Distal-less expression, embryos, 375; Ubx/abd-A expression, embryos, **166**, 167
acceleration, 365, **365**
Achaeta domesticus (house cricket): eve expression, 159; homeotic gene expression, embryos, 167; number of molts, 265; number of sperm transferred, 76; pseudocleavage, 136; sperm entry into egg, 77; superficial anatrepsis, 127
achaete (ac). See genes, neuronal
achaete scute Complex (AS-C). See genes, neuronal
acroblast, 18, **19**
acron, 122, **130**, 134, 148, 151–153, **152**, **153**, **166**, 172–174, 218–219, **219**, **371**
acrosomal vesicle, 18, **19**
acrosome: development, 18, **19**; reaction, 78, **80**; structure, 8, **10**
Acyrthosiphon pisum (aphid), 346
adaptive zone, 354
adrenocorticotropin, **63**
Aedes aegypti (yellowfever mosquito): endocrine control, vitellogenesis, **60**; vitellogenin genes, expression, 39; mouthparts, **297**; vitellogenin receptor, 38
Aedes stimulans (mosquito): temperature effects on, 94–95
aeropyle: deposition, **36**, 41; structure, 111, **112**
Afrocimex constrictus (cimicid): homosexual mating, 75

Afrocimex leleupi (cimicid): homosexual mating, 75
Ageratum houstonianum (composite): molting, effects on, 341; vitellogenesis, effects on, 62
Agrin, 204
Agrotis segetum (noctuid): neurons, posterior migration, 280
air sacs, **300**, 301
alate ametaboly, 241, **242**, **243**, 253
Aleochara curtula (staphylinid): sperm displacement, **73**, 85; spermatheca, 73, **73**
alimentary canal: embryogenesis of, **130**, 134, 152, 184–186, **185**; genes and, Drosophila, 152, 184–186; metamorphosis, 277–278, **279**
allatostatin, **338**, 340–341
allatotropin, **59**, 60, 337, **338**
Allocorrhynchus flavipes (nabid): traumatic insemination, 74
Allocorrhynchus plebejus (nabid): traumatic insemination, 74
Allometabola, 238, **240**
allometric variation: experimental analysis, 261–262, 264, 345–346; sources, 260–261, **263**; types, 259
allometry, 259–262, **260–263**
amber, Cretaceous, 361–362
Ametabola (Apterygota), 236–237, **237**, **238**
amnion, 127–129, 131
amnioserosa, 131, **132**
amphioxus (lancelet): gene expression, embryos, 160, 168, **169**, **170**
amphitoky, 104, **104**
Amphoraphora tuberculata (aphid): spermatogenesis, **106**

anachronism (ana), 188, 279
Anagasta kühniella (Mediterranean flour moth): blood-spermatocyst barrier, 22; circadian sperm release, 65–66
Anajapyx sp. (japygid dipluran): ovary structure, **56**
anamorphosis, 135
Anastrepha fraterculus (tephritid): yolk extrusion and consumption, 131, 134
anatrepsis, 127, 128, **128**, **129**
anatrepsis center, **128**
anautogeny: oogenesis, 37; spermatogenesis, 22
angiosperms: insect-plant coevolution, 361–362, **362**; origin and diversification, 361, 362
Aniridia, 209
Anisolabis annulipes (earwig): eve expression, embryos, 159
Anomalocaris nathorsti†, **377**
Anopheles gambiae (malaria mosquito): eve expression, embryos, 159
Anoplognathus pindarus (scarab beetle): larva, **244**
antennae: embryogenesis, **130**, 134, 172–174, **173**; genes and, Drosophila, 172–174, **173**, 316, **316**; metamorphosis, 283–284, 286, **286**; sense organs/projection patterns, 196, 202, 218, 280, 283–284, 286, **286**
Antennagalea (Ag5), 172
antennal lobe: embryogenesis, 196; metamorphosis, **281**, 283–284
antennal segment: genes and, Drosophila, 172–174, **173**
Antennapedia (Antp). See genes, homeotic

Antennapedia Complex (ANT-C). See genes, homeotic
antennomaxillary complex, 218, **219**
anterior midgut rudiment, 184, **185**
Antheraea pernyi (saturniid): eclosion hormone, 341; sperm activator, 66
Antheraea polyphemus (polyphemus moth): choriogenesis, 41
Anthocoris sp. (minute pirate bug), 75
Anthonomus grandis (cotton boll weevil): sperm movement, spermatophore to spermatheca, 72, **72**; spermatogenesis rate, 20
antisymmetry, 262, **264**
aorta. See dorsal vessel
Aphidius ervi (braconid): gene expression, embryos, 160, **161**
Aphis fabae (bean aphid): life history, 105, **106**
apical complex, 9, **13**, 13–14
apical tissue (testis rudiment), 95–96, **96**
Apis mellifera (honeybee): allatostatins, 341; bicoid expression, embryo, 153; blastokinesis, **133**; brain metamorphosis, 283; caste formation, 95, 349–350, **350**; differentiation center, embryos, 137; egg production, inhibitory pheromones, 62; en expression, embryos, 160; eve expression, embryos, 159; fecundity, 54; fertilization, 79; gastrulation, 126; homeotic gene expression, embryos, 168; mesoderm, induction, 229; midgut-hindgut junction, 278; salivary glands, 215; sperm movement, 72; stinger, 87

apodemes: ecdysis during molt, 225, 267; embryogenesis, **181**, 225, **225**

apoptosis in: *Drosophila* embryogenesis, 220, 228; *Drosophila* head involution, 220; metamorphosis, 273–301; pretarsal metamorphosis, higher flies, 298–299; right mandible degeneration, thrips embryos, 220, **221**, 228; wing development, plume moths, 298; wing development, polymorphic moths, 295

appendages: embryogenesis, **130**, **132**, 134–135, 179, **181**, **185**, **188**, **194**, **202**, 202–203, **204**, **205**, 235; evolution, 355, **356**, 358, 373–375, **374**, 377–382, **377**, **380–382**; genes and, *Drosophila*, 215–218, **216**, **217**, 307–312, **308–311**; metamorphosis, 270–273, **272**, 294–299, **295–299**

apterous (*ap*): expression, wing disc, 309, **309**, **310**

Apterygota (Ametabola), 237

Archaearanea tepidariorum (spider): homeotic gene expression, embryos, 167, 172

Archaefructus liaoningensis†, 362

Archaeognatha (bristletails): diversity, **4**; fossil history, **356**, 359, **362**; life history, 236–237, **238**, **365**; number of molts, 264

Archegozetes longisetosus (oribatid): gene expression, embryos, **166**, 167, 172, 375

Archichauliodes guttiferus (megalopteran): pupa, **246**

aristaless (*al*), 307, 308

arrhenotoky, 101–103, **102**, 104, 109

Artemia franciscana (brine shrimp): gene expression, embryos, 160, **166**, 167, **371**, 375, 377; insect wings, evolution, 380–382, **381**

Ascogaster quadridentata (braconid): host spermatogenesis, effects on, 25–26, **26**

asense (*ase*). See genes, neuronal

Asobara citri (braconid), 62–63

Asobara tabida (braconid): *Wolbachia* infection, 62 asymmetry, 262–265, **264**

Athalia infumata (tenthredinid): ovariole transplant/artificial fertilization, 80

Athalia rosae (tenthredinid): ovariole transplant/ artificial fertilization, 80

atonal (*ato*): compound eye development, 287–289, **290**

Atrichopogon polydactylus (ceratopogonid): larva, **245**

attachment fibers, 269–270, **270**

Aulocara ellioti (acridid): parent condition and eggs, 232

autogeny: oogenesis, 37; spermatogenesis, 22

Autographa californica (alfalfa looper): nuclear polyhedrosis virus, 335

autosomes, 91

autotomy, 302

auxetic growth, 256

axillary sclerites, 355–356, 382

axis specification: appendages, 307–312, **308–311**; eggs, 49–52, **51**

axonal transport, 323

axoneme: development, 18, **19**; diversity, 11–13, **12**; function, 11, **11**; genes and, *Drosophila*, 17; structure, 9–10, **10**

Azadirachta indica (neem tree), 335

azadirachtin, **328**, 335

B tubule, 9–10, **10**, **11**

Bacillus rossius (stick insect): chromosome doubling, 103

Bactrothrips brevitubus (phlaeothripid): pole plasm, 39

bag-of-marbles (*bam*), 14, **17**

Balbiani ring, 331

Bambara invisibilis (ptiliid): number of sperm transferred, 76; sperm/ spermatheca coadaptation, 83, **84**

Barathra brassicae (noctuid): sensory continuity during molt, 269, **269**

behavior: adult eclosion, 247, **247**, 271–273; hatching, 235–236; mating, 65; molting, 236, **236**, 266–270, **267–269**, 339, 341–345, **342**

benign gonial cell neoplasm (*bcgn*), 14, **17**

Bibio hortulanus (bibionid): larval/adult CNS, **280**

bicaudal (*bic*), 141, 182

Bicaudal-C (*Bic-C*): microtubule organizing center, 44

Bicaudal-D (*Bic-D*): oocyte specification, 43

bicoid (*bcd*): anteroposterior axis, eggs, 50, 51; anteroposterior axis, embryos, 152–155, **153–155**; *eve* stripe 2, specification, 169–170, **171**; head segments, specification, 172–174, **173**

Bicyclus anynana (nymphalid): eyespot specification, 315–316

Bilobella grassei (collembolan): polytene chromosomes, 331

bisacylhydrazine insecticides, **328**, 335

Bithorax Complex (*BX-C*). See genes, homeotic

Blaberus craniifer (death's head cockroach): antigenic differences, leg segments, 307; ootheca and vitellogenesis, 61

blastoderm: formation, 115, 117, **117**; mitotic domains, 118–119, **119**

blastodermal cuticle, 129

blastokinesis, 126–131, **128**, **129**, **132**

blastomeres: Collembola, 112; polyembryonic embryos, 112, 123–125, **124–126**

Blattella germanica (German cockroach): 20E and choriogenesis, 61; accessory hermaphroditism, 96; allatostatins, 341; homeotic gene expression, embryos, 168; molting/ regeneration, 269; yolk mobilization, embryos, 115

Blattodea (cockroaches): diversity, **4**; fossil history, **356**, **362**; number of molts, 264

blood meal: *Aedes aegypti*, **60**; and fecundity, 62; and metamorphosis, 321–322, **321**, **322**; and molting, 256, **257**, 319–320, **320**; *Rhodnius prolixus*, 22–24, 59, **59**, **60**; and spermatogenesis, 22–24

blood-testis barrier, 22, **23**, **24**

body cavity, embryonic origin, **178**, 182–183, **185**

body plan: genes and, *Drosophila*, 146–174, 306–312, 316–318; and phylotypic stage, **130**, 135, 137, 364, 369–373, **372**

Bolwig's organ (*Drosophila*), 208, 218

Bombus terrestris (bumblebee), 350

Bombyx mandarina (bombycid): artificial polyploid bisexual species, 104

Bombyx mori (commercial silkworm): bombyxin, **328**; choriogenesis, 41; diapause hormone, 231–232, **232**; ecdysone, 324; ecdysteroid receptors, 331; eclosion hormone, **328**, 341; prothoracic gland, function, 320–321; prothoracicotropic hormone, 336

bombyxin, **328**

Bone morphogenetic protein- 4 (*BMP-4*), 151

border cells: egg termini, specification, 51, 52

Brachydanio rerio (zebrafish): *en* expression, embryos, 160, **162**; *Hox* gene expression, embryos, 168

Bracon hebetor (braconid): gene expression, embryos, 160, **161**

brain: embryogenesis, 195–200, **196**; genes and, *Drosophila*, 197; insect versus vertebrate, 199–200; metamorphosis, 280–285, **281–285**; structure, 195, **195**

branched (*bch*), 214

Branchiostoma sp. (lancelet): gene expression, embryos, 160, 168, **169**, **170**

branchless (*bnl*), 214

breathless (*btl*), 214

Brevicoryne brassicae (cabbage aphid): corpora allata in juveniles, 346

Brevipalpus phoenicis (false spider mite): haploid females, 101

bride-of-sevenless (*boss*): photoreceptor R7, specification, 289, **291**

Broad Complex, 333–335, **335**, **336**

bromodeoxyuridine (BrdU), 186, 195, 197, 273, 279, 281, 283, 285

Bruchidius sp. (bruchine weevil): physiological centers, embryos, 140

brustia, 235

bursicon, 268, **339**, **342**, 342–343

buttonhead (*btd*): head segment, specification, **173**, 174

cactus (*cact*), 50, 52

Caenorhabditis elegans (nematode): apoptosis, 228; computer models, embryogenesis, 234;

ecdysteroids, 345; heterochrony genes, 366; homeotic genes, embryos, 168, **169**; neural genes, 201; neuron growth across midline, 198

Calliphora erythrocephala (blowfly): bursicon, 342; ectodermal induction, embryonic mesoderm, 229; nurse cell ploidy, 47, 55; polyploidy, larval epidermis, 257; proliferation, abdominal histoblasts, 297

Callosobruchus maculatus (cowpea weevil): number of sperm transferred, 76; segmentation gene expression, embryos, 159

Caloneurodea†: fossil history, **356**, **362**

calyx: lateral oviduct, 30, **30**, 326; corpora pedunculata, **195**, **281**, **283**, 283

Cambrian "explosion," 357, 369

Cambrian Period, 357–358, **358**, 375

Campodea montis (campodeid dipluran): adult, **244**

Campodea sp. (campodeid dipluran): ovaries, 56, **56**

campodeiform larva, 243–244, **244**

Camponotus laevigatus (carpenter ant): *eve* expression, embryos, 159

cappuccino (*capu*): anteroposterior axis, 152; pole plasm, 39, 121, **121**

Carausius morosus (stick insect): corpora pedunculata, 197; follicle cell ploidy, 40; germ cells, 121; metaphase I block, 40; ommatidial embryogenesis, 291; parthenogenesis, 103; substitutive embryogenesis, 121; temperature and sex determination, 94

Carboniferous Period, **356**, **358**, 359, **362**, 376

cardioblasts: embryonic origin, 176, **178**; genes and, *Drosophila*, 183–184, **183**

cartridges, optic, 293, **293**

caste determination, 347–351, **348–350**

caterpillars: diet effects, 260, 262; prolegs, 134, 165, 243, **245**, 248, 319, 378; why not short and fat? 259

caudal (*cad*), 152–153, 185, 317; *Tribolium castaneum* embryos, 155

caudal visceral mesoderm, 182

Cecidomyiidae (gall midges): chromosome elimination, cleavage, 113–114; meiotic block, oocytes, 40; sperm motility, 11

Cediopsylla simplex (cottontail rabbit flea): host hormones, gametogenesis, 63

cell death program. *See* apoptosis

cell types: border, 51 (*Drosophila* ovary), 52, 179, 203, **204** (locust embryonic legs); cap, 42, **43**; cyst, 6, **9**, **13**, 13–18, 22–24, 65, 66; follicular, 29, **31**, 33, 34, **36**, 37–39, **38**, **42**, **44**, **46–51**; ganglion, 187; ganglion mother, 187–190, **188**, **189**, **191**, **192**, 193, **194**, 195–202, **196**, **201**; glial, 191, 193, **193**, 200, **201**, 279, 284, 293; Inka, 267, 341–342, **342**; muscle pioneer, 179–182, **180**, **181**; neuroblast, 187–190, **188**, **190–192**, 193, **194**, 195–197, **196**, **201**, **202**, 278–284, **282**, **283**; neuron, **188**, 190–191, **190–194**, 196; neurosecretory, 60, 95–96, **96**, **195**, 198–199, 284–285, **284**, **285**, 323–324, **325–327**; nurse, 29, **31**, 41–52, **42–49**, 51, 55–56, 356, **356**, 382; pioneer neuron, **202**, 202–203, **204**, **205**; sense organ precursor, 204–208, **205**, **207**, 286; stem, female, 42–43, **42**, **43**; stem, male, **13**, 13–14, **27**; terminal filament, 42, **43**; tip, 134 (hatching thread), 185–186 (Malpighian tubule)

cement layer, 168, **168**

cenogenesis, 364

Cenozoic Era, 362–363

centers: activation, 136–137, **137**; cleavage, 136–137; commitment, 137; germinal, 259, **261**; germinal proliferation, 6, 13–14

central doublet, 10, **10**

central nervous system: embryogenesis, 187–202; genes and, *Drosophila*, 188–190, 200–202, **201**; structure, 186, **186**

centrin, 82

centriolar adjunct, 9, **19**

centrioles in: cleavage, 101, 113, **115**; fertilization, 79–82, **81**; meiosis, 15, 17;

oogenesis, **43**, 44, 47; sperm, 8–9, 18, **19**

centrolecithal, 1, 111

centrosomes in: cleavage, 101, 113, **115**; fertilization, 79–82, **81**

centrosomin (*cnn^{mfs}*), 82, 113

cephalic furrow, **132**, 174

cephalic horns, 265, 345–346

cercus: embryogenesis, 134, 204, **205**, **206**; hatching threads, 134, 235–236; sensillar differentiation, 204, **205**, **206**

Chaetopterus variopedatus (annelid): *Distal-less* expression, parapodia, 375

cheliceral segment: homeotic gene expression, **166**, **371**

Chelicerata, 166, 167, 308, 331, 381

chemoreceptors, gustatory: cuticle continuity during molt, 269; embryogenesis, 204, 207

chemoreceptors, olfactory: cuticle continuity during molt, 269; differentiation, 269, **270**; embryogenesis, 204, 207; projection pattern, 196, 283–284; structure, **270**

Cherax destructor (decapod crustacean): NB pattern, 193

chiasmata, 15, 16, 17–18

Chicxulub crater, 362

Chironomus dorsalis (chironomid): anteroposterior axis in embryos, 141–143, **142**; embryogenesis promoting factor, 230–231

Chironomus hyperboreus (chironomid): mermithid parasitism effects, 98

Chironomus samoensis (chironomid): anteroposterior axis in embryos, 141–143, **142**

Chironomus tentans (chironomid): chromosome puffs, tissue specificity, 331; ecdysteroids, chromosome puffing, 332, **333**; EcR ortholog, 331

cholesterol: ecdysone biosynthesis, 324

chordoblast, 46, **47**

chordocyte, 46, **47**

chordotonal organs, 202, 204, 205

choriogenesis, 36, 41

chorion: deposition, 36, 41; structure, 111, **112**

Choristoneura fumiferana (spruce budworm): EcR ortholog, 331

Chortophaga australior (grasshopper): mitosis without chromosomes, 14

chromatin condensation (spermiogenesis), 18, **19**

chromosomes, types: autosomes, 91, **92**; lampbrush, 34, **35**, 37; polytene, 296, 298–299, **298**, **299**, 331–335, **332**, **333**, **335**; sex, 91–92, **92**

chromosome elimination in: cleavage, 113–115; gynandromorphs, 97, **98**; male meiosis, aphids, 106, **106**; sex determination, 93–94; symphypleon collembolans, 93–94

chromosome puffs in: pretarsal pulvillar cells, 296, 298–299, **298**, **299**; salivary glands, 331–335, **332**, **333**, **335**; socket- and hair-forming cells, 299

Chrysopa perla (chrysopid): blastokinesis, **133**; mesoderm differentiation, **178**; mesoderm induction by ectoderm, 228–229, **229**

Cimex lectularius (bed bug): growth, 257; mating stimuli/vitellogenesis, 62; sperm activation, 66; traumatic insemination, 74–76, **75**

circular rDNA: previtellogenesis, 34, **36**

c-Jun amino-terminal kinase (JNK): dorsal closure, 182

clathrin: yolk uptake, 38, **38**

cleavage, 111–115, **113–116**

cleavage center, 111, 136–137

cleavage energid, 111, **113**

climate/Earth history, 357–363

Clitumnus extradentatus (stick insect): ovulation, 77

clypeolabral disc, **216**, 218–220, **219**, 294, **295**

Cnidaria (jellyfish): homeotic gene expression, embryos, 168, **170**

coated pit: yolk uptake, 37, **38**

coated vesicle: yolk uptake, 37, **38**

Coccoidea (scales): life history, 238–239, 241–242, **241**; mouthpart embryogenesis, 222, **224**; sex determination, 93; sperm, 11, **12**, 13

cocoon: endopterygotes, 246; male scales, 238

coelom, 176, **178**, 182

coelomic sacs, 176, **178**, 182

Coleoptera (beetles): diversity, **4**; fossil history, **356**, **362**; number of molts, 264

Collembola (springtails): diversity, **4**; fossil history, **356**, **362**; number of molts, 264

commissureless (*comm*), 191, 193, **193**, 197

commitment center, 137

compartment boundary: demonstration, 144–146, **147–149**; genes and, *Drosophila* embryos, 147–160, **160**, **166**, 170; genes and, *Drosophila* imaginal discs, 306–311, **308–310**

compartment hypothesis, 144–147, **145–149**

complete circle rule, 304, **305**

compound eyes. *See* eyes, compound

Comstockiella chromosome system, 93

Comstock-Needham system, 311

Confirm (RH-5992), **328**, 335

continental drift, 357–363, **358**, **361**

Copidosoma floridanum (encyrtid wasp): gene expression, embryos, 160, **161**; germ cells, 121–122; host hormone effects, 232–233; polyembryony, 124, **124**

copulation, 65, **66**, **67**, **85**

copulatory courtship, 86

Cordulia aenea (dragonfly): male grasping female, **71**

coremata: development, 301

Coriophagus rieki (strepsipteran): adult male, **382**

cornichon (*cni*), 50

corpora allata: embryogenesis, 209, 211; juvenile hormone and, 22–23, **25**, **58–60**, 58–63, 319–324, **320**, **324**, **327**, 337–340, **338**, **339**, 343; metamorphosis, **284**, 284–285; structure, **195**, 209, **210**, 320, 324

corpora cardiaca: embryogenesis, 211–212; hormone release, 323; metamorphosis, 284–285; structure, **195**, 209, **210**, **320**, 324, **325–327**

corpus cardiacum stimulating factor, 60, **60**

corpus luteum, 77

corpus pedunculatum: embryogenesis, 197; metamorphosis, 283, **283**; structure, **195**, **281**

corticosterone, 63, **63**

cortisol, 63, **63**

Cosmoconus meridionator (ichneumonid): egg axis markers, 49; pole plasm formation, 39

Cossus cossus (cossid): growth, 256

Cossus ligniperda (cossid): prothoracic glands, 320

Crambus mutabilis (pyralid): Dyar's law, **258**

Creatonotos transiens: pheromone gland development, 301

Cressida cressida (papilionid): sperm plugs, 83

Cretaceous Period, **361**, 361–362

crooked legs (*crol*), 334

Crumbs (Crb), 186

Crustacea: compound eye development, 291; *Distal-less* expression, embryonic appendages, 174; *en* expression, embryos, 160, **162**; fossil history, 357, 375–376, **375**, **376**; homeotic gene expression, embryos, 166, 167–168, **170**, **371**, 373, 375; hormones, 345; NB distribution, embryos, 193; wing/distal epipodite homology, 380–382, **381**

crustacean cardioactive peptide, 342, **342**

cryptobiosis, 231

cryptonephridia: metamorphosis, 278

crystal cell, 182, **183**

crystallomitin, 9

Ctenolepisma longicaudata (long-tailed silverfish): JH effects, embryogenesis, 235; ventral nerve cord embryogenesis, 187–188, 193

Ctenophalides felis (cat flea): embryonic head segments, 172; homeotic gene expression, embryos, 167

Cubitermes fungifaber (termite): physogastry, 259

Culicoides melleus (ceratopogonid): number of sperm transferred, 77; sperm movement, spermatophore to spermatheca, 72, 73

Cupiennius salei (spider): homeotic gene expression, embryos, **166**, 167

cuticle: blastodermal, 129; chemical composition, 166; deposition, 266–270, **267–270**; embryonic, 229–230, **230**; muscular continuity during molt, 269–270; sclerotization, 268, **268**; sensory continuity during molt, 269, **269**; serosal, 129, 229–230, **230**; structure, 266, **267**, **268**

cuticulin envelope, 266

Cyclorrhapha: metamorphosis, 270–273, **272**, **273**

Cydia pomonella (codling moth): apyrene spermatogenesis, 25; parasitoid effects on spermatogenesis, 25–26, **26**

Cynipidae (gall wasps): cyclical parthenogenesis, 105, **105**

cyst, 9, 13, 14

cystocytes, 42–44, **42**, **43**, 382

cytoplasmic incompatability, 102

cytoplasmic streaming during: cleavage, 111, **113**; fertilization, 79, **81**

dachshund (*dac*): proximal distal domains, leg discs, 307

Dacus tryoni (olive fly): embryogenesis, summary, 233, **233**

Danaus plexippus (monarch butterfly): brain metamorphosis, 280–283, **281**, **282**

daughterless (*da*): sex determination, 92–93

decalin fragment (azadirachtin), 323, 335

decapentaplegic (*dpp*): abdominal mesoderm, 179; abdominal terga/sterna, 317, **317**

Deilephila spp. (moth): prothoracic gland? 320

Delta (*Dl*). *See* genes, neuronal

denticle belt, larva, 157, **159**, **245**

Dermaptera (earwigs): diversity, **4**; fossil history, **356**, **362**; number of molts, 264

Dermestes frischi (dermestid beetle): *eve* expression, embryos, 159

Dermestes lardarius (dermestid beetle): mesoderm induction, embryos, 229

Deuterostomia, 199–200, **200**

deutocerebrum. *See* brain

developmental genetics, 144

developmental stability, 262

Devonian Period, **356**, **358**, 358–359

diapause, 231

diapause hormone, 231–232, **232**

Diaphanopterodea†, 356, **356**, **362**, 382

diaphanous (*dia*), 14

dicephalic (*dic*), 141

Dicyrtomina ornata (collembolan): chromosome elimination/sex determination, 93–94

diet and: allometry, 260; developmental polymorphism, 262, **264**

Diopsidae (stalk-eyed flies), 84

Diploptera punctata (cockroach): allatostatins, 340; hormones/ vitellogenesis, 61; mating/ vitellogenesis, 62

Diplura (diplurans): abdominal appendages, 134, 378; adult, **244**; blastodermal cuticle, 129; blastokinesis, 129; diversity, **4**; fossil history, **362**; long germ embryo, 122, 135; number of molts, 264; oogonial clones, **42**; predation/small size, 253; sperm, 10, **12**; sperm tubes, 8; trachea, 213

Diptera (2-winged flies): diversity, **4**; fossil history, **356**, **362**; number of molts, 264

disjunction-defective, 27

distal epipodite: and wing origin, 218, 379, **380**, **381**, 380–382

Distal-less (*Dll*): expression in leg/wing discs, embryos, 217, **217**; expression in leg/wing discs, larvae, 307–309, **308**, **309**; mouthpart embryogenesis, **173**, 174; wing color pattern, 314, 314–315

division of labor: during metamorphosis, 247–248, 254–256, **255**, 382–383; in social insects, 347–351

DNA microarrays, 334–335

DNA-binding proteins: *Drosophila* embryos, 146–174; *even-skipped* stripe 2 expression, 169–172, **171**; evolution, 369–373, **370**; pattern-control gene expression, 146–174

Dorcadia ioffi (tunga flea): neosomy, 259, **259**

dorsal (*dl*): dorsoventral axis, embryos, 149–151, **151**; dorsoventral axis, oocytes, 50, 52 dorsal closure: embryogenesis, 182–183, **185**; metamorphosis, 295

dorsal ocelli, 294

dorsal ridge, **132**, 218

dorsal vessel: embryogenesis, **178**, **179**, **183**, 183–184; genes and, *Drosophila*, 183–184; metamorphosis, 275; structure, **177**, **178**

dorsoventral axis, *Drosophila* eggs/embryos, 50–52, **51**, 149–151, **151**

dorsoventral axis inversion, 151

dosage compensation, 93

double abdomen phenotype, 141–143, **142**, 152, **152**

double head phenotype, 140, **141**, 152, **152**

double sex (*dsx*): male/female promoridia, genital disc, 318; sex determination, **92**, 92–93; sexual differences, ventral nerve cord, 280

Downstream receptor kinases (*Drk*), 289, **291**

Drepanosiphum platanoides (aphid): oogonial clone, **42**

Drosophila acanthoptera: number of sperm transferred, 77

Drosophila bifurca: number of sperm transferred, 77; sperm entry, egg, 78, 79; sperm length, 12; sperm mitochondrial derivatives, persistence, 89

Drosophila bucksii: sperm length, 79

Drosophila ezoana: sperm length, 79

Drosophila hydei: sperm length, 79; spermatocyte lampbrush Y-chromosome, 16

Drosophila mauritiana: sperm competition in heterospecific crosses, 83; *Odysseus* homeobox divergence, 353

Drosophila melanogaster, 147, 150. *See also entries for specific genes, processes, and structures*

Drosophila pachea: sperm configuration within egg, 79

Drosophila pseudoobscura: sperm configuration within egg, 79

Drosophila sechellia: sperm competition in heterospecific crosses, 83; *Odysseus* homeobox divergence, 353

Drosophila simulans: gynandromorph fate mapping, 144; *Odysseus* homeobox divergence, 353

Drosophila subobscura: female choice of polymorphic sperm, 86

Dugesia trigrina (flatworm): *Pax-6* expression, 209

dumbfounded (*duf*), 181

duplication, **303**

Dysdercus fasciatus (pyrrhocorid): allometric shape change during postembryogenesis, 262, **263**

Dysdercus intermedius (pyrrhocorid): electrophoresis, nurse cell-oocyte transfer, **49**

E74A, 332–333, **333**

E74B, 332–333, **333**

E75, 332–333, **333**

E78B, 332–333, **333**

Earth history, 356, 357–363, **358**, 361

ecdysial canal, 269

ecdysial line, 267, 338

ecdysial membrane, 266, **267**

ecdysis, 236, **236**, 267, **267**

Ecdysone Receptor (*EcR*), 329–331

ecdysone response element, 329–331, **330**, 332

ecdysoneless (*ecd*), 289

ecdysones (ecdysteroids): adult commitment peak, 338, **339**, 343; biosynthesis in follicle cells, 39, **59**, 60, **60**; chemical structure/synthesis, 324–325, **328**

Ecdysozoa, 345

ecdysteroid inducible genes, 331–335, **333**, **335**, **336**

ecdysteroid pool, 326–328, **330**

E-chromosomes: elimination during cleavage, 113–115

eclosion, adult, 247

EcR-A, 329–331

EcR-B1, 329–331

EcR-B2, 329–331

ectadenes: development, 275, **276**; function, 66, 68–69

ectoderm: derivatives, 186; embryogenesis, 3, 125–126, **127**, 149–151, **151**; genes and, *Drosophila*, 149–151, **151**; metamorphosis, 270–301

egg burster, 235

egg development neurosecretory hormone, 60, **60**

egg envelopes, **33**, **36**, 41, 111

egg size: adaptive significance, 54; miniaturization, effects on, 367

egg structure, **33**, **36**, 111, **112**

ejaculatory duct: embryogenesis, **227**, 228; function, 65–66, **68**; metamorphosis, 275–276, **276**; structure, 7, **7**

Eldana saccharina (pyralid): stases, 243

electrophoresis, 48, **49**

Embioptera (embiids): diversity, **4**; fossil history, 356, 362; number of molts, 264

embryogenesis, early: blastoderm formation, **113–115**, 115–117, **117–119**; blastokinesis/embryonic membranes, 126–131, **128**, **129**, **132**, **133**; cleavage, 111–115, **113–116**; DNA, RNA, protein synthesis, 117–118, 118; gastrulation/germ layer formation, **119**, 125–126, **127**; germ band formation, **121**, 122–125, **123–126**; mitotic domains, 118–119, **119**; pole cell formation, **114**, 119–122, **120–121**; polyembryony, 121–122, 123–125, **124–126**; segmentation/appendage formation, 130, **132**, 134–135; yolk extrusion/consumption, 131, 133, 254

embryogenesis, genes and, *Drosophila*: apoptosis, 228; body pattern, 146–174, **145–155**, **157–160**, **162–166**, **169–171**, **173**; Bolwig's organ, 208–209; brain development, 196–197; cleavage rate, 112–113, **115**; compartment boundaries, 144–146, **145–148**; computer simulation, 233–234; endoderm, 184–185; ganglion mother cell identity, 188–190, **189**; gastrulation, 126, **127**; germ band shortening/dorsal closure, **128**, 128–129, 131, **132**; glial cells, 200–201, **201**; gonads, 226–228; head involution, 218–220, **219**; insect versus vertebrate brain, 199; larval mesoderm, 176–179, **179**; larval muscles, hemocytes, dorsal vessel and body cavity, 180–184, **183**; leg/wing specification, **217**, 217–218; mesodermal induction, 228–229, **229**; neuronal crossing of midline, 191, 193, **193**; pole cells, 120–121, **121**; salivary glands, 215; sense organs, 205–207, **207**; synapse formation, 204; tracheal system, 212–215, **214**

embryogenesis, phylogeny, **133**

embryogenesis, summary, **233**

embryonic membranes: adaptive significance, 129; amnion, 126–128, **128**, **129**; serosa, 2, **3**, **121**, 122, 126–129, **128**, 131

embryos, stages: oligopod, **128**, 135, 243; percentage total development time, 111; polypod (phylotypic), **130**, 135, 243; protopod, **121**, 122, **128–130**, 134, 135

Empididae: male nuptial gifts, 85

empty spiracles (*ems*): anterior head segmentation, 172–174, 197, 377; as gap gene, **155**, 156

endo cell cycle in: follicle cells, 40; larval cells (*Drosophila*), 216–217, 256, **257**; nurse cells, 47; prothoracic gland cells (*Drosophila*), 212; serosal cells, 122; vitellophages, 115

endocuticle: in ectospermalege, 74–75, **75**; in exoskeleton, 266, **267**

endoderm: embryonic origin, 125–126, 184–185, **185**; genes and, *Drosophila*, 126, 152, 184–185; metamorphosis, 277–278, **279**

Endopterygota (Holometabola): characteristics, 239–240; evolution, 236, **237**, **239**, **242**, 243, 248–256, **249**, **251**, **255**, 356–357, 382–383; fossil history, **356**, 359

endosome: yolk uptake, 37–38

engrailed (*eng*). See genes, segmentation: pair-rule

Enhancer-of-split (*E[spl]*). See genes, neuronal

enteric nervous system. See stomatogastric nervous system

enteric placode, 210–211, **210**, **211**

Entwicklungsmechanik, 136

environmental effects: on embryogenesis, 231–233; on vitellogenesis, 61–63, **63**

enzymes, molting, 266–270, **267**

Eocene Epoch, **358**, 363

Ephemeroptera (mayflies): diversity, **4**; fossil history, **253**, **356**, **362**; life history, 237, **238**; number of molts, 264; surface skimming, 379

Ephestia sp. (Mediterranean flour moth): blastokinesis, **133**

epicuticle, 266, **267**, **268**, 270

epidermal feet, 271

epitracheal gland, 341

escargot (*esg*): diploidy, maintenance, imaginal disc cells, 217; leg disc domains, maintenance, 307

Ethmostigmus rubripes (centipede): *Ubx/abd-A* expression, embryos, **166**, 167, **371**; ventral nerve cord embryogenesis, 193–194

Eublaberus posticus (cockroach): ootheca, inhibition of JH secretion, 61

Eudalia macleayi (carabid beetle): larva, **244**

Euscelis plebejus (cicadellid): anteroposterior axis, embryos, 139–141, **141**; previtellogenic transcripts in egg periplasm, 117

even-skipped (*eve*). *See* genes, segmentation

evolution of: complete metamorphosis, 248–256, **249**, **251**, **252**, **255**, **356**, 356–357, 382–383; di-condylic mandibles, 355, **356**, 378; internal fertilization, 355, **356**, 376–377; nurse cells, **356**, 356, 382; ovipositor, 355, **356**, 378; six legs, 355, **356**, **377**, 377–378; wing flexing, 355, **356**, 382; wings, 355, **356**, 378–382

exaptation. *See* key innovations

Exopterygota (Hemimetabola): characteristics, 237–239; fossil history, **356**, **362**; phylogeny, **239**

extradenticle (*exd*): and antenna development, 316; and leg disc domains, specification, 307

eye-antennal disc: embryogenesis, 215–217, **216**, 218–220, **219**; metamorphosis, 286–291, **287–298**, 294–296, **295**

eyegone (*eyg*), 209

eyeless (*ey*), 209

eyelisch, 209

eyes, compound: embryogenesis, 208; gene expression during metamorphosis, 286–291, **288–292**; innervation, 293, **293**, **294**; metamorphosis, not *Drosophila*, 291–293; structure, 286–287, **287**

eyes, larval. *See* Bolwig's organ; stemmata

eyes absent, 209

eyespot determination in butterflies, 312–316, **312–315**

farnesoic acid, 337, 345

Fasciclin I–IV (fasI–IV), 190, **192**

fat body: CNS and hormones, influence on metamorphosis, 275; embryogenesis, **178**, 184; genes and, *Drosophila*, **179**; vitellogenin synthesis, 59–60

fate maps: abdominal segment, 316–317, **317**; blastoderm, **143**, 143, 144; constructing, 143, 144; genital disc, **317**, 317–318; leg disc, 307–308, 308, **309**; mitotic domains, 118, 119, **119**; segmental mesoderm, 176–179, **179**, **180**; wing disc, 308–312, **310**, **311**

fear of intimacy (*foi*), 226

fecundity, 54

fenoxycarb, **328**

fertilization, 78–82, **80**, **81**

fertilization chamber, 31, **78**

filing cabinet model, 250–251, **252**

flagellum: male intromittent organ, 74; sperm, 9–10, **9**, **10**, 18, **19**, 20

flight, evolution, 354–355, **356**, 378–382

follicle cells: ecdysteroid production by, 39; egg envelope deposition by, 40–41; genes and, *Drosophila*, 41–44, **43**, 49–52, **51**; metaphase I block, 39–40; origin, **35**, **36**, 40, **43**

foregut: embryogenesis, 130, 134, 185, **185**; genes and, *Drosophila*, 185; metamorphosis, 277–278, **279**; segmental origin, 134, **130**, **185**

forkhead (*fkh*): acron, foregut and AMR specification, 152; head segment specification, 172, **173**; midline NB, 199

fossils: formation, 357; in geological record, **356**, 357–368, **362**

Frankliniella occidentalis (western flower thrips): sperm, 11–12, **12**

Frieseomelitta varia (bee): ovariolar degeneration, 62

fringe (*frg*): dorsoventral axis, compound eye, 291, **292**; wing pattern, 310

frizzled (*fz*), **160**

frontal ganglion: embryogenesis, 209–210, **210**

fushi tarazu (*ftz*). *See* genes, segmentation: pair-rule

fusomes: in oogenesis, 43, **43**; in spermatogenesis, **13**, 14

Galleria mellonella (wax moth): integumental transplants and hormonal effects, 321–322, **323**

gametogenesis, female, **31**, 32–54, **33**, **35–36**, **37**, **42–49**, **51**, 58–64, **58–60**, **63**

gametogenesis, male, **9**, **13**, 13–28, **15**, **19–21**, **25–27**

ganglion mother cells: brain, 195–197, 280–284; embryogenesis, 187–190, **188**, **189**, **191**, **192**; genes and, *Drosophila*, 188–190, **189**; optic lobes, 197, 280–283, **282**; ventral nerve cord, 187–190, **188**, **189**, **191**, **192**, 278–280

gap genes. *See* genes, segmentation

gap phenomenon, 140–142, **141**, **142**

Gastroneuralia, 151

gastrulation, **3**, 125–126, **127**, 185, **185**, 199, **200**

gastrulation arrested (*ga*), 229

genes, dorsoventral: 50–52, **51**, 149–151, **151**

genes, ecdysteroid-inducible: *Broad Complex* (*BX-C*), **333**, 333–334, **335**, **336**; *E74A*, 333; *E74B*, 333; *E75A*, 333; *E75C*, 333; *E75G*, 333; temporal expression, *Drosophila* metamorphosis, 334–335, **335**, **336**

genes, homeotic: evolution, 168–169, **170**, 369–373, **370**, **371**, **374**, **377**; expression, *Drosophila* embryos, **159**, 161–165, **163–166**, **169**, **170**, **173**, **370–372**, 377; expression, *Drosophila* imaginal discs, 163; expression, other metazoan embryos, **161**, 165–169, **166**; regulation of expression, 162–165

genes, maternal effect and induction of: anteroposterior axis, 30, **51**, 152–155, **153–155**; dorsoventral axis, 50–52, **51**, 149–150; egg termini, **51**, 52

genes, neuronal: CNS, 200–202, **201**; PNS, 205–207

genes, pattern-control, 147–175

genes, segmentation: in embryos of other animals, 160–161, **161**, **162**; gap, 156, **155**, **157–159**; homeotic gene expression, effects on, **159**, 162; pair-rule, **154**, 156, **157–159**, **161**, **171**; regulation, *Drosophila*, **159**; segment-polarity, **150**, 156–158, 157, **159–162**

genetic address, 146, **149**

genetic dissection, 144

genitalia, external: accessory, 70–71, **71**; female, 32, **32**; male, 7, 74, **85**; postembryogenesis, 275–275, **276**, **278**

genomic conflict, 99

geological record and hexapod evolution, **356**, 357–363, **358**, **361**, **362**

germ band: blastokinesis, 126–131, **128**, **129**, **133**; elongation, **121**, 122, 127, **130**, **131**, **132**, **133**, **134**; extension/retraction of in *Drosophila*, 131, **132**, **133**; formation, **121**, 122; intermediate, 122, **128**, **133**; long, 122, **132**, **133**; phylogenetic distribution, **133**; in polyembryonic insects, 123–125, **124**, **125**, 126; position of, pentatomid bug eggs, 122, **123**, **133**; short, **121**, 122, **130**, 133

germ cells: embryonic origin, **114**, **119**, 119–122, **121**; encapsulation, 226; genes and, *Drosophila*, 120–121, **121**; migration to gonadal rudiments, 225–226; numbers of, 120; polyembryonic insect embryos, origin, 121–122; substitutive embryogeny, origin, 121

germarium: ovariole, 29, **30**, **31**, 32, **33**, 34, **35**, 41–45, **43**, **47**, 53, 56; sperm tube, 6, **9**, **13**, 13–14

germinal vesicle: **31**, **33**, 34–35, **35**, **36**, **37**, 46, 55

Gerris paludum insularis (waterstrider): anatrepsis center, 128

giant pulvillar nuclei, 296, **298**, 298–299

glands, accessory reproductive: development, 275–277, **276**, **277**; ectadenes, 7; female, **30**, 31, 78; male, 7, **7**, **68**; mesadenes, 7, **7**, 65, 66, **68**; products, 66, 78, **80**; spermathecal, **72**, **73**, **78**

glial cell deficient (glide), 200

glial cells: brain embryogenesis, role, 195; embryonic origin, 177, 200, **201**, 208; function, 190, 191, 193, 195, 200; in ganglia of VNC, larval *Drosophila*, 200, **201**; genes and, *Drosophila*, 182, 200; in honeybee brain, 284; metamorphosis, 279, 280, 284, 293; in stomatogastric nervous system, 211

glial cells missing (gcm), 182, 200

global climate change, 357–363

glomeruli: antennal lobe, 196, 283–284; brain, 195; embryogenesis, 195, 196; metamorphosis, 283–284

Glossina palpalis gambiensis (tsetse fly): fluctuating asymmetry, 263, 265

Glossina spp. (tsetse fly): uterus, 31

glycoproteins: growth cone guidance by, 180, **192**, 203

Golgi in: acrosome formation, 18, **19**; egg envelope deposition, 41; neurohormone synthesis in neurosecretory cells, 323

gonads: embryogenesis, 226–227; female, 29, **30**; male, 6, **7**; metamorphosis, 277

gooseberry (gsb). *See* genes, segmentation: pair-rule

gradients: in appendages, 302–306, **303–306**, 307, **308**, 309, **310**, 311; of Bicoid, 50, **51**, 153, **154**, **155**; of Caudal, 153; classic experiments, insect embryos, 137–143, **139–142**, 149; of Decapentaplegic, 150, **151**, 307, **308**, 309, **310**, 311; of Dorsal, 150, **151**; of gap proteins, **155**, 156, **159**; of maternal Hunchback, 153, **155**; of Nanos, 50, **51**, 153, **155**; of Snail, 150, **151**; of Spätzle, 150; of Twist, 150,

151; in wing pattern specification, butterflies, 314, **314**; of Wingless, 307, **308**, 309, **310**, 311

gradients/pattern formation, 137–143, **138–142**

gradualism versus saltationism, 353–354

Great Chain of Being, 363

gregarization factor, 347

grim, 228

Gromphadorhina portentosa (cockroach): interommatidial chimeras, 192

growth: accretionary, 256; allometric (disproportionate), 259–262, **260–263**; auxetic, 256; metamorphosis, 344–345, **344**, **345**; molting, 256–258, **257**, **258**, 343–344; multiplicative, 256; and sclerotization, 258–259, **258**

growth cones in: crossing midline, 191, **192**, 193, **193**, 195–196, **196**; epidermal feet, 271; glycoproteins and pathfinding, **188**, 190–191, **191**, **192**, **194**, 201–205, **202**, **204**, **205**, 209–211, **210**, **211**; growing neuronal axons, **188**, 190–191, **192**, 193, **194**, 196, 201–205, **202**, **204**, **205**, 209–211, **210**, **211**, 280–286, **282–286**, 293, **294**; muscle pioneers, 179–182, **180**; tracheoblasts, 213

growth laws: allometry, 259, **260**, 261; Dyar's, 258–259; geometric progression, 258; linear progression, 258

growth zone, 121, 122, **130**, 134

Grylloblattodea (ice worms): diversity, 4; fossil history, **356**, 362; number of molts, 264

Gryllodes supplicans (cricket): number of sperm transferred, 76

Gryllus bimaculatus (cricket): oviposition, 88

Gryllus rubens (cricket): JH and wing polymorphism, 346

guide post cells, embryonic: brain, 196; cerci, 205; legs, **202**, 203, 204

gurken (grk): anteroposterior axis, egg, 50, **51**; dorsoventral axis, egg, 50, **51**, 52

gynandromorph, 97, **98**

gynatrial complex: and selection for small size,

367; sperm transfer, role, **85**

gynogenesis, 104

Haematopinus suis (hog louse): mouthpart embryogenesis, 222, 225

Hairless (H). *See* genes, neuronal

hairy (h). *See* genes, segmentation: pair-rule

Halictoxenos simplicis (strepsipteran): polyembryony, 125, **126**

halofenozide, **328**, 335

haltere disc: genes and, *Drosophila*, **166**, 167, **371**, 379

Haplothrips verbasci (mullein thrips): germ cells, 120; male external genitalia, postembryogenesis, 276; mouthpart embryogenesis, 220–221, **221**; mouthpart structure, 220, **220**; pole plasm, 39

harlequin lobe, 13

hatching, 235–236

hatching thread, 134–135, 235–236

head eversion, *Drosophila*, 271, **272**, 294, 334

head involution, *Drosophila*, **132**, 218–220, **219**

head involution deficient (hid), 334

head segmentation: genes and, *Drosophila*, **166**, 172–174, **173**, **371**; in other arthropod embryos, **166**, 172, **371**; morphological evidence for, **130**, **132**, 134, 171, 225, **225**

heart. *See* dorsal vessel

heartless (htl), 226

Hebrus pusillus (velvet water bug): external genitalia, female, 87, **87**; external genitalia, male, **85**; gynatrial complex, **85**, 367; miniaturization, 367; oviposition, 87, **87**; sexual selection, effects, 85; sperm transfer, 74, 83, 85, **85**

hedgehog (hh). *See* genes, segmentation: segment-polarity

Heliconius spp. (nymphalid): genes/wing color pattern, 314

Heliothis virescens (tobacco budworm): 20E in testes, 24; allatostatin, effects, 340; JH synthesis by male accessory glands, 60; polydnavirus and failure to pupate, 326

Hemimetabola (Exopterygota): characteristics, 237–239, **237–241**

Hemimetabola (*sensu stricto*), 237, **238**

Hemiptera (bugs): diversity, **4**; fossil record, **356**, 362; number of molts, 264

hemocoel: formation, **178**, 182–183, **185**

hemocytes: basal lamina deposition by, 182; embryonic origin, **178**, 182, **183**; genes and, *Drosophila*, 182; metamorphosis, 274–275, **275**

hermaphrodites, 96–97

Hesperoctenes sp. (polyctenid [bat] bug): viviparity, 90

heterochrony, **365**, 365–366

heterogony. *See* allometry

Heteropeza pygmaea (cecidomyiid): cyclical parthenogenesis, 107–108, **107**, **108**; maternal hemolymph/sex determination, 96, **97**

hexapods, evolution of: complete metamorphosis, 356, **356**, 382–383; dicondylic mandibles, 355, **356**, 378; internal fertilization, 355, **356**, 376–377; nurse cells, 356, **356**, 382; ovipositor, 355, **356**, 378; six legs, 355, **356**, 377–378; wing flexion, 355–356, **356**, 381; wings and active flight, 355, **356**, 378–382, **380**, **381**

higher taxa, origin by: adaptive zones/key innovations, 354–363, **356**; homeotic gene expression, changes in and effects on modularity, 369–373, **370–374**; hormonal changes and abnormal metamorphosis, 366–367; punctuated equilibrium, 354; saltation versus gradualism, 353–354

hindgut: embryogenesis, 185, **185**; metamorphosis, 277–278, 279

hindsight (hnt): germ band shortening and, 131

Holocene Epoch, 363

Holometabola (Endopterygota), **238**, **239**, 239–240, **242**, **243**, 243–256, **244–247**, **249**, **251**, **252**, **255**

HOM Complex (HOM C). *See* genes, homeotic homeodomain, 169–172, **171**

homology, 364

homothorax (hth) and: antennal development, 316; hinge specification in wing discs, 310; leg disc, domain specification, 307, **309**

hormones: adrenocorticotropin, **63**; allatostatins, **338**, 340–341; allatotropin, **59**, 60, 61, 337, **338**; bursicon, 230, 268, **339**, 342–343; corticosterone, 63, **63**; cortisol, 63, **63**; crustacean cardioactive peptide, 342, **342**; diapause, 231–232, **232**, 343; ecdysis- and pre-ecdysis-triggering, 267, 341–342, **342**; eclosion, 266, **328**, **339**, 341, 343; egg development neurosecretory, 60, **60**; 20-hydroxyecdysone, 58–60, **58–60**, 229–230, **230**, 324–336, **324**, **328**, 330–333, **335**, **336**, **339**, **342**, 343; insulin-like, 230; juvenile, 22, 23–24, **25**, 58, **58**, 59–60, **59**, **60**, 61, 230, **230**, 232, **328**, 337–340, **338–340**, 345–350, **347**, **349**, **350**; luteinizing, 63, **63**; oostatic, 60–61; ovulation, 62; pheromone biosynthesis activating neuropeptide, 60; precocenes, 62, **328**, 341; prothoracicotropic, **324**, **328**, 329, 336, **337**, **339**, **342**; somatotropin, 63, **63**

hormones, classic experiments on: metamorphosis, 321–323, **321–323**; molting, 319–321, **320**; spermatogenesis, 22, **23**; vitellogenesis, 59

hormones, roles in: complete metamorphosis, evolution, 254–256, **255**; embryogenesis, 229–230, **230**; macroevolution, 366–367; metamorphosis, **324**, 324, **339**, **340**, **342**; molting, 266–269, **267**, **324**, **330**, **339**, **342**, 343; polymorphism/polyphenism, 345–351, **347–350**; previtellogenesis, 58, **58**; spermatogenesis, 22–24, **25**; vitellogenesis, 59–61, **59**, **60**

HOX Complex (HOX C). See genes, homeotic

huckebein (hkb). See genes, segmentation: gap

hu-li tai shao (hts): fusome, ringcanals, spectrosome, 14, 42, **43**, **44**

hunchback (hb). See genes, segmentation: gap

Hyalophora cecropia (cecropia moth): 20E and spermatogenesis, 22, **23**; autogeny and vitellogenesis in pupae, 37, 59; chemical identity of JH, 337; classic experiments on metamorphosis, 321; cuticular proteins, 250, 252; eclosion hormone, 341; pupal prothoracic gland, **320**; yolk uptake by oocytes, 38

Hydra sp.: Hox genes, 168, **170**

hydroxy furan fragment, azadirachtin, **328**, 335

Hylemya antiqua (onion maggot): spermatogenesis, rate, 20

Hymenoptera (sawflies, wasps, bees): diversity, **4**; fossil history, **356**, **362**; number of molts, 264

hypermorphosis, **245**, 245–246

Hypogastrura lapponica (collembolan): synchronized molting, 269

Hypogastrura socialis (collembolan): synchronized molting, 269

Hypoperlida†: fossil history, **356**, **362**

Hystrichopsylla talpae (flea): nurse cells, 44

Icerya bimaculata (margarodid scale), 96

Icerya purchasi (cottony cushion scale): hermaphroditism, 96

Icerya zeteki (margarodid scale), 96

ichno (trace) fossil, 357

imaginal discs: adaptive significance, 239–240, 247–248; axis specification in, **306**, 307–311, **308–310**; compartment boundaries, 144–146, **145–149**, 160, 163, **166**, **306**, 306–311, **308–310**; competition between, 265; embryogenesis, 215–218, **216**, **217**; eversion/ differentiation, 271, **272**, 294–296, **296**; genes and, Drosophila, 215–218, **217**, 306–312, **308–311**, 316–318, **316–317**; hormonal effects, 343; identity of, Drosophila, **216**, 294, **295**; pattern specification and signal proteins, 217, **217**,

306–311, **308–310**, 316–318, **316–317**

induction: assimilatory (intradermal), 138, **139**; definition, 228–229; gonadal sheath cells by germ cells, 226; mesoderm by ectoderm, 228–229, **229**; midgut epithelium by visceral mesoderm, 184–185

Indy, 235

infra-abdominal 2–5 (iab-2, iab-3, iab-4, iab-5): in bithorax complex, **165**

initiatorin, 66

Inka cells, 267, 341

inscuteable (ins), 189, **189**

insect diversity, **4**

insect growth regulators, **328**, 339–340

insect ontogeny, summary, 1–4, **3**

insect structure, **177**

insemination, artificial, 80

insulin-like peptide, 280

intercalary segment: appendages, 134, 172; genes and, Drosophila, **166**, 167, 172–174, **173**, **371**

intersex (ix), 92–93, **92**

intersexes, 97–98, **99**

intertrepsis, 2, 126–128, **128**

intromittent organ, 7, **68**, **70**, 74, **85**

invected (inv): imaginal disc expression, 307, **308**, 309, **310**, 317

Isoptera (termites): caste formation, 347–348, **349**; diversity, **4**; fecundity, 54; fossil history, **356**, **362**; number of molts, 264; physogastry, 259

Japyx sp. (japygid dipluran): ovaries, **56**

JH sensitive periods in: caste determination, 347–351, **349**, **350**; development of exaggerated structures, 265, 345–346; M. sexta metamorphosis, **339**, **340**, 343; metamorphosis of exopterygotes, **340**, 343; molting, **340**

Jurassic Period, 361, **361**

juvenile hormone, roles in: complete metamorphosis, origin, 254–256, **255**, 383; development of exaggerated structures, 265, 345–346; embryogenesis, 230, **230**; macroevolution, 366–367; metamorphosis, 321–323, **321**, **322**, 329, 337–340, **338–340**; molting, **339**, **340**, 343; phase change in

locusts, 346–347, **347**; polymorphism, 347–351, **349**, **350**; polyphenism, 345–347, **347**; spermatogenesis, 22–24, **25**; vitellogenesis, 59–63, **59**, **60**

K10, 50

Kalotermes flavicollis (kalotermitid): blastokinesis, **133**; caste determination, 347–348, **348**, **349**

Kapala terminalis (eucharitid), 54

karyosome, 57

katatrepsis, **128**, 128–129, 131

Keilin's organ, 285

Kenyon cells: embryogenesis, 197; metamorphosis, **281**, 283; structure, **195**

key innovations: and adaptive zones, 354–355; complete metamorphosis, 356, 382–383; dicondylic mandibles, 355, 378; and hexapod adaptive radiation, 355–357, **356**; and insect fossil record, **356**, 357–363, **362**; internal fertilization, 355, 376–377; nurse cell origin, 356, 382; ovipositor, 355, 378; six legs, 355, 377–378; wing flexing, 355–356, 382; wings/active flight, 355, 378–382, **380–382**

kinetochore, 14, **15**, 16–17

knirps (kni). See genes, segmentation: gap

knot (kn): hypopharynx, 174

Krüppel (Kr). See genes, segmentation: gap

K-selection, 365

labial (lab). See genes, homeotic

labial segment: genes and, Drosophila, **166**, 167, **170**, **371**

labral segment: reality of? 137, 172

ladybird early (lbe), 184

ladybird late (lbl), 184

lampbrush chromosomes, 34–35, 37, **36**

Lampyris noctiluca (firefly): androgens and sex determination, 95–96, **96**, **97**

Laphria sp. (asilid): larva, larval head, 245

larvae, types, 243–246, **238**, **244**, **245**

larval body plan, specification of: A-P pattern, 49–50, **51**,

138–143, **139–142**; determinants and, 136; D-V pattern, 49– 50, **51**, 52, 137–138, **139**; homeotic genes and, 161–168, **166**, **371**; maternal effect genes and, 148–151; P-D pattern (limbs), 307–308, **308**, **309**; physiological centers and, 136–137, **137**; segmentation genes and, **154**, **155**, 155–161, **157–162**

laws, growth: biogenetic, 363–364; Dyar's, 258–259; Halley's (1886), 49; von Baer's: 364

leg segments: antigenetic differences within, 307; genes and, *Drosophila*, 307–308, **309**

Lepidoptera (butterflies and moths): diversity, **4**; evolution, 361–362; fossil history, **356**, **362**; metamorphosis, 239; number of molts, 264

Lepinotis spp. (trogid psocid): Pthiraptera, and evolution, 366–367

Lepisma sp. (lepismatid): intertrepsis, **133**

Lepismodes inquilinus (lepismatid): axoneme structure in sperm, **10**

Leptinotarsa decemlineata (Colorado potato beetle): ectoderm induces mesoderm, 229; gonadal mesoderm, 226; photoperiod and vitellogenesis, 61; polar granules in egg, 136

Leucophaea maderae (Madeira cockroach): number of follicle cells/oocyte, 40; ommatidial chimeras, 292

Liassothrips longipennis†, 361

life histories of: apterygotes, 236–237, **238**, 264; endopterygotes, **238**, 239–240, 264; exopterygotes, 237–239, **238**, **240**, **241**, 264

life history, concepts of, 240–243, **243**

life history theory, 235

limb branching: *Distal-less* expression/evolution of arthropods, 373, 375, **371**, **377**

Limulus polyphemus (horseshoe crab): homeotic gene expression, embryos, 167, 168

Lineus sanguineus (nemertean worm), 209

Liposcelis spp. (liposcelid psocid): Pthiraptera, and evolution, 366–367

Lissonota catenator (ichneumonid): pole plasm formation in egg, 39

Locusta migratoria migratorioides (African migratory locust): 20E and meiotic reinitiation, 40; 20E and spermatogonial mitosis, 23; 20E in embryos, 229–230, **230**, 256; 20E production by follicle cells, 39; growth of hind femur and abdomen, **258**; hemocytes deposits basal lamina in embryo, 182; JH in embryos, 230, **230**; midline precursors in embryos, 193, **194**; origin/projection of G neuron, **191**; oviposition, 88; peripheral pioneer neurons in embryos, 202; phase change, 346–347, **347**; sperm movement to seminal vesicles, 65; spermatophore formation, 68; subesophageal body in embryo, 184; Vg receptor, 38; weight change during postembryogenesis, **257**

Locustana pardalina (grasshopper): ventral glands in embryo, 229–230

lozenge (*lz*), 182

Lyctocoris campestris (lyctocorid): traumatic insemination, 74

Lyctocoris dorini (lyctocorid): traumatic insemination, 74

Lygaeus simulans (lygaeid): time to insert phallus, 74

Lymantria dispar (gypsy moth): circadian sperm release by, 65–66; hindgut ecdysiotropin, 336–337; intersexes, 97; Kopéc's experiments on "brain factor," 319; ovary transplants into males, 38; testis ecdysiotropin, 24

lymph glands, 182, **183**

Lytta viridana (caragana blister beetle): cephalic apodemes, embryogenesis, 225, **225**; cleavage, 112, **113**; embryogenesis of optic lobe, 197, **198**; embryogenesis of stemmata, **198**, 208; embryogenesis of subesophageal body, 184

Mach 2 (RH-0345), **328**, 335

Machilis sp. (bristletail): intertrepsis, **133**

macroevolution, role of: conserved cellular processes, 367–368; developmental constraints, 368–369; heterochrony, 365–366; homeotic and *Distal-less* genes, **370–374**, 373–375; hormones, 366–367; JH sensitive periods, 369; miniaturization, 367; modularity, 369–373, **370–372**; phenotypic plasticity, 373

macroglomerular complex, 284

macromolecular factor, 284

macromutation, 353

macrophages: functions of during metamorphosis, 274–275; origin of, *Drosophila* embryos, 182, **183**

Macrotermes bellicosus (termite): physogastry in, 259

mago nashi (*mago*): pole cell specification, 121; pole plasm formation, 39, 50, 152

Malpighian tubules, 185

Manduca sexta (tobacco hornworm): critical size, molting and metamorphosis, 343–345; ecdysteroids and spermatogenesis, 24; homeotic gene expression, embryos, 160, 165, 168; hormones and metamorphosis, 273–274, 281–282, 292–293, 323–343, **324**, **327–329**, **337**, **339**, **340**, **342**; Inka cells, 266, 341–342; larva, **245**; lipophorins in vitellogenesis, 37; macroglomerular complex, 284; neuroblasts, 193; neurons, posterior migration of during metamorphosis, 280; neurons and adult muscle differentiation, 273; neurosecretory cells, embryogenesis, 198–199; neurosecretory cells, structure, **327**; peripheral nerve plexus, 205; prothoracic glands, degeneration, 285; prothoracic glands, embryogenesis, 212, **212**; reproductive system, embryogenesis, 226, 227–228, **227**, 376; sensory remodelling during metamorphosis, 285; stomatogastric nervous system, embryogenesis, 209–211, **210**, **211**; tracheal nodes, 215

Mantodea (preying mantids): diversity, **4**; fossil history, **356**, 362, **362**; hatching threads in embryos, 134–135, 235–236; number of molts, 264

MAPK, 50

Margosan-O (azadirachtin), **328**, 335

Mastotermes darwiniensis (termite): sperm, 13

mate copying, 105

material compensation, 366

Mayetiola destructor (Hessian fly): chromosome elimination during cleavage, 113–115

mechanoreceptors: cuticle continuity during molt, 269, **269**; embryogenesis, **202**, **204**, 204–207, **205–207**; genes and, *Drosophila* embryos, 205–207, **207**; genes and, *Drosophila* larvae, 285–286, **287**; metamorphosis, 285–286; structure, **205**, **206**, 269, **269**

meconium, 247, 278

Mecoptera (scorpionflies): diversity, **4**; fossil history, **356**, 362; number of molts, 264

Megaloptera (alder and dobsonflies): diversity, **4**; fossil history, **356**, 362; number of molts, 264

Megasecoptera†: fossil history, **356**, 362

Megaselia abdita (phorid): derivation of *bicoid* from *zerknüllt*, 154–155

Megaselia scalaris (phorid): sperm, **9**, 11

Megoura viciae (aphid): life history, **106**, **107**, 346

meiosis: adaptive significance, 17–18; centriole allocation during, **15**, 17; summary, **15**

meiosis, stages, 14–17, **15**, **92**

meiosis in: amphitoky, 104, **104**; arrhenotoky, 101, **102**; cyclical parthenogenesis, 105–109, **105**, **106**, **107**; females, **3**, 33–37, **33**, **35**, **36**, 39–40; males, **3**, **9**, 14–18, **15**, **92**; thelytoky, 103–104, **103**, **104**

Melanoplus differentialis (differential grasshopper): CA and vitellogenesis, 59; overlapping cycles of vitellogenesis, 61; spermatogenesis, rate, 20

Melanoplus sanguinipes (migratory grasshopper): allatostatins, 340; microtubules in spermatogonial mitosis, 14

Meroista, 41

mesectoderm, genes and, *Drosophila*, 126, 151, **151**

mesoderm: embryogenesis, **3**, 137–138, **138**, **139**, 176, **178**, **179**, 226–227; genes and, *Drosophila* embryos, 52, 176–179, **179**, 226–227, 229; gonadal, 226–227; induction by ectoderm, 228–229, **229**; visceral, **178**, **179**, 184–185, **185**

Mesozoic Era, **356**, 360–362, **361**, **362**

metamorphosis, ecdysteroid influence on: compound eye morphogenesis, 292–293; nurse cell cytoplasm transfer into oocyte, 48; optic lobe metamorphosis, 281–282; skeletal muscle metamorphosis, 273; ventral diaphragm metamorphosis, 273–274

metamorphosis, structural changes in: abdominal histoblasts, 295–296, **297**; alimentary canal, 277–278, **279**; antennal lobes, 281, 283–284; brain, 280–285, **281–285**; compound eyes, *Drosophila*, 286–291, **287–292**; compound eyes, other arthropods, 291–293; corpora pedunculata, **281**, 283, **283**; cryptonephridia, 283; dorsal ocelli, *Drosophila*, 294; dorsal vessel, 275; fat body, 275; giant pulvillar nuclei, 296, 298–299, **298**, **299**; glia, 284; hemocytes, 274–275, **275**; higher flies, 270–273, **272**, **273**; imaginal discs and epidermis, 294–296, **295**, **296**; Malpighian tubules, 278; mouthparts, mosquitoes, 296, **297**; musculature, 273–274, **274**; neuroendocrine system, 284–285, **284**, **285**; optic lobes, 280–283, **281**, **282**, 293, **293**, **294**; peripheral nervous system, 285–293, **286–294**; pheromone glands, 301; reproductive systems, 275–277, **276–278**; salivary glands, 301; sense organs other than compound eyes, 285–286, **286**; tracheal system, 299–301, **300**; ventral nervous system, 278–280,

280; wing dimorphism, 295

Metaplastes ornatus (bush cricket): number of sperm transferred, 76; removal of sperm of previous male, 84

methoprene, effects on: cephalic horn growth, 345; metamorphosis, 339; muscle metamorphosis, 274; nurse cell transport, 48; wing disc growth, 344

methoprene, structure, **328**

methuselah (*mth*), 235

methyl farnesoate: in crustaceans, 337, 345

Miastor sp. (cecidomyiid): cyclical parthenogenesis, **107**

Micromalthus debilis (micromalthid): cyclical parthenogenesis, **108**, 108–109

Micropteryx aruncella (micropterygid): no apyrene sperm, 12

Micropteryx calthella (micropterygid): no apyrene sperm, 12

microtubules, function in: axonal transport, 323; blastoderm formation, 115, 117, **117**; cleavage, 113, **116**; fertilization, 78–83, **81**; nurse cell cytoplasm transfer into oocyte, 47–49, **49**; oocyte axis specification, 49–52, **51**; oocyte specification, *Drosophila*, 43–44; sperm motility, 11, **11**; spermatogonial mitosis and meiosis, 14, **15**, 16–17; spermiogenesis, 18, **19**

microtubules, types, 113, **116**

midblastula transition, 118

midgut: changes in larva, 277–278; embryogenesis, 184, **185**; genes and, *Drosophila*, 184–185; metamorphosis, 278, **279**

midgut rudiments: embryogenesis, 184, **185**

midline precursors, **188**, **189**, **191**, **192**, 193, **194**

migration cytaster, **81**

Mimic (RH-5992), **328**, 335

miniaturization, 367

minute, 147

Miocene-Pliocene Epochs, 363

Miomoptera†: fossil history, **356**, 360, **362**

miranda, 189

mirror (*mirr*): D-V axis in compound eye, 291, **292**

Mississippian Period, 359

mitochondrial derivative: genes and, **27**, 28; during

spermiogenesis, 18, **19**, **20**; structure, 9, **10**

mitotic domains, 118–119, **119**,

mitotic oscillator, 112–113, **115**

mitotic recombination, 145, **145**

models: filing cabinet, 250–251, **252**; French flag, 139, **140**; homeotic take-over, 372–373, **373**, **374**

models of head segmentation: L-/bent Y-, 172; linear, **130**, 172

modularity and: gene expression, 373–375, **370–372**; macroevolution, 369–373; mosaic evolution, 369

molecular genetics and pattern formation: abdominal segments, 316–317, **317**; antennae, 316; eggs, 49–52, **51**; embryos, 146–159, 150–156, **157–160**, 161–165, **162–166**, 169–172, **171**, 172–174, **173**; genital discs, **317**, 317–318; leg discs, 307–308, **308**, **309**; lepidopteran wings, **312–315**, 312–316; *Oncopeltus* embryonic head, 222, **372**; wing discs, 308–316, **310–315**

molting: critical weight, 343–344; endocrine control, 266–269, 319–321, 320, 321, **324**, 324–343, **329–333**, **335**, **336**, **339**, **340**, **342**, 343; other arthropods, 345; sclerotization effects, **258**, 258–259; stages, 266–269, **267**, **268**

molting space, 266, **267**, **269**

molts, numbers, **264**, 265

monsters, "hopeful," 353

Monura†: fossil history, **356**, 359, **362**

morphogenetic furrow (compound eye): *Drosophila*, 287–293, **288–290**

morula (*mr*), 47, 212

mouthparts: embryogenesis, **130**, **132**, 134, **218**, **219**, 218–222, **221**, **223**, **224**, 225; genes and, *Drosophila*, 172–174, **173**; genes and, *Oncopeltus*, 222, **372**; metamorphosis, 294–295, 296, **297**; types, **218**, 218–222, **220**, **222**, **297**

Muellerianella brevipennis (delphacid): pseudogamy, 104

Muellerianella fairmaireii (delphacid): pseudogamy, 104–105

multiple wing hair (*mwh*), **120**, 145, **145**

Murgantia histrionica (harlequin bug): germ band position in egg batch, **123**

Mus musculus (house mouse): *en* expression, embryos, 160–161; *Hox* gene expression, embryos, 168–169, **170**

Musca domestica (house fly): *bicoid* expression, embryos, 153; determinants, embryonic, 136; egg position in ovariole, 49; *even-skipped* expression, embryos, 159; fluctuating asymmetry in wings, 263, 265; oostatic hormone, 61; sex determination, 93; sperm entry into egg, 78, **80**

Muscidifurax uniraptor (pteromalid): cortical centrosomes in egg, 101

muscle pioneers: in appendages, 179, **181**; in body segments, 179–**180**; genes and, *Drosophila*, 176–182

muscles: continuity with cuticle during molt, 269–270, **271**; eclosion, 271–273, 274; embryogenesis, 176–182, **178–181**; genes and, *Drosophila*, 176–182, **179**; hatching, 235; metamorphosis, 273–274, **274**; molting, 267–270, **271**

mushroom body (corpus pedunculatum): embryogenesis, 197; function, 280; metamorphosis, **281**, 283, **283**; structure, **195**

Mycophila speryeri (cecidomyiid): ecdysteroids and paedogenesis, 108

myoblasts. See myogenesis

myogenesis: metamorphic, 273–274, **274**; embryonic, 176–182, **178–181**

Myriapoda (myriapods): *Ubx/abd-A* expression, embryos, 166, 167–168, **371**; ventral nervous system embryogenesis, 193–194

Myrmica rubra (ant): corpora allata, 321

Myrmica sp. (ant): caste formation, 349

Mysidopsis bahia (crustacean): *Distal-less* expression, embryonic limbs, 375

Myzus persicae (green peach aphid): parthenogenesis, **106**, 107

naked (*nkd*), 157

nanos (*nos*), role in: egg, 50, **51**; embryo, **121**, 152–153, **155**; germ cell movement, 226

Nasonia vitripennis (pteromalid): cortical centrosomes, eggs, 101; paternal sex ratio chromosome, 102

Naturphilosophie, 363

nautilus (*nau*), 181

Nebenkern, 18, **19**, **20**, 27

Necrophorus vespilloides (silphid carrion beetle): chemoreceptor development, 269, **270**

Nemoria arizonaria (geometrid): diet and developmental polymorphism, 262, **264**

Neobellieria bullata (sarcophagid): bursicon, 342–343; eclosion muscles, 274; number of ovarioles, 29; number of sperm, males, 21; number of sperm transferred to female, 77; pretarsus, metamorphosis, 296, 298–299, **298**

Neogene Epoch, 363

Neometabola, **238**, 239, **239**, **243**

Neoproterozoic Era, 357

Neoptera, **4**, 355, **356**, 382

neosomy, 259, **259**

neoteny, 365, **365**

Neotermes zuluensis (termite): accessory hermaphroditism, 96

nervous system, central: embryogenesis, 186–191, **186–196**, 193–202, **200**, **201**; genes and, *Drosophila*, 200–202, **201**; metamorphosis, 278–285, **280–285**

nervous system, peripheral: embryogenesis, **198**, 202–208, **202**, **204**, **205–207**; genes and, *Drosophila*, 205–207, **207**; metamorphosis, 285–294, **286–295**

netrin, 191, 193, **193**

neuroblasts: embryonic origin, 186–190, **188**, **189**, **191**, **192**, **194**, 195–197, **196**; genes and, *Drosophila* (embryonic), 200–202, **201**;

genes and, *Drosophila* (imaginal), 188, 279–284, **282**; in other arthropod embryos, 193–195

neuroectoderm: genes and, *Drosophila*, 149–151, **151**

neuroendocrine system: embryogenesis, 198–199, 209–212, **210–212**; metamorphosis, 284–285, **284**, **285**; structure, 195, **195**, **284**, **285**, 323–324, **325–327**

neurogenic zones Z1–Z3, 209–212, **210**

neuromere, 186, **186**, **188**, **189**, **191**, **192**, **194**

Neuronice sp. (caddisfly): spermatozoon, **10**

neurons, role in: inducing muscle differentiation during regeneration, 302; synapse formation, 204

neurons, types, 190–191, **190**

Neuroptera (lacewings): diversity, **4**; fossil history, **356**, **362**; metamorphosis, **239**, 241; number of molts, 264

neurosecretory cells, roles in: metamorphosis, **339**, **342**, 343; molting, 336–337, **339**, 340–343, **342**; previtellogenesis, 258, **258**; spermatogenesis, 22–24, **23**, **25**; vitellogenesis, 60, **60**, 61–63

neurosecretory cells, types, 323–324, **324–327**

neurosecretory granules, 323, **325**

Nezara viridula (southern green stink bug): oviposition and germ band formation, 222–223, **223**

nitric oxide and: R1–R6, projection to optic cartridge, 293; subepidermal nerve plexus, growth, 205

nodal (critical) point, 355

Nosopsyllus fasciatus (European rat flea), 63

Notch (*N*). See genes, neuronal

Notiophilus biguttatus (carabid): photoperiod and eggs, 61–62

Notonecta glauca (notonectid): blastokinesis, **133**; microtubules in trophic cords, **49**; weight change during life history, **257**

Notoneuralia, 151

Numb: ganglion mother cell specificity, 188–189, **189**, 193, 207

nurse cells: adaptive significance, 55–56; evolution, 54–55, **55**, 356, **356**, 382; function, 41, 50, 55–56, 382; origin, 41–46, **42**, **43**, **45**, **47**; polyploidy, 41, 47

Nymphalidae (four-footed butterflies): wing color pattern, **312–315**, 312–316

ocular segment: genes and, *Drosophila*, **166**, 167, **173**, 174

odd-paired (*opa*). See genes, segmentation: pair-rule

odd-skipped (*odd*). See genes, segmentation: pair-rule

Odonata (damsel and dragonflies): diversity, **4**; fossil history, 355, **356**, **362**; number of molts, 264; postembryogenesis, 237, **239**

Odontotermes badius (termite): RNA synthesis and katatrepsis, 129

Oecanthus pellucens (tree cricket): oviposition, 88

Oidaematophorus hirosakianus (pterophorid moth): wing development and apoptosis, 298

Oligocene Epoch, 363

ommatidium: development, 287–291, **288–292**; structure, 286–287, **287**

Oncopeltus fasciatus (milkweed bug): critical weight and molting, 344; insemination, 74; mouthpart embryogenesis, 222, **223**, 372; prothoracic gland degeneration, 326

Onthophagus taurus (scarab beetle): cephalic horn development, 265, 345

ontogeny: and hexapod evolution, **356**, 363–383, **365**, **370–374**

Onychophora (velvet worms): *Distal-less* expression, appendages, 373; homeotic gene expression, embryos, 165, **166**, 167–168, **371**

oogenesis, 32–64, **33**, **35**, **36**, **38**, **42–49**, **51**, **54–56**, **58–60**, 63

oogenesis-flight syndrome, 61–62

oostatic hormone, **59**, 60–62

Opabinia regalis†: legs, 377, **377**

optic cartridge, 293, **293**

optic lobe: embryogenesis, 197, **198**; hormones and metamorphosis, 281–282; photoreceptor axons and

metamorphosis, 293, **293**, **294**; structure, 195, **195**

optomotor blind (*omb*), 309

orders, insect: evolution, 352–383; fossil history, **356**, 358–363, **362**; known diversity, **4**; number of molts in members, 264; phylogeny, **4**, **356**

Ordovician Period, 358, **358**

organogenesis, ectodermal structures: brain, 195–199, **196**; compound eyes, 208; endocrine system, 198–199; epidermis and imaginal discs, 215–218, **216**, **217**; fore- and hindguts, 185, **185**; Malpighian tubules, 185–186; peripheral nervous system, 202–207, **202**, **204–207**; sense organs, **202**, 204–207, **204–207**; stemmata, 198, 208; stomatogastric nervous system, 209–212, **210**, **211**; tracheal system, 212–215, **213**, **214**; ventral nerve cord, 187–193, **188**, **189**, **191–194**

organogenesis, mesodermal structures: body cavity, **178**, 182–183, **185**; dorsal vessel, 178, **183**, 183–184; fat body, 184; gonadal sheath, 226–227; hemocytes, **178**, 182; primary exit system, **227**, 227–228; somatic muscles, 176–182, **178–181**; subesophageal body, 184; visceral muscles, 176–182, **178**, **179**, **185**

organogenesis, organs of mixed origin: gut, 184–186, **185**; reproductive system, 225–228, **227**

Ornithoderus parkeri (tick): ecdysone, 345

orthodenticle (*otd*): expression of, preblastoderm, **155**, 156; expression of, other arthropod embryos, **166**, 167, **371**; head segmentation, role, 172–174, **173**

Orthoptera (grasshoppers and crickets): diversity, **4**; fossil history, **356**, **362**; number of molts, 264; postembryogenesis, 237, **238**, **239**

oskar (*osk*): anteroposterior axis, egg, 50; anteroposterior axis, embryo, 152–153, **153**; pole cell specification, 120–121, **120**, **121**

Ostrinia nubialis (European corn borer): ecdysiotropin, 336–337

ovarioles: embryogenesis, 227; evolution, **54–56**, 54–57; meroistic: 29–30, **31**, 41–49, **41–49**; panoistic: 29, **31**, 33, 34–41, **35**, **36**, 38; polytrophic: 29, **31**, **42–44**, 42–45; postembryogenesis, **46**; telotrophic: 29–30, **31**, 45–46, **45–49**

oviposition, 86–89, **87**, **88**

ovipositors: adaptive significance, 89; function, 86–88, **87**, **88**; innervation, 32, **79**; phylogenetic diversity, **89**; structure, 32, **32**, **79**, **87**

ovotestis, 96

ovulation, 77, **78**

ovulation hormone, 62, 77

Oxidus gracilis (millipede): homeotic gene expression, embryos, **166**, 167–168, **371**

Pacifastacus leniusculus (crayfish): wing and distal epipodite, homology of? 380–382, **381**

paired (*prd*). *See* genes, segmentation: pair-rule

Paleocene-Eocene Epochs, 363

Paleodictyoptera†: fossil history, **356**, 359–360, **362**; wings, evolutionary origin, 241–242, **243**, **251**, 253–254, 378–379

palingenesis, 364

Panorpa japonica (scorpionfly): male symmetry, effects, 83

Panstrongylus megistus (reduviid bug): endocrine control of previtellogenesis, 58, **58**; ovariolar differentiation, 45, **46**

Parametabola, 238, **241**

parasegment, 147, **150**, 160, **170**

Paropsisterna beata (chrysomelid): larva, **244**

parthenogenesis: adaptive significance, 101, **102**, 104–105; amphitoky, 104, **104**; arrhenotoky, 101–103, **102**; cyclical, **105–108**, 105–109; microorganisms and, 102, 109–110; phylogenetic distribution, **109**; thelytoky, apomictic, 103–104, **104**; thelytoky, automictic, 103, **103**; tychoparthenogenesis, 101

parthenogenesis, cyclical: aphids, 105–107, **106**; gall midges, 107–108, **107**, **108**; gall wasps, 105, **105**; *Micromalthus debilis*, 108, 108–109

Parthenothrips dracaenae (palm thrips): oogonial clone formation, 45

partner of inscuteable (*pins*): GMC specification, 189

partner-of-numb (*pon*): GMC specification, 189

patched (*ptc*), **160**

pattern formation: genes and, *Drosophila* embryos, 146–174, **150–155**, **157–160**, **162–166**, **169–171**, **173**; genes and, *Drosophila* imaginal discs, 306–312, **308–311**, **316**, **317**, 316–318

Paurometabola, 237, **238**

Pax-6: and eye evolution, 208–209

pdm-1: expression in distal epipodite/wing primordia, 380–381, **380**, **381**

pdm-2: expression in distal epipodite/wing primordia, 380–381, **380**, **381**

Pedetontus unimaculatus (machilid jumping bristletail): mandible-maxilla, homology, 378

Pediculus sp. (head louse): blastokinesis, **133**

peduncle. *See* mushroom body

Pelecorhynchus fulvus (pelecorhynchid fly): pupa, **246**

Pennsylvanian Period, 359

peramorphosis in: macroevolution, 365, **365**; ovary evolution, 56

Peripatopsis capensism (onychophoran): *Distal-less* expression, embryos, 375

peripheral nerve plexus, 205

peripheral nervous system: embryogenesis, **198**, **202**, 202–209, **204–207**; evolution, 207–208; genes and, *Drosophila*, 205–207, **207**; structure, **190**; synapse formation, 204

peripheral pioneer neurons, **202**, 202–203, **204**

Periplaneta americana (American cockroach): antennal glomeruli, embryogenesis, 196; compound eye development, 291; corpora pedunculata, embryogenesis, 197; melatonin and PTTH release, 336; oocyte

growth, 33, **35**; optic lobe, postembryogenesis, 282; sperm, 11; vitellogenesis, overlapping cycles, 61

periplasm, 2, 111, **112**

peripodial membrane: Hedgehog signaling from, 288; structure/function, 271, 294, **296**

perisympathetic organs, 323, **325**, **326**

Perla marginata (perlid stonefly): hermaphroditism, accessory, 96

Permian Period, **356**, **358**, 359–360, **362**

Permothemistida†: fossil history, **356**, **362**

Phaeophilacris spectrum (cricket): oviposition by, 88

Phanerozoic Era, 357–363

pharate stage, 236, **236**

phase change, locusts, 346–347, **347**

Phasmatodea (stick insects): diversity, 4; fossil history, **356**, 360, **362**; number of molts, 264; postembryogenesis, 237, **239**

Pheidole bicarinata (ant): caste determination, 349, **349**

phenotypic plasticity: definition, 345; macroevolution, role of, 372–373

pheromone biosynthesis activating neuropeptide, 60

pheromone systems, 301

pheromones, role in: caste formation, 347–351; inducing embryogenesis, 230–231, **231**

Philosamia ricini (saturniid moth): Dyar's law, **258**

Phormia regina (black blowfly): endocrine control of vitellogenesis, 61; ovarioles, number/ovary, 29

photoreceptor cells: in compound eyes, 286–287, **287**; in dorsal ocelli, 294; genes and, *Drosophila*, 289–290, **291**, 293–294, **293**, **294**; in stemmata, **198**, 208

Phryna vanessae (muscid fly): central nervous system, metamorphosis, **280**

Phthiraptera (lice): diversity, **4**; evolution, 366–367; fossil history, **356**, **362**; life history, 237; number of molts, 264

phyletic phenocopy paradigm, 312

phylogeny, reconstruction, 4–5

phylotypic stage, 135, 369–373

Physarum polycephalum (slime mold): mating types, 99

physiological centers, role in embryogenesis, 136, 137, **137**

physogastry, 259, **259**

piercing-sucking mouthparts, ontogeny: bug embryos, 222, **222–224**, 372; genes and, *Oncopeltus fasciatus* embryos, 222, **372**; lice embryos, 222, 225; mosquito larvae and pupae, 296, **297**; scale embryos, 222, **224**; thrips embryos, 220–221, **220**, **221**

Pieris napi (cabbage white butterfly): apyrene sperm and mating delay, 86; apyrene/eupyrene sperm ratio, 24

Pimpla turionellae (ichneumon wasp): egg activation, **102**; egg gradients, 140

planidium, **245**, 245–246

Platycnemis pennipes (damselfly): blastokinesis, **133**; egg regulation, 136; mesoderm induction by ectoderm, 228; physiological centers in embryo, 136, 137, **137**

Plecoptera (stoneflies): diversity, **4**; fossil history, **356**, **362**; life history, 237, 239; number of molts, 264

Pleistocene Epoch, 263

pleuropodia: ecdysteroid synthesis by, 134; embryogenesis, 134, 165, 167; function, 134; genes and, 165; position, **130**

Plodia interpunctella (Indian meal moth): apyrene/eupyrene sperm ratio, 24; diet and eupyrene sperm, 77

Poecilia formosa (sailfin molly): mate copying, 105

Poecilia latipinna (sailfin molly): mate copying, 105

pointed (*pnt*), 200, 214

polar bodies in: female meiosis, 3, 33, **33**; gynandromorphs, 97, **98**; parthenogenesis, 101–107, **102–105**, **107**; polyembryony, 121–122, 124–125

polar granules: components, **120**; formation, 36, 39, **112**, **114**, 119; function, 120, **120** pole cells: embryogenesis, **114**, 119–122; genes and, *Drosophila*: 120–121, **121**

pole plasm: components, 120; formation, **36**, 39; function, 120; genes and, *Drosophila*, 120–121, **120**, **121**

Polistes annularis (vespid wasp): caste formation, 350

Polycomb (*Pc*), 164

polyembryony, 121–122, 123–125, **124–126**

polyfusome, 13, 14, 43

polymorphism, 345, 347–351, **348–350**

polymorula: primary, 124, **124**, 125; secondary, 124, **124**, 125

polyphenism, 345–347

polytene chromosomes: in pulvillar nuclei, 296, 298–299, **298**, **299**; in salivary glands, 331–332, **332**, **333**

Porcellio scaber (isopod crustacean): genes and mouthpart embryogenesis, 167, 372–373

positional information, 139, **140**

postembryogenesis, concepts of, 240–243

posterior midgut rudiment, 184–185, **185**

postvitellogenesis, **36**, 40–41

pox-neuro (*pox-n*). *See* genes, neuronal

pre-adaptation, 354–355

Precambrian Era, 357

Precis coenia (buckeye butterfly): *en* expression, embryos, 160; homeotic gene expression and prolegs, 165, 378; JH, wing disc, and larval growth, 344; wing color pattern, specification, 218, **312–315**, 312–316

precocenes, 62, 341

pretarsus, metamorphosis, **272**, 296, **298**, 298–299

previtellogenesis, 34–37, **35**, **36**, **42–49**, 58, **58**

Primicimex cavernis (cimicid bug): traumatic insemination, 74–75

proboscipedia (*pb*). *See* genes, homeotic

Procambarus clarkii (crayfish): methyl farnesoate, farnesoic acid, 337

Procrustes analysis, 261

proctodaeum, embryogenesis: **130**, 134, 185, **185**

progenesis, 365, **365**

programmed cell death. *See* apoptosis

prolarva, 235, 243, 254–256

prolegs: genes and, butterfly embryos, 165, 378

proliferation clusters, 195, **196**

Prometabola, 237, 238, 242, **243**

Prospero (Pros): ganglion mother cell specification, 188–189, **189**, 207

prospero (*pros*): ganglion mother cell specification, 207

Protelytropterat†: fossil history, **356**, 362

Protereismatidae†: fossil history, **356**, 362

prothoracic glands: embryogenesis, 212, **212**; function, 320–321, **324**, 324–335, **329**, **330–333**, **335**, **336**; imaginal degeneration, 326; structure, **212**, 320

prothoracicotropic hormone, **324**, 336, **339**, 342, 343

protocephalon, 121, 122, **130**, **132**, 134, 172, **173**

protocerebrum, **186**, **195**, **281**, **284**

protocorm, 121, 122, **130**, 134

Protodonatat†: fossil history, **356**, 362

Protoelytopterat†: fossil history, **356**, 362

Protophormia terrae novae (blowfly): embryonic gradients, 140; regulation, embryonic, 136

"Protorthoptera"†: fossil history, **356**, 362

Protostomia, 151, 185, 199–200, **200**, 371

Protura: diversity, **4**; fossil history, **356**, 359; life history, 236–237, **239**; number of molts, 264

PS integrins, 233

PS integrins: during embryogenesis, 182, 233

Pseudaletia separata (noctuid): number of sperm in spermatheca, 76

Pseudaletia unipuncta (armyworm): diet and allometry, 260, **262**; feeding efficiency and growth, 251

pseudocleavage, 136

pseudogamy, 104

Pseudometabola, 237

Psilopsocus mimulus (psocid): wood boring nymphs/adults, 248

Psocoptera (psocids): diversity, **4**; fossil history, **365**, 362; life history, 237, **239**; number of molts, 264; paraphyletic with Pthiraptera? 366–367

Ptenothrix italica (collembolan): chromosome elimination and sex determination, 98–94

Pterygota: Neoptera, 355–356; Paleoptera, 355

ptilinum, 272, **272**, 274

puff formation, 298–299, **298**, **299**, 331–332, **332**, **333**, 334–335, **335**, **336**

pulvillus. *See* pretarsus

Pulvinaria hydrangeae (cottony hydrangea scale): diploid chromosome number, reconstitution, 103, **103**

pumilio (*pum*): anteroposterior axis, 50, 152

punctuated equilibrium, 354

pupa, models of evolutionary origin: Berlese-Jeschikov, 248, 249; Bernays, 251–252; Bryant, 251; Downes, **249**, 252–253; Heslop-Harrison, **249**, 249–250; Hinton, **249**, 250; Kukalová-Peck, **251**, 253–254; Poyarkoff-Hinton, 248–249, **249**; Sehnal, Svácha, and Zrzavy, 242, **243**, 254; Truman and Riddiford, 254–256, **255**; Wigglesworth, 250–251, **251**, **252**

pupae: adult eclosion, 247; functional significance, 247–248; models of origin, 248–256, **249**, **251**, **252**, **255**; protection, 246

pupae, types of: **238**, **239**, **245–247**, 270–273, **272**, **273**

puparium: adult eclosion from, 271–273; formation, 271; structure, **272**

Quadraspidiotus perniciosus (San José scale): life history, **241**

Quarternary Epoch, 363

queen substance (*Apis mellifera*): in caste formation, 350, **350**; in inhibition of worker vitellogenesis, 62, 95

radial component. *See* regeneration: polar coordinate hypothesis

Raphidioptera (snakeflies): diversity, **4**; fossil history, **356**, 362; life history, **238**; number of molts, 264

reaction norms. *See* growth, allometric

reaction-diffusion hypothesis: and eyespot formation in butterfly wings, 313–314, **314**; and limb pattern formation, 304–306, **306**

reaper (*rpr*): and apoptosis, 220, 228, **334**

recombination, **15**, 16, 17–18

reduced bristles on the palpus (*rbp*), 334

reductionist versus evolutionist approaches to ontogeny, 1–2

regeneration: of appendages, 302; Cartesian coordinate hypothesis, 306; morphogen gradients and, 302–304, **303**; of ommatidia in compound eyes, 292; polar coordinate hypothesis, 304, **304**, **305**; reaction-diffusion hypothesis, 304–306, **306**; of tracheal system, 303

regeneration field, 302

regenerative blastema, 302

Remetabola, 237–238, **237**, **238**

reproductive system, female: components, 29–32, **30**, **42**; embryogenesis, 225–228, **227**; number of ovarioles, **30**; ovaries, 29–30, **30**, **42**; ovipositor, 31, **32**, **79**, **86–82**, **87**; postembryogenesis, 276–277, **278**; primary exit system, 30, **30**, **42**, 227; secondary exit system, 30–32, **30**, **42**

reproductive system, male: components, 6–7, **7**; embryogenesis, 225–228, **227**; external genitalia, **7**, **68**, 74, **85**, 86; postembryogenesis, 275–276, **276**; primary exit system, 7, **7**, **68**; secondary exit system, 7, **7**, **68**; sperm tubes, 6–7, **7**, **8**, **9**

resegmentation, 161

retardation, 365, **365**

RH-0345: **328**, 335

RH-2485: **328**, 335

RH-5849: **328**, 335

RH-5992: **328**, 335

rhabdom, 287, 290

Rhagoletis pomonella (apple maggot): metamorphosis, **272**

Rhodnius prolixus (reduviid): ecdysteroid secretion, regulation, 325; JH and vitellogenesis, 58, **59**; Malpighian tubules, embryogenesis, 186; meal size and molting, 343–344; metamorphosis, classic experiments, 321–322, **321**, **322**; molting, classic experiments, 319–320, **320**; muscles, molting, 267; oostatic hormone, 60–61; ovariolar differentiation, 45;

Rhodnius prolixus (reduviid) (continued)
ovulation hormone, 77; prothoracic glands, degeneration, 326; sensillar postembryogenesis, 285; sperm activation, 66; sperm tube and ovarioles, embryogenesis, **128**, 227; spermatogenesis, control, 22–24; stylet-secreting organs, 225

rhomboid (*rho*): and dorsoventral axis, embryo, 150, **151**; and ommatidial spacing, 291; and wing vein specification, 311

Rhyacophila dorsalis (rhyacophilid): mandible, pharate adult, **247**

Rhynchosciara americana (sciarid): mitotic domains in embryo, 119

Rhyniella praecursor† (collembolan), 359

ring canals in: oogenesis, 41–45, **42–45**; spermatogenesis, 13, 14

ring gland: embryogenesis, 212; metamorphosis, 285

Romdan (RH-5992), **328**, 335

rotund (*rn*): expression pattern, leg disc, **309**

roundabout (*robo*): and crossing midline in ventral nerve cord, 191, 193, **193**

royal jelly (*Apis mellifera*): and queen development, 62, 95, 350

r-selection, 365

runt (*run*). See genes, segmentation: pair-rule

salivary glands: embryogenesis, 215; genes and, *Drosophila*, 215; metamorphosis, 301; polytene chromosomes, 331–332, **332**, **333**, 334–335, **335**, **336**

saltationism versus gradualism, 353–354

Samia cynthia (cynthia moth): bombyxin and prothoracic glands, 337

Samia walkeri (saturniid): ecdysteroids and spermatogenesis, 22, **23**

Sarcophaga barbata (sarcophagid): polytene chromosomes in hair and socket cells, 299

scaling relationships: cell size/body size, 261, 265, 311; and exaggerated structures, 260, 345–346

Scatophaga stercoraria (dung fly): sperm, differential storage by female, 85

Schistocerca americana (American bird grasshopper): *even-skipped* expression, embryos, 159; homeotic gene expression, embryos, 168; muscle pioneers, embryos, 179, **180**, **181**; neuroblast 7-4 lineage, embryos, **188**; neuroblast map, **189**; peripheral pioneer neurons, embryos, 202

Schistocerca gregaria (desert locust): brain embryogenesis, 195–196, **196**, 197; cleavage cells, 117; compound eye development, 289, 291–292; gene expression, embryonic legs, 308; hatching, 235; homeotic gene expression, embryos, 168; JH secretion by corpora allata, 337; male pheromones and vitellogenesis, 62; neuroblast mitosis, 187; optic lobe metamorphosis, 282; ovipositor motorneurons, 88; phase change, 347; pleuropodia function, 134, 230; *zerknüllt* expression, embryos, 169

Schistocerca nitens (vagrant grasshopper): neuroblast lineage and neuronal pathfinding, experiments on embryos, **192**; peripheral pioneer neurons, **202**, 202–203; staging embryogenesis, 52

schizocoelomate, 176

Schizolachnus piniradiatae (aphid): fecundity, 54

S-chromosomes, 113–114

sclerotization (tanning): biosynthetic pathway, 268, **268**; bursicon and, 268, 342–343, **339**, **342**; and ecdysis, 268, **268**; muscular continuity during molt, 270, **271**; and puparium formation, 271, **272**; sensory continuity during molt, 269, **269**

secondary dorsal organ, 128, **128**

segmentation and appendage formation: long germ embryos, **132**, 135; short germ embryos, **130**, 134, 135

segments, gnathal: genes and, *Drosophila* embryos, 161–163, **163**, 167, **167**, **170**, 172–174, **173**, **371**, **372**; in other arthropod embryos, **166**, 167, 168–169, **170**, **371**, **372**

segments, pre-oral: genes and, *Drosophila* embryos, 172–174, **173**, **371**, **372**; in other arthropod embryos, 168–169, **170**, **371**, **372**

selection: natural, 352; sexual, 82

selector genes. See genes, homeotic

semaphorin 1 (*sema 1*): and leg segment boundaries in locust embryo, 203, **204**

sense organs: cuticular continuity during molt, **206**, 269, **269**; embryogenesis, **189**, **202**, **204–207**, 204–209; evolution, 207–208; genes and, *Drosophila*, 205–207, **207**; metamorphosis, 285–294, **288–292**; projection to CNS, **202**, 203–204, **204**, **205**

sense organs, types: Bolwig's organ, 208; chordotonal, 202, 204, 205; compound eyes, 208–209; dorsal ocelli, 294; gustatory, 269; Keilin's organ, 285; mechanosensory, 204–205, **205**, **206**; olfactory, 269, **270**; stemmata, **198**, 208

serosa, 121, 122, 126–129, **128**, 131

serosal cuticle, 129

serpent (*srp*): fat body specification, 178, 184; germ cell migration, 226

Serrate (*Ser*): dorsoventral axis, eye disc, 291, **292**; dorsoventral axis, wing disc, 309, **310**; leg joints, 307, **309**

Sesamia nonagrioides (saturniid): diapause, facultative, 265

sevenless (*sev*): photoreceptor R7, 289–290, **291**

seven-up (*svp*): photoreceptor R1/R6, 290

Sex combs reduced (*Scr*). See genes, homeotic

sex chromosomes, 91, **92**

sex determination, 91–100, **92**, **94**, **96–99**; environmental effects on, 94–96, **94**, **96–98**

sexes, why two? 99–100

Sex-lethal (*Sxl*), **92**, 92–98

short gastrulation (*sog*): dorsoventral axis specification, embryo, 150–151; and brain embryogenesis, 199

short germ embryo: blastokinesis, 126–129, **129**; characteristics, **121**, 122; midblastula transition, 118; phylogenetic distribution, **132**; segmentation and appendage formation, **130**, 134–135; stages, 135

shortest intercalation rule. *See* regeneration: polar coordinate hypothesis

Silurian Period, 353, **358**

simulation, computer; embryogenesis, 233–234; wing color pattern, **313**, 313–314

Simulium decorum (black fly): number of sperm/spermatophore, 77

Simulium ornatum (black fly): flight muscle development, **247**, 248–250

sine oculis (*so*), 209

single-minded (*sim*), 193

Siphlonurus armatus (siphlonurid mayfly): telotrophic ovarioles, **48**

Siphonaptera (fleas): diversity, **4**; fossil history, **356**, 362, **362**; number of molts, 264; postembryogenesis, **239**

sisterless-a (*sis-a*), 92, **92**

sisterless-b (*sis-b*), 92, **92**

size, critical and: metamorphosis, 344–345, **344**, **345**; molting, 343–344; wing imaginal discs, caterpillars, 344

skeleton, cephalopharyngeal: embryogenesis, **132**, 218–220, **219**; fate during metamorphosis, 271; genes and, *Drosophila* embryos, 172–174, **173**; structure, **218**

skeleton, hemostatic in: caterpillars, 259; hatching, 235; head eversion in higher flies, 271; leg expansion, 267–268; mating, 74, **85**; molting, 267–268; wing expansion, 267–268

slit blastopore, 185, **185**, 199, **200**

sloppy-paired (*slp*). See genes, segmentation: pair-rule

small eye (*Pax-6*), 209

Smittia parthenogenetica (chironomid): anteroposterior axis specification, 141–143, **142**

snail (*sna*): dorsoventral axis, egg, 52; dorsoventral axis, embryo, 150–151, **151**; visceral muscle, 184

Solenobia triquetrella (psychid): intersexes, 97, **98**

Solenopsis invicta (red imported fire ant): number

of sperm in spermatheca, 77
solitaria phase, 346–347
spanandric males, 103
spätzle (*spz*): dorsoventral axis, egg, 50, **51**, 51, 150
speciation, 352–353
spectrosome in: stem oogonia, 42–43, **43**; stem spermatogonia, 14
sperm transfer: circadian release from sperm tubes, 65–66; direct, **70**, 74, **85**; in hexapods, 68–77, **70**; in marine arthropods, 67, **70**; number transferred, 76–77; in odonates, 70–71, **71**; phylogenetic diversity, **76**; seminal vesicles to spermatophore, 65, **68**; sperm tube to seminal vesicles, 65–66, **68**; spermatheca to egg, 71, 77–78, **78**; spermatophore to spermatheca, **72**, 72–73; in terrestrial arthropods, 67–72, **70**; by traumatic insemination, **70**, 74–76, **75**
sperm tubes: embryogenesis, 227; evolutionary origin, 56; number of/testis, 6, **8**; structure, 6–7, **7**, **9**, 22, **23**, **24**
spermalege phase, 74, **75**
spermatheca: development, 276, 277, **277**; function, 31, **72**, 72–73, **78**; structure, 31, **72**, **73**, **84**, **85**
spermathecal factor, 62
spermatogenesis: apyrene/eupyrene, 24–25, **26**; blood-testis barrier and, 22, **23**, 24; diapause and, 25; endocrine control, 21–25, **23**, **25**; genes and, *Drosophila*, 26–28, **27**; mitotic synchrony, **13**, 14; parasitoid effects on, 25–26, **26**; productivity, 20–21; rate, 18–20, **21**
spermatophores: contributions of to vitellogenesis, 74; fate, 73–74; formation, **68**, 68–69; types, 69–72, **70**, **72**, **73**
spermatophylax, 72
spermatozoa: activation, 66, **68**; capacitation, 65, **69**; competition between, 82–86; motility, 11, **11**; numbers produced, 20–21; phylogenetic diversity, 11–13, **12**; structure, 8–11, **9**, **10**, **12**, 19, 69, **80**
spermiogenesis: characteristics, **9**, 18, **19–21**, 20; endocrine control, 22–24; genes and,

Drosophila, 26–28, **27**; rate, 18–20, **21**
Sphodromantis bioculata (preying mantid): Dyar's law, **258**
Sphodromantis centralis (preying mantid): hatching threads, 134–135, 235–236
Spilopsylla cuniculi (rabbit flea): host hormones and reproduction, 63, **63**
Spodoptera frugiperda (fall armyworm): *Autographa* NPV and molting, 335–336
stage and instar, 236
stase, 242–243
staufen (*stau*) and: anteroposterior axis, egg, 50; anteroposterior axis, embryo, 152–153; pole cells, 121, **121**; pole plasm, 39
Steatoda triangulosa (spider): homeotic gene expression, embryos, 167, 171
stemmata: embryogenesis, *Drosophila*, 208; embryogenesis, *Lytta viridana*, 198, 208; fate of at metamorphosis, 208; genes and, *Drosophila* embryos, 208
stomatogastric (enteric) nervous system: embryogenesis, 209–211, **210**, **211**; function, 209; genes and, *Drosophila* embryos, 211; metamorphosis, **284**, 284–285; structure, **195**, 209, **210**
stomodaeum: embryogenesis, **130**, 134, 185, **185**
Strepsiptera (twisted wings): diversity, **4**; fossil history, **356**, **362**; number of molts, 264; wings and halteres, 382, **382**
Stricticimex brevispinosus (cimicid): traumatic insemination, 75
string (*stg*): maternal, **115**, 118; zygotic, **115**, 118
Stylops sp. (strepsipteran): embryogenesis, 125, **125**
subesophageal body, 184
subimago, 237, **238**
Suppressor hairless (*Su[H]*). See genes, neuronal switch points, in: caste development, 347–351, **349**, **350**; cleavage divisions, 112–113, **115**; development of exaggerated structures, 345–346; postembryogenesis of endopterygotes, 339, 343; postembryogenesis of

exopterygotes, **340**, 343; postembryogenesis of polyphenic insects, 345–347; spermatogenesis, **27**, 27–28
Sympetrum sanguineum (dragonfly): tandem oviposition, **71**
synapse formation, 204
synaptonemal complex, **15**, 16

Tachycines asynomorus (longhorn grasshopper): blastokinesis by, **133**, 138, **138**; dorsoventral axis specification in embryos, 137–138, **138**, **139**
tailless (*tll*): and head segments, 173, 174; and termini, 52, 152, **155**, 156, **159**
tail-up (*tup*): and germ band shortening/amnioserosa, 131
Talaeporia sp. (psychid moth): temperature and sex determination, 94, **94**
taxa, origin of: adaptive zones and key innovations, 354–355; saltation/gradualism, 353–354; speciation, 352–353; variation and natural selection, 352
teashirt (*tsh*): and leg discs, **309**; and wing discs, 310
tebufenozide, **328**, 335
Teleogryllus commodus (bush cricket): sperm entry into eggs, 77–78, **79**
telomerase, 265
telomeres: in meiosis, 16; and number of molts, 265
Tenebrio molitor (yellow mealworm beetle): blastokinesis, embryos, **133**; homeotic gene expression, embryos, 222; leg/wing disc development, 295; water loss during molt, 266
tenent hairs, 296, 298
tentorium, 225
terminal epithelium (sperm tube), 66
termini: derivatives, **130**, 134, **143**, 185, 185–186; early experiments on, 139–143, **141**, **142**; genes and, *Drosophila* embryos, 151–152, **152–155**, 159, 172–174, **173**; genes and, *Drosophila* oocytes, **51**, 52
Tertiary Period, 362–363
testis: embryogenesis, 227; number of sperm tubes, **8**; structure, 6–7, **7**, **9**, **13**, 21, **23**

testis ecdysiotropin, 24
Tetrodontophora bielanensis (collembolan): polyspermy, 80
thelytoky, **103**, 103–105
Thermobia domestica (firebrat): *en* expression, embryos, 160, 172; homeotic gene expression, embryos, 167, 168, 222, **372**, 376, 378, 379; mating and vitellogenesis, 62
thin plate spline method, 261
third axillary sclerite: development, 382; function, 355
thorax closure, 295
Thysanoptera (thrips): diversity, **4**; fossil history, **356**, 360, 361, **362**; life history, 237–238, **238**; number of molts, 264
Tineola bisselliella (webbing clothes moth): starvation and molting, 265
tinman (*tin*): and dorsal vessel, 178, 183–184; and germ cells alignment, 226
tip cells in: hatching threads, 134; Malpighian tubules, 186
Titanoptera†: fossil history, **356**, 360, **362**
Toll (*Tl*): and dorsoventral axis in egg, 50, 52
tonofibrillae, 269–270, **271**
torpedo (*top*): and anteroposterior axis of egg, 50
torso (*tor*): and termini, 151, **153**
torsolike (*tsl*): and termini of egg, **51**, 52; termini, embryos, 152
tracheal system: embryogenesis, 212–215, **213**, **214**; genes and, *Drosophila* embryos, **214**, 214–215; marine versus terrestrial origin, hexapods, 214; metamorphosis, 299, 301, **300**; regeneration, 303; structure, **177**, 212, **300**
trachealess (*trh*), **214**, 214–215
transcription factors, 169–172, **171**, 369–373, **370**
transformer (*tra*), 92–93
transformer-2 (*tra-2*), 92–93
transplantation of: embryonic nuclei, 120, **120**, 143, 144, **144**; endocrine organs, 319–323, **320**, **322**, **323**; integument, 152, 303, 321–322, **323**; ovaries (between species), 38; pole cells, 120, **120**; posterior pole plasm, 140, **141**, 152, **152**;

transplantation of (continued)
stylet-secreting organs,
225
transport and reutilization
theory (ecdysteroids), 328,
329
traumatic insemination,
74–76, **75**
Trialeurodes vaporariorum
(greenhouse whitefly): life
history, 238, **240**
Triassic Period, **356**, 358,
360–361
Tribolium castaneum
(red flour beetle):
Antennagalea, mutant
effects on embryo, 172;
compound eye
development, 289, 292; *eve*
expression, embryos, 159;
homeotic gene deletion,
effects, 162, 377; homeotic
gene expression, embryos,
167; re-use of displaced
sperm by males, 84; *Scr*
expression in labium,
embryonic, 376; *Tc-cad*
expression, embryos, 155;
Ubx expression and
pleuropodia, 165; *zerknüllt*
expression and embryonic
function, 169
Tribolium confusum
(confused flour beetle):
homeotic gene expression,
embryos, 168; *Wolbachia*
infection, effects, 86
Trichoplusia ni (cabbage
looper): as host of
Copidosoma, 122, **124**,
124–125, 232; midgut
endoderm, larval
replacement, 277–278;
spermatophore formation,
68
Trichoptera (caddisflies):
diversity, **4**; fossil history,
356, 362; number of molts,
264
Trigona sp. (sweat bee): caste
formation, 350
trigonotarbid, 358
Trigonotarsus rugosus
(weevil): larva, **244**
trilobites: anamorphosis, 135;
extinction, 360
Triops cancriformes
(crustacean): ommatidial
assembly in embryos, 291

trithorax (*trx*), 226
triungulin larva: in meloid
beetles, **245**, 245–246; in
Micromalthus debilis, **108**,
108–109
trophallaxis in: honeybees,
62, 350; termites, 347–348;
wasps, 350
twist (*twi*): and mesoderm
specification, 52, 150, **151**,
176–182, **179**, 184
two sexes: why? 99–100

Uca pugilator (decapod
crustacean): ecdysteroid
receptor, 331
Ultrabithorax (*Ubx*). *See*
genes, homeotic
Urbilateria: and CNS in
insect and vertebrate
embryos, 199–200, **200**;
definition, 161; segmented?
161
Urochordata (tunicates), 168
u-shaped (*ush*): and germ
band shortening, 131

valois (*val*): and
anteroposterior axis of egg,
50; and anteroposterior axis
of embryo, 152; and pole
cells, 121, **121**
Vanessa io (nymphalid):
appendage transplants,
319
Vanessa urticae (nymphalid):
appendage transplants,
319
vasa (*vas*): and
anteroposterior axis of egg,
50; and anteroposterior axis
of embryo, 152; and pole
cells, 121, **121**
ventral glands. *See*
prothoracic glands
ventral nerve cord:
embryogenesis, 187–195,
188, **189**, **191–194**; genes
and, *Drosophila* embryo,
188–190, **189**; glycogens
and neuronal pathfinding,
190–193, **192**, **193**;
interneurons, posterior
migration of during
metamorphosis, 280;
metamorphosis, 278–280,
280; midline precursors,
188, **189**, **191**, 193, **194**;
neuroblasts, 187–190, 188,

189, **191**, **192**, **194**;
neurones, **190**, 190–191;
phyletic variation,
193–195; sexual
differences, 280; structure,
186, **186**, **190**
vermiform larva, 235
Vespula spp. (vespid wasps):
caste formation, 350
vestigial (*vsg*): expression of,
wing disc, 309–310
visual system. *See* eyes
vitelline membrane:
deposition, **33**, 36, 41;
function, 49–52, **51**
vitellins, 37–38
vitellogenesis: endocrine
control, 59–61, **59**, **60**;
extrinsic factors, influence
on, 61–63, **63**; genes and,
Drosophila, 63–64;
male proteins and, 38;
overlapping cycles, 61;
termination, 38; yolk
uptake, specificity, 38
vitellogenin receptors, 38
vitellogenins: synthesis,
59–60, **59**, **60**; uptake of by
oocyte, 37–38, **38**
vitellophages, 115
viviparity: phylogenetic
distribution, **89**; types,
89–90

weight, critical: and
metamorphosis, 344–345,
344, **345**; and molting,
343–344
wing color pattern,
butterflies, **312–315**,
312–316
wing discs: in *Drosophila*
larvae, 294–296, **295**, **296**;
genes and, *Drosophila*
embryos, 215–218, **216**,
217; genes and, *Drosophila*
larvae, 308–312, **310**, **311**;
scaling of with body size,
311
wing evolution: fossil
evidence for, 355, **356**, 359,
378–382, **380–382**; genes
and, *Drosophila*, 215–218,
216, **217**, 308–312, **310**,
311; homology, 378–382,
380–382; leg exite theory
on origin, 217–218, **218**,
379; paranotal theory on
origin, 379; surface

skimming and origin,
379
wing flexing: adaptive
significance, 355–356, **356**;
evolutionary orgin, **356**,
359, 382; structural basis,
355
wing polymorphism: in
aphids, 105–107, **106**, 346;
in crickets, 346; and
evolutionary origin of
Phthiraptera, 366–367;
JH effects on, 346; in
lepidopterans, 295; in
psocids, 366
wingless (*wg*). *See* genes,
segmentation: pair-rule
Wolbachia pipientis:
fertilization and sperm
competition, influence on,
86
Wolbachia sp.:
parthenogenesis, induction,
109–110
worker jelly, 95, 350
wound healing, 302, **303**

X/A ratio, 92–93
X-chromosome, 91–92, **92**
Xenopsylla cheopis (oriental
rat flea), 63

Y-chromosome, 91–92
yolk: carbohydrate, 37; lipid,
37; protein, 37; synthesis of
in fat body, 59; uptake by
oocytes, **36**, 37–38, **38**
yolk consumption: in
Anastrepha embryos, 131,
134; complete
metamorphosis, role in
evolution, 254–256, **255**; by
lepidopteran protolarvae,
254
yp1–yp3, 39

zerknüllt (*zen*): dorsoventral
axis specification in
embryo, 150–151, **151**
zones, neurogenic, 209–211,
210
zootype, 168, **169**
Zoraptera: diversity, **4**; fossil
history, **356**, 362; number
of molts, 264
Zygentoma (silverfish):
fossil history, **356**, 362;
life history, 236–237;
number of molts, 264
zygogenetic, 101